W9-ATB-299

4521

Selected Works in Probability and Statistics

For other titles published in this series, go to
www.springer.com/series/8556

Evarist Giné • Vladimir Koltchinskii
Rimas Norvaiša
Editors

Selected Works of R.M. Dudley

Editors
Evarist Giné
University of Connecticut
Department of Mathematics
Auditorium Road 196
06269-3009 Storrs Connecticut U-9
USA
gine@math.uconn.edu

Vladimir Koltchinskii
Georgia Institute of Technology
School of Mathematics
30332-0160 Atlanta
USA
vlad@math.gatech.edu

Rimas Norvaiša
Institute of Mathematics and Informatics
Akademijos str. 4
LT-08663 Vilnius
Lithuania
norvaisa@ktl.mii.lt

ISBN 978-1-4419-5820-4 e-ISBN 978-1-4419-5821-1
DOI 10.1007/978-1-4419-5821-1
Springer New York Dordrecht Heidelberg London

Library of Congress Control Number: 2010932623

Printed on acid-free paper

Springer is part of Springer Science+Business Media (www.springer.com)

Preface to the Series

Springer's Selected Works in Probability and Statistics series offers scientists and scholars the opportunity of assembling and commenting upon major classical works in statistics, and honors the work of distinguished scholars in probability and statistics. Each volume contains the original papers, original commentary by experts on the subject's papers, and relevant biographies and bibliographies.

Springer is committed to maintaining the volumes in the series with free access on SpringerLink, as well as to the distribution of print volumes. The full text of the volumes is available on SpringerLink with the exception of a small number of articles for which links to their original publisher is included instead. These publishers have graciously agreed to make the articles freely available on their websites. The goal is maximum dissemination of this material.

The subjects of the volumes have been selected by an editorial board consisting of Anirban DasGupta, Peter Hall, Jim Pitman, Michael Sörensen, and Jon Wellner.

Richard Mansfield Dudley

Preface

It is almost impossible to describe all the research of Richard Mansfield Dudley, as it comprises more than 100 articles and spans over more than 45 years. The three co-editors of this volume have selected what we think are the most influential and representative of his articles and have shared the task of offering a few comments on them. Consistent with the overview of his work that follows, these Selected Works of Richard M. Dudley are divided into six chapters, each preceded by a short note on its content.

Dudley has been extremely influential in the development of Probability and Mathematical Statistics during the second half of the last century (or more exactly, between the nineteen sixties and the present). The two subjects on which he has left the deepest mark so far, in our view, are Gaussian processes and empirical processes. Succinctly, on the first: his research changed the framework of study of Gaussian processes by highlighting the intrinsic metric structure on the parameter space, and provided one of the main tools of their study, the famous metric entropy bound. And on the second: Vapnik and Červonenkis, in connection with their work on machine learning, initiated the modern view of empirical processes as processes indexed by general classes of sets and functions and proved uniform laws of large numbers for them, but it was the results of Dudley on the uniform central limit theorem that a) made this theory so useful and pervasive in asymptotic statistics, and b) initiated a vigorous development by many authors, that in turn made this theory even more useful. We may well say that in the first case Dudley changed the direction of the field in a way that led to the solution of some of its outstanding questions and in the second he created (or co-created with Vapnik and Červonenkis) a whole new field.

There are at least three more quite large fields of research where Richard M. Dudley's achievements are crucial and manifold. When Dudley started his career, the works of Prokhorov, Skorokhod, Varadarajan and LeCam on the general theory of weak convergence were very recent, and Dudley made substantial contributions to this theory in several ways, in particular, to complete the proofs of results of Donsker on the empirical process, and to study metrizability of this convergence in metric spaces. He also wrote, between 1965 and 1973, a series of articles on 'relativistic Markov processes', where he introduced Lorentz invariant diffusions as the only possible analogues in Lorentz space of Brownian motion and (in 3 dimensions) rotationally invariant Lévy processes, gave their properties, including asymptotics, and pointed at possible applications in Cosmology. Empirical processes can be thought of as the 'linear term' in Taylor type developments of statistical functionals, and several researchers in the seventies and eighties used compact differentiation of functionals defined on $D[0, 1]$ endowed with the supremum norm; Dudley advocates replacing the sup norm by the p-variation norm and compact differentiability by the more regular Fréchet differentiability, and shows this is possible (and desirable) by using empirical process techniques and by proving the Fréchet differentiability of some of the most usual operators like composition and inverse; this theory is not only quite elegant but it leads to optimality results.

This classification of Dudley's research into large areas leaves out many other remarkable works that touch upon approximation theory, learnability, Wiener functionals, singularity of measures, prediction theory, mathematical statistics, IQ and heredity, ... Some of them are commented upon below.

If we try to think about the common attributes of Dudley's works, three come immediately to mind: *good taste* in the choice of subjects by their relevance and beauty, *rigor* and *scholarship*. These characteristics make of Richard Dudley an excellent adviser, as one of us directly experienced, and this may help explain why he has had as many as thirty-one PhD students, while fairness and generosity are some of his other personal attributes that help further explain why he has been such a successful adviser.

It has been both an honor and a pleasure to select and comment on the works of a scholar of Richard M. Dudley's quality, and even more so as he has had a deep influence on the work (and on more than just the work) of each of us. We are thus grateful to the Editors of the Springer Selected Works in Probability and Statistics for the opportunity to assemble this volume. We also thank Edith Dudley Sylla for correspondence on Dudley's biography. Finally, we wish to acknowledge John Kimmel's and Jim Pitman's support in the preparation of this volume.

Evarist Giné
Vladimir Koltchinskii
Rimas Norvaiša

Acknowledgements

This series of selected works is possible only because of the efforts and cooperation of many people, societies, and publishers. The series editors originated the series and directed its development. The volume editors spent a great deal of time organizing the volumes and compiling the previously published material. The contributors provided comments on the significance of the papers. The societies and publishers who own the copyright to the original material made the volumes possible and affordable by their generous cooperation:

American Institute of Physics
American Mathematical Society
American Statistical Association
Applied Probability Trust
Bernoulli Society
Cambridge University Press
Canadian Mathematical Society
Danish Society for Theoretical Statistics
Elsevier
Finnish Statistical Society
Indian Statistical Institute
Institute of Mathematical Statistics
International Chinese Statistical Association
International Statistical Institute
John Wiley and Sons
New Zealand Statistical Association
Norwegian Statistical Society
Oxford University Press
Princeton University and the Institute for Advanced Studies
Royal Statistical Society
Statistical Society of Australia
Swedish Statistical Society
University of California Press
University of Illinois, Department of Mathematics
University of North Carolina Press

Contents

Part 3: Gaussian Processes

Part 4: Empirical Processes

Part 5: Nonlinear functionals and p-variation

Part 6: Miscellanea

A Biographical Sketch of Richard M. Dudley

Richard Mansfield Dudley was born on July 28, 1938, in Cleveland, Ohio, into a family with a strong university tradition, some of it mathematical. He graduated as valedictorian from Bloom Township High School in Chicago Heights, Illinois in 1955, attended Harvard University from 1955 to 1959, where he graduated Summa cum Laude as Bachelor of Arts, and obtained the PhD in Mathematics from Princeton University in 1962. Richard Dudley had two advisers, Gilbert A. Hunt and Edward Nelson, and wrote a thesis entitled 'Lorentz invariant random distributions'. Professor Dudley's professional career has taken place at the University of California at Berkeley, from 1962 to 1966, and at the Massachusetts Institute of Technology since 1967. He was a Fellow of the A. P. Sloan Foundation in 1966–68 and of the Guggenheim Foundation in 1991.

At Berkeley, Dudley coincided with Volker Strassen, who introduced him to Sudakov's idea of using metric entropy for the study of sample Gaussian paths and with whom he also shared an interest in probability distances that metrize weak convergence. He also worked with two other Berkeley colleagues, Lucien Le Cam and Jack Feldman, on different aspects of Gaussian processes. While at Berkeley, Dudley published work on Lorentz-invariant Markov processes, singularity of measures on linear spaces, random walks on groups, sequential convergence, prediction theory for non-stationary processes (the subject of the first PhD thesis he supervised), and on weak convergence of probabilities on nonseparable metric spaces (having to do with empirical processes) and Baire measures, among other topics.

Dudley completed his deep and extensive study of Gaussian processes when he was already at M.I.T., and, although he also started working on empirical processes at Berkeley (empirical measures on Euclidean spaces), he developed this theory at M.I.T., in about twenty publications including his landmark 1978 paper 'Central limit theorems for empirical processes' in Annals of Probability, that basically created a new subfield within Mathematical Statistics, the very influential Saint-Flour lecture notes (1984), and the book 'Uniform Central Limit Theorems' (Cambridge Studies in Advanced Mathematics). Empirical process theory is pervasive in modern Mathematical Statistics and has had a deep impact in the field of machine learning. Moreover, Dudley has worked on measurability problems, testing hypotheses, Wiener functionals, and approximation theory (in connection with metric entropy, Gaussian processes and empirical processes), to name a few subjects. During the last twenty years, he has also been interested in

p-variation and in differentiable functionals, in connection with the delta method for statistical functionals. Most of the work of Richard Dudley is single authored but at the same time he has collaborated with several colleagues and with a few of his thirty-one PhD students. Aside from the already mentioned Strassen, Le Cam and Feldman, the list of collaborators includes S. L. Cook, M. Durst, J. M. González-Barrios, S. Gutmann, D. Haughton, J. Hoffmann-Jørgensen, Y.-C. Huang, M. Kanter, J. Kuelbs, S. R. Kulkarni, J. Llopis, L. Pakula, N. P. Peng, P. Perkins, W. Philipp, A. Quiroz, B. Randol, T. J. Richardson, L. Shepp, S. Sidenko, D. Smith, D. Stroock, F. Topsøe, Z. Wang, R. S. Wenocur, O. Zeitouni, J. Zinn, and the editors of this volume.

Richard Dudley was visiting professor at the University of Aarhus, Denmark, in the spring of 1976, and, while there, he wrote a very nice set of lectures for a graduate course, Probabilities and Metrics. Most of these notes were later adapted and incorporated in the second part of his deservedly successful graduate text Real Analysis and Probability (1989, 2002). This book is special in particular because the author looked for the best and shortest proofs available and each chapter ends with very accurate and complete historical notes. He has also written monographs on two of the subjects on which he has recently worked most: on empirical processes, he has written the book Uniform Central Limit Theorems, which also originated in lecture notes, in this case the above mentioned Saint-Flour lectures, 'A course on empirical Processes', and on p-variation and related matters, jointly with Rimas Norvaiša, he has written two sets of lecture notes (Differentiability of Six Operators on Non-Smooth Functions and p-variation, and An Introduction to p-variation and Young Integrals), and a forthcoming book, Concrete Functional Calculus. Beside the above, his scholarly service to the profession also includes having been the Editor of the Annals of Probability, 1979–1981, after six years of being an Associate Editor, and having served on the Editorial Board of the Wadsworth Advanced Series in Statistics/Probability, 1982–1992. He has talked and talks about his work at many scientific meetings, and in particular he has been an invited speaker at the 1974 International Congress of Mathematicians, at American Mathematical Society, Institute of Mathematical Statistics and Bernoulli Society meetings, at Vilnius Conferences in Probability Theory and Mathematical Statistics, at Saint-Flour, at many Probability in Banach Spaces meetings (some of which he helped organize), etc.

Dudley has participated at different times in very concrete and effective ways in civic activities related to his convictions or to his interests, so, for instance, he was a volunteer news writer and broadcaster one afternoon a week in 1963–1966 for public radio station KPFA in Berkeley, and in 1979 he was the Editor of the Appalachian Mountain Club White Mountain Guide. He has hiked up all the mountains in New England over 4000 feet, besides Mont Blanc and other mountains in the Alps and elsewhere.

Professor Dudley has received several honors, in particular, he has served in the (honorary) Advisory Board of Stochastic Processes and their Applications, 1987–2001 and is a Fellow of: the Institute of Mathematical Statistics, the American Statistical Association, and the American Association for the Advancement of Science; and an elected Member of the International Statistical Institute.

Richard Dudley lives in Newton, Massachusetts, with Liza Martin, his wife of thirty-one years.

Author's Bibliography 1961–2009

[1] R. M. Dudley. Continuity of homomorphisms. *Duke Math. J.*, 28:587–594, 1961.

[2] R. M. Dudley and B. Randol. Implications of pointwise bounds on polynomials. *Duke Math. J.*, 29:455–458, 1962.

[3] R. M. Dudley. *Lorentz invariant random distributions*. PhD thesis, Princeton, 1962. Mathematics.

[4] R. M. Dudley. Random walks on abelian groups. *Proc. Amer. Math. Soc.*, 13:447–450, 1962.

[5] R. M. Dudley. A property of a class of distributions associated with the Minkowski metric. *Illinois J. Math.*, 8:169–174, 1964.

[6] R. M. Dudley. On sequential convergence. *Trans. Amer. Math. Soc.*, 112:483–507, 1964. Corrections *ibid.* 148:623–624, 1970.

🕮 [7] R. M. Dudley. Pathological topologies and random walks on abelian groups. *Proc. Amer. Math. Soc.*, 15:231–238, 1964.

[8] R. M. Dudley. Singular translates of measures on linear spaces. *Z. Wahrscheinlichkeitstheorie und Verw. Gebiete*, 3:128–137, 1964.

[9] R. M. Dudley. Fourier analysis of sub-stationary processes with a finite moment. *Trans. Amer. Math. Soc.*, 118:360–375, 1965.

[10] R. M. Dudley. Gaussian processes on several parameters. *Ann. Math. Statist.*, 36:771–788, 1965.

[11] R. M. Dudley. Convergence of Baire measures. *Studia Math.*, 27:251–268, 1966.

🕮 [12] R. M. Dudley. Lorentz-invariant Markov processes in relativistic phase space. *Ark. Mat.*, 6:241–268, 1966.

[13] R. M. Dudley. Singularity of measures on linear spaces. *Z. Wahrscheinlichkeitstheorie und Verw. Gebiete*, 6:129–132, 1966.

🕮 [14] R. M. Dudley. Weak convergences of probabilities on nonseparable metric spaces and empirical measures on Euclidean spaces. *Illinois J. Math.*, 10:109–126, 1966.

🕮 [15] R. M. Dudley. A note on Lorentz-invariant Markov processes. *Ark. Mat.*, 6:575–581, 1967.

🕮 [16] R. M. Dudley. Measures on non-separable metric spaces. *Illinois J. Math.*, 11:449–453, 1967.

[17] R. M. Dudley. On prediction theory for nonstationary sequences. In *Proc. Fifth Berkeley Sympos. Math. Statist. and Probability (Berkeley, Calif., 1965/66)*, pages Vol. II: Contributions to Probability Theory, Part 1, pp. 223–234. Univ. California Press, Berkeley, Calif., 1967.

[18] R. M. Dudley. Sub-stationary processes. *Pacific J. Math.*, 20:207–215, 1967.

🕮 [19] R. M. Dudley. The sizes of compact subsets of Hilbert space and continuity of Gaussian processes. *J. Functional Analysis*, 1:290–330, 1967.

🕮 [20] R. M. Dudley. Distances of probability measures and random variables. *Ann. Math. Statist.*, 39:1563–1572, 1968.

🕮 Indicates article found in this volume

[21] R. M. Dudley. The speed of mean Glivenko-Cantelli convergence. *Ann. Math. Statist.*, 40:40–50, 1968.

[22] R. M. Dudley. Fourier transforms of stationary processes are not functions. *J. Math. Mech.*, 19:309–316, 1969.

[23] R. M. Dudley. Random linear functionals. *Trans. Amer. Math. Soc.*, 136:1–24, 1969.

[24] V. Strassen and R. M. Dudley. The central limit theorem and ε-entropy. In *Probability and Information Theory (Proc. Internat. Sympos., McMaster Univ., Hamilton, Ont., 1968)*, pages 224–231. Springer, Berlin, 1969.

[25] R. M. Dudley. Random linear functionals: Some recent results. In C. T. Taam, editor, *Lectures in Modern Analysis and Applications, III*, Lecture Notes in Mathematics, Vol. 170, pages 62–70. Springer, Berlin, 1970.

[26] R. M. Dudley. Small operators between Banach and Hilbert spaces. *Studia Math.*, 38:35–41, 1970.

[27] R. M. Dudley. Some uses of ε-entropy in probability theory. In *Les probabilités sur les structures algébriques (Actes Colloq. Internat. CNRS, No. 186, (Clermont-Ferrand, 1969)*, pages 113–122. Éditions Centre Nat. Recherche Sci., Paris, 1970. With discussion by L. Le Cam, Ju. V. Prohorov, A. Badrikian and R. M. Dudley.

[28] R. M. Dudley. Convergence of sequences of distributions. *Proc. Amer. Math. Soc.*, 27:531–534, 1971.

[29] R. M. Dudley. Non-linear equivalence transformations of Brownian motion. *Z. Wahrscheinlichkeitstheorie und Verw. Gebiete*, 20:249–258, 1971.

[30] R. M. Dudley. On measurability over product spaces. *Bull. Amer. Math. Soc.*, 77:271–274, 1971.

[31] R. M. Dudley, J. Feldman, and L. Le Cam. On seminorms and probabilities, and abstract Wiener spaces. *Ann. of Math.*, 93:390–408, 1971. Corrections *ibid.* 104:391, 1976.

[32] R. M. Dudley and L. Pakula. A counter-example on the inner product of measures. *Indiana Univ. Math. J.*, 21:843–845, 1972.

[33] R. M. Dudley. A counterexample on measurable processes. In *Proceedings of the Sixth Berkeley Symposium on Mathematical Statistics and Probability (Univ. California, Berkeley, Calif., 1970/1971), Vol. II: Probability theory*, pages 57–66, Berkeley, Calif., 1972. Univ. California Press. Corrections *ibid.* 1:191–192, 1973.

[34] R. M. Dudley. Speeds of metric probability convergence. *Z. Wahrscheinlichkeitstheorie und Verw. Gebiete*, 22:323–332, 1972.

[35] R. M. Dudley. A note on products of spectral measures. In *Vector and operator valued measures and applications (Proc. Sympos., Alta, Utah, 1972)*, pages 125–126. Academic Press, New York, 1973.

[36] R. M. Dudley. Asymptotics of some relativistic Markov processes. *Proc. Nat. Acad. Sci. U.S.A.*, 70:3551–3555, 1973.

[37] R. M. Dudley. Sample functions of the Gaussian process. *Ann. Probability*, 1:66–103, 1973.

[38] R. M. Dudley. Metric entropy and the central limit theorem in *C(S)*. *Ann. Inst. Fourier (Grenoble)*, 24:49–60, 1974. Colloque International sur les Processus Gaussiens et les Distributions Aléatoires (Colloque Internat. du CNRS, No. 222, Strasbourg, 1973).

[39] R. M. Dudley. Metric entropy of some classes of sets with differentiable boundaries. *J. Approximation Theory*, 10:227–236, 1974.

[40] R. M. Dudley. Recession of some relativistic Markov processes. *Rocky Mountain J. Math.*, 4:401–406, 1974. Papers arising from a Conference on Stochastic Differential Equations (Univ. Alberta, 1972).

[41] R. M. Dudley and M. Kanter. Zero-one laws for stable measures. *Proc. Amer. Math. Soc.*, 45:245–252, 1974. Corrections *ibid.* 88:689–690, 1983.

[42] R. M. Dudley, P. Perkins, and E. Giné. Statistical tests for preferred orientation. *J. Geology*, 83:685–705, 1975.

[43] R. M. Dudley. The Gaussian process and how to approach it. In *Proceedings of the International Congress of Mathematicians (Vancouver, B.C., 1974), Vol. 2*, pages 143–146. Canad. Math. Congress, Montreal, Que., 1975.

[44] R. M. Dudley. *Probabilities and Metrics: Convergence of Laws on Metric Spaces with a View to Statistical Testing*, volume 45 of *Lecture Notes Series*. Matematisk Institut, Aarhus Universitet, Aarhus, 1976.

[45] F. Topsøe, R. M. Dudley, and J. Hoffmann-Jørgensen. Two examples concerning uniform convergence of measures w.r.t. balls in Banach spaces. In P. Gaenssler and P. Révész, editors, *Empirical distributions and processes (Selected Papers, Meeting on Math. Stochastics, Oberwolfach, 1976)*, pages 141–146. Lecture Notes in Math., Vol. 566. Springer, Berlin, 1976.

[46] R. M. Dudley. Unsolvable problems in mathematics. *Science*, 191:807–808, 1976.

[47] R. M. Dudley. A survey on central limit theorems in Banach spaces. In *Symposia Mathematica, Vol. XXI (Convegno sulle Misure su Gruppi e su Spazi Vettoriali, Convegno sui Gruppi e Anelli Ordinati, INDAM, Rome, 1975)*, pages 11–14. Academic Press, London, 1977.

[48] R. M. Dudley. On second derivatives of convex functions. *Math. Scand.*, 41:159–174, 1977. Acknowledgement of priority, *ibid.* 46:6, 1980.

[49] R. M. Dudley and S. Gutmann. Stopping times with given laws. In *Séminaire de Probabilités, XI (Univ. Strasbourg, Strasbourg, 1975/1976)*, pages 51–58. Lecture Notes in Math., Vol. 581. Springer, Berlin, 1977.

[50] R. M. Dudley. Wiener functionals as Itô integrals. *Ann. Probability*, 5:140–141, 1977.

[51] R. M. Dudley. Central limit theorems for empirical measures. *Ann. Probab.*, 6:899–929, 1978. Corrections *ibid.* 7:909–911, 1979.

[52] R. M. Dudley. Balls in $\mathbf{R}^k$ do not cut all subsets of $k+2$ points. *Adv. in Math.*, 31:306–308, 1979.

[53] R. M. Dudley. Lower layers in R^2 and convex sets in R^3 are not GB classes. In *Probability in Banach spaces, II (Proc. Second Internat. Conf., Oberwolfach, 1978)*, volume 709 of *Lecture Notes in Math.*, pages 97–102. Springer, Berlin, 1979.

[54] J. Hoffmann-Jørgensen, L. A. Shepp, and R. M. Dudley. On the lower tail of Gaussian seminorms. *Ann. Probab.*, 7:319–342, 1979.

[55] R. M. Dudley. On χ^2 tests of composite hypotheses. In *Probability theory (Papers, VIIth Semester, Stefan Banach Internat. Math. Center, Warsaw, 1976)*, volume 5 of *Banach Center Publ.*, pages 75–87. PWN, Warsaw, 1979.

[56] J. Kuelbs and R. M. Dudley. Log log laws for empirical measures. *Ann. Probab.*, 8:405–418, 1980.

[57] R. M. Dudley. Donsker classes of functions. In M. Csörgő et al., editors, *Statistics and related topics (Ottawa, Ont., 1980)*, pages 341–352. North-Holland, Amsterdam, 1981.

[58] M. Durst and R. M. Dudley. Empirical processes, Vapnik-Chervonenkis classes and Poisson processes. *Probab. Math. Statist.*, 1:109–115, 1981.

[59] R. M. Dudley. Some recent results on empirical processes. In *Probability in Banach spaces, III (Medford, Mass., 1980)*, volume 860 of *Lecture Notes in Math.*, pages 107–123. Springer, Berlin, 1981.

[60] R. S. Wenocur and R. M. Dudley. Some special Vapnik-Chervonenkis classes. *Discrete Math.*, 33:313–318, 1981.

[61] R. M. Dudley. Vapnik-Červonenkis Donsker classes of functions. In *Aspects Statistiques et Aspects Physiques des Processus Gaussiens: Saint-Flour, 1980*, volume 307 of *Colloq. Internat. CNRS*, pages 251–269. CNRS, Paris, 1981.

[62] R. M. Dudley. Empirical and Poisson processes on classes of sets or functions too large for central limit theorems. *Z. Wahrscheinlichkeitstheorie und Verw. Gebiete*, 61:355–368, 1982.

[63] R. M. Dudley, P. Humblet, and M. Goldring. Appendix to: Single photon transduction in *Limulus* photoreceptors and the Borsellino-Fuortes model, by M. A. Goldring and J. E. Lisman. *IEEE Trans. Systems, Man, Cybernet*, SMC-13:727–731, 1983.

[64] R. M. Dudley. Heritability under genotype-environment interaction and dependence. *Bull. Inst. Math. Statist.*, 12:152, 1983. Abstract.

📖 [65] R. M. Dudley and W. Philipp. Invariance principles for sums of Banach space valued random elements and empirical processes. *Z. Wahrscheinlichkeitstheorie und Verw. Gebiete*, 62:509–552, 1983.

[66] R. M. Dudley. A course on empirical processes. In *École d'été de probabilités de Saint-Flour, XII—1982*, volume 1097 of *Lecture Notes in Math.*, pages 1–142. Springer, Berlin, 1984.

[67] R. M. Dudley. Discussion of Special Invited Paper by E. Giné and J. Zinn "Some limit theorems for empirical processes". *Ann. Probab.*, 12:991–992, 1984.

📖 [68] R. M. Dudley. An extended Wichura theorem, definitions of Donsker class, and weighted empirical distributions. In A. Beck, R. Dudley, M. Hahn, J. Kuelbs, and M. Marcus, editors, *Probability in Banach spaces, V (Medford, Mass., 1984)*, volume 1153 of *Lecture Notes in Math.*, pages 141–178. Springer, Berlin, 1985.

[69] R. M. Dudley. Manifolds. In *Encyclopedia of Statistical Sciences*, volume 5, pages 198–201. Wiley, New York, 1985.

[70] R. M. Dudley. The structure of some Vapnik-Červonenkis classes. In *Proceedings of the Berkeley conference in honor of Jerzy Neyman and Jack Kiefer, Vol. II (Berkeley, Calif., 1983)*, Wadsworth Statist./Probab. Ser., pages 495–508, Belmont, CA, 1985. Wadsworth.

[71] R. Dudley and D. W. Stroock. Slepian's inequality and commuting semigroups. In *Séminaire de Probabilités, XXI*, volume 1247 of *Lecture Notes in Math.*, pages 574–578. Springer, Berlin, 1987.

[72] R. M. Dudley. Some inequalities for continued fractions. *Math. Comp.*, 49:585–593, 1987.

[73] R. M. Dudley. Some universal Donsker classes of functions. In Yu. V. Prohorov, V. A. Statulevičius, V. V. Sazonov, and B. Grigelionis, editors, *Probability theory and mathematical statistics, Vol. I (Vilnius, 1985)*, pages 433–438. VNU Sci. Press, Utrecht, 1987.

📖 [74] R. M. Dudley. Universal Donsker classes and metric entropy. *Ann. Probab.*, 15:1306–1326, 1987.

[75] R. M. Dudley. Comment on D. Pollard, "Asymptotics via empirical processes". *Statistical Science*, 4:354, 1989.

[76] R. M. Dudley. *Real Analysis and Probability*. The Wadsworth & Brooks/Cole Mathematics Series. Wadsworth & Brooks/Cole Advanced Books & Software, Pacific Grove, CA, 1989.

[77] R. M. Dudley, S. L. Cook, J. Llopis, and N. P. Peng. A. N. Kolmogorov and statistics: a citation bibliography. *Ann. Statist.*, 18:1017–1031, 1990.

[78] R. M. Dudley. Interaction and dependence prevent estimation (Commentary). *Behavioral and Brain Sciences*, 13:132–133, 1990.

[79] R. M. Dudley. Nonlinear functionals of empirical measures and the bootstrap. In E. Eberlein, J. Kuelbs, and M. B. Marcus, editors, *Probability in Banach Spaces, 7 (Oberwolfach, 1988)*, volume 21 of *Progr. Probab.*, pages 63–82. Birkhäuser Boston, Boston, MA, 1990.

[80] R. M. Dudley. Nonmetric compact spaces and nonmeasurable processes. *Proc. Amer. Math. Soc.*, 108:1001–1005, 1990.

[81] R. M. Dudley. Program verification. *Notices Amer. Math. Soc.*, 37:123–124, 1990.

[82] R. M. Dudley. IQ and heredity (letter). *Science*, 252:191, 1991.

[83] A. J. Quiroz and R. M. Dudley. Some new tests for multivariate normality. *Probab. Theory Related Fields*, 87:521–546, 1991.

[84] R. M. Dudley, E. Giné, and J. Zinn. Uniform and universal Glivenko-Cantelli classes. *J. Theoret. Probab.*, 4:485–510, 1991.

[85] R. M. Dudley. Why are adoptees so similar in IQ? (Commentary.). *Behavioral and Brain Sciences*, 14:336, 1991.

[86] D. L. Smith and R. M. Dudley. Exponential bounds in Vapnik-Červonenkis classes of index 1. In R. M. Dudley, M. G. Hahn, and J. Kuelbs, editors, *Probability in Banach spaces, 8 (Brunswick, ME, 1991)*, volume 30 of *Progr. Probab.*, pages 451–465. Birkhäuser Boston, Boston, MA, 1992.

[87] R. M. Dudley. Fréchet differentiability, p-variation and uniform Donsker classes. *Ann. Probab.*, 20:1968–1982, 1992.

[88] R. M. Dudley. Nonlinear functionals of empirical measures. In R. M. Dudley, M. G. Hahn, and J. Kuelbs, editors, *Probability in Banach spaces, 8 (Brunswick, ME, 1991)*, volume 30 of *Progr. Probab.*, pages 403–410. Birkhäuser Boston, Boston, MA, 1992.

[89] J. M. González-Barrios and R. M. Dudley. Metric entropy conditions for an operator to be of trace class. *Proc. Amer. Math. Soc.*, 118:175–180, 1993.

[90] R. M. Dudley, S. R. Kulkarni, T. Richardson, and O. Zeitouni. A metric entropy bound is not sufficient for learnability. *IEEE Trans. Inform. Theory*, 40:883–885, 1994.

[91] R. M. Dudley and V. I. Koltchinskii. Envelope moment conditions and Donsker classes. *Teor. Ĭmovīr. Mat. Stat.*, 51:39–49, 1994.

[92] R. M. Dudley. Metric marginal problems for set-valued or non-measurable variables. *Probab. Theory Related Fields*, 100:175–189, 1994.

[93] R. M. Dudley and J. González-Barrios. On extensions of Mercer's theorem. In M. E. Caballero and L. G. Gorostiza, editors, *III Simposio de Probabilidad y Procesos Estocásticos*, volume 11 of *Aportaciones Matemáticas*, pages 91–97, 1994.

[94] R. M. Dudley. The order of the remainder in derivatives of composition and inverse operators for p-variation norms. *Ann. Statist.*, 22:1–20, 1994.

[95] R. M. Dudley. Empirical processes and p-variation. In D. Pollard, E. Torgersen, and G. L. Yang, editors, *Festschrift for Lucien Le Cam*, pages 219–233. Springer, New York, 1997.

[96] R. M. Dudley and D. Haughton. Information criteria for multiple data sets and restricted parameters. *Statist. Sinica*, 7:265–284, 1997.

[97] R. M. Dudley and R. Norvaiša. *An Introduction to p-Variation and Young Integrals, with Emphasis on Sample Functions of Stochastic Processes*, volume 1 of *MaPhySto Lecture Notes*. Aarhus, Denmark, 1998.

[98] R. M. Dudley and J. M. González-Barrios. Conditions for integral and other operators to be of trace class. *Bol. Soc. Mat. Mexicana*, 4:105–114, 1998.

[99] R. M. Dudley. Consistency of M-estimators and one-sided bracketing. In E. Eberlein, M. Hahn, and M. Talagrand, editors, *High dimensional probability (Oberwolfach, 1996)*, volume 43 of *Progr. Probab.*, pages 33–58. Birkhäuser, Basel, 1998.

[100] R. M. Dudley and R. Norvaiša. *Differentiability of Six Operators on Nonsmooth Functions and p-Variation*, volume 1703 of *Lecture Notes in Mathematics*. Springer-Verlag, Berlin, 1999. With the collaboration of Jinghua Qian.

[101] R. M. Dudley. *Notes on Empirical Processes*, volume 4 of *MaPhySto Lecture Notes*. Aarhus, Denmark, 1999.

[102] R. M. Dudley. *Uniform Central Limit Theorems*, volume 63 of *Cambridge Studies in Advanced Mathematics*. Cambridge University Press, Cambridge, 1999.

[103] V. Koltchinskii and R. M. Dudley. On spatial quantiles. In V. Korolyuk, N. Portenko, and H. Syta, editors, *Skorokhod's Ideas in Probability Theory*, pages 195–210. Inst. Math. Nat. Acad. Sci. Ukraine, Kyiv, 2000.

[104] Y.-C. Huang and R. M. Dudley. Speed of convergence of classical empirical processes in p-variation norm. *Ann. Probab.*, 29:1625–1636, 2001.

[105] R. M. Dudley and D. Haughton. Asymptotic normality with small relative errors of posterior probabilities of half-spaces. *Ann. Statist.*, 30:1311–1344, 2002.

[106] R. M. Dudley and D. M. Haughton. One-sided hypotheses in a multinomial model. In C. Huber-Carol, N. Balakrishnan, M. S. Nikulin, and M. Mesbah, editors, *Goodness-of-fit Tests and Model Validity (Paris, 2000)*, Stat. Ind. Technol., pages 387–399. Birkhäuser, Boston, MA, 2002.

[107] R. M. Dudley. *Real Analysis and Probability*, volume 74 of *Cambridge Studies in Advanced Mathematics*. Cambridge University Press, Cambridge, 2002. 2d. Edition.

[108] R. M. Dudley. Statistical nearly universal Glivenko-Cantelli classes. In J. Hoffmann-Jørgensen, M. B. Marcus, and J. A. Wellner, editors, *High Dimensional Probability, III (Sandbjerg, 2002)*, volume 55 of *Progr. Probab.*, pages 295–312. Birkhäuser, Basel, 2003.

[109] R. M. Dudley. Some facts about functionals of location and scatter. In E. Giné, V. Koltchinskii, W. Li, and J. Zinn, editors, *High Dimensional Probability*, volume 51 of *IMS Lecture Notes Monogr. Ser.*, pages 207–219. Inst. Math. Statist., Beachwood, OH, 2006.

[110] R. M. Dudley, S. Sidenko, and Z. Wang. Differentiability of t-functionals of location and scatter. *Ann. Statist.*, 37:939–960, 2009.

[111] R. M. Dudley and R. Norvaiša. Concrete Functional Calculus, to appear. Research monograph in preparation, about 750 pp., under contract for publication.

PhD students of Richard M. Dudley

Chandrakant Deo, 'Prediction Theory of Non-stationary Processes,' University of California, Berkeley, 1965.

José Abreu Leon, 'Smoothing, Filtering and Prediction of Generalized Stochastic Processes,' Massachusetts Institute of Technology, 1970.

Lewis Pakula, 'Covariances of Generalized Stochastic Processes,' Massachusetts Institute of Technology, 1972.

Evarist Giné-Masdéu, 'Invariant Tests for Uniformity on Compact Riemannian Manifolds Based on Sobolev Norms,' Massachusetts Institute of Technology, 1973.

Ruben Klein, 'Topics on Gaussian Sample Functions,' Massachusetts Institute of Technology, 1974 .

Michael Zuker, 'Speeds of Convergence of Random Probability Measures,' Massachusetts Institute of Technology, 1974.

Marjorie Hahn, 'Central Limit Theorems for $D[0,1]$-Valued Random Variables,' Massachusetts Institute of Technology, 1975.

Donald Cohn, 'Topics in Liftings and Stochastic Processes,' Harvard University, 1975.

Eric Slud, 'Inequalities for Binomial, Normal, and Hypergeometric Tail Probabilities,' Massachusetts Institute of Technology, 1976.

Samuel Gutmann, 'Non-Stationary Markov Transitions,' Massachusetts Institute of Technology, 1977.

Steven Pincus, 'Strong Laws of Large Numbers for Products of Random Matrices,' Massachusetts Institute of Technology, 1980.

Mark Durst, 'Donsker Classes, Vapnik-Chervonenkis Classes, and Chi-Squared Tests of Fit with Random Cells,' Massachusetts Institute of Technology, 1980.

Zakhar Maymin, 'Minimax Estimation on Subsets of Parameters,' Massachusetts Institute of Technology, 1981.

Joseph Yukich, 'Convergence in Empirical Probability Measures, Massachusetts Institute of Technology,' 1982.

Kenneth Alexander, 'Some Limit Theorems and Inequalities for Weighted and Non-Identically Distributed Empirical Processes,' Massachusetts Institute of Technology, 1982.

Rae Michael Shortt, 'Existence of Laws with Given Marginals and Specified Support,' Massachusetts Institute of Technology, 1982.

David Marcus, 'Non-Stable Laws with All Projections Stable and Relationships Between Donsker Classes and Sobolev Spaces,' Massachusetts Institute of Technology, 1983.

Richard Dante DeBlassie, 'Hitting Times of Brownian Motion' , Massachusetts Institute of Technology, 1984.

Dominique Haughton, 'On the choice of a model to fit data from an exponential family,' Massachusetts Institute of Technology, 1984.

Joseph Fu, 'Tubular Neighborhoods of Planar Sets' (with J. Almgren), Massachusetts Institute of Technology, 1984.

Daphne Smith, 'Vapnik-Červonenkis Classes and the Supremum Distribution of a Gaussian Process, Massachusetts Institute of Technology,' 1985.

Adolfo Quiroz Salazar, 'On Donsker Classes of Functions and their Application to Tests for Goodness of Fit,' Massachusetts Institute of Technology, 1986.

Robert Holt, 'Computation of Gamma Tail Probabilities,' Massachusetts Institute of Technology, 1986.

Michael Schmidt, 'Optimal Rates of Convergence for Nonparametric Regression Function Estimators,' Massachusetts Institute of Technology, 1988.

José González-Barrios, 'On Von Mises Functionals with Emphasis on Trace Class Kernels,' Massachusetts Institute of Technology, 1990.

Evangelos Tabakis, 'Asymptotic and Computational Problems in Single-Link Clustering,' Massachusetts Institute of Technology, 1992 .

Yen-Chin Huang, 'Empirical Distribution Function Statistics, Speed of Convergence, and p-Variation,' Massachusetts Institute of Technology, 1994.

Jinghua Qian, 'The p-Variation of Partial Sum Processes and the Empirical Process,' Tufts University 1997.

Li He, 'Modeling and Prediction of Sunspot Cycles,' Massachusetts Institute of Technology, 2001.

Martynas Manstavičius, 'The p-variation of Strong Markov Processes,' Tufts University, 2003.

Xia Hua, 'Testing regression models with residuals as data', Massachusetts Institute of Technology, 2010

Part 1

Convergence in Law

Introduction

The papers in this chapter deal with important properties of weak convergence of probability measures on metric spaces. Most of them are motivated by, and applied to, the question of establishing convergence in law of empirical processes, a basic topic of statistics.

The problem considered in the first paper comes from the fact that, due to certain set-theoretic assumptions, a finite, countably additive measure defined on all Borel sets of a metric space is concentrated in a separable subspace whereas, on the other hand, almost all sample paths of empirical processes are not elements of a separable subset of a metric space. Dudley extended the notion of weak convergence to countably additive probability measures defined on σ-algebras of a metric space not necessarily related to its metric topology. Namely, the weak* convergence of measures β_n defined on σ-algebras $\mathcal{B}_n$ of subsets of a metric space S to a Borel measure β_0 on S means that

$$\lim_{n\to\infty} \int^* f\,d\beta_n = \lim_{n\to\infty} \int_* f\,d\beta_n = \int f\,d\beta_0 \tag{1}$$

for every bounded continuous real-valued function f on S, where $\int^*$ and $\int_*$ are upper and lower integrals, respectively. Results were obtained for this convergence in case each $\mathcal{B}_n$ includes the smallest σ-algebra $\mathcal{U}$ generated by all open balls of S. A probability measure on $\mathcal{U}$ does not need to be confined to a separable subspace of S. Dudley proved the weak* convergence of measures α_n, $n \geq 1$, on a suitable function space J with the uniform metric when α_n is the probability distribution of normalized empirical distribution functions induced by a sequence of independent identically distributed $\mathbf{R}^k$-valued random variables. The suggested solution of the problem corrected the main results of M. D. Donsker (Mem. Amer. Math. Soc., 1951, No. 6 and Ann. Math. Statist., 1952, **23**, 277-281) for real-valued random variables and generalized them to random variables with values in a Euclidean space.

The second paper continued the subject of the first one by giving a more general definition of weak* convergence. Let $\mathcal{U}$ be the σ-algebra on a metric space S generated by open balls as before. Let $M(S, \mathcal{U})$ be the set of all finite, countably additive, real-valued set functions on $\mathcal{U}$, and let $C(S, \mathcal{U})$ be the closed linear subspace of $\mathcal{U}$-measurable elements of the Banach space $C(S)$ of all bounded, continuous, real-valued functions on S with the supremum norm. Then any $\mu \in M(S, \mathcal{U})$ defines a bounded linear functional $f \to \int f\,d\mu$ on $C(S, \mathcal{U})$ and we have the weak* topology of pointwise convergence on $C(S, \mathcal{U})$. Let (β_n) be a sequence of nonnegative elements of $M(S, \mathcal{U})$ and let β_0 be a nonnegative element of $M(S, \mathcal{U})$ concentrated in a separable subspace of S. Under the hypothesis that the metric space S is complete Dudley proved that $\beta_n \to \beta_0$ for the weak* topology if and only if (1) holds for every f in $C(S)$.

The third paper in this chapter compares various metrics on the set of all probability measures of a metric space, and relates the weak* convergence of probability measures with almost surely convergent realizations. Let $\mathcal{P}(S)$ be the set of all Borel probability measures on a separable metric space S, endowed with the weak* topology. For S complete and $\mu, \nu \in \mathcal{P}(S)$, V. Strassen (Ann. Math. Statist., 1965, **36**, 423-439) proved that the Prokhorov distance $\rho(\mu, \nu)$ is the minimum distance in probability between random variables distributed according to μ and ν. Dudley generalized this result without assuming completeness of S and by using the finite combinatorial "marriage lemma". Useful bounds for the Prokhorov and the bounded Lipschitz metrics are given in this paper. Also Dudley proved that if $\beta_n \to \beta_0$ in $\mathcal{P}(S)$ then there exist random variables X_n with distributions β_n such that $X_n \to X_0$ almost surely. This was proved by A. V. Skorokhod (Theor. Prob. Appl., 1956, **1**, 261-290) to hold if S is complete. Later M. J. Wichura (Ann. Math. Statist., 1970, **41**, 284-291) and P. J. Fernandez (Bol. Soc. Brasil. Math., 1974, **5**, 51-61) proved another extension of Skorokhod's result when the metric space S may be non-separable. In this case probability measures β_n, $n \geq 1$, are defined on the σ-algebra $\mathcal{U}$ generated by all open balls of the metric space S, β_0 is a Borel probability as before, and β_n converges to β_0 in the sense Dudley defined.

Further improvement on almost surely convergent realizations of probability measures was made in the last paper of this chapter. Let $(\Omega_n, \mathcal{A}_n, P_n)$ be probability spaces for $n = 0, 1, 2, \ldots$, and let X_n a

E. Giné et al. (eds.), *Selected Works of R.M. Dudley*, Selected Works in Probability and Statistics,
DOI 10.1007/978-1-4419-5821-1_1, © Springer Science+Business Media, LLC 2010

function from Ω_n into S, where X_0 takes values in some separable subset of S and is measurable for the Borel sets on its range. Following J. Hoffmann-Jørgensen (Various Publication Series no. 39, Matematisk Institut, Aarhus Universitet, 1991) one says that X_n converges to X_0 in law if

$$\lim_{n\to\infty} \int^* f(X_n)\, dP_n = \int f(X_0)\, dP_0$$

for every bounded continuous real-valued function f on S, where the upper integral and integral are taken over Ω_n, not S as in (1), so that the laws of X_n for $n \geq 1$ need not be defined on any particular σ-algebra of the metric space S. Dudley proved that X_n converges in law to X_0 if and only if one can redefine each X_n on a new probability space in such a way that the new sequence of S-valued random variables converges almost surely (in fact, almost uniformly). He uses this theorem in particular to show that the empirical process based on a sample from P and indexed by a class of functions converges in law in the sense of Hoffmann-Jørgensen to the corresponding 'P-Brownian bridge' if and only if the class of functions is functional P-Donsker, as previously defined by Dudley (Lect. Notes in Math., 1984, **1097**, 1-142) and by Dudley and Philipp (Z. Wahrsch. verw. Geb., 1983, **62**, 509-552).

Reprinted from ILLINOIS JOURNAL OF MATHEMATICS
Vol. 10, No. 1, March 1966
Printed in U.S.A.

WEAK CONVERGENCE OF PROBABILITIES ON NONSEPARABLE METRIC SPACES AND EMPIRICAL MEASURES ON EUCLIDEAN SPACES

BY
R. M. DUDLEY[1]

1. Introduction

It is known that under certain mild set-theoretic assumptions, a finite, countably additive measure defined on all Borel sets of a metric space is concentrated in a separable subspace (Marczewski and Sikorski [8]). However, there are interesting probability measures on metric spaces not concentrated in separable subspaces. In this paper, we consider countably additive probability measures on the smallest σ-field containing the open balls of a metric space. This σ-field is the Borel field for a separable space, but is smaller in general. A probability measure on it need not be confined to a separable subspace.

A sequence of such measures will be said to converge weak* to a Borel measure if the upper and lower integrals of each bounded continuous real function converge. Some abstract results on this convergence, similar to those in Prokhorov [9] for separable metric spaces, will be given in §2.

The rest of the paper deals with "empirical measures" on Euclidean spaces, whose study motivated the abstract results and provides an application of them. Two of the main results of Donsker [3], [4] for measures on the real line will be generalized to arbitrary Euclidean spaces. At the same time, his results are corrected by replacing some integrals, which may not be defined, by upper and lower integrals.

I discovered after writing most of the rest of this paper that a generalization of Donsker's work to multidimensional spaces was proved several years ago by L. LeCam, who is now revising a paper embodying his results for the Illinois Journal of Mathematics. I shall try to explain what seem to be the main differences between our approaches.

While my abstract results in §2 are for metric spaces and guided by those in Prokhorov [9] for the separable case, LeCam uses a more elaborate abstract apparatus involving the second dual spaces of topological linear spaces and nonmetric topologies; the place of upper and lower integrals is taken by integrals with respect to finitely additive extensions of a measure.

With regard to the more concrete equicontinuity properties of empirical distribution functions, my approach in §4 below uses a sort of Markov property for the random empirical measure μ_n, namely given $\mu_n(E)$ for a set E, the

Received September 28, 1964.

[1] Partially supported by a National Science Foundation Grant and by the Office of Naval Research.

E. Giné et al. (eds.), *Selected Works of R.M. Dudley*, Selected Works in Probability and Statistics,
DOI 10.1007/978-1-4419-5821-1_2, © Springer Science+Business Media, LLC 2010

values of μ_n on subsets of E are independent of its values on sets disjoint from E. LeCam instead lets n be a random variable $n(\ )$ with a Poisson distribution and obtains a random measure with independent values on disjoint sets.

Acknowledgments. I am grateful both to Prof. LeCam for providing the above information and to V. Strassen for earlier conversations about the Skorokhod topology, etc.

2. Weak*-convergence in nonseparable metric spaces

Suppose β is a measure on a σ-field $\mathcal{S}_1$ in a space S and let F be any real-valued function on S. Then the *lower integral*

$$\int_* F(s)\, d\beta(s)$$

is defined as the supremum of all integrals

$$\int f(s)\, d\beta(s)$$

where $f \leq F$ on S, f is $\mathcal{S}_1$-measurable, and the integral of f is defined. Similarly, the upper integral

$$\int^* F\, d\beta$$

is defined as the infimum of $\int f\, d\beta$ for $f \geq F$ on S and $\int f\, d\beta$ defined. Clearly

$$\int_* F\, d\beta \leq \int^* F\, d\beta$$

for any F. If χ_A is the indicator function of a set A, let

$$\beta_*(A) = \int_* \chi_A\, d\beta, \qquad \beta^*(A) = \int^* \chi_A\, d\beta;$$

β_* and β^* are clearly the usual inner and outer measures for β.

A *Baire* set in a topological space is a member of the smallest σ-field with respect to which all continuous functions are measurable. In a metric space, the Baire sets are precisely the Borel sets. Now if β_n are measures on a topological space S (not necessarily defined on all Baire sets), and β is a Baire measure on S, we say

$$\beta_n \rightarrow \beta \quad (\text{weak}^*)$$

if for every bounded continuous real function F on S,

$$\lim_{n\to\infty} \int^* F\, d\beta_n = \lim_{n\to\infty} \int_* F\, d\beta_n = \int F\, d\beta.$$

Let (S, ρ) be a metric space, and let $\mathcal{S}$ be the σ-field generated by the balls

$$[y \in S : \rho(x, y) < \varepsilon]$$

for any $x \in S$ and $\varepsilon > 0$. Let $\mathcal{B}$ be the Borel σ-field generated by the open sets of S. In the cases of interest here, S will be non-separable and $\mathcal{B}$ strictly larger than $\mathcal{S}$.

Let $(\mathcal{C}, \| \ \|)$ be the Banach space of bounded real continuous functions on S with supremum norm. For any subset A of S and $\varepsilon > 0$ let

$$A^{\varepsilon} = [x \in S : \rho(x, y) < \varepsilon \quad \text{for some} \quad y \in A].$$

We need the following well-known fact:

LEMMA 1. *If F is a continuous real-valued function on a metric space (S, ρ), K is a compact subset of S, and $\varepsilon > 0$, then there is a $\delta > 0$ such that if $x \in K$, $y \in S$, and $\rho(x, y) < \delta$ then*

$$|F(x) - F(y)| < \varepsilon.$$

Proof. If the conclusion is false there are $x_n \in K$ and $y_n \in S$, $n = 1, 2, \cdots$, with $\rho(x_n, y_n) < 1/n$ and

$$|F(x_n) - F(y_n)| \geq \varepsilon.$$

A subsequence of the x_n converges to an $x \in K$ at which F is not continuous, a contradiction which completes the proof.

We call a set $\mathcal{K}$ of measures on S weak*-*precompact* if any sequence $[\beta_n]$ of *distinct* elements of $\mathcal{K}$ has a subsequence which is weak*-convergent (to a Borel measure on S).

THEOREM 1. *If (S, ρ) is a metric space and $\mathcal{K}$ is a set of probability measures each defined at least on $\mathcal{S}$ in S, then $\mathcal{K}$ is weak*-precompact if for every $\varepsilon > 0$ there is a compact set $K \subset S$ such that for every $\delta > 0$,*

$$\beta(K^{\delta}) \geq 1 - \varepsilon$$

for all but finitely many $\beta \in \mathcal{K}$.

Proof. Note that K^{δ} is a countable union of open balls and hence is in $\mathcal{S}$. For each positive integer N, let K_N be a compact set in S such that for any $\delta > 0$,

$$\beta(K_N^{\delta}) \geq 1 - 1/N$$

for all but finitely many $\beta \in \mathcal{K}$. Let $\{F_n\}_{n=1}^{\infty}$ be a countable set of continuous functions on S with $\| F_n \| \leq 1$ for all n, uniformly dense on K_N for each N in the continuous functions F with $\| F \| \leq 1$ (such F_n exist since $\mathcal{C}(K_N)$ is a separable Banach space for each N and we can use the Tietze extension theorem).

Given N and n, let $\delta > 0$ be such that $\rho(x, y) < \delta$ and $x \in K_N$ imply

$$|F_n(x) - F_n(y)| < 1/N.$$

Let $x_1, \cdots, x_r$ be points of K_N such that for each $x \in K_N$,

$$\rho(x_j, x) < \delta$$

for some j. For $j = 1, \cdots, r$ let A_j be the set of all $x \in S$ such that $\rho(x_i, x) \geq \delta$ for $i < j$ and $\rho(x_j, x) < \delta$. Then the sets A_j are disjoint and belong to $\mathcal{S}$; if A is their union,

$$K_N \subset A \subset K_N^\delta.$$

Since A is open, $K_N^\gamma \subset A$ for some $\gamma > 0$, so that $\beta(A) > 1 - 1/N$ for all but finitely many $\beta \in \mathcal{K}$. Let $\varepsilon = 1/N$,

$$\begin{aligned} G(x) &= F_n(x_j) - \varepsilon, & x \in A_j, j = 1, \cdots, r \\ &= -1, & x \notin A \\ H(x) &= F_n(x_j) + \varepsilon, & x \in A_j, j = 1, \cdots, r \\ &= 1, & x \notin A. \end{aligned}$$

Then $G \leq F_n \leq H$, G and H are $\mathcal{S}$-measurable, and

$$\int (H - G)\, d\beta \leq 2\varepsilon + 2\varepsilon = 4\varepsilon$$

for any $\beta \in \mathcal{K}$ with $\beta(A) > 1 - \varepsilon$.

If $\{\beta_m\}$ is any sequence of distinct elements of $\mathcal{K}$, we can find a subsequence $\{\beta_{m_r}\}$ such that

$$\int_* F_n \, d\beta_{m_r}$$

is convergent for a given n, so that

$$\limsup \int^* F_n \, d\beta_{m_r} - \lim \int_* F_n \, d\beta_{m_r} \leq 4\varepsilon.$$

Taking further subsequences and diagonalizing, we can assume this holds for all n. Letting ε tend to zero through some sequence and diagonalizing again, we get a subsequence $\{\gamma_q\}$ of $\{\beta_m\}$ such that

$$\liminf_{q\to\infty} \int_* F_n \, d\gamma_q = \limsup_{q\to\infty} \int^* F_n \, d\gamma_q$$

for all n, so that lim inf and lim sup can be replaced by lim.

Now let F be a bounded continuous function on S with $\| F \| \leq 1$. Given N, choose n so that

$$|(F - F_n)(x)| < 1/N$$

for all $x \in K_N$; then this will hold for all $x \in K_N^\delta$ for some $\delta > 0$. Then except for finitely many $\beta \in \mathcal{K}$,

$$\int_* F \, d\beta \geq \int_* F_n \, d\beta - 3/N,$$

$$\int^* F \, d\beta \leq \int^* F_n \, d\beta + 3/N.$$

Thus

$$\limsup_{q\to\infty} \int^* F \, d\gamma_q - \liminf_{q\to\infty} \int_* F \, d\gamma_q \leq 6/N.$$

Since this holds for all N, the limit $M(F)$ defined by

$$M(F) = \lim_{q\to\infty} \int^* F \, d\gamma_q = \lim_{q\to\infty} \int_* F \, d\gamma_q$$

exists. Then this clearly holds for all $F \in \mathcal{C}$ without the restriction $\| F \| \leq 1$. Clearly M is linear on $\mathcal{C}$, $M(1) = 1$, and $M(F) \geq 0$ if $F \geq 0$. If $F_n \in \mathcal{C}$ and $F_n \downarrow 0$ pointwise, then $F_n \downarrow 0$ uniformly on compact sets. Given $F \in \mathcal{C}$ with $\| F \| \leq 1$ and

$$\sup_{x \in K_N} | F(x)| \leq 1/N,$$

there is a $\delta > 0$ such that

$$\sup_{x \in K_N{}^\delta} | F(x)| \leq 2/N$$

so that

$$M(F) \leq 3/N.$$

Thus $M(F_n) \downarrow 0$.

Hence there is a nonnegative, countably additive probability measure P on S such that each $F \in \mathcal{C}$ is P-measurable with

$$M(F) = \int F \, dP.$$

Since S is a metric space, P is a Borel measure on S. Now $\gamma_q \to P$ (weak*), and the proof is complete.

PROPOSITION 1. *Suppose (S, ρ) is a complete metric space, α is a finite Borel measure on S, concentrated in a separable subspace, the α_n are defined on $\mathcal{S}$, and $\alpha_n \to \alpha$ (weak*). Then for any bounded real function F on S which is continuous almost everywhere with respect to α,*

$$\lim_{n\to\infty} \int^* F \, d\alpha_n = \lim_{n\to\infty} \int_* F \, d\alpha_n = \int F \, d\alpha.$$

Proof. We can assume that the α_n and α are all probability measures and $\| F \| = 1$. Given $\varepsilon > 0$, there is a compact set K on which F is continuous with $\alpha(K) \geq 1 - \varepsilon$. Take $\delta > 0$ such that whenever $\rho(x, y) < \delta$ and $x \in K$, $| F(x) - F(y)| < \varepsilon$.

Let G be F restricted to K. Then by the Tietze extension theorem G can be extended to a continuous function on all of S with $\| G \| \leq 1$. There is a $\gamma > 0$ such that $\rho(x, y) < \gamma$ and $x \in K$ imply

$$| G(x) - G(y)| < \varepsilon;$$

we can assume $\gamma < \delta$. For $x \in K^\gamma$,

$$| F(x) - G(x)| \leq 2\varepsilon.$$

Let f be a continuous function on S with $0 \leq f(x) \leq 1$ for all x, $f(x) = 1$ for $x \in K$, and $f(x) = 0$ for $x \notin K^\gamma$. Then

$$\int_* f\, d\alpha_n \to \int f\, d\alpha \geq 1 - \varepsilon$$

so that $\alpha_n(K^\gamma) \geq 1 - 2\varepsilon$ for n large enough, say $n \geq n_0$. Now

$$\lim_{n\to\infty} \int_* G\, d\alpha_n = \lim_{n\to\infty} \int^* G\, d\alpha_n = \int G\, d\alpha,$$

$$\left| \int F\, d\alpha - \int G\, d\alpha \right| \leq \int | F - G |\, d\alpha \leq 2\varepsilon.$$

For $n \geq n_0$,

$$\int_* F\, d\alpha_n \geq \int_* (G - 2\varepsilon)\, d\alpha_n - \alpha_n(S \sim K^\gamma)$$

$$\geq \int_* G\, d\alpha_n - 4\varepsilon,$$

$$\int^* F\, d\alpha_n \leq \int^* G\, d\alpha_n + 4\varepsilon.$$

Thus

$$\lim_{n\to\infty} \int^* F\, d\alpha_n = \lim_{n\to\infty} \int_* F\, d\alpha_n = \int F\, d\alpha, \qquad \text{Q.E.D.}$$

In proving weak*-precompactness using Theorem 1, the following is useful: suppose (C, ρ) is a metric space, let $(B, \| \ \|)$ be the Banach space of all bounded real-valued functions on C, and for ε and $\delta > 0$ let B_ε^δ be the set of all $f \in B$ such that for some $x, y \in C$, $\rho(x, y) < \delta$ and $|f(x) - f(y)| \geq \varepsilon$.

PROPOSITION 2. *Suppose (C, ρ) is compact and μ_n, $n = 1, 2, \cdots$, are probability measures on the σ-field $\mathcal{S}$ in B such that for any $\varepsilon > 0$ there are $\delta > 0$, n_0, and M such that*

$$(\mu_n)^*(B_\varepsilon^\delta) < \varepsilon \qquad \textit{for} \quad n \geq n_0,$$

and

$$\mu_n\{f : \| f \| \geq M\} < \varepsilon \qquad \textit{for all} \quad n.$$

Then for any $\varepsilon > 0$ there is a compact set K in B, consisting entirely of continuous functions, such that for any $\gamma > 0$,

$$\mu_n(K^\gamma) \geq 1 - \varepsilon$$

for n sufficiently large.

Proof. We may assume $0 < \varepsilon < 1$. Choose $M > 1$ such that

$$\mu_n\{f : \| f \| \geq M\} < \varepsilon/2$$

for all n, and for $m = 1, 2, \cdots$, $\beta > 0$, let

$$A_m^\beta = B_{\varepsilon/2^m}^\beta .$$

We choose a decreasing sequence $\{\beta_m\}$ of positive numbers satisfying the following two conditions:

(I) If $\delta_m = \beta_m \varepsilon/2^{m+1}M$, then $\delta_{m+1} < \delta_m/4$ for all m.
(II) For some sequence $\{n_0(m)\}$, $(\mu_n)^*(A_m^{\beta_m}) < \varepsilon/2^m$ for $n \geq n_0(m)$.

Then let $A_m = A_m^{\beta_m}$. Now if $\| f \| \leq M, f \notin A_j$, and $\rho(x, y) \geq \beta_j$, then

$$| f(x) - f(y) | \leq 2M \leq \frac{\varepsilon\beta_j/2^j}{\varepsilon\beta_j/2^{j+1}M} \leq \frac{\varepsilon\rho(x, y)}{2^j\delta_j},$$

while if $\rho(x, y) < \beta_j$, $|f(x) - f(y)| \leq \varepsilon/2^j$. Thus for any $x, y \in C$,

$$(*) \qquad | f(x) - f(y) | \leq \frac{\varepsilon}{2^j} \max\left(1, \frac{\rho(x, y)}{\delta_j}\right).$$

Let B_m be the set of $f \in B$ such that $\| f \| \leq M$ and $(*)$ holds for $j = 2, \cdots, m$ and all $x, y \in C$. Then

$$(\mu_n)_*(B_m) \geq 1 - \varepsilon \qquad \text{for} \quad n \geq N = N(m).$$

Now let K be the set of all $g \in B$ such that $\| g \| \leq M$ and for all j,

$$\rho(x, y) < \delta_j/2 \quad \text{implies} \quad |g(x) - g(y)| < 3\varepsilon/2^j.$$

K is a set of continuous functions, compact by Ascoli's theorem.

Given a $\gamma > 0$, choose an integer $m > 1$ such that $\varepsilon/2^m < \gamma/2$. Let us show that $B_m \subset K^\gamma$. Choose a finite set C_m of points of C such that $\rho(x, y) \geq \delta_m$ for any distinct $x, y \in C_m$, and such that for any $z \in C$, $\rho(x, z) < \delta_m$ for some $x \in C_m$ (choosing points one by one to satisfy the first condition, we end with a finite set satisfying the second condition).

If $f \in B_m$ and $x, y \in C_m$,

$$| f(x) - f(y) | \leq \frac{\varepsilon\rho(x, y)}{2^m\delta_m}.$$

Let f_m be f restricted to C_m. Then f_m can be extended to a function g on C satisfying

$$| g(x) - g(y)| \leq \varepsilon\rho(x, y)/2^m\delta_m$$

for all $x, y \in C$ (Czipszer and Geher [2]). We can assume $\| g \| \leq M$. Let us show that $g \in K$. For $j \geq m$, since $\varepsilon/2^m\delta_m \leq \varepsilon/2^j\delta_j$,

$$| g(x) - g(y)| \leq \varepsilon/2^j \qquad \text{for} \quad \rho(x, y) < \delta_j.$$

For $j < m$, given $x, y \in C$ with $\rho(x, y) < \delta_j/2$, choose $x_m, y_m \in C_m$ with $\rho(x, x_m) < \delta_m$ and $\rho(y, y_m) < \delta_m$. Then

$$\rho(x_m, y_m) < \delta_j/2 + 2\delta_m < \delta_j,$$

$$| g(x) - g(y)| \leq | g(x) - g(x_m)| + | f(x_m) - f(y_m)| + | g(y_m) - g(y)|$$
$$\leq \varepsilon/2^m + \varepsilon/2^j + \varepsilon/2^m \leq 3\varepsilon/2^j.$$

Thus $g \in K$. Finally $\| f - g \| < \gamma$ since for any $x \in C$ and $x_m \in C_m$ with

$\rho(x, x_m) < \delta_m$,

$$|f(x) - g(x)| \leq |f(x) - f(x_m)| + |g(x_m) - g(x)| \leq 2\varepsilon/2^m < \gamma.$$

Thus indeed $B_m \subset K^\gamma$, so that $\mu_n(K^\gamma) \geq 1 - \varepsilon$ for n sufficiently large, Q.E.D.

3. Empirical measures on Euclidean spaces

Let R^k be the Cartesian space of ordered k-tuples $t = \langle t_1, \cdots, t_k\rangle$ of real numbers. Suppose μ is a Borel probability measure on R^k. Let $X_1, X_2, \cdots,$ be independent R^k-valued random variables with distribution μ; specifically, let Ω be a countably infinite product of probability spaces isomorphic to (R^k, μ), with X_i as coordinate functions, and call the product measure Pr. For any $t \in R^k$, let δ_t be the unit measure at t, and let μ_n be the measure

$$(\delta_{X_1} + \cdots + \delta_{X_n})/n, \qquad n = 1, 2, \cdots.$$

Then the μ_n will be called "empirical measures" for μ. They may be thought of as approximations to μ given by a series of independent trials. Let

$$\mu^n = \sqrt{n}(\mu_n - \mu).$$

Let H be the Hilbert space $L^2(R^k, \mu)$. For any $f_1, \cdots, f_m \in H$ the multi-dimensional central limit theorem implies that the joint distribution of

$$\int f_1 \, d\mu^n, \cdots, \int f_m \, d\mu^n$$

converges as $n \to \infty$ to that of

$$L(f_1), \cdots, L(f_m)$$

where L is a linear map of H into a space of Gaussian random variables with mean zero and

$$E(L(f)L(g)) = \int \left(f - \int f \, d\mu\right)\left(g - \int g \, d\mu\right) d\mu,$$

$f, g \in H$. L is the "centered noise" r.l.f. with spectral measure μ as defined in [5, §7].

Given $t = \langle t_1, \cdots, t_k\rangle \in R^k$ let A_t be the indicator function of the set B_t of all $s \in R^k$ such that for each j,

$$t_j \leq s_j < 0 \quad \text{or} \quad 0 \leq s_j < t_j.$$

(B_t is empty if any t_j is zero.)

For $k = 1$, if μ is a nonatomic measure, the celebrated Kolmogorov-Smirnov theorems give information on the limiting behavior as $n \to \infty$ of the distribution of

$$\sup_t |\mu^n(B_t)|$$

and related quantities (see e.g. Fortet [6, Chapter 5]). It is not hard to see that, still for $k = 1$, the distributions are the same for any nonatomic μ on the half-line $t > 0$. One approach to the Kolmogorov-Smirnov results is to show that the limiting distribution is that of

$$\sup_t | L(A_t)|,$$

and that this is true not only for the supremum but for a large class of other functionals (Donsker [3], [4]). In this paper I shall extend this last result to $k > 1$ although the "invariance principle" no longer holds, i.e. the distributions depend on what continuous measure μ is chosen.

Let J be the space of all bounded real-valued functions f on R^k such that for any $t \in R^k$, $f(s) \to f(t)$ if $s_j \uparrow t_j$ for $j = 1, \cdots, k$. Then it is easy to verify that J is a Banach space with the supremum norm

$$\| f \| = \sup_{t \in R^k} | f(t)|,$$

and that if Q is any countable dense set in R^k and $f \in J$,

$$\| f \| = \sup_{t \in Q} | f(t)|.$$

It is clear that the functions V_n^μ or V_n,

$$V_n(t) = \mu^n(B_t),$$

belong to J, and that the set of such functions includes no countable dense subset for the supremum norm, even if a set of zero probability is removed, unless μ is purely atomic.

I want to show that for any real-valued function F on J which is continuous for the given norm, the limiting distribution of $F(\mu^n(B_t))$ is that of $F(L(A_t))$. The formulation requires special attention since the distribution of $\mu^n(B_t)$ will in general *not* be a Borel measure on J, and F may not be measurable for this distribution, so that $F(\mu^n(B_t))$ will not have a well-defined distribution. Nevertheless its distribution is defined with increasing precision as $n \to \infty$ and does approach that of $F(L(A_t))$ in a sense to be explained below.

The functions $\mu^n(B_t)$ for $k = 1$ have only jump discontinuities, and by introducing the "Skorokhod topology" (Skorokhod [10] and [11], Kolmogorov [7]) on the space of such functions and considering only functionals F continuous for this topology, one can avoid the imprecision in the definition of the distribution of $F(\mu^n(B_t))$ (Prokhorov [9, Theorem 2.4]). The method used here yields a larger class of functionals and easily implies the results using the Skorokhod topology. Also, no useful generalization of the Skorokhod topology to functions on R^k seems to be known.

Let Q_μ be as defined in [5, §4], i.e. the set of real-valued functions on R^k, continuous except on hyperplanes t_j=constant having positive μ-measure, and there being continuous from below with limits from above, and with limits at $\pm\infty$. Then $Q_\mu \subset J$, and Q_μ is a separable Banach space.

Let L or L^μ be the centered noise r.l.f. with spectral measure μ. Then by [5, Theorem 4.2] the functions

$$t \to L^\mu(A_t)$$

may be taken in Q_μ with probability 1. Since $(Q_\mu, \| \quad \|)$ is a separable Banach space, and a function in Q_μ is determined, for purposes of membership in an open ball and hence any set in $\mathcal{S}$, by its values on any countable dense set in R^k, there is a Borel measure α or α^μ on Q_μ such that for any $t(1), \cdots, t(n) \in R^k$ the joint probability law of $f(t(1)), \cdots, f(t(n))$, where f has distribution α, is the same as that of

$$L(A_{t(1)}), \cdots, L(A_{t(n)}).$$

α can also be regarded as a Borel measure on J since Q_μ is a closed subspace.

Now let $\alpha_n = \alpha_n^\mu$ be the probability distribution of the function $V_n^\mu = \mu^n(B_{(\quad)})$ in J for $n = 1, 2, \cdots$. Then the α_n are, under certain mild assumptions, *not* definable as (countably additive) Borel measures on J (or any other space containing all the V_n^μ, with supremum norm). For example, if μ is not purely atomic, α_1 gives positive measure to an uncountable set $\mathfrak{N}$ such that $\| f - g \| = 1$ for any $f, g \in \mathfrak{N}$ with $f \neq g$. If α_1 had a Borel extension, we would have a countably additive probability measure on all subsets of $\mathfrak{N}$ (since every subset is closed), giving points measure zero, which is impossible assuming the continuum hypothesis (Banach and Kuratowski [1]).

What we have, then, is the following: each α_n, $n = 1, 2, \cdots$, is defined by mapping W_n of an n-fold product of R^k's into J, and α_n is defined exactly on those sets A such that $W_n^{-1}(A)$ is a measurable set in the product.

Here is my main theorem on empirical measures, whose proof will be completed in §5:

THEOREM 2. *If F is a bounded real-valued function on J, continuous almost everywhere with respect to α, then*

$$\lim_{n\to\infty} \int_* F\, d\alpha_n = \lim_{n\to\infty} \int^* F\, d\alpha_n = \int F\, d\alpha.$$

In particular, $\alpha_n \to \alpha$ (weak).*

For $k = 1$, Donsker [4] asserts that under the same hypotheses

$$\int F\, d\alpha_n \to \int F\, d\alpha.$$

There is not even a measurability assumption on F away from the support of α, and an examination of his proof indicates that one has only convergence of upper and lower integrals.

In order to treat unbounded functionals such as the supremum, we have

THEOREM 3. *If F is a real-valued function on J, continuous for the supremum* *norm almost everywhere with respect to* α, *and* b *is a real number such that* $\alpha(f : F(f) = b) = 0$, *then*

$$\lim_{n\to\infty} \alpha_n^*(f : F(f) < b) = \lim_{n\to\infty} (\alpha_n)_*(f : F(f) < b) = \alpha(f : F(f) < b).$$

Proof. We use Theorem 2. Given $\varepsilon > 0$ take continuous functions g and h from the real line to the unit interval with

$$\begin{aligned} g(x) &= 1, & x &\le b - \varepsilon \\ g(x) &= 0, & x &\ge b; \\ h(x) &= 1, & x &\le b \\ h(x) &= 0, & x &\ge b + \varepsilon. \end{aligned}$$

Then we apply Theorem 2 to $g \circ F$ and $h \circ F$. Noting that

$$\int_* g \circ F \, d\alpha_n \le (\alpha_n)_*(F < b) \le \alpha_n^*(F < b)$$
$$\le \int^* h \circ F \, d\alpha_n ,$$

and

$$\int (h - g) \circ F \, d\alpha \to 0 \quad \text{as} \quad \varepsilon \downarrow 0,$$

the proof is complete.

Theorem 2 will be proved first for "continuous" measures on the unit cube. In this case, we have an "equicontinuity" result (Theorem 4, §4) which implies, using the abstract results in §2, that the α_n form a "weak*-precompact" set. The final details and the passage from a general probability measure on R^k to a continuous one on the unit cube will be given in §5.

4. "Equicontinuity"

A measure μ on R^k will be called *continuous* if each hyperplane $t_j =$ constant has measure zero. Let C be the unit cube

$$\{t : 0 \le t_j < 1, j = 1, \cdots, k\}.$$

We use the notation of Proposition 2 for this choice of C.

THEOREM 4. *Suppose* μ *is a continuous Borel probability measure on* C. *Then for any* $\varepsilon > 0$ *there is a* $\delta > 0$ *such that*

$$\Pr(V_n \in B_\varepsilon^\delta) \le \varepsilon$$

for all large enough n.

For the proof, we first mention a sort of Markov property of the random measures μ^n. Note that μ^n and μ_n are functions of each other and of μ, and for a given measurable set E, $\mu^n(E)$ has only finitely many possible values, so that the definition of conditional probabilities is elementary.

PROPOSITION 3. *Given a probability space* (S, μ) *and measurable sets*

$$E_i \subset F \subset G \subset S$$

for finitely many values of i, $\mu^n(G)$ *is conditionally independent of the* $\mu^n(E_i)$ *given* $\mu^n(F)$. *The conditional distribution of* $\mu_n(G \sim F)$ *given that* $\mu_n(F) = r/n$, *and hence also given any consistent values of* $\mu_n(E_i)$, *is exactly that of*

$$\frac{n-r}{n} \nu_{n-r}(G \sim F)$$

where ν *is* μ *restricted to* $C \sim F$ *and then normalized to total mass* 1 (*or, if* $\mu(C \sim F) = 0$, $\mu_n(G \sim F)$ *is almost surely zero*).

Proof. It suffices to show that the conditional distribution of $\mu_n(G \sim F)$, given $n\mu_n(F) = r$ and given the $\mu_n(E_i)$, is as indicated. Taking independent random variables $X_1, \cdots, X_n$ with distribution μ defining μ_n, the conditional distribution clearly does not depend on which set of r of the X_i is included in F. For a given set, since the $\mu_n(E_i)$ are independent of the complementary set, the conclusion follows by an elementary calculation.

Now to prove Theorem 4 we use induction on k. For $k = 1$, given μ, let F be its distribution function:

$$F(x) = \mu([0, x)).$$

Let λ be Lebesgue measure on $[0, 1) = I$,

$$W_n(t) = \lambda^n(B_t), \qquad V_n(t) = \mu^n(B_t).$$

Then we can put

$$V_n(t) = W_n(F(t)).$$

Since F is uniformly continuous, it suffices to prove our assertion where μ is Lebesgue measure. Here, Prokhorov [9, Lemmas 2.8 and 2.9, pp. 209–210 (original), pp. 187–188 (translation)] has proved the following: given $\beta > 0$, $\varepsilon > 0$, $s, t \in I$, let $A(\beta, \varepsilon, s, t)$ be the set of all $\omega \in \Omega$ such that

$$\min (| V_n(s - \beta) - V_n(t) |, | V_n(s + \beta) - V_n(t) |) \geq \varepsilon.$$

Then there is a $\delta > 0$ such that

$$\Pr (\omega \in A(\beta, \varepsilon, s, t) \text{ for some } \beta, s, t \text{ with } | s - t | \leq \beta \leq \delta) \leq \varepsilon.$$

Now if $\omega \notin A(\beta, \varepsilon, s, t)$ whenever $| s - t | \leq \beta \leq \delta$, and there are $s \in I$ and $\beta, 0 < \beta \leq \delta$, such that

$$| V_n(s + \beta) - V_n(s - \beta) | \geq 3\varepsilon,$$

then there is at least one jump of height ε or more in the graph of V_n between $s - \beta$ and $s + \beta$. However, for $n > 1/\varepsilon^2$, since μ_n is concentrated in n distinct points with probability one, the probability of such a jump is zero. Thus

$$P_r(V_n \in B_{3\varepsilon}^{2\delta}) \leq \varepsilon$$

for $n > 1/\varepsilon^2$, and the theorem is proved for $k = 1$.

Now suppose the conclusion is true for $1, \cdots, k - 1$. Let L_m be the cubical lattice of all points

$$\langle r_1/2^m, \cdots, r_k/2^m \rangle,$$

$r_j = 0, 1, \cdots, 2^m, j = 1, \cdots, k$. We put the lexical ordering on L_m :

$$\langle a_1, \cdots, a_k \rangle < \langle b_1, \cdots, b_k \rangle$$

if and only if there is a j such that $a_i = b_i$ for $i < j$ and $a_j < b_j$.

It suffices to prove the theorem inserting the condition that $s_j = t_j$ for all $j \neq i$ for some i, say $i = k$. We can also assume that s and t both belong to L_m for some m (of course, δ must not depend on m). This shows in particular that the probability which appears in the theorem is well-defined. Let

$$\phi_n(x) = V_n(x, 1, \cdots, 1), \qquad 0 \leq x \leq 1.$$

Let $\varepsilon > 0$ be given. By the result for $k = 1$, there is a positive integer M such that

$$\Pr(\phi_n \in B_{\varepsilon/4}^{1/2M}) \leq \varepsilon/4$$

for $n > N_1 = 144/\varepsilon^2$.

For $j = 0, 1, \cdots, 2^M - 1$ let S_j be the slab

$$j/2^M < t_1 \leq (j + 1)/2^M.$$

We assume M is large enough so that $\mu(S_j) < 1$ for all j.

By the result for $k - 1$ there are a $\delta_2 > 0$ and an $N_2 > 0$ such that

$$\Pr(|V_n(s) - V_n(t)| \geq \varepsilon/2^{M+2} \text{ for some } s, t \text{ with } |s - t| < \delta_2 \text{ and } s_1 = t_1 = 1) \leq \varepsilon/2^{M+2}$$

if $n \geq N_2$. Also, there is a δ_3, $0 < \delta_3 < \frac{1}{2}^M$, such that if λ is an arbitrary probability measure on a measurable space $(S, \mathfrak{J})$, $A \in \mathfrak{J}$, and $\lambda(A) < \delta_3$, then for all r,

$$\Pr(|\lambda^r(A)| > \varepsilon/2^{M+2}) < \varepsilon/2^{M+2}$$

(this follows e.g. from Chebyshev's inequality). Let

$$b = \min_j \mu(C \sim S_j);$$

then $b > 0$ by assumption. Choose $\delta_4 > 0$ so that

$$\sup_a \mu\{s : a \leq s_k \leq a + \delta_4\} \leq b\delta_3.$$

Let $\delta = \min(\delta_2, \delta_4)$. We shall show that this δ satisfies the desired condition. For the rest of the proof, not only ε and δ but m will be fixed, $m \geq M$.

Let $\Gamma(j, n)$ be the set of all ω such that for some s, $t \in S_j \cap L_m$,

17

$$|s - t| < \delta, \qquad s_i = t_i \qquad \text{for } i < k,$$

and

$$|\mu^n(B_s \cap S_j) - \mu^n(B_t \cap S_j)| \geq \varepsilon/2^M.$$

For $\omega \in \Gamma(j, n)$ let $s(\omega) = s(j, n, \omega)$ be the least s (lexically) in $S_j \cap L_m$ for which a t satisfying the above three conditions exists, and then let $t(\omega)$ be the least such t. Let

$$E = E(j, n, \omega) = \{u \in C : j/2^M < u_1 < s(\omega)_1\},$$

and for any $s \in C$ let

$$P(s) = (1, s_2, \cdots, s_k).$$

Then, since $\delta \leq \delta_2$,

$$\Pr[|V_n(P(s(\omega))) - V_n(P(t(\omega)))| \geq \varepsilon/2^{M+2}] \leq \varepsilon/2^{M+2}$$

for any j and n (a set defined by a condition on $s(\omega)$ and $t(\omega)$ will be regarded as a subset of $\Gamma(j, n)$, their domain of definition).

Let $F = F(j, n, \omega)$ be the set

$$(B_{P(t(\omega))} \sim B_{P(s(\omega))}) \sim (B_{t(\omega)} \cap S_j).$$

Then F is disjoint from E, and E is determined by F, $E = E(F)$.

The conditional distribution of $\mu^n(F)$ given a value F_0 of F and given $\mu^n(E)$ is the same as the conditional distribution of $\mu^n(F_0)$ given $\mu^n(E(F_0))$, by Proposition 3, since knowing that $F = F_0$ in addition yields information only on values of μ_n on subsets of $E(F_0)$.

Let $\Omega_1 = \Omega_1(n)$ be the set of ω with ($\omega \in \Gamma(j, n)$ and)

$$|\mu^n(E(j, n, \omega))| > \varepsilon/4$$

for some j. Then for $n \geq N_1$, $\Pr(\Omega_1(n)) \leq \varepsilon/4$. Let $G = C \sim E$ and $r = n\mu_n(G)$.

For given values of j, m, n, F and r, the conditional distribution of $n\mu_n(F)$ is exactly that of $r\lambda_r(F)$, where λ is μ restricted to G and normalized to mass 1, by Proposition 3. Since $\mu(F)/\mu(G) \leq \delta_3$,

$$\Pr\left\{\left|\frac{n\mu_n(F)}{\sqrt{r}} - \sqrt{r}\,\frac{\mu(F)}{\mu(G)}\right| \geq \varepsilon/2^{M+2} \,\middle|\, F, r\right\} \leq \varepsilon/2^{M+2}.$$

Thus if $\Omega_2(j, n)$ is the set of $\omega \in \Gamma(j, n)$ with

$$\left|\frac{n\mu_n(F)}{\sqrt{r}} - \frac{\sqrt{r}\,\mu(F)}{\mu(G)}\right| \geq \varepsilon/2^{M+2},$$

then $\Pr(\Omega_2(j, n)) \leq \varepsilon/2^{M+2}$.

Let $N = \max(N_1, N_2)$. We fix $n \geq N$ and suppress "n" in some notations.

Now note that

$$\mu^n(F) = \sqrt{n}(\mu_n(F) - \mu(F))$$
$$= \frac{\sqrt{r}}{\sqrt{n}}\left(\frac{n\mu_n(F)}{\sqrt{r}} - \frac{\sqrt{r}\mu(F)}{\mu(G)}\right) + \left(\frac{r\mu(F)}{\sqrt{n}\,\mu(G)} - \sqrt{n}\,\mu(F)\right),$$

and

$$\frac{r\mu(F)}{\sqrt{n}\,\mu(G)} - \sqrt{n}\,\mu(F) = \frac{\mu(F)}{\mu(G)}\,\mu^n(G) = -\frac{\mu(F)}{\mu(G)}\,\mu^n(E).$$

For $\omega \in \Gamma(j)$ but neither in Ω_1 nor in $\Omega_2(j)$, $r \leq n$ and $\mu(F)/\mu(G) \leq \frac{1}{2}^M$ imply

$$|\mu^n(F)| \leq 2\varepsilon/2^{M+2} = \varepsilon/2^{M+1},$$

and hence

$$|V_n(P(t(\omega))) - V_n(P(s(\omega)))| \geq \varepsilon/2^{M+1}.$$

If $\Omega_3(j)$ is the set of $\omega \in \Gamma(j)$ for which the latter inequality holds, then since $\delta \leq \delta_2$,

$$\Pr(\Omega_3(j)) < \varepsilon/2^{M+2}.$$

Thus for each j,

$$\Pr(\Gamma(j) \sim \Omega_1) \leq \Pr(\Omega_2(j)) + \Pr(\Omega_3(j)) \leq \varepsilon/2^{M+1}.$$

Now if $s, t \in C$, $|s - t| < \delta$, and $s_j = t_j$ for $j < k$, then $V_n(s) - V_n(t)$ is the sum of at most 2^M terms

$$\mu^n(B_{f(s,q)} \cap S_q) - \mu^n(B_{f(t,q)} \cap S_q)$$

where for any $u \in C$, $f(u, q) \in S_q$,

$$\bar{B}_u \cap S_q = \bar{B}_{f(u,q)} \cap S_q,$$

and for $u \in L_m$ and $m \geq M$, $f(u, q) \in L_m$. Thus

$$|f(s, q) - f(t, q)| < \delta$$

for each q, and the probability that at least one of the 2^M terms exceeds $\varepsilon/2^{M+1}$ in absolute value is less than

$$\Pr(\Omega_1) + \varepsilon/2 < \varepsilon.$$

Thus for $n \geq N$,

$$\Pr(|V_n(s) - V_n(t)| \geq \varepsilon \text{ for some } s, t \in C \text{ with } s_j = t_j \text{ for } j < k \text{ and } |s - t| < \delta) < \varepsilon.$$

For any $\delta > 0$ and $s, t \in R^k$ with $|s - t| < \delta$, there are $u, v \in C$ with $|u - v| < \delta$ and $V_n(s) \equiv V_n(u)$, $V_n(t) \equiv V_n(v)$. Thus the condition "$s, t \in C$" can be removed and the proof of Theorem 4 is complete.

5. Proof of Theorem 1

We first complete the proof assuming μ is continuous and concentrated in C. In this case, Theorems 1 and 4 and Proposition 2 show that the measures α_n

on J form a weak*-precompact set. Now suppose

$$\alpha_{n_m} \to \beta \quad (\text{weak}^*)$$

for some subsequence $\{\alpha_{n_m}\}$ of $\{\alpha_n\}$. Then for any $t^{(1)}, \cdots, t^{(n)} \in C$ the joint distribution of $f(t^{(1)}), \cdots, f(t^{(n)})$ for f distributed according to β is the same as for f distributed according to α, by the central limit theorem convergence mentioned in §3. Thus $\beta = \alpha$.

Suppose there is a bounded continuous function F on $(J, \| \quad \|)$ and an $\varepsilon > 0$ such that

$$\max \left(\left| \int^* F \, d\alpha_n - F \, d\alpha \right|, \quad \left| \int F \, d\alpha - \int_* F \, d\alpha_n \right| \right) \geq \varepsilon$$

for an infinite sequence of values of n. Then taking a subsequence n_m such that α_{n_m} is weak* convergent, we have a contradiction. Thus

$$\alpha_n \to \alpha \quad (\text{weak}^*).$$

Applying Proposition 1, §2 the proof for μ continuous on C is complete.

The general case of a probability measure on R^k is easily reduced to that of a measure on the cube

$$-1 < t_j < 1, \qquad j = 1, \cdots, k,$$

by the transformation

$$t_j \to (2/\pi) \text{ arc tan } t_j, \qquad j = 1, \cdots, k,$$

which preserves all the structure we need.

Let E_t be the indicator function of the set

$$\{s : -1 \leq s_j < t_j, j = 1, \cdots, k\}$$

Then for $t_j \geq -1, j = 1, \cdots, k$,

$$A_t = \textstyle\sum_F S(F) E_{P(t,F)}$$

where F runs over all finite subsets of $(1, \cdots, k)$, $S(F) = (-1)^r$ where $r = k - \#(F)$ and $\#(F)$ is the number of elements in F, and $P(t, F) = s$ with $s_j = \max(t_j, 0)$ for $j \in F$ and $s_j = \min(t_j, 0)$ otherwise. Since the functions $t \to P(t, F)$ are uniformly continuous for each F, results for the E_t, obtained by an obvious linear transformation, yield corresponding results for the A_t. Thus we have reduced to the case of a measure on C.

In [5, §4] the following was proved: if μ is any probability measure on C, there is a continuous measure ν on some rectangular solid

$$C_1 : 0 \leq t_j < a_j, \qquad j = 1, \cdots, k,$$

and a continuous mapping G of the closure $\bar{C}_1$ of C_1 onto $\bar{C}$, of the form

$$G(t_1, \cdots, t_k) = \langle G_1(t_1), \cdots, G_k(t_k) \rangle,$$

with each G_j (weakly) monotone increasing, such that $\nu \circ G^{-1} = \mu$. Clearly we can assume $a_j \equiv 1$, $C_1 = C$, and extend each G_j to a monotone function from $[0, \infty)$ *onto* itself.

If X_n, $n = 1, 2, \cdots$, are independent with distribution ν then $G(X_n)$ are independent with distribution μ, so we can put

$$\mu^n = \nu^n \circ G^{-1}.$$

For $j = 1, \cdots, k$, $0 \leq t_j \leq 1$, let

$$F_j(t_j) = \inf (s : G_j(s) = t_j),$$

and let $F(t) = \langle F_1(t_1), \cdots, F_k(t_k)\rangle$.

Now for any j, s_j, t_j we have $G_j(s_j) < t_j$ if and only if $s_j < F_j(t_j)$. Thus $A_t = A_s \circ G$ if and only if $F(s) = t$, and for any $s \in \bar{C}$,

$$V_n^{\mu}(s) = \int A_s(v)\, d\mu^n(v) = \int A_s(G(u))\, d\nu^n(u)$$

$$= \int A_{F(s)}(\mu)\, d\nu^n(u) = V_n^{\nu}(F(s)).$$

Now for any continuous function f on $\bar{C}$, $F^*(f) = f \circ F \in Q_\mu$ since $F_j \in Q_\mu$ for each j. The map $F^* : f \to f \circ F$ is linear with norm one. Since

$$\alpha_n^{\mu} = \alpha_n^{\nu} \circ (F^*)^{-1} \quad \text{and} \quad \alpha_n^{\nu} \to \alpha^{\nu}\ (\text{weak}^*),$$

we have $\alpha_n^{\mu} \to \alpha^{\mu}$ (weak*). Then applying Proposition 1 of §2, the proof of Theorem 2 is complete.

Note. The supremum norm $N = \| \ \|$ is an $\mathcal{S}$-measurable function. Letting $s \in C_t$ if and only if $s_j \leq t_j$, $j = 1, \cdots, k$, and letting $f_n(t) = \mu^n(C_t)$, it is clear that $N(f_n)$ has a well-defined distribution for each n. For $k = 1$ and μ continuous, it was shown by Kolmogorov (see e.g. [6, Chapter 5]) that the limiting distribution of $N(f_n)$ as $n \to \infty$ is

$$\Pr (N < z) = K(z) = 1 + 2\textstyle\sum_{r=1}^{\infty}(-1)^r \exp(-2r^2z^2).$$

On a k-fold product of real lines for $k > 1$,

$$R^k = \textstyle\prod_{j=1}^{k} R_j,$$

the limiting distribution of N will not be the same for all continuous measures μ. However, it will be the same for all product measures

$$\mu = \textstyle\prod_{j=1}^{k} \mu_j,$$

where μ_j is continuous on R_j for each j. In this case,

$$\mu^n(C_t) = \textstyle\prod_{j=1}^{k} \mu_j^n(C_{t_j}),$$

the μ_j^n are independent for different j, and if f_j is a function of t_j only,

$$N(\textstyle\prod_{j=1}^{k} f_j) = \textstyle\prod_{j=1}^{k} N(f_j).$$

Thus we have a product of n independent random variables with the same distribution function K. I have not found the distribution of this product.

Although the function(al) S defined by

$$S(f) = \sup_t f(t)$$

has a simpler limiting distribution for a continuous measure in one dimension,

$$\Pr(S < z) = 1 - \exp(-2z^2)$$

[6, Chapter 5], the situation is more complicated in several dimensions because

$$S\left(\prod_{j=1}^{k} f_j(t_j)\right)$$

depends on the infima as well as the suprema of the f_j.

References

1. S. Banach and C. Kuratowski, *Sur une géneralisation du problème de la mesure*, Fund. Math., vol. 14 (1929), pp. 127–131.
2. J. Czipszer and L. Geher, *Extension of functions satisfying a Lipschitz condition*, Acta Math. Acad. Sci. Hungar., vol. 6 (1955), pp. 213–220.
3. M. D. Donsker, *An invariance principle for certain probability limit theorems*, Mem. Amer. Math. Soc., no. 6 (1951).
4. ———, *Justification and extension of Doob's heuristic approach to the Kolmogorov-Smirnov theorems*, Ann. Math. Statist., vol. 23 (1952), pp. 277–281.
5. R. M. Dudley, *Gaussian processes on several parameters*, Ann. Math. Statist., vol. 36 (1965), pp. 771–788.
6. R. Fortet, *Recent advances in probability theory*, Some aspects of analysis and probability, New York, Wiley, 1958, pp. 171–240.
7. A. N. Kolmogorov, *On the Skorokhod convergence*, Theory of probability and its applications, vol. 1 (1956), pp. 239–247 (Russian), pp. 215–222 (English translation).
8. E. Marczewski and P. Sikorski, *Measures in nonseparable metric spaces*, Colloq. Math., vol. 1 (1948), pp. 133–139.
9. Yu. V. Prokhorov, *Convergence of random processes and limit theorems in probability*, Theory of probability and its applications, vol. 1, (1956), pp. 177–238 (Russian), pp. 151–214 (English translation).
10. A. V. Skorokhod, *On passage to the limit from sequences of sums independent random variables to a homogeneous random process with independent increments*, Dokl. Akad. Nauk SSR, vol. 104 (1955), pp. 364–367 (in Russian).
11. ———, *On a class of limit theorems for Markov chains*, Dokl. Akad. Nauk SSR, vol. 106 (1956), pp. 781–784 (in Russian).

University of California
Berkeley, California

Reprinted from ILLINOIS JOURNAL OF MATHEMATICS
Vol. 11, No. 3, September 1967
Printed in U.S.A.

MEASURES ON NON-SEPARABLE METRIC SPACES

BY
R. M. DUDLEY[1]

1. Introduction

The main purpose of this note is to give a simpler and more general definition of "weak" or "weak-star" convergence of certain measures on non-separable metric spaces, and to prove its equivalence with the convergence introduced in [1] for the cases considered there.

Let (S, d) be a metric space. Let $\mathcal{B}$ or $\mathcal{B}(S)$ be the class of all Borel sets in S, i.e. the smallest σ-algebra containing all the open sets. One can safely assume that a finite, countably additive measure on $\mathcal{B}$ is concentrated in a separable subset [2]. It has seemed useful to consider finite, countably additive measures on metric spaces, not concentrated in separable subsets, defined on some, but not all, Borel sets [1]. Specifically, one can use the σ-algebra $\mathcal{U}$ or $\mathcal{U}(S)$ generated by the open balls

$$B(x, \varepsilon) = \{y \in S : d(x, y) < \varepsilon\}$$

for arbitrary x in S and $\varepsilon > 0$. Examples of finite measures on $\mathcal{U}$ not concentrated in separable subsets are the probability distributions of distribution functions of "empirical measures" [1]. For a simpler example, let S be uncountable and $d(x, y) = 1$ for $x \neq y$. Then $\mathcal{U}$ consists of countable sets, which we give measure 0, and sets with countable complement, which we give measure 1.

If S is separable, then all open sets are in $\mathcal{U}$ by the Lindelöf theorem, hence $\mathcal{U} = \mathcal{B}$. I don't know whether $\mathcal{U}$ is always strictly included in $\mathcal{B}$ for S non-separable, but it is in the cases mentioned above, and under the following conditions:

PROPOSITION. *Suppose that the smallest cardinal of a dense set in S is c (cardinal of the continuum). Then $\mathcal{U}$ has cardinal c and $\mathcal{B}$ has cardinal 2^c. Hence $\mathcal{U}$ is strictly included in $\mathcal{B}$.*

Proof. Let A be a dense set in S of cardinal c. Let G be the class of balls $B(x, r)$ with x in A and r (positive) rational. We show that G generates $\mathcal{U}$. Let $x \in S$, $r > 0$. Let $x_n \in A$, $x_n \to x$. We can assume $d(x_n, x) < r$ for all n. Let r_n be positive rational numbers such that $r_n \to r$ and $r_n < r - d(x_n, x)$ for all n. Then

$$B(x, r) = \bigcup_{n=1}^{\infty} B(x_n, r_n),$$

showing that G generates $\mathcal{U}$.

Received July 1, 1966.

[1] This research was supported in part by a National Science Foundation grant, and was presented to the International Congress of Mathematicians in Moscow, August 1966.

E. Giné et al. (eds.), *Selected Works of R.M. Dudley*, Selected Works in Probability and Statistics,
DOI 10.1007/978-1-4419-5821-1_3, © Springer Science+Business Media, LLC 2010

Let ω be the cardinal of the set of all integers. Then G has cardinal at most ωc, and $\omega c = c$. Hence the class of complements of sets in G has cardinal at most c. The class of countable unions of elements of G has cardinal at most equal to c^ω, and

$$c^\omega = (2^\omega)^\omega = 2^{\omega^2} = 2^\omega = c.$$

Using transfinite induction, we obtain that the cardinal of $\mathfrak{U}$ is at most $\aleph_1 c$ (where $\aleph_1$ is the least uncountable cardinal; we are assuming the axiom of choice, but not the continuum hypothesis). Now $\aleph_1 c = c$. Since each one-point set in S clearly belongs to $\mathfrak{U}$, the cardinal of $\mathfrak{U}$ is exactly c. The cardinal of $\mathfrak{B}$ is exactly 2^c [3, Remark 3.7 p. 106] and $c < 2^c$. Thus $\mathfrak{U}$ is properly included in $\mathfrak{B}$, q.e.d.

If in the statement of the above proposition we replace c by another uncountable cardinal α, then the proof goes through except that possibly $\alpha < \alpha^\omega$, which will happen e.g. if $\alpha = \aleph_\omega$ [3, p. 100], but not if $\alpha = 2^\beta$ for some (infinite) β. When $\mathfrak{U}$ and $\mathfrak{B}$ have the same cardinal, it remains unclear whether they are equal.

It should be noted that the σ-algebras $\mathfrak{U}$ in non-separable metric spaces have certain unpleasant properties. For example, they are not always preserved by homeomorphisms or even by uniform isomorphisms. Also, they are not always preserved by "relativization" to a subset of S with the same metric. Finally, if one takes a cartesian product of two metric spaces S and T, with any of the usual metrics for the product topology, $\mathfrak{U}(S \times T)$ may not even contain all "rectangles" $A \times B$ where $A \in \mathfrak{U}(S)$, $B \in \mathfrak{U}(T)$.

The Borel σ-algebras are superior in all these respects, although $\mathfrak{B}(S \times T)$ may not be generated by the rectangles whose sides are Borel sets. Of course, the Borel σ-algebras are generally too large to carry a finite measure with non-separable support. One might hope for a σ-algebra which, like $\mathfrak{U}$, would allow such measures, but which had better "functorial" properties.

2. Measures on $\mathfrak{U}$

Let $M(S, \mathfrak{U})$ be the set of all finite, countably additive, real-valued set functions (signed measures) on $\mathfrak{U}$, $M^+(S, \mathfrak{U})$ the set of elements of $M(S, \mathfrak{U})$ with nonnegative values, and $P(S, \mathfrak{U})$ the set of elements of $M^+(S, \mathfrak{U})$ with total mass 1 (probability measures).

In [1], "weak-star" convergence of a sequence in $M^+(S, \mathfrak{U})$ to a Borel measure μ was defined as convergence of the upper and lower integrals of every bounded continuous function f to $\int f\, d\mu$. Here we define a natural convergence in $M(S, \mathfrak{U})$ and prove that if S is complete, the new convergence agrees with the old one whenever the latter is defined (if μ has separable support, which, as noted above, practically follows from μ being a Borel measure).

Let $\mathfrak{C}(S)$ be the Banach space of all bounded, continuous, real-valued functions on S with supremum norm $\| \quad \|_\infty$. Let $C(S, \mathfrak{U})$ be the closed linear subspace of $\mathcal{U}$-measurable elements of $\mathcal{C}(S)$. Then any μ in $M(S, \mathcal{U})$ defines a bounded linear functional

$$f \to \int f\, d\mu$$

on $\mathcal{C}(S, \mathcal{U})$. Then on $M(S, \mathcal{U})$, we have the "weak-star" topology of pointwise convergence on $\mathcal{C}(S, \mathcal{U})$. (Note that $M(S, \mathcal{U})$ is a proper subset of the dual space $\mathcal{C}(S, \mathcal{U})^*$ unless S is compact.)

Given a real-valued function f and a measure μ we define the usual upper and lower integrals:

$$\int^* f\, d\mu = \inf\left\{\int h\, d\mu : h \geq f, \int h\, d\mu \text{ defined}\right\},$$

$$\int_* f\, d\mu = \sup\left\{\int g\, d\mu : g \leq f, \int g\, d\mu \text{ defined}\right\}.$$

THEOREM. *Suppose (S, d) is a complete metric space, $\{\mu_n\}$ is a sequence of elements of $M^+(S, \mathcal{U})$ and μ in $M^+(S, \mathcal{U})$ is concentrated in a separable subspace. Then $\mu_n \to \mu$ for the weak-star topology on $M(S, \mathcal{U})$ if and only if*

$$\lim_{n\to\infty} \int^* f\, d\mu_n = \lim_{n\to\infty} \int_* f\, d\mu_n = \int f\, d\mu$$

for every f in $\mathcal{C}(S)$.

Proof. "If" holds since the upper and lower integrals of functions in $\mathcal{C}(S, \mathcal{U})$ are integrals.

To prove "only if", suppose $\mu_n \to \mu$ on $\mathcal{C}(S, \mathcal{U})$ and f is in $\mathcal{C}(S)$. Since μ has separable support it has a natural extension to all Borel sets. We may assume $\| f \|_\infty \leq 1$ and $\mu_n(S) \leq 1$ for all n. Let ε be given, $0 < \varepsilon < 1$. By Ulam's theorem [4], there is a compact set K such that $\mu(S \sim K) < \varepsilon$. Choose $\delta > 0$ so that $d(x, y) < \delta$ and x in K imply $| f(x) - f(y) | < \varepsilon$. Let C be countable and dense in K. Let

$$d(y, K) = \inf_{x \in K} d(x, y) = \inf_{x \in C} d(x, y).$$

Then $d(\cdot, K)$ is $\mathcal{U}$-measurable and continuous (in fact,

$$| d(y, K) - d(z, K) | \leq d(y, z)$$

for all y and z). Let

$$g(y) = \min(1, 4d(y, K)/\delta).$$

Then $g \in \mathcal{C}(S, \mathcal{U})$, so

$$\int g\, d\mu_n \to \int g\, d\mu < \varepsilon.$$

Let F be a finite subset of K such that for any x in K, $d(x, z) < \delta/4$ for some z in F. Let

$$\phi(t) = \varepsilon t/\delta, \quad 0 \leq t \leq \delta/2$$

$$= 2, \qquad t \geq \delta$$

and let ϕ also be linear in the interval $[\delta/2, \delta]$. Let

$$u(x) = \min\,(1, \min\,(f(z) + \varepsilon + \phi(d(x, z)) : z \in F)),$$
$$v(x) = \max\,(-1, \max\,(f(z) - \varepsilon - \phi(d(x, z)) : z \in F)).$$

Then clearly $u, v \in \mathcal{C}(S, \mathfrak{U})$. Let

$$W = \{x : d(x, w) < \delta/4 \text{ for some } w \text{ in } K\}.$$

For any x in W, $d(x, z) < \delta/2$ for some z in F, so

$$|f(x) - f(z)| < \varepsilon \quad \text{and} \quad \phi(d(x, z)) < \varepsilon.$$

Thus

$$u(x) \leq f(z) + 2\varepsilon \leq f(x) + 3\varepsilon.$$

Given x, let G_x be the set of all z in F such that $d(x, z) < \delta$. Then $f(x) \leq f(z) + \varepsilon$ for all z in G_x, while for z in $F \sim G_x$, $\phi(d(x, z)) = 2$. Thus $f(x) \leq u(x)$ for all x in W. Likewise

$$f(x) \geq v(x) \geq f(x) - 3\varepsilon$$

for all x in W. Now since $W \in \mathfrak{U}$,

$$\int^* f\, d\mu_n \leq \int_W u\, d\mu_n + \mu_n(S \sim W),$$

$$\int_* f\, d\mu_n \geq \int_W v\, d\mu_n - \mu_n(S \sim W),$$

$$\limsup \int^* f\, d\mu_n \leq \limsup \int_W u\, d\mu_n + \varepsilon,$$

$$\liminf \int_* f\, d\mu_n \geq \liminf \int_W v\, d\mu_n - \varepsilon,$$

and

$$\limsup \int_W (u - v)\, d\mu_n \leq 6\varepsilon,$$

so

$$\limsup \int^* f\, d\mu_n - \liminf \int_* f\, d\mu_n \leq 8\varepsilon.$$

Since an upper integral is greater than a lower integral of the same function, the limits of $\int^* f\, d\mu_n$ and $\int_* f\, d\mu_n$ exist and are equal. These limits are also approached by $\int_W f\, d\mu$ as $\varepsilon \to 0$ (of course, W depends on ε), thus they equal $\int f\, d\mu$, q.e.d.

References

1. R. M. Dudley, *Weak convergence of probabilities on nonseparable metric spaces and empirical measures on Euclidean spaces*, Illinois J. Math., vol. 10 (1966), pp. 109–126.

2. E. Marczewski and P. Sikorski, *Measures in nonseparable metric spaces*, Colloq. Math., vol. 1 (1948), pp. 133–139.
3. A. H. Stone, *Cardinals of closed sets*, Mathematika, vol. 6 (1959), pp. 99–107.
4. S. Ulam and J. C. Oxtoby, *On the existence of a measure invariant under a transformation*, Ann. of Math., vol. 40 (1939), pp. 560–566.

University of California
Berkeley, California
Massachusetts Institute of Technology
Cambridge, Massachusetts

The Annals of Mathematical Statistics
1968, Vol. 39, No. 5, 1563–1572

DISTANCES OF PROBABILITY MEASURES AND RANDOM VARIABLES

BY R. M. DUDLEY[1]

Massachusetts Institute of Technology

1. Introduction. Let (S, d) be a separable metric space. Let $\mathcal{P}(S)$ be the set of Borel probability measures on S. $\mathcal{C}(S)$ denotes the Banach space of bounded continuous real-valued functions on S, with norm

$$\|f\|_\infty = \sup \{|f(x)| : x \in S\}.$$

On $\mathcal{P}(S)$ we put the usual weak-star topology TW^*, the weakest such that

$$P \to \int f\, dP, \quad P \in \mathcal{P}(S)$$

is continuous for each $f \in \mathcal{C}(S)$.

It is known ([8], [11], [1]) that TW^* on $\mathcal{P}(S)$ is metrizable. The main purpose of this paper is to discuss and compare various metrics and uniformities on $\mathcal{P}(S)$ which yield the topology TW^*.

For S complete, V. Strassen [10] proved the striking and important result that if $\mu, \nu \in \mathcal{P}(S)$, the Prokhorov distance $\rho(\mu, \nu)$ is exactly the minimum distance "in probability" between random variables distributed according to μ and ν. Theorems 1 and 2 of this paper extend Strassen's result to the case where S is measurable in its completion, and, with "minimum" replaced by "infimum", to an arbitrary separable metric space S. We use the finite combinatorial "marriage lemma" at the crucial step in the proof rather than the separation of convex sets (Hahn-Banach theorem) as in [10]. This offers the possibility of a constructive method of finding random variables as close as possible with the given distributions.

For S complete, V. Skorokhod ([9], Theorem 3.1.1, p. 281) proved the related result that if $\mu_n \to \mu_0$ for TW^* there exist random variables X_n with distributions μ_n such that $X_n \to X_0$ almost surely. This is proved in Section 3 below for a general separable S. Note that it is not sufficient to establish consistent finite-dimensional joint distributions for the X_n ; the Kolmogorov existence theorem for stochastic processes is not available in this generality. Instead we construct the joint distribution of $\{X_n\}_{n=0}^\infty$ out of suitable infinite Cartesian product measures.

When S is the real line R, various special constructions involving distribution and characteristic functions are known. In Section 4, we compare some of these uniformities on $\mathcal{P}(R)$.

2. Strassen's theorem. The metric of Prokhorov [8] is defined as follows. For any $x \in S$ and $T \subset S$ let

$$d(x, T) = \inf (d(x, y) : y \in T),$$

Received 11 January 1968.
[1] Fellow of the A. P. Sloan Foundation.

E. Giné et al. (eds.), *Selected Works of R.M. Dudley*, Selected Works in Probability and Statistics, DOI 10.1007/978-1-4419-5821-1_4, © Springer Science+Business Media, LLC 2010

and for $\delta \geqq 0$ let

$$T^{\delta} = \{x \,\varepsilon\, S: d(x, T) < \delta\},$$
$$T^{\delta]} = \{x \,\varepsilon\, S: d(x, T) \leqq \delta\}.$$

Given P and Q in $\mathcal{P}(S)$ let

$$(1)\quad \begin{aligned} \sigma(P, Q) &= \inf\,(\epsilon > 0: P(F) \leqq Q(F^{\epsilon}) + \epsilon \quad \text{for all closed} \quad F \subset S), \\ \rho(P, Q) &= \max\,(\sigma(P, Q), \sigma(Q, P)). \end{aligned}$$

Then ρ is a metric and metrizes TW^* on $\mathcal{P}(S)$ (this was proved in [8], Section 1.4, for S complete and is established for general separable S by results toward the end of this section).

We may replace F^{ϵ} by $F^{\epsilon]}$ in the definition of σ without changing its value. Also we may replace "all closed F" by "all Borel sets B" since if F is the closure of B, $F^{\epsilon} = B^{\epsilon}$ and $F^{\epsilon]} = B^{\epsilon]}$.

PROPOSITION 1 (known to Strassen [10]). *If P, $Q \,\varepsilon\, \mathcal{P}(S)$ and α, $\beta > 0$, then $P(F) \leqq Q(F^{\alpha}) + \beta$ for all closed F if and only if the same conditions hold with P and Q interchanged. Thus $\sigma(P, Q) = \sigma(Q, P) = \rho(P, Q)$.*

PROOF. Suppose $P(F) \leqq Q(F^{\alpha}) + \beta$ for all closed F and let T be closed. Then T^{α} is open,

$$T \subset (S \sim (S \sim T^{\alpha})^{\alpha}), \quad \text{and}$$
$$P(S \sim T^{\alpha}) \leqq Q((S \sim T^{\alpha})^{\alpha}) + \beta, \quad \text{so}$$
$$Q(T) \leqq Q(S \sim (S \sim T^{\alpha})^{\alpha}) \leqq P(T^{\alpha}) + \beta.$$

The conclusions follow.

Let $(\Omega, \mathcal{B}, \Pr)$ be a probability space and let $\mathfrak{F}(\Omega, S)$ be the set of S-valued random variables over Ω, i.e. functions from Ω to S, measurable from $\mathcal{B}$ to the Borel σ-algebra in S, modulo functions vanishing with probability 1.

Then the natural topology of convergence in probability in $\mathfrak{F}(\Omega, S)$ is metrized by the metric

$$d_{\Pr}(f, g) = \inf\,(\epsilon > 0: \Pr\,(d(f(\omega), g(\omega)) \geqq \epsilon) < \epsilon).$$

Now $f \times g: \omega \to (f(\omega), g(\omega))$ maps Ω measurably into $S \times S$ and defines an element $\Pr \circ (f \times g)^{-1}$ of $\mathcal{P}(S \times S)$. On $S \times S$ let π_1 and π_2 be the natural projections onto S:

$$\pi_1(x, y) \equiv x, \; \pi_2(x, y) \equiv y.$$

THEOREM 1. *Let S be a separable metric space, P, $Q \,\varepsilon\, \mathcal{P}(S)$, $\alpha \geqq 0$ and $\beta \geqq 0$. Then the following are equivalent:*

(I) *$P(T) \leqq Q(T^{\alpha]}) + \beta$ for all closed $T \subset S$;*

(II) *For any $\epsilon > 0$ there is a μ in $\mathcal{P}(S \times S)$ with $\mu \circ \pi_1^{-1} = P$, $\mu \circ \pi_2^{-1} = Q$, and $\mu(d(x, y) > \alpha + \epsilon) \leqq \beta + \epsilon$.*

PROOF. First assume (II). Then for any $\epsilon > 0$ and closed $T \subset S$, $P(T) \leqq Q(T^{\alpha+\epsilon}) + \beta + \epsilon$. Letting $\epsilon \downarrow 0$, this yields (I).

Conversely, assume (I). Given $\epsilon > 0$, let $\gamma = \epsilon/9$.

Let $\{x_n\}$ be a dense sequence in S. For any $x \varepsilon S$ let $f(x) = x_n$ for the least n such that $d(x, x_n) < \gamma$. Let $P_\gamma = P \circ f^{-1}$, $Q_\gamma = Q \circ f^{-1}$. Let $H_n = \{x_1, x_2, \cdots, x_n\}$ and choose n so that

$$\min (P_\gamma(H_{n-1}), Q_\gamma(H_{n-1})) > 1 - \gamma.$$

Then choose an integer m so that $n < m\gamma$. Let $P' \varepsilon \mathcal{P}(H_n)$ be such that for $i = 1, \cdots, n-1$, $mP'(x_i)$ is the largest integer $\leqq mP_\gamma(x_i)$. Likewise construct Q' from Q_γ. Then for any set $T \subset S$,

$$\max (|(Q' - Q_\gamma)(T)|, |(P' - P_\gamma)(T)|) \leqq 2\gamma,$$

$$P'(T) \leqq P_\gamma(T) + 2\gamma \leqq P(T^\gamma) + 2\gamma \leqq Q(T^{\gamma+\alpha]}) + 2\gamma + \beta$$

$$\leqq Q_\gamma(T^{2\gamma+\alpha}) + 2\gamma + \beta \leqq Q'(T^{2\gamma+\alpha}) + 4\gamma + \beta,$$

$$P'(T) \leqq Q'(T^{2\gamma+\alpha}) + r/m,$$

where r is the largest integer $\leqq m(4\gamma + \beta)$.

Let I be the unit interval [0, 1] with Lebesgue measure λ. On the Cartesian product $S \times I$ we form the product measures $P \times \lambda$ and $Q \times \lambda$. Let X be the natural projection of $S \times I$ on S.

For $i = 1, \cdots, n-1$ we select measurable subsets E_i and F_i of $(f \circ X)^{-1}(x_i)$ such that

$$(P \times \lambda)(E_i) = P'(x_i), \qquad (Q \times \lambda)(F_i) = Q'(x_i).$$

Let

$$E_n = (S \times I) \sim (E_1 \cup \cdots \cup E_{n-1}),$$

$$F_n = (S \times I) \sim (F_1 \cup \cdots \cup F_{n-1}).$$

We divide each E_i into $mP'(x_i)$ sets E_{ij} with $(P \times \lambda)(E_{ij}) = 1/m$; likewise each F_i into $mQ'(x_i)$ sets F_{ij} with $(Q \times \lambda)(F_{ij}) = 1/m$. We call the E_{ij} "boys" and the F_{ij} "girls". For ω in E_{ij} let $b(\omega) = x_i$; we say the boy E_{ij} "lives at" x_i. Likewise on F_{ij} let $g(\omega) = x_i$. Let B (resp. G) be the set of boys (resp. girls) so far defined, m of each. Let U (resp V) be a new disjoint set of r elements called boys (resp. girls).

We say a boy b *knows* a girl g if they live at points less than $2\gamma + \alpha$ apart or if $b \varepsilon U$ or $g \varepsilon V$. Then for any set $A \subset B$, with k members, living on a set $T \subset H_n$,

$$k \leqq mP'(T) \leqq mQ'(T^{2\lambda+\alpha}) + r$$

$\leqq$ the numbers of girls known by the k boys.

Thus any set of boys in $B \cup U$ knows at least as many girls. Hence by the marriage lemma (Philip Hall [5]; cf. also [4], p. 60) there is a function M from $B \cup U$ onto $G \cup V$ such that b knows $M(b)$ for each b. Hence there is a function h from B onto G such that b knows $h(b)$ except for at most r boys in B.

Now for each boy $b = E_{ij}$, let

$$p_b(A) = (P \times \lambda)(A \cap b), \qquad q_b(A) = (Q \times \lambda)(A \cap h(b))$$

for any measurable set $A \subset S \times I$. Let μ_b be the product measure

$$m(p_b \circ X^{-1}) \times (q_b \circ X^{-1}) \quad \text{on} \quad S \times S, \quad \text{and} \quad \sum_{b \varepsilon B} \mu_b = \mu.$$

Then since the $b \varepsilon B$ are disjoint with union $S \times I$, as are the $h(b)$, and $p_b(S \times I) = q_b(S \times I) = 1/m$, we have $\mu \circ \pi_1^{-1} = P$ and $\mu \circ \pi_2^{-1} = Q$.

All but at most $2m\gamma$ of the boys in B are subsets each of some $(f \circ X)^{-1}(x_i)$, $i = 1, \cdots, n-1$, all but at most r of them know $h(b)$, and likewise for the girls in G. Thus except for at most $4m\gamma + r$ of the boys b in B, the following three statements all hold:

$$b \subset (f \circ X)^{-1}(x_i) \quad \text{for some} \quad i,$$

$$h(b) \subset (f \circ X)^{-1}(x_j) \quad \text{for some} \quad j,$$

and $d(x_i, x_j) < 2\gamma + \alpha$.
Thus

$$\mu(d(x, y) > \alpha + \epsilon) = \sum_b \mu_b(d(x, y) > \alpha + \epsilon) < 4\gamma + r/m < 8\gamma + \beta < \beta + \epsilon.$$

This completes the proof.

A separable metric space (S, d) is called *inner regular* if for every Borel probability measure ν on S and Borel set $A \subset S$,

$$\nu(A) = \sup (\nu(K): K \subset A, K \quad \text{compact}).$$

Then S is inner regular if it is complete, or a Borel subset of its completion $\bar{S}$, or if (and only if) it is P-measurable for every $P \varepsilon \mathcal{P}(\bar{S})$ (Varadarajan [11], b, p. 224).

THEOREM 2. *If in addition to the hypotheses of Theorem* 1 *S is inner regular, then* (I) *is equivalent to*

(II′) *There is a* μ *in* $\mathcal{P}(S \times S)$ *with*

$$\mu \circ \pi_1^{-1} = P, \qquad \mu \circ \pi_2^{-1} = Q, \qquad \textit{and} \qquad \mu(d(x, y) > \alpha) \leqq \beta.$$

PROOF. Clearly (II′) $\Rightarrow$ (II) $\Rightarrow$ (I). Assuming (I) let $\epsilon = \epsilon_k \downarrow 0$ in (II) and let μ_k be corresponding measures on $S \times S$. For any $\delta > 0$ there is a compact $K \subset S$ such that

$$P(S \sim K) < \delta/2, \qquad Q(S \sim K) < \delta/2,$$

so

$$\mu_k((S \times S) \sim (K \times K)) < \delta.$$

Thus the sequence $\{\mu_k\}$ is "tight" and has a TW^*-convergent sub-sequence (Varadarajan [11], Appendix, p. 223, Theorem 2; Part II, p. 202, Theorem 27). Thus we may assume $\mu_k \to \mu$ (TW^*) for some μ in $P(S)$. Then $\mu \circ \pi_1^{-1} = P$,

$\mu \circ \pi_2^{-1} = Q$, and

$$\mu(d(x, y) > \alpha) = \lim_{c \downarrow 0} \mu(d(x, y) > c + \alpha)$$
$$\leqq \lim_{c \downarrow 0} \liminf_{k \to \infty} \mu_k(d(x, y) > c + \alpha) \quad \text{([8], Theorem 1.2,)}$$
$$\leqq \liminf (\beta + \epsilon_k) = \beta. \qquad \text{q.e.d.}$$

We shall see in a moment that Theorem 2 cannot be proved under the hypotheses of Theorem 1 only. The following holds by definition of ρ, Proposition 1, and a passage to the limit:

COROLLARY 1. *Under the hypotheses of Theorems* 1 *or* 2, (I) *holds* (*hence* (II) *or* (II′) *respectively*) *when*

$$\alpha = \beta = \rho(P, Q).$$

In Theorems 1 and 2, μ depends on α and β. Now (I) will hold for different pairs (α, β) yet it may be impossible to obtain (II) for two different pairs simultaneously. For example let $S = R$, $P(0) = P(\frac{3}{2}) = \frac{1}{2} = Q(1) = Q(\frac{5}{2})$. Then (I) holds for $\alpha = \beta = \frac{1}{2}$ and for $\alpha = 1, \beta = 0$. If μ satisfied (II) for both these pairs then $\mu(x = \frac{3}{2}, y = 1) = \frac{1}{2}$ and $\mu(x = 0, y = 1) = \mu(x = \frac{3}{2}, y = \frac{5}{2}) = \frac{1}{2}$, a contradiction.

Note that Theorem 1 yields an independent proof of Proposition 1.

Now we give an example showing that the hypothesis of inner regularity cannot simply be dropped from Theorem 2. Let λ be Lebesgue measure on the real line. Then there is a subset A of the interval [0, 3] whose outer measure $\lambda^*(A)$ is 3, and such that A and $A + 1$ are disjoint (Halmos [6], Theorem E, p. 70). (Then A is not Lebesgue measurable and hence not inner regular.) Let $S = A$ and for any Borel set B in S let

$$P(B) = \lambda^*(B \cap [0, 2]/2),$$
$$Q(B) = \lambda^*(B \cap [1, 3])/2.$$

Then $P, Q \in \mathcal{P}(S)$ ([6], p. 75 Theorem A), and for any Bore lset B in S, $P(B) \leqq Q(B^1)$. Suppose

$$\mu \in \mathcal{P}(S \times S), \qquad \mu \circ \pi_1^{-1} = P, \qquad \mu \circ \pi_2^{-1} = Q, \quad \text{and} \quad \mu(|x - y| > 1) = 0.$$

Then $y \leqq x + 1$ almost surely, and

$$\int y \, d\mu = 2 = \int x \, d\mu + 1 = \int (x + 1) \, d\mu, \quad \text{so } y = x + 1$$

almost surely, contradicting disjointness of A and $A + 1$.

We shall use Theorem 1 to compare ρ with another metrization of TW^*[1]. Let $BL(S, d)$ denote the set of all bounded real-valued functions f on S which are Lipschitzian, i.e.

$$\|f\|_L \equiv \sup_{x \neq y} |f(x) - f(y)|/d(x, y) < \infty.$$

We let $\|f\|_{BL} = \|f\|_\infty + \|f\|_L$. (The use of Lipschitzian functions has been suggested in the excellent survey by Fortet [2], p. 191, and the boundedness assumption assures integrability for each probability measure. Cf. also Fortet and Mourier [2a].)

Now $(BL(S, d), \|\cdot\|_{BL})$ is a Banach space. If

$$\|\mu\|_{BL}^* = \sup \{|\int f\, d\mu| : \|f\|_{BL} \leqq 1\}$$

then the metric $\|\mu - \nu\|_{BL}^*$ metrizes TW^* on $\mathcal{P}(S)$ ([1], Theorems 6, 8, and 18).

Proposition 2. *If the hypotheses of Theorem* 1 *and* (I) *hold then* $\|P - Q\|_{BL}^* \leqq 2 \max(\alpha, \beta)$.

Proof. By (II), given $\epsilon > 0$ we take random variables X with distribution P and Y with distribution Q such that

$$P(d(X, Y) > \alpha + \epsilon) \leqq \beta + \epsilon.$$

Then for any f in $BL(S, d)$,

$$|\int f\, d(P - Q)| = |E(f(X) - f(Y))| \leqq (\alpha + \epsilon)\|f\|_L + 2(\beta + \epsilon)\|f\|_\infty.$$

Letting $\epsilon \downarrow 0$ we get the desired conclusion.

Corollary 2. *For S separable metric and $P, Q \,\varepsilon\, \mathcal{P}(S)$,*

$$\|P - Q\|_{BL}^* \leqq 2\rho(P, Q).$$

Proposition 3. *If $P, Q \,\varepsilon\, \mathcal{P}(S)$, F is a closed set in the metric space S, $\alpha \geqq 0$, $\beta > 0$, and $P(F) > Q(F^\beta) + \alpha$, then*

$$\|P - Q\|_{BL}^* \geqq 2\alpha\beta/(2 + \beta).$$

Proof. We deffine a function f in $BL(S)$ such that $f = 1$ on F, $f = -1$ on $S \sim F^\beta$, $\|f\|_\infty = 1$, and $\|f\|_{BL} \leqq 1 + 2/\beta$ ([1], Lemma 5)[2]. Then

$$\begin{aligned}(1 + 2/\beta)\|P - Q\|_{BL}^* &\geqq \int f\, d(P - Q) \\ &= \int (f + 1)\, d(P - Q) \geqq 2(P(F) - Q(F^\beta)) \geqq 2\alpha,\end{aligned}$$

and the conclusion follows.

Now note that

$$f(x) \equiv 2x^2/(2 + x) = 2/[2/x^2 + 1/x]$$

is an increasing function of x for $x > 0$. Thus if $x \geqq 0$, $0 \leqq f(x) \leqq \frac{2}{3}$ if and only if $x \leqq 1$, and for $0 \leqq x \leqq 1$, $2x^2/3 \leqq f(x)$.

Corollary 3. *For S metric and $P, Q \,\varepsilon\, \mathcal{P}(S)$, $f(\rho(P, Q)) \leqq \|P - Q\|_{BL}^*$. Thus if $\rho(P, Q) \leqq 1$ or $\|P - Q\|_{BL}^* \leqq \frac{2}{3}$,*

$$\|P - Q\|_{BL}^* \geqq \tfrac{2}{3}\rho(P, Q)^2, \qquad \rho(P, Q) \leqq (\tfrac{3}{2}\|P - Q\|_{BL}^*)^{\frac{1}{2}}.$$

Corollaries 2 and 3 together imply that if S is separable (metric), $\|\cdot\|_{BL}^*$

[2] The extension of a Lipschitzian function f from $A \subset S$ to S without increasing $\|f\|_L$ was reportedly first shown by Banach (unpublished); cf. also McShane, E. J., "Extension of range of functions," *Bull. Amer. Math. Soc.* **40** (1934) 837–842.

and ρ define the same uniformity on $\mathcal{P}(S)$ (but this is not the weak-star uniformity, defined by all pseudo-metrics $|\int f\, d(P - Q)|$, $f \in \mathcal{C}(S)$, which indeed is not metrizable, unless S is compact ([1], Theorem 13)).

Here are examples showing that the inequalities in Corollaries 2 and 3 can be improved at most by a factor of 2. Let $d(p, q) = 1/n$, and let μ be a point mass 1 at p, and ν at q. Then

$$\rho(\mu, \nu) = 1/n, \qquad \|\mu - \nu\|_{BL}^{*} = 2/(2n + 1),$$

and the two distances are asymptotic as $n \to \infty$. On the other hand let

$$\sigma(p) = \sigma(q) = \tfrac{1}{2}, \qquad \tau(p) = \tfrac{1}{2} + 1/n, \qquad \tau(q) = \tfrac{1}{2} - 1/n.$$

Then $\rho(\sigma, \tau) = 1/n$, $\|\sigma - \tau\|_{BL}^{*} = \|\mu - \nu\|_{BL}^{*}/n$, asymptotic to $1/n^2$ as $n \to \infty$.

3. Almost sure convergence. A set A in a topological space S is called a *continuity set* of a measure $\mu \geqq 0$ if the boundary of A has μ-measure 0. If S is metrizable and $P_n \to P_0$ for TW^* in $\mathcal{P}(S)$, then $P_n(A) \to P_0(A)$ for every continuity set A of P_0([11], Theorem 2(IV), p. 182). The continuity sets of P_0 form an algebra ([8], Lemma 1.1) the proof does not use completeness of S.

Given $x \in S$, the balls

$$\{y \in S: d(x, y) < \epsilon\}$$

are continuity sets of P_0 except for at most countably many values of ϵ. Thus if S is separable, given $\delta > 0$ we can find finitely many disjoint continuity sets of P_0, each of diameter less than δ, and with total P_0-measure at least $1 - \delta$ (cf. [9], p. 281).

THEOREM 3. *Let S be a separable metric space, $P_n \in \mathcal{P}(S)$, $n = 0, 1, \cdots$, and $P_n \to P_0$ weak-star. Then there is a probability space $(\Omega, \mathcal{B}, \mu)$ with S-valued random variables X_n, $X_n \to X_0$ almost surely, and $\mu \circ X_n^{-1} \equiv P_n$.*

PROOF. For each $k = 1, 2, \cdots$, we take finitely many disjoint continuity sets of P_0, called $A(k, j)$, $j = 1, 2, \cdots, J_k$, each of diameter less than $1/k$, and satisfying

$$\textstyle\sum_j P_0(A(k, j)) \geqq 1 - 2^{-k}.$$

We may assume each term in the above sum is positive. Then for each k there is an n_k such that for all $n \geqq n_k$

$$\textstyle\sum_j |(P_n - P_0)(A(k, j))| < 2^{-k} \min_j P_0(A(k, j)).$$

We may assume $n_1 < n_2 < \cdots$.

Now for each n let S_n be a copy of S and I_n of the unit interval [0, 1] with Lebesgue measure λ_n. Let $\Omega_n = S_n \times I_n$ and let P_n' be the product measure $P_n \times \lambda_n$ on Ω_n. We define countable Cartesian products

$$\Omega_* = \textstyle\prod_{n=1}^{\infty} \Omega_n, \qquad \Omega = \Omega_0 \times \Omega_*.$$

Let X_n be the natural projection of Ω onto S_n and π the projection of Ω_n onto S_n for each n.

For each k, $j \leqq J_k$, and $n \geqq n_k$ we let

$$B(n, k, j) = A(k, j) \times [0, \delta(n, k, j)] \subset \Omega_n,$$
$$C(n, k, j) = A(k, j) \times [0, \gamma(n, k, j)] \subset \Omega_0,$$

choosing δ and γ so that

$$P_n'(B(n, k, j)) = P_0'(C(n, k, j)) = \min(P_n(A(k, j)), P_0(A(k, j))).$$

Then one of δ and γ is 1 and the other is at least $1 - 2^{-k}$. Let $B(n, k, 0) = \Omega_n \sim \bigcup_{j \geqq 1} B(n, k, j)$, $C(n, k, 0) = \Omega_0 \sim \bigcup_{j \geqq 1} C(n, k, j)$.

Let $n_0 = 1$ and for each n let $k(n)$ be the unique k such that $n_k \leqq n < n_{k+1}$.

For each n, Ω_0 is the disjoint union of finitely many sets $E(n, j) = C(n, k(n), j)$, $j = 0, 1, \cdots, J_{k(n)}$. For $j \geqq 1$ the $E(n, j)$ have diameters less than $1/k(n)$, and if also $n \geqq n_2$, $P_0(E(n, j)) > 0$. Likewise Ω_n is the disjoint union of finitely many sets

$D(n, j) = B(n, k(n), j)$, $j = 0, 1, \cdots, J_{k(n)}$, with the same properties.

For each n, and x in Ω_0, let $j(n, x)$ be the j such that $x \in E(n, j)$. Let

$$A = \{x \in \Omega_0 : P_0'(E(n, j(n, x))) > 0 \quad \text{for all } n\}.$$

Then clearly $P_0'(\Omega_0 \sim A) = 0$. For x in A let $P(n, x)$ be the measure P_n' restricted to measurable subsets of $D(n, j(n, x))$ in Ω_n, then normalized to mass 1 (i.e. divided by $P_0'(E(n, j(n, x)))$). Let μ_x be the product measure

$$\prod_{n=1}^{\infty} P(n, x) \text{ on } \Omega_*$$

(Halmos [6], Section 38, Theorem B, p. 157). Now I claim that for any measurable subset F of Ω_*, $x \to \mu_x(F)$ is a measurable function on Ω_0. In fact, for a given n, $P(n, x)$ has only finitely many possible values, each for x in a measurable set, and hence so does

$$\prod_{n=1}^{N} P(n, x), \quad N \text{ finite.}$$

Thus the claim is true for sets $F = Y_N^{-1}(G)$ where Y_N is the projection of Ω_* on, and G is measurable in,

$$\prod_{n=1}^{N} \Omega_n.$$

But the algebra of such sets generates the σ-algebra of measurable sets in Ω_*, and the class of sets F for which the claim holds is closed under countable monotone increasing and decreasing limits. Thus the claim holds for all measurable F ([6], Section 6, Theorem B, p. 27).

For any measurable $H \subset \Omega$, and $x \in \Omega_0$, let

$$H_x = \{y: (x, y) \in H\}, \qquad \text{and} \qquad \mu(H) = \int \mu_x(H_x)\, dP_0(x).$$

Note that $x \to \mu_x(H_x)$ is measurable if H is a finite union of measurable "rectangles" $A \times B$, $A \subset \Omega_0$, $B \subset \Omega_*$. Hence by monotone convergence it is measurable for any measurable $H \subset \Omega$, and μ is a countably additive probability measure on W. The distribution of X_n for μ is

$$[\sum_j P_0'(E(n, j))P(n, x)_{x \varepsilon E(n,j)}] \circ \pi^{-1} = P_n' \circ \pi^{-1} = P_n .$$

Since $\sum_k P_0(S \sim \bigcup_j A(k, j)) < \infty$, P_0-almost every point of S_0 belongs to $\bigcup_j A(k(n), j)$ for all large enough n. Also if $t \varepsilon I_0$ and $t < 1$, then $t < \gamma(n, k(n), j)$ for all j if n is large enough. Thus P_0'-almost all x belong to an $E(n, j)$ with $j \geqq 1$ for n large enough, and then

$$d(X_0, X_n) \leqq 1/k(n) \to 0$$

so $X_n \to X_0$. Thus $\mu(X_n \to X_0) = 1$, q.e.d.

4. The real line R. If $S = R$, the proof of Skorokhod ([9], Theorem 3.1) reduces naturally to the following. Let $P_n \varepsilon \mathcal{P}(R)$ and let F_n be their distribution functions

$$F_n(x) = P_n((-\infty, x]).$$

Let Ω be the unit interval [0, 1] with Lebesgue measure λ and for y in Ω let $X_n(y)$ be any x such that $F_n(x) = y$, or $F_n(x^-) \leqq y \leqq F_n(x)$. X_n is well-defined except for at most countably many values of y and hence is a well-defined random variable. If $P_n \to P_0$ for TW^*, then $F_n(x) \to F_0(x)$ whenever F_0 is continuous at x, and $X_n(y) \to X_0(y)$ except on the possible countable set of y where X_0 is not well-defined. Thus $X_n \to X$ almost surely. Clearly $\lambda \circ X_n^{-1} \equiv P_n$.

The above method seems unsuited to proving Theorem 1 on R. Let

$$P_n(j) = Q_n(j + 1) = 1/n, \quad j = 0, 1, \cdots, n - 1, P_n, Q_n \varepsilon \mathcal{P}(R).$$

$P_n - Q_n \to 0$ (even in total variation), but if X_n and Y_n are random variables on (Ω, λ) constructed from P_n and Q_n as above, then $X_n + 1 \equiv Y_n$.

For P in $\mathcal{P}(R)$ we introduce the usual characteristic function

$$\hat{P}(t) = \int_{-\infty}^{\infty} e^{ixt} \, dP(X).$$

On $\mathcal{P}(R)$ let UC be the uniformity of uniform convergence of $\hat{P}$ on compact sets, with a base given by the vicinities $\{(P, Q): |\hat{P}(t) - \hat{Q}(t)| \leqq 1/n$ whenever $|t| \leqq n\}$. Clearly the identity on $\mathcal{P}(R)$ is uniformly continuous from the $BL^*(=$ Prokhorov) uniformity to UC. We do not have uniform continuity in the converse direction, as is shown by the following stronger result:

PROPOSITION 4. *For any $\delta > 0$ there exist $P, Q \varepsilon \mathcal{P}(R)$ with $\|P - Q\|_{BL}^* \geqq 1$ (in fact P concentrated in $x \geqq 1$ and Q in $x \leqq -1$) and $|\hat{P}(t) - \hat{Q}(t)| < \delta$ for all t.*

PROOF. For each $n = 1, 2, \cdots$, let

$$C_n = \sum_{k=1}^{n} 1/k$$

and let P_n have mass $1/kC_n$ at $k = 1, \cdots, n$, with $Q_n(A) \equiv P_n(-A)$. Then clearly $\|P_n - Q_n\|_{BL}^* \geqq 1$. Also $C_n|\hat{P}_n(t) - \hat{Q}_n(t)|$ is bounded uniformly in n and t (see e.g. Zygmund [12], volume 1, II, 9, p. 61), so $\hat{P}_n(t) - \hat{Q}_n(t) \to 0$ uniformly in t as $n \to \infty$, q.e.d.

The central limit theorem is generally proved using characteristic functions, and as long as one considers convergence $P_n \to P$ for a specific limit P, it is a question of topology rather than uniformity on $\mathcal{P}(R)$. But it is notable, and not surprising given Proposition 4, that in order to prove uniform closeness of n-fold convolutions $P*P* \cdots *P$, $P \in \mathcal{P}(R)$, to infinitely divisible distributions, one does not use characteristic functions (Kolmogorov [7]).

For $P, Q \in \mathcal{P}(R)$, Paul Lévy's metric $\rho_L(P, Q)$ may be defined by replacing, in the definition of Prokhorov's metric ρ, closed sets F by semi-infinite intervals $(-\infty, x]$. Now let $P_n, Q_n \in \mathcal{P}(R)$ where

$$P_n(2j) = Q_n(2j+1) = 1/n, \qquad j = 1, \cdots, n.$$

Then $\rho_L(P_n, Q_n) \equiv 1/n$, while $\|P_n - Q_n\|_{BL}^* \geqq \frac{1}{2}$, so the uniformity of Lévy's metric is strictly weaker than that of $\|\cdot\|_{BC}^*$ and ρ. ρ_L metrizes TW^* on $\mathcal{P}(R)$ ([3], p. 33, Theorem 1).

REFERENCES

[1] Dudley, R. M. (1966). Convergence of Baire measures. *Studia Math.* **27** 251–268.

[2] Fortet, Robert (1958). Recent advances in probability theory. *Some Aspects of Analysis and Probability*. Wiley, New York.

[2a] Fortet, Robert and E. Mourier (1953). Convergence de la répartition empirique vers la répartition théorique. *Ann. Sci. Ecole Norm. Sup.* **70** 266–285.

[3] Gnedenko, B. V., and A. N. Kolmogorov (1954). *Limit Distributions for Sums of Independent Random Variables* (transl. by K. L. Chung). Addison-Wesley, Reading, Mass.

[4] Hall, M. (1958). A survey of combinatorial analysis. *Some Aspects of Analysis and Probability*. Wiley, New York.

[5] Hall, P. (1935). On representatives of subsets. *J. London Math. Soc.* **10** 26–30.

[6] Halmos, P. (1950). *Measure Theory*. Van Nostrand, Princeton.

[7] Kolmogorov, M. N. (1963). Approximation to distributions of sums of independent terms by means of infinitely divisible distributions. *Trans. Moscow Math. Soc.* **12** 492–509.

[8] Prokhorov, Yu. V. (1956). Convergence of random processes and limit theorems in probability theory. *Theor. Prob. Appl.* **1** 157–214.

[9] Skorokhod, A. V. (1956). Limit theorems for stochastic processes. *Theor. Prob. Appl.* **1** 261–290.

[10] Strassen, V. (1965). The existence of probability measures with given marginals. *Ann. Math. Statist.* **36** 423–439.

[11] Varadarajan, V. S. (1961). Measures on topological spaces. *Amer. Math. Soc. Transl.* Series 2, **48** 161–228.

[12] Zygmund, Antoni (1959). *Trigonometrical Series*. Cambridge Univ. Press.

An extended Wichura theorem, definitions of Donsker class, and weighted empirical distributions

by R. M. Dudley[1]

Abstract. Let $(X, \mathcal{A}, P)$ be a probability space, $x(1), x(2), \cdots$ coordinates for the countable product of copies of X, and $\mathcal{F} \subset \mathcal{L}^2(X, \mathcal{A}, P)$. Let ν_n be the normalized empirical measure $n^{-1/2}(\delta_{x(1)} + \cdots + \delta_{x(n)} - nP)$. Let $\ell^\infty(\mathcal{F})$ be the set of all bounded real functions G on $\mathcal{F}$ with norm $\|G\|_{\mathcal{F}} := \sup_{f \in \mathcal{F}} |G(f)|$. Let $\mu(f) := \int f d\mu$ for any measure μ. We are interested in central limit theorems where ν_n converges in law for $\|\cdot\|_{\mathcal{F}}$. The limit is a Gaussian process G_P with mean 0 and covariance

$$EG_P(f)G_P(g) = P(fg) - P(f)P(g) := (f,g)_{P,0},$$

$f, g \in \mathcal{F}$. Say $\mathcal{F}$ is G_PBUC if G_P can be chosen to have each sample function bounded and uniformly continuous for $(\cdot,\cdot)_{P,0}$. Say the central limit theorem holds for $\mathcal{F}$ if $\mathcal{F}$ is G_PBUC and we have convergence of upper integrals $\lim_{n\to\infty} \int^* H(\nu_n) dP^n = EH(G_P)$ for every bounded continuous real function H on $\ell^\infty(\mathcal{F})$. (Thus, as noted by J. Hoffmann-Jørgensen, convergence "in law" does not require definition of laws.) In Sec. 4 an extended Wichura's theorem is proved: given such convergence in law, there exist almost surely convergent realizations. Sec. 5 shows that the new definition of central limit theorem holding is equivalent to the previous definition of "functional Donsker class." Secs. 6-7 treat the case where $\mathcal{F} = \{M1_{h \geq M}: 0 < M < \infty\}$ for a random variable h. This case reduces to the study of weighted empirical distribution functions. Conditions for the central limit theorem are collected and made precise.

1. **Partially supported by National Science Foundation Grant MCS-8202122**

E. Giné et al. (eds.), *Selected Works of R.M. Dudley*, Selected Works in Probability and Statistics, DOI 10.1007/978-1-4419-5821-1_5,

1. <u>Introduction</u>. Let (S,d) be a metric space. If (S,d) is separable and complete, Skorohod (1956) shows that if Y_n are S-valued random variables, $n = 0,1,\cdots$, which converge in law to Y_0 as $n \to \infty$, then on some probability space there exist S-valued random variables X_n such that for each n, X_n and Y_n have the same distribution, and $X_n \to X_0$ almost surely. The author (1968) removed the completeness assumption. M. Wichura (1970) and P. J. Fernandez (1974) proved an extension of the theorem where S may be non-separable, only Y_0 is required to have separable range, and the Y_n are measurable with respect to the σ-algebra $\mathcal{B}_b$ generated by all balls $\{y: d(x,y) < r\}$, $x \in S, r > 0$. Here convergence in law is defined as convergence of upper integrals

$$\int^* F(Y_n)dP \longrightarrow \int F(Y_0)dP$$

for every bounded continuous real F on S (Dudley, 1966). Also, Wichura replaced almost sure convergence by almost uniform convergence (see below), which is stronger and more useful in the non-separable case. Hoffmann-Jørgensen (1984) defined convergence of random elements $Y_n \longrightarrow Y_0$ in law as above without requiring laws of Y_n to be defined on any non-trivial σ-algebra. One main result of the paper, stated and proved in Sec. 4, is that, essentially, Wichura's theorem can be extended to hold under Hoffmann's definition. The extension uses the notion of "perfect" function (Sec. 2 below), also due to Hoffmann-Jørgensen.

Let $(X, \mathcal{A}, P)$ be a probability space and f any real-valued function on X (not measurable, in general). Let

$$\begin{aligned}\int^* f dP &:= \inf\{\int h dP: h \geq f, \ h \text{ measurable}\} \\ &= \int f^* dP, \text{ where} \\ f^* &:= \text{ess. inf}\{g: g \geq f, \ g \text{ measurable}\} \geq f\end{aligned}$$

(Dudley and Philipp, 1983, Lemma 2.1). Also let $P^*(B) := \int^* 1_B dP$ for any set $B \subset X$ and

$$\int_* f dP := -\int^* -f dP = \int f_* dP$$
$$= \sup\{\int g dP: g \le f, \quad g \text{ measurable}\},$$

where $f_* := -((-f)^*)$.

Let $(S,\mathcal{T})$ be any topological space and β_n probability measures defined on σ-algebras $\mathcal{B}_n$ of subsets of S (not necessarily related to $\mathcal{T}$ for $n > 0$), $n = 0,1,\cdots$, $\beta := \beta_0$.

In 1966 I defined

"$\beta_n \to \beta$ (weak*) if for every bounded continuous real function F on S, $\lim_{n\to\infty} \int^* F d\beta_n = \lim_{n\to\infty} \int_* F d\beta_n = \int F d\beta$."

Results were obtained for this convergence mainly in case $\mathcal{T}$ is metrizable by a metric d and each $\mathcal{B}_n$ includes $\mathcal{B}_b$ (Dudley 1966, 1967a, 1978; Wichura 1970; Fernandez 1974). The following will help to elucidate the notion of almost uniform convergence. It is not claimed as new, but a proof will be given for completeness.

1.1 Proposition. Let $(\Omega,\mathcal{S},P)$ be a probability space, (S,d) a metric space, and X_n any functions from Ω into S, $n = 0,1,\cdots$. Then the following are equivalent:

A) $d(X_n,X_0)^* \to 0$ almost surely;

B) for any $\varepsilon > 0$,
$P^*\{\sup_{n\ge m} d(X_n,X_0) > \varepsilon\} \downarrow 0$ as $m \to \infty$;

C) For each $\delta > 0$, there is some $B \in \mathcal{S}$ with $P(B) > 1 - \delta$ such that $X_n \to X_0$ uniformly on B.

D) There exist measurable $h_n \ge d(X_n,X_0)$ with $h_n \to 0$ a.s.

Proof. A) implies that

$$(\sup_{n\ge m} d(X_n,X_0))^* \le \sup_{n\ge m}(d(X_n,X_0)^*) \downarrow 0$$

a.s. as $m \to \infty$, which implies B).

Assuming B), for $k = 1,2,\cdots$ take $m(k)$ such that $P^*(\sup_{n\ge m(k)} d(X_n,X_0) > 1/k) < 2^{-k}$. Take measurable covers B_k,

$$B_k \supset \{\sup_{n\ge m(k)} d(X_n,X_0) > 1/k\}$$

with $B_k \in \mathcal{S}$, $P(B_k) < 2^{-k}$. For $r = 1,2,\cdots$, let $A_r := \Omega \setminus \bigcup_{k>r} B_k$. Then $P(A_r) > 1 - 2^{-r}$ and $X_n \to X_0$ uniformly on A_r, so C) holds.

Now assume C). Take $C_k \in \mathcal{S}$, $k = 1,2,\cdots$, such that $P(C_k) \uparrow 1$ and $X_n \to X_0$ uniformly on C_k. We can take $C_1 \subset C_2 \subset \cdots$. Take m_k such that $d(X_n,X_0) < 1/k$ on C_k for all $n \geq m_k$. Then $d(X_n,X_0)^* \leq 1/k$ on C_k, so $d(X_n,X_0)^* \to 0$ a.s., giving A). Clearly, A) and D) are equivalent, Q.E.D.

A sequence X_n satisfying, say, A) in Prop. 1.1 will be said to converge <u>almost uniformly</u> to X_0. This agrees with the definition of Halmos (1950, p.89) for functions such that $d(X_n,X_0)$ is measurable. Then, almost uniform convergence is equivalent to almost sure convergence.

On the other hand, in $[0,1]$ with P = Lebesgue measure let $A(1) \supset A(2) \supset \cdots$ be sets with $P^*(A(n)) = 1$ for all n and $\bigcap_{n=1}^{\infty} A(n) = \emptyset$ (e.g. Cohn, 1980, p.35). Then $1_{A(n)} \to 0$ everywhere and, in that sense, almost surely, but not almost uniformly. To avoid such pathology it will be useful to obtain almost uniform convergence.

Wichura (1970) proved (results stronger than) the following:

<u>Theorem A</u>. Let (S,d) be any metric space and μ_n probability measures on $\mathcal{B}_b$, $n = 0,1,\cdots$. Suppose $\mu_0(S_0) = 1$ for some separable $S_0 \subset S$. If $\mu_n \to \mu_0$ (weak*) then there exist a probability space $(\Omega,\mathcal{S},P)$ and functions X_n from Ω to S, measurable from $\mathcal{S}$ to $\mathcal{B}_b$, such that $P \circ X_n^{-1} = \mu_n$ for each n and $X_n \to X_0$ almost uniformly.

The measurability for $\mathcal{B}_b$ does not always hold in cases of interest. To obtain it may require replacing S by a suitable subset and/or giving a non-trivial proof. To avoid these difficulties we have the following, due to J. Hoffmann-Jørgensen (1984).

<u>Definition</u>. Let $(S,\mathcal{T})$ be any topological space, $(X_\alpha, \mathcal{A}_\alpha, P_\alpha)_{\alpha\in J}$

probability spaces, where J is a directed index set, $0 \in J$, and f_α are any functions from X_α into S. Say that $f_\alpha \xrightarrow{\mathcal{L}} f_0$, or $f_\alpha \to f_0$ <u>in law</u>, iff for every bounded continuous real-valued function G on S, $\int G(f_0)dP_0$ is defined and

$$\int^* G(f_\alpha)dP_\alpha \longrightarrow \int G(f_0)dP_0$$

as $\alpha \to \infty$.

Taking $-G$, we will also have

$$\int_* G(f_\alpha)dP_\alpha \longrightarrow \int G(f_0)dP_0 .$$

If we let $\mathcal{B}_\alpha := \{C \subset S: f_\alpha^{-1}(C) \in \mathcal{Q}_\alpha\}$, a σ-algebra, and set $\beta_\alpha := P \circ f_\alpha^{-1}$ on $\mathcal{B}_\alpha$, Hoffmann's definition is not equivalent to $\beta_\alpha \to \beta_0$ (weak*), as the following shows.

1.2 <u>Example</u>. Let $(X_n, \mathcal{Q}_n, Q_n) = ([0,1], \mathcal{B}, \lambda)$ for all n (λ = Lebesgue measure, $\mathcal{B}$ = Borel σ-algebra). Take sets $C(n) \subset [0,1]$ with $0 = \lambda_*(C(n)) < \lambda^*(C(n)) = 1/n^2$ (Halmos, 1950, p.70). Let S be the two-point space $\{0,1\}$ with usual metric. Then $f_n := 1_{C(n)} \to 0$ in law and almost uniformly, but each $\beta_n := Q_n \circ f_n^{-1}$, $n > 0$, is only defined on the trivial σ-algebra $\{\emptyset, S\}$ and $\beta_n \not\to \delta_0$ (weak*). The σ-algebra $\mathcal{B}_b$ generated by balls in this case is 2^S, the only σ-algebra larger than $\{\emptyset, S\}$, but no β_n for $n > 0$ is defined on 2^S.

1.3 <u>Remark</u>. To show that $f_\alpha \xrightarrow{\mathcal{L}} f_0$ it is enough to show that

$$\limsup_{\alpha\to\infty} \int^* G(f_\alpha)dP_\alpha \le \int G(f_0)dP_0$$

for every bounded continuous G. For then

$$\begin{aligned}
&\limsup \int^* -G(f_\alpha)dP_\alpha \le \int -G(f_0)dP_0, \text{ so} \\
&\int G(f_0)dP_0 \ge \limsup \int^* G(f_\alpha)dP_\alpha \\
&\ge \liminf \int^* G(f_\alpha)dP_\alpha \ge \liminf \int_* G(f_\alpha)dP_\alpha \\
&\ge \int G(f_0)dP_0,
\end{aligned}$$

so these terms are all equal.

2. Perfect functions. It will be useful that under some conditions on a measurable function g and general real-valued f, $(f \circ g)^* = f^* \circ g$. Here are some conditions, first found by Hoffmann-Jørgensen and Andersen (1984), given here for completeness.

2.1 Theorem. Let $(X, \mathcal{A}, P)$ be a probability space, $(Y, \mathcal{B})$ any measurable space, and g a measurable function from X to Y. Let Q be the restriction of $P \circ g^{-1}$ to $\mathcal{B}$. For any real-valued function f on Y, define f^* for Q. Then the following are equivalent:

1) for any $A \in \mathcal{A}$ there is a $B \in \mathcal{B}$ with $B \subset g(A)$ and $Q(B) \geq P(A)$;

2) for any $A \in \mathcal{A}$ with $P(A) > 0$ there is a $B \in \mathcal{B}$ with $B \subset g(A)$ and $Q(B) > 0$;

3) for any real function f on Y, $(f \circ g)^* = f^* \circ g$ a.s.;

4) for any $D \subset Y$, $(1_D \circ g)^* = 1_D^* \circ g$ a.s.

Proof. 1) implies 2), clearly.

2) implies 3): note that always $(f \circ g)^* \leq f^* \circ g$. Suppose $(f \circ g)^* < f^* \circ g$ on a set of positive probability. Then for some rational r, $(f \circ g)^* < r < f^* \circ g$ on a set $A \in \mathcal{A}$ with $P(A) > 0$. Let $g(A) \supset B \in \mathcal{B}$ with $Q(B) > 0$. Then $f \circ g < r$ on A implies $f < r$ on B, so $f^* \leq r$ on B a.s., contradicting $f^* \circ g > r$ on A.

3) implies 4), clearly.

4) implies 1): given $A \in \mathcal{A}$, let $D := Y \setminus g(A)$. Then we can take $1_D^* = 1_C$ for some $C \in \mathcal{B}$ ($C = \{1_D^* \geq 1\}$), $D \subset C$, and $1_D \circ g = (1_D \circ g)^* = 0$ a.s. on A. Let $B := Y \setminus C$. Then $B \subset g(A)$, and

$$Q(B) = 1 - Q(C) = 1 - \int 1_D^* d(P \circ g^{-1})$$
$$= 1 - \int 1_D^* \circ g dP = 1 - \int (1_D \circ g)^* dP \geq P(A), \quad \text{Q.E.D.}$$

Following Hoffmann-Jørgensen and Andersen (1984), a function g satisfying any of the four conditions in 2.1 is called perfect or P-perfect.

Call g quasiperfect or P-quasiperfect iff for every $C \subset Y$ with $g^{-1}(C) \in \mathcal{A}$, C is Q-completion measurable. Then $(X, \mathcal{A}, P)$ is called perfect iff every real-valued function g on X, measurable (for the usual Borel σ-algebra on $\mathbb{R}$) is quasiperfect.

2.2 Example. A measurable, quasiperfect function g on a finite set X need not be perfect: let $X = \{a,b,c,d,e,j\}$, $U := \{a,b\}$, $V := \{c,d\}$, $W := \{e,j\}$, $\mathcal{A} := \{\varphi, U, V, W, U \cup V, U \cup W, V \cup W, X\}$, $P(U) = P(V) = P(W) = 1/3$, $Y := \{0,1,2\}$, $g(a) = g(c) = 0$, $g(b) = g(e) = 1$, $g(d) = g(j) = 2$. Let $\mathcal{B} = \{\varphi, Y\}$. For $C \subset Y$, $g^{-1}(C) \in \mathcal{A}$ iff $C \in \mathcal{B}$, so g is quasiperfect. But, $P(U) > 0$ and $g(U)$ does not include any non-empty set in $\mathcal{B}$, so g is not perfect.

2.3 Proposition. Any perfect function g (as in 2.1) is quasiperfect.

Proof. Let $C \subset Y$, $A := g^{-1}(C) \in \mathcal{A}$. By Theorem 2.1 take $B \subset g(A)$ with $B \in \mathcal{B}$, $Q(B) \geq P(A)$. Then $B \subset C$, so $Q(B) = P(g^{-1}B) \leq P(g^{-1}C) = P(A)$, and $Q(B) = P(A)$. Thus the inner measure $Q_*(C) = (P \circ g^{-1})(C)$. Likewise, $Q_*(Y \setminus C) = (P \circ g^{-1})(Y \setminus C)$, so $Q^*(C) = (P \circ g^{-1})(C)$ and C is Q-completion measurable, Q.E.D.

2.4 Example. Let $C \subset [0,1]$ satisfy $0 = \lambda_*(C) < \lambda^*(C) = 1$ (cf. Example 1.2). Let $X = C$, $Y = [0,1]$ with Borel σ-algebra, and $P = \lambda^*$ giving a probability measure on the Borel sets of X. Let g be the identity from X into Y. Then as the range C of g is not Q-completion measurable, g is not quasiperfect, although it is a 1-1, Borel measurable function between metric spaces. If, instead, $X = [0,1]$ with the same P on C and $P(X \setminus C) = 0$, with g the identity, then g is onto $[0,1]$ but still not quasiperfect. (Such examples, if not the terminology, are well known.)

2.5 Proposition. Suppose $A = X \times Y$, P is a product law $v \times m$, and g is the natural projection of A onto Y. Then g is P-perfect.

Proof. Here $P \circ g^{-1} = m$. For any $B \subset A$ let $B_y := \{x: \langle x,y \rangle \in B\}$, $y \in Y$. If B is measurable, then by the Tonelli-Fubini theorem, for $C := \{y: v(B_y) > 0\}$, C is measurable, $C \subset g(B)$, and $P(B) \leq m(C)$, Q.E.D.

3. Convergence in outer probability. Let $(X, \mathcal{A}, P)$ be any probability space, (S,d) a metric space, and f_n functions from X into S. Say that $f_n \to f_0$ in outer probability if $d(f_n,f_0)^* \to 0$ in probability as $n \to \infty$, or equivalently, for every $\varepsilon > 0$,

$$\lim_{n\to\infty} P^*(d(f_n,f_0) > \varepsilon) = 0.$$

We have, clearly:

3.1 Proposition. For any probability space $(X, \mathcal{A}, P)$, metric space (S,d), and functions f_n from X into S, $n = 0,1,\cdots$, almost uniform convergence $f_n \to f_0$ implies convergence in outer probability.

The example after Prop. 1.1 shows that $f_n \to f_0$ everywhere pointwise does not imply convergence in outer probability.

3.2 Proposition. Let (S,d) and (Y,e) be two metric spaces and $(X, \mathcal{A}, P)$ a probability space. Let f_n be functions from X into S, $n = 0,1,2,\cdots$, such that $f_n \to f_0$ in outer probability and f_0 has separable range and is Borel measurable. Let g be any continuous function from S into Y. Then $g(f_n) \to g(f_0)$ in outer probability.

Note. If $\int G(f)dP$ is defined for all bounded continuous real G (as it must be if $f_n \xrightarrow[\mathcal{L}]{} f$, by definition) then $P \circ f^{-1}$ is defined on all Borel subsets of S. Such a law does have a separable support except perhaps in some set-theoretically pathological cases (Marczewski and Sikorski, 1948). Indeed, it is consistent with the usual axioms of set theory (including the axiom of choice) that such pathology never arises (e.g. Drake, 1974, pp.67-68, 177-178). It is apparently unknown whether it can ever, consistently, arise (Drake, 1974, pp.185-186).

Proof of Prop. 3.2. Given $\varepsilon > 0$, $k = 1,2,\cdots$, let $B_k := \{x \in S\colon d(x,y) < 1/k$ implies $e(g(x),g(y)) \le \varepsilon$, $y \in S\}$. Then each B_k is closed and $B_k \uparrow S$ as $k \to \infty$. Fix k large enough so that $P(f_0^{-1}(B_k)) > 1 - \varepsilon$. Then

$$\{e(g(f_n),g(f_0)) > \varepsilon\} \cap f_0^{-1}(B_k) \subset \{d(f_n,f_0) \ge \tfrac{1}{k}\}.$$

Thus

$$P^*\{e(g(f_n),g(f_0)) > \varepsilon\} < \varepsilon + P^*\{d(f_n,f_0) \ge 1/k\} < 2\varepsilon$$

for n large enough, Q.E.D.

On any metric space, the σ-algebra will be the Borel σ-algebra unless stated otherwise.

3.3 Lemma. Let $(X,\mathcal{A},P)$ be a probability space and $\{g_n\}_{n=0}^{\infty}$ a uniformly bounded sequence of real-valued functions on X such that g_0 is measurable. If $g_n \to g_0$ in outer probability then

$$\limsup_{n\to\infty} \int^* g_n dP \le \int g_0 dP.$$

Proof. Let $|g_n(x)| \le M < \infty$ for all n and all $x \in X$. We may assume $M = 1$. Given $\varepsilon > 0$, for n large enough $P^*(|g_n-g_0| > \varepsilon) < \varepsilon$. Let $A(n)$ be a measurable set on which $|g_n-g_0| \le \varepsilon$ with $P(X\setminus A(n)) < \varepsilon$. Then

$$\int g_n^* dP \le \varepsilon + \int^*_{A(n)} g_n dP \le 2\varepsilon + \int_{A(n)} g_0 dP \le 3\varepsilon + \int g_0 dP.$$

Letting $\varepsilon \downarrow 0$ completes the proof. □

3.4 Corollary. If $f_n \to f_0$ in outer probability and f_0 is measurable with separable range then $f_n \xrightarrow[\mathcal{L}]{} f_0$.

Proof. Apply Prop. 3.2, Lemma 3.3 and Remark 1.3.

3.5 Theorem. Let $(X, \mathcal{A}, P)$ be a probability space, (S,d) a metric space. Suppose that for $n = 0,1,\cdots$, $(Y_n, \mathcal{B}_n)$ is a measurable space, g_n a perfect measurable function from X into Y_n, and f_n a function from Y_n into S, where f_0 has separable range and is measurable. Let $Q_n := P \circ g_n^{-1}$ on $\mathcal{B}_n$ and $f_n \circ g_n \to f_0 \circ g_0$ in outer probability. Then $f_n \xrightarrow{\mathcal{L}} f_0$ as $n \to \infty$.

Before proving this, here is:

3.6 Example. Theorem 3.5 can fail without the hypothesis that g_n be perfect. In Example 2.4 let $X = C$, $Y_n := [0,1]$, and let g_n be the identity for all n. Let $S = \{0,1\}$, $f_0 = 0$, and $f_n := 1_{[0,1]\setminus C}$. Then $f_n \circ g_n \equiv 0$ for all n, so $f_n \circ g_n \to f_0 \circ g_0$ in outer probability (and any other sense). Let $\mathcal{B}_n$ be the Borel σ-algebra for each n. Then for G the identity,

$$\int^* G(f_n)dQ_n = \int^* f_n d\lambda = 1 \quad \text{for } n \geq 1,$$

while $\int G(f_0)dQ_0 = 0$, so $f_n \not\xrightarrow{\mathcal{L}} f_0$.

Proof of Theorem 3.5. By Cor. 3.4, $f_n \circ g_n \xrightarrow{\mathcal{L}} f_0 \circ g_0$. Let G be any bounded, continuous, real-valued function on S. Then

$$\int^* G(f_n(g_n))dP \to \int G(f_0(g_0))dP = \int G(f_0)dQ_0 .$$

Also,

$$\begin{aligned} \int^* G(f_n(g_n))dP &= \int G(f_n(g_n))^* dP \\ &= \int (G \circ f_n)^*(g_n)dP \qquad \text{by Th. 2.1} \\ &= \int (G \circ f_n)^* dQ_n \\ &= \int^* G(f_n)dQ_n \end{aligned}$$

and the result follows. □

4. An extended Wichura theorem. Here is one of the main results of the paper:

4.1 <u>Theorem</u>. Let (S,d) be any metric space, $(X_n, \mathcal{A}_n, Q_n)$ any probability spaces, and f_n a function from X_n into S for each $n = 0,1,\cdots$. Suppose f_0 has separable range S' and is measurable. Then $f_n \xrightarrow[\mathcal{L}]{} f_0$ if and only if there exists a probability space $(\Omega, \mathcal{S}, Q)$ and perfect measurable functions g_n from $(\Omega, \mathcal{S})$ to $(X_n, \mathcal{A}_n)$ for each $n = 0,1,\cdots$, such that $Q \circ g_n^{-1} = Q_n$ on $\mathcal{A}_n$ for each n and $f_n(g_n) \to f_0(g_0)$ almost uniformly, $n \to \infty$.

<u>Proof</u>. "If" follows from Prop. 3.1 and Theorem 3.5. "Only if" will be proved largely as in Dudley (1968, Theorem 3) and Fernandez (1974). Let $f_n \xrightarrow[\mathcal{L}]{} f_0$.

The space Ω will be taken as the Cartesian product $\prod_{n=0}^{\infty} X_n \times I_n$ where each I_n is a copy of $[0,1]$. Here g_n will be the natural projection of Ω onto X_n for each n.

Let $P := Q_0 \circ f_0^{-1}$ on the Borel σ-algebra of S, concentrated in the separable subset S'. A set $B \subset S'$ will be called a <u>continuity set</u> in S' if $P(\partial B) = 0$ where ∂B is the boundary of B in S'. We then have:

4.2 <u>Lemma</u>. For any $\varepsilon > 0$ there exist $J < \infty$ and disjoint open continuity sets U_j, $j = 1,\cdots,J$, each with diameter

$$\operatorname{diam} U_j := \sup\{d(x,y) : x,y \in U_j\} < \varepsilon,$$

and with $\Sigma_{j=1}^{J} P(U_j) > 1 - \varepsilon$.

<u>Proof</u>. This is proved as in Skorohod (1956). Let $\{x_j\}_{j=1}^{\infty}$ be dense in S'. Let $B(x,r) := \{y \in s' : d(x,y) < r\}$, $0 < r < \infty$, $x \in S'$. Then $B(x_j,r)$ is a continuity set of P for all but at most countably many values of r. Choose r_j with $\varepsilon/3 < r_j < \varepsilon/2$ such that $B(x_j,r_j)$ is a continuity set of P for each j. The continuity sets form an algebra. Let

$$U_j := B(x_j,r_j) \setminus \cup_{i<j} \{y : d(x_i,y) \le r_i\} .$$

Then U_j are disjoint open continuity sets of diameters $< \varepsilon$

with $\Sigma_{j=1}^{\infty} P(U_j) = 1$, so there is a $J < \infty$ with $\Sigma_{j=1}^{J} P(U_j) > 1 - \varepsilon$, Q.E.D.

Now for each $k = 1,2,\cdots$, by Lemma 4.2 take disjoint open continuity sets $U_{kj} := U(k,j)$ of P, $j = 1,2,\cdots, J(k) < \infty$, with $\mathrm{diam}(U_{kj}) < 1/k$, $P(U_{kj}) > 0$, and

$$(4.3) \qquad \Sigma_{j=1}^{J(k)} P(U_{kj}) > 1 - 2^{-k}.$$

For any open set U in S with complement F, let $d(x,F) := \inf\{d(x,y): y \in F\}$. For $r = 1,2,\cdots$, let $F_r := \{x: d(x,F) \geq 1/r\}$. Then F_r is closed and $F_r \uparrow U$ as $r \to \infty$. There is a continuous h_r on S with $0 \leq h_r \leq 1$, $h_r = 1$ on F_r and $h_r = 0$ outside F_{2r}: take $h_r(x) := \min(1, \max(0, 2rd(x,F) - 1))$.

For each j and k, setting $F(k,j) := S \setminus U_{kj}$, take $r = r(k,j)$ large enough so that

$$P(F(k,j)_r) > (1-2^{-k})P(U_{kj}).$$

Let h_{kj} be the h_r above for such an r and H_{kj} the h_{2r}.

For n large enough, say for $n \geq n_k$, we have

$$\int_* h_{kj}(f_n)dQ_n > (1-2^{-k})P(U_{kj}) \quad \text{and}$$

$$\int^* H_{kj}(f_n)dQ_n < (1+2^{-k})P(U_{kj})$$

for all $j = 1,\cdots,J(k)$. We may assume $n_1 \leq n_2 \leq \cdots$.

For every $n = 0,1,\cdots$, let $f_{kjn} := (h_{kj} \circ f_n)_*$ for Q_n, so that

$$\int_* h_{kj}(f_n)dQ_n = \int f_{kjn}dQ_n, \quad 0 \leq f_{kjn} \leq h_{kj}(f_n),$$

and f_{kjn} is $\mathcal{A}_n$-measurable. For $n \geq 1$ let $B_{kjn} := \{f_{kjn} > 0\} \in \mathcal{A}_n$. Let $B_{kj0} := f_0^{-1}(U_{kj}) \in \mathcal{A}_0$. For each k and n, the $B_{kjn} \subset U_{kj}$ are disjoint, $j = 1,\cdots,J(k)$, and $H_{kj}(f_n) = 1$ on B_{kjn}, so

$$(4.4) \qquad (1-2^{-k})P(U_{kj}) < Q_n(B_{kjn}) < (1+2^{-k})P(U_{kj}).$$

Let $T_n := X_n \times I_n$. Let μ_n be the product law $Q_n \times \lambda$ on T_n where λ is Lebesgue measure on I_n. For each $k \geq 1$, $n \geq n_k$, and $j = 1, \cdots, J_k$, let

$$C_{kjn} := B_{kjn} \times [0, f(k,j,n)] \subset T_n ,$$

$$D_{kjn} := B_{kj0} \times [0, g(k,j,n)] \subset T_0 ,$$

defining f and g here so that

$$\mu_n(C_{kjn}) = \mu_0(D_{kjn}) = \min(Q_n(B_{kjn}), Q_0(B_{kj0})) .$$

Then for each k, j, and $n \geq n_k$, we have

$$\max(f,g)(k,j,n) = 1 \quad \text{and}$$

$$(4.5) \qquad \min(f,g)(k,j,n) > 1-2^{-k}, \text{ by } (4.4).$$

Let $C_{k0n} := T_n \setminus \bigcup_{j=1}^{J(k)} C_{kjn}$, $D_{k0n} := T_0 \setminus \bigcup_{j=1}^{J(k)} D_{kjn}$.

For $k = 0$ let $J_0 := 0$, $C_{00n} := T_n$, $D_{00n} := T_0$, $n_0 := 0$.

For each $n = 1, 2, \cdots$, let $k(n)$ be the unique k such that $n_k \leq n < n_{k+1}$. Then for $n \geq 1$, T_n is the disjoint union of sets $W_{nj} := C_{k(n)jn}$, $j = 0, 1, \cdots, J_{k(n)}$. Also, T_0 is the disjoint union of sets $E_{nj} := D_{k(n)jn}$. If $j \geq 1$ or $k(n) = 0$, then $\mu_n(W_{nj}) = \mu_0(E_{nj}) > 0$.

For x in T_0, and each n, let $j(n,x)$ be the j such that $x \in E_{nj}$. Let

$$L := \{x \in T_0 : \mu_0(E_{nj(n,x)}) > 0 \text{ for all } n\} .$$

Then $T_0 \setminus L \subset \bigcup_i E_{n(i)0}$ for some (possibly empty or finite) sequence $n(i)$ such that $\mu_0(E_{n(i)0}) = 0$ for all i. Thus $\mu_0(L) = 1$.

For $x \in L$ and any measurable set $B \subset T_n$ $(B \in \mathcal{A}_n \otimes \mathcal{B})$, let

$$P_{nj}(B) := \mu_n(B \cap W_{nj})/\mu_0(E_{nj}), \quad P_{nx} := P_{nj} ,$$

where $j := j(n,x)$. Then P_{nx} is a probability measure on $\mathcal{A}_n \otimes \mathcal{B}$. Let ρ_x be the product measure $\prod_{n=1}^{\infty} P_{nx}$ on $T := \prod_{n=1}^{\infty} T_n$ (Halmos, 1950, p. 157).

4.6 Lemma. For every measurable set $H \subset T$ ($H \in \times_{n=1}^{\infty}(\mathcal{A}_n \otimes \mathcal{B})$), $x \to \rho_x(H)$ is measurable on $(T_0, \mathcal{A}_0 \otimes \mathcal{B})$.

Proof. Let $\mathcal{H}$ be the collection of all H for which the assertion holds. Given n, P_{nx} is one of finitely many laws, each obtained for x in a measurable set E_{nj}. Thus if Y_n is the natural projection of T onto T_n and $H = Y_n^{-1}(B)$ for some $B \in \mathcal{A}_n \otimes \mathcal{B}$ then $H \in \mathcal{H}$.

If $H = \cap_{i \in F} Y_{m(i)}^{-1}(B_i)$, where $B_i \in \mathcal{A}_{m(i)} \otimes \mathcal{B}$ and F is finite, we may assume the $m(i)$ are distinct. Then

$$\rho_x(H) = \prod_{i \in F} \rho_x(Y_{m(i)}^{-1}(B_i)) ,$$

so $H \in \mathcal{H}$. Then, any finite, disjoint union of such intersections is in $\mathcal{H}$. Such unions form an algebra. If $H_n \in \mathcal{H}$ and $H_n \uparrow H$ or $H_n \downarrow H$, then $H \in \mathcal{H}$. As the smallest monotone class including an algebra is a σ-algebra (Halmos, 1950, p. 27) the Lemma follows.

Now $\Omega = T_0 \times T$. For any product measurable set $C \subset \Omega$ and $x \in T_0$, let $C_x := \{y \in T: \langle x,y\rangle \in C\}$, and $Q(C) := \int \mu_x(C_x)d\mu_0(x)$. Here $x \to \mu_x(C_x)$ is measurable if H is a finite union of products $A_i \times F_i$ where $A_i \in \mathcal{A}_0 \otimes \mathcal{B}$ and F_i is product measurable in T. Thus by monotone convergence, $x \to \mu_x(H_x)$ is measurable for any product measurable set $H \subset \Omega$. Thus Q is defined. It is then clearly a countably additive probability measure.

Let p be the natural projection of T_n onto X_n. Recall that $P_{nx} = P_{nj}$ for all $x \in E_{nj}$. The marginal of Q on X_n, or $Q \circ g_n^{-1}$, is

$$\left(\int P_{nx} d\mu_0(x)\right) \circ p^{-1} = \Sigma_{j=0}^{J(k(n))} \mu_0(E_{nj}) P_{nj} \circ p^{-1} = \mu_n \circ p^{-1} = Q_n ,$$

where $J(k(n)) := J_{k(n)}$. Thus Q has marginal Q_n on X_n for each n as desired.

By (4.3), $\Sigma_{k=1}^{\infty} Q_0(X_0 \setminus \cup_{j=1}^{J(k)} f_0^{-1}(U_{kj})) < \Sigma\, 2^{-k} < \infty$. Thus Q_0-almost every $y \in X_0$ belongs to $\cup_{j=1}^{J(k)} f_0^{-1}(U_{kj})$ for all large enough

k. Also if $t \in I_0$ and $t < 1$, then by (4.5), $t < g(k,n,j)$ for all $j \geq 1$ as soon as $1 - 2^{-k} > t$. Thus for μ_0-almost all $\langle y,t\rangle$, there is an m such that

$$\langle y,t\rangle \in \cup_{j=1}^{J(k(n))} E_{nj}$$

for all $n \geq m$. Since $\operatorname{diam}(U_{kj}) < 1/k$ for each j,

$$Q^*(d(f_n(g_n),f_0(g_0)) > 1/k(n) \quad \text{for some} \quad n \geq m)$$
$$\leq \mu_0(\{\langle g_0,t\rangle \in E_{n0} \quad \text{for some} \quad n \geq m\}) \to 0$$

as $m \to \infty$, so $f_n(g_n) \to f_0(g_0)$ almost uniformly.

Lastly, let us see that the g_n are perfect. Suppose $Q(A) > 0$ for some A. Now

$$Q(A) = \int \mu_x(A_x)d\mu_0(x).$$

Then for some x, $\mu_x(A_x) > 0$. Given $n \geq 1$, if $\mu_0(E_{n0}) = 0$, we take $x \notin E_{n0}$. Now $T = T_n \times \mathfrak{T}^{(n)}$ where $T^{(n)} := \prod_{1\leq i\neq n} T_i$. Then on T, $\mu_x = P_{nx} \times Q_{nx}$ for some law $Q_{nx} = \prod_{m\neq n} P_{mx}$ on $T^{(n)}$. Let $A(x) := A_x$. By the Tonelli-Fubini theorem,

$$\mu_x(A_x) = \iint 1_{A(x)}(u,v)dP_{nx}(u)dQ_{nx}(v).$$

Thus for some v, $\int 1_{A(x)}(u,v)dP_{nx}(u) > 0$. Choose and fix such a v as well as x. Now $P_{nx} = P_{nj}$ for $j = j(n,x)$ with $\mu_0(E_{nj}) > 0$. Let $u = (s,t)$, $s \in X_n$, $t \in I_n$. Then since $P_{nj} = Q_n \times \lambda$ restricted to a set of positive measure and normalized,

$$0 < \iint 1_{A(x)}(s,t,v)dQ_n(s)dt.$$

Choose and fix a t with

$$0 < \int 1_{A(x)}(s,t,v)dQ_n(s).$$

Let $C := \{s \in X_n: (s,t,v) \in A_x\}$. So $Q_n(C) > 0$. Clearly $C \subset g_n(A)$, so g_n is perfect, $n \geq 1$, by Theorem 2.1, 1), finishing the proof of Theorem 4.1.

5. Definitions and stability of Donsker classes. Recall the notations of the Abstract. Let $\rho_P(f,g) := (f-g,f-g)_{P,0}^{1/2}$. First, some properties of $G_P BUC$ classes will be developed. A process $(\omega,f) \longrightarrow Y(f)(\omega)$, $f \in \mathcal{F}$, $\omega \in \Omega$, for some probability space Ω, which has the finite-dimensional distributions of G_P, will be called a suitable G_P iff for all ω, $f \longrightarrow Y(f)(\omega)$ is bounded and uniformly continuous for ρ_P. Let $Pf := \int f dP$, $f \in \mathcal{L}^1(P)$.

5.1 Theorem. Let Y be a suitable G_P on $\mathcal{F}$. Then a) for almost all ω, the function defined as Y on $\mathcal{F}$ and 0 on constant functions is well-defined and extends uniquely to a linear functional on the linear span of $\mathcal{F}$ and constant functions, ρ_P-uniformly continuous on the symmetric convex hull of $\mathcal{F}$; b) there is a complete separable linear subspace M of $\ell^\infty(\mathcal{F})$ and a Borel probability measure γ on M such that every suitable G_P has its sample functions in M and has distribution γ there.

Proof. First, $\mathcal{F}$ is totally bounded for ρ_P (Dudley, 1967b, Prop. 3.4, p. 295). Let M be the set of all real functions on $\mathcal{F}$ uniformly continuous for ρ_P. Then $Y(\cdot)(\omega) \in M$ for each ω. The functions in M are bounded and it is separable and complete in $\ell^\infty(\mathcal{F})$. Let $\mathcal{H} := \{h_k\}_{k\geq 1}$ be a countable dense set in $\mathcal{F}$ for ρ_P. On M, $\|\cdot\|_{\mathcal{H}} = \|\cdot\|_{\mathcal{F}}$. For each $g \in M$, $\omega \longrightarrow \|Y-g\|_{\mathcal{H}}$ is measurable. Thus for any Borel set $B \subset M$, $\gamma(B) := \Pr(Y \in B)$ is defined, and does not depend on which suitable G_P was chosen, proving b).

Next, for any $f \in \mathcal{L}^2(P)$ let $\pi f := f - Pf$, $H_0 := \{f \in \mathcal{L}^2(P): Pf = 0\}$, so that π is the usual projection from $\mathcal{L}^2(P)$ onto H_0. For the usual $\mathcal{L}^2$ metric $e_P(\pi u,\pi v) = \rho_P(u,v)$ for any $u,v \in \mathcal{L}^2(P)$. Let $g_k := \pi h_k$, $k = 1,2,\cdots$. Then $\{g_k\}_{k\geq 1}$ is dense in $\pi\mathcal{F}$ for e_P. By Gram-Schmidt orthonormalization, let $\{\varphi_m\}_{m\geq 1}$ be an orthonormal basis of the linear span of $\pi\mathcal{F} \subset H_0$, where for a subsequence $\{j_m\}$ of $\{g_k\}$ and some real a_{mi},

$$\varphi_m = \Sigma_{i=1}^{m} a_{mi} j_i$$

for each m. Let $W_m := \Sigma_{i=1}^{m} a_{mi} Y(j_i)$. Then the W_m are i.i.d. $N(0,1)$ ("orthoGaussian") variables. For each $f \in \mathcal{L}^2(P)$,

$$f \sim Pf + \Sigma_{m\geq 1} (f,\varphi_m)\varphi_m,$$

convergent in $\mathcal{L}^2$, and

$$Y(f) = \Sigma_{m\geq 1} (f,\varphi_m)W_m \quad \text{a.s.}$$

by the three-series theorem.

If $\pi f = \pi h$, $f,h \in \mathcal{F}$, then $\rho_P(f,h) = 0$ so $Y(f) \equiv Y(h)$. Thus we can set $W(\pi f) := Y(f)$ for all $f \in \mathcal{F}$ and W is well-defined on $\pi\mathcal{F}$. Since $f \longrightarrow \pi f$ is an isometry for ρ_P, W has each sample function bounded and uniformly continuous on $\pi\mathcal{F}$ for ρ_P. Thus each $W(\cdot)(\omega)$ extends uniquely to the ρ_P-closure B of $\pi\mathcal{F}$, remaining ρ_P-uniformly continuous and bounded. Now B is compact for ρ_P. On B, W is isonormal, i.e. a Gaussian process with mean 0 and covariance equal to the inner product. Now B is a GC-set in H_0 (Dudley 1967b, 1973). It follows that almost surely the series $\Sigma_m (h,\varphi_m)W_m$ converges uniformly in $h \in \pi\mathcal{F}$ (Feldman, 1971; Dudley, 1973, Theorem 0.3), hence to the ρ_P-uniformly continuous function $h \longrightarrow W(h)(\omega)$. So almost surely

$$Y(f) = \Sigma_m (\pi f,\varphi_m)W_m$$

for $f = h_k$ for all k and thus, by ρ_P-uniform continuity of both sides, for all $f \in \mathcal{F}$. The right side is linear in f and convergent for all f in the linear span of $\mathcal{F}$, except on an event of probability 0 not depending on f. The right side is 0 on constants. It follows that it is uniformly continuous for ρ_P on the closed (for ρ_P), symmetric, convex hull of $\mathcal{F}$ (Feldman, 1971, Theorem 3; Dudley, 1967, Theorem 4.6). This proves a), Q.E.D.

We have the countable product $(X^\infty, \mathcal{A}^\infty, P^\infty)$ and let

$$(\Omega, \mathcal{S}, \Pr) := (X^\infty, \mathcal{A}^\infty, P^\infty) \times ([0,1], \mathcal{B}, \lambda)$$

where $\mathcal{B}$ is the Borel σ-algebra. $\mathcal{F}$ is called a <u>functional Donsker class</u> (for P) iff it is G_PB UC and there exist processes $Y_j(f,\omega)$, $f \in \mathcal{F}$, $\omega \in \Omega$, where Y_j are independent, suitable G_P processes, such that for every $\varepsilon > 0$

$$\lim_{n\to\infty} \Pr^*\{n^{-1/2}\max_{m\le n}\|\Sigma_{j=1}^m(\delta_{x(j)}-P-Y_j\|_{\mathcal{F}} > \varepsilon\} = 0$$

(Dudley and Philipp, 1983; Dudley, 1984).

5.2 <u>Theorem</u>. Let $\mathcal{F} \subset \mathcal{L}^2(X,\mathcal{A},P)$ be G_PB UC . Then the following are equivalent:

1) $\mathcal{F}$ satisfies the central limit theorem as in Sec. 1, i.e. for every bounded real function H on $\ell^\infty(\mathcal{F})$ continuous for $\|\cdot\|_{\mathcal{F}}$, $\int^* H(\nu_n)dP^n \to EH(G_P)$ as $n \to \infty$;

2) for every $\varepsilon > 0$ there is a $\delta > 0$ and an N such that for $n \ge N$

$$\Pr^*\{\sup[|\nu_n(f-g)|: f,g \in \mathcal{F}, \rho_P(f,g) < \delta] > \varepsilon\} < \varepsilon.$$

3) $\mathcal{F}$ is a functional Donsker class.

<u>Proof</u>. 1) implies 2): given $\varepsilon > 0$, take $0 < \delta \le \varepsilon/3$ such that for any suitable G_P,

$$\Pr\{\sup\{|G_P(f)-G_P(g)|: \rho_P(f,g) < \delta\} > \varepsilon/3\} < \varepsilon/2 .$$

Note that such events are measurable by Theorem 5.1, corresponding to open sets in M. By the extended Wichura theorem (4.1), for $n \ge N$ large enough we may assume $\Pr^*\{\|\nu_n-G_P\|_{\mathcal{F}} \ge \varepsilon/3\} < \varepsilon/2$.
If $\|\nu_n-G_P\|_{\mathcal{F}} < \varepsilon/3$ and $|G_P(f)-G_P(g)| \le \varepsilon/3$ then $|\nu_n(f)-\nu_n(g)| < \varepsilon$.
Thus 2) holds with the δ and N chosen.

2) implies 3): as noted, the G_PBUC assumption implies $\mathcal{F}$ is totally bounded for ρ_P. Then 2) is equivalent to 3)

(Dudley, 1984, Theorem 4.1.1).

3) implies 1): given $\varepsilon > 0$, by Theorem 5.1 and Ulam's theorem take a compact $K \subset M$ with $\gamma(K) > 1-\varepsilon$. Let H be bounded and continuous on $\ell^\infty(\mathcal{F})$. For some $\delta > 0$, if $\|u-v\|_{\mathcal{F}} < \delta$, $u \in K$, and $v \in \ell^\infty(\mathcal{F})$, then $|H(u)-H(v)| < \varepsilon$ (e.g. Dudley, 1966, Lemma 1). Take $n_0 = n_0(\varepsilon)$ such that for $n \geq n_0$,

$$\Pr^*(\|\nu_n - T_n\|_{\mathcal{F}} \geq \delta) < \varepsilon,$$

where $\nu_n := n^{-1/2} \Sigma_{j=1}^n (\delta_{x(j)} - P)$, $T_n := n^{-1/2} \Sigma_{j=1}^n Y_j$. Then,

$$\Pr^*(|H(\nu_n) - H(T_n)| > \varepsilon) < \varepsilon,$$
$$\int^* H(\nu_n)\, dP^n \leq \varepsilon(1+\sup|H|) + EH(T_n)\ .$$

Now T_n is a suitable G_P, so

$$\limsup_{n\to\infty} \int^* H(\nu_n)\, dP^n \leq EH(G_P) + \varepsilon(1+\sup|H|).$$

Letting $\varepsilon \downarrow 0$ gives 1), Q.E.D.

Next is a stability result.

5.3 <u>Theorem</u>. Let $\mathcal{F}$ be a functional Donsker class. Let $\mathcal{G}$ be the symmetric convex hull of $\mathcal{F}$, and $\mathcal{H}$ the set of all $g \in \mathcal{L}^2(P)$ such that for some $g_m \in \mathcal{G}$, $g_m(x) \to g(x)$ for all x and $\int (g_m - g)^2 dP \to 0$ as $m \to \infty$. Then $\mathcal{H}$ is a functional Donsker class.

<u>Proof</u>. Take $Y_1, Y_2, \cdots$ as in the definition of functional Donsker class. For any $x(1), \cdots, x(n)$,

$$\|\Sigma_{j=1}^n \delta_{x(j)} - P - Y_j\|_{\mathcal{G}} = \|\Sigma_{j=1}^n \delta_{x(j)} - P - Y_j\|_{\mathcal{F}}\ ,$$

using the fact that the $\delta_{x(j)}$, P, and Y_j are all linear on the linear span of $\mathcal{F}$, by Theorem 5.1(a) for the Y_j. Now if $g_m \to g$ as assumed, then $\delta_x(g_m - g) \to 0$ for all x, $P(g_m) \to P(g)$, and

$\rho_P(g_m, g) \to 0$, so $Y_j(g_m)(\omega) \to Y_j(g)(\omega)$ for all j and ω. Thus

$$\|\Sigma_{j=1}^n \delta_{x(j)} - P - Y_j\|_{\mathcal{H}} = \|\Sigma_{j=1}^n \delta_{x(j)} - P - Y_j\|_{\mathcal{G}},$$

and the result follows. □

6. Weighted empirical distribution functions and related families of functions. Let $(A, \mathcal{A}, P)$ be any probability space and X a nonnegative random variable on A. Let X have distribution $\mathcal{L}(X) = Q$ on $[0, \infty[$. Let

$$\mathcal{G}_X := \{M 1_{X \geq M} : 0 < M < \infty\}.$$

Let $G(t) := P(X \leq t)$ and

$$G^{-1}(y) := \inf\{t: G(t) \geq y\}, \quad 0 < y < 1.$$

It is well known that for the uniform distribution λ on $]0,1[$, G^{-1} has distribution Q. Let

$$h_X(t) := \lim_{u \uparrow 1-t} G^{-1}(u) := G^{-1}((1-t)^-), \quad 0 < t < 1.$$

Then h_X is a non-increasing, nonnegative function on $]0,1[$, continuous from the left, whose distribution for λ is also Q.

To deal with classes $\mathcal{G}_X$, classes with an exponent on M will also be helpful.

6.1 Lemma. For any $X \geq 0$ in $\mathcal{L}^p(P)$, $2 \leq p < \infty$, let $\mathcal{F}_{X,p} := \{X M^{(p-2)/2} 1_{X \geq M}\}_{M > 0}$. Then $\mathcal{F}_{X,p}$ is a functional Donsker class.

Proof. For $p = 2$ this follows directly from a central limit theorem of Pollard (1982), see also Dudley (1984, Theorems 11.3.1, 11.1.2). Thus $\{X^{p/2} 1_{X \geq K}\}_{K > 0}$ is a functional Donsker class. Then by Theorem 5.3, so is

$$\{X^{p/2} \Sigma_{i \geq 1} \lambda_i 1_{X \geq N(i)} : \lambda_i \geq 0, \Sigma \lambda_i = 1\},$$

in particular if $0 \le N(1) < N(2) < \cdots$. Then $\Sigma_{i\ge 0} \lambda_i 1_{X\ge N(i)} = \Sigma_{i\ge 0} c_i 1_{N(i)\le X<N(i+1)}$ where $N(0) := 0$ and $0 \le c_0 < c_1 < \cdots \le 1$. Bounded pointwise limits of such sums give all functions $G(X)$ where G is non-decreasing with $0 \le G \le 1$. Then by Theorem 5.3,

$$\{X^{p/2}G(X): G(0) = 0,\ \text{tot. var}(G) \le 1\},$$

where "tot. var" denotes total variation, is a functional Donsker class.

If $\mathcal{F}$ is a functional Donsker class, so is $\{cf: f \in \mathcal{F}\}$ for any constant c, clearly. Now

$$XM^{(p-2)/2} 1_{X\ge M} = X^{p/2}G(X)$$

where $G(x) := (M/x)^{(p-2)/2} 1_{x\ge M}$, $0 < x < +\infty$. Then G has total variation 2, and the result follows.

6.2 <u>Theorem</u>. For any probability space $(A, \mathcal{A}, P)$ and nonnegative random variable X on it, the following are equivalent:

a) $\mathcal{G}_X$ is a functional P-Donsker class on A;

b) $\mathcal{G}_{id}$ is a functional Q-Donsker class on $[0,\infty[$, where id is the identity function, $Q = \mathcal{L}(X)$;

c) $\mathcal{G}_h$ is a functional λ-Donsker class on $]0,1[$, $h := h_X$.

<u>Proof</u>. As above, let $x(1), x(2), \cdots$, be coordinates on A^∞. Then $X(x(j))$, $j = 1,2,\cdots$, are i.i.d. (Q) in $\mathbb{R}$. For any $M \ge 0$ and rational $q(j) \uparrow M$, $q(j)1_{X\ge q(j)}$ converge boundedly and pointwise, everywhere, to $M1_{X\ge M}$. Thus the supremum over $\mathcal{G}_X$ of any finite signed measure on $(A, \mathcal{A})$ equals the sup over the countable set $\{q1_{X\ge q}: q \text{ rational}, q \ge 0\}$. The same holds for id on $\mathbb{R}$ or h on $]0,1[$. Thus there is no difficulty about measurability of such suprema for empirical measures.

For any $\omega \in A^\infty$, $n = 1,2,\cdots$, and $0 \le M < \infty$, $P_n(X \ge M) = Q_n(\text{id} \ge M)$ where $Q_n := n^{-1} \Sigma_{j=1}^n \delta_{X(x(j))}$ serves as an empirical measure for Q. Also, $P(X \ge M) = Q(\text{id} \ge M)$ for all M. Distances

ρ_P between functions $M1_{X\geq M}$ equal ρ_Q distances between the corresponding functions $M1_{id\geq M}$. Thus the asymptotic equicontinuity criterion (Theorem 5.2,(2)) holds for $\mathcal{G}_X$ if and only if for $\mathcal{G}_{id}$, so a) and b) are equivalent. Applying this to $A =]0,1[$, $P = \lambda$ shows that c) is also equivalent, Q.E.D.

Condition c) is (in effect) a condition on weighted empirical distribution functions (for λ, near 0). An integral condition, f) below, was shown equivalent to a central limit theorem (and thus to (c) in the present formulation) by Chibisov (1964) if h is regularly varying, then by O'Reilly (1974) if h is continuous. (Always, h is non-decreasing at least on some interval $(0,\delta)$, $\delta > 0$.) The following is an effort to put the Chibisov-O'Reilly result in a somewhat more final form by

i) removing the continuity assumption,

ii) adding several more equivalent conditions, all also considered explicitly or implicitly by other previous authors, but perhaps not in the present combinations ((g1) and (g2)),

iii) collecting a more complete and self-contained proof rather than, e.g., citing the "arguments of O'Reilly" who cites "arguments of Chibisov." There has been some confusion in the literature as noted by M. Csörgő (1984).

The past results have allowed a singularity $h \to \infty$ as $t \uparrow 1$ as well as $t \downarrow 0$. This is natural for empirical distribution functions. The two endpoints are symmetrical via $t \leftrightarrow 1-t$. The present formulation, with h non-decreasing on $]0,1[$, is natural in the context of Theorem 6.2. Results for $h \uparrow +\infty$ as $t \uparrow 1$ and/or $t \downarrow 0$ can be written down easily if desired.

Let $W(t)$ be a standard Wiener process.

6.3 <u>Theorem</u> (O'Reilly, Chibisov, et al.). For any non-increasing function $h \geq 0$ on $]0,1[$ the following are equivalent:

c) $\mathcal{G}_h$ is a functional λ-Donsker class;

d) $\{h(t)1_{[0,t]}(\cdot): 0 \leq t \leq 1\}$ is a functional λ-Donsker class;

e) $h(t)W(t) \to 0$ a.s. as $t \downarrow 0$;

e') $h(2^{-k})W(2^{-k}) \to 0$ a.s. as $k \to \infty$;

f') for every $\delta > 0$, $\Sigma_{k=1}^{\infty} \exp(-2^k\delta/h^2(2^{-k})) < \infty$;

f) for every $\varepsilon > 0$, $\int_0^1 t^{-1}\exp(-\varepsilon/(th^2(t)))dt < \infty$;

g) both

g1) $t^{1/2}h(t) \to 0$ as $t \downarrow 0$, and

g2) for every $\varepsilon > 0$,

$$\int_0^1 t^{-3/2}h(t)^{-1}\cdot\exp(-\varepsilon/(th^2(t)))dt < \infty .$$

Proof. First, c) implies d): if $u_1, u_2, \cdots$ are i.i.d. (λ), then almost surely no u_i falls at any jump of h (there are at most countably many jumps). Thus we may assume h is left continuous in this step. The only difficulty is with possible atoms of Q. There are at most countably many such atoms M_k, with $h(x) = M_k$ for $a_k < x \leq b_k$ (possibly also at a_k) where we can take a_k minimal, and b_k maximal by left continuity. A function $h(t)1_{[0,t]}(\cdot)$, if not of the form $M1_{h\geq M}$, has $a_k \leq t < b_k$ for some k.

Taking finitely many of the intervals $]a_k, b_k]$, say for $k = 1,\cdots,m$, we may number them so that

$$a_1 < b_1 \leq a_2 < b_2 \leq \cdots \leq b_m ,$$
$$M_1 > M_2 > \cdots > M_m .$$

Now

$$M\lambda(h\geq M)^{1/2} \to 0 \quad \text{as} \quad M \to \infty , \tag{6.4}$$

for otherwise the set $\mathcal{G}_h$ is not totally bounded for ρ_λ, contrary to c) (Dudley, 1984, Theorem 4.1.1). Thus, given $\varepsilon > 0$, there is an $M_0 < \infty$ such that $M\lambda(h \geq M)^{1/2} < \varepsilon/16$ for $M \geq M_0$. We may and do choose $M_0 > h(1/2)$. Also, by c) and

Theorem 5.2, let $M_0 := M(0)$ be large enough so that

$$(6.5) \qquad \Pr\{\sup_{M \geq M(0)} |\nu_n(M1_{h \geq M})| \geq \varepsilon/8\} < \varepsilon/4 .$$

For any constant c, $\{c1_{[0,t]}: 0 \leq t \leq 1\}$ is a functional Donsker class (essentially by the classical theorem of Donsker (1952), cf. Theorem 5.2 above and Theorem 7.4 of Dudley and Philipp (1983)). Thus by Theorem 5.3, $\{h(t)1_{[0,t]}: \gamma \leq t \leq 1\} := \mathcal{G}_\gamma$ is a functional Donsker class for any $\gamma > 0$, or for $\gamma = 0$ if h is bounded. So we may assume h unbounded and choose $\gamma > 0$ with $h(\gamma) > h(t) \geq M_0$ as $t \downarrow \gamma$. By Theorem 5.2, take $\delta > 0$ and n_1 such that for $n \geq n_1$,

$$(6.6) \qquad \Pr\{\sup\{|\nu_n(f-g)|: f,g \in \mathcal{G}_\gamma, \operatorname{Var}(f-g) < \delta^2\} > \tfrac{\varepsilon}{2}\} < \tfrac{\varepsilon}{2} .$$

Now let us restrict to intervals with $b_k \leq \gamma$, so that $b_m \leq \gamma$ and $M_m \geq M_0$. For any given n and ω, let τ be the least t such that for some $k \leq m$, $t \in [a_k,b_k]$ and $M_k|\nu_n([0,t])| \geq \varepsilon/2$, if such a t exists. Note that $(t,\omega) \longrightarrow \nu_n([0,t])(\omega)$ is a strong Markov process. Given τ and $\nu_n([0,\tau])$, the conditional distribution of $\nu_n([0,b_k])$ (an affine function of a binomial variable) has mean $\nu_n([0,\tau])(1-b_k)/(1-\tau)$, whose absolute value is at least $|\nu_n(0,\tau)|/2$. Note that $|\nu_n(M_k1_{[0,\tau]})|/2 \geq \varepsilon/4$. The conditional variance of $M_k\nu_n([0,b_k])$ is

$$M_k^2 n^{-1}(b_k-\tau)(1-b_k)(1-\tau)^{-2}\, n(1-\lambda_n([0,\tau])) \leq M_k^2 b_k .$$

Thus by Chebyshev's inequality there is a conditional probability at least $3/4$ that

$$M_k|\nu_n(\{h \geq M_k\})| = |M_k\nu_n([0,b_k])| \geq \frac{\varepsilon}{4} - 2M_k b_k^{1/2} \geq \varepsilon/8$$

by choice of $M(0)$ and γ. Then by (6.5),

$$\Pr\{\text{for some } k = 1,\cdots,m \text{ and } t \in [a_k,b_k],$$
$$M_k|\nu_n([0,t])| \geq \varepsilon/2\} = \Pr\{\tau \text{ exists}\}$$

$$\leq \frac{4}{3} \Pr\{M_k |\nu_n([0,b_k])| \geq \varepsilon/8 \quad \text{for some} \quad k\}$$
$$\leq \frac{4}{3} \cdot \frac{\varepsilon}{4} = \varepsilon/3 .$$

Now let the number m of intervals increase to obtain all intervals $]a_j,b_j] \subset [0,\gamma]$. (Note that if for some j, $a_j < \gamma$, then $b_j \leq \gamma$ by choice of γ.) Then another use of (6.5) yields

$$\Pr\{|\nu_n(h(t)1_{[0,t]})| \geq \varepsilon/2 \quad \text{for some} \quad t \leq \gamma\} \leq \varepsilon/3 .$$

Then using (6.6),

$$\text{(6.7)} \quad \begin{array}{l} \Pr\{|\nu_n(h(t)1_{[0,t]} - h(s)1_{[0,s]})| > \varepsilon \ \text{for some} \\ 0 \leq s < t \leq 1 \ \text{ with } \ \mathrm{Var}(h(t)1_{[0,t]} - h(s)1_{[0,s]}) < \delta^2\} < \varepsilon, \end{array}$$

noting that if $s < \gamma < t$ then

$$\mathrm{Var}(h(\gamma)1_{[0,\gamma]} - h(t)1_{[0,t]}) < \delta^2 .$$

The function $t \longmapsto h(t)1_{[0,t]}(\cdot)$ in $\mathcal{L}^2([0,1],\lambda)$ is left continuous and has limits from the right, with a right limit 0 at 0 by (6.4). Thus $\mathcal{G}_h$ is totally bounded in $\mathcal{L}^2$. This and (6.7), with Theorem 4.1.1 of Dudley (1984), imply d).

Next, d) implies e): clearly d) implies c), giving (6.4) in the last step, which implies g1) in the statement of Theorem 6.3. Since the class in d) is G_PBUC, we have $h(t)B_t \to 0$ a.s. as $t \downarrow 0$, where $B_t := G_P(1_{[0,t]})$ is a standard Brownian bridge process. We can write $W_t = B_t + tG$ where G $(= W_1)$ is a standard normal variable independent of B_t, so e) follows. Clearly e) implies e').

Next, e') implies f'): since $h(2^{-k}) \leq h(2^{-k-1})$, we have $h(2^{-k})W(2^{-k-1}) \to 0$ a.s., so $h(2^{-k})(W(2^{-k})-W(2^{-k-1})) \to 0$ a.s. Since $W(2^{-k}) - W(2^{-k-1})$ are independent and equal to $2^{-(k+1)/2}G_k$ for i.i.d. standard normal variables G_k, this means that for every $\varepsilon > 0$, by the Borel-Cantelli lemma

$$\Sigma_{k=1}^{\infty} \Pr(|G_k| > 2^{(k+1)/2}\varepsilon/h(2^{-k})) < \infty .$$

As $x \to \infty$,

$$\Pr(|G_1|>x) \sim (2/\pi)^{1/2} x^{-1} \cdot \exp(-x^2/2) \geq \exp(-x^2) .$$

Thus

$$\Sigma_{k=1}^{\infty} \exp(-2^{k+1} \varepsilon^2 / h^2(2^{-k})) < \infty .$$

Letting $\varepsilon = (\delta/2)^{1/2}$ proves f').

Now f') implies f): given $\varepsilon > 0$, if $\delta := \varepsilon/2$,

$$\begin{aligned} \infty &> \Sigma_{k=1}^{\infty} \exp(-\delta 2^k / h^2(2^{-k})) \\ &= \Sigma_{j=0}^{\infty} (2^{-j} - 2^{-j-1}) 2^{j+1} \cdot \exp(-\varepsilon 2^j / h^2(2^{-j-1})) \\ &\geq \Sigma_{j=0}^{\infty} \int_{2^{-j-1}}^{2^{-j}} \frac{1}{t} \exp(-\varepsilon/(th^2(t)))\, dt \\ &= \int_0^1 t^{-1} \exp(-\varepsilon/(th^2(t)))\, dt . \end{aligned}$$

Conversely, f) implies f'): given $\delta > 0$, if $\varepsilon := \delta/2$,

$$\begin{aligned} \infty &> \int_0^1 \frac{1}{t} \exp(-\varepsilon/(th^2(t)))\, dt \\ &= \Sigma_{k=0}^{\infty} \int_{2^{-k-1}}^{2^{-k}} \frac{1}{t} \exp(-\varepsilon/(th^2(t)))\, dt \\ &\geq \Sigma_{k=0}^{\infty} (2^{-k} - 2^{-k-1}) 2^k \exp(-\varepsilon 2^{k+1} / h^2(2^{-k})) \\ &= \frac{1}{2} \Sigma_{k=0}^{\infty} \exp(-\delta 2^k / h^2(2^{-k})) . \end{aligned}$$

Also, f) implies g): let $q(t) := 1/h(t)$. For any $t > 0$,

$$\begin{aligned} &\int_{t/2}^{t} s^{-1} \exp(-q^2(s)/(2s))\, ds \\ &\qquad \geq \int_{t/2}^{t} s^{-1} \exp(-q^2(t)/t)\, ds \\ &\qquad = (\log 2) \exp(-q^2(t)/t) . \end{aligned}$$

Letting $t \downarrow 0$ gives g1). So for any $\varepsilon > 0$ and t small enough,

$$q(t)/t^{1/2} < \exp(\varepsilon q^2(t)/(2t)) .$$

Then the integrand in g2) is smaller than that of f) for $\varepsilon/2$, so g2) holds.

Conversely, g) implies f): if g1) holds then as $t \downarrow 0$, $t^{-1} < t^{-3/2}/h(t)$ and f) follows. So f), f') and g) are all equivalent.

It remains to show that these conditions imply c). Assume f), f') and g). Let

$$g(t) := M_k := \max(h(2^{-k-1}),1) \quad \text{for} \quad 2^{-k-1} < t \leq 2^{-k},$$
$$k = 0,1,\cdots .$$

Then $h(t) \leq g(t)$, $0 < t \leq 1$. If c) holds for g, then so does d), then d) for h by Theorem 5.3, then c) for h (recall that d) implies c) directly). So it is enough to prove that

$$\{M_k 1_{[0,2^{-k}]}\}_{k\geq 0}$$

is a functional Donsker class, given that $M_k \leq M_{k+1}$ for all k and

$$(6.8) \qquad \Sigma_{j=0}^{\infty} \exp(-2^j \alpha / M_j^2) < \infty$$

for every α $(= \delta/2) > 0$. Then clearly $2^j/M_j^2 \to \infty$ as $j \to \infty$. So for some j_1,

$$(6.9) \qquad 2^j \geq M_j^2, \ j \geq j_1.$$

The following is essentially Lemma 4 of Chibisov (1964). Let $\nu_n(u) := \nu_n([0,u])$, $0 \leq u \leq 1$.

6.10 Lemma. If f) holds then for any $a < \infty$ and $\varepsilon > 0$,

$$\lim_{n\to\infty} \Pr\{\sup_{0<u\leq a/n} |\nu_n(h(u)1_{[0,u]})| \geq \varepsilon\} = 0 .$$

Proof. Let $P_{a,n} := \Pr\{|\nu_n(u)| < \varepsilon/h(u), \ 0 < u \leq a/n\}$

$$= \Pr\{|\nu_n(v/n)| < \varepsilon/h(v/n), \ 0 < v \leq a\}$$

$$= \Pr\{|n\lambda_n(v/n)-v| < \varepsilon n^{1/2}/h(v/n)), \quad 0 < v \leq a\}.$$

Let $c_n := \inf_{0<u\leq a/n} u^{-1/2}/h(u) \to \infty$ as $n \to \infty$ by g1). Note that $\varepsilon n^{1/2}/h(v/n) = \varepsilon v^{1/2}/(h(v/n)(v/n)^{1/2}) \geq \varepsilon v^{1/2}c_n$. Thus

$$P_{a,n} \geq \Pr(|n\lambda_n(v/n)-v| < \varepsilon v^{1/2}c_n, \; 0 < v \leq a).$$

Let $m_n := n\lambda_n(a/n)$. Then as $n \to \infty$, $m := m_n$ converges in law to a Poisson variable with parameter a. In particular, m is bounded in probability.

Let μ be the uniform distribution on $[0,a]$. Then we can write

$$n\lambda_n(v/n) = m\mu_m(v), \quad 0 \leq v \leq a.$$

Given m,

$$\Pr\{|m\mu_m(v)-v| < \varepsilon c_n v^{1/2}, \quad 0 < v \leq a\} \to 1$$

as $n \to \infty$ because a.s. $\mu_m(\{0\}) = 0$ and then $|m\mu_m(v) - v|$ is bounded and equal to v in a neighborhood of 0, so less than $c_n \varepsilon v^{1/2}$ for n large. Thus $P_{a,n} \to 1$, proving Lemma 6.10.

To continue proving c), for $0 \leq p \leq 1$ let

$$B(x,n,p) := \Sigma_{0\leq k\leq x}\binom{n}{k}p^k(1-p)^{n-k},$$

$$E(x,n,p) := \Sigma_{x\leq k\leq n}\binom{n}{k}p^k(1-p)^{n-k},$$

where k runs through integers but x need not be an integer. Then for any $\gamma > 0$,

$$\Pr\{|\nu_n([0,t])| > \gamma) = E(nt+n^{1/2}\gamma,n,t) + B(nt-n^{1/2}\gamma,n,t).$$

For E and B there are Chernoff-Okamoto bounds (e.g. Dudley, 1984, 2.2.7, 2.2.8): for $0 < t \leq 1/2$,

$$(6.11)\qquad B(nt-n^{1/2}\gamma,n,t) \le \exp(-\gamma^2/(2t(1-t))) \le \exp(-\gamma^2/(2t)),$$

$$(6.12)\qquad E(nt+n^{1/2}\gamma,n,t) \le (nt/(nt+n^{1/2}\gamma))^{nt+n^{1/2}\gamma}e^{n^{1/2}\gamma}$$

$$= ((1+y)^{1+y}e^{-y})^{-nt}$$

where $y := \gamma/(n^{1/2}t)$. (Note: Chibisov [1964, Lemma 5] gave a bound of the form (6.12) for Poisson probabilities with $\lambda = nt$, $z = n^{1/2}\gamma$.)

Now for $0 < \beta < 1$ there is a largest $y_\beta > 0$ such that $\log(1+y) \ge \beta y$ for $0 \le y \le y_\beta$. As $\beta \downarrow 0$,

$$(6.13)\qquad y_\beta \uparrow +\infty \quad\text{and}\quad \log(1+y_\beta) = \beta y_\beta \to +\infty .$$

Following Chibisov [1964, after Lemma 5],

$$(1+y)\cdot\log(1+y) - y = \int_0^y \log(1+x)dx \ge \beta y^2/2,\ 0 \le y \le y_\beta .$$

Then

$$(6.14)\qquad E(nt+n^{1/2}\gamma,n,t) \le \exp(-nt\beta y^2/2) = \exp(-\beta\gamma^2/(2t)) .$$

Next,

$$(6.15)\qquad (1+y)\cdot\log(1+y) - y > (1+y)\cdot\log(1+y_\beta) - y$$

$$> y(\beta y_\beta - 1),\ y > y_\beta .$$

Given $\varepsilon > 0$, choose and fix a $\beta > 0$ small enough so that by (6.13)

$$(6.16)\qquad (\beta y_\beta - 1)\varepsilon > 1.$$

Setting $t = 2^{-k}$ and $\gamma = \varepsilon/M_k$, $k = 1,2,\cdots$, we have by (6.11) and (6.8) for some $j(0)$ large enough

$$(6.17)\qquad \begin{aligned}&\Sigma_{k\ge j(0)} \Pr\{M_k\nu_n([0,2^{-k}]) < -\varepsilon\}\\ &= \Sigma_{k\ge j(0)} B(nt-n^{1/2}\varepsilon/M_k,n,2^{-k})\\ &\le \Sigma_{k\ge j(0)} \exp(-\varepsilon^2 2^{k-1}/M_k^2) < \varepsilon/4 .\end{aligned}$$

For the upper tail terms in (6.14) with $y = \gamma/(n^{1/2}t) = 2^k \varepsilon/(n^{1/2}M_k) \le y_\beta$, we also have by (6.8)

$$(6.18) \qquad \Sigma_{k\ge j', y\le y_\beta} \exp(-\beta\varepsilon^2 2^{k-1}/M_k^2) < \varepsilon/4$$

for some $j' \ge \max(j(0), j_1)$.

It remains to treat the upper tail terms E with $y > y_\beta$. Choose $a < \infty$ large enough so that

$$(6.19) \qquad \exp(-a^{1/2})/(1-\exp(-a^{1/2}(\log 2)/2)) < \varepsilon/4 .$$

By Lemma 6.10, we may restrict to those k such that $2^{-k} \ge a/n$, or $k \le r := r(n) := [\log_2(n/a)]$. Let $M(n)$ be the set of all k with $j' \le k \le r(n)$ for which $y \equiv 2^k\varepsilon/(n^{1/2}M_k) > y_\beta$. Then by (6.12) and (6.15) it suffices to bound

$$T_n := \Sigma_{k\in M(n)} \exp(-n2^{-k}(2^k\varepsilon/(n^{1/2}M_k))(\beta y_\beta-1)).$$

Then

$$\begin{aligned} T_n &\le \Sigma_{k\in M(n)} \exp(-n^{1/2}/M_k) && \text{by (6.16)} \\ &\le \Sigma_{k\in M(n)} \exp(-2^{-k/2}n^{1/2}) && \text{by (6.9)} \\ &\le \Sigma_{s=0}^{\infty} \exp(-2^{(s-r)/2}n^{1/2}) \\ &\le \Sigma_{s=0}^{\infty} \exp(-2^{s/2}a^{1/2}) \end{aligned}$$

since $2^{-r}n \ge a$ by choice of $r := r(n)$. Now $2^{s/2} = \exp(s(\log 2)/2) \ge 1 + s(\log 2)/2$ for all s, so

$$T_n \le \exp(-a^{1/2})/(1-\exp(-a^{1/2}(\log 2)/2)) < \varepsilon/4$$

by a geometric series and (6.19). Take n_0 such that the probability in Lemma 6.10 is less than $\varepsilon/4$ for $n \ge n_0$. Combining with (6.17) and (6.18) then gives

$$(6.20) \qquad \Sigma_{k\ge j'} \Pr(M_k|\nu_n([0,2^{-k}])| > \varepsilon) < \varepsilon, \quad n \ge n_0 .$$

Let $\delta := 2^{-(j'+3)/2}$, $f_k := M_k 1_{[0,2^{-k}]}$.
If $i < k$, then

$$\rho_P(f_i,f_k)^2 = \int_0^1 (f_i - M_i 2^{-i} - f_k + M_k 2^{-k})^2\,dx$$
$$\geq \int_{2^{-k}}^{2^{-i}} (M_i - 2^{-i}M_i + M_k 2^{-k})^2\,dx$$
$$\geq 2^{-i-3}M_i^2 \geq 2^{-j'-3} = \delta^2$$

if $i \leq j'$, as $M_i \geq 1$ for all i. Thus if $\rho_P(f_i,f_k) < \delta$, then $i > j'$. So for $n \geq n_0$, by (6.20),

$$\Pr\{\sup\{|\nu_n(f_i - f_k)|: \rho_P(f_i,f_k) < \delta\} > 2\varepsilon\} < \varepsilon.$$

As $i \to \infty$, $\rho_P(f_i,0) \to 0$ by (6.8) and the sentence after it. Thus $\{f_i\}$ is totally bounded for ρ_P, and is a functional Donsker class (Dudley, 1984, Theorem 4.1.1). So c) holds and Theorem 6.3 is proved.

7. Corollaries and remarks on Sec. 6.

7.1 Corollary. If $h \geq 0$ is any non-increasing function on $]0,1[$ with $h(t) = o((t \log\log \frac{1}{t})^{-1/2})$ then the conditions of Theorem 6.3 all hold.

Proof. It is easy to verify f).

7.2 Example. It will be shown that for a positive decreasing h, the sufficient condition in Cor. 7.1 is not necessary for its conclusion. This will be a counter-example to Shorack [1979, (3), (4)] and thus to Shorack and Wellner (1982, Theorem 1.1). A review of the latter by M. Csörgő (1984) also noted the error, stating that S. Csörgő and D. M. Mason also found such examples.

Let $t_k := \exp(-e^k)$, $k = 1,2,\cdots$, and $M_k := (t_k \cdot \log|\log t_k|)^{-1/2} = (kt_k)^{-1/2}$. Let $h(t) := M_k$ for $t_{k+1} < t \leq t_k$, $k = 1,2,\cdots$, $h(t) := M_0 := 1$ for $t_1 < t < t_0 := 1$. Then h is non-increasing and $h(t) \neq o((t \log \log \frac{1}{t})^{-1/2})$ as $t \downarrow 0$.

For any $\mathcal{E} > 0$,

$$\int_0^1 \frac{1}{t} \exp(-\mathcal{E}/(th^2(t)))\, dt =$$

$$\Sigma_{k=0}^{\infty} \int_{t_{k+1}}^{t_k} \frac{1}{t} \exp(-\mathcal{E}/(tM_k^2))\, dt$$

$$= \Sigma_{k=0}^{\infty} \int_{t_{k+1}}^{t_k} \frac{1}{t} \exp(-\mathcal{E}\, k \cdot t_k/t)\, dt .$$

Each term in the sum is clearly finite. Note that $\frac{d}{dt}(\frac{1}{t} \exp(-C/t)) \geq 0$ if $C \geq t$. Thus the sum, for $k \geq 1/\mathcal{E}$, is bounded above by

$$\Sigma_{k \geq 1/\mathcal{E}} (t_k - t_{k+1}) t_k^{-1} e^{-k\mathcal{E}}$$

$$\leq \Sigma_{k \geq 1/\mathcal{E}} e^{-k\mathcal{E}} < \infty .$$

So f) does hold, as do the other conditions of Theorem 6.3.

Nevertheless, Cor. 7.1 is sharp in at least two senses, shown by the next two results.

7.3 Corollary. If for some $\delta > 0$, $h(t) \geq \delta(t \log|\log t|)^{-1/2}$ for $0 < t \leq \delta$, then $\mathcal{G}_h$ is not a functional Donsker class.

Proof. If $\mathcal{G}_h$ is a functional Donsker class, then so is $\mathcal{G}_j$ where $j(t) := \delta(t \log|\log t|)^{-1/2}$, $0 < t \leq \delta$; $j(t) := 0$, $\delta < t \leq 1$, by Theorem 5.3 (one can take $\delta < e^{-2}$ so that j is non-increasing). But this contradicts Theorem 6.3f) for $0 < \mathcal{E} \leq \delta^2$, Q.E.D.

7.4 Proposition. If $h \geq 0$, $h \downarrow$ and $t \longrightarrow t^{1/2}h(t) \uparrow$ (is non-decreasing) for $0 < t \leq 1$, then the conditions of Theorem 6.3 are equivalent to $h(t) = o((t \log \log \frac{1}{t})^{-1/2})$ as $t \downarrow 0$.

Proof. Cor. 7.1 gives one direction. For the other, suppose that for some $\delta > 0$, and $t_k \downarrow 0$, $h(t_k) \geq \delta(t_k \log|\log t_k|)^{-1/2}$ for $k = 1,2,\cdots$. Taking a subsequence, it can be assumed that

$t_{k+1} < t_k^2$ for all k. For $t_k \le t \le t_{k-1}$, since $t^{1/2}h \uparrow$, $th^2(t) \ge t_k h^2(t_k) \ge \delta^2/(\log|\log t_k|)$. Then for $\mathcal{E} := \delta^2$,

$$\int_{t_k}^{t_{k-1}} \frac{1}{t} \exp(-\mathcal{E}/(th^2(t)))\, dt$$

$$\ge \int_{t_k}^{t_{k-1}} \frac{1}{t} \exp(-\log|\log t_k|)$$

$$= (\log t_{k-1} - \log t_k)/|\log t_k|$$

$$= 1 - |\log t_{k-1}|/|\log t_k| \ge \frac{1}{2} .$$

Thus the sum of these numbers over k diverges and f) fails, Q.E.D.

7.5 Remarks. Note that in Theorem 6.3 g2) does not imply g1): let $h(t) := 1/t$. Nor does g1) imply g2): let $h(t) := 1/(tLLL(1/t))^{1/2}$ as $t \downarrow 0$ where $Lx := \max(1, \log x)$.

If $t^{1/2}h(t)$ is non-decreasing, then g2) implies e) by the classical Kolmogorov-Petrovskii test, see Petrovskii (1935), Erdős (1942), and Itô and McKean (1974, pp. 33-35). O'Reilly (1974, p.644) states that $t^{1/2}h$ non-decreasing is "not needed for the relevant half of the test". This is correct in the context there, where g1) has been proved, but not in general: if $h(t) = 1/t$, g2) holds but not e).

Itô and McKean (1974, pp.33-35) prove that g) implies e), in effect, although they do not explicitly make the conjunction of g1) with g2). At any rate, integral condition f) seems preferable to g2), as f) is simpler and does not require an extra condition such as g1).

O'Reilly's (1974) assumption that h is continuous would also follow from the two assumptions $h\downarrow$ and $t^{1/2}h\uparrow$. At any rate, continuity is now seen to be unnecessary.

7.6 Corollary. On any probability space $(A, \mathcal{a}, P)$, for any function

$h \in \mathcal{L}^2(P)$, $\mathcal{G}_h$ is a functional Donsker class.

Proof. By Theorem 6.2, one can take $(A, \mathcal{A}, P) = (]0,1[, \mathcal{B}, \lambda)$, and h non-increasing. Then

$$\Sigma_{k=1}^{\infty} 2^{-k-1} h(2^{-k})^2 \leq \int_0^1 h^2(t)\, dt < \infty .$$

Let $a_k := 2^k/h(2^{-k})^2$. Then $\Sigma_{k=1}^{\infty} 1/a_k < \infty$. Thus for every $\varepsilon > 0$, $\Sigma_k \exp(-\varepsilon a_k) < \infty$, which gives f') of Theorem 6.3, hence c), the conclusion.

Note. Cor. 7.6 is not surprising since by a theorem of D. Pollard (1982), $\{h1_{h\geq M}: 0 \leq M < \infty\}$ is a functional Donsker class and $M1_{h\geq M} \leq h1_{h\geq M}$. Still, Cor. 7.6 is sharp in the following senses:

7.7 Proposition. For any sequence $t_k \to 0$ (however slowly) as $k \to \infty$, there exist $M_k \uparrow +\infty$ as $k \to \infty$ such that $\Sigma_{k=1}^{\infty} t_k M_k^2/2^k < \infty$ but $M_k^2/2^k \not\to 0$ as $k \to \infty$, so $\mathcal{F} := \{M_k 1_{[0,2^{-k}]}\}_{k\geq 1}$ is not a functional Donsker class.

For any function $\varphi(x) \downarrow 0$ (however slowly) as $x \to +\infty$, there is a function $h\downarrow$ such that

$$\int_0^1 \varphi(1/t) h^2(t) dt < \infty \tag{7.8}$$

and $\mathcal{G}_h$ is not a functional Donsker class. Also, there is an $h\downarrow$ such that

$$\int_0^1 \varphi(h(t)) h^2(t) dt < \infty \tag{7.9}$$

and $\mathcal{G}_h$ is not a functional Donsker class.

Proof. Let $s_k := \sup_{j\geq k} t_j \downarrow 0$ as $k \to \infty$, with $s_k \geq t_k$, so we may assume $t_k \downarrow 0$. Take a subsequence $k(i) \uparrow +\infty$ such that $\Sigma_i t_{k(i)} < \infty$, $k(0) := 1$. Let $M_j := 2^{k(i)/2}$ for $k(i) \leq j < k(i+1)$. Then $M_{k(i)}^2/2^{k(i)} = 1 \not\to 0$. So by g1), $\mathcal{F}$ is not a functional Donsker class. Also

$$\Sigma_k \, t_k M_k^2/2^k \leq \Sigma_{i=0}^{\infty} 2^{k(i)} t_{k(i)} \, \Sigma_{k \geq k(i)} 2^{-k}$$
$$\leq 2\Sigma_{i=0}^{\infty} t_{k(i)} < \infty,$$

as stated.

For (7.8) let $t_k := \varphi(2^k)$, define the M_j as above, and let

$$h(t) := M_j, \; 2^{-j-1} < t \leq 2^{-j}, \; j = 0,1,\cdots .$$

Then $\mathcal{G}_h$ is not functional Donsker class and

$$\int_0^1 \varphi(1/t)h^2(t)dt = \Sigma_{k=0}^{\infty} \int_{2^{-k-1}}^{2^{-k}} \varphi(1/t)M_k^2 \, dt$$
$$\leq \Sigma_{k=0}^{\infty} M_k^2 2^{-k-1} t_k < \infty,$$

so (7.8) holds.

For (7.9), let $t_k := \varphi(2^{k/2})$ and apply the previous arguments.

Thus each of the sufficient conditions in Corollaries 7.1 and 7.6 is sharp in its own ways. Neither implies the other: if $h(t) = (t \log(3/t))^{-1/2}$ then h satisfies the conditions of 7.1 but not 7.6. Conversely, let $s_k := \exp(-\exp(e^k))$, $k = 1,2,\cdots,$ $M_k := (s_k \log|\log s_k|)^{-1/2} = (e^k s_k)^{-1/2}$. Let $h(t) := M_k$ for $s_{k+1} < t \leq s_k$, $k = 1,2,\cdots, h(t) := M_0 := 1$ for $s_1 < t < s_0 := 1$. Then h is non-increasing and $h(t) \neq o((t \log|\log t|)^{-1/2})$ as $t \downarrow 0$. Also

$$\int_0^1 h^2(t)dt = \Sigma_{k=0}^{\infty} M_k^2(s_k - s_{k+1}) \leq \Sigma_{k=0}^{\infty} e^{-k} < \infty .$$

For other recent work related to the equivalent conditions in Sec. 6 see Stute (1982), but in light of M. Csörgő (1984).

This paper does not treat laws of the iterated logarithm. See Dudley and Philipp (1983, Theorem 7.5) and references there.

Acknowledgment. I am grateful to M. Talagrand, J. Hoffmann-Jørgensen and N. T. Andersen for communicating to me some of their unpublished work. Talagrand (1984) stated a result close to Theorem 5.2 above.

REFERENCES

Chibisov, D. M. (1964). Some theorems on the limiting behavior of the empirical distribution function. Trudy Mat. Inst. Steklov (Moscow) 71 104-112; Selected Transls. Math. Statist. Prob. 6 147-156.

Cohn, Donald L. (1980). Measure Theory. Birkhäuser, Boston.

Csörgő, M. (1984). Review of Shorack and Wellner (1982). Math. Revs. 84f:60041.

Donsker, M. D. (1952). Justification and extension of Doob's heuristic approach to the Kolmogorov-Smirnov theorems, Ann. Math. Statist. 23 pp. 277-281.

Drake, F. R. (1974). Set Theory: An Introduction to Large Cardinals. North-Holland, Amsterdam.

Dudley, R. M. (1966). Weak convergence of probabilities on nonseparable metric spaces and empirical measures on Euclidean spaces. Illinois J. Math. 10 109-126.

____________ (1967a). Measures on non-separable metric spaces. Ibid. 11 449-453.

____________ (1967b). The sizes of compact subsets of Hilbert spaces and continuity of Gaussian processes. J. Functional Analysis 1 290-330.

____________ (1968). Distances of probability measures and random variables. Ann. Math. Statist. 39 1563-1572.

____________ (1973). Sample functions of the Gaussian process. Ann. Probab. 1 66-103.

____________ (1978). Central limit theorems for empirical measures. Ibid. 6 899-929; Correction, ibid. 7 (1979) 909-911.

____________ (1984). A course on empirical processes. Ecole d'été de probabilités de St.-Flour, 1982. Lecture Notes in Math. 1097 2-142.

____________, and Walter Philipp (1983). Invariance principles for sums of Banach space valued random elements and empirical processes. Z. Wahrsch. verw. Geb. 62 509-552.

Erdős, P. (1942). On the law of the iterated logarithm. Ann. Math. 43 419-436.

Feldman, Jacob (1971). Sets of boundedness and continuity for the canonical normal process. Proc. Sixth Berkeley Symp. Math. Statist. Prob. 2, 357-368. Univ. Calif. Press.

Fernandez, Pedro J. (1974). Almost surely convergent versions of sequences which converge weakly. Bol. Soc. Brasil. Math. 5 51-61.

Halmos, P. (1950). Measure Theory. Princeton, Van Nostrand. 2d. printing, Springer, N. Y. 1974.

Hoffmann-Jørgensen, J., and Niels Trolle Andersen (1984). Personal communication.

________________ (1984). Envelopes and perfect random variables (preprint, section of forthcoming book).

Itô, K., and H. P. McKean Jr. (1974). Diffusion processes and their sample paths. Springer, N. Y. (2d. printing, corrected).

Marczewski, E., and R. Sikorski (1948). Measures in nonseparable metric spaces. Colloq. Math. 1 133-139.

O'Reilly, N. E. (1974). On the weak convergence of empirical processes in sup-norm metrics. Ann. Probab. 2 642-651.

Petrovskii, I. G. (1935). Zur ersten Randwertaufgabe der Wärmeleitungsgleichung. Compositio Math. 1 383-419.

Pollard, D. B. (1982). A central limit theorem for empirical processes. J. Austral. Math. Soc. Ser. A 33 235-248.

Pyke, R. (1969). Applications of almost surely convergent constructions of weakly convergent processes. Lecture Notes in Math. 89 187-200.

Shorack, G. R. (1979). Weak convergence of empirical and quantile processes in sup-norm metrics via KMT-constructions. Stochastic Processes Applics. 9 95-98.

_____________, and J. Wellner (1982). Limit theorems and inequalities for the uniform empirical process indexed by intervals. Ann. Probab. 10 639-652.

Skorohod, A. V. (1956). Limit theorems for stochastic processes. Theor. Prob. Appls. 1 261-290 (English), 289-319 (Russian).

Stute, W. (1982). The oscillation behavior of empirical processes, Ann. Probab. 10 86-107.

Talagrand, M. (1984). The Glivenko-Cantelli problem (preprint).

Wichura, M. J. (1970). On the construction of almost uniformly convergent random variables with given weakly convergent image laws. Ann. Math. Statist. 41 284-291.

Room 2-245, MIT
Cambridge, MA 02139
USA

Part 2

Markov Processes

Introduction

The papers in this chapter deal with the development of a relativistic version of Brownian motion. There were difficulties with the extension of classical Brownian motion to the relativistic framework related to the instantaneous propagation of the heat flow in the classical case. Dudley developed a Lorentz invariant version of Brownian motion (this definition happened to be the only possible one under the assumption of Lorentz invariance).

The first paper studies Markov processes with sample paths $t \mapsto (x(t), \dot{x}(t)) \in \mathbf{R}^6$, where $x(t)$ is a Lipschitz curve in $\mathbf{R}^3$ and $\dot{x}(t)$ is its velocity such that $|\dot{x}(t)| < c$, the speed of light. Here $\dot{x}(t)$ is a Lévy process in hyperbolic space. The orthochronous Lorentz group in $\mathbf{R}^4$ (i.e., the group of linear transformations preserving the form $c^2t^2 - |x|^2$, $(t,x) \in \mathbf{R}^4 = \mathbf{R} \times \mathbf{R}^3$, with determinant $+1$) induces transformations of smooth paths $x(t)$ and their velocities. The goal is to characterize Markov processes that are temporally homogeneous and, at the same time, spatially homogeneous with respect to the action of the Lorentz group. The paper provides a complete solution of this problem (some earlier work was done by V. N. Tutubalin and by R. Gangolli).

In the follow up note that is also included in the volume, several supplements and improvements were added to the first paper.

The last paper deals with asymptotic properties of strongly Markov processes $(x(t), \dot{x}(t))$ in relativistic phase space (whose velocity is continuous from the right and has limits from the left). Specifically, the existence of the limit $\lim_{t\to\infty} \dot{x}(t)$ was studied both for processes invariant with respect to the Lorentz group and for some processes that are not invariant.

The problems studied in these papers have remained a topic of research both in the mathematics and in the physics literature. In fact, an extension of Dudley's results to the context of general relativity was developed only recently by J. Franchi and Y. Le Jan, Relativistic diffusions and Schwarzschild geometry, Comm. Pure Appl. Math, 2007, 60, 2, 187–251 (see also M. Émery, On some relativistic-covariant stochastic processes in Lorentzian space-times, C.R. Acad. Sci. Paris, Ser I, 2009, 347).

E. Giné et al. (eds.), *Selected Works of R.M. Dudley*, Selected Works in Probability and Statistics,
DOI 10.1007/978-1-4419-5821-1_6, © Springer Science+Business Media, LLC 2010

1. 65 02 1 Communicated 27 January 1965 by LARS GÅRDING and LENNART CARLESON

Lorentz-invariant Markov processes in relativistic phase space

By R. M. DUDLEY

1. Introduction

This paper deals with certain "random motions" permitted by the special theory of relativity; that is, with probability measures on sets of trajectories on which speeds are less than or equal to the speed c of light (which we take equal to 1 throughout). We deal with classes of such measures indexed by possible initial states, related to each other by the Lorentz group (implying certain invariance conditions for individual measures).

The definition of "state" as mentioned above is governed by the Markov property, i.e. that given the present state, further knowledge of past states should be irrelevant to the prediction of future states. It is known that position is insufficient for this purpose (see e.g. Theorem 11.3 below), and it seems natural to include the velocity in specifying the state. Indeed, velocities must exist at almost all times since position is a Lipschitzian function of time, and the existence of velocities is generally incompatible with the Markov property of a position process.

We distinguish between speeds strictly less than 1 and those equal to 1, and do not consider processes in which both occur (except in Theorem 11.3). This corresponds to the physical distinction between particles of positive or zero rest mass. Invariant processes of speed 1 turn out to be essentially uninteresting (see § 11) in that they cannot change direction.

We are left, then, with processes of speeds almost always strictly less than 1. For these processes, we can introduce the relativistic "proper time" on each trajectory (see § 6 below). The possible "4-velocities", i.e. derivatives of space-time position with respect to proper time, then lie in a three-dimensional hyperboloid $\mathcal{U}$ with a symmetric, Lorentz-invariant Riemannian structure (see § 7). Because of our invariance assumptions, the velocity process is itself Markovian (Theorem 3.2) and defined by a "convolution" semigroup on $\mathcal{U}$.

Such convolution semigroups have fortunately been completely classified by Tutubalin [1]. We thus arrive at an explicit description of all the processes which concern us, in Theorem 8.2, the main theorem of the paper. There are two extreme possibilities: on the one hand, "Brownian motions" in $\mathcal{U}$ (a one-parameter family indexed by a diffusion constant), which yield (the only) processes in which velocity is a continuous (but, of course, not differentiable) function of time; see § 10. On the other hand, there are "Poisson" processes in which the velocity changes only by jumps. Finally, there are (roughly speaking) mixtures of the two.

E. Giné et al. (eds.), *Selected Works of R.M. Dudley*, Selected Works in Probability and Statistics, DOI 10.1007/978-1-4419-5821-1_7, © Springer Science+Business Media, LLC 2010

In order to make the change from coordinate time to proper time as parameter, it seems necessary to make an assumption of continuity in probability. One would like to infer this from the other assumptions through a sort of "infinite divisibility," but in our situation this seems difficult before the change to proper time, since one has a convolution semigroup on a symmetric space only after the change.

Sections 2 and 3 deal with generalities about Markov processes. Section 4 introduces Lorentz-invariant Markov processes of speeds less than 1. Sections 5–10 carry through the characterization of these processes. In section 11, we establish the triviality of processes with speeds equal to 1 or states given by position only. In section 12 some unsolved problems are mentioned.

For Markov processes I have used specializations of the definitions in Dynkin's books [1] and [2], since several results are also quoted. It is worth noting, however, that this leads to at least one unsatisfying situation. In the usual approach to the strong Markov property one demands right continuity for sample functions. Here, where a sample function is (in part) a derivative, it becomes a right derivative after imposition of right continuity. It may seem unnatural to consider the right derivative of a function at time t as known at that time, if the left derivative is different. The way out of this difficulty, if it is one, will be left for other researches.

The "diffusion" or "Brownian motion" processes studied in § 10 below have been worked on previously by G. Schay [1] and H. Dinges. Both the latter and R. Hermann advised me against beginning with the proper time (i.e., essentially starting in the middle of § 6 below), as I did in an earlier draft, and I am now glad to have followed their advice. An exchange of correspondence with H. Dinges on this subject in general has been most helpful.

2. Starting probabilities and Markov processes

First we review some measure-theoretic notation. If $(X, \mathcal{S}, \mu)$ and $(Y, \mathcal{T}, \nu)$ are σ-finite measure spaces (which for our purposes may as well be finite), then $\mathcal{S}\times\mathcal{T}$ denotes the σ-field of subsets of the Cartesian product $X\times Y$ generated by all sets $A\times B$, $A\in\mathcal{S}$, $B\in\mathcal{T}$, and $\mu\times\nu$ the product measure. If F maps X onto another set S, then $F(\mathcal{S})$ denotes the class of all $C\subset S$ such that $F^{-1}(C)=\{x: F(x)\in C\}\in\mathcal{S}$. We let

$$(\mu\circ F^{-1})(C)=\mu(F^{-1}(C)),\ C\in F(\mathcal{S}).$$

Finally if f is μ-integrable and $\mathcal{H}$ is a sub-σ-field of $\mathcal{S}$, then $E_\mu(f\,|\,\mathcal{H})$ denotes the μ-conditional expectation of f given $\mathcal{H}$, and if f is the indicator function of a set A, we let

$$E_\mu(f\,|\,\mathcal{H})=\mu(A\,|\,\mathcal{H}).$$

To clarify the argument of a function $F=\mu(A\,|\,\mathcal{H})$ we may write

$$F(x)=\mu(x: x\in A\,|\,\mathcal{H}),$$

where "$x\in A$" will be replaced by a defining condition.

Throughout this paper, R will denote the real line (with its usual topology), R^n its n-fold Cartesian product, and R^+ the nonnegative axis $[0,\infty)$. If S is any topological space, $B(S)$ will denote the Borel σ-field generated by the open sets in S.

We shall consider certain function spaces defined as follows:

Definition. If S is a set, a *path space* for S is a pair $(\mathcal{A}, \zeta)$ satisfying the following conditions:

(1) ζ is a function on $\mathcal{A}$ with $0 \leqslant \zeta(f) \leqslant +\infty, f \in \mathcal{A}$.
(2) Each $f \in \mathcal{A}$ is a function from the interval $[0, \zeta(f))$ to S.
(3) For each $s \geqslant 0$ and $x \in S$ there is an $f \in \mathcal{A}$ with $f(s)=x$.

If S is any set, the set $\mathcal{A}$ of all functions f from intervals $[0, \zeta(f))$ to S will be called the *maximal* path space for S.

Given a path space $(\mathcal{A}, \zeta)$ and $s \geqslant 0$, let

$$\mathcal{A}^s = \{f \in \mathcal{A} : \zeta(f) > s\}.$$

Suppose $\mathcal{S}$ is a σ-field of subsets of S. If $0 \leqslant s \leqslant t$, $B^s_t(\mathcal{A}, \mathcal{S})$ will denote the σ-field of subsets of $\mathcal{A}^s$ generated by all sets of the form

$$\{f \in \mathcal{A}^s : f(r) \in A\}$$

for $s \leqslant r \leqslant t$ and $A \in \mathcal{S}$ (here and throughout $f(r) \in A$ implies $\zeta(f) > r$). $B^s(\mathcal{A}, \mathcal{S})$ will be the σ-field generated by all the $B^s_t(\mathcal{A}, \mathcal{S})$ for $t > s$. Context permitting, $B^s_t(\mathcal{A}, \mathcal{S})$ will be written $B^s_t(\mathcal{A})$ or B^s_t, and likewise for B^s. We call $(\mathcal{A}, \zeta, S, \mathcal{S})$ a *measurable path space.*

We shall need the following fact, proved in Dynkin [1] Lemma 5.9:

Lemma 2.1. *Suppose S is a metric space and $(\mathcal{A}, \zeta)$ is a path space for S such that each $f \in \mathcal{A}$ is continuous from the right wherever it is defined. For $0 \leqslant r \leqslant s$ let M^r_s be the product σ-field*

$$B^r_s(\mathcal{A}, B(S)) \times B(R^+)$$

in $\mathcal{A} \times R^+$. Let $F(f, t) = f(t)$ for $r \leqslant t \leqslant s$. Then the domain D of F belongs to M^r_s and F is M^r_s-measurable on D.

For $h \geqslant 0$ let θ_h be the transformation of functions defined by

$$(\theta_h f)(t) = f(t+h), \quad t \geqslant 0.$$

A path space $(\mathcal{A}, \zeta)$ will be called *stationary* if for each $h \geqslant 0$, θ_h takes $\mathcal{A}$ into itself. If $(\mathcal{A}, \zeta, S, \mathcal{S})$ is a stationary measurable path space, clearly θ_h takes B^{s+h}_{t+h} onto B^s_t for any $h \geqslant 0$, $t \geqslant s \geqslant 0$.

A Markov process will be defined below by a class of measures on a path space corresponding to different initial states. The restriction to path spaces is a specialization of the definitions in Dynkin [1] and [2]. The definition in [2] requires stationary transition probabilities, which are sufficient for our purposes. It will be reproduced after our own is completed.

Definition. Suppose $(\mathcal{A}, \zeta, S, \mathcal{S})$ is a measurable path space in which $\mathcal{S}$ contains all one-point sets. Given $x \in S$, a probability measure P on $B^0(\mathcal{A})$ will be called a *starting probability* at x on $(\mathcal{A}, \zeta, S, \mathcal{S})$ if

$$P(f : f(0) = x) = 1.$$

Definition. Suppose $(\mathcal{A}, \zeta, S, \mathcal{S})$ is stationary and for each $x \in S$, P_x is a starting probability at x on $\mathcal{A}$. Then $(\{P_x\}, \mathcal{A}, \zeta, S, \mathcal{S})$ or, briefly, $(\{P_x\}, \mathcal{A})$ or $\{P_x\}$, is a *Markov process* if

(1) For any $t \geq 0$ and $A \in \mathcal{S}$, $P(t, x, A) = P_x(f : f(t) \in A)$ is $\mathcal{S}$-measurable in x.
(2) Whenever $0 \leq t \leq u$, $x \in S$, and $A \in \mathcal{S}$, $P_x(f : f(u) \in A \mid B_t^0) = P(u - t, f(t), A)$ almost everywhere with respect to P_x on $\mathcal{A}^s$.

The following is known (Dynkin [1], Lemma 2.2):

Lemma 2.2. *If* $(\{P_x\}, \mathcal{A}, \zeta, S, \mathcal{S})$ *is a Markov process and* $A \in B^0(\mathcal{A}, \mathcal{S})$, *then* $P_x(A)$ *is* $\mathcal{S}$*-measurable in* x.

The definition of Markov process in Dynkin [2] can be reformulated in our terms as follows:

Definition. Suppose given a stationary measurable path space $(\mathcal{A}, \zeta, S, \mathcal{S})$, a set Ω, a mapping

$$X : \omega \to x(\ \ , \omega)$$

of Ω onto $\mathcal{A}$, for each $t \geq 0$ a σ-field $\mathcal{M}_t$ of subsets of

$$\Omega_t = \{\omega \in \Omega : \zeta(x(\ \ , \omega)) > t\},$$

a σ-field $\mathcal{M}^\circ$ of subsets of Ω including all the $\mathcal{M}_t$, and a set $\{Q_x, x \in S\}$ of probability measures on $\mathcal{M}^\circ$. Then $(x(\ \ ,\ \ ), \zeta, \{\mathcal{M}_t\}, \{Q_x\})$ is a *fields Markov process* if:

(A) For $0 \leq t \leq u$ and $A \in \mathcal{M}_t$, or $A \in X^{-1}(B_u^0(S, \mathcal{S}))$,
$$A \cap \Omega_u \in \mathcal{M}_u.$$
(B) If P_x is $Q_x \circ X^{-1}$ restricted to $B^0(\mathcal{A}, \mathcal{S})$, then $(\{P_x\}, \mathcal{A}, \zeta, S, \mathcal{S})$ is a Markov process (in the path space sense), and (2) holds with B_t^0 replaced by $X(\mathcal{M}_t)$.

The word "fields" has been added to clarify the difference between the two definitions. If the Q_x are defined only on the minimal σ-field $X^{-1}(B^0)$, we have an isomorphism.

It is clear that a Markov process on a path space defines a fields Markov process, and conversely a fields Markov process defines a path space Markov process. However, non-isomorphic fields Markov processes may define the same path space Markov process.

A Markov process on a maximal path space is called "canonical" by Dynkin [1, 2.11]. Any path space Markov process extends naturally to a canonical one. Conversely, a canonical Markov process $(\{P_x\}, \mathcal{B})$ can be restricted to a path space $\mathcal{A} \subset \mathcal{B}$ if and only if $\mathcal{A}$ has outer measure 1 for all the P_x (Dynkin [1] Theorem 2.5).

We shall also use the strong Markov property, which we proceed to define. Suppose $(\mathcal{A}, \zeta, X, \mathcal{S})$ is a measurable path space. A *stopping time* on the path space is a $B^0(\mathcal{A})$-measurable function τ from $\mathcal{A}$ to R^+ such that $\tau(f) \leq \zeta(f)$ for all $f \in \mathcal{A}$ and, for any $t \geq 0$,

$$\{f \in \mathcal{A} : \tau(f) \leq t < \zeta(f)\} \in B_t^0(\mathcal{A}).$$

If τ is a stopping time, let

$$\mathcal{A}^\tau = \{f \in \mathcal{A} : \zeta(f) > \tau(f)\}$$

and let B_τ be the σ-field of sets $A \subset \mathcal{A}$ such that for any $t \geq 0$,

$$A \cap \{f : \tau(f) \leqslant t < \zeta(f)\} \in B_t^0(\mathcal{A}).$$

Then a Markov process $\{P_x\}$ on $\mathcal{A}$ is called *strongly Markovian* if:
$SM(1)$ Given $B \in \mathcal{S}$,

$$\langle t, x \rangle \to P_x\{f : f(t) \in B\}$$

is $B(R^+) \times \mathcal{S}$-measurable,
$SM(2)$ for any stopping time τ, $B \in \mathcal{S}$, and B_τ-measurable function $\eta \geqslant \tau$,

$$P_x(f : f(\eta) \in B \mid B_\tau) = P_{f(\tau)}(g : g(\eta - \tau) \in B)$$

almost everywhere on $\mathcal{A}^\tau$ with respect to P_x, and $\{x : f(\tau) = x \text{ for some } f \in \mathcal{A}\} \in \mathcal{S}$.

3. Markov processes in product spaces

Suppose $(\mathcal{A}, \xi, X, \mathcal{S})$ and $(\mathcal{B}, \eta, Y, \mathcal{T})$ are stationary measurable path spaces with $\xi \equiv \eta \equiv +\infty$, and for each $x \in X$ $(y \in Y)$, P_x (Q_y) is a starting probability on $\mathcal{A}$ $(\mathcal{B})$.

Let
$$Z = X \times Y, \ \mathcal{U} = \mathcal{S} \times \mathcal{T}, \ \mathcal{C} = \mathcal{A} \times \mathcal{B},$$

and $\zeta \equiv +\infty$. For $z = \langle x, y \rangle \in Z$ let R_z be the measure $P_x \times Q_y$ on $\mathcal{C}$. Regard $h = \langle f, g \rangle \in \mathcal{C}$ as the function

$$h(t) = \langle f(t), g(t) \rangle \in Z, \quad t \geqslant 0.$$

Theorem 3.1. $(\{R_z\}, \mathcal{C}, \zeta, Z, \mathcal{U})$ *is a Markov process if and only if both* $\{P_x\}$ *and* $\{Q_y\}$ *are Markov processes.*

Proof. Clearly $(\mathcal{C}, \zeta, Z, \mathcal{U})$ is a measurable path space and each R_z is a starting probability at z. First suppose $\{P_x\}$ and $\{Q_y\}$ are both Markovian. The class of sets $C \in \mathcal{U}$ for which both 1) and 2) in the definition of Markov process hold is closed under finite, disjoint unions and countable increasing unions and decreasing intersections. Thus we can put $C = A \times B$, $A \in \mathcal{S}$, $B \in \mathcal{T}$, and the conclusion follows easily.

Now suppose $\{R_z\}$ is Markovian; let us show that $\{P_x\}$ is Markovian. The measurability condition 1) is clear. Given $0 \leqslant t \leqslant u$, $x \in X$, and $A \in \mathcal{S}$, let $z = \langle x, y \rangle$ for some $y \in Y$ and note that the conditional probability of a "rectangle" for a product measure is itself a product, so that if $\mathbf{1}_g = 1$ for all g,

$$\begin{aligned} &P_x(f: f(u) \in A \mid B_t^0(\mathcal{A})) \cdot \mathbf{1}_g \\ &= R_z(\langle f, g \rangle: f(u) \in A \mid B_t^0(\mathcal{C})) \\ &= R_{\langle f(t), g(t) \rangle}(\langle \phi, \psi \rangle: \phi(u-t) \in A) \\ &= P_{f(t)}(\phi: \phi(u-t) \in A) \cdot \mathbf{1}_g, \end{aligned}$$

almost everywhere for R_z. Choosing g suitably, we obtain that $\{P_x\}$ is Markovian, q.e.d.

Now suppose $(X, \mathcal{S})$ and $(Y, \mathcal{T})$ are measurable spaces, $Z = X \times Y$, $\mathcal{U} = \mathcal{S} \times \mathcal{T}$, and $(\{R_z\}, \mathcal{C}, \zeta, Z, \mathcal{U})$ is a Markov process. Suppose G is a transitive group of automorphisms of the measurable space $(X, \mathcal{S})$, and each $\gamma \in G$ acts on Z by $\gamma\langle x, y \rangle = \langle \gamma(x), y \rangle$ and on $\mathcal{C}$ in the obvious way. Let K be the natural projection of Z onto Y and $K(\langle f, g \rangle) = g$, $\langle f, g \rangle \in \mathcal{C}$, where f and g are both defined on the interval $0 \leqslant t < \zeta(\langle f, g \rangle)$;

we can let $\eta(g)=\zeta(\langle f,g\rangle)$. Let $\mathcal{B}$ be the set of all $K(h)$ for $h\in\mathcal{C}$. Clearly $(\mathcal{B},\eta,Y,\mathcal{J})$ is a stationary, measurable path space.

Theorem 3.2. *Suppose $\{R_z\}$ is G-invariant, i.e. for any $z\in Z$, $A\in B^0(\mathcal{C})$, and $\gamma\in G$,*

$$R_z(A)=R_{\gamma(z)}(\gamma(A)).$$

Then for each $y\in Y$ the measure $R_{\langle x,y\rangle}\circ K^{-1}$ is independent of x. Calling it Q_y,

$$(\{Q_y\},\mathcal{B},\eta,Y,\mathcal{J})$$

is a Markov process.

Proof. Each Q_y is well-defined by G-invariance and is clearly a starting probability on $\mathcal{B}$. For the Markov property, first given $A\in\mathcal{J}$ we have

$$Q_y(g\colon g(t)\in A)=R_{\langle x,y\rangle}(\langle f,g\rangle\colon g(t)\in A)$$

for any fixed $x\in X$, and this is $\mathcal{J}$-measurable in y. Second, suppose $y\in Y$, $0<t<u$, $A\in\mathcal{J}$. For any $x\in X$,

$$R_{x,y}(\langle f,g\rangle\colon g(u)\in A\mid B_t^0(\mathcal{C}))$$
$$=R_{f(t),g(t)}(\langle\phi,\psi\rangle\colon\psi(u-t)\in A)$$

almost everywhere for $R_{x,y}$ in $\mathcal{C}^t$. Transforming both sides by K we get

$$Q_y(g\colon g(u)\in A\mid B_t^0(\mathcal{B}))=Q_{g(t)}(\psi\colon\psi(u-t)\in A)$$

almost everywhere for Q_y on $\mathcal{B}^t$, q.e.d.

4. Lorentz-invariant random motions

Let X be a three-dimensional Euclidean space R^3 of ordered triples

$$x=(x_1,x_2,x_3)$$

of real numbers and let

$$|x|=(x_1^2+x_2^2+x_3^2)^{1/2}.$$

Let M be the space R^4 of pairs $z=\langle x,t\rangle$, $x\in X$, $t\in R$, with

$$\|z\|^2=t^2-|x|^2.$$

$\mathcal{L}$ will denote the group of linear transformations of M into itself which preserve $\|\;\|^2$, have determinant 1, and do not change the sign of t (the proper, orthochronous Lorentz group).

M_o will denote the universal "tangent space" of M, or space of derivatives of functions from the real line to M at points of M. M_o is naturally isomorphic to M, but for some purposes the two spaces will be distinguished.

Let $\mathcal{A}$ be the space of functions f from R^+ to X such that

(1) $|f(s)-f(t)|\leqslant|s-t|$ for $s,t\geqslant 0$
(2) If $f'(s)$ is defined (as it is for almost all s by (1)), $|f'(s)|<1$.

Let $I(f) \equiv +\infty$. Then $(\mathcal{A}, I, X, B(X))$ is a measurable path space. We shall need the following fact:

Lemma 4.1. *Suppose U is an open subset of R^k, $S = B(U)$, and $(\mathfrak{B}, \zeta, U, S)$ is a measurable path space consisting of continuous functions. Let $F = \mathfrak{B} \times U$ and*

$$\mathfrak{J} = B^0(\mathfrak{B}) \times S.$$

Then the function

$$\langle f, t\rangle \to df(t)/dt$$

is defined on a set in $\mathfrak{J}$ and $\mathfrak{J}$-measurable there.

Proof. By Lemma 2.1,

$$\langle f, t\rangle \to f(t) \text{ and } \langle f, t\rangle \to (f(t+r) - f(t))/r$$

for any $r \neq 0$ have the desired property. For any open set $V \subset R^k$, let

$Z(V) = \{\langle f, t\rangle : (f(t+r) - f(t))/r \in V$ for small enough rational $r \neq 0\}$. Clearly $Z(V) \in \mathfrak{J}$.

For each $n = 1, 2, \ldots$, let $\{U_{mn}\}_{m=1}^{\infty}$ be a locally finite open cover of R^k by sets of diameter less than $1/n$. Then

$$A = \{\langle f, t\rangle : f'(t) \text{ exists}\} = \bigcap_n \bigcup_m Z(U_{mn}).$$

Now any open $V \subset R^k$ is the union of the U_{mn} with $\bar{U}_{mn} \subset V$, where the bar denotes closure. Thus

$$\{\langle f, t\rangle : f'(t) \in V\} = \bigcup_{m,n} \{Z(U_{mn}) : \bar{U}_{mn} \subset V\} \cap A.$$

This completes the proof.

Now let V be the open unit ball $\{v : |v| < 1\}$ in R^3. V is the space of admissible velocities. We shall consider families $\{P_x^v\}$ of measures, where for each $x \in X$ and $v \in V$ P_x^v is a starting probability at x on $(\mathcal{A}, I, X, B(X))$. The family $\{P_x^v\}$ will be assumed to satisfy conditions $A) - E)$ to be formulated below.

(A) For any $n = 1, 2, \ldots, t_i \geqslant 0, A_i \in B(X), x \in X$, and $v \in V$,

$$P_x^v(f : f(t_i) \in A_i, i = 1, \ldots, n) = P_0^v(f : f(t_i) \in A_i - x, i = 1, \ldots, n).$$

(B) For any $x \in X$ and $v \in V$,

$$P_x^v(f : f'(0^+) = v) = 1.$$

Since each $f \in \mathcal{A}$ is differentiable almost everywhere and $\{\langle f, t\rangle : f'(t) \text{ exists}\}$ is product measurable,

$$\{s > 0 : P_x^v(f'(s) \text{ exists}) < 1\}$$

has Lebesgue measure zero by the Fubini theorem for each x and v. We assume this set is actually empty, and formulate a Markov property:

(C) If $t > 0$, $x \in X$, and $v \in V$, then

$$P_x^v(f'(t) \text{ exists}) = 1$$

and for any $A \in B^t(\mathcal{A})$,

$$P_x^v(A \mid B_t^0(\mathcal{A})) = P_{f(t)}^{f'(t)}(\theta_t A)$$

almost everywhere for P_x^v (where θ_t is the translation by t, defined in §2 above).

We next formulate a Lorentz-invariance condition. Each $L \in \mathfrak{L}$ defines a transformation L^* of functions $f \in \mathcal{A}$ by

$$L\langle f(t), t\rangle = \langle L_{(1)}\langle f(t), t\rangle, L_{(2)}\langle f(t), t\rangle\rangle,$$
$$L_{(1)}\langle f(t), t\rangle = (L^* f)(L_{(2)}\langle f(t), t\rangle).$$

Clearly L^* is measurable from $(\mathcal{A}, B^0(\mathcal{A}))$ to itself. L also defines a transformation V_L of velocities by

$$V_L(f'(t)) = (L^* f)'(L_{(2)}\langle f(t), t\rangle).$$

Our Lorentz-invariance assumption is

(D) $P_0^v \circ (L^*)^{-1} = P_0^{V_L(v)}$ for any $L \in \mathfrak{L}$ and $v \in V$.

Let W be the "phase space" $X \times V$. Let d be the metric on W defined by

$$d(\langle x, u\rangle, \langle y, v\rangle) = |x - y| + |u - v|.$$

The last assumption is that derivatives are continuous in probability:

(E) $\lim_{t \downarrow 0} P_0^v(d(\langle f(t), f'(t)\rangle, \langle 0, v\rangle) \geqslant \varepsilon) = 0$

uniformly in v, for any $\varepsilon > 0$.

I don't know whether (E) follows from the preceding assumptions.

Let Q be the set of all functions from R^+ to W. Then $(Q, I, W, B(W))$ is a stationary, measurable path space. Each measure P_x^v on sets

$$\{f : \langle f(t_i), f'(t_i^+)\rangle \in B_i, i = 1, \ldots, n\}, t_i \geqslant 0, B_i \in B(W),$$

extends uniquely by Kolmogorov's theorem (Loève [1] p. 93) to a probability measure $Q_{x,v}$ on $B^0(Q)$. For $t_i > 0$, t_i^+ can be replaced by t_i. Of course, there may exist with positive probability points t, depending on f, at which $f'(t)$ does not exist.

5. Properties of the $Q_{x,v}$.

In this section we infer from $(A) - (E)$ that $\{Q_{x,v}\}$ is a "Fellerian" Markov process, and discuss other continuity properties.

We consider the transition measures

$$Q(h, x, v, A) = Q_{x,v}(f : f(h) \in A)$$

defined for $\langle x, v\rangle \in W$, $A \in B(W)$, $h \geqslant 0$. Each $Q(h, x, v)$ is a probability measure on W. We consider weak* convergence of such measures:

$$\mu_n \to \mu \text{ (weak*)}$$

if for every bounded continuous real-valued function f on W,

$$\int f\, d\mu_n \to \int f\, d\mu.$$

Let $BL(W)$ be the class of bounded Lipschitzian real-valued functions on W, i.e. functions f with

$$\|f\|_{BL} = \sup_w |f(w)| + \sup_{u \neq w} \frac{|f(u) - f(w)|}{d(u, w)} < \infty.$$

Then $(BL(W), \| \|_{BL})$ is a Banach space, and any probability measure μ on W defines an element of the dual space $BL^*(W)$, with the dual norm

$$\|\mu\|^*_{BL} = \sup\left(\int f\, d\mu : \|f\|_{BL} = 1\right).$$

We have $\mu_n \to \mu$ (weak*) if and only if

$$\|\mu_n - \mu\|^*_{BL} \to 0;$$

see e.g. R. R. Rao [1] Theorem 3.1. (For "if", nonnegativity of the measures is essential.)

Theorem 5.1. *Under assumptions* $(A) - (E)$ *of* § 4, $Q(h, x, v, \ )$ *is jointly weak* continuous in* $h, x,$ *and* v.

Proof. We shall prove continuity in h and x uniformly on the triple product, then simple continuity in v.

We have continuity in h, uniformly in all x, v and h, by assumptions (A) and (E). For continuity in x, let

$$T_y\langle x, v\rangle = \langle x + y, v\rangle,$$

and note that

$$Q(h, x + y, v, \ ) = Q(h, x, v, \ ) \circ T_y^{-1}$$

by (A). Letting $\mu = Q(h, x, v, \ )$, we have

$$\|\mu - \mu \circ T_y^{-1}\|^*_{BL} = \sup_{\|f\|_{BL}=1} \left| \int f\, d\mu - \int f \circ T_y\, d\mu \right|.$$

Since $|(f - f \circ T_y)(w)| \leqslant |y|$ if $w \in W, y \in X$, and $\|f\|_{BL} = 1$, we have continuity in x, uniformly in h and v.

It remains to prove continuity in v for fixed x and h, and we can take $x = 0$. Suppose $v_n \to v$ in V. There are $L_n \in \mathfrak{L}$ such that $v = V_{L_n}(v_n)$ for all n, with L_n converging to the identity. We also choose numbers $\varepsilon_n \downarrow 0$ such that

$$L_n\langle x, h + \varepsilon_n\rangle = \langle x', t'\rangle \quad \text{and} \quad (t')^2 - |x'|^2 \geqslant 0 \quad \text{imply} \quad t' > h.$$

For $f \in \mathcal{A}$, $f(0)=0$, let $t_n(f)$ be the unique value of s such that

$$\langle f(s), s\rangle = L_n\langle x, h+\varepsilon_n\rangle$$

for some x. (Then $t_n(f) > h$ for all $f \in \mathcal{A}$ with $f(0)=0$.) We need the following:

Lemma 5.2. *$f \to \langle f(t_n(f)), V_{L_n} f'(t_n(f))\rangle$ is B^h-measurable on $\mathcal{A}$.*

Proof. Clearly $f \to t_n(f)$ is B^h-measurable, and so is $f \to \langle f, t_n(f)\rangle$, (from B^h to $B^h \times B((h, \infty))$. The map

$$\langle f, s\rangle \to \langle f(s), V_{L_n} f'(s)\rangle, \quad s > h,$$

is $B^h \times B(R^+)$-measurable by Lemmas 2.1 and 4.1 and continuity of V_{L_n}. Composing the last two mappings finishes the proof of the Lemma.

Now we apply assumption (C) to obtain, given $\varepsilon > 0$,

$$(^*) \qquad \begin{aligned} &P_0^v(f : d(\langle f(t_n(f)), V_{L_n} f'(t_n(f))\rangle, \langle f(h), f'(h)\rangle) \geqslant \varepsilon \,|\, B_h^0) \\ &= P_{f(h)}^{f'(h)}(g : d(\langle f(h), f'(h)\rangle, \langle g(s_n(g)), V_{L_n} g'(s_n(g))\rangle) \geqslant \varepsilon) \end{aligned}$$

almost everywhere for P_0^v, where $s_n(g)$ is the unique value of s such that

$$\langle g(s), s+h\rangle = L_n\langle x, h+\varepsilon_n\rangle$$

for some x. The set of $\langle y, s\rangle \in M$ such that

$$\langle y, s+h\rangle = L_n\langle x, h+\varepsilon_n\rangle$$

for some $x \in X$ is the set of points of the form $L_n\langle z, \alpha_n\rangle$ for some fixed $\alpha_n > 0$ and arbitrary $z \in X$, since L_n^{-1} takes parallel hyperplanes into parallel hyperplanes.

For $y \in X$ the translation

$$T_y: \langle \xi, t\rangle \to \langle \xi - y, t\rangle$$

of M onto itself takes points of the form $L_n\langle a, \alpha_n\rangle$ onto points of the form $L_n\langle b, \alpha_n(y)\rangle$ for some $\alpha_n(y)$. If $|y| \leqslant h$, then $\alpha_n(y) > 0$, and as $n \to \infty$, $\alpha_n(y) \to 0$ uniformly for $|y| \leqslant h$.

For any $y \in X$ and $u \in V$, let

$$A_{n,y,u} = P_y^u(g : d(\langle y, u\rangle, \langle g(s_n(g)), V_{L_n} g'(s_n(g))\rangle) \geqslant \varepsilon.$$

Applying first T_y and then L_n^{-1}, letting $u_n = V_{L_n}^{-1} u$, we obtain by (A) and (D)

$$A_{n,y,u} = P_0^{u_n}(\phi : d(\langle 0, u_n\rangle, \langle \phi(\alpha_n(y)), \phi'(\alpha_n(y))\rangle) \geqslant \varepsilon).$$

Of course, $0 \leqslant A_{n,y,u} \leqslant 1$ for all n, y and u. As $n \to \infty$, since $\alpha_n(y) \to 0$, and the sequence u_n and its limit u form a compact set for any u, we have by (E) that $A_{n,y,u} \to 0$ for any u and y with $|y| \leqslant h$. Integrating (*) with respect to P_0^v over all of $\mathcal{A}$, we conclude by the bounded convergence theorem that

$$P_0^v(f : d(\langle f(t_n(f)), V_{L_n} f'(t_n(f))\rangle, \langle f(h), f'(h)\rangle \geqslant \varepsilon)$$

tends to 0 as $n \to \infty$. Now if $B \in B(W)$ let

$$P_n(B) = P_0^v(f : \langle f(t_n(f)), V_{L_n} f'(t_n(f))\rangle \in B),$$

defined by Lemma 5.2. We have shown that

$$P_n \to Q(h, 0, v) \text{ weak}^* \text{ as } n \to \infty. \text{ By (D),}$$

$$P_n(B) = P_0^{v_n}(g : \langle g(h+\varepsilon_n), g'(h+\varepsilon_n)\rangle \in B) = Q(h+\varepsilon_n, 0, v_n, B).$$

We also know that

$$\| Q(h+\varepsilon_n, 0, v_n, \ ) - Q(h, 0, v_n, \ ) \|_{BL}^* \to 0$$

as $n \to \infty$ since $\varepsilon_n \to 0$ and the v_n lie in a compact set. Hence $Q(h, 0, v_n, \ ) \to Q(h, 0, v, \ )$ weak*, and the proof is complete.

It is now easy to prove

Lemma 5.3. *Under assumptions* (A) *through* (E), $(\{Q_{x,v}\}, Q)$ *is a Markov process.*

Proof. We know that $(Q, I, W, B(W))$ is a measurable path space. The $Q_{x,v}$ are probability measures on $B^0(Q, B(W))$ and are starting probabilities by assumption (B). For the Markov property, we must first show that given $t \geq 0$ and $A \in B(W)$, $Q(t, x, v, A)$ is jointly measurable in x and v. If A is open, $x_n \to x$, and $v_n \to v$,

$$Q(t, x, v, A) \leq \limsup Q(t, x_n, v_n, A)$$

by Theorem 5.1. This implies joint measurability in x and v, and then A can be replaced by an arbitrary Borel set. The Markov property itself then holds by assumption (C), q.e.d.

(The following section, which presents Lemma 5.4, was received as manuscript added to proof on 15 September 1965.—*Editor*)

We next verify Dynkin's condition "$L(\Gamma)$" for compact sets Γ, which says in our case:

Lemma 5.4. For any compact set $K \subset X \times V$ and $u \geq 0$,

$$\lim_{\langle x,v\rangle \to \infty} \sup_{0 \leq t \leq u} Q(t, x, v, K) = 0.$$

(Here $\langle x, v\rangle \to \infty$ means $|x| \to \infty$ and/or $|v| \to 1$.)

Proof. For $|x| \to \infty$ the result is clear. Thus it suffices to show that for any compact set $C \subset V$,

$$\lim_{|v|\uparrow 1} \sup_{0 \leq t \leq u} Q(t, 0, v, X \times C) = 0.$$

We may assume that C is a closed ball centered at 0, and then that $v = (z, 0, 0)$ with $0 \leq z \uparrow 1$. For $\langle x, s\rangle \in M$ let

$$L(v)\langle x, s\rangle = \langle (x_1 + zs)/\sqrt{1-z^2}, x_2, x_3, (s + zx_1)/\sqrt{1-z^2}\rangle.$$

Then $L(v) \in \mathfrak{L}$ and $V_{L(v)}(0) = v$. Also for any $t \geq 0$, by (*D*),

$$Q(t, 0, v, X \times C) = P(f : V_{L(v)} f'(s(v)) \in C),$$

where $P = P_0^0$ and $s(v) = s(v, t, f)$ is the unique number s such that

$$\mathring{f}(s) = \langle f(s), s\rangle \in H = L(v)^{-1}\{\langle y, t\rangle : y \in X\}.$$

Let C_δ be the cone $|x| \leqslant \delta s$. By assumption (B), for any $\delta > 0$ $\mathring{f}(s)$ will P-almost surely lie in C_δ for s small enough. In $H \cap C_\delta$, $s \geqslant \alpha(v) > 0$, where

$$\alpha(v) = \alpha(v, t, \delta) = t\sqrt{1 - z^2}/(1 + z\delta),$$

so that $\alpha(v) \to 0$ as $z \uparrow 1$. Thus if $D(\delta, v)$ is the set of all $\langle x, \alpha(v) \rangle$ with $|x| \leqslant \delta\alpha(v)$, and

$$S(v) = \{f : \mathring{f}(\alpha(v)) \in D(\delta, v)\},$$

then $P(S(v)) \to 1$ as $z \uparrow 1$. Also, by (E), for any $\varepsilon > 0$

$$\lim_{z \uparrow 1} P(f : |f'(\alpha(v))| > \varepsilon) = 0.$$

Let $u(f, v) = V_{L(v)} f'(\alpha(v))$. We can apply the Markov property at time $\alpha(v)$ to obtain

$$(*) \qquad \begin{aligned} &P(f : f \in S(v) \quad \text{and} \quad |V_{L(v)} f'(s(v)) - u(f, v)| \leqslant \varepsilon \,|\, B^0_{\alpha(v)}) \\ &= \prod(f, v, \delta) = P^{f'(\alpha(v))}_{f(\alpha(v))}(g : |V_{L(v)} g'(\sigma(v)) - u(f, v)| \leqslant \varepsilon), \; f \in S(v) \\ &= 0 \text{ otherwise,} \end{aligned}$$

P-almost everywhere; for $f \in S(v)$ we have let $\sigma(v) = \sigma(v, g, \delta, t)$ be the unique s such that

$$\langle g(s), s + \alpha(v) \rangle \in H.$$

Translating by $-f(\alpha(v))$, using (A), we obtain

$$\prod(f, v, \delta) = P_0^{f'(\alpha(v))}(h : |V_{L(v)} h'(\varkappa(v)) - u(f, v)| \leqslant \varepsilon),$$

where $\varkappa(v) = \varkappa(v, h, f, \delta, t)$ is the unique $\varkappa$ such that

$$\langle h(\varkappa) + f(\alpha(v)), \varkappa + \alpha(v) \rangle \in H.$$

Now we transform by $L(v)$, using (D), to get

$$\prod(f, v, \delta) = P_0^{u(f,v)}(j : |j'(\lambda(v)) - u(f, v)| \leqslant \varepsilon),$$

where $$\lambda(v) = t - L(v)_{(2)} \mathring{f}(\alpha(v)).$$

Note that $\lambda(v)$ does not depend on j, only on v, t, δ, and $f(\alpha(v))$. Now

$$L(v)\langle x, \alpha(v) \rangle = \langle (x_1 + z\alpha(v))/\sqrt{1 - z^2}, x_2, x_3, (\alpha(v) + zx_1)/\sqrt{1 - z^2} \rangle = \langle \; , x_2, x_3, T \rangle,$$

where $$T = \frac{t}{1 + z\delta} + \frac{zx_1}{\sqrt{1 - z^2}}.$$

For $\langle x, \alpha(v) \rangle \in D(\delta, v)$, T is smallest when $x_1 = -\delta\alpha(v)$.

Then $$T = t(1 - z\delta)/(1 + z\delta).$$

Thus $$\lambda(v) \leqslant 2z\delta t/(1 + z\delta) \leqslant 2\delta t.$$

Hence $$\prod(f, v, \delta) \geqslant \inf_{0 \leqslant s \leqslant 2\delta t} P_0^{u(f,v)}(j : |j'(s) - u(f, v)| \leqslant \varepsilon).$$

Thus by (E), $$\lim_{\delta\downarrow 0} \prod(f, v, \delta) = 1,$$

uniformly in f and v. Thus, integrating our Markov equation $(^*)$ with respect to P, we have

$$\lim_{\delta\downarrow 0} \lim_{|v|\uparrow 1} P(f : f \in S(v) \quad \text{and} \quad |V_{L(v)} f'(s(v)) - u(f, v)| \leqslant \varepsilon) = 1.$$

For $(a, b, c) \in V$ we have

$$V_{L(v)}(a, b, c) = (a+z, b\sqrt{1-z^2}, c\sqrt{1-z^2})/(1+za),$$

$$|V_{L(v)}(a, b, c) - v|^2 \leqslant (a^2(1-z^2)^2 + b^2 + c^2)/(1+za)^2$$

$$\leqslant (a^2 + b^2 + c^2)/(1-|a|)^2.$$

Thus for $\gamma \leqslant \frac{1}{2}$, $|(a, b, c)| \leqslant \gamma$ implies

$$|V_{L(v)}(a, b, c) - v| \leqslant 2\gamma.$$

Hence for any $\varepsilon > 0$ and $0 < \delta < 1$,

$$\lim_{|v|\uparrow 1} P(f : |V_{L(v)} f'(\alpha(v)) - v| \leqslant \varepsilon) = 1.$$

Thus, letting $\delta \to 0$, we get

$$\lim_{|v|\uparrow 1} P(f : |V_{L(v)} f'(s(v)) - v| \leqslant \varepsilon) = 1.$$

Now for any compact $C \subset V$, there is an $\varepsilon > 0$ such that $|v| > 1 - \varepsilon$ and $|w - v| > \varepsilon$ imply $w \notin C$. Then

$$\lim_{|v|\uparrow 1} P(f : V_{L(v)} f'(s(v)) \in C) = 0$$

This limit is uniform in $0 \leqslant t \leqslant u$ since

$$\lim_{|v|\uparrow 1} \alpha(v) = \lim_{\delta\downarrow 0} \lambda(v) = 0$$

uniformly for $0 \leqslant t \leqslant u$. The proof is complete.

Under assumptions (A) – (E), each measure $Q_{x,v}$ gives outer measure 1 to the class $\mathcal{G}$ of functions in Q having limits from the left and continuous from the right for each $t \geqslant 0$ (Dynkin [2], Theorem 3.6).

The set of $\langle x(\ ), v(\ )\rangle \in \mathcal{G}$ such that $|x(s) - x(t)| \leqslant |s - t|$ for all rational $s, t \geqslant 0$ is $B^0(\mathcal{G})$-measurable and has $Q_{y,u}$-measure 1 for each $\langle y, u\rangle \in W$ since P_y^u is concentrated in $\mathcal{A}$. Thus by definition of $\mathcal{G}$, $|x(s) - x(t)| \leqslant |s - t|$ for all $s, t \geqslant 0$ with $Q_{y,u}$-probability 1 on $\mathcal{G}$. The map

$$\langle t, \langle x(\ ), v(\ )\rangle\rangle \to x'(t)$$

is $B(R^+) \times B^0(\mathcal{G})$-measurable by Lemma 4.1. For each t, $x'(t) = v(t)$ with $Q_{y,u}$-probability one. Hence by the Fubini theorem, for $Q_{y,u}$-almost all $\langle x(\ ), v(\ )\rangle \in \mathcal{G}$ we have $x'(t) = v(t)$ for almost all t, so that since $v(\ )$ is locally bounded and $x(\ )$ is Lipschitzian, $x(\ )$ is an indefinite integral of $v(\ )$ and $x'(t^+) = v(t)$ for all $t \geqslant 0$. Let $\mathcal{W}$ be the set of $\langle x(\ ), v(\ )\rangle \in \mathcal{G}$ with $x'(t^+) = v(t)$ for all $t \geqslant 0$. Then we have a Markov process

$$(\{Q_{x,v}\}, \mathcal{W}, I, W, B(W)).$$

Let $\overline{\mathcal{W}}$ be the set of functions $t \to \langle c+t, f(t)\rangle, f \in \mathcal{W}, c, t \in R$. Let $\overline{W}$ be the product space $R \times W$, and let R_c be the unit mass concentrated in the function $t \to c+t$. Let $Q_{c,x,v}$ be the product measure $R_c \times Q_{x,v}$. Then by Theorem 3.1,

$$(\{Q_{c,x,v}\}, \overline{\mathcal{W}}, I, \overline{W}, B(\overline{W}))$$

is a Markov process. This process is "Fellerian" by Theorem 5.1 and right continuous, hence strongly Markovian (Dynkin [1], Theorem 5.10).

6. The proper time

For any $f \in \mathcal{A}$ (defined early in § 4) the proper time $\tau(s, f)$ is defined (cf. Møller [1] §§ 36, 37) by

$$\tau(s, f) = \int_0^s \left(1 - \left|\frac{df}{dt}\right|^2\right)^{1/2} dt$$

where the integrand is clearly bounded and measurable. By assumption on $\mathcal{A}$, it is strictly positive almost everywhere (Lebesgue measure). For $F \in \mathcal{W}, F(t) = \langle f(t), v(t)\rangle$, or for $F \in \overline{\mathcal{W}}, F(t) = \langle f(t), c+t, v(t)\rangle$, we let $\tau(s, F) = \tau(s, f)$. Then let

$$\varphi_t^s(f) = \tau(t, f) - \tau(s, f)$$

for $0 \leqslant s \leqslant t, f \in \overline{\mathcal{W}}$. Clearly $\varphi_s^s(f) = 0$ for any $s \geqslant 0$, and $\varphi_t^s(f) > 0$ if $0 \leqslant s < t$. We have

$$\varphi_t^s(f) + \varphi_u^t(f) = \varphi_u^s(f)$$

for $0 \leqslant s \leqslant t \leqslant u, f \in \overline{\mathcal{W}}$, and $\varphi_t^s(f)$ is continuous in s and t. The functional φ is stationary in the sense that

$$\varphi_t^s(\theta_h f) = \varphi_{t+h}^{s+h}(f)$$

for $h \geqslant 0, 0 \leqslant s \leqslant t, f \in \overline{\mathcal{W}}$.

The strong Markov process $(\{Q_{c,x,v}\}, \overline{\mathcal{W}}, I, \overline{W}, B(\overline{W}))$ is "strongly" measurable as defined by Dynkin [2], 3.17, i.e. the conclusion of Lemma 2.1 above holds, since the hypothesis (right continuity) does. We can extend each of the σ-fields $B_t^s(\overline{\mathcal{W}})$ by the sets which are subsets of sets of measure zero for all the $Q_{c,x,v}$, thus obtaining a "complete" fields Markov process as defined by Dynkin [2], 3.6.

Thus we have established all the hypotheses of a theorem on "random change of time": Dynkin [2], Theorem 10.10. To state the theorem in our case, we let $\tau \to t(\tau, f)$ be the inverse of the increasing function $t \to \tau(t, f)$ for any $f \in \overline{\mathcal{W}}$. Then for any fixed $\tau > 0, t(\tau, \ )$ is a stopping time. For $f \in \overline{\mathcal{W}}$ let

$$(Af)(\tau) = f(t(\tau, f)), \zeta'(Af) = \sup(\tau(t, f) : t \geqslant 0), x(\tau, f) = (Af)(\tau).$$

Theorem 6.1. *$(x(\tau, \ ), \zeta' \circ A, B_{t(\tau)}, Q_{c,x,v})$ is a fields Markov process.*

Corollary. *$(\{Q_{c,x,v} \circ A^{-1}\}, A(\overline{\mathcal{W}}), \zeta', \overline{W}, B(\overline{W}))$ is a (path space) Markov process.*

We have the inclusion

$$A(B_{t(\tau, \ )}) \supset B_\tau^0;$$

I don't know whether the converse inclusion holds (given right continuity, in general).

The "velocity component" of a function in $A(\overline{\mathfrak{W}})$ is not the derivative of the "space component". To remedy this situation, let U be the transformation of V into M_0 defined by

$$U(v) = \langle v, 1\rangle / \sqrt{1 - |v|^2}.$$

Then $\| U(v) \|^2 = 1$ for any $v \in V$, and the last component of $U(v)$ is positive. $U(v)$ is the four-velocity associated with the "velocity" v (cf. Møller [1] § 37). Let $\mathfrak{U}$ be the set of all $U(v)$, $v \in V$. $\mathfrak{U}$ is one nappe of a three-dimensional hyperboloid. For

$$\xi = \langle x, c, v\rangle \in M \times V,$$

let $U_2(\xi) = \langle x, c, U(v)\rangle \in M \times \mathfrak{U}$. Then U_2 is a homeomorphism of $M \times V$ onto $M \times \mathfrak{U}$, and it defines a transformation U_2 of $A(\overline{\mathfrak{W}})$ onto another function space $\mathfrak{V}$. For $\langle x, c, U\rangle \in M \times \mathfrak{U}$, $z = \langle x, c\rangle$, $U = U(v)$, let

$$\textstyle\prod_{z,U} = Q_{c,x,v} \circ A^{-1} \circ U_2^{-1}, \zeta(U_2 f) = \zeta'(f)$$

(U_2 is one-to-one). Then by a standard result on transformation of phase spaces (Dynkin [2] Theorem 10.13) we have

Theorem 6.2. *$(\{\prod_{z,U}\}, \mathfrak{V}, \zeta, M \times \mathfrak{U}, B(M \times \mathfrak{U}))$ is a Markov process.*

We now have that for $f = \langle \phi, \psi\rangle \in \mathfrak{V}$, where ϕ has values in M and ψ in $\mathfrak{U}$, $\phi'(\tau^+) = \psi(\tau)$ for all $\tau < \zeta(f)$, since the derivative with respect to τ is the 4-velocity (Møller [1] § 37, equation (38)). Also ψ has a limit from the left for all $\tau > 0$. Thus ψ is bounded on any bounded closed subinterval of $[0, \zeta(f))$, and ϕ is locally Lipschitzian and an indefinite integral of ψ.

Assumption (A) of § 4 (spatial homogeneity) and the same condition for the measures R_c of § 5 imply the corresponding condition for the $\prod_{z,U}$:

(I) If $y, z \in M$, $U \in \mathfrak{U}$, $\tau \geq 0$, and $A \in B(M)$,

$$\textstyle\prod_{z,U}(\langle \phi, \psi\rangle : \phi(\tau) \in A) = \prod_{z+y,U}(\langle \phi, \psi\rangle : \phi(\tau) \in A + y).$$

Also, assumption (D) of § 4 yields

(II) For any $L \in \mathfrak{L}$, $A \in B^0(\mathfrak{V})$, and $U \in \mathfrak{U}$,

$$\textstyle\prod_{0,U}(A) = \prod_{0,L(U)}(L(A)).$$

An $L \in \mathfrak{L}$ acts on the tangent space M_0 of M through the natural isomorphism of M and M_0, or, equivalently here, through its differential dL, so that if $U \in \mathfrak{U}$ and $\phi'(\tau) = U$, then

$$L(U) = dL(\phi(\tau))/d\tau.$$

Now let P be the natural projection of $M \times \mathfrak{U}$ onto $\mathfrak{U}$. By (I) and Theorem 3.2,

$$(\{\textstyle\prod_{z,U} \circ P^{-1}\}, P(\mathfrak{V}), \eta, \mathfrak{U}, B(\mathfrak{U}))$$

is a Markov process independent of z, where $\eta(P(f)) = \zeta(f)$ for any $f \in \mathfrak{V}$. P is one-to-one on $\mathfrak{V}$, so that this process determines $\{\prod_{z,U}\}$ (by integration).

For $\sigma \geq 0$, $U \in \mathfrak{U}$, and $A \in B(\mathfrak{U})$, let

$$P_\sigma(U, A) = \textstyle\prod_{z,U}(\langle \phi, \psi\rangle \in \mathfrak{V} : \psi(\sigma) \in A)$$

(which is independent of z). By the Markov property, we have the Chapman–Kolmogorov equation (semigroup property)

$$P_{\sigma+\tau}(U,B)=\int_{\mathcal{U}} P_\sigma(U,dV)\,P_\tau(V,B)$$

for any $\sigma,\tau\geqslant 0$, $U\in\mathcal{U}$, $B\in B(\mathcal{U})$. By (II), we have

$$P_\sigma(U,B)=P_\sigma(L(U),L(B)),\ L\in\mathfrak{L}.$$

In § 8, we shall reproduce the known classification of all semigroups $\{P_\sigma\}$ satisfying our conditions.

Lemma 6.3. *For any* $\{\prod_{z,U}\}$ *satisfying our conditions and* $\tau>0$,

$$\textstyle\prod_{z,U}\{\langle\phi,\psi\rangle:\phi'(\tau)\ \textit{exists}\}=\prod_{z,U}\{f:\zeta(f)>\tau\}.$$

Proof. Let $\mathcal{N}$ be the set of all $\tau>0$ for which the conclusion does not hold. For $f=\langle\phi,\psi\rangle\in\mathcal{V}$, $\phi'(\tau)$ exists for almost all τ in the interval $(0,\zeta(f))$. Thus by Lemma 4.1 and the Fubini theorem, $\mathcal{N}$ has Lebesgue measure 0. Thus for any $\tau>0$ there exist $t>0$ and $u>0$, neither of them in $\mathcal{N}$, such that $t+u=\tau$. For any $\sigma>0$,

$$\textstyle\prod_{0,U}\{\langle\phi,\psi\rangle:\phi'(\sigma)\ \text{exists}\}$$

is independent of U by condition (II). Letting $\sigma=u$ and applying the Markov property at time t, we conclude that $\tau\notin\mathcal{N}$, q.e.d.

Of course, for $\langle\phi,\psi\rangle\in\mathcal{V}$, $\phi'(\tau)$ is equal to $\psi(\tau)$ if the former is defined.

7. The space $\mathcal{U}$

$\mathcal{U}$ is well-known as "hyperbolic space" or "Lobachevsky space" (Gelfand and Berezin [1]). It is acted on transitively by the proper Lorentz group $\mathfrak{L}$. The subgroup K_p of $\mathfrak{L}$ leaving a point p of $\mathcal{U}$ fixed is isomorphic to the orthogonal group K on three-dimensional Euclidean space, and is a maximal compact subgroup of $\mathfrak{L}$. $\mathcal{U}$ can be regarded as the homogeneous space $\mathfrak{L}/K_p$ of right cosets of K_p in $\mathfrak{L}$. (Of course, K_p is not a normal subgroup of $\mathfrak{L}$.)

$\mathcal{U}$ has a natural $\mathfrak{L}$-invariant Riemannian structure, inherited from the "pseudo-Riemannian" or Lorentz form $\|\ \|^2$ on M_o by restriction to the tangent spaces of $\mathcal{U}$.

We put geodesic polar (or spherical) coordinates on $\mathcal{U}$ (see Helgason [1] p. 401) as follows: we take p as $(0,0,0,1)$ for a given Lorentz coordinate system (x,y,z,t) on M_o. Given any point U in $\mathcal{U}$ we let $\varrho(U)$ be the Riemannian distance from p to U in $\mathcal{U}$. Now each surface $\varrho=\varrho_0>0$ in $\mathcal{U}$ is isometric to a Euclidean sphere of some radius $f(\varrho_0)$. (K_p acts transitively on these surfaces, i.e. $\mathcal{U}$ is a "two-point homogeneous" Riemannian manifold as defined by Helgason [1] p. 335.) in each such sphere we choose standard spherical coordinates θ and ϕ, $0\leqslant\theta<2\pi$, $-\pi/2\leqslant\varphi\leqslant\pi/2$, constant on each geodesic emanating from p. The functions (ϱ,θ,ϕ) make up a regular local coordinate system except where ϱ or θ is zero or $\phi=\pm\pi/2$.

Now let $r=(x^2+y^2+z^2)^{1/2}$. Then ϱ is a function of r in $\mathcal{U}$, and $f(\varrho(r))=r$, $0<r<\infty$, since the induced Riemann metric on a sphere $\varrho=$ constant can also be induced from a 3-plane $t=$ constant, which has the standard Euclidean structure.

To find the function $\varrho(r)$, and hence its inverse f, it suffices to find the Riemann structure induced on the hyperbola H: $t^2-x^2=1$, $y=z=0$. Here $r=|x|$. x is a coordinate function on H, and the tangent vector d/dx is $(1, 0, 0, x/\sqrt{1+x^2})$ at the point $(x, 0, 0, \sqrt{1+x^2})$ of H. This vector has "magnitude" $1/\sqrt{1+x^2}$ for the Riemann metric induced on H. Thus a unit vector $d/d\varrho$ for the Riemann metric is $\sqrt{1+x^2}\, d/dx$ for $x>0$, so that $dx/d\varrho = \sqrt{1+x^2}$, $\varrho = \log(x+\sqrt{1+x^2})$ (since $\varrho=0$ when $x=0$), $|x| = \sinh \varrho$ on H, and $r = \sinh \varrho$ throughout $\mathcal{U}$.

Now let g be the induced Riemann form on $\mathcal{U}$:

$$g_{ij} = g(\partial/\partial x_i, \partial/\partial x_j), \quad \text{where} \quad x_1 = \varrho,\, x_2 = \theta,\, x_3 = \phi.$$

Then $g_{ij}=0$ for $i \neq j$, $g_{11}=1$, $g_{22} = \sinh^2 \varrho$, and $g_{33} = \sinh^2 \varrho \sec^2 \phi$ (except at the exceptional points $\theta = 0$ etc.). Now the "Laplace-Beltrami" operator Δ on $\mathcal{U}$ is (see Helgason [1] p. 386)

$$\frac{1}{\sqrt{\bar{g}}} \sum_k \frac{\partial}{\partial x_k} \sum_i g^{ik} \sqrt{\bar{g}} \frac{\partial}{\partial x_i}$$

($\bar{g}$ = determinant of (g_{ij}), g^{ik} = inverse matrix)

$$= \partial^2/\partial\varrho^2 + 2\coth\varrho\, \partial/\partial\varrho + \operatorname{csch}^2\varrho\, \partial^2/\partial\phi^2 - \tan\phi \operatorname{csch}^2\varrho\, \partial/\partial\phi + \operatorname{csch}^2\varrho \sec^2\phi\, \partial^2/\partial\theta^2$$

(wherever ϱ, θ, ϕ form a coordinate system).

8. Semigroups of measures on $\mathcal{U}$

At the end of § 6 we had obtained a collection of nonnegative, K_p-invariant measures $P_t(p, \;)$ on $B(\mathcal{U})$ for $p \in \mathcal{U}$, $t \geq 0$, of total mass at most 1. For a fixed p, we call finite, K_p-invariant, nonnegative measures on $\mathcal{U}$ "radial". The "convolution" of two radial measures μ and ν on $\mathcal{U}$ is given by

$$(\mu * \nu)(A) = \int_{\mathcal{U}} \nu(A_x)\, \mu(dx)$$

for any $A \in B(\mathcal{U})$, where A_x is a set $T_x(A)$ and T_x is any element of $\mathcal{L}$ taking x into p. By K_p-invariance, it is irrelevant which such element is chosen, and we can take T_x continuous in x to insure measurability.

Since $P_t(U, B) = P_t(L(U), L(B))$ for all $L \in \mathcal{L}$, $B \in B(\mathcal{U})$, we have

$$P_s * P_t = P_{s+t} \quad \text{for all} \quad s, t \geq 0.$$

Thus the P_t form a semigroup under convolution. Now each radial measure on $\mathcal{U}$ has a "Fourier transform", defined in terms of the eigenfunctions of Δ which are functions of ϱ alone ("spherical functions": see Helgason [1] Chap. X § 3 p. 398). As given by Tutubalin, Karpelevich and Shur [1] (hereafter called "TKS") the nonconstant spherical functions on $\mathcal{U}$, normalized to the value 1 at p, are

$$F_\lambda(\varrho) = \sin \lambda\varrho / \lambda \sinh \varrho,$$

where λ is any complex number and $\Delta F_\lambda = (-1-\lambda^2) F_\lambda$. For λ or ϱ equal to 0, $F_\lambda(\varrho)$ is defined by continuity. These functions are bounded for $|Im\, \lambda| \leq 1$. The Fourier transform of a radial measure μ is given by

$$\hat{\mu}(\lambda) = \int_{\mathcal{U}} F_\lambda(x)\, d\mu(x),$$

defined at least for $|Im\,\lambda| \leqslant 1$. For radial measures convolution is commutative, and

$$(\mu * \nu)^{\wedge}(\lambda) = \hat{\mu}(\lambda)\,\hat{\nu}(\lambda)$$

(see TKS [1]).

A radial measure μ on $\mathcal{U}$ such that for all $n = 1, 2, \ldots, \mu$ is the n-fold convolution $(\mu_n)^n$ of a radial measure μ_n with itself, is called "infinitely divisible". Clearly the measures P_t have this property. Such measures have been characterized by Tutubalin [1] and Gangolli [1]: μ is infinitely divisible of total mass $\beta > 0$ if and only if

$$(^*) \qquad \hat{\mu}(\lambda) = \beta \exp\left((-1-\lambda^2)\,\alpha - \int_{\mathcal{U}} (1 - F_\lambda(u))\, dL(u)\right)$$

where $\alpha \geqslant 0$ and L is a nonnegative, K_p-invariant measure on $\mathcal{U}$ satisfying

$$\int_0^\infty \frac{\varrho(u)^2}{1+\varrho(u)^2}\, dL(u) < \infty$$

(so that the integral of $1 - F_\lambda$ is absolutely convergent). $\alpha = \alpha(\mu)$ and $L = L(\mu)$ are uniquely determined. Conversely, for any such measure L and numbers α and $\beta \geqslant 0$, there is a radial, infinitely divisible measure μ satisfying (*).

Tutubalin [1] proved the representation formula (*) under a definition of infinite divisibility requiring that for every $\varepsilon > 0$, μ is a convolution of possibly distinct measures each concentrated within Riemann distance ε of p, except for mass ε. An argument in his proof, partly reproduced below, shows that this is equivalent to the definition given above. A proof under our definition was also given by R. Gangolli [1] for general symmetric spaces of non-compact type.

We say a radial measure μ *divides* another one, ν, if $\mu * \varrho = \nu$ for some radial ϱ. Clearly this implies

$$\alpha(\mu) + \alpha(\varrho) = \alpha(\nu) \quad \text{and} \quad L(\mu) + L(\varrho) = L(\nu),$$

so that μ divides ν if and only if $\alpha(\mu) \leqslant \alpha(\nu)$ and $L(\mu) \leqslant L(\nu)$.

Now given a convolution semigroup P_t, we have $L(P_{rt}) = rL(P_t)$ and $\alpha(P_{rt}) = r\alpha(P_t)$ for all rational $r > 0$ by uniqueness, and hence for all $r > 0$ by the ordering just mentioned. Thus the semigroup is completely determined by $L(P_1)$ and $\alpha(P_1)$.

$P_t(p, \ )$ converges weak* to the unit mass concentrated at p as $t \downarrow 0$, since $\hat{P}_t$ converges to 1 pointwise and each $|F_\lambda|$ for λ real is bounded below 1 outside any neighborhood of p. Note that the original continuity assumption (E) was used only to obtain a Markov process with proper time as parameter; for the latter, the continuity properties follow from the invariance conditions, and the first part of assumption (C) is taken care of by Lemma 6.3.

We now show that ζ is actually infinite:

Lemma 8.1. *For any* $z \in M$ *and* $U \in \mathcal{U}$, $\prod_{z,U}(\zeta = +\infty) = 1$.

Proof. The Markov process

$$(\{\textstyle\prod_{z,U} \circ P^{-1}\}, P(\mathcal{V}), \eta, \mathcal{U}, B(\mathcal{U}))$$

is right continuous and Fellerian, hence strongly Markovian. Given $f \in P(\mathcal{V})$, let $t_0(f) = 0$. Given $t_0(f), \ldots, t_n(f)$, let $t_{n+1}(f)$ be the least t, if any, such that the Riemann distance from

$$f(t_0(f)+\ldots+t_n(f)) \quad \text{to} \quad f(t_0(f)+\ldots+t_n(f)+t)$$

is greater than or equal to 1 (if such a t exists, $t_{n+1}(f)$ is well-defined by right continuity). Clearly $t_1(f)+\ldots+t_n(f)$ is a stopping time for each n.

If $\eta(f)$ is finite, then all the $t_n(f)$ are defined and their sum converges. In any case, each t_n is defined with $\prod_{z,U}\circ P^{-1}$ probability one. The random variables t_n for $n\geqslant 1$ are independent by the strong Markov property and have the same probability distribution on the nonnegative real axis. Each is almost surely strictly positive by right continuity. Thus the probability of their sum converging is zero, q.e.d.

It follows that under our assumptions the measures $P_t(p,\ )$ are all probability measures. We summarize our results as follows:

Theorem 8.2. *There is a natural* $1-1$ *correspondence between sets* $\{P_x^v\}$ *of measures on* $\mathcal{A}$ *satisfying* (A)–(E) *of* § 4, *and infinitely divisible radial probability measures* P_1 *on* $\mathcal{U}$, *and hence with pairs* (L,α) *as described earlier in this section. Each* P_x^v *gives outer measure* 1 *to the set of functions* $f\in\mathcal{A}$ *having left and right derivatives at all points* (*except left derivatives at* 0), *equal except at countably many points, with right derivatives continuous from the right and left derivatives from the left.*

9. Further discussion of radial semigroups on $\mathcal{U}$

In this section we discuss further facts which were not needed for the proof of Theorem 8.2, but may well be of interest.

Suppose μ is an infinitely divisible radial probability measure on $\mathcal{U}$, and $\{\prod_{z,U}\}$ the corresponding Markov process on $\mathcal{V}$. It follows from results of Gangolli [3] that if $n(M,f,t)$ for $M,t\geqslant 0$ and $f=\langle\phi,\psi\rangle\in\mathcal{V}$ is the number of values of $\tau<t$ such that the Riemann distance from $\phi'(\tau-)$ to $\phi'(\tau+)$ is at least M, then

$$\int n(M,f,t)\,d\textstyle\prod_{z,U}(f)=tL(\mu)(U:\varrho(U)\geqslant M).$$

Thus if $L=0$, $\alpha>0$, ψ is almost surely continuous. These processes ("Brownian motions") will be considered in § 10.

On the other hand, we have so-called "Poisson" processes defined as follows: let μ_σ be normalized K_p-invariant "surface area" measure on the sphere $\varrho=\sigma$ in $\mathcal{U}$, for any $\sigma>0$. Then the "Poisson measure" $\pi(\sigma,c)$ is defined by

$$\pi(\sigma,c)=e^{-c}\sum_{n=0}^{\infty}(c\mu_\sigma)^n/n!,$$

where the powers represent convolutions. It is easy to verify that $P_t=\pi(\sigma,ct)$ gives a convolution semigroup. The associated process in $\mathcal{U}$ will remain at one point for some time, then jump to one of the points at Riemann distance σ, with uniform probability distribution over the sphere of such points. The probability of such a jump during an interval $t_0<t<t_0+\varepsilon$ is $1-e^{-\varepsilon c}$ (see Loève [1] p. 548), which is asymptotic to $c\varepsilon$ as $\varepsilon\downarrow 0$.

A Poisson process seems very natural as a velocity process, where a particle undergoes collisions at various times which cause discontinuities in its velocity, constant between collisions.

If I is the identity operator on bounded continuous functions on $\mathcal{U}$, then

$$\lim_{t \downarrow 0} (P_t - I)/t$$

will be, in general, an unbounded operator defined on a subspace, called the "generator" of the semigroup P_t. Transferring our semigroups from $\mathcal{U}$ to $\mathfrak{L}$ in a rather obvious way, we obtain information on their generators from Hunt [1] (see also Wehn [1]).

Since any radial measure μ on $\mathcal{U}$ is an average

$$\mu = \int \mu_\sigma \, d\nu(\sigma)$$

for some finite measure ν on R^+, we can explicitly calculate any convolution of radial measures if we can find $\mu_a * \mu_b$ for $a, b > 0$. In doing this, we can assume $b \leqslant a$. Suppose given $c \geqslant 0$; let us find

$$f(a,b,c) = (\mu_a * \mu_b)(\varrho \leqslant c).$$

Clearly

$$f(a,b,c) = 0 \quad \text{if} \quad c < a - b,$$
$$= 1 \quad \text{if} \quad c > a + b.$$

For any $u \in \mathcal{U}$ and $t > 0$ let $S(u,t)$ be the set of points of $\mathcal{U}$ at Riemann distance t from u. Let

$$q = (\sinh a, 0, 0, \cosh a);$$

then $\varrho(q) = a$. Also $L_a \in \mathfrak{L}$ takes $p = (0,0,0,1)$ into q, where

$$L_a(x,y,z,t) = (x \cosh a + t \sinh a, y, z, t \cosh a + x \sinh a).$$

$S(p,b)$ is the set of $(X, y, z, T) \in \mathcal{U}$ with $T = \cosh b$ and

$$X^2 + y^2 + z^2 = \sinh^2 b.$$

Thus $S(q,b)$ is the set of points

$$(X \cosh a + \cosh b \sinh a, y, z, \cosh b \cosh a + X \sinh a)$$

with $X^2 + y^2 + z^2 = \sinh^2 b$. Then $S(p,c) \cap S(q,b)$ is the set of (x,y,z,t) with $t = \cosh c$, $x = X \cosh a + \cosh b \sinh a$ where $X = (\cosh c - \cosh b \cosh a)/\sinh a$, and $y^2 + z^2 = \sinh^2 b - X^2$. Applying L_a^{-1} we obtain a circle of radius $\sinh^2 b - X^2$ in $S(p,b)$, a sphere of radius $\sinh b$, in the given coordinate system for M_0. We now want to find the areas of the two zones demarcated by this circle.

For c small enough so that $\cosh c \leqslant \cosh b \cosh a$, the area of the zone which concerns us is

$$2\pi \sinh b \left(\sinh b - \frac{\cosh b \cosh a - \cosh c}{\sinh a} \right).$$

For $\cosh c > \cosh b \cosh a$ we want the area of the larger of the two zones marked off by the circle, which is given by the same formula. The total area of the sphere being $4\pi \sinh^2 b$, we obtain

$$f(a, b, c) = \tfrac{1}{2}\left[1 + \frac{\cosh c - \cosh b \cosh a}{\sinh b \sinh a}\right],$$

$$df(a, b, c)/dc = \sinh c/2 \sinh a \sinh b.$$

Proposition 9.1. For any $a, b, \geq 0$,

$$\mu_a * \mu_b = \int_{|a-b|}^{a+b} \frac{\sinh x}{2 \sinh a \sinh b} \mu_x \, dx.$$

We now turn to proving "transience" or "non-recurrence" of our invariant processes in $\mathcal{U}$. Proposition 9.1 will not be used.

Theorem 9.2. *For any invariant Markov process* $\{\prod_{z,U}\}$ *as described in* § 6, *with* $\prod_{z,U}$ *not concentrated in one function, and any compact subset* K of $\mathcal{U}$,

$$\textstyle\prod_{z,U}\{\langle\phi, \psi\rangle : \psi(\tau) \notin K \text{ for all sufficiently large } \tau\} = 1.$$

We first prove the following

Lemma 9.3. *For any radial probability measure* μ *on* $\mathcal{U}$ *not concentrated at* p, $0 < \hat{\mu}(0) < 1$ *and for any* $R \geq 0$ *there is a* $K > 0$ *such that*

$$\mu^n(\varrho \leq R) \leq K[\hat{\mu}(0)]^n.$$

Proof. Since μ is not concentrated at p,

$$0 < \hat{\mu}(0) = \int_{\mathcal{U}} \frac{\varrho}{\sinh \varrho} \, d\mu < 1.$$

Then
$$\int \frac{\varrho}{\sinh \varrho} \, d\mu^n = (\mu^n)^\wedge(0) = [\hat{\mu}(0)]^n.$$

Thus given $R \geq 0$,

$$\mu^n(\varrho \leq R) < \frac{\sinh R}{R} [\hat{\mu}(0)]^n, \text{ q.e.d.}$$

Now to prove Theorem 9.2, suppose for some z, U, and compact $K \subset \mathcal{U}$,

$$P_K = \textstyle\prod_{z,U}(\langle\phi, \psi\rangle \in \mathcal{V} : \psi(t) \in K \text{ for } t \text{ arbitrarily large}) > 0.$$

For $k = 1, 2, \ldots, f = \langle\phi, \psi\rangle \in \mathcal{V}$, let $n_k(f)$ be the kth integer n such that $\psi(t) \in K$ for some t in the interval $[n, n+1)$, or $+\infty$ if there is no such n; let $t_k(f)$ be the least such t, or $+\infty$ if $n_k = +\infty$. Then each t_k is a stopping time.

Let d be the Riemann distance in $\mathcal{U}$. There is an $R > 0$ such that

$$\textstyle\prod_{z,U}(\langle\phi, \psi\rangle : d(\psi(t), \psi(0)) \leq R, 0 \leq t \leq 1) = \varepsilon > 0.$$

ε is clearly independent of z and U. Let A_k be the set of all $f \in \mathcal{V}$ such that $n_k(f)$ and $t_k(f)$ are finite, and let C_k be the set of $\langle\phi, \psi\rangle \in A_k$ such that

$$d(\psi(t_k), \psi(t_k + t)) \leq R, 0 \leq t \leq 1.$$

Then by the the strong Markov property of $\{\prod_{z,U}\}$

$$\prod_{z,U}(C_k \mid B_{t_k}) = \varepsilon$$

almost everywhere on A_k. Thus

$$\prod_{z,U}(C_k) = \varepsilon \prod_{z,U}(A_k) \geq \varepsilon P_K > 0 \quad \text{for} \quad k = 1, 2, \ldots.$$

Thus $$\prod_{z,U}\{f : f \in C_k \text{ for infinitely many } k\} \geq \varepsilon P_K.$$

Hence if K_R is the set of $U \in \mathcal{U}$ such that $d(U, V) \leq R$ for some $V \in K$,

$$\prod_{z,U}\{\langle \phi, \psi \rangle : \psi(n_k + 1) \in K_R \text{ for infinitely many } k\} > 0.$$

This contradicts Lemma 9.3, so the proof of Theorem 9.2 is complete.

10. Diffusion processes in $\mathcal{U}$ and $M \times \mathcal{U}$

A convolution semigroup P_t in a symmetric space, e.g. in $\mathcal{U}$, is called a *diffusion* semigroup if

$$\lim_{t \downarrow 0} P_t(\varrho \geq t)/t = 0,$$

where ϱ is the Riemannian distance from the fixed point used in defining convolution. The theory of such semigroups tells us that the generator A is defined at least on all functions which, together with their first- and second-order partial derivatives, are bounded and uniformly continuous on $\mathcal{U}$, and is equal on these functions to a second-order differential operator (Yosida [1]).

On $\mathcal{U}$, or other "two-point homogeneous" symmetric spaces G/K, G-invariance then implies that $A = m\Delta + b$ where Δ is the Laplace–Beltrami operator. By definition of A it is zero on constants and $Af \leq 0$ where f has a relative maximum, so that $b = 0$ and $m \geq 0$.

Thus we have the "Brownian motion" semigroup with parameter m as defined by Yosida [1]. On $\mathcal{U}$, the P_t for this semigroup are given explicitly by Tutubalin [1] as

$$P_t = (4\pi m t)^{-3/2} \frac{\varrho}{\sinh \varrho} \exp\left(-mt - \frac{\varrho^2}{4mt}\right) \cdot N,$$

where N is the $\mathfrak{L}$-invariant measure on $\mathcal{U}$ given by $4\pi(\sinh^2 \varrho)\, d\varrho\, d\Omega$ and Ω is normalized orthogonally invariant surface area measure on a sphere ($d\Omega = \cos\phi\, d\theta\, d\phi / 4\pi$). In this case the measures P_x^v are concentrated in the set of functions f having continuous first derivatives $f'(t)$, since the P_t are a diffusion semigroup (Yosida [1], Dynkin [1] Theorem 6.5).

Now, the space $M \times \mathcal{U}$ is a homogeneous space under the action of the "Poincaré group" $\bar{\mathfrak{L}}$ generated by $\mathfrak{L}$ (acting on M and $\mathcal{U}$ together) and by translations of M which leave $\mathcal{U}$ pointwise fixed. The subgroup of $\bar{\mathfrak{L}}$ leaving the point $(0, p)$ of $M \times \mathcal{U}$ fixed remains equal to K_p, so that $M \times \mathcal{U}$ can be regarded as the space $\bar{\mathfrak{L}}/K_p$ of right cosets of K_p in $\bar{\mathfrak{L}}$.

One can define a convolution for finite, (left) K_p-invariant measures on $M \times \mathcal{U}$, as before, by

$$(\mu * \nu)(A) = \int_{M \times \mathcal{U}} \mu(T_z(A))\,\nu(dz), \quad A \in B(M \times \mathcal{U}),$$

where $T_z \in \mathcal{L}$ takes z into $(0, p)$ and μ and ν are invariant under K_p.

Our conditions (I) and (II) of § 6 can be regarded as conditions of $\bar{\mathcal{L}}$-invariance of transition probabilities

$$Q_t((x, p); B) = \textstyle\prod_{x,p}(f : f(t) \in B),$$

$t \geqslant 0, x \in M, p \in \mathcal{U}, B \in B(\mathrm{M} \times \mathcal{U})$. Our assumptions imply, letting

$$Q_t = R_t((0, p);\ ),$$

that $Q_s * Q_t = Q_{s+t}$ on $M \times \mathcal{U}$. If P_t is a diffusion semigroup on $\mathcal{U}$ with generator $m\Delta$, the generator of Q_t is the differential operator whose value at the point (x, U) of $M \times \mathcal{U}$ is $-U + M\Delta$, where $m\Delta$ acts on $\mathcal{U}$ and U acts as a first-order differential operator on M at x.

G. Schay [1] also studied $\bar{\mathcal{L}}$-invariant diffusion processes in $M \times M_0$, and arrived at essentially the same generator or "diffusion equation" (Schay [1] Theorem 4 p. 39, Equation 3.33). Our use of the proper time permits a considerable simplification of the result.

11. Non-existence results

In § 8 above we proved the existence of a class of Lorentz-invariant processes with speeds less than 1 in $M \times V$. In this section we show that such processes with speeds equal to 1 (in $M \times S^2$, where S^2 is the sphere $|x| = 1$ in R^3) are deterministic (trivial), and that this remains essentially true if we allow states to be specified by momenta rather than velocities only. We also prove the non-existence of invariant processes in M itself. The results in this section do not require the Markov property or continuity in probability.

On a trajectory with speed almost everywhere equal to 1 the proper time τ is a constant and hence not suitable as a parameter.

Let $\mathcal{D}$ he the set of all functions f from R^+ to X such that for $0 \leqslant s \leqslant t$,

$$|f(t) - f(s)| \leqslant t - s,$$

and if $f'(t)$ is defined, $|f'(t)| = 1$.

Now suppose given starting probabilities P_x^v on the stationary, measurable path space $(\mathcal{D}, I, X, B(X))$, satisfying (A), (B) and (D) of § 4, with V replaced by S^2.

A collection $\{P_x^v\}$ will be called *deterministic* if each P_x^v gives mass 1 to a set containing only one function.

For any $f \in \mathcal{D}$ with f' non-constant there is a unique $t = t_0(f) > 0$ such that

$$t^2 - f(t)^2 = 1$$

(i.e., $\langle f(t), t \rangle \in \mathcal{U}$). Clearly $t_0(\ )$ is $B^0(\mathcal{D})$-measurable.

Suppose $\{P_x^v\}$ is non-deterministic. Then by (A) and (D) no P_x^v gives mass 1 to one function. Let $x=0$ and fix $v \in S^2$. Then there exist Lorentz transformations L_n, $n=1,2,\ldots$, such that $V_{L_n}(v)=v$ for all n and for any compact subset C of $\mathcal{U}$, $L_n(C)$ and C are disjoint for n large enough (we take L_n as Lorentz transformations defined by relative velocities v_n parallel to v with $|v_n| \uparrow 1$). For $A \in B(\mathcal{U})$, let

$$P(A) = P_x^v\{f : \langle f(t_0(f)), t_0(f)\rangle \in A\}.$$

Then P is a finite, non-zero Borel measure on $\mathcal{U}$, invariant under all the L_n, which is impossible. We have proved

Theorem 11.1. *If* $(\{P_x^v\}, \mathcal{D})$ *satisfies* (A), (B) *and* (D) *of* § 4, *then each* P_x^v *is concentrated in the function* $f \in \mathcal{D}$ *with* $f(0)=x$ *and* $f'(t)=v$ *for all* $t \geqslant 0$. *Conversely, such* P_x^v *satisfy* (A) – (E).

We can try to avoid the paucity of processes in Theorem 11.1 by introducing different possible "states" of a particle moving with speed 1 in a given direction. One possibility is to use an "energy" analogous to the energy of a photon or other particle of zero rest mass. We take such an energy as defined by a function

$$E : (\mathsf{C}, s, f,) \to E(\mathsf{C}, s, f),$$

where $f \in \mathcal{D}$, $s \geqslant 0$, and C is an arbitrary Lorentz coordinate system $\langle x(\ ), t(\ )\rangle$, where $x(\ )$ and $t(\ )$ take M onto X and R respectively. Then the trajectory defined by f in the original coordinate system of M is defined in C by another function $f_{\mathsf{C}} \in \mathcal{D}$ with

$$x(\langle f(s), s\rangle) = f_{\mathsf{C}}(t(\langle f(s), s\rangle)), \; s \geqslant 0.$$

We assume that $E(\mathsf{C}, s, f)$ is defined if and only if $f'_{\mathsf{C}}(s)$ is defined.

We require that

$$\langle E(\mathsf{C}, t, f)\, f'_{\mathsf{C}}(t), E(\mathsf{C}, t, f)\rangle$$

transform as an "energy-momentum vector" under Lorentz transformations L (in $\mathfrak{L}$) of C (see Møller [1] p. 72). This is simply the natural action of $\mathfrak{L}$ on M_0, the tangent space of M, through its isomorphism with M. The space of possible energy-momentum vectors is the open half-cone Q of all points $\langle \xi, E\rangle$, $\xi \in R^3$, $E>0$, with $|\xi|^2 = E^2$. Q is acted on transitively by $\mathfrak{L}$.

The subgroup $\mathcal{K}$ of $\mathfrak{L}$ leaving the point $(1,0,0,1)$ fixed contains all transformations K_{bc} with matrices

$$\begin{pmatrix} 1-a & b & c & a \\ -b & 1 & 0 & b \\ -c & 0 & 1 & c \\ -a & b & c & 1+a \end{pmatrix}$$

where $a=(b^2+c^2)/2$ and b and c are arbitrary real numbers. Thus (in its relative topology from the general linear group) $\mathcal{K}$ is not compact.

For $k \geqslant 0$ let $\mathcal{U}_k$ be the set of $\langle x_1, x_2, x_3, t\rangle \in \mathcal{U} \subset M$ with $x_2 \geqslant 0$, $x_3 \geqslant 0$, and $t \leqslant k$. In $\mathcal{U}_k$, $t - x_1 > 0$ (since $t>0$ and $t^2 > x_1^2$) so by compactness there is an $\varepsilon > 0$ such that $t + x_1 \geqslant \varepsilon$ for all $\langle x, t\rangle \in \mathcal{U}_k$. Now

$$K_{bc}\langle x, t\rangle = \langle\ ,\ -ax_1 + bx_2 + cx_3 + (1+a)t\rangle.$$

Thus for $\langle x, t\rangle \in \mathcal{U}_k$ and $b, c \geqslant 0$ the time component of $K_{bc}\langle x, t\rangle$ it at least $a\varepsilon$. Hence for b and c sufficiently large, $K_{bc}(\mathcal{U}_k)$ is disjoint from $\mathcal{U}_k$, but, on the other hand, it is included in a larger set $\mathcal{U}_{k'}$. Thus we can find an infinite sequence of disjoint sets in $\mathcal{U}$ taken into $\mathcal{U}_k$ by transformations in $\mathcal{K}$. Thus for the initial energy-momentum $(1, 0, 0, 1)$ the probability that the trajectory passes through $\mathcal{U}_k$ is zero. We can apply the same argument to the cases where $y \geqslant 0, z \geqslant 0$ in the definition of $\mathcal{U}_k$ is replaced by $y \geqslant 0, z \leqslant 0$ or $y \leqslant 0, z \geqslant 0$ or $y \leqslant 0, z \leqslant 0$ (letting b and c have the same signs as y and z respectively). Thus there is probability zero of passing through $\mathcal{U}$.

Let $\mathcal{M}$ be the set of all functions $f = \langle g, h\rangle$ from R^+ to $M \times Q$ such that $g \in \mathcal{D}$ and $g'(t)$, where defined, is proportional to $h(t)$ (where the constant may vary with t). We have proved

Theorem 11.2. *Suppose starting probabilities* $\{P_{z,\langle\xi,E\rangle}\}$ *on* $(\mathcal{M}, I, M \times Q, B(M \times Q))$ *are homogeneous in z and Lorentz-invariant in the sense that for any* $L \in \mathcal{L}$, *defining a map* $L \times L$ *of* $M \times Q$ *onto itself and hence of* $\mathcal{M}$ *onto itself,*

$$P_{z,\langle\xi,E\rangle} \circ (L \times L)^{-1} = P_{L(z), L\langle\xi,E\rangle}.$$

Then for each $z \in M$ *and* $\langle\xi, E\rangle \in Q$, $P_{z,\langle\xi,E\rangle}$ *is concentrated in the set of functions* $f = \langle g, h\rangle$ *where g defines a straight half-line in M and* $h(t)$ *is proportional to* $\langle\xi, E\rangle$ *for almost all t (Lebesgue measure), possibly with a varying proportionality factor.*

Thus the energy-momentum approach does not yield any essentially non-trivial processes either. We do not consider here the possibility of allowing still more information, e.g. a "polarization", in the definition of states.

Random processes with speed 1 have been considered by Rudberg [1] in one or two space dimensions. His approach is different from ours; to help clarify the situation we now consider the cases of one and two space dimensions.

For the "velocity" approach (as in Theorem 11.1) in any number of space dimensions, the same arguments as above yield the conclusion that invariant processes do not exist. This seems to be reflected in Rudberg's condition (p. 12, above Equ. (19),and p. 28, A) that there is a distinguished time-axis direction for which the probabilities of scattering in all directions are equal.

For the "momentum" approach, apparently not treated by Rudberg, the situation is as follows: in two or more space dimensions, the situation for invariant processes is essentially the same; in the argument in two dimensions $(1, 0, 0, 1)$ and K_{bc} can be replaced respectively by $(1, 0, 1)$ and

$$K_b = \begin{pmatrix} 1-a & b & a \\ -b & 1 & b \\ -a & b & 1+a \end{pmatrix}, a = b^2/2.$$

In one space dimension, the argument which proved Theorem 11.2 does *not* apply, since the group $\mathcal{K}$ reduces to the identity. The question of the truth of Theorem 11.2 in one space dimension will be left open here.

Now we turn to processes in M itself. Let $\mathcal{J}$ be the class of all functions f from R^+ to M with

$$|f(s) - f(t)| \leqslant |s-t|, s, t \geqslant 0.$$

Suppose given starting probabilities $\{P_z\}$ on

$$(\mathcal{J}, I, M, B(M))$$

which satisfy (A) and (B) below:

(A) For $t_i \geqslant 0$, $A_i \in B(M), z \in M$,

$$P_0\{f: f(t_i) \in A_i, i=1, \ldots, n\} = P_z\{f: f(t_i) \in A_i + z, i=1, \ldots, n\}$$

(B) For any $L \in \mathcal{L}$ and $A \in B^0(\mathcal{J})$,

$$P_0(A) = P_0(L(A)).$$

As before, for those $f \in \mathcal{J}$ with $\langle f(t), t\rangle \in \mathcal{U}$ for some t we let $t_0(f)$ be the unique such t. Assuming P_0 gives positive probability to the set of such f,

$$P(A) = P_0(f: \langle f(t_0(f)), t_0(f)\rangle \in A)$$

is a finite, $\mathcal{L}$-invariant Borel measure on $\mathcal{U}$, which is impossible. Thus P_0 is concentrated in the set of $f \in \mathcal{J}$ with $f(0)=0$ and f' equal to a constant $v \in S_2$. (This result was obtained under additional hypotheses by G. Schay [1] Theorem 1 pp. 17–18.) Since orthogonal transformations of X are in $\mathcal{L}$, the distribution of v is an orthogonally invariant probability measure, hence the standard surface area measure divided by 4π.

Let $\mathcal{D}$ be the set of half-lines

$$t \to (at, bt, ct, t), t \geqslant 0$$

on which $a^2+b^2+c^2=1$ and $a, b, c \geqslant 0$, so that $P_0(\mathcal{D}) = \frac{1}{8}$. If $w > 0$, the Lorentz transformation X_w:

$$\langle x, y, z, t\rangle \to \langle \frac{x+wt}{\sqrt{1-w^2}}, y, z, \frac{t+wx}{\sqrt{1-w^2}}\rangle$$

takes the half-line defined by $v=(a, b, c)$ into the half-line of all points

$$\langle \frac{a+w}{1+wa}t', b\frac{\sqrt{1-w^2}}{1+wa}t', c\frac{\sqrt{1-w^2}}{1+wa}t', t'\rangle, t' \geqslant 0.$$

Thus $\mathcal{D}$ is taken into itself. The second components of the velocities associated with elements of $X_w(\mathcal{D})$ are at most equal to $\sqrt{1-w^2}$. Thus an open subset of S^2, the (a, b, c) with a and c sufficiently small and positive, has measure zero for the distribution of v in S^2, a contradiction. Thus we have

Theorem 11.3. *No starting probabilities* $\{P_z\}$ *satisfying* (A) *and* (B) (*of this section*) *exist, i.e. there are no Lorentz-invariant processes in space-time* M.

12. Unsolved problems

A first set of problems is the explicit calculation of transition probabilities for the processes of Theorem 8.2, such as the following: the distribution of velocity in $\mathcal{U}$ at proper time τ for nondiffusion processes; the distribution of position at proper time τ for all processes; and the distribution of position and velocity at coordinate time t for all processes. The latter would be of special interest for diffusion processes; to simplify it, one can first seek only the distribution of velocity at time t. For a diffusion process with parameter m this leads to a parabolic partial differential equation in $\mathcal{U}$:

$$\cosh \varrho \, \partial f/\partial t = m\Delta f$$

(cf. Schay [1] equation 3.60 for the case of one space dimension).

Also of interest are the "relativistic Maxwell(–Boltzmann)" distributions originally defined by Jüttner [1] (see also Synge [1] equation (118) p. 36). These are radial measures on $\mathcal{U}$ of the form

$$M_{\alpha\beta} = \alpha \exp(-\beta \cosh \varrho) \cosh \varrho \cdot N$$

where N is the $\mathcal{L}$-invariant measure

$$4\pi \sinh^2\varrho \, d\varrho \, d\Omega$$

and $\alpha, \beta > 0$. (Note that if $\mathcal{U}$ is projected into the spacelike hyperplane $t=0$ perpendicular to $p=(0,0,0,1) \in \mathcal{U}$, $\cosh \varrho dN$ goes into Lebesgue measure.)

Given $\beta > 0$, $M_{\alpha\beta}$ is a probability measure if and only if

$$\alpha = 1/4\pi \int_0^\infty \exp(-\beta \cosh \varrho) \cosh \varrho \sinh^2\varrho \, d\varrho = \beta/4\pi K_2(\beta),$$

where K_2 is a Bessel function (see e.g. Synge [1] § 14). Let M_β be $M_{\alpha\beta}$ for this value of α.

Schay [1] asserts that M_β for fixed β is a "steady-state" solution of a diffusion equation with a term representing "internal friction." Thus it may be irrelevant to ask whether M_β is infinitely divisible, etc., but it seems that Schay's result should be followed up.

Thirdly, it would be interesting to move from the "special relativity" assumptions of this paper to the case of general relativity, in which M is no longer a vector space but a 4-dimensional manifold with a Lorentz quadratic form on its tangent spaces. The proper time is still available, but spatial homogeneity and Lorentz-invariance require reformulation. Instead of the product $M \times \mathcal{U}$ one has a subset $\mathcal{U}(M)$ of the "tangent bundle" $T(M)$: $\mathcal{U}(M)$ consists of all "forward" timelike vectors of unit magnitude at all points of M. (We assume it is possible to choose a "forward" direction at all points in a continuous way.)

In particular, there should be diffusion processes in $\mathcal{U}(M)$, generated by the differential operators which, roughly speaking, have the form $-U + m\Delta$ at a point (x, U) of $\mathcal{U}(M)$, where $m > 0$, $x \in M$, U belongs to the tangent space at x, and Δ is the Laplace-Beltrami operator in a hyperboloid in this tangent space. U may be regarded as a first-order partial derivation, or tangent vector, to $\mathcal{U}(M)$ by way of the pseudo-Riemannian "affine connection" (or "parallel displacement": see Helgason [1] Chapter 1 §§ 4–6). Diffusions in sufficient generality to cover this case have been considered by Nelson [1] and Gangolli [2], but their results are not as complete as might be desired for our purposes. For example, it apparently is not known whether the semigroup generated by the operator mentioned above on bounded measurable functions actually takes bounded continuous functions into bounded continuous functions. This may require supplementary, but physically reasonable, hypotheses on M.

Finally, given a "Brownian motion" diffusion in $\mathcal{U}(M)$, even where M is a vector space, one might let the parameter m in the generator approach infinity for use in defining a sort of "Feynman integral" as in Ito's approach [1] to the non-relativistic Feynman integral. Of course, there is also the problem of finding a suitable replacement for the classical Lagrangian function.

ACKNOWLEDGEMENT

This research was partially supported by National Science Foundation grant GP-2 and ONR contract 222 (60).

Department of Mathematics, University of California, Berkeley, Cal., U.S.A.

REFERENCES

DYNKIN, E. B., *Osnovanya Teorii Markovskikh Protsessov* (Moscow, Fizmatgis, 1959). English translation; *Theory of Markov Processes* (New York, Pergamon, 1960). There are also French and German translations.

— *Markovskie Protsessi* (Moscow, Gostekhizdat, 1963).

FEYNMAN, R. P., "Space–time approach to nonrelativistic quantum mechanics", *Reviews of Modern Physics 20*, 368–387 (1948).

GELFAND, I. M., and F. A. BEREZIN. "Some remarks on the theory of spherical functions" (in Russian), *Trudy Moskovskovo Matematicheskovo Obshchestva 5*, 311–351 (1956).

GANGOLLI, R., "Isotropic infinitely divisible measures on symmetric spaces", *Acta Mathematica* (Uppsala) *111*, 213–246 (1964).

— "On the construction of certain diffusions on a differentiable manifold", *Zeitschrift für Wahrscheinlichkeitstheorie 2*, 406–419 (1964).

— "Sample functions of certain differential processes on symmetric spaces" (preprint).

HUNT, G. A., "Semi-groups of measures on Lie groups", *Transactions of the American Mathematical Society 81*, 264–293 (1956).

ITO, K., "Wiener integral and Feynman integral", *Proceedings of the Fourth Berkeley Symposium on Mathematical Statistics and Probability* (Berkeley and Los Angeles, University of California Press, 227–238 (1961).

JÜTTNER, F., "Das Maxwellsche Gesetz der Geschwindigkeitsverteilung in der Relativtheorie", *Annalen der Physik 34*, 856–882 (1911).

LOÈVE, M., *Probability Theory* (second edition) (Princeton, Van Nostrand, 1960).

NELSON, E., "An existence theorem for second-order parabolic equations", *Transactions of the American Mathematical Society 88*, 414–429 (1958).

RAO, R. RANGA, "Relations between weak and uniform convergence of measures with applications", *Annals of Mathematical Statistics 33*, 659–680 (1962).

RUDBERG, H., *On the Theory of Relativistic Diffusion* (Uppsala, Almqvist and Wiksells, 1957).

SCHAY, G., "The equations of diffusion in the special theory of relativity" (physics Ph. D. thesis Princeton University, 1961; available on Xerox through University Microfilms, Ann Arbor, Michigan).

SYNGE, J. L., *The Relativistic Gas* (North-Holland, Amsterdam, 1957).

TUTUBALIN, V. N., "On the limit behavior of convolutions of measures in the Lobachevsky plane and space" (in Russian), *Teoriya Veroyatnosti i eyo Primeneniya 7*, 197–204 (1962). English translation in *Theory of Probability and its Applications 7*, 189–196.

TUTUBALIN, V. N., KARPELEVICH, F. I., and SHUR, M. G., "Limit theorems for convolutions of distributions in the Lobachevsky plane and space", *Teoriya Veroyatnosti i eyo Primeneniya 4*, 432–436 (1959). English translation: *Theory of Probability and its Applications 4*, 399–402.

WEHN, D., "Probabilities on Lie groups", *Proceedings of the National Academy of Science of the U.S.A. 48*, 791–795 (1962).

YOSIDA, K., "On Brownian motion in a homogeneous Riemannian space", *Pacific Journal of Mathematics 2*, 263–270 (1952).

Tryckt den 15 december 1965

Uppsala 1965. Almqvist & Wiksells Boktryckeri AB

ARKIV FÖR MATEMATIK Band 6 nr 30

1.66 11 3 Communicated 14 September 1966 by L. Carleson

A note on Lorentz-invariant Markov processes

By R. M. Dudley

1. Introduction

This note will remove some unnecessary regularity assumptions from my first paper on the subject [1]. There, beyond the conditions of spatial and temporal homogeneity, Lorentz-invariance, the Markov property, and the requirement that speeds be less than that of light, there were additional assumptions: continuity in probability ((E), p. 248) and existence of derivatives with probability one at each time (first half of (C), p. 248). Both these assumptions will be shown to follow from the others. This makes easier the proof of the converse part of the main theorem of [1], Theorem 8.2, as will be indicated below.

It was assumed in [1] that speeds remain less than (resp. equal to) 1, taken as the speed of light, if the initial speed is less than (resp. equal to) 1. This follows the physical distinction between particles of positive and zero rest mass. In section 3 we show that this assumption is unnecessary for initial speed 1, but that there are processes with initial speeds less than 1, satisfying all our other assumptions, which jump to speed 1 at later times.

2. Removing regularity assumptions

We shall use freely the notation and terminology of [1, sections 2 and 4]. Also, for any function f from the real line to three-space X let

$$\hat{f}(t) = \langle f(t), t \rangle \in M.$$

As in [1], we consider families of measures $\{P_x^v\}$ where for each x in X and v in V, P_x^v is a starting probability at x on $(\mathcal{A}, I, X, B(X))$. We retain assumptions (A) (spatial homogeneity), (B) (initial velocity v), and (D) (Lorentz-invariance) of [1, section 4], and a weakened form of the Markov property (C) as follows.

Each f in $\mathcal{A}$ is differentiable almost everywhere (Lebesgue) and $f'(t) \in V$ whenever it is defined. By [1, Lemma 4.1]

$$\{\langle f, t \rangle : f'(t) \text{ exists}\}$$

is product measurable for $B^0(\mathcal{A}) \times B(R^+)$. Let

$$T(v) = \{t > 0 : P_x^v(f'(t) \text{ exists}) = 1\}.$$

E. Giné et al. (eds.), *Selected Works of R.M. Dudley*, Selected Works in Probability and Statistics, DOI 10.1007/978-1-4419-5821-1_8, © Springer Science+Business Media, LLC 2010

(Clearly $T(v)$ is independent of x.) We know that the complement of $T(v)$ has measure 0 in R^+ for each v in V. Now rather than assuming $T(v) = (0, \infty)$ we simply assume the stationary Markov property wherever it makes sense:

(C′) For any v in V, x in X, t in $T(v)$, and A in $B^t(\mathcal{A})$, $P_x^v(f \in A \mid B_t^0(\mathcal{A})) = P_{f(t)}^{f'(t)}(\theta_t A)$ almost everywhere for P_x^v.

Theorem 1. *Let* $\{P_x^v\}$ *on* $\mathcal{A}$ *satisfy conditions* (A), (B), (C′), *and* (D). *Then they also satisfy* (C) *and* (E) [1, *section* 4], *thus define a process given by an infinitely divisible radial probability on Lobachevsky space* [1, *Theorem* 8.2], *and conversely.*

Proof. We first establish continuity in probability at 0 through $T(v)$ (a weakened form of (E)). We put a natural Lorentz-invariant metric on V. Let U be the transformation

$$U(v) = \langle v, 1\rangle / (1 - |v|^2)^{1/2}.$$

Then U transforms V one-to-one onto the hyperbolic space $\mathcal{U}$ (three-velocity to four-velocity). On $\mathcal{U}$, there is a natural Lorentz-invariant Riemannian metric ϱ [1, section 7]. Let e be the metric on V defined by

$$e(v, w) = \varrho(U(v), U(w)).$$

For any L in $\mathcal{L}$, $V_L = U^{-1} L U$. Hence e is V_L-invariant. For any v in V, let

$$L_v\langle x, t\rangle = \langle x + vt, t + v\cdot x\rangle / (1 - |v|^2)^{1/2}.$$

Then $L_v \in \mathcal{L}$ and if $L = L_v$, $V_L(0) = v$.

Lemma 1. *For any* v *and* w *in* V, (*a*) $\sinh e(v,w) = |v - w| / ((1 - v\cdot w)^2 - |v - w|^2)^{1/2}$. (*b*) $e(v, w) \geq |v - w|$.

Proof. Let $L = L_{-v}$. Then

$$\begin{aligned} e(v, w) &= e(0, V_L(w)) = e(0, (w - v)/(1 - v\cdot w)) \\ &= \varrho(U(0), \langle (w-v)/(1 - v\cdot w), 1\rangle / (1 - |v - w|^2/(1 - v\cdot w)^2)^{1/2}) \end{aligned}$$

and (*a*) follows [1, section 7].

To prove (*b*), note that $v \cdot w \geq -|v - w|^2/4$. Letting $x = |v - w|$, $y = e(v, w)$, this yields

$$\sinh y \geq x/((1 + x^2/4)^2 - x^2)^{1/2} \geq x/(1 - x^2/4) = 4x/(4 - x^2),$$

$$\begin{aligned} y &\geq \operatorname{arg\,sinh}\,(4x/(4 - x^2)) = \ln\,(4x/(4 - x^2) + (1 + 16x^2/(4 - x^2)^2)^{1/2}) \\ &= \ln\,((4x + (x^4 + 8x^2 + 16)^{1/2})/(4 - x^2)) = \ln\,((2 + x)/(2 - x)) = f(x). \end{aligned}$$

Now $f(0) = 0$ and $f'(x) \geq 1$ for $0 \leq x < 2$, so $f(x) \geq x$ and $y \geq x$, q.e.d.

Lemma 2. *For any $\varepsilon > 0$ there is a $\delta > 0$ such that for all v in V, $0 \leqslant t \leqslant \delta$ and t in $T(v)$,*

$$P_0^v(f : e(f'(t), v) > \varepsilon) < \varepsilon.$$

Proof. We may assume $\varepsilon \leqslant 1$. Take $\alpha > 0$ such that $|v| \leqslant 5\alpha$ implies $e(0, v) \leqslant \varepsilon$. Thus $\alpha < 1/5$. Since

$$P_0^0(f'(0^+) = 0) = 1,$$

there is a $\beta > 0$ such that

$$P_0^0(|f(t)| > \alpha t \quad \text{for some} \quad t \leqslant 2\beta) < \varepsilon/2.$$

(The set of all f in $\mathcal{A}$ satisfying the given condition is in $B^0(\mathcal{A})$ since it suffices to consider rational t.) Let

$$K = \{\langle x, t\rangle : |x| \leqslant \alpha t \leqslant 2\alpha\beta\}, \quad D = \{\langle x, 2\beta\rangle \in K\}.$$

The idea of the rest of the proof is that a large derivative for small t would cause departure from the cone K too soon.

We have for any w in V

$$\begin{aligned} 1 - \varepsilon/2 &\leqslant P_0^0(f : \hat{f}(t) \text{ first leaves } K \text{ through } D) \\ &\leqslant P_0^w(g : \hat{g}(s) \text{ first leaves } L_w(K) \text{ through } L_w(D)) \\ &\leqslant P_0^w(g : \hat{g}(s) \in L_w(K) \quad \text{for all} \quad s \leqslant \beta), \end{aligned}$$

since if $L = L_w$ and $\langle x, t\rangle \in D, L_{(2)}\langle x, t\rangle \geqslant \beta$.

Now we let $\delta = \beta/2$. Given v in V, let $L = L_v, t \in T(v)$, and $t \leqslant \delta$. Then

$$P_0^v(e(g'(t), v) \geqslant \varepsilon) = P_0^0(e(f'(\sigma), 0) \geqslant \varepsilon),$$

where $\sigma = \sigma(f, t) = (L^{-1})_{(2)}(L^* f)\hat{}(t)$, i.e.

$L\hat{f}(\sigma) = (L^* f)\hat{}(t), \sigma + v \cdot f(\sigma) = t(1 - |v|^2)^{1/2}$. Thus if $\hat{f}(\sigma) \in K, \sigma \leqslant 2t \leqslant \beta$. Now

$$\begin{aligned} 1 - \varepsilon/2 &\leqslant \textstyle\int_{\hat{g}(t) \in L(K)} P_{g(t)}^{g'(t)}(h : \langle h(s), s + t\rangle \text{ first leaves } L(K) \text{ through } L(D))\, dP_0^v(g) \\ &= \textstyle\int_{\hat{g}(t) \in L(K)} P_0^{g'(t)}(h : \hat{h}(s) + \hat{g}(t) \text{ first leaves } L(K) \text{ through } L(D))\, dP_0^v(g) \\ &= \textstyle\int_{\hat{f}(\sigma) \in K} P_0^{f'(\sigma)}(j : \hat{j}(r) + \hat{f}(\sigma) \text{ first leaves } K \text{ through } D)\, dP_0^0(f) \\ &\leqslant \textstyle\int_{\hat{f}(\sigma) \in K} P_0^{f'(\sigma)}(j : \hat{j}(\beta) + \hat{f}(\sigma) \in K)\, dP_0^0(f). \end{aligned}$$

Now let $w = w(f) = f'(\sigma)$. If $\hat{f}(\sigma) \in K, |w| > 5\alpha$, and $\hat{j}(\beta) \in \mathrm{L}_w(K)$, then $|j(\beta) - w\beta| \leqslant \alpha(\beta - w \cdot j(\beta)) \leqslant 2\alpha\beta$,

$$|j(\beta)| > 3\alpha\beta, |j(\beta) + f(\sigma)| > 3\alpha\beta - \alpha\beta = 2\alpha\beta, \hat{j}(\beta) + \hat{f}(\sigma) \notin K.$$

Thus $\hat{f}(\sigma) \in K$ and $|w| > 5\alpha$ imply

$$P_0^w(j : \hat{j}(\beta) + \hat{f}(\sigma) \in K) \leqslant P_0^w(j : \hat{j}(\beta) \notin L_w(K)) \leqslant \varepsilon/2 \leqslant 1/2.$$

Thus $$P_0^0(e(f'(\sigma), 0) > \varepsilon) \leqslant P_0^0(|f'(\sigma)| > 5\alpha) \leqslant \varepsilon, \text{ q.e.d.}$$

Next, let $v \in V$ and let W be a countable dense subset of $T(v) \cup \{0\}$ with $0 \in W$. Given $\varepsilon > 0$ and $\delta > 0$ let

$$\alpha(\varepsilon, \delta) = \sup [P_0^w(e(f'(t), w) \geqslant \varepsilon : w \in V, t \in T(w), 0 \leqslant t \leqslant \delta].$$

For each positive integer k and $a \geqslant 0$ let
$A(k, a, \varepsilon, \delta) = \{f \in \mathcal{A}:$ for some s_i in W with $a \leqslant s_1 < s_2 < \ldots < s_{2k} \leqslant a + \delta, e(f'(s_{2i-1}), f'(s_{2i})) \geqslant 4\varepsilon$ for all $i = l, \ldots, k\}$.

Lemma 3. *For any $\varepsilon > 0, \delta > 0$, positive integer $k, a \geqslant 0, x \in X$, and $v \in V, P_x^v(A(k, a, \varepsilon, \delta)) \leqslant [2\alpha(\varepsilon, \delta)]^k$.*

The statement and proof of Lemma 3 are essentially those of Lemma 6.4 of Dynkin [2]. We make only the following remarks: application of the strong Markov property is permissible since W is countable, as mentioned at the end of section 5.8 of [2]. We need only the Markov property at each t in W. We cannot formally apply the statement of [2, Lemma 6.4] since we have the Markov property only on a set of times $T(v)$, depending on the initial v. Our assumptions imply that for any $s \leqslant t$ in $T(v), t - s \in T(f'(s))$ for P_x^v-almost all f in $\mathcal{A}$.

Lemma 2 implies that for every $\varepsilon > 0, \lim_{\delta \to 0} \alpha(\varepsilon, \delta) = 0$. Thus for any $M > 0$, we can take k large enough so that $\alpha(\varepsilon, M/k) < 1/2$ and obtain from Lemma 3

$$\lim_{n \to \infty} P_x^v(A(n, 0, \varepsilon, M)) \leqslant \lim_{n \to \infty} kP_x^v(A([(n-k)/k], 0, \varepsilon, M/k)) = 0,$$

where $[x]$ is the largest integer $\leqslant x$.

Let $G = \bigcup_{m=1}^{\infty} \bigcap_{n=1}^{\infty} A(n, 0, 1/m, m)$. Then $P_x^v(G) = 0$. For $f \notin G$, the limits

$$w(t) = \lim_{s \downarrow t,\, s \in W} f'(s), \; t \geqslant 0, \text{ and}$$

$$w(t^-) = \lim_{s \uparrow t,\, s \in W} f'(s), \; t > 0$$

always exist, w is continuous from the right, and for all $t > 0$

$$\lim_{s \uparrow t} w(s) = w(t^-).$$

For each fixed t in $T(v), w(s) \to f'(t)$ in P_x^v-probability as $s \downarrow t$ through W, by the Markov property at t and Lemma 2. Thus

$$P_x^v(w(t) = f'(t)) = 1.$$

Now since the maps $\langle f, t \rangle \to f'(t)$ and $\langle f, t \rangle \to w(t)$ are both measurable [1, Lemmas 4.1 and 2.1], we have $P_x^v(f'(t) = w(t)$ for (Lebesgue) almost all $t \geqslant 0) = 1$. Since f is Lipschitzian, $f' = w$ almost everywhere implies that f is the indefinite integral of w. Then by the left and right limit properties of w,

$$P_x^v(f'(t^+) = w(t) \quad \text{for all } t \geqslant 0)$$
$$= P_x^v(f'(t^-) = w(t-) \quad \text{for all } t > 0) = 1.$$

If $t \notin T(v)$, we take $s_n \uparrow t$ and $t_n \downarrow t$, $s_n, t_n \in T(v)$. For any $\varepsilon > 0$, $\lim_{n\to\infty} P_x^v(e(f'(s_n), f'(t_n)) > \varepsilon) = 0$, by Lemma 2 and the Markov property at s_n. Thus

$$P_x^v(f'(t^+) \neq f'(t^-)) = 0,$$

and in fact $T(v) = (0, \infty)$, i.e. assumption (C) holds. We have

$$P_x^v(\lim_{t\downarrow 0} |f(t) - x| = 0) = 1$$

since the functions in $\mathcal{A}$ are Lipschitzian. Thus assumption (E) holds by Lemmas 1(*b*) and 2. We have now obtained the full assumptions (A) through (E) of [1], and can draw all the inferences of [1, sections 4–8]. (In particular, we have by [1, Lemma 5.3] that for any $t \geq 0$ and A in $B(X \times V)$,

$$\langle x, v\rangle \to P_x^v(\langle f(t), f'(t)\rangle \in A),$$

is $B(X \times V)$-measurable. Such a condition is made part of the definition of Markov process by Dynkin [2], but not by all authors in the field. It is not needed for the earlier results in this paper (even for Lemma 3), since the weaker measurability implied by (C′) suffices.)

To finish the proof of Theorem 1 we now sketch a proof of the converse part of Theorem 8.2 of [1], which was not proved there. Let P_1 be an infinitely divisible, radial probability measure on the Lobachevsky space $\mathcal{U}$. Then P_1 can be imbedded in a convolution semigroup $\{P_\tau, \tau \geq 0\}$ of such measures, as is obvious from the "Lévy–Khinchin" representation formula [1, section 8]. Such a semigroup is always weak-star continuous, hence defines a strong Markov process $\{x_\tau, \tau \geq 0\}$ having sample functions continuous from the right with left limits everywhere (see the remarks before [1, Lemma 8.1]). We take an "indefinite integral" of the process, obtaining paths in M. Then we can invert the arguments of [1, section 6], making a "random change of time" from proper time τ to co-ordinate time t. We thus obtain a process which satisfies (A), (B), (C′), and (D), as desired.

3. Mixing speeds less than 1 and equal to 1

Let $\mathcal{J}$ be the set of functions from R^+ to X satisfying

$$|f(s) - f(t)| \leq |s - t| \quad \text{for all } s, t \geq 0.$$

Each f in $\mathcal{J}$ is differentiable almost everywhere and $|f'(t)| \leq 1$ whenever $f'(t)$ is defined.

Theorem 2. *Suppose* $\{P_x^v\}$ *for* x *in* X *and* v *in* S^2 (i.e. $v \in X, |v| = 1$) *is a family of probability measures on* $\mathcal{J}$ *satisfying assumptions* (A), (B), *and* (D). *Then each* P_x^v *is concentrated in the one function* $f(t) = x + vt, t \geq 0$.

Proof. The proof of [1, Theorem 11.1] is sufficient since it only uses the fact that initial speeds are 1, not that speeds remain equal to 1.

Next, we describe informally a class of invariant Markov processes for which transitions *from* speeds less than 1 *to* speeds equal to 1 *do* occur.

Suppose $\{P_x^v\}$ satisfy conditions (A), (B), (C′), and (D) on $\mathcal{A}$, as in section 2. We construct a new process on $\mathcal{J}$ as follows. Let τ_0 be a random variable with a distribution given, for some $k>0$, by

$$P(\tau_0 > c) = e^{-kc} \quad \text{for all} \quad c \geqslant 0.$$

We assume τ_0 is independent of the P_x^v. Now let a particle move according to P_x^v until proper time τ_0 has passed, say at time t. From then on, let it move in a straight line with velocity w, where w is uniformly distributed over the sphere $\{w: |w|=1\}$ in a co-ordinate system where the particle has zero velocity at time t^- (and w is independent of τ_0 and other events before time t). This defines a set $\{R_x^v\}$ of probability measures on $\mathcal{J}$ (using the usual deterministic ones for $|v|=1$) which have all the properties (A) through (E) of the P_x^v without being concentrated in $\mathcal{A}$ for $|v|<1$.

4. Diffusion processes

Here we make some additions and corrections to [1, section 10]. First, the definition of diffusion semigroup given there is too strong. It should read: for every $\varepsilon > 0$,

$$\lim_{t \downarrow 0} P_t(\varrho \geqslant \varepsilon)/t = 0.$$

The phase space $M \times \mathfrak{U}$ is acted on transitively by the Poincaré group (inhomogeneous Lorentz group) $\bar{\mathfrak{L}}$. I thank C. C. Moore for the following construction of $\bar{\mathfrak{L}}$-invariant Riemannian metrics. Take a tangent space V to $M \times \mathfrak{U}$ at one point, say $(0,p)$. Then V is a seven-dimensional real vector space $R^3 \times R \times R^3$. It suffices to find the positive definite quadratic froms B on V such that

$$B(\langle x,t,y\rangle) = B(\langle A(x), t, A(y)\rangle)$$

for all orthogonal transformations A of R^3 with determinant 1 (the latter restriction is also needed in the first paragraph of [1, section 7]). B satisfies the above conditions if and only if

$$B(\langle x,t,y\rangle) = a_1 t^2 + a_2 |x|^2 + 2a_3 x\cdot y + a_4 |y|^2,$$

where $$a_1 > 0, a_2 > 0, a_3^2 < a_2 a_4.$$

The definition of diffusion process depends only superficially on the metric. For a homogeneous Riemannian manifold, as here, it clearly depends only on the original topology of the manifold. On a general differentiable manifold, Dynkin [3] defines a diffusion process as one whose generator is a second-order differential operator.

In any case, the diffusion processes in $M \times \mathfrak{U}$ which are also Lorentz-invariant Markov processes as in Theorem 1 above are those defined by "Brownian motions" in $\mathfrak{U}$ [1, section 10].

5. Note

In [1, section 5], use was made of the relations between weak-star convergence of probability measures and convergence in the dual space of the bounded Lipschitzian functions. Details of these relations are given in a paper of mine, "Convergence of Baire measures", to appear in *Studia Mathematica*.

University of California, Berkeley, Cal. U.S.A.

REFERENCES

1. DUDLEY, R. M., Lorentz-invariant Markov processes in relativistic phase space, *Arkiv för Matematik 6*, 241–268 (1965).
2. DYNKIN, E. B., Theory of Markov Processes (Moscow, 1959; English translation, London, Pergamon, 1960).
3. DYNKIN, E. B., Markov Processes (Moscow, 1963; English translation, Berlin, Springer, 1965).

Tryckt den 14 mars 1967

Uppsala 1967. Almqvist & Wiksells Boktryckeri AB

Reprinted from

Proc. Nat. Acad. Sci. USA
Vol. 70, No. 12, Part I, pp. 3551–3555, December 1973

Asymptotics of Some Relativistic Markov Processes

(four-velocities/galactic recession)

R. M. DUDLEY

Room 2-245, M.I.T., Cambridge, Massachusetts 02139

Communicated by E. J. McShane, August 3, 1973

ABSTRACT Markov processes in special-relativistic position-velocity phase space are proved to have converging velocities as $t \to +\infty$ under some mild assumptions on transition probabilities. This offers a possible mechanism to explain the recession of galaxies.

For a large class $\mathfrak{M}$ of strong Markov processes in special-relativistic phase space it will be proved that velocities converge almost surely as $t \to +\infty$. The assumptions defining $\mathfrak{M}$ will require that transition probabilities for velocities not be too badly skewed toward or away from any direction. The Lorentz-invariant processes described in refs. 4 and 5 all belong to $\mathfrak{M}$. For them, the limiting velocities have the speed of light (except in a trivial case). The relative motions of galaxies do not form a Lorentz-invariant process but may form a process in $\mathfrak{M}$. Galactic motions will be discussed in section 2 below. The main difference between the relativistic and classical cases is that in the hyperboloid $\mathfrak{U}$ of relativistic four-velocities, a sphere of radius x (for the natural Lorentz-invariant metric) has circumference $2\pi\sinh x$, which goes to ∞ exponentially as $x \to \infty$. It follows, as will be seen, that a point in $\mathfrak{U}$ which moves very far randomly tends strongly to move out to ∞ in some limiting direction. A randomly moving point in Newtonian velocity space, i.e., ordinary Euclidean space, is more likely to go to ∞ in a spiraling way with recurring major changes of direction.

We will obtain convergence of velocities under several sets of hypotheses, ranging from weak but complicated ones in *Theorem 1* to strong but simple ones in *Theorem 3*. The intermediate *Theorem 2* may be the most informative.

1. Formulation of results

On Euclidean space, R^3, let $|\cdot|$ denote the usual norm. On $R^4 \equiv \{\langle x,t\rangle: x \in R^3, t \in R\}$ we have the Lorentz–Minkowski quadratic form $||\langle x,t\rangle||^2 = |x|^2 - t^2$. Here we assume units chosen so that the speed of light $c = 1$.

Let $V = \{v \in R^3: |v| < 1\}$. Then V is the set of relativistically possible velocities for particles with positive rest mass. (In this paper, speeds $|v| = 1$, such as for photons, will not be considered.)

For general information on Markov processes, see Dynkin (6, 7). In particular we recall that a *stopping time* for a process $x(t,\omega)$ is a random time $t(\omega)$ such that for any fixed t_0, the question whether $t(\omega) \le t_0$ can be answered by observing $x(t,\omega)$ for $t \le t_0$. The process is a *strong* Markov process iff the Markov property holds at every stopping time $t(\cdot)$, i.e., $\{x(s,\omega): s \ge t(\omega)\}$ and $\{x(s,\omega): s \le t(\omega)\}$ are independent given $x(t(\omega),\omega)$. Clearly this implies the ordinary Markov property since fixed times are stopping times.

For a process $x(t,\omega)$ in R^3, the Markov property is somewhat incompatible with existence of velocities $dx(t,\omega)/dt$, since given the present position, past and future positions are dependent through the present velocity. (Nonexistence of Lorentz-*invariant* Markov processes in space-time was noted in Theorem 11.3 of ref. 4 and Proposition 2 of ref. 8.) To allow a reasonably large class of relativistic Markov processes, we shall consider trajectories $\langle x(t,\omega), dx(t,\omega)/dt\rangle$ in phase space.

Relativity theory also postulates equivalence of different uniformly moving coordinate systems, so that simultaneity is relative. Time, t, has meaning only in a coordinate system whose choice is somewhat arbitrary. Therefore, it is convenient to introduce an intrinsic or *proper time* τ on each trajectory. This τ is the time shown by a clock carried along the trajectory in space-time. Given a coordinate system $\langle x_1,x_2,x_3,t\rangle$ we have

$$d\tau = (dt^2 - dx_1^2 - dx_2^2 - dx_3^2)^{1/2}.$$

Thus $d\tau \le dt$, with equality for particles at rest. The proper time for relativistic Markov processes was used previously in refs. 4 and 5.

"Velocities" $d\langle x,t\rangle/d\tau$, where $|dx/dt| < 1$, are called "four-velocities" and form a hyperboloid, defined as follows:

$$\mathfrak{U} \equiv \{\langle x,t\rangle \in R^4: t > 0, |x|^2 - t^2 = -1\}.$$

Let $u(v) \equiv (1 - |v|^2)^{-1/2}\langle v,1\rangle$. Then $u(\cdot)$ is C^∞ from V onto $\mathfrak{U}$ and has a C^∞ inverse v given by

$$v(u) \equiv v(\langle x,t\rangle) = x(1 + |x|^2)^{-1/2}.$$

This is the usual correspondence between 3-velocity v and 4-velocity u.

Let $\mathcal{P}$ be the relativistic phase space $R^4 \times \mathfrak{U}$.

Definition: Let $\mathcal{R}$ be the class of all strong Markov processes $\tau \to \langle x(\tau,\omega),t(\tau,\omega),u(\tau,\omega)\rangle \in \mathcal{P}$, $0 \le \tau < +\infty$, such that the following conditions hold:

(1) For almost all ω, $u(\tau,\omega)$ is continuous from the right and has (finite) limits from the left in τ.

(2) For all such ω, $\langle x(\tau,\omega),t(\tau,\omega)\rangle = \int_0^\tau u(\sigma,\omega)d\sigma$ for $0 \le \tau < +\infty$.

(3) $u(0,\omega) = p$ almost surely, where $p = u(0) = \langle 0,0,0,1\rangle$.

There exist trajectories on which $t \to +\infty$ while τ remains bounded. Such trajectories are excluded, however, for processes in $\mathcal{R}$ since we assume $\tau \to \infty$ and t continuous.

It is well known (Kinney's theorem) that condition 1 and the strong Markov property hold for a suitable version of each Markov process satisfying certain mild conditions on its

E. Giné et al. (eds.), *Selected Works of R.M. Dudley*, Selected Works in Probability and Statistics, DOI 10.1007/978-1-4419-5821-1_9, © Springer Science+Business Media, LLC 2010

ansition probabilities (see Kapitel 6, section 3 of ref. 6). In rticular, all the Markov processes in $\mathcal{P}$ with Lorentz-variant transition probabilities as defined in refs. 4 and have such versions.

To formulate our assumptions on transition probabilities shall use the following notions.

Definition: A *half-space* in $\mathfrak{U}$ is a set $H \cap \mathfrak{U}$ where H is closed half-space in R^4 such that the boundary K of H a three-dimensional linear subspace of R^4 and intersects , i.e., $0 \in K$ and $K \cap \mathfrak{U} \neq \phi$.

A subset $C \subset \mathfrak{U}$ will be called *convex* iff for all $q \in \mathfrak{U} \sim C$, ere is a half-space J in $\mathfrak{U}$ containing q and disjoint from C. Clearly $C \subset \mathfrak{U}$ is convex iff $C = B \cap \mathfrak{U}$ where B is a conx cone in R^4 with apex at the origin. It follows that C is nvex iff for every pair of points $p, q, \in C$, the geodesic ining p to q is included in C.

In $\mathfrak{U}$ we have a natural Lorentz-invariant Riemannian stance $\rho(\cdot,\cdot)$ induced by $\|\cdot\|^2$. It is easily seen that the lls $\{u: \rho(u,q) < r\}$ are convex in $\mathfrak{U}$ for any $r > 0$ and $q \in \mathfrak{U}$.

Associated with our Markov processes are transition probilities $P_\tau(\xi,u;B,A)$ where $\tau \geq 0$, $\xi \in R^4$, $u \in \mathfrak{U}$, B is a orel set in R^4, and A is a Borel set in $\mathfrak{U}$. $P_\tau(\xi,u;B,A)$ is the obability, starting at space-time point ξ with 4-velocity that after τ units of proper time have passed the moving int will have space-time position in B and 4-velocity in A. he Markov property corresponds as usual to the Chapman–olmogorov equation: if $0 \leq \sigma \leq \tau$,

$$P_\tau(\xi,u;B,A) = \int P_{\tau-\sigma}(\eta,w;B,A)P_\sigma(\xi,u;d\eta,dw).$$

We shall consider marginal transition probabilities in $\mathfrak{U}$, efined by

$$P_\tau(\xi,u;C) \equiv P_\tau(\xi,u;R^4,C)$$

r $\xi \in R^4$, $\tau \geq 0$, $u \in \mathfrak{U}$, and C a Borel set in $\mathfrak{U}$.

Let $p \equiv \langle 0,1\rangle \in \mathfrak{U}$. Let φ be the map from $\mathfrak{U} \sim \{p\}$ onto e unit sphere $S^2 \subset R^3$ defined by $\psi\langle x,t\rangle \equiv x/|x|$. For any $\in \mathfrak{U}$ let $\rho \equiv \rho(u) \equiv \rho(p,u)$. Let d be the geodesic (greatrcle, angle) distance in S^2. Some sets in $\mathfrak{U}$ will now be dened for use as neighborhoods of points at ∞, corresponding neighborhoods in V of points in its boundary S^2.

Given $u \in \mathfrak{U}$, $u \neq p$, and $n > 1$, let $J(n,u)$ be the threeimensional subspace of R^4 spanned by the plane orthogonal both p and u and by a point $\zeta \in \mathfrak{U}$ with $\rho(\zeta) = n$ such hat ζ is in the linear span of p and u, with $\rho(\zeta,u) < \rho(p,u)$. et $H(n,u)$ be the closed half-space of R^4 containing p with oundary $J(n,u)$. Given also $\gamma > 0$, let

$$C(u,\gamma,n) = \{q \in \mathfrak{U}: d(\psi(u),\psi(q)) < \gamma \text{ and } q \notin H(n,u)\}.$$

Note that for any $q \in C(u,\gamma,n)$, $\rho(q) \geq n$.

Let $K = K(n,u,\gamma) = \{q: d(\psi(q),\psi(u)) < \gamma \text{ and } q \in H(n, u)n\ \mathfrak{U}\}$. [Note: all these sets depend on u only through $\psi(u)$.] I claim that if $0 < \gamma < \pi/2$, then K is bounded. To rove this we can assume $u = (x,0,0,(1 + x^2)^{1/2})$ for some > 0. Then $\zeta = (a,0,0,(1 + a^2)^{1/2})$ for some $a > 0$, and $K = \langle X,Y,Z,t\rangle \in \mathfrak{U}: 0 \leq X \leq a,\ Y^2 + Z^2 \leq X^2\tan^2\gamma\}$, which s clearly bounded. Thus, $B_n(\gamma) \equiv \sup\{\rho(q): q \in K\} < \infty$. Let $B_n \equiv B_n(1/n)$. Then, since $\tan {}^1/_2 < 1$, we have for $n \geq 2$, $B_n \leq \arg\sinh(2 \sinh n) \leq 2n \arg \sinh (2 \sinh n)$.

Given a process in $\mathfrak{R}$, let $u_\tau \equiv u(\tau,\omega)$, $\rho_\tau \equiv \rho(u_\tau)$. Now, we re ready to state the hypotheses on transition probabilities or our theorem.

Definition: $\mathfrak{M}$ denotes the class of strong Markov processes in $\mathfrak{R}$ (satisfying conditions *1–3* above), having transition probabilities for which the following four conditions also hold:

(4) For any $B \geq 0$ there is an $\alpha = \alpha(B) > 0$ such that for any half-space H with boundary containing $u \in \mathfrak{U}$ with $\rho(u) \leq B$, we have for any $\xi \in R^4$ and $\tau \geq 0$ that $P_\tau(\xi,u;H) \geq \alpha$.

(5) There is a $\beta > 0$ such that whenever H is a half-space with boundary containing both u and p, we have $P_\tau(\xi,u;H) \geq \beta$ for any $\xi \in R^4$ and $\tau \geq 0$.

(6) Let $D_\delta(u) = \{w: \rho(u,w) \geq \delta\}$. For any $B \geq 0$ there is a $\kappa = \kappa(B) > 0$ such that whenever $0 < \delta < \kappa$ there is an $\epsilon = \epsilon(\delta,B) > 0$ such that whenever $\tau \geq 0$, $\xi \in R^4$, $u \in \mathfrak{U}$, and $\rho(u) \leq B$, we have

$$P_\tau(\xi,u;D_\delta(u) \cap \{w: \rho(w) > \rho(u) + \epsilon\}) \geq \delta P_\tau(\xi,u;D_\delta(u)).$$

(7) Given any $n > 0$, there is an $M = M(n)$ such that whenever $u \in \mathfrak{U}$ with $\rho(u) \geq M$, $\tau \geq 0$, and $\xi \in R^4$, we have

$$P_\tau(\xi,u;C(u,1/n,n)^c) < \alpha(B_n)/n$$

where $C^c \equiv \mathfrak{U} \sim C$.

THEOREM 1. *For any process in $\mathfrak{M}$, $v(\tau,\omega)$ converges almost surely as $\tau \to +\infty$ to some $v(\omega)$ with $|v(\omega)| \leq 1$.*

In order to formulate the next theorem, we decompose each transition probability as an integral of probabilities on spheres, letting

$$S_\lambda(u) \equiv \{w \in \mathfrak{U}: \rho(u,w) = \lambda\},$$

(8) $P_\tau(\xi,u,A) \equiv \int_0^\infty Q_\tau(\xi,u,\lambda,A)\, P_\tau'(\xi,u,d\lambda)$, where $P_\tau'(\xi,u,\cdot)$ is the probability measure on $[0,\infty)$ defined by $P_\tau'(\xi,u,[a,b)) \equiv P_\tau(\xi,u,\{w: a \leq \rho(u,w) < b\})$, $0 \leq a < b < \infty$, and each $Q_\tau(\xi,u,\lambda,\cdot)$ is a probability measure on $S_\lambda(u)$.

Since $\mathfrak{U}$ is homeomorphic to $\mathbf{R}^3$, every Borel probability measure on it has a decomposition (*8*) (see ref. 3, Theorem I.9.5).

THEOREM 2. *Suppose given a process in $\mathfrak{R}$ satisfying condition 5 and the following: for all $\xi \in R^4$ and $u \in \mathfrak{U}$ and $P_\tau'(\xi, u,\cdot)$ almost all λ, $Q_\tau(\xi,u,\lambda,\cdot)$ has a density $q_\tau(\xi,u,\lambda,\sigma)$ with respect to the normalized rotationally invariant surface measure $d\sigma$ on the sphere $S_\lambda(u)$. Assume also:*

(I) *For some function $q(\cdot) > 0$, $q_\tau(\xi,u,\lambda,\sigma) \geq q(B)$ whenever $\rho(u) \leq B$, i.e., q_τ is bounded away from 0 for u in bounded sets.*

(II) $\lim_{\rho(u)\to\infty} e^{-\rho(u)} \sup\{q_\tau(\xi,u,\lambda,\sigma): \tau \geq 0, \xi \in R^4, u \in \mathfrak{U}, \lambda \geq 0, \sigma \in S_\lambda(u)\} = 0$. *Then the process is in $\mathfrak{M}$.*

COROLLARY. *Suppose a process in $\mathfrak{R}$ has transition probabilities satisfying $P_\tau(\xi,u;A) = P_\tau(\xi,u;LA)$ for any Lorentz transformation L leaving u fixed, $\tau \geq 0$, and $\xi \in R^4$. Then the process is in $\mathfrak{M}$.*

If we choose a coordinate system in which we are at rest, so that u becomes $p = \langle 0,0,0,1\rangle$, then the Lorentz transformations L leaving u fixed are just orthogonal transformations of the space R^3, leaving the time axis fixed.

The next theorem concerns the processes treated in refs. 4 and 5.

THEOREM 3. *The Lorentz-invariant Markov processes all satisfy the assumptions of Theorem 2 and, hence, of Theorem 1 and for them, except in the trivial case of constant velocities, we have $|v(\omega)| = 1$ almost surely.*

Proof of Theorem 1: Let $\Omega_1 = \{\omega: \limsup \rho_\tau(\omega) < \infty\}$, $\Omega_2 =$

$\Omega \sim \Omega_1$. We shall use conditions **4** and **6** on Ω_1 and conditions **4**, **5**, and **7** on Ω_2.

If $Pr(\Omega_1) > 0$, take $B > 0$ such that $Pr\{\limsup \rho_\tau(\omega) < B\} > 0$. In condition **6** take $\kappa = \kappa(B)$, any $\delta \in (0,\kappa)$ and $\epsilon = \epsilon(\delta,B)$. Suppose $0 < A < A + \epsilon < B$ and $Pr\{A < \limsup \rho_\tau < A + \epsilon\} > 0$. Such an A exists if B is chosen large enough and ϵ small enough, as they can be.

We choose an increasing sequence of stopping times $\tau_n(\omega)$ as follows. Let $\tau_1(\omega)$ be the least τ such that $A \leq \rho_\tau \leq A + \epsilon$, or $+\infty$ if there is no such τ. Suppose given $\tau_1, \ldots, \tau_n(\omega)$ such that $A \leq \rho(u(\tau_j)) \leq A + \epsilon, j = 1, \ldots, n$. Since τ_n is a stopping time we can apply the strong Markov property at τ_n. Let T_n be the least $\tau > \tau_n$ such that $\rho(u(\tau), u(\tau_n)) \geq \delta$, or $T_n = +\infty$ if there is no such τ. Then T_n is a stopping time. Let $D = D_\delta(u(\tau_n))$ as in condition **6**. For any $S > 0$, by the strong Markov property at T_n we have $Pr\{\rho(u(\tau_n + S), u(\tau_n)) \geq \delta\} \geq Pr\{T_n \leq \tau_n + S, \rho(u(T_n)) \leq B\} \inf P_{\tau_n+S-T}(\xi,\zeta;D)$ where the infimum extends over all $\xi \in R^4$, $T \leq \tau_n + S$, and $\zeta \in D$ with $\rho(\zeta) \leq B$. (The values $T = T_n, \zeta = u(T_n)$ are relevant.) Now take $\alpha = \alpha(B)$ in condition **4**. Since $\mathfrak{U} \sim D$ is convex, condition **4** gives that the infimum is $\geq \alpha$.

Next we use condition **6** to get

$$Pr\{\rho(u(\tau_n + S)) > \rho(u(\tau_n)) + \epsilon\}$$
$$\geq \delta Pr\{\rho(u(\tau_n),u(\tau_n + S)) \geq \delta\}$$
$$\geq \alpha\delta Pr\{T_n \leq \tau_n + S, \rho(u(T_n)) \leq B\}.$$

These inequalities remain valid conditional on any event up to time τ_n. Given the space-time position and velocity at $\tau_n(\omega)$, we choose $S \equiv S_n(\omega) \equiv S_n\{u(\tau,\omega): \tau \leq \tau_n(\omega)\}$ large enough so that

$$Pr\{T_n \leq \tau_n + S, \rho(u(T_n)) \leq B\}$$
$$\geq {}^1/_2 Pr\{T_n < \infty, \rho(u(T_n)) \leq B\}.$$

We can assume $S_n \geq 1$. Let τ_{n+1} be the least $\tau \geq \tau_n + S_n$ such that $A \leq \rho(u(\tau)) \leq A + \epsilon$, or $\tau_{n+1} = +\infty$ if there is no such τ.

Now we use the strong Markov property to conclude that if A_n is any event measurable for the σ-algebra of events up to time τ_n, then

$$Pr\{\rho(u(\tau_n + S)) > \rho(u(\tau_n)) + \epsilon | A_n\}$$
$$\geq {}^1/_2\alpha\delta\, Pr\{T_n < \infty, \rho(u(T_n)) \leq B | A_n\}.$$

Let E_n be the event

$$\{\tau_n < \infty \text{ and } \rho(u(\tau)) \leq \rho(u(\tau_n)) + \epsilon \text{ for all } \tau \in [\tau_n,\tau_{n+1}]\}.$$

Let $A_n{}^M = \bigcap_{M \leq m < n} E_m$. Suppose that $Pr\{T_n < \infty, \rho(u(T_n)) \leq B | A_n{}^M\} \geq \sigma > 0$ for infinitely many values of n. Then for such n, $Pr(E_n | A_n{}^M) \leq 1 - {}^1/_2\, \alpha\delta\sigma$. Hence, by induction $Pr(A_n{}^M) \to 0$ as $n \to \infty$, contrary to assumption for M large enough. Thus

$$(9) \quad \lim_{n\to\infty} Pr\{T_n < \infty, \rho(u(T_n)) \leq B | A_n{}^M\} = 0.$$

Let C be the event $A < \limsup \rho_\tau < A + \epsilon$. Recall that we assumed $Pr(C) > 0$. Let $A^M = \bigcap_{n > M} A_n{}^M$. Then

$$\lim_{M\to\infty} Pr\{A^M | C\} = 1, A_n{}^M \downarrow A^M,$$

and

$$\lim_{n\to\infty} Pr\{\rho(u(T_n)) < B | \{T_n < \infty\} \cap C\} = 1.$$

Hence by Eq. **9**,

$$\lim_{n\to\infty} Pr\{T_n < \infty | C\} = 0.$$

So for almost all $\omega \in C$ we have $\rho(u_\sigma,u_\tau) \leq 2\delta$ for all σ and large enough. Letting A vary, we get the same result fo almost all ω such that $\limsup \rho_\tau < B$. Now letting $B \to \infty$ and $\delta \downarrow 0$, we have u_τ converging in $\mathfrak{U}$, and hence $v(\tau,\omega)$ converging in V, for almost all $\omega \in \Omega_1$.

Now we treat Ω_2. Suppose $Pr(\Omega_2) > 0$. Given a positive integer n we take $M(n)$ from condition **7**. We can assume $M(n)\uparrow$ as $n\uparrow$. For each n let t_n be the least τ such that $\rho(u(\tau)) \geq M(n)$, or $+\infty$ if there is no such τ. For any $S >$ we have, since each $C(u,\gamma,n)$ is convex, $Pr\{u(t_n + S) \notin C(u(t_n),1/n,n)\} \geq Pr\{$for some T with $t_n \leq T \leq t_n + S$, $u(T) \notin C(u(t_n),1/n,n)\}\min(\beta,\alpha(B_n))$, where we use the strong Markov property at the least such T; for $d(\psi(u(T)), \psi(u(t_n))) \geq 1/n$ we use condition **5** and a factor of β, and otherwise we use condition **4** and $\alpha(B_n)$. By condition **7**, letting $S \to \infty$, we have

$$\lim_{n\to\infty} Pr\{\text{for some } T \geq t_n, u(T) \notin C(u(t_n),1/n,n)\} = 0.$$

Hence as $\tau \to \infty$, for almost all $\omega \in \Omega_2$, $v(u(\tau))$ converge to some $v(\omega)$ with $|v| = I$ where $\psi(v(\omega)) = \lim_{n\to\infty} \psi(u(t_n))$. Thus *Theorem 1* is proved.

Proof of Theorem 2: We shall prove that if densities q_τ exist and satisfy **I** and **II**, then conditions **4**, **6**, and **7** also hold.

I implies **4** immediately with $\alpha(B) = {}^1/_2 q(B)$.

For **6**, we can replace our transition probability by probability $q_\tau d\sigma$ on some $S_\lambda(u)$, $\lambda \geq \delta$. We can assume $u = \langle w,0,0,(1 + w^2)^{1/2}\rangle$ for some $w \geq 0$. Using a formula from ref. 4 (pp. 256–257), we see that such a sphere is the transform of an ordinary sphere

$$(10) \quad \{q = \langle x,y,z,t\rangle: (x^2 + y^2 + z^2)^{1/2} = r, t = (r^2 + 1)^{1/2}\}$$

by the Lorentz transformation

$$(11) \quad T(q) \equiv \langle (x + vt)(1 - v^2)^{-1/2}, y, z, (t + vx)(1 - v^2)^{-1/2}\rangle$$

where $v = w(1 + w^2)^{-1/2}$. Then if $x \geq 0$, we have

$$\rho(T(q)) = \sinh^{-1}[(x^2 + 2vxt + v^2t^2)(1 - v^2)^{-1} + y^2 + z^2]^{1/2}$$
$$\geq \sinh^{-1}[w^2(r^2 + 1) + r^2]^{1/2}$$
$$\geq \sinh^{-1}[w + {}^1/_2 r^2(w^2 + 1)(w^2r^2 + w^2 + r^2)^{-1/2}]$$
$$\geq \sinh^{-1} w + r^2/[3(1 + r^2)^{1/2} + r^2]$$
$$\geq \sinh^{-1} w + r^2/(4r^2 + 3).$$

We have $r = \sinh \lambda \geq \sinh \delta$. We now take $\kappa(B) \equiv {}^1/_2 q(B)$ and $\epsilon(\delta,B) \equiv \epsilon(\delta) \equiv (\sinh^2 \delta)/(3 + 4\sinh^2 \delta)$. Since $x^2(3 + x^2)$ is an increasing function for $x > 0$, it is clear that condition **6** holds.

Now to prove condition **7**, we can again assume $u = \langle w,0,0,(1 + w^2)^{1/2}\rangle$, $w \geq 0$, assume P_τ is concentrated in some sphere **10** and transform it by T in **11**.

The angle between $\psi(T(q))$ and $\psi(T(p)) = \psi(u)$ is

$$\theta \equiv \arctan[(y^2 + z^2)^{1/2}(1 - v^2)^{1/2}/|x + vt|].$$

As $v \uparrow 1$, $\theta \to 0$ uniformly on sets $x \geq (-1+\epsilon)r$, $\epsilon > 0$, since then for $v \geq 1 - {}^1/_2\epsilon$, we have

$$\theta \leq \text{arc tan } [2(1-v^2)^{1/2}/\epsilon].$$

Now, the set C_ϵ of points of the sphere with $x < (\epsilon - 1)r$ is a zone of height ϵ and, hence, $\sigma(C_\epsilon) = {}^1/_2\epsilon$. Hence, $P_\tau(T(C_\epsilon))$ is at most ${}^1/_2\epsilon$ times the supremum which appears in **II**; call it $\bar{q}(\rho(u))$. Hence,

$$P_\tau(T(C_\epsilon)) = o(\epsilon e^{\rho(u)}) \text{ as } \rho(u) \to \infty.$$

Here $e^{\rho(u)}$ can be replaced by $\cosh \rho(u) = (1-v^2)^{-1/2}$.

Also, for points ζ of **10** not in C_ϵ, we have $\sinh \rho(T(\zeta)) \geq |x+vt|(1-v^2)^{-1/2} \geq {}^1/_2\epsilon(1-v^2)^{-1/2}$ if $v \geq 1 - {}^1/_2\epsilon$. Thus, given n, we wish to choose ϵ and v_0 to satisfy the following four conditions: $v_0 \geq 1 - {}^1/_2\epsilon$, $[2(1-v_0^2)^{1/2}/\epsilon] \leq \tan(1/n)$, ${}^1/_2\epsilon(1-v_0^2)^{-1/2} \geq \sinh B_n$, and $P_\tau(T(C_\epsilon)) \leq q(B_n)/2n$. Letting $\delta = (1-v_0^2)^{1/2}$, a sufficient set of conditions is that $\delta^2 \leq \epsilon - \epsilon^2/4$, $2\delta \leq \epsilon/n$, $\delta \leq \epsilon/2\sinh(3n^2)$, and $\epsilon o(1/\delta) \leq q(3n^2)/2n$. For n large enough, the first three of these conditions will be satisfied if $\epsilon = \delta \exp(4n^2)$, and $\delta \leq \exp(-5n^2)$. Then for fixed n and $\delta \downarrow 0$,

$$\epsilon o(1/\delta) = o(1) \exp(4n^2) \leq q(3n^2)/2n$$

for δ small enough. Q.E.D.

The corollary follows directly from *Theorem 2* since in condition **5** we can take $\beta = {}^1/_2$ and we have $q_\tau \equiv 1$ for orthogonally invariant probabilities, so **I** and **II** hold.

Proof of Theorem 3: We use *Theorem 2* and note that by the proof of Lemma 9.3, p. 261, of ref. 4, either velocities remain constant or we have, for proper times $\tau = n$ and any $K > 0$,

$$Pr\{\rho(u(n)) < K\} \leq c^n(\sinh K)/K$$

for some c with $0 < c < 1$. Choosing $\delta > 0$ small enough so that $ce^\delta < 1$ and using $\sinh K < e^K$, we have

$$\sum_{n=1}^{\infty} Pr\{\rho(u(n)) \leq \delta n\} < \infty.$$

Thus, almost surely $\rho(u(n)) > \delta n$ for all large enough n. Hence in the proof of *Theorem 1*, above, $Pr(\Omega_1) = 0$, and on Ω_2 with $Pr(\Omega_2) = 1$ we have $|v(\omega)| = 1$ almost surely. In refs. 4 and 5 it is proved that these processes are in $\mathcal{R}$. Q.E.D.

2. Possible applications to gases and galaxies

Given a gas of many interacting particles with individual identities (Boltzons), we may consider the motion of one of them as a stochastic process $x(t,\omega)$ with velocity $v(t,\omega)$. For a gas enclosed in a vessel, at points x near the sides of the vessel the transition probabilities to near futures will give small probability to half-spaces of velocities pointing outside the vessel. Thus the assumptions of *Theorem 1* will not hold, and clearly the conclusion also does not.

Suppose we consider, then, an unconfined gas. Such a gas, not in equilibrium, has velocity distributions satisfying a relativistic Boltzmann equation, but exact solution of such an integro-differential-functional equation is difficult (see refs. 2 and 14) as it is in the classical case. (The more special diffusion or Fokker–Planck equations do not hold for most processes in $\mathfrak{M}$ nor even for most Lorentz-invariant processes; see ref. 4 section 10, and ref. 5, section 4.) If we consider stars or galaxies as the "molecules" in our gas, we presumably cannot expect elastic collisions, but, in any case, certain qualitative features seem reasonably clear. We consider some alternate possibilities.

(*I*) A finite gas, with finite total energy, expands into an infinite surrounding empty space. Suppose that the initial probability distributions of position and velocity are spherically symmetric about some central point, p. Suppose also that the particles only interact by collisions or similar short-range repulsive forces, i.e., we neglect long-range gravitational interactions. Then after a long time passes, the mean velocities will point away from p and the variance of velocities at any given space point, corresponding to temperature, will approach 0. These conclusions hold for special-relativistic statistical mechanics (ref. 11, section 21) as well as for Newtonian mechanics.

(*II*) Suppose the gas is distributed throughout an infinite space with positive density. Then after equilibrium is reached, velocities will have a Maxwell–Boltzmann distribution whose relativistic form is

$$a \exp(-b(1-v^2)^{-1/2})dv \text{ on } V,$$

where a and b are constants depending on the state of the gas (ref. 9). In this case assumptions **4**, **5**, and **6** (section **1**) hold but not **7**. For widely separated times (enough for one particle to undergo many collisions between them), velocities are nearly independent and identically distributed and, hence, do not converge. For a general-relativistic treatment of this case see Tolman (ref. 12, p. 321).

(*III*) A final alternative possibility is that although the total mass of the gas may be infinite, its average density approaches zero when larger and larger regions (of reasonable shape, such as balls or cubes) are considered. Zero density has been suggested by de Vaucouleurs (ref. 13).

In this case it appears that conditions **4**, **5**, and **6** hold, while **7** may or may not, depending on further conditions.

Red-shifts of galaxies do not increase in strict proportion to their distance from us. There are major differences in the red-shifts of galaxies which appear to be interacting strongly and close to each other (see ref. 1). This is not surprising in a model for which galactic velocities and interactions are random.

The conclusion of *Theorem 1* above appears to hold for the known galaxies. It remains to be seen whether a more precise quantitative model satisfying the hypotheses of *Theorem 1* or *2* would fit the gas of galaxies. (The simpler conditions in *Theorem 3* almost surely would not hold for a real system.) At any rate the usual theory of a "big bang" from a highly concentrated initial "fireball" has not been established with great detail or precision either. A further possibility worth noting is that while galaxies may have been closer together in the past, they may also have been smaller and have grown in the interim by accretion of matter from intergalactic or extragalactic space. Such random accretion, as well as interaction of large galaxies, provides a possible mechanism, in view of the above theorems, to increase and directionally stabilize velocities of separation of galaxies.

Several years ago, Edward Nelson conjectured that Lorentz-invariant processes have limiting velocities with the speed of light, as proved in *Theorem 3* of this paper. Geza Schay also made a useful remark leading to the use of proper time rather than coordinate time. This research was partially supported by National Science Foundation Grant GP-29072.

1. Arp, H. (1971) "Observational paradoxes in extragalactic astronomy," *Science* **174,** 1189–1200.
2. Bancel, D. (1970) "Sur le problème de Cauchy pour l'équation de Boltzmann relativiste," *C. R. Acad. Sci.* **271,** A694–A696.
3. Doob, J. L. (1953) *Stochastic Processes* (New York, Wiley).
4. Dudley, R. M. (1965) "Lorentz-invariant Markov processes in relativistic phase space," *Ark. Mat. Astron. Fys.* **6,** 241–268.
5. Dudley, R. M. (1967) "A note on Lorentz-invariant Markov processes," *Ark. Mat. Astron. Fys.* **6,** 575–581.
6. Dynkin, E. B. (1959) *Die Grundlagen der Theorie der Markoffschen Prozesse* (German translation: Springer, Berlin, 1961).
7. Dynkin, E. B. (1963) *Markov Processes* (English translation: Academic Press and Springer-Verlag, New York and Berlin, 1965), in two volumes.
8. Hakim, R. (1968) "Relativistic stochastic processes," *J. Math. Phys.* **9,** 1805–1818.
9. Jüttner, F. (1911) "Das Maxwellsche Gesetz der Geschwindigkeitsverteilung in der Relativtheorie," *Ann. Phys.* (4. *Folge*) **34,** 856–882.
10. Schay, G. (1961) *The Equations of Diffusion in the Special Theory of Relativity,* Ph.D. thesis in physics, Princeton University.
11. Synge, J. L. (1957) *The Relativistic Gas* (Amsterdam, North-Holland).
12. Tolman, R. C. (1934) *Relativity, Thermodynamics and Cosmology* (Oxford, Clarendon Press).
13. de Vaucouleurs, G. (1970) "The case for a hierarchical cosmology," *Science* **167,** 1203–1213.
14. Vignon, B. (1969), "Une méthode de résolution approchée de l'équation de Boltzmann relativiste," *Ann. Inst. Henri Poincaré Sec. A* **10,** 31–66.

Part 3

Gaussian Processes

Introduction

The theory of Gaussian processes underwent a spectacular development during the last century. In one of its main directions, starting with the formulation of Brownian motion (Bachelier in 1900, Einstein in 1905), it continued with the proof of its existence (Wiener in 1923) and ended with the characterization of all Gaussian processes X_t, $t \in T$, that admit versions with bounded sample paths and all those that admit versions with bounded uniformly continuous (with respect to the intrinsic L_2 metric) sample paths, solely in terms of properties of the metric space (T,d), $d(s,t) = (E(X_t - X_s)^2)^{1/2}$, by Talagrand in 1987 (Acta Math. 159, 99-149). R. M. Dudley's contribution to this extraordinary achievement was of paramount importance, and is better described in Talagrand's (1987) Introduction: '... a Gaussian process is determined by its covariance structure, which has no reason to be closely related to the structure of the index set as a subset of $\mathbf{R}^n$. It should then be expected that a more intrinsic point of view would yield better results. This was achieved in the landmark paper of R. M. Dudley...' (the first one in this chapter). '...One major result of R. M. Dudley is that the metric entropy condition implies the boundedness of the process. More precisely,

$$E \sup_T X_t \leq K \int_0^\infty (\log N_\varepsilon)^{1/2} \, d\varepsilon \tag{1}$$

for some constant K.... the inequality is in some sense sharp' as V. N. Sudakov (1969) proved

$$\sup_{\varepsilon > 0} \varepsilon (\log N_\varepsilon)^{1/2} \leq KE \sup_T X_t$$

for some universal constant K (here and in (1), N_ε is the smallest number of d-balls of radius at most ε needed to cover T).

The first paper of this section contains many results, examples and conjectures regarding several ways of measuring the size of subsets of a Hilbert space and their relations: whether they are GB or GC (meaning whether the isonormal Gaussian process restricted to them has sample bounded or sample continuous paths), their metric entropy, or their exponent of volume, and in particular it contains the famous metric entropy bound (1) (more precisely, its original version, which proves sample continuity when the right side of (1) is finite). The third article is a survey about sample paths of Gaussian processes, with new results about GB and GC sets, the modulus of continuity of Gaussian processes, quadratic variation of Brownian motion, GC and GB-sets consisting of classes of sets in $\mathbf{R}^n$ with differentiable boundaries, and others. These two articles had a tremendous influence in the development of Gaussian process theory.

Dudley worked on other questions related to Gaussian processes, and the second and fourth papers in this chapter are excellent further instances of his work. In the second, R. M. Dudley, J. Feldman and L. Le Cam obtain very general and complete results on Gross' 'abstract Wiener spaces', in particular showing that the canonical normal distribution n on H gives rise to a countably additive Gaussian measure on the Banach space obtained from H by completing with respect to a seminorm $|\cdot|$ if and only if this seminorm is n-measurable in the sense of Gross; and that Gaussian Borel measures on locally convex spaces satisfying very mild conditions can be realized as continuous images of the canonical Gaussian measure n from some H. The last article, with J. Hoffmann-Jørgensen and L. Shepp, considers the 'small balls' problem: to study the probabilities of small balls for various norms, with respect to Gaussian measures on an infinite dimensional space, a problem that has recently attracted considerable interest.

E. Giné et al. (eds.), *Selected Works of R.M. Dudley*, Selected Works in Probability and Statistics,
DOI 10.1007/978-1-4419-5821-1_10, © Springer Science+Business Media, LLC 2010

Reprinted from JOURNAL OF FUNCTIONAL ANALYSIS Vol. 1, No. 3, October 1967

Printed in Belgium

The Sizes of Compact Subsets of Hilbert Space and Continuity of Gaussian Processes

R. M. DUDLEY*

Department of Mathematics, Massachusetts Institute of Technology, Cambridge, Massachusetts 02139

Communicated by Irving E. Segal

Received April 18, 1967

1. THE SIZES OF COMPACT SETS

The first two sections of this paper are introductory and correspond to the two halves of the title.

As is well known, there is no complete analog of Lebesgue or Haar measure in an infinite-dimensional Hilbert space H, but there is a need for some measure of the sizes of subsets of H. In this paper we shall study subsets C of H which are closed, bounded, convex and symmetric ($-x \in C$ if $x \in C$). Such a set C will be called a *Banach ball*, since it is the unit ball of a complete Banach norm on its linear span. In most cases in this paper C will be compact.

We use three main measures of the size of C. One is as follows. Let $V_n = V_n(C)$ be the supremum of (n-dimensional Lebesgue) volumes of projections $P_n(C)$ where P_n is any orthogonal projection with n-dimensional range. Then we define the *exponent of volume* of C, $EV(C)$, by

$$EV(C) = \limsup_{n\to\infty} \frac{\log V_n}{n \log n}.$$

Another numerical measure of the size of C involves the notion of ϵ-entropy [12]. Let (S, d) be a metric space. The *diameter* of a set $T \subset S$ is defined as

$$\sup \{d(x, y) : x, y \in T\}.$$

Given $\epsilon > 0$, one defines $N(S, \epsilon)$ as the minimal number of sets of diameter at most 2ϵ which cover S. Then the *ϵ-entropy* of S, $H(S, \epsilon)$,

* Fellow of the Alfred P. Sloan Foundation.

E. Giné et al. (eds.), *Selected Works of R.M. Dudley*, Selected Works in Probability and Statistics, DOI 10.1007/978-1-4419-5821-1_11, © Springer Science+Business Media, LLC 2010

is defined as $\log N(S, \epsilon)$. (The logarithm is taken to the base e. The ideas of information theory and thermodynamics play no explicit role in this paper.) Finally, we define the *exponent of entropy* r by

$$r = r(S) \equiv \limsup_{\epsilon \downarrow 0} \frac{\log H(S, \epsilon)}{\log (1/\epsilon)}.$$

(In case the lim sup is equal to a limit, r has been called the *metric order* of S—see [*12*], p. 22.)

We prove below (Proposition 5.8) that if $EV(C) \geqslant -1/2$, then $r(C) = +\infty$, while if $EV(C) < -1/2$, then

$$r(C) \geqslant -\frac{2}{1 + 2EV(C)}.$$

If the above inequality becomes an equality C will be called *volumetric*. In Section 6 we prove that ellipsoids, rectangular solids, certain "full approximation sets", and, if $EV(C) < -1$, octahedra, are volumetric. The question is left open for $-1 \leqslant EV(C) < -1/2$, but I conjecture (5.9) only that a Banach ball C with $EV(C) < -1$ is volumetric.

Our third general measure of the size of a Banach ball C involves the canonical "normal distribution" L on H ([*18*], pp. 116-119; [*9*]). L is a linear mapping of H into a set of Gaussian random variables with mean 0, which preserves inner products. Let A be a countable dense subset of C and

$$\bar{L}(C) = \sup \{|L(x)| : x \in A\}.$$

Then $\bar{L}(C)$ is a well-defined functionoid; i.e., a different choice of A will affect $\bar{L}(C)$ only on a set of zero probability.

For any $k > 0$, $r(kC) = r(C)$ and $EV(kC) = EV(C)$, but the random variable $\bar{L}(C)$ does not have this homothetic invariance. We call C a *GB-set* if $\bar{L}(C)$ is finite with probability one. This property is homothetically invariant, and for other reasons which will become clearer in the next section, we study mainly the GB-property rather than the entire random variable $\bar{L}(C)$. To relate this property to r and EV we have the following main results: if $r(C) < 2$ then C is a GB-set (V. Strassen (unpublished) and Corollary 3.2 below). If $r(C) = 2$, C need not be a GB-set (Section 6) and I conjecture (3.3) that if $r(C) > 2$ it never is. If $EV(C) > -1$, C is not a GB-set (Theorem 5.3); I conjecture (5.4) that C is a GB-set if $EV(C) < -1$, and prove this for $EV(C) < -3/2$ (Proposition 5.5). The conjectures are proved in all four classes of special cases considered in Section 6.

However, at $r = 2$ and $EV = -1$ there is some "overlap" and the GB-property is not a monotone function of the $H(S, \epsilon)$ as $\epsilon \downarrow 0$ nor of V_n as $n \to \infty$ (Proposition 6.10).

2. Continuity of Gaussian Processes

We shall study sample function continuity and boundedness of Gaussian processes from a general viewpoint. Let (S, d) be a metric space and let $\{x_t, t \in S\}$ be a real-valued Gaussian stochastic process over S (for definitions see, e.g., [5], p. 72). Then the x_t are all elements of a Hilbert space $H = L^2(\Omega, \mathscr{B}, \Pr)$ over some probability space $(\Omega, \mathscr{B}, \Pr)$. ($\Omega$ is a set, $\mathscr{B}$ a σ-algebra of subsets, and Pr a probability measure on $\mathscr{B}$). If two Gaussian processes over the same S have the same mean and covariance functions Ex_t and $Ex_s x_t$, then they have the same joint probability distributions for $\{x_t, t \in F\}$ for any finite or countable subset F of S ([5], p. 72, (3.3)). Such processes will be called "versions" of each other. We say that a process is *sample-continuous* if it has a version $\{x_t, t \in S\}$ such that for almost all ω in Ω, $t \to x_t(\omega)$ is continuous on S. (In case S is e.g. the real line it is well known that not all versions of a process will be continuous.)

Sequential convergence of functions on Ω almost everywhere implies convergence in measure and then, for the Gaussian case, convergence in H. Thus since S is metric, sample continuity implies that $t \to x_t$ is continuous from S into H and we can and will restrict ourselves to this case. Then, Ex_t is continuous on S, and x_t is sample-continuous if and only if $x_t - Ex_t$ is, so we may and do assume $Ex_t \equiv 0$.

A subset C of an abstract Hilbert space H_1 is realized as a Gaussian process $\{x_t, t \in C\}$ with $Ex_t \equiv 0$ and $Ex_s x_t = (s, t)$ by letting $x_t = L(t)$ where L is the "normal" random li ear functional or weak distribution mentioned in Section 1. We callnC a *GC-set* (Gaussian continuity set) if L is sample-continuous on C. Thus if $\{x_t, t \in S\}$ is a (Gaussian) process with $Ex_t \equiv 0$, the function $t \to x_t$ is continuous from S into H, and its range is a GC-set, then the process is sample-continuous.

For any set $A \subset H$ there is a sample-continuous process $\{x_t, t \in S\}$ whose range is A, letting S be A with discrete topology, but such examples are rather artificial and much of the study of sample-continuous Gaussian processes reduces to the study of GC-sets. (See e.g. the end of Section 4.)

A process $\{x_t, t \in S\}$ will be called *sample-bounded* if it has a version such that the sample functions $t \to x_t(\omega)$ are bounded uniformly on S, for each ω. Here we have a perfect correspondence: a Gaussian process is sample-bounded if and only if its range is a GB-set. The convex, closed, symmetric hull of a GB-set is a GB-set and is compact (Proposition 3.4 below). We shall on the whole restrict ourselves to compact sets, and a compact GC-set is a GB-set. Conversely, most, but not all, GB-sets are GC-sets. Sample continuity and boundedness are equivalent for ellipsoids and rectangular blocks (Propositions 6.3 and 6.6 below) and stationary processes on a finite interval ([2], Theorem 1). A narrow class of GB-sets which are not GC-sets appears among octahedra (Propositions 6.7 and 6.9 below), and other examples can be constructed by the law of the iterated logarithm. We shall prove severe narrowness of the class of GB-sets which are not GC-sets in general (Theorem 4.7).

V. Strassen proved (unpublished) in 1963 or 1964 that, if S is a set of Gaussian random variables with $r(S) < 2$, then (in our terminology) it is a GC-set. Strassen's result is sharpened somewhat (Theorem 3.1 below) to include some sets with $r(S) = 2$ and to yield a result of Fernique [7], [7a] for processes over the unit cube as a corollary (Theorem 7.1 below).

Conjecture 3.3 (if $r(S) > 2$ S is not a GB-set) is verified for certain random Fourier series with independent Gaussian coefficients, both those covered by a result of Kahane [10] and some others (Propositions 7.2 and 7.3 below).

In Section 4 we give some general results about $\bar{L}(C)$, convergence of series defining L, etc. Among other things, we establish an exact natural correspondence between GC-sets and the "measurable pseudo-norms" of L. Gross [9] (see Theorem 4.6 below).

Section 8 gives some brief comments on possible methods of attack in proving the conjectures.

3. Sample Continuity and ϵ-Entropy

Here is a sufficient condition for sample continuity in terms of ϵ-entropy:

THEOREM 3.1. *Suppose S is a subset of a Hilbert space and*

$$\sum_{n=1}^{\infty} \frac{H(S, 1/2^n)^{1/2}}{2^n} < \infty. \tag{1}$$

Then S is a GC-*set.*

Proof. Given a positive integer n, we decompose S into $N(1/2^{n+4})$ sets of diameter at most $1/2^{n+3}$, and choose one point from each set, forming a set A_n. Let G_n be the set of all random variables $L(x - y)$ for x and y in $A_{n-1} \cup A_n$ and $\| x - y \| \leqslant 1/2^n$. Then the cardinality of G_n is at most $4N(2^{-n-4})^2$.

We shall use below the well-known estimate, for $a > 0$,

$$\int_a^\infty e^{-x^2/2}\,dx \leqslant \int_a^\infty \frac{xe^{-x^2/2}\,dx}{a} = \frac{e^{-a^2/2}}{a}.$$

Let $\{b_n\}$ be any sequence of positive real numbers. Let

$$P_n = \Pr(\max\{|(z)| : z \in G_n\} \geqslant b_n) \leqslant 4N(2^{-n-4})^2 (\exp[-4^n b_n^2/2])2^n/b_n.$$

Thus $P_n \leqslant b_n$ if $n \geqslant 2$ and $-4^n b_n^2/2 + 2H(2^{-n-4}) \leqslant 2\log b_n$, or

$$[H(2^{-n-4}) - \log b_n]/4^{n-1} \leqslant b_n^2.$$

Let $a_n^2 = H(2^{-n-4})/4^{n-1}$. Then $\sum a_n < \infty$ by (1), and a_n is independent of b_n. But now we specify b_n, letting $b_n = \max(2a_n, 1/n^2)$. Then $a_n^2 \leqslant b_n^2/2$ and $\log b_n \geqslant -2\log n$, so for n large enough

$$(-4\log b_n)/4^n \leqslant 1/2n^4 \leqslant b_n^2/2,$$

and then $P_n \leqslant b_n$. Since $\sum b_n < \infty$, we have $\sum P_n < \infty$.

Thus for almost all ω there is an $n_0(\omega)$ such that $|z| < b_n$ for all $n \geqslant n_0(\omega)$ and all z in G_n.

Now let T be any countable dense subset of S. We shall show that on T, L is uniformly continuous with probability one. Its extension to S is then a version of L with continuous sample functions, as desired.

Given $\delta > 0$, we choose n_0 so that

$$3\sum_{n=n_0}^{\infty} b_n < \delta \qquad \text{and} \qquad P(\Omega(n_0)) < \delta$$

where

$$\Omega(n_0) = \{\omega : n_0(\omega) > n_0\}$$

For any s in T, we choose points $A_n(s)$ in A_n such that $\| s - A_n(s) \| < 1/2^{n+3}$. Now if $n \geqslant n_0$, $s, t \in A$, and $\| s - t \| \leqslant 1/2^{n+3}$, then $\| A_n(s) - A_n(t) \| \leqslant 1/2^n$. Thus $L(A_n(s) - A_n(t)) \in G_n$. Also, $L(A_n(s) - A_{n+1}(s)) \in G_{n+1}$. Thus for $\omega \notin \Omega(n_0)$, $L(A_n(s))(\omega) \to L(s)(\omega)$

for all s in T, and for any t in T such that $d(s, t) \leqslant 1/2^{n_0+3}$, we have

$$|L(s)(\omega) - L(t)(\omega)| \leqslant \delta.$$

Letting $\delta \downarrow 0$, we see that L is uniformly continuous on T with probability 1. Q.E.D.

COROLLARY 3.2. *If S is a subset of a Hilbert space and $r(S) < 2$, then S is a* GC-*set.*

There are numerous examples of sets S with $r(S) = 2$ which are neither GC- nor GB-sets; see, e.g., Section 6 below. Moreover, Theorem 7.1 below and its partial converse, due to Fernique [7], indicate that, even when specialized to stochastic processes on the real line, Theorem 3.1 is essentially the best possible result of its kind.

However, we prove in Proposition 6.10(a) below that *no* sufficient condition for the GC-property of a Banach ball in terms of $H(S, \epsilon)$ is necessary, i.e., the GC-property is not a "monotone function" of the function $\epsilon \to H(S, \epsilon)$ as $\epsilon \downarrow 0$. Yet I make

Conjecture 3.3. If S is a GB-set (and hence if S is a compact GC-set), then $r(S) \leqslant 2$.

In Sections 6 and 7 below, Conjecture 3.3 is proved in a number of special cases. In the general case, I shall prove at present only the following:

PROPOSITION 3.4. *If S is a* GB-*set then S is totally bounded (i.e., its closure is compact).*

Proof. If S is a GB-set, it is certainly bounded. Suppose it is not totally bounded. Then for some $\epsilon > 0$ there is an infinite sequence $\{f_j\}_{j=1}^{\infty}$ in S such that the distance of f_{j+1} from the linear span F_j of $f_1, \dots, f_j$ is at least ϵ for all j. Let

$$f_{n+1} = g_n + \sum_{j=1}^{n} a_{nj} f_j$$

where $\| g_n \| \geqslant \epsilon$ and $g_n \perp F_n$. Given $M > 0$, let

$$A_n = \{\omega : \max\{|L(f_j)| : 1 \leqslant j \leqslant n\} < M\}.$$

Then

$$\Pr(A_n \cap \{\omega : |L(f_{n+1})| \geqslant M\})$$

$$\geqslant \Pr\left(A_n \text{ and } L(g_n) \geqslant M \text{ and } L\left(\sum_{j=1}^{n} a_{nj} f_j\right) \geqslant 0\right)$$

$$= \Pr(L(g_n) \geqslant M) \Pr(A_n)/2.$$

Now for some $\delta > 0$, we have, for all n,

$$\Pr(L(g_n) \geqslant M) \geqslant 2\delta; \qquad \text{so} \qquad \Pr(A_n) \leqslant (1-\delta)^{n-1}$$

by induction. This contradicts the fact that S is a GB-set and completes the proof.

The method of proof just used will yield a stronger result. Using also (5.2) and Lemma 5.6 (cf. also Proposition 6.9), it can be shown that, if S is a GB-set, then for any $\delta > 0$

$$N(S, \epsilon) \leqslant \exp(\exp(1/\epsilon^{2+\delta}))$$

for ϵ sufficiently small. Since the examples in Section 6 indicate that this inequality has an unnecessary extra exponentiation, no further details will be given.

4. Pseudo-Norms

Let V be a real linear space and let W be a linear space of linear functionals on V. Then for any set $C \subset V$, the *polar* C^1 is defined by

$$C^1 = \{w \in W : w(x) \leqslant 1 \text{ for all } x \text{ in } C\}.$$

When C is symmetric,

$$C^1 = \{w \in W : |\, w(x)\,| \leqslant 1 \text{ for all } x \text{ in } C\}.$$

If A is a linear transformation of V into itself and W is closed under the adjoint A^* (i.e., composition with A), then for any $C \subset V$,

$$A(C)^1 = (A^*)^{-1}(C^1).$$

(Here $(A^*)^{-1}$ is a set mapping and A^* need not be invertible.) In particular V may be a Hilbert space and W its dual space, possibly identified with V.

On k-dimensional Euclidean space R^k, let λ or λ_k be Lebesgue measure and let G be the standard Gaussian probability measure;

$$dG = (2\pi)^{-k/2} \exp(-r^2/2)\, d\lambda,$$

where r is the distance from the origin.

Proposition 4.0 (Gross [9]). *Let A be a linear transformation* *from R^k into itself with norm $\| A \| \leqslant 1$ and let C be a convex symmetric set in R^k. Then*

$$G(A(C)^1) \geqslant G(C^1).$$

Proof. This follows directly from [9], Theorem 5, stated in different language. For A symmetric and invertible it is Lemma 5.2 of [9], and arguments to reduce to this case are given in the proof of Theorem 5. Q.E.D.

Now as usual, let H be a separable, infinite-dimensional Hilbert space. Every GB-set in H is included in some Banach ball which is still a GB-set.

For any subset C of H, let $s(C)$ be its linear span. Then if C is convex and symmetric, it is the unit ball of a norm $\| \cdot \|_C$ on $s(C)$. If C is a Banach ball, then $(s(C), \| \cdot \|_C)$ is a Banach space and its natural injection into H is continuous.

Let H^* be the dual space of H. (For clarity, we do not identify the two.) For each φ in H^*, let

$$\| \varphi \|_C' = \sup \{| \varphi(\psi) | : \psi \in C\}.$$

Then if C is a Banach ball in H, $\| \cdot \|_C'$ is the dual norm to $\| \cdot \|_C$ (composed with the natural map of H^* into the dual space $(s(C)', \| \cdot \|_C')$ of $(s(C), \| \cdot \|_C)$).

L is an assignment of random variables to elements of H, or equivalently to continuous linear functionals on H^*. The assignment can be extended to some nonlinear functionals in various ways. For example, if φ is a Borel measurable function on R^k and $f_1, \ldots, f_n \in H$, then $\varphi(f_1, \ldots, f_n)$ defines by composition a function on H^*. (Such a function is called "tame.") The assignment

$$L(\varphi(f_1, \ldots, f_n)) = \varphi(L(f_1), \ldots, L(f_n))$$

is well-defined, as is well known [9], [18]. Thus, e.g., we let $L(| f |) = | L(f) |, f \in H$.

Now in general, an assignment such as

$$L(\sup g_n) = \sup (L(g_n))$$

will not be well-defined, but if $g_n = | f_n |$, $f_n \in H = (H^*)^*$, then $\sup g_n = \| \cdot \|_C'$ where C is the closed symmetric convex hull of the f_n. Also

$$\sup L(g_n) = \sup (| L(f_n) |) = \bar{L}(C),$$

and the assignment

$$L(\| \cdot \|_C) = \bar{L}(C)(\cdot)$$

is well-defined.

By f.d.p. (finite-dimensional projection) we shall mean an orthogonal projection of H onto a finite-dimensional subspace. For projections P and Q, one says $P \leqslant Q$ if the range of P is included in that of Q, and $P_n \uparrow I$ if $P_1 \leqslant P_2 \leqslant \cdots$ and $P_n(f) \to f$ in (Hilbert) norm for each f in H. Also $P \perp Q$ means the ranges of P and Q are orthogonal. If $\{f_n\}$ is an orthonormal basis of H, g_n are independent, normalized Gaussian random variables, and $L_n(f) = (f, f_n)\, g_n$, then the series

$$\sum_{n=1}^{\infty} L_n(\cdot) \tag{1'}$$

is a version of L. If $P_n \uparrow I$ (and the P_n are f.d.p.'s) then there is an orthonormal basis $\{f_j\}$ of H such that for each n, $\{f_1, \ldots, f_{k_n}\}$ is a basis of the range of P_n for some k_n, $k_n \uparrow \infty$. The convergence of $L \circ P_n$ to L is equivalent to convergence of a certain sequence of partial sums of (1′).

We shall need an infinite-dimensional form of Proposition 4.0.

PROPOSITION 4.1. *Let A be a linear operator from H into itself with $\| A \| \leqslant 1$, and let $C \subset H$. Then for any $t \geqslant 0$,*

$$\Pr(\bar{L}(AC) \leqslant t) \geqslant \Pr(\bar{L}(C) \leqslant t).$$

Proof. If C is finite, the result follows immediately from Proposition 4.0. In general, let C_n be finite sets which increase up to a dense set in C. Then

$$\Pr(\bar{L}(C) \leqslant t) = \lim_{n\to\infty} \Pr(\bar{L}(C_n) \leqslant t) \leqslant \lim_{n\to\infty} \Pr(\bar{L}(AC_n) \leqslant t)$$

$$= \Pr(\bar{L}(AC) \leqslant t), \qquad \text{Q.E.D.}$$

PROPOSITION 4.2. *If P_n are* f.d.p.'*s*, $P_n \uparrow I$, $C \subset H$, *and* $t \geqslant 0$, *then*

$$\Pr(\bar{L}(C) \leqslant t) = \lim_{n\to\infty} \Pr(\bar{L}(P_n C) \leqslant t).$$

Proof. Let A be countable and dense in C. Then $L(P_n f) \to L(f)$ as $n \to \infty$ for all f in A, with probability 1. Hence

$$\Pr(\bar{L}(C) \leqslant t) \leqslant \liminf_{n\to\infty} \Pr(\bar{L}(P_n C) \leqslant t).$$

On the other hand Prnposition 4.1 yields

$$\limsup_{n\to\infty} \Pr(\bar{L}(P_nC) \leqslant t) \leqslant \Pr(\bar{L}(C) \leqslant t),$$

completing the proof.

The following definition is essentially that of Gross [9].

DEFINITION. A pseudo-norm $\|\cdot\|$ on H^* is *measurable* (for L) if for every $\epsilon > 0$ there is a f.d.p. P_0 such that, for every f.d.p. $P \perp P_0$,

$$\Pr(L(\|\cdot\| \circ P) > \epsilon) < \epsilon.$$

Note that $\|\cdot\| \circ P$ is a tame function on H^* so that L of it is defined. If $C \subset H$ then $\|\cdot\|'_C$ is measurable if and only if for every $\epsilon > 0$ there is a f.d.p. P_0 such that, for every f.d.p. $P \perp P_0$,

$$\Pr(\bar{L}(PC) > \epsilon) < \epsilon.$$

It then follows by Propositions 4.1 and 4.2 that

$$\Pr(\bar{L}P_0^{\perp}C) > \epsilon) \leqslant \epsilon.$$

(For any projection P, $P^{\perp} = I - P$ where I is the identity operator.)

THEOREM 4.3. *If C is a Banach ball in H, the following are equivalent:*

(a) *C is a* GB-*set*, i.e., $\Pr(\bar{L}C) < \infty) = 1$.

(b) $\Pr(\bar{L}(C) < \infty) > 0$

(c) [resp. (d)] *$\bar{L}(P_nC)$ converges in law for some* (resp. *every*) *sequence of* f.d.p.'s *$P_n \uparrow I$.*

(e) *L restricted to $s(C)$ has a version linear and continuous with probability 1 for $\|\cdot\|_C$.*

Proof. Let $\{f_n\}$ be an orthonormal basis of H. For each f in H, the series (1′) converges almost everywhere on Ω and in $L^2(\Omega)$. For any finite N,

$$\sum_{n=1}^{N} L_n(\cdot)(\omega)$$

is bounded on C for each ω in Ω, and finiteness of $\bar{L}(C)(\omega)$ thus depends on the g_n for $n > N$. Thus by the zero-one law ([*13*], B, p. 229), $\Pr(\bar{L}(C) < \infty) = 0$ or 1, and (a) is equivalent to (b).

(a) is equivalent to (c) and (d) by Proposition 4.2.

(a) is equivalent to (e) since a linear functional on a normed space is continuous if and only if it is bounded on the unit ball. The proof is complete.

Before treating GC-sets, we introduce some facts we need about function-valued random variables. Let S be a metric space with a countable dense subset $A = \{x_j\}_{j=1}^{\infty}$. Let $\mathscr{C}(S)$ be the Banach space of bounded continuous real-valued functions on S, with supremum norm $\| \cdot \|_\infty$. We say $X_1, X_2, \ldots$ are given as a set of $\mathscr{C}(S)$-valued random variables if probabilities

$$\Pr(X_i(t_j) \in A_{ij}, i, j = 1, 2, \ldots)$$

are defined for any points $t_1, t_2, \ldots,$ in S and Borel sets A_{ij} in the real line. Then the norms

$$\| X_i \|_\infty = \sup \{| X_i(t) | : t \in A\}$$

are measurable. Note, however, that $\mathscr{C}(S)$ will not be separable if S is not compact. Then, the distributions of the X_i will not be expected to be defined on all open sets in $\mathscr{C}(S)$ for the supremum norm topology (cf. [*6*]).

A random variable X in $\mathscr{C}(S)$ will be called *symmetric* if $-X$ has the same distribution as X. Independence of random variables X_i in $\mathscr{C}(S)$ is defined also, naturally, to mean that the sets of real random variables

$$A_i = \{X_i(t) : t \in S\}$$

are independent for different values of i.

Let X_i be independent and symmetric in $\mathscr{C}(S)$ and

$$S_n = X_1 + \cdots + X_n .$$

The following generalization of a Lemma of P. Lévy is proved much like the classical version (Loève [*13*], p. 247).

LEMMA 4.4. *For any* $\alpha > 0$,

$$\Pr(\max\{\| S_k \| : k = 1, \ldots, m\} > \alpha) \leqslant 2 \Pr(\| S_m \| > \alpha).$$

Proof. For each $k = 1, \ldots, m$, $j = 1, 2, \ldots,$ and $s = \pm 1$, let

$$A(k, j, s) = \{\omega : \| S_i \| \leqslant \alpha, i = 1, \ldots, k-1, | S_k(x_q) | \leqslant \alpha,$$
$$q = 1, \ldots, j-1, sS_k(x_j) > \alpha\}.$$

Then $\{\omega : \| S_k \| > \alpha$ for some k, $1 \leqslant k \leqslant m\}$ is the disjoint union of the $A(k, j, s)$. We also have for each k, j and s,

$$\Pr(A(k, j, s) \text{ and } \| S_m \| > \alpha) \geqslant \Pr(A(k, j, s) \text{ and } s(S_m - S_k)(x_j) \geqslant 0)$$
$$\geqslant \Pr(A(k, j, s))/2.$$

Hence

$$2 \Pr(\| S_m \| > \alpha) \geqslant \sum_{k,j,s} \Pr(A(k, j, s)) = \Pr(\max\{\| S_k \| : k = 1,..., m\} > \alpha);$$

Q.E.D.

PROPOSITION 4.5. *The series* $\sum_{n=1}^{\infty} X_n$ *of independent symmetric* $\mathscr{C}(S)$*-valued random variable converges in* $\mathscr{C}(S)$ *(i.e., uniformly on* S*) with probability 1 if and only if it converges (uniformly) in probability.*

Proof. "Only if" is obvious. "If" is proved from Lemma 4.4 just as in the classical case where S has only one point: see [*13*], p. 249.

THEOREM 4.6. *For any compact Banach ball* C *in* H, *the following are equivalent:*

(a) *for any* $\epsilon > 0$, $\Pr(\bar{L}C) < \epsilon) > 0$;

(b) C *is a* GC-*set;*

(c) [resp. (d)] $L \circ P_n$ *converges uniformly on* C *in probability for some* (resp. *all*) *sequences of* f.d.p.'*s* $P_n \uparrow I$;

(c′) [resp. (d′)] *replace "in probability" by "with probability 1" in* (c) [resp. (d)];

(e) $\| \cdot \|_C$ *is a measurable pseudo-norm on* H^*.

Proof. Throughout let A be a countable dense subset of C.

(a) $\Rightarrow$ (b): Let P_n be f.d.p.'s and $P_n \uparrow I$. Given $\epsilon > 0$, let

$$C_n(\epsilon) = \{\omega : \bar{L}(P_n^{\perp}C) < \epsilon/3\},$$
$$K(\epsilon) = \limsup C_n(\epsilon)$$
$$= \{\omega : C_n(\epsilon) \text{ holds for arbitrarily large } n\}.$$

Then $K(\epsilon)$ is a tail event, having a probability 0 or 1.

By (a) and Proposition 4.1,

$$0 < \Pr(\bar{L}(C) < \epsilon/3) \leqslant \Pr(\bar{L}P_n^{\perp}C < \epsilon/3)$$

for all n, where $P_n^{\perp} = I - P_n$. Thus $K(\epsilon)$ has positive probability, hence probability 1. Hence almost every ω belongs to $C_n(\epsilon)$ for some n. Then since $L \circ P_n$ is continuous, there is an $\alpha > 0$ such that if $x, y \in A$ and $\| x - y \| < \alpha$, then

$$| L(x) - L(y) | (\omega) \leqslant | L(P_n(x - y)) (\omega) | + 2\bar{L}P_n{}^{\perp}C < \epsilon.$$

Since ϵ was an arbitrary positive number, (b) is proved.

(b) $\Rightarrow$ (c): given $\epsilon > 0$ we use uniform continuity on C with probability 1 to infer that for some $\delta > 0$,

$$\Pr (\sup \{| L(x - y) | : x, y \in A, \| x - y \| < \delta\} \geqslant \epsilon) < \epsilon.$$

We choose a finite-dimensional subspace F such that $F \cap C$ is within δ of every point of C. Let P be the projection onto F. Then by Proposition 4.1

$$\begin{aligned} \epsilon &> \Pr (\sup \{| L(P^{\perp} (x - y)) | : x, y \in C, \| x - y \| \leqslant \delta\} \geqslant \epsilon) \\ &\geqslant \Pr (\sup \{| LP^{\perp}x | : x \in C\} \geqslant \epsilon) \end{aligned}$$

since, for any x in C, there is a y in $F \cap C$ with $\| x - y \| \leqslant \delta$ and $P^{\perp}y = 0$. Thus

$$\Pr (| (L - L \circ P)^{-} (C) | \geqslant \epsilon) < \epsilon$$

as desired, so (c) holds.

(c) $\Rightarrow$ (c′) by Proposition 4.5.

(c′) $\Rightarrow$ (d): let $L \circ Q_n \to L$ uniformly on C with probability 1 and $Q_n \uparrow I$, $P_m \uparrow I$, where P_m and Q_n are f.d.p.'s. Then given $\epsilon > 0$ there is an n such that

$$\Pr (\bar{L}(Q_n{}^{\perp}C) > \epsilon/2) < \epsilon/2.$$

Now the operator norm $\| P_m{}^{\perp}Q_n \| \to 0$ as $m \to \infty$ since Q_n has finite-dimensional range and $P_m{}^{\perp} \to 0$ pointwise. Hence $\bar{L}(P_m{}^{\perp}Q_nC) \to 0$ in probability as $m \to \infty$. Also

$$\begin{gathered} \Pr (\bar{L}(P_m{}^{\perp}Q_n{}^{\perp}C) \leqslant \epsilon/2) \geqslant \Pr (\bar{L}(Q_n{}^{\perp}C) \leqslant \epsilon/2) \geqslant 1 - \epsilon/2, \\ \bar{L}(P_m{}^{\perp}C) \leqslant \bar{L}(P_m{}^{\perp}Q_nC) + \bar{L}(P_m{}^{\perp}Q_n{}^{\perp}C); \end{gathered}$$

so, for m large enough,

$$\Pr (\bar{L}(P_m{}^{\perp}C) \geqslant \epsilon) \leqslant \epsilon$$

and (d) holds.

(d) $\Rightarrow$ (d') by Proposition 4.5.

(d') $\Rightarrow$ (e): clearly (d') $\Rightarrow$ (c), and (c) $\Rightarrow$ (e) by Proposition 4.1.

(e) $\Rightarrow$ (a): given $\epsilon > 0$, we choose a f.d.p. P such that

$$\Pr(\bar{L}(P^{\perp}C) < \epsilon/2) > 0.$$

Then also

$$\Pr(\bar{L}(PC) < \epsilon/2) > 0$$

and since $\bar{L}(P^{\perp}C)$ and $\bar{L}(PC)$ are independent, we have

$$\Pr(\bar{L}(C) < \epsilon) \geqslant \Pr(\bar{L}(PC) < \epsilon/2 \text{ and } \bar{L}(P^{\perp}C) < \epsilon/2) > 0.$$

Q.E.D.

Not every GB-set is a GC-set, as we shall see below (Propositions 6.7 and 6.9). Thus all possible implications among the conditions listed in Theorems 4.3 and 4.6 are settled. However, these conditions suggest others, e.g., replacing "in law" in (c) and (d) of Theorem 4.3 by "in probability" or "with probability one". If $P_nC \subset C$ for all n, then $\bar{L}(P_nC)$ is nondecreasing, so the different forms of (c) are equivalent in this case. In Section 6 we present a GB-set (octahedron with axes $(\log n)^{-1/2}$), which is not a GC-set, and for which $P_nC \to C$ for certain natural projections $P_n \uparrow I$. Thus the stronger forms of (c) do not imply that C is a GC-set, but other possible implications are not settled.

We shall conclude this section with a result showing that the class of GB-sets which are not GC-sets is quite narrow.

If B and C are Banach balls in H, we shall say B is *C-compact* if $B \subset s(C)$ and B is compact for $\|\cdot\|_C$. If B is a GB-set, we call it *maximal* if whenever B is C-compact, C is not a GB-set. (No GB-set A is maximal in a strict set-theoretic sense since $2A$ includes A strictly and $2A$ is a GB-set.)

THEOREM 4.7. *Every* GB-*set is either maximal or a* GC-*set.*

Proof. Suppose B is C-compact where C is a Banach ball and a GB-set. Then L restricted to $s(C)$ has a version which is linear and continuous for $\|\cdot\|_C$. The $\|\cdot\|_C$ topology is stronger than the original Hilbert topology on $s(C)$ since C is bounded, hence these two topologies are equal on the compact set B ([*11*], Theorem 8, p. 141). Thus B is a GB-set. Q.E.D.

If $\|\cdot\|$ is a measurable pseudo-norm on H^*, then L is defined by a countably additive probability measure on the completion of H^* for $\|\cdot\|$.[1] At the moment the converse seems to be an open question.

Suppose $(x_t, t \in S)$ is a sample-continuous Gaussian process over a compact metric space (S, d). Then $t \to x_t$ is continuous from S into H, and

$$e(s, t) = (E(x_s - x_t)^2)^{1/2}$$

defines a pseudo-metric e on S which is continuous for d and hence defines a weaker topology. If (S, e) is Hausdorff, i.e., if $x_s \neq x_t$ for $s \neq t$, then the d and e topologies on S are equal, and hence the range of the process in H is a GC-set; its closed convex symmetric hull is a GB-set, which, by Theorem 4.7, is not much worse.

5. Sequences of Volumes

We shall need the volumes of certain simple sets in R^k. First, suppose A is a simplex, i.e., a convex hull of $(k + 1)$ points $x_0, \ldots, x_k$, having an interior. Let F be a *face* of A, i.e., a convex hull of k of its vertices. Let λ or λ_k be Lebesgue measure on R^k and S or S_{k-1} the $(k - 1)$-dimensional Lebesgue "surface" or "area" measure. Then

$$\lambda(A) = S(F)\, d/k,$$

where d is the distance from the vertex not in F to the hyperplane through F. Now suppose $x_0 = 0$ and let d_j be the distance from x_j to the linear span of $x_0, \ldots, x_{j-1}$, $j = 1, \ldots, k$. Then

$$\lambda(A) = \Big(\prod_{j=1}^{k} d_j\Big)\Big/ k!.$$

Now, recalling the definitions of $V_n(C)$ and $EV(C)$ given is Section 1, we have the following fact (a stronger statement is given as Proposition 5.10 below):

Lemma 5.0. *If C is a convex set in H and $EV(C) < -1$, then C is totally bounded.*

[1] For the proof, see L. Gross, Abstract Wiener spaces, in *Proceedings of the Fifth Berkeley Symposium on Mathematical Statistics and Probability (1964)*. University of California Press, Berkeley, 1967.

Proof. If C is not totally bounded we make the same construction as in the proof of Proposition 2.4. Then for some $\epsilon > 0$, $V_n(C)$ is greater than or equal to the volume of the convex hull of $0, f_1, \ldots, f_n$, so

$$V_n(C) \geqslant \epsilon^n/n! \qquad \text{for all} \qquad n.$$

By Stirling's formula, this contradicts the hypothesis. Q.E.D.

Next, let $c_k = \lambda_k(B)$ where B is a ball of radius 1 in R^k. Then it can be shown by induction that, for any positive integer k,

$$c_{2k+1} = 2^{2k+1}\pi^k k!/(2k+1)!,$$
$$c_{2k} = \pi^k/k!.$$

Thus by Stirling's formula we have the following estimate:

$$\lim_{j\to\infty} c_j(\pi j)^{1/2}\,(j/2\pi e)^{j/2} = 1. \tag{5.1}$$

We shall also need the following fact. Let $\{a_n\}$ be a sequence of positive real numbers such that $a_n \downarrow 0$ as $n \to \infty$. For such a sequence and $\epsilon > 0$ we define

$$n(\epsilon) = n(\{a_n\}, \epsilon) = \max\,(n : a_n \geqslant \epsilon),$$

$$\lambda = \lambda(\{a_n\}) = \inf\left(\alpha : \sum_{n=1}^{\infty} a_n^{\alpha} < \infty\right).$$

Then it is known, and easy to prove, that

$$\lambda = \limsup_{\epsilon\downarrow 0} \log n(\epsilon)/\log(1/\epsilon). \tag{5.2}$$

Now let C be a convex symmetric set in H.

THEOREM 5.3. *If C is a* GB-*set, then*

$$\sup_n\,[n^{-1}\log V_n + \log n] < \infty.$$

Hence $EV(C) \leqslant -1$.

Proof. Since C is a GB-set there is an $M > 0$ such that

$$\Pr(\bar{L}(C) \leqslant M) > \gamma > 0.$$

C may be replaced in the above inequality by any orthogonal projection $P(C)$, according to Proposition 4.1. Multiplying C by a positive number leaves the relevant properties unchanged, so we may assume $M = 1$. Suppose the first conclusion is false. Then for any $K > 0$ there is an n such that $V_n \geqslant (K/n)^n$.

Let P_n be a projection with n-dimensional range F. Then

$$\gamma \leqslant \Pr(\bar{L}(P_n C) \leqslant 1) = G((P_n C)^1),$$

where the polar is taken in the dual of F and G is normalized Gaussian probability measure. We use the general inequality

$$\lambda_n(B)\,\lambda_n(B^1) \leqslant c_n{}^2$$

where B is any convex symmetric set in R^n (due to Santalò [17]). (Later work by Bambah [1] on a lower bound for $\lambda_n(B)\,\lambda_n(B^1)$ may also be noted.) For any $\beta > 0$ there is a P_n such that

$$\lambda_n(P_n C) \geqslant (n\beta)^{-n}, \qquad \text{so} \qquad \lambda_n((P_n C)^1) \leqslant c_n{}^2(n\beta)^n.$$

Using (5.1) we obtain for any $\alpha > 0$

$$\lambda_n((P_n C)^1) \leqslant c_n(\alpha n)^{n/2}$$

for certain arbitrarily large n. Now, given $\lambda_n(A)$ for a set A, $G(A)$ is clearly maximized when A is a ball $E(r)$ centered at 0, say of radius r. Hence

$$G((P_n C)^1) \leqslant G(E(r_n)),$$

where $r_n \leqslant (\alpha n)^{1/2}$. Then

$$G(E(r_n)) \leqslant \int_0^{(\alpha n)^{1/2}} r^{n-1} e^{-r^2/2}\, dr/I_n\,,$$

where

$$I_n = \int_0^\infty r^{n-1} e^{-r^2/2}\, dr.$$

The integrand increases for $0 < r < (n-1)^{1/2}$. But $(\alpha n/(n-1))^{1/2} \to 0$ as $n \to \infty$ and $\alpha \downarrow 0$, so $G(E(r_n)) \to 0$ as $n \to \infty$ through a suitable sequence, contradicting the fact that $G((P_n C)^1) \geqslant \gamma > 0$.
Thus for some $M > 0$,

$$\frac{\log V_n}{n \log n} \leqslant -1 + \frac{M}{\log n}$$

for all n, so $EV(C) \leqslant -1$. Q.E.D.

Conjecture. 5.4. If C is a Banach ball and $EV(C) < -1$, then C is a GC-set.

The above conjecture may be made plausible by a supporting conjecture (5.9 below) and proofs of both conjectures in four classes of special cases (Section 6). In the general case, I can prove the following.

PROPOSITION 5.5. *If C is a Banach ball and $EV(C) < -\frac{3}{2}$, then C is a GC-set.*

Before proving Proposition 5.5 we introduce another construction and some other facts. Given a compact Banach ball C in H and an orthonormal basis $\{\varphi_j\}_{j=1}^{\infty}$ of H, let F_n be the linear span of $\varphi_1, \ldots, \varphi_n$, and

$$C_n = C \cap F_n \qquad (C_0 = F_0 = \{0\}).$$

Given two sets A and B in H we define their distance as usual,

$$e(A, B) = \sup_{x \in A} \inf_{y \in B} \| x - y \| ,$$
$$d(A, B) = e(A, B) + e(B, A).$$

We shall say the basis $\{\varphi_j\}$ is *adapted* to C if

$$d(C_{n-1}, C) = d(C_{n-1}, C_n)$$

for $n = 1, 2, \ldots$. Since C is compact, a basis adapted to C always exists. Then the sequence $\{F_n\}$ of subspaces will also be called *adapted* to C.

If there is a sequence $G_0 \subset G_1 \subset \cdots$ of subspaces of H with each G_n n-dimensional and $d(C \cap G_n, C) \leqslant a_n$ for all n, $a_n \downarrow 0$, then the sequence $\{a_n\}$ will be called *adapted* to C (whether or not the G_n are).

In order to find an upper bound for ϵ-entropies of sets with a given adapted sequence $\{a_n\}$ we use the following result.

LEMMA 5.6. *Let $B(\{a_j\}_{j=1}^{n})$ be a rectangular n-dimensional block of sides $2a_j$, $0 < a_j \leqslant 1$,*

$$B = B(\{a_j\}) = \left\{ \sum_{i=1}^{n} x_i \varphi_i : | x_i | \leqslant a_i , i = 1, \ldots, n \right\} ,$$

where the φ_i are orthonormal. Let $0 < \epsilon < 1$. Then

$$N(B, \epsilon) \leqslant \prod_{j=1}^{n} (2 + n^{1/2} a_j / \epsilon) \leqslant 3^n n^{n/2} / \epsilon^n .$$

Proof. We consider the cubes of side $2\epsilon/n^{1/2}$ whose vertices are of the form

$$\sum_{j=1}^{n} 2m_j\epsilon/n^{1/2}, \qquad |m_j| \leqslant 1 + n^{1/2}a_j/2\epsilon,$$

and the m_j are integers. B is included in the union of these cubes, their diameters are 2ϵ, and the number of them is bounded as indicated. Q.E.D.

The latter, cruder estimate in the above Lemma is sufficient for its applications below except for one rather delicate one (Proposition 6.10).

PROPOSITION 5.7. *Let C be a compact Banach ball in H and $\{a_n\}$ adapted to C. Then*

$$r(C) \leqslant \lambda(\{a_n\}).$$

Proof. Let $s = \lambda(\{a_n\})$ and let $F_0 \subset F_1 \subset F_2 \cdots$ be subspaces of H, F_n n-dimensional, such that for all n, $d(C_n, C) \leqslant a_n$ where $C_n = C \cap F_n$.

If $\beta > \alpha > s$ then for small enough $\epsilon > 0$,

$$n(\{a_n\}, \epsilon/2) \leqslant 1/\epsilon^{\alpha}$$

by (5.2). For such an $\epsilon < 1$ and $n = n(\epsilon/2)$,

$$N(C, \epsilon) \leqslant N(C_n, \epsilon/2).$$

Since r and s are homothetically invariant we can assume $a_1 \leqslant 1$. Clearly C_n is included in the block $B(\{a_j\}_{j=1}^{n})$ of Lemma 5.6, so for ϵ small enough

$$N(C, \epsilon) \leqslant \exp\left(n\left(\log 3 + \tfrac{1}{2}\log n + \log(1/\epsilon)\right)\right) \leqslant \exp(\epsilon^{-\beta}).$$

Thus $r(C) \leqslant \beta$ for all $\beta > s$ and $r(C) \leqslant s$. Q.E.D.

Proof of Proposition 5.5. By Lemma 5.0, C is compact. There is a $c > \frac{3}{2}$ such that $V_n(C) \leqslant n^{-nc}$ for n large enough. We choose a basis $\{\varphi_n\}$ adapted to C and v_n in C_{n+1} such that $e(v_n, C_n) = a_n = d(C, C_n)$, $n = 0, 1, \ldots$. Then C includes the symmetric convex hull of the v_n, so

$$V_n \geqslant \left(\prod_{j=1}^{n} 2a_j\right)\Big/ n!$$

Then by Stirling's formula there is a $\beta > \frac{1}{2}$ such that

$$a_n{}^n < n^{n-nc} = n^{-\beta n}$$

for n large enough, and $a_n \leqslant n^{-\beta}$. Thus 5.7 and 3.2 imply that C is a GC-set. Q.E.D.

Suppose given a Banach ball (= convex symmetric bounded closed set) C in H. Suppose also that $\{F_n\}$ is a sequence of subspaces adapted to C. Given $F_1, \ldots, F_{n-1}$, we assume F_n can be and is chosen among its possible values so as to minimize $\lambda_n(F_n \cap C)$. Then we define

$$W_n = \lambda_n(F_n \cap C),$$

$$EW(C) = \limsup_{n\to\infty} (\log W_n)/(n \log n).$$

For a sufficiently "smooth" set C, e.g., an ellipsoid, we shall have $EV(C) = EW(C)$ and even $V_n \equiv W_n$ (see Proposition 6.1 below). At the end of Section 6 we show that $EW(C) < EV(C)$ is possible.

Next we obtain a lower bound for $r(C)$ in terms of $EV(C)$. In each of the four classes of examples treated in Section 6, it becomes an equality at least for $EV(C) < -1$.

PROPOSITION 5.8. *For any convex symmetric set C in H,*

(a) $r(C) \geqslant -2/(2EV(C) + 1)$ *if* $EV(C) < -\frac{1}{2}$

(b) $r(C) = +\infty$ *if* $EV(C) \geqslant -\frac{1}{2}$.

Proof. If C is covered by m sets, each of diameter at most ϵ, then any n-dimensional projection P_nC is covered by m balls of radius ϵ, and

$$mc_n\epsilon^n \geqslant V_n, \qquad \text{so} \qquad N(C, \epsilon/2) \geqslant V_n/c_n\epsilon^n$$

for all n. Let $EV(C) = -b > -c$ and $c > \frac{1}{2}$. Then for n large enough

$$V_n/c_n\epsilon^n \geqslant n^{n/2}(\pi n)^{1/2}/[(2\pi e)^{n/2}\, \epsilon^n n^{nc}] = k_n,$$

say. The following paragraph gives motivation only.

To maximize k_n, we note that

$$k_{n+1}/k_n = ((n+1)/n)^{n[(1/2)-c]}\,(n+1)^{(1-c)}/\epsilon(2\pi en)^{1/2},$$

which is asymptotic as $n \to \infty$ to

$$e^{-c}n^{(1/2)-c}/\epsilon(2\pi)^{1/2}.$$

At any rate, as $\epsilon \downarrow 0$ we choose $m = m(\epsilon)$ so that $m^{(1/2)-c}$ is asymptotic to $e^c\epsilon(2\pi)^{1/2}$, as is clearly possible. Then for any $\delta > 0$, and ϵ small enough,

$$\begin{aligned} k_m &= (m^{(1/2)-c}/\epsilon(2\pi e)^{1/2})^m (\pi m)^{1/2} \\ &\geqslant \exp(m(c - \tfrac{1}{2} - \delta)) \\ &\geqslant \exp((c - \tfrac{1}{2} - \delta)[(1-\delta)/e^c\epsilon(2\pi)^{1/2}]^{2/(2c-1)}. \end{aligned}$$

Hence for some constant $\gamma > 0$,

$$N(C, \epsilon) \geqslant \exp(\gamma\epsilon^{2/(1-2c)})$$

for ϵ small enough. If $-b < -\frac{1}{2}$ we let c approach b and obtain (a). If $-b \geqslant -\frac{1}{2}$, we let c approach $\frac{1}{2}$ and obtain (b). Q.E.D.

Definition. A Banach ball C is *volumetric* if $EV(C) < -\frac{1}{2}$ and

$$r(C) = -2/(2EV(C) + 1).$$

Conjecture. 5.9. If C is a Banach ball and $EV(C) < -1$, then C is volumetric (hence $r(C) < 2$ and C is a GC-set).

A weaker inequality in the direction converse to 5.8 (a) is easily proved. Let $\{F_n\}$ be adapted to C and $T_n = \lambda_n(F_n \cap C)$. If $T_n \leqslant n^{-n(1+\delta)}$ for n large enough, $\delta > 0$, then since $a_1 \cdots a_n/n! \leqslant T_n$, we have $a_n \leqslant n^{-\delta}$ for n large enough; hence, by Proposition 5.7, $r(C) \leqslant 1/\delta$. Thus:

Proposition 5.10. *If* $\beta = EV(C)$ *or* $\beta = EW(C)$, $\beta < -1$, *then*

$$r(C) \leqslant -1/(\beta + 1).$$

Suppose given a compact Banach ball C in H for which $EW(C)$ is defined and equals

$$\lim_{n\to\infty} (\log W_n)/(n \log n)$$

(not just lim sup). Let $\{F_n\}$ and $\{a_n\}$ be adapted sequences of subspaces and numbers, respectively, and $\{\varphi_n\}$ an adapted orthonormal basis. Let A be the linear transformation such that

$$A(\varphi_n) = b_n\varphi_n \qquad \text{for all } n,\ b_n \downarrow 0.$$

Then the F_n and φ_n are adapted to $A(C)$, F_n now being uniquely determined, and

$$\begin{aligned} a_n(A(C)) &= b_n a_n, \\ W_n(A(C)) &= b_1 \cdots b_n W_n(C). \end{aligned}$$

Thus

$$EW(A(C)) = EW(C) + ew(\{b_j\}) \tag{5.11}$$

where

$$ew(\{b_j\}) = \limsup_{n\to\infty} \left(\sum_{j=1}^{n} \log b_j\right)\Big/ n \log n.$$

Thus the following is useful.

PROPOSITION 5.12. *If* $b_j \downarrow 0$,

$$ew(\{b_j\}) = -1/\lambda(\{b_j\}).$$

Proof. Given $\delta > 0$, we have by (5.2):

$$n(\epsilon) = n(\{b_j\}, \epsilon) \leqslant 1/\epsilon^{\lambda+\delta}$$

for ϵ small enough, and $n(\epsilon) \geqslant 1/\epsilon^{\lambda-\delta}$ for arbitrarily small $\epsilon > 0$. Now if $n = n(\epsilon)$,

$$\left(\sum_{j=1}^{n} \log b_j\right)\Big/ n \log n \geqslant (\log \epsilon)/(\log n).$$

When $n \geqslant 1/\epsilon^{\lambda-\delta}$ and $0 < \epsilon < 1$,

$$\log n \geqslant (\lambda - \delta) \log (1/\epsilon) \qquad \text{and} \qquad (\log \epsilon)/\log n \geqslant -1/(\lambda - \delta).$$

Thus letting $\delta \downarrow 0$ we have

$$ew(\{b_j\}) \geqslant -1/\lambda.$$

For the converse inequality, we can assume $b_1 < 1$. For any positive integer m let $\epsilon = \epsilon(m)$ satisfy $m = \epsilon^{-\lambda-2\delta}$. Then as $m \to \infty$, $\epsilon \downarrow 0$. Since

$$n(\epsilon) \leqslant 1/\epsilon^{\lambda+\delta} < 1/\epsilon^{\lambda+2\delta}$$

for ϵ small enough,

$$\begin{aligned}\left(\sum_{j=1}^{m} \log b_j\right)\Big/ m \log m &< (m - n(\epsilon(m)))\,(\log \epsilon)\,\epsilon^{\lambda+2\delta}/(\lambda + 2\delta) \log (1/\epsilon)\\ &\leqslant (1 - \epsilon^{\delta})\,(\log \epsilon)/(\lambda + 2\delta) \log (1/\epsilon)\\ &= (-1 + \epsilon^{\delta})/(\lambda + 2\delta) \to -1/(\lambda + 2\delta),\end{aligned}$$

where $\epsilon = \epsilon(m)$, $m \to \infty$. Thus, letting $\delta \downarrow 0$, we have

$$ew(\{b_j\}) \leqslant -1/\lambda.$$

Q.E.D.

6. Simple Subsets of Hilbert Space

In this section we study symmetric rectangular solids, ellipsoids, and "octahedra" and determine when they are GC- and GB-sets. We also study certain "full approximation sets" (see [14]), which are maximal sets with a given adapted sequence $\{a_n\}$, while octahedra are (among the) minimal sets.

For each class we shall have sequences $\{b_n\}$ of real numbers, $b_n \downarrow 0$, related to an orthonormal set $\{\varphi_n\}$ in H, usually complete. Let F_n be the subspace spanned by $\varphi_1, \dots, \varphi_n$. For any orthonormal set $\{\varphi_n\}$ and any $b_n \geqslant 0$ we define the ellipsoid

$$E = E(\{b_n\}) = E(\{b_n\}, \{\varphi_n\}) = \left\{ \sum_{b_n > 0} x_n \varphi_n : \sum_{b_n > 0} x_n^2 / b_n^2 \leqslant 1 \right\}.$$

Clearly, E is compact if and only if the b_n for $b_n > 0$ can be arranged into a sequence $b_n \downarrow 0$. Then the $\{\varphi_n\}$, $\{F_n\}$ and $\{b_n\}$ are adapted to E. (The F_n are uniquely determined unless some positive b_n are equal, and the b_n are unique.)

More abstractly, we can define a compact ellipsoid as an image $A(B_1)$ of the unit ball $B_1 = \{x : \| x \| \leqslant 1\}$ in H under a compact operator A.[2]

It follows that if E is a compact ellipsoid and S is a bounded linear transformation from H into itself, then $S(E)$ is a compact ellipsoid.

Lemma 6.0. *If $E = E(\{b_n\}, \{\varphi_n\})$ is a compact ellipsoid and P is a* f.d.p.,

$$P(E) = E(\{\beta_n\}, \{\psi_n\}), \qquad b_n \downarrow 0, \qquad \beta_n \downarrow 0,$$

then $\beta_n \leqslant b_n$ for all n.

Proof. We may assume the $\{\varphi_n\}$ are complete. Given n let G_n be the linear span of $\psi_1, \dots, \psi_n$. G_n has at least one-dimensional intersection with the set of vectors u orthogonal to $P(\varphi_j)$, $j = 1, \dots, n-1$. If also $u \in P(E)$ then $\| u \| \leqslant \| Pv \|$ for some $v \in E(\{b_j\}, \{\varphi_j\}_{j \geqslant n})$, so $\| u \| \leqslant b_n$ and hence $\beta_n \leqslant b_n$. Q.E.D.

Now we find the exponents of volume of ellipsoids.

Proposition 6.1.

$$EV(E) = EW(E) = -\frac{1}{2} - \frac{1}{\lambda(\{b_n\})}.$$

[2] For the equivalence of the definitions, see R. T. Prosser. The ϵ-entropy and ϵ-capacity of certain time-varying channels. *J. Math. Anal. Appl.* **16** (1966), 553–573.

Proof. Lemma 6.0 implies that

$$V_n = W_n = c_n b_1 b_2 \cdots b_n \qquad \text{for all } n.$$

Let B be the unit ball $E(\{1\})$ in H. Then

$$V_n(B) \equiv W_n(B) \equiv c_n .$$

Using (5.11) and (5.12) the proof is complete.

PROPOSITION 6.2. *For any compact ellipsoid $E = E(\{b_n\})$,*

$$r(E) = \lambda(\{b_n\}).$$

Thus if $EV(E) < -\frac{1}{2}$, E is volumetric.

Proof.[3] We have $r \geqslant \lambda$ by Propositions 5.8 and 6.1, and $r \leqslant \lambda$ by Proposition 5.7. The second conclusion follows then from 6.1 and the definition of "volumetric" (just before Conjecture 5.9).

PROPOSITION 6.3. *The following are equivalent:*

(a) $E = E(\{b_n\})$ *is a* GC-*set*

(b) E *is a* GB-*set*

(c) $\sum_{n=1}^{\infty} b_n{}^2 < \infty$ *(E is a "Schmidt ellipsoid").*

Proof. (a) implies (b) clearly if E is compact; if not, both fail. If (b) holds, and A is the linear operator such that $A(\varphi_n) = b_n \varphi_n$, $L \circ A$ has a version continuous on H (Theorem 4.3(e) above). It is known that this is true if and only if A is a Hilbert-Schmidt operator (see [*8*], Lemma 4, p. 344). Thus (b) and (c) are equivalent. Next, assume (c). Then for some $k_n \uparrow \infty$, $\sum k_n{}^2 b_n{}^2 < \infty$. Let $E^1 = E\,\{k_n b_n\})$. Then E is E^1-compact and not maximal, so by Theorem 4.7, E is a GC-set. Q.E.D.

It follows immediately from the above results that Conjectures 3.3, 5.4, and 5.9 all hold for ellipsoids.

Now we turn to our second class of examples. Let $\{F_n\}$ be an increasing sequence of subspaces of H with F_n n-dimensional, $n = 0, 1, 2, \ldots$. Let $b_n \downarrow 0$. Specializing [*14*], we define the *full approximation set* $A = A(\{b_n\})$ as

$$\{x : \text{for all } n,\ \| x - y_n \| \leqslant b_n \text{ for some } y_n \text{ in } F_n\}.$$

[3] $r = \lambda$ is also proved by Prosser; see *op. cit.* in previous footnote.

It is easy to see that $A \supset E(\{b_n\})$. Also we can choose y_n in $A \cap F_n \equiv A_n$. Hence $\{b_n\}$ is adapted to A and A is simply a maximal set having $\{b_n\}$ as an adapted sequence.

PROPOSITION 6.4. *$r(A) = \lambda(\{b_n\})$. If $EV(A) < -\frac{1}{2}$ then A is volumetric.*

Proof. Since $A \supset E(\{b_n\})$, we have $r(A) \geqslant r(E) = \lambda(\{b_n\})$ by Proposition 6.2. $r \leqslant \lambda$ by Proposition 5.7, so $r = \lambda$.

If $EV(A) = -\frac{1}{2} - \delta$, $\delta > 0$, then $EV(E(\{b_n\})) \leqslant -\frac{1}{2} - \delta$ so

$$\begin{aligned} r(A) &= \lambda(\{b_n\}) = r(E) = -2/(1 + 2EV(E)) \leqslant 1/\delta \\ &= 1/(-\tfrac{1}{2} - EV(A)) = -2/(1 + 2EV(A)). \end{aligned}$$

The converse inequality holds by Proposition 5.8 (a), so A is volumetric. Q.E.D.

Note that the ellipsoid E with same parameters $\{b_n\}$, included in A, also has the same exponent of entropy and the same exponent of volume if that of either is less than $-\frac{1}{2}$. We have proved Conjectures 5.9 and (hence) 5.4 for A. Conjecture 3.3 also holds since if $r(A) > 2$ then $r(E) > 2$ and 3.3 holds for ellipsoids.

The condition $\sum b_n^2 < \infty$ is clearly necessary for $A(\{b_n\})$ to be a GB-set but I don't know whether it is sufficient for A to be a GC-set or GB-set.

Next we consider the rectangular solid or "block"

$$B = B(\{b_n\}) = \left\{ \sum_{n=1}^{\infty} x_n \varphi_n : | x_n | \leqslant b_n , n = 1, 2, \ldots \right\} .$$

We assume as usual $b_n \downarrow 0$ and, to assure $B \subset H$, $\sum b_n^2 < \infty$. (Since $B(\{b_n\}) \supset E(\{b_n\})$, no GB-sets are lost here.) For blocks we shall not find adapted subspaces, but we shall characterize GC-blocks and GB-blocks and verify the three conjectures.

PROPOSITION 6.5. *If $\lambda = \lambda(\{b_n\})$, $t = EV(B(\{b_n\}))$, and $r = r(B)$, then $t = -1/\lambda = -\frac{1}{2} - 1/r$ if any of these terms is less than $-\frac{1}{2}$ (i.e., if $t < -\frac{1}{2}$, $\lambda < 2$ or $r < \infty$). Thus under these conditions B is volumetric.*

Proof. Given $\delta > 0$, we have for n large enough

$$b_n \leqslant V_n^{1/n} < n^{t+\delta},$$

so $\lambda \leqslant -1/t$ if $t < 0$. Thus by 5.8 (b), any of our hypotheses implies $\lambda < 2$.

Then by 5.2 we have for n large enough

$$b_n \leqslant n^{-1/(\lambda+\delta)}$$

so for $\delta < 2 - \lambda$ we have for k large

$$\left(\sum_{n=k}^{\infty} b_n^2\right)^{1/2} \leqslant k^{1/2-1/(\lambda+\delta)}$$

so letting $\delta \downarrow 0$ we have by Proposition 5.7

$$r \leqslant 1/(-\tfrac{1}{2} + 1/\lambda) < \infty$$

so by 5.8 $t < -\frac{1}{2}$ and $r \geqslant -2/(1+2t)$, so

$$t \leqslant -\frac{1}{2} - \frac{1}{r} \leqslant -\frac{1}{\lambda} \leqslant t. \qquad \text{Q.E.D.}$$

Thus Conjectures 5.4 and 5.9 hold for blocks.

PROPOSITION 6.6. *The following are equivalent:*

(a) $\sum b_n \mid L(\varphi_n) \mid$ *converges with probability 1;*

(b) $\sum b_n < \infty$*;*

(c) $B = B(\{b_n\})$ *is included in some* GC-*ellipsoid;*

(d) B *is a* GC-*set;*

(e) B *is a* GB-*set.*

Proof. (a) implies (b) by an application of the three-series theorem ([*13*], p. 237).

If $\sum b_n < \infty$, we let

$$a_n = \left(b_n \sum_{j=1}^{\infty} b_j\right)^{1/2}.$$

Then $E(\{a_n\})$ is a GC-ellipsoid by 6.3, and $B \subset E$, so (b) implies (c). Clearly (c) implies (d) which implies (e).

If B is a GB-set, then for almost every ω, there is an $M < \infty$ such that

$$\sum_{j=1}^{\infty} s_j b_j L(\varphi_j)(\omega) \leqslant M$$

for all possible choices of $s_j = \pm 1$. Hence (a) holds, and the proof is complete.

Now if a block B is a GB-set, then $r(B) \leqslant 2$ by (c) so Conjecture 3.3 holds for blocks.

If $r(B) < 2$ and E is the ellipsoid of (c), then it is easily shown that $r(B) < r(E) < 2$.

Next we discuss some other classes of subsets of H: orthogonal sets $S(\{b_n\})$ and their closed symmetric convex hulls, octahedra $\mathrm{Oc}\,(\{b_n\})$. These sets refute a number of conjectures which up to now might have seemed plausible (cf. Propositions 6.7, 6.9, 6.10, and the remarks between and after them) while satisfying Conjectures 3.3, 5.4, and 5.9.

Given $b_n \downarrow 0$ and $\{\varphi_n\}$ an orthonormal basis let

$$S = S(\{b_n\}) = \{0\} \cup \{b_n \varphi_n\}_{n=1}^{\infty},$$

$$\mathrm{Oc} = \mathrm{Oc}\,(\{b_n\}) = \text{symmetric closed convex hull of } S$$

$$= \left\{ \sum_{n=1}^{\infty} b_n x_n \varphi_n : \sum_{n=1}^{\infty} | x_n | \leqslant 1 \right\}.$$

In this case, as for ellipsoids but not blocks, the $\{\varphi_n\}$ and $\{b_n\}$ are adapted to $\mathrm{Oc}\,(\{b_n\})$. It is easy to see that

$$N(\mathrm{Oc}, \epsilon) \geqslant N(S, \epsilon) = n(\{b_n\}, \sqrt{2}\epsilon) \pm \tfrac{1}{2} + \tfrac{1}{2}$$

for all $\epsilon > 0$ such that $b_n = \epsilon$ for at most one value of n.

Proposition 6.7. *The following are equivalent:*

(a) $\mathrm{Oc}\,(\{b_n\})$ *is a* GC-*set;*

(b) $S(\{b_n\})$ *is a* GC-*set;*

(c) $b_n = o\,(\log n)^{-1/2}$.

Proof. Clearly (a) $\Rightarrow$ (b). To prove the converse, note that (b) is equivalent to $b_n L(\varphi_n) \to 0$ as $n \to \infty$ with probability one. Given $\epsilon > 0$, for almost all ω there is an N such that for all $n > N$,

$$| b_n L(\varphi_n)(\omega) | < \epsilon/4,$$

and there is a $\delta > 0$ such that whenever $x, y \in \mathrm{Oc}\,(\{b_n\})$ and $\| x - y \| < \delta$,

$$\left| L\left(\sum_{n=1}^{N} (x_n - y_n) \varphi_n \right) \right| < \epsilon/2,$$

and we infer (a).

Now (b) is equivalent by the zero-one law ([*13*] A, p. 228) to the following: for any $\epsilon > 0$,

$$\sum_{n=1}^{\infty} \Pr (b_n \mid L(\varphi_n) \mid \geqslant \epsilon) < \infty,$$

or

$$\sum_{n=1}^{\infty} \int_{\epsilon/b_n}^{\infty} e^{-x^2/2}\, dx < \infty.$$

As is well known, an integration by parts shows that as $M \to \infty$,

$$\int_M^{\infty} e^{-x^2/2}\, dx \qquad \text{is asymptotic to} \qquad e^{-M^2/2}/M.$$

Thus (b) is equivalent to

$$\sum_{n=1}^{\infty} b_n \exp (- \epsilon^2/2b_n^2) < \infty.$$

Letting $b_n = \alpha_n (\log n)^{-1/2}$, $n \geqslant 2$, we obtain the series

$$\sum_{n=2}^{\infty} \alpha_n (\log n)^{-1/2}\, n^{-\epsilon^2/2\alpha_n^2}. \tag{6.8}$$

If (c) holds, i.e., $\alpha_n \to 0$, then the terms of (6.8) become less than n^{-2} for n large, so (b) holds.

Conversely suppose (c) is false, so that for some $\delta > 0$, $\alpha_n \geqslant \delta$ for arbitrarily large values of n. For such an n and $n^{1/2} \leqslant j \leqslant n$, we have

$$\begin{aligned} \alpha_j &= b_j (\log j)^{1/2} \geqslant b_n (\log j)^{1/2} \\ &= \alpha_n (\log j/\log n)^{1/2} \geqslant \alpha_n/2. \end{aligned}$$

Letting $\epsilon = \delta/2$ we then have

$$\epsilon^2/2\alpha_j^2 \leqslant 4\epsilon^2/2\alpha_n^2 \leqslant \tfrac{1}{2},$$

$$\begin{aligned} \sum_{n^{1/2} \leqslant k \leqslant n} b_j\, j^{-\epsilon^2/2\alpha_j^2} &\geqslant \sum_{n^{1/2} \leqslant j \leqslant n} \delta (j \log n)^{-1/2}/2 \\ &\geqslant \delta (n - n^{1/2} - 1)/2n^{1/2} \log n \to \infty \end{aligned}$$

as $n \to \infty$ (recall that δ is independent of n). Thus (6.8) diverges and (b) fails, so (b) $\Rightarrow$ (c). Q.E.D.

PROPOSITION 6.9. *The following are equivalent:*

(a) $\mathrm{Oc}\,(\{b_n\})$ *is a* GB-*set;*

(b) $S(\{b_n\})$ *is a* GB-*set;*

(c) $b_n = O((\log n)^{-1/2})$.

Proof. We use some notation and results of the previous proof. (By the way, note that Theorem 4.7 and either of 6.7 and 6.9 make the other at least very plausible.) Here the equivalence of (a) and (b) is obvious. (b) is equivalent to the statement that for some $M > 0$, $b_n \mid L(\varphi_n) \mid < M$ for n sufficiently large, with probability 1, or that (6.8) converges for $\epsilon = M$. If (c) holds, i.e., if for some $N > 0$, $\mid \alpha_n \mid \leqslant N$ for all n, we can let $M = 2N$ and infer (b). If $\{\alpha_n\}$ is unbounded, then given M we choose n so that $\alpha_n \geqslant 2M$. Then $\alpha_j \geqslant M$ for $n^{1/2} \leqslant j \leqslant n$,

$$\sum_{n^{1/2} \leqslant j \leqslant n} M(n \log n)^{-1/2} \geqslant (n - n^{1/2} - 1)\, M(n \log n)^{-1/2}.$$

Since n can be chosen arbitrarily large, (6.8) diverges for $\epsilon = M$ for all $M > 0$. Thus (b) implies (c). Q.E.D.

We infer from Propositions 6.3 and 6.7 that a GC-set, $\mathrm{Oc}\,(\{1/\log n\})$, is not included in any GB-ellipsoid, since

$$\sum_{n=2}^{\infty} 1/(\log n)^2 = +\infty$$

(see [*15*], Lemma 2).

We next show that the GC- and GB-properties are not monotone functions of the "size" of a set as measured by volumes V_n or by ϵ-entropy.

PROPOSITION 6.10. *There exist a* GC-*set* $\mathrm{Oc} = \mathrm{Oc}\,(\{a_n\})$ *and a non*-GB-*ellipsoid* $E = E(\{b_n\})$ *such that*

(a) $H(E, \epsilon)/H(\mathrm{Oc}, \epsilon) \to 0$ *as* $\epsilon \downarrow 0$,

(b) $V_n(E)/V_n(\mathrm{Oc}) \to 0$ *as* $n \to \infty$.

Proof. We let $a_n = \alpha_n(\log n)^{-1/2}$, $n \geqslant 2$, where $\alpha_n \downarrow 0$ sufficiently slowly; for definiteness we can let $\alpha_n = (\log\log n)^{-1/4}$, $n \geqslant 3$. Let

$$b_n = (n \log n \log\log n)^{-1/2}, \qquad n \geqslant 3.$$

Then Oc is a GC-set and E is not a GB-set. $V_n(E)$ is asymptotic to a constant times

$$\left(\frac{2\pi e}{n}\right)^{n/2} n^{-1/2} \left(\frac{e}{n}\right)^{n/2} n^{-1/4} \prod_{j=3}^{n} (\log j \log \log j)^{-1/2},$$

and for n large,

$$V_n(\mathrm{Oc}) \geqslant \left(\frac{e}{n}\right)^n (3\pi n)^{-1/2} \prod_{j=2}^{n} \alpha_j (\log j)^{-1/2},$$

Thus there is a $K > 0$ such that for n large,

$$V_n(E)/V_n(\mathrm{Oc}) \leqslant K \prod_{j=3}^{n} (4\pi^2/\log \log j)^{1/4},$$

which implies (b).

To prove (a) it suffices to show that

$$H(E(\{(n \log n)^{-1/2}\}), \epsilon)/H(S(\{a_n\}), \epsilon) \to 0$$

as $\epsilon \downarrow 0$. Let $S = S(\{a_n\})$.

Given $\epsilon > 0$, let

$$N(S, \epsilon) = n = n(\{a_j\}, \sqrt{2}\epsilon) + \tfrac{1}{2} \pm \tfrac{1}{2}\,.$$

Because of the slow growth of the logarithms, this implies that, for ϵ small enough,

$$1/9\epsilon^4 \leqslant (\log n)^2 \log \log n \leqslant 1/\epsilon^4,$$
$$\log n \leqslant 1/\epsilon^2, \qquad \log \log n \leqslant 2 \log (1/\epsilon),$$
$$H(S, \epsilon) = \log n \geqslant 1/5\epsilon^2(\log (1/\epsilon))^{1/2}.$$

To estimate $N(E, \epsilon)$ from above we take the smallest integer n such that

$$(n \log n)^{-1/2} \leqslant \epsilon/2, \qquad \text{i.e.,} \qquad n \log n \geqslant 4/\epsilon^2.$$

For ϵ small enough this implies $n \log n < 5/\epsilon^2$. Now

$$N(E, \epsilon) \leqslant N(E_n\,, \epsilon/2) \leqslant N(B_n\,, \epsilon/2)$$

where

$$E_n = E(\{\beta_j\}), \qquad B_n = B(\{\beta_j\}),$$

$$\beta_j = (j \log j)^{-1/2}, \qquad j = 2, \ldots, n, \qquad \beta_j = 0, \qquad j > n.$$

By Lemma 5.6, for ϵ small and hence for n large enough,

$$N(B_n, \epsilon/2) \leqslant \prod_{j=2}^{n} (2 + n^{1/2}(j \log j)^{-1/2}/\epsilon)$$

$$\leqslant 3^n n^{n/2} \epsilon^{-n} (n!)^{-1/2}.$$

(*Note:* the logarithms have served to make n smaller, but they are no longer needed.)
For n large we have $n! \geqslant (n/e)^n$, so

$$N(E, \epsilon) \leqslant (3e/\epsilon n^{1/2})^n = \exp \{n[\log 3 + 1 + \log (1/\epsilon) - \tfrac{1}{2} \log n]\}.$$

Since $n < 5/\epsilon^2$, we have, for ϵ small enough,

$$\log n < \log 5 + 2 \log (1/\epsilon) < 3 \log (1/\epsilon),$$
$$n \geqslant 4/\epsilon^2 \log n > 4/3\epsilon^2 \log (1/\epsilon),$$
$$\log n > \log (4/3) + 2 \log (1/\epsilon) - \log \log (1/\epsilon),$$
$$H(E, \epsilon) \leqslant 5[3 + \log \log (1/\epsilon)]/\epsilon^2 \log (1/\epsilon).$$

Thus $H(E, \epsilon)/H(S, \epsilon) \to 0$ as $\epsilon \downarrow 0$. Q.E.D.

Suppose given a sufficient condition that a set C be a GC-set, asserting that $H(C, \epsilon)$ is sufficiently small (e.g., Theorem 2.1) or that the $V_n(C)$ are sufficiently small (e.g., Proposition 5.5, Conjecture 5.4). Then the GC-octahedron of Proposition 6.10 will never satisfy such a condition since the ellipsoid does not. Hence no such sufficient conditon can be necessary.

In the converse direction, likewise, a sufficient condition for a Banach ball *not* to be a GB-set such as Theorem 5.3 or Conjecture 3.3 cannot be necessary.

One may, however, seek "best possible" conditions of the given kinds. In the four cases, Theorem 3.1 has a fairly strong claim to be best (see the next section). Theorem 5.3 has a weaker claim. Conjectures 3.3 and 5.4, if they are true, could probably be improved upon.

The volume of the n-dimensional octahedron

$$\left\{\sum_{j=1}^{n} x_j \varphi_j : \sum_{j=1}^{n} | x_j | \leqslant 1\right\}$$

is $2^n/n!$, which is asymptotic to $(2e)^n/n^n(2\pi n)^{1/2}$ by Stirling's formula. Thus by 5.11 and 5.12

$$EW(\text{Oc}(\{b_n\})) = -1 - 1/\lambda(\{b_n\}). \tag{6.11}$$

A sequence $b_n \downarrow 0$ such that $\lambda(\{b_n\}) < \infty$ is $o((\log n)^{-1/2})$ (cf. end of the proof of Proposition 6.7). Thus conjecture 5.4 holds for octahedra. The next proposition implies that conjecture 5.9 also holds for octahedra.

PROPOSITION 6.12. *Let* $\lambda = \lambda(\{b_n\})$, $s = EV(\text{Oc}(\{b_n\}))$, $r = r(\text{Oc})$. *Then* $r = -2/(2s+1) = 2\lambda/(2+\lambda)$ *if any of these terms is less than 2 (i.e., if* $r < 2$, $s < -1$, *or* $\lambda < \infty$*). Thus under these conditions* Oc *is volumetric.*

Proof. $r \geqslant -2/(2s+1)$ in general by 5.8 (a). If $s < 1$, then $\lambda < \infty$ and $s \geqslant -1 - 1/\lambda$ by 6.11. Thus any of the hypotheses implies $\lambda < \infty$, and then $1/\lambda \geqslant -1 - s$,

$$2\lambda/(2+\lambda) = 2/((2/\lambda)+1) \leqslant -2/(2s+1)$$

if $s < -\frac{1}{2}$. It will now suffice to show that if $\lambda < \infty$,

$$r \leqslant 2/((2/\lambda)+1)$$

(since then $r < 2$ and $s < -1 < -\frac{1}{2}$).

Let $0 < \gamma < 1/\lambda$. Then for n large enough, $b_n < 1/n^\gamma$ by 5.2. Thus for some $K > 0$, $\text{Oc}(\{b_n\}) \subset KC_\gamma$ where $C_\gamma = \text{Oc}(\{1/n^\gamma\})$, and $r(\text{Oc}) \leqslant r(KC_\gamma) = r(C_\gamma)$. Thus it is enough to prove that $r(C_\gamma) \leqslant 2/(1+2\gamma)$.

For x in C_γ, we have

$$x = \sum x_j \varphi_j / j^\gamma, \qquad \sum |x_j| \leqslant 1.$$

Given $\epsilon > 0$, let $A(x)$ be the set of all j such that

$$|x_j| > j^{2\gamma}\epsilon^2/4.$$

Then the number m of integers in $A(x)$ satisfies

$$m^{1+2\gamma}/(1+2\gamma) = \int_0^m x^{2\gamma}\, dx \leqslant \sum_{j=1}^m j^{2\gamma} \leqslant 4/\epsilon^2.$$

Let $\alpha < \beta < \gamma$. Then for some $c(\gamma)$,

$$m \leqslant c(\gamma)/\epsilon^{2/(1+2\gamma)} < \epsilon^{-2/(1+2\beta)}$$

for ϵ small enough. (Of course m depends on γ and ϵ).

The largest integer N in $A(x)$ is at most $(2/\epsilon)^{1/\gamma}$. Thus the number of possible choices of $A(x)$ is at most

$$\binom{N}{m} < N^m < \exp\left(c(\gamma)\log(2/\epsilon)/\gamma\epsilon^{2/(1+2\gamma)}\right)$$
$$< \exp\left(\epsilon^{-2/(1+2\beta)}\right)$$

for ϵ small enough. For any x in C_γ,

$$\left\| x - \sum_{j\in A(x)} x_j\varphi_j/j^\gamma \right\| = \Big(\sum_{j\notin A(x)} x_j^2/j^{2\gamma}\Big)^{1/2} \leqslant \max\{|x_j|/j^{2\gamma} : j \notin A(x)\}^{1/2} \leqslant \epsilon/2.$$

Thus

$$N(C_\gamma, \epsilon) \leqslant \sum_A N(C_\gamma(A), \epsilon/2)$$

where the sum is over the possible sets $A = A(x)$ and $C_\gamma(A)$ is the set of all sums

$$\sum_{j\in A(x)} x_j\varphi_j/j^\gamma, \qquad \sum |x_j| \leqslant 1.$$

Here we use a crude estimate from Lemma 5.6 to obtain for ϵ small enough

$$N(C_\gamma, \epsilon) \leqslant \exp\left(\epsilon^{-2/(1+2\beta)}\right)(3m^{1/2}/\epsilon)^m$$
$$\leqslant \exp\left(\epsilon^{-2/(1+2\alpha)}\right).$$

Thus $r(C_\gamma) \leqslant 2/(1+2\alpha)$. Letting $\alpha \uparrow \beta \uparrow \gamma$ we infer $r(C_\gamma) \leqslant 2/(1+2\gamma)$. Q.E.D.

By Proposition 6.9, to prove Conjecture 3.3 for octahedra its uffices to prove the following, where $a_n = (\log n)^{-1/2}$, $n \geqslant 2$.

PROPOSITION 6.13. $r(\mathrm{Oc}(\{a_n\})) = 2$.

Proof. $r \geqslant 2$ since this Oc is not a GC-set (Corollary 3.2, Proposition 6.7), or by volumes (5.8 (a) and 6.11).

To prove $r \leqslant 2$ we shall use the method of the previous proof with some additional complications. Let $\epsilon > 0$ and $\delta > 0$. Given x in Oc let $A(x)$ be the set of all j such that

$$|x_j| > \epsilon^2/4a_j^2 = (\epsilon^2 \log j)/4, \qquad j \geqslant 2.$$

Then (for ϵ small enough) $A(x)$ has at most $4/\epsilon^2$ elements. The largest possible integer n in $A(x)$ satisfies $n \leqslant \exp(4/\epsilon^2)$. For any x in Oc

$$\left\| x - \sum_{j \in A(x)} x_j a_j \varphi_j \right\| \leqslant (\max_{j \notin A(x)} |x_j| a_j^2)^{1/2} \leqslant \epsilon/2.$$

Let $Q(\epsilon)$ be the number of possible sets $A(x)$ for a given $\epsilon > 0$. Then by Lemma 5.6,

$$N(\mathrm{Oc}, \epsilon) \leqslant Q(\epsilon)\,(6/\epsilon^2)^{4/\epsilon^2} < Q(\epsilon) \exp(\epsilon^{-2-\delta})$$

for ϵ small enough.

(The estimate $Q(\epsilon) \leqslant n^{4/\epsilon^2} \leqslant \exp(16/\epsilon^4)$ is clearly inadequate.) Let s be a positive integer such that $1/s < \delta$. For $r = 0, 1, \ldots, s - 1$, let

$$Z_{rs} = \{ j : 4\epsilon^{-2r/s} \leqslant \log j < 4\epsilon^{-2(r+1)/s} \}.$$

If $j \in A(x) \cap Z_{rs}$, then

$$| x_j | \geqslant \epsilon^{2(s-r)/s},$$

so the number of elements of $A(x) \cap Z_{rs}$ is at most $\epsilon^{2(r-s)/s}$. Thus the number of ways of choosing $A(x) \cap Z_{rs}$ is at most

$$[\exp(4\epsilon^{-2(r+1)/s})]^{\epsilon^{2(r-s)/s}} = \exp[4\epsilon^{-2(r+1)/s}\,\epsilon^{2(r-s)/s}] \leqslant \exp(\epsilon^{-2(1+\delta)}).$$

Thus for ϵ small enough

$$Q(\epsilon) \leqslant 2e^4 \exp(s\epsilon^{-2(1+\delta)}) \leqslant \exp(\epsilon^{-2-3\delta}),$$

and

$$N(\mathrm{Oc}, \epsilon) \leqslant \exp(\epsilon^{-2-5\delta}).$$

Letting $\delta \downarrow 0$ we get $r(\mathrm{Oc}) \leqslant 2$. Q.E.D.

Next we show that $EW(C)$ may be strictly smaller than $EV(C)$. Let

$$C = \mathrm{Oc}(\{2/(2n + 1)\}) \times E(\{1/n\}),$$

a Banach ball in $H \times H$ which of course is a separable Hilbert space. Then subspaces adapted to C are uniquely determined, with

$$a_{2n} = 2/(2n + 1), \qquad a_{2n+1} = 1/(n + 1).$$

It follows easily that $EW(C) = -7/4$. Taking projections of the ellipsoid only we get $EV(C) \geqslant -3/2$. By 5.8 (a), 6.2, and 6.12 we obtain $r(C) = 1$, $EV(C) = -3/2$. Thus in measuring volumes it seems better to use EV primarily, as we have done, rather than EW, since, e.g., Conjecture 5.9 is false if EV is replaced by EW, and r and EW are no functions of each other over a reasonable range.

We have not evaluated $EV(\mathrm{Oc}\,\{b_n\})$ if $\lambda(\{b_n\}) = +\infty$, although then for $\{b_n\}$ bounded we have $EW(\mathrm{Oc}) = -1$. Thus it is conceivable that Conjecture 5.9 could hold even for $EV < -\frac{1}{2}$, but it seems unlikely.

7. Processes on Euclidean Spaces

In this section we apply Theorem 3.1 to Gaussian processes over a finite-dimensional Euclidean parameter set, e.g., the usual one dimensional "time". Conjecture 3.3 is also verified in certain cases. Since any compact Banach ball is a continuous image of the unit interval,[4] our hypotheses in general do not restrict the geometry of the Banach balls in H which arise, and we do not try to evaluate their volumes.

THEOREM 7.1 (Fernique [7], [7a] for $T =$ cube). *Suppose* $\{x_t, t \in T\}$ *is a Gaussian process where* T *is a bounded subset of* R^k. *Suppose* φ *is a nonnegative real-valued function such that*

(a) $E\,|\,x_s - x_t\,|^2 \leqslant \varphi(|\,s - t\,|)^2$ *for all* $s, t \in T$,

(b) $\varphi(u)$ *is monotone-increasing on some interval* $0 < u < \alpha$,

(c) $\int_M^\infty \varphi(e^{-x^2})\,dx < \infty$ *for some* $M < \infty$.

Then x_t *is sample-continuous.*

Proof. Let C be the set of all x_t, $t \in T$, $C \subset H$. We shall prove that C is a GC-set and hence, since x_t is continuous from T into H by (a) and (c), that x_t has a continuous version.

Since T is bounded, there is an $A > 0$ such that

$$N(T, \delta) \leqslant A/\delta^k \qquad \text{for all} \qquad \delta > 0$$

(see [*12*], Section 3, I, p. 20; cf. also Lemma 5.6 above). (b) and (c) imply $\varphi(\delta) \downarrow 0$ as $\delta \downarrow 0$.

For any $\epsilon > 0$ let

$$\delta \equiv \Psi(\epsilon) \equiv \sup\{t : \varphi(t) < \epsilon\}.$$

(If φ is continuous and ϵ is small enough, $\delta = \varphi^{-1}(\epsilon)$.)

---

[4] See, for example, K. Kuratowski, "Topologie" (3rd ed.), Vol. II, Chapter VII, Section 45. Warszawa, 1961.

Let $\delta_n = \Psi(1/2^n)$, defined and positive for n large enough (unless $\varphi \equiv 0$, in which case the conclusion is trivial). Then $\delta_n \downarrow 0$ as $n \to \infty$. Now

$$N(C, 1/2^n) \leqslant A/\delta_n{}^k,$$

so

$$H(C, 1/2^n) \leqslant \log A + k \log (1/\delta_n).$$

Let $x_n = (\log (1/\delta_n))^{1/2}$. By Theorem 3.1 it suffices to prove that

$$\sum_{n=1}^{\infty} x_n/2^n < \infty.$$

(Note how the dimension becomes irrelevant.)
Now

$$\varphi(e^{-x^2}) \geqslant 1/2^n \qquad \text{for} \qquad x_{n-1} \leqslant x < x_n\,,$$

so

$$\int_{x_N}^{\infty} \varphi(e^{-x^2})\, dx \geqslant \sum_{n=N}^{\infty} (x_{n+1} - x_n)/2^{n+1} = \sum_{n=N+1}^{\infty} x_n/2^n - \sum_{m=N}^{\infty} x_m/2^{m+1}$$

$$= \tfrac{1}{2} \sum_{n=N+1}^{\infty} x_n/2^n - x_N/2^{N+1},$$

so the required series converges. Q.E.D.

Fernique [7] shows that Theorem 7.1 is optimal of its kind in a sense, even for $k = 1$, since if

$$\int^{\infty} \varphi(e^{-x^2})\, dx = +\infty$$

and φ satisfies some additional mild monotonicity assumptions, then counterexamples to sample continuity exist. However, note that we may take a process x_t on $T = [0, 1]$ satisfying the hypotheses of Theorem 7.1 and transform it by a "steep" homeomorphism f of T, e.g. $f(t) = 1/\log(1/t)$, into a process $x_{f(t)}$ which may no longer satisfy 7.1. (c) but of course is still sample-continuous. The ϵ-entropy of the range is unchanged, so Theorem 3.1 applies to $x_{f(t)}$ and has a broader range of applications. Note however that such a transformation destroys stationarity of the process, and for stationary processes Theorem 7.1 may be essentially the best possible.

It has been shown [4] that for T an interval, hypothesis (c) of Theorem 7.1 can be replaced by any of several conditions, of which the best ([4], p. 186, 3°) seems to be

$$\sum_{k=1}^{\infty} 2^{k/2}[\varphi(1/2^{2^k})]^{1/2} < \infty.$$

But this condition is easily shown to imply Fernique's.

Next we discuss random Fourier series and the work of Kahane [10]. Let $\{x_t, t \in R\}$ be a Gaussian process, stationary and periodic of period 2π. [*Note:* Fernique's counterexamples showing that Theorem 7.1 (c) cannot be improved are all of this type, so the additional hypotheses do not change that situation.) We assume x_t is continuous in probability and that $Ex_t \equiv 0$. It is then well known and not hard to prove that a version of x_t is given by

$$x_t(\omega) = \frac{\beta_0 \xi_0}{2} + \sum_{n=1}^{\infty} \beta_n(\xi_n \sin nt + \eta_n \cos nt), \tag{1''}$$

where the $\xi_i(\omega)$ and $\eta_j(\omega)$ are all independent, normalized Gaussian random variables and the β_n are nonnegative constants, $\sum \beta_n^2 < \infty$. (Conversely, any such series (1″) defines a process of the given type.) Kahane [10] assumes $\beta_0 = 0$, which does not affect the sample continuity.

Let

$$t_i^2 = \sum_{n=2^i+1}^{n=2^{i+1}} \beta_n^2.$$

(*Note:* t_i are not values of t!) Kahane ([10], p. 2, Théorèmes 3, 4) proves the

THEOREM. *The condition* $\sum_{i=1}^{\infty} t_i < \infty$ *is necessary for sample continuity or boundedness of* x_t *and, if the* t_i *are decreasing, also sufficient (even for almost sure uniform convergence of* (1″)).

Neither half of the above theorem will be proved here, and I doubt that the methods of this paper would give such a complete result. However, it will be shown that Conjecture 3.3 holds to the extent that Kahane's rather sharp result applies. Also we shall treat some additional cases where Kahane's theorem does not apply but the conjecture still holds.

PROPOSITION 7.2. *Suppose* $t_1 \geqslant t_2 \geqslant \cdots$ *and* $\sum t_i < \infty$. *Let* S *be the set of all* x_t *in* H. *Then* $r(S) \leqslant 2$.

Proof. We can restrict ourselves to $0 \leqslant t < 2\pi$. For any s and t in $(0, 2\pi)$,

$$E((x_s - x_t)^2) = E\left\{\left[\sum_{n=1}^{\infty} \beta_n \xi_n (\cos ns - \cos nt) + \beta_n \eta_n (\sin ns - \sin nt)\right]^2\right\}$$

$$= 2 \sum_{n=1}^{\infty} \beta_n^2 (1 - \cos (n(s - t))).$$

Let

$$\beta^2 = \sum_{n=1}^{\infty} \beta_n^2 = \sum_{i=1}^{\infty} t_i^2, \qquad b = \sum_{i=1}^{\infty} t_i .$$

Given $\epsilon > 0$, we choose a minimal $M(\epsilon)$ such that

$$\sum_{n=M+1}^{\infty} \beta_n^2 \leqslant \epsilon^2/8.$$

For all x, $1 - \cos x \leqslant x^2$, so if

$$| s - t | \leqslant \epsilon/2\sqrt{2}\beta M,$$

then

$$2 \sum_{n=1}^{M} \beta_n^2 (1 - \cos (n(s - t))) \leqslant \epsilon^2/4$$

Hence

$$N(S, \epsilon) \leqslant 2 \sqrt{2}\beta M/\epsilon + 1.$$

Now $M(\epsilon) \leqslant 2^i$ for the least i such that

$$\sum_{j=1}^{\infty} t_j^2 \leqslant t_i b \leqslant \epsilon^2/8.$$

For any $\delta > 0$,

$$n(\{t_i\}, \epsilon^2/8b) < (1/\epsilon^2)^{1+\delta}$$

for ϵ small enough by (5.2). Thus

$$M(\epsilon) \leqslant 2\epsilon^{-2(1+\delta)}$$

for ϵ small enough. Hence $r(S) \leqslant 2$. Q.E.D.

PROPOSITION 7.3. *If* $\sum \beta_n < \infty$, *then series* (1″) *converges uni-*

formly in t with probability 1, so x_t is sample-continuous. Then if X is the set of all x_t in H, $r(X) \leqslant 2$.

Proof. $\sum \beta_n(|\xi_n| + |\eta_n|)$ and hence (1″) converge uniformly in t by the three-series theorem. We represent X in H as follows: let $\{\varphi_n\}$ be an orthonormal basis, and

$$X = \left\{\sum_{n=1}^{\infty} \beta_n[(\cos nt)\,\varphi_{2n} + (\sin nt)\,\varphi_{2n-1}] : 0 \leqslant t < 2\pi\right\}.$$

Then $X \subset B(\{b_n\})$ where $\beta_n = b_{2n-1} = b_{2n}$, and $\sum b_n < \infty$. As remarked after Proposition 6.6, $r(B) \leqslant 2$, so $r(X) \leqslant 2$. Q.E.D.

For "lacunary" random Fourier series of the form

$$x_t(\omega) = \sum_{k=1}^{\infty} \beta_k \xi_k(\omega) \cos(n_k t),$$

where $n_{k+1}/n_k \geqslant \gamma > 1$ for all k, $\beta_k \geqslant 0$ and the ξ_k are independent normalized Gaussian random variables, it is easy to see that $\sum t_i < \infty$ implies $\sum \beta_k < \infty$. Thus Kahane's theorem and Proposition 7.3 together imply that Conjecture 3.3 holds for lacunary series (without any further monotonicity assumptions).

8. Comments on the Conjectures

Of the three Conjectures 3.3, 5.4, and 5.9, Conjecture 5.4 is supported by 5.9 which has nothing *a priori* to do with Gaussian processes. One might seek similar support for 3.3. But $r(C) > 2$ does not imply (for octahedra) that the $W_n(C)$ are too large to satisfy Theorem 5.3. However, Theorem 5.3 may not be the best possible, and the largest $V_n(C)$ and $W_n(C)$ I know for a GB-set C are those of $\mathrm{Oc}(\{K(\log n)^{-1/2}\})$, $K > 0$, namely

$$V_n(C) \geqslant W_n(C) = \frac{K^n}{n!} \prod_{j=2}^{n} (\log j)^{-1/2}.$$

The following approach to some of our problems might seem natural. Given $\epsilon > 0$ and $C_n \subset R^n$, let C_n^{ϵ} be the set of points within ϵ of C_n. Then

$$N(C_n, 2\epsilon) \leqslant \lambda_n(C_n^{\epsilon})/c_n \epsilon^n.$$

For C_n convex, $\lambda_n(C_n^{\epsilon})$ can be expressed in terms of "mixed volumes" of C_n (Bonnesen and Fenchel [3], Paragraph 29, p. 38; Paragraph 32, p. 49; Paragraph 38, p. 61). I believe some estimates of Santalò [16] are of this sort.

However, such estimates do not seem adequate for our purposes. Consider for example Oc $(\{\log n)^{-1/2}\})$. To estimate $N(\mathrm{Oc}, \epsilon)$ is more or less equivalent to estimating $N(\mathrm{Oc}_n, \epsilon/2)$ where Oc_n is the intersection of Oc with the span of $\varphi_1, \ldots, \varphi_n$, and n is approximately e^{4/ϵ^2}. Then every point of the boundary of Oc_n is at least $(n \log n)^{-1/2}$ from the origin, so

$$\lambda_n(\mathrm{Oc}_n^{\epsilon}) \geqslant c_n \epsilon^n (1 + 1/2n^{1/2})^n,$$

$$\lambda_n(\mathrm{Oc}_n^{\epsilon})/c_n \epsilon^n \geqslant \exp(\gamma \exp(2/\epsilon^2))$$

if $\gamma < \frac{1}{2}$ and n is large enough. Thus this method seems quite inferior to that used to prove Proposition 6.13, in this case, since it produces an extra exponentiation.

Acknowledgments

I am greatly indebted to Volker Strassen for the idea of introducing ϵ-entropy into the study of sample continuity of Gaussian processes, and for the statement of the result which now appears as Corollary 3.2.

Another main result, Theorem 5.3, is proved using L.A. Santalò's theorem [17] on volumes of convex symmetric sets and their polars. I thank G. D. Chakerian for telling me of Santalò's result via a network of mutual friends.

References

1. Bambah, R. P., Polar reciprocal convex bodies. *Proc. Cambridge Phil. Soc.* **51** (1955), 377–378.
2. Belyaev, Yu. K., Continuity and Hölder's conditions for sample functions of stationary Gaussian processes. *Proc. Fourth Berkeley Symp. Math. Stat. Prob.* **2** (1961), 23–34.
3. Bonnesen, T. and Fenchel, W., "Theorie der Konvexen Körper." Springer, Berlin, 1934.
4. Delporte, J., Fonctions aléatoires presque sûrement continues sur un intervalle fermé. *Ann. Inst. Henri Poincaré.* **B.I** (1964), 111–215.
5. Doob, J. L., "Stochastic Processes." Wiley, New York, 1953.
6. Dudley, R. M., Weak convergence of probabilities on non-separable metric spaces and empirical measures on Euclidean spaces. *Ill. J. Math.* **10** (1966), 109–126.
7. Fernique, Xavier, Continuité des processes Gaussiens, *Compt. Rend. Acad. Sci Paris* **258** (1964), 6058–60.

7a. Fernique, Xavier, Continuité de certains processus Gaussiens. Sém. R. Fortet, Inst. Henri Poincaré, Paris, 1965.

8. Gelfand, I. M. and Vilenkin, N. Ya., "Generalized Functions, Vol. 4: *Applications of Harmonic Analysis* (translated by Amiel Feinstein). Academic Press, New York, 1964.
9. Gross, L., Measurable functions on Hilbert space. *Trans. Am. Math. Soc.* **105** (1962), 372–390.
10. Kahane, J.-P., Propriétés locales des fonctions à séries de Fourier aléatoires, *Studia Math.* **19** (1960), 1–25.
11. Kelley, J. L., "General Topology." Van Nostrand, Princeton, New Jersey, 1955.
12. Kolmogorov, A. N. and Tikhomirov, V. M., ϵ-entropy and ϵ-capacity of sets in function spaces (in Russian), *Usp. Mat. Nauk* **14** (1959), 1–86. [English transl.: *Am. Math. Soc. Transl.* **17** (1961), 277–364.]
13. Loève, M., "Probability Theory" (2nd ed.). Van Nostrand, Princeton, New Jersey, 1960.
14. Lorentz, G. G., Metric entropy and approximation. *Bull. Am. Math. Soc.* **72** (1966), 903–937.
15. Minlos, R. A., Generalized random processes and their extension to measures, *Trudy Moskovsk. Mat. Obsc.* **8** (1959), 497–518. [English transl.: *Selected Transl. Math. Stat. Prob.* **3** (1963), 291–314.
16. Santalò, L. A., Acotaciones para la longitud de una curva o para el numero de puntos necesarios para cubrir approximadente un dominio. *An. Acad. Brasil. Ciencias* **16** (1944), 111–121.
17. Santalò, L. A., Un invariante afin para los cuerpos convexos del espacio de n dimensiones. *Portugal. Math.* **8** (1950), 155–161.
18. Segal, I. E., Tensor algebras over Hilbert spaces, I. *Trans. Am. Math. Soc.* **81** (1956), 106–134.

PRINTED IN BRUGES, BELGIUM, BY THE ST CATHERINE PRESS, LTD.

On seminorms and probabilities, and abstract Wiener spaces

By R. M. Dudley, Jacob Feldman, and L. Le Cam

Reprinted from Annals of Mathematics
Vol. 93, No. 2, March, 1971, pp. 390-408
Printed in Japan

E. Giné et al. (eds.), *Selected Works of R.M. Dudley*, Selected Works in Probability and Statistics,
DOI 10.1007/978-1-4419-5821-1_12, © Springer Science+Business Media, LLC 2010

On seminorms and probabilities, and abstract Wiener spaces

By R. M. DUDLEY*, JACOB FELDMAN*, and L. LE CAM*

1. Introduction

In real Hilbert space H there is a finitely additive measure n on the ring of sets defined by finitely many linear conditions, which is analogous to the normal distribution in the finite-dimensional case. This has been examined from various points of view in Gelfand and Vilenkin [11], Gross [12], Segal [19], as well as by many earlier authors. Now, if the Hilbert space is "completed" in any of number of ways, this "cylinder set measure" extends to an actual Borel measure on the completed space. For example, if we define $|x|_T = \|Tx\|$, where T is a Hilbert-Schmidt operator with trivial nullspace, then the completion of H with respect to the norm $|\cdot|_T$ is such a completion. A more subtle example is the following. Let H be realized explicitly as $L_2(0, 1)$. Define a norm on H by $|f| = \sup_{0 \leqq t \leqq 1} \left| \int_0^t f(s)ds \right|$. Then the completion of H with respect to this norm is isomorphic to the continuous functions on $[0, 1]$ with sup norm and vanishing at zero, via the map which sends f to the function g whose value at t is $\int_0^t f(s)ds$, $0 \leqq t \leqq 1$. The finitely additive measure n is then realized as Wiener measure. Other natural norms give rise to realizations of n as Wiener measure on the Hölder-continuous function, exponent α, $0 < \alpha < 1/2$. This sort of phenomenon has been investigated abstractly by L. Gross in [14]. He shows there (Theorem 4) that if H is completed with respect to any of his measurable seminorms, as defined in Gross [13], then n gives rise to a countably additive Borel measure on the Banach space obtained from H by means of the seminorm.

Here we shall prove the converse of Gross' result just mentioned (Theorem 3 below) and supply a new proof of the theorem itself. The equivalence thus obtained is extended to general cylinder set measures on linear spaces in duality (Theorem 2). The generalization of "measurable seminorm" to linear spaces in duality which we use is equivalent, but not trivially so, to Gross' definition for the normal distribution on Hilbert space. Orthogonal projections in Hilbert space provide some structure not available in the gen-

* Research partially supported by National Science Foundation Grants GP-9141, GP-3977 and GP-7176, GP-15283. respectively.

eral case, which requires some special treatment.

Theorems 2 and 3 appeared in preprint form in Feldman [10] (in a weaker form) and in Dudley-Feldman [8]. The present proofs are considerably simpler, however. Theorem 3 has also been announced by S. Chevet [5].

In Lemma 5 we show that the maximum (or the sum) of two measurable, Mackey-continuous seminorms is still measurable. Then, Corollary 2.1 shows that a cylinder set measure m on a linear space Y in duality with some X yields a countably additive measure on the completion of Y for the topology defined by any countable family of Mackey-continuous, m-measurable seminorms.

H. Sato [1] and J. Kuelbs [15] have proved that every Gaussian Borel measure μ on a separable Banach space B arises from the completion of a Hilbert space H and the standard Gaussian measure n on H with respect to an n-measurable seminorm. In Theorem 4 below we extend this result, replacing B by a general locally convex space, provided μ is nearly concentrated in convex, weakly compact sets (as is always true in a separable Banach space). This is another kind of converse to the result of Gross [14, Th. 4]. The final example in Sato [18] (which would contradict our Theorem 3) is incorrect.

The present form of the proof that (a) $\Rightarrow$ (b) in Theorem 2 was suggested by A. D. de Acosta.

In an appendix we give simple proofs of some inequalities for Gaussian measures, known from Gross [13], which play an important role in the present circle of ideas.

2. Preliminaries

Let X and Y be real linear spaces, put in duality by a bilinear form $\langle \cdot, \cdot \rangle : X \times Y \to R$, which separates points of X and of Y. Let $w(X, Y)$ be the weakest topology on X making each linear form $\langle \cdot, Y \rangle$ continuous, $y \in Y$. Likewise one defines $w(Y, X)$ on Y. These topologies may be called "weak" topologies where that appears unambiguous.

The *Mackey topology* on Y is the strongest locally convex topology for which its dual space is X (see Schaefer [18, Ch. 4] for further details). As usual, x, in X, is identified with the linear form $\langle x, \cdot \rangle$ on Y.

For any vector space Z let $FD(Z)$ be the set of all finite-dimensional subspaces of Z. If E and F are linear subspaces of a vector space Z, then we shall call E and F *supplementary* if and only if $E \cap F = \{0\}$ and $E + F = Z$. We shall only need the fact that if $E + G = Z$ for some $G \in FD(Z)$, then E has a supplement $F \in FD(Z)$.

Given X and Y in duality and a linear subspace F of X, let $\mathfrak{M}(Y, F)$ be the smallest σ-algebra of subsets of Y for which the functionals in F are all measurable. If $F \in FD(X)$, elements of $\mathfrak{M}(Y, F)$ will be called *cylinder sets, based on F*. Let $\mathcal{C}(Y,X)$ denote the algebra of all cylinder sets in Y, $\mathcal{C}(Y, X) = \bigcup \{\mathfrak{M}(Y, F): F \in FD(X)\}$. A non-negative, finitely additive set function m on $\mathcal{C}(Y, X)$, with $m(Y) = 1$, countably additive on $\mathfrak{M}(Y, F)$ for each fixed $F \in FD(X)$, will be called a *cylinder set measure on Y*.

For any $A \in X$, let

$$A^{\perp} = \{y \in Y: \ \langle x, y\rangle = 0 \quad \text{for all } y \in A\} ,$$

$$A^{0} = \{y \in Y: |\langle x, y\rangle| \leqq 1 \quad \text{for all } y \in A\} .$$

If A is a linear subspace, $A^{\perp} = A^0$. If $B \subset A^{\perp}$, we say $A \perp B$.

For any real vector space Z, let Z^a denote the algebraic dual space of A, i.e. the vector space of all linear forms $Z \to R$. Z and Z^a are naturally in duality. For any linear subspace W of Z, the quotient vector space $Z^a/W^{\perp}$ is naturally isomorphic to W^a, $W^{\perp}$ being the kernel of the natural map of Z^a onto W^a. Likewise $W^{\perp}$ is isomorphic to $(Z/W)^a$.

For any X and Y in duality there is a natural map of Y into X^a and of Y onto F^a if $F \in FD(X)$. Thus $Y/F^{\perp}$ is isomorphic to F^a. For any linear subspace J of X, let π_J be the natural map of X^a onto $X^a/J^{\perp}$ and the induced map of Y onto $Y/J^{\perp}$, $\pi_J(y) \equiv y + J^{\perp}$.

LEMMA 1. *Given X and Y in duality and $G \in FD(Y)$, then $G^{\perp\perp} = G$, even in X^a. Similarly, if $G = F^{\perp}$ for $F \in FD(X)$, then $G^{\perp\perp} = G$ in Y.*

Proof. If $G \in FD(Y)$ there is a natural one-to-one linear map of G into $G^{\perp\perp}$ ($G^{\perp} \subset X$, $G^{\perp\perp} \subset X^a$). $G^{\perp\perp}$ is isomorphic to $(X/G^{\perp})^a$, and $X/G^{\perp}$ to G^a, so these spaces all have the same (finite) dimension as G, and $G = G^{\perp\perp}$. If $F \in FD(X)$ and $G = F^{\perp}$, then $F^{\perp\perp} = F$ in X so $G^{\perp\perp} = G$ in Y. □

LEMMA 2. *Given X and Y in duality, $F \in FD(X)$, $G \in FD(Y)$, we have $(F \cap G^{\perp})^{\perp} = F^{\perp} + G$.*

Proof. $(F^{\perp} + G)^{\perp} = F^{\perp\perp} \cap G^{\perp} = F \cap G^{\perp} \in FD(X)$. Thus by the bipolar theorem (Schaefer [18, p. 126]) it suffices to show $F^{\perp} + G$ is weakly closed. $F^{\perp}$ is closed, so $Y/F^{\perp}$ is a Hausdorff finite-dimensional space under the quotient of the weak topology [ibid., 2.3 p. 20], so $F^{\perp} + G$ is the inverse image of a finite-dimensional subspace in $Y/F^{\perp}$ and hence closed. □

For any $F \in FD(X)$, the finite-dimensional vector space $Y/F^{\perp}$ with its (unique) Hausdorff vector topology has a natural class $\mathcal{B}(F)$ of Borel sets, and

$$\{\pi_F^{-1}(B): B \in \mathcal{B}(F)\} = \mathfrak{M}(Y, F) .$$

A cylinder set $\pi_F^{-1}(U)$ is open for $w(Y, X)$ if and only if U is open in $Y/F^\perp$. Such a cylinder set will be called *open* and its complement *closed*.

A cylinder set measure m on Y defines a countably additive Borel probability measure m_F on $Y/F^\perp$ or F^a for each $F \in FD(X)$, and conversely a class $\{m_F: F \in FD(X)\}$ of such Borel measures satisfying a natural consistency condition is defined by a cylinder set measure m on Y (cf. Dudley [6, Th. 1.4]). Given a cylinder set measure m on Y, there also exists a unique countably additive probability measure P_m on $\mathfrak{M}(X^a, X)$ in X^a with $(P_m)_F = m_F$ on F^a for each $F \in FD(X)$.

Definition. A finite non-negative Borel measure μ is called *regular* if for each $\varepsilon > 0$ and each Borel set A there is some compact $K \subset A$ with $\mu(A \subset K^c) < \varepsilon$.

It is known that any finite non-negative Borel measure on a complete separable metric space is regular, and that conversely any regular Borel finite measure gives full measure to some separable subspace; see, for example, Billingsley [1, p. 9].

The following fact is well-known. Since the proof is so simple, we include it.

LEMMA 3. *The cylinder set measure m on $\mathcal{C}(Y, X)$ is countably additive $\Leftrightarrow$ whenever A_j is an ascending sequence of open sets in $\mathcal{C}(Y, X)$ with $\bigcup_j A_j = Y$, we have $m(A_j) \uparrow 1$.*

Proof. $\Rightarrow$: clear. $\Leftarrow$: suppose $B_j \in \mathcal{C}(Y, X)$ and $B_j \downarrow \varnothing$, but $m(B_j)$ always $\geqq \alpha > 0$. Choose C_j *closed* $\subset B_j$ with $m(B_j \sim C_j) < \alpha/2^{j+1}$, since each m_F is regular. Let $D_j = C_1 \cap \cdots \cap C_j$. Then

$$m(B_j \sim D_j) \leqq m\left(\bigcup_{i=1}^{j} (B_i \sim C_i)\right) < \alpha/2 .$$

Let $A_j = D_j^c$. Then $A_j \uparrow$, $\bigcup_j A_j = Y$, but $m(A_j) < 1 - \alpha/2$. □

Definition. For any finitely additive non-negative set function m on a family $\mathcal{A}$ of subsets of S, we define $m^*(B)$, for each $B \subset S$, as $\inf \{m(A): A \in \mathcal{A}, A \supset B\}$.

The following theorem is a variant of results of Prokhorov [16]; it can be deduced by reexamining the proofs there. We shall prove it directly.

THEOREM 1. *Let Y be a locally convex Hausdorff topological vector space, and m a cylinder set measure on $\mathcal{C}(Y, Y')$. Then a necessary and sufficient condition for the existence of a regular Borel measure μ on Y extending m is*

$$\sup \{P_m^*(K): K \text{ compact} \subset Y\} = 1 .$$

If such a μ exists, it is unique, and $\mu(K) = P_m^*(K)$ for all compact K.

Proof. Necessity of the condition $\sup\{P_m^*(K): K \text{ compact} \subset Y\} = 1$ is evident. We show sufficiency. Let $X = Y'$. Then X and Y are in duality. A vertical line denotes *restriction*, e.g. $X|K$ is the set of all restrictions to K of the linear functions on Y defined by X. Let $\mathfrak{M}(K, X|K)$ be the smallest σ-algebra of subsets of K making all these functions measurable. Then $\mathfrak{M}(K, X|K) \subset \mathcal{B}_0(K)$, the class of Baire subsets of K. But X separates K, so the algebra of functions on K generated by $X|K$ is uniformly dense in the continuous functions on K, by Stone-Weierstrass. Then $\mathfrak{M}(K, X|K) = \mathcal{B}_0(K)$.

There exists $L \supset K$, $L \in \mathfrak{M}(X^a, X)$, such that $P_m(L) = P_m^*(K)$. Let φ be a bounded and continuous function of k real variables, let $\{x_1, \cdots, x_k\} \subset X$, and let f be defined on X^a by $f(x^a) = \varphi(\langle x_1, x^a\rangle, \cdots, \langle x_k, x^a\rangle)$. Then consider $\int_L f dP_m$. This is independent of L: that is, if L_1 also $\supset K$ and $P_m(L_1) = P_m^*(K)$, then $\int_{L_1} f dP_m = \int_L f dP_m$. If $f|K = g|K$, where g is also of the above form, then $\{x^a: f(x^a) - g(x^a) = 0\}$ is in $\mathfrak{M}(X^a, X)$ and includes K, hence $\int_L f dP_m = \int_L g dP_m$. Clearly $\int_L 1 dP_m = P_m^*(K)$, and if $f|K \geqq 0$ then $\{x^a: f(x^a) \geqq 0\} \in \mathfrak{M}(X^a, X)$ and $\supset K$, so $\int_L f dP_m \geqq 0$. Consequently a non-negative linear functional Φ_K is defined on the algebra $\mathcal{A}_K = \{f|K: f = \varphi(\langle x_1, \cdot\rangle, \cdots, \langle x_k, \cdot\rangle) \text{ for some } \varphi, x_1, \cdots, x_k\}$. Now, $\mathcal{A}_K$ is a uniformly dense subalgebra of the continuous functions on K, so there is a unique measure μ_K^0 on $\mathcal{B}^0(K)$, satisfying $\int_K f|K d\mu_K^0 = \int_L f dP_m$ for f of the form $\varphi(\langle x_1, \cdot\rangle, \cdots, \langle x_k, \cdot\rangle)$, φ bounded and continuous. The same formula then holds for any $\mathfrak{M}(X^a, X)$-measurable, bounded function on X^a. Furthermore: $\exists$ a unique regular Borel measure μ_K on $\mathcal{B}(K)$, the Borel sets of K, which agrees with μ_K^0 on $\mathcal{B}_0(K)$. μ_K may be extended to a measure on $\mathcal{B}(Y)$, since $K \in \mathcal{B}(Y)$; and regularity is retained, as is easily seen. The extension to $\mathcal{B}(Y)$ will also be called μ_K. It is uniquely characterized by regularity, agreeing with μ_K^0 on $\mathcal{B}_0(K)$, and giving measure zero to K^c.

Next, we see that if $K_2 \subset K_1$ then $\mu_{K_1}|K_2 = \mu_{K_2}$. Given $A \in \mathcal{B}(K_2)$, choose $B \in \mathfrak{M}(X^a, X)$ such that $\mu_{K_1}((B \cap Y)\Delta A)$ and $\mu_{K_2}((B \cap Y)\Delta A)$ are zero. This can be done, since $\mu_{K_1} + \mu_{K_2}$ is regular. Choose $L_j \supset K_j$, $L_j \in \mathfrak{M}(X^a, X)$, $P_m(L_j) = P_m^*(K_j)$. We may take $L_1 \supset L_2 \supset B$. Then:

$$\begin{aligned}\mu_{K_1}(A) &= \mu_{K_1}(B \cap Y) = \mu_{K_1}^0(B \cap Y) = P_m(L_1 \cap B) = P_m(B) \\ &= P_m(L_2 \cap B) = \mu_{K_2}(B \cap Y) = \mu_{K_2}(A)\,.\end{aligned}$$

So $\mu_K \uparrow$ with K, and now we may define μ by $\mu(A) = \sup_K \mu_K(A)$. If

$A_j \downarrow \emptyset$, then $\mu(A_j) = \mu(A_j) - \mu_K(A_j) + \mu_K(A_j)$. But $\mu(A_j) - \mu_K(A_j) \leqq \mu(Y) - \mu_K(Y)$, so $\mu(A_j) \downarrow 0$. Thus μ is a Borel measure, and $\mu(Y) = \sup_K P_m^*(K)$, which we are assuming to be 1.

μ is regular: for choose $\varepsilon > 0$ and $A \in \mathcal{B}(Y)$. Choose K such that $\mu(A) - \mu_K(A) < \varepsilon/2$. Choose $K_1 \subset A$ such that $\mu_K(A) - \mu_K(K_1) < \varepsilon/2$. Then

$$\begin{aligned}&\mu(A) - \mu(K_1)\\ &\leqq \big(\mu(A) - \mu_K(A)\big) + \big(\mu_K(A) - \mu_K(K_1)\big) + \big(\mu_K(K_1) - \mu(A_1)\big) < \varepsilon/2 + \varepsilon/2 = \varepsilon\,.\end{aligned}$$

If K is compact, then we claim $\mu(K) = P_m^*(K)$. For: if L is compact then $\mu_L(K) \leqq \mu_{L\cup K}(K) = \mu_K(K) = P_m^*(K)$, so $\sup_L \mu_L(K) = P_m^*(K)$.

As for the uniqueness, let μ and ν be regular Borel measures on Y which extend m. Then since $\mu+\nu$ is regular, for any $A \in \mathcal{B}(Y)$ there is $B \in \mathfrak{M}(X^a, X)$ with $\mu\big((B \cap Y)\Delta A\big) = \nu\big((B \cap Y)\Delta A\big) = 0$. Then

$$\mu(A) = \mu(B \cap Y) = P_m(B) = \nu(B \cap Y) = \nu(A)\,. \qquad \square$$

COROLLARY 1.1. *If Y is a σ-compact locally convex topological vector space, and m is a countably additive cylinder set measure on $\mathcal{C}(Y, Y')$, then m extends to a regular Borel measure on Y.*

Proof. Obvious. $\square$

This corollary applies to the important case where X is a Banach space, and $Y = X'$ in the weak* topology, so that $X = Y'$.

3. m-measurability and regularity

LEMMA 4. *The following conditions are equivalent for a cylinder set measure m on $\mathcal{C}(Y, X)$, seminorm $|\cdot|$ on Y, and $\varepsilon > 0$:*

(a) *$\exists G \in FD(Y)$ such that $m^*\big(\{y\colon |y - G| \leqq \varepsilon\}\big) \geqq 1 - \varepsilon$*

(b) *$\exists G \in FD(Y)$ such that if $F \in FD(X)$ and $F \perp G$ then $m\big(\{y\colon |y - F^\perp| < \varepsilon\}\big) \geqq 1 - \varepsilon$.*

Proof. Note first that $\{y\colon |y - F^\perp| \leqq \varepsilon\}$ is a closed cylinder set based on F. Clearly (a) $\Rightarrow$ (b). As for the converse: choose G as in (b). Suppose $A \in \mathcal{C}(Y, X)$ and $\{y\colon |y - G| \leqq \varepsilon\} \subset A$. Since $A = A + J^\perp$ for some $J \in FD(X)$, $A \supset \{y\colon |y - G| \leqq \varepsilon\} + J^\perp = \{y\colon |y - (G + J^\perp)| \leqq \varepsilon\}$. But $G + J^\perp = (J \cap G^\perp)^\perp$ by Lemma 2. Setting $F = J \cap G^\perp$, we then have $F \perp G$ and thus $m(A) \geqq m\big(\{y\colon |y - F^\perp| \leqq \varepsilon\}\big) \geqq 1 - \varepsilon$. $\square$

Definition. A seminorm satisfying the above equivalent conditions with respect to m will be called *m-measurable.*

THEOREM 2. *The following are equivalent for a cylinder set measure m on $\mathcal{C}(Y, X)$ and a Mackey-continuous seminorm $|\cdot|$ on Y:*

(a) *the cylinder set measure induced by m on the Banach space obtained from Y via the seminorm $|\cdot|$ extends to a regular Borel measure.*

(b) *for each $\varepsilon > 0$, $\exists$ finite $A \subset Y$ such that $m^*(\{y: |y - A| \leqq \varepsilon\}) \geqq 1 - \varepsilon$.*

(c) *$|\cdot|$ is m-measurable.*

Proof. Let W be the Banach space obtained from Y by means of $|\cdot|$ (the completion modulo the null space). Let θ be the canonical map: $Y \to W$. S_r will denote the closed ball of radius r about 0 in W, and likewise S'_r in W' and S''_r in W''. For any set $A \subset W$, let $S_r(A) = S_r + A$, and analogously in W' and W''.

(a)$\Rightarrow$(b): Given $\varepsilon > 0$, choose K compact $\subset W$ such that $P^*_{m\circ\theta^{-1}}(K) > 1 - \varepsilon$. Choose A_0 finite $\subset W$ such that $S_{\varepsilon/4}(A_0) \supset K$. By density of $\theta(Y)$ in W, $\exists$ finite $A \subset Y$ such that $S_{\varepsilon/5}(\theta(A)) \supset A_0$ and so K is included in the interior of $S_{\varepsilon/2}(\theta(A))$. Now, suppose we are given $B \in \mathcal{C}(Y, X)$, B disjoint from $S_{\varepsilon/2}(A)$ in Y, with $m(B) > \varepsilon$. Let B be based on $F \in FD(X)$, so that $B = B + F^\perp$. Let G be a supplement to $F^\perp$, $G \in FD(Y)$. Since m_F is regular we may assume $B = C + F^\perp$ for some compact $C \subset G$. Now: $\theta(B)$ is disjoint from $S_{2\varepsilon/5}(\theta(A))$, so the closure $\theta(B)^-$ is disjoint from K. But $\theta(B)^- = \theta(C + F^\perp)^- = \theta(C) + \theta(F^\perp)^-$. Also, $\theta(F^\perp)^-$ is of finite codimension since

$$W = \theta(G + F^\perp)^- = [\theta(G) + \theta(F^\perp)]^- \subset [\theta(G) + \theta(F^\perp)^-]^- = \theta(G) + \theta(F^\perp)^- .$$

(A finite-dimensional extension of a closed subspace is closed, as in the proof of Lemma 2 above.) Hence $\theta(B)^- \in \mathcal{C}(W, W')$, and

$$(m \circ \theta^{-1})(\theta(B)^-) \geqq m(B) > \varepsilon ,$$

contradicting the assumption.

(b) $\Rightarrow$ (c): obvious.

(c) $\Rightarrow$ (a): first observe that condition (c) on the system $\{X, Y, m, |\cdot|\}$ immediately implies the same condition on $\{W', W, m\circ\theta^{-1}, \|\cdot\|\}$. Write n for $m \circ \theta^{-1}$.

Next: there exists some $r(\varepsilon)$ such that $n^*(S_{r(\varepsilon)}) \geqq 1 - \varepsilon$. To see this, choose $G_\varepsilon \in FD(W)$ such that $n^*(S_{\varepsilon/2}(G_\varepsilon)) > 1 - \varepsilon/2$. Let F be supplementary to $G_\varepsilon^\perp$ in W'. Choose t so large that $n(S_t(F^\perp)) > 1 - \varepsilon/2$. This can be done because $n(S_t(F^\perp)) = n_F(\pi_F(S_t)) = n_F(t\pi_F(S_1))$, and $\pi_F(S_1)$ is a neighborhood of 0 in $Y/F^\perp$. Then $n^*(S_{\varepsilon/2}(G_\varepsilon) \cap S_t(F^\perp)) > 1 - \varepsilon$. Now: if $w \in S_{\varepsilon/2}(G_\varepsilon) \cap S_t(F^\perp)$, then choose w' in W' and write it as $w'_1 + w'_2$ with $w'_1 \in G_\varepsilon^\perp$ and w'_2 in $F = F^{\perp\perp}$. Then

$$|\langle w, w' \rangle| \leqq |\langle w, w'_1 \rangle| + |\langle w, w'_2 \rangle| \leqq (\varepsilon/2) \|w'_1\| + t\|w'_2\| .$$

So $S_{\varepsilon/2}(G_\varepsilon) \cap S_t(F^\perp)$ is weakly bounded, hence strongly bounded (by Banach-

Steinhaus), i.e. included in some S_r.

Now consider W as embedded in W''. There is a natural map: $\mathcal{C}(W'', W') \to \mathcal{C}(W, W')$ gotten by sending $A \in \mathcal{C}(W'', W')$ to $A \cap W$. Define n'' on $\mathcal{C}(W'', W')$ by $n''(A) = n(A \cap W)$. We next show that n'' is countably additive on $\mathcal{C}(W'', W')$. To see this, let A_j be open in $\mathcal{C}(W'', W)$, with $A_j \uparrow$ and $\bigcup_j A_j = W''$. Since the sphere $S''_{r(\varepsilon)}$ in W'' is weak* compact, $\exists j(\varepsilon)$ such that $A_{j(\varepsilon)} \supset S''_{r(\varepsilon)}$. So

$$n''(A_{j(\varepsilon)}) = n(A_{j(\varepsilon)} \cap W) \geqq n^*(S_{r(\varepsilon)}) \geqq 1 - \varepsilon .$$

Hence $n(A_j) \uparrow 1$, and we apply Lemma 3.

Thus n'' extends to a regular Borel measure ρ for the weak* topology on W'', by Corollary 1.1. Notice that $S''_\varepsilon(G_\varepsilon)$ is closed in the weak* topology, since it is the sum of the closed set G_ε and the compact set S''_ε.

We now claim that $\rho(S''_\varepsilon(G_\varepsilon)) \geqq 1 - \varepsilon$. For suppose not; then by regularity of ρ, there is some weak* compact K disjoint from $S''_\varepsilon(G_\varepsilon)$ with $\rho(K) > \varepsilon$. Now: $S''_\varepsilon(G_\varepsilon)^c = \bigcup_\alpha V_\alpha$ for certain open $V_\alpha \in \mathcal{C}(W'', W')$, since such open sets form a base for the weak* topology. So $\exists \alpha_1, \cdots, \alpha_k$ such that $K \subset V_{\alpha_1} \cup \cdots \cup V_{\alpha_k} = V_0 \in \mathcal{C}(W'', W')$. But $V_0 \cap S''_\varepsilon(G_\varepsilon) = \varnothing$, while $n(V_0 \cap W) = n''(V_0) = \rho(V_0) \geqq \rho(K) > \varepsilon$. This contradicts the choice of G_ε. So $\rho(S''_\varepsilon(G_\varepsilon)) \geqq 1 - \varepsilon$.

Let W_0 be the norm-closed linear subspace of W'' spanned by $\bigcup_{j=1}^\infty G_{1/j}$. Notice that W_0 is equal to $\bigcap_{k=1}^\infty S'_{1/k}(W_0)$. W_0 is clearly norm-separable. Choose a norm-dense countable subset $A \subset W$; then also $W_0 = \bigcap_{k=1}^\infty S''_{1/k}(A)$. This shows that W_0 is weak* Borel. Finally: $S''_{1/k}(W_0) \supset S''_{1/k}(G_{1/k})$, so $\rho(W_0) = 1$.

Let $\mathcal{B}''$ be the weak* Borel sets of W''. Then $\mathcal{B}'' \mid W_0 \supset \mathfrak{M}(W_0, W') = \mathfrak{M}(W_0, W_0') = \mathcal{B}(W_0)$, because of separability of W_0. (Actually the opposite inclusion also holds, because the map $W_0 \to W''$ is continuous from the norm topology on W_0 into the weak* topology on W''.) So $\rho \mid W_0$ is a norm-Borel measure on W_0. It is regular, since W_0 is separable. This gives rise to a norm-Borel measure on W, also regular, call it μ, by the definition $\mu(A) = \rho(A \cap W_0)$ for $A \in \mathcal{B}(W)$. It remains only to see that $\mu \mid \mathcal{C}(W, W') = n$. But if $A \in \mathcal{C}(W, W')$, then A is of the form $B \cap W$, $B \in \mathcal{C}(W'', W')$, and $\mu(B \cap W) = \rho(B \cap W_0) = \rho(B) = n''(B) = n(B \cap W)$. □

The implication (b) $\Rightarrow$ (c) may be strengthened as follows. If Y is a linear space, and $\{|\cdot|_\alpha: \alpha \in A\}$ are seminorms on Y, let $Y_0 = \{y: |y|_\alpha = 0 \text{ for all } \alpha\}$. Then the $|\cdot|_\alpha$ give rise to seminorms $\|\cdot\|_\alpha$ on Y/Y_0, so that one obtains a map $\theta: Y \to W$, where W is the completion of Y/Y_0 with respect to the topology obtained from the seminorms. If A is countable, then W is a Fréchet space.

LEMMA 5. *Let m be a cylinder set measure on $\mathcal{C}(Y, X)$ and let $|\cdot|_j$,*

$j = 1, 2$, be m-measurable Mackey-continuous seminorms on Y; then so is $|\cdot|_0$, where $|y|_0 = |y|_1 \vee |y|_2$.

Proof. Mackey continuity is obvious. In view of Theorem 2, one may reduce the theorem to the following assertion. Let W be a Banach space with a cylinder set measure m on $\mathcal{C}(W, W')$. For $j = 1, 2$ let ψ_j be a continuous linear map of W onto a dense subset of a Banach space $(W_j, \|\cdot\|_j)$. Assume that $\|w\| = \|\psi_1(w)\|_1 \vee \|\psi_2(w)\|_2$. Finally, assume that $m_j = m \circ \psi_j^{-1}$ extends to a regular measure μ_j on $\mathcal{B}(W_j)$. Then m extends to a regular measure on $\mathcal{B}(W)$.

To prove this, let $W_1 \times W_2$ be the product space with the norm $\|(w_1, w_2)\| = \|w_1\|_1 \vee \|w_2\|_2$. Let ψ be the linear map defined by $\psi(w) = (\psi_1(w), \psi_2(w))$. This is an isometry of W onto a closed subspace $W_0 = \psi(W)$ of $W_1 \times W_2$. Thus ψ is a homeomorphism for the norms and also for the corresponding weak topologies. Since the weak topology of $W_1 \times W_2$ is the product of the weak topologies of W_1 and W_2, a filter in W converges weakly if and only if its image by ψ_j, $j = 1, 2$ converges weakly.

Let K_j be compact and convex in W_j, $j = 1, 2$, with $\mu_j(K_j^c) < \varepsilon/2$. Let $K = \psi^{-1}(K_1 \times K_2) = (\psi_1^{-1}(K_1)) \cap (\psi_2^{-1}(K_2))$. Suppose that C is a cylinder set with $K \subset C$; we claim $m(C) \geqq 1 - \varepsilon$. For suppose $m(C) < 1 - \varepsilon$. Then, because of the regularity of the finite-dimensional images of m, there is an *open* cylinder set V with $K \subset C \subset V$ and $m(V) < 1 - \varepsilon$. Now we show that there are weakly open cylinder sets V_j with $K_j \subset V_j \subset W_j$ and $K \subset \psi^{-1}(V_1 \times V_2) \subset V$. This will be a contradiction, since then

$$\begin{aligned}\varepsilon \leqq m(V^c) \leqq m(\psi^{-1}(V_1 \times V_2)^c) &\leqq m(\psi_1^{-1}(V_1)^c \cup \psi_2^{-1}(V_2)^c) \\ \leqq m \circ \psi_1^{-1}(V_1^c) + m \circ \psi_2^{-1}(V_2^c) &\leqq \mu_1(K_1^c) + \mu_2(K_2^c) < \varepsilon .\end{aligned}$$

Suppose no such V_j exists. Then, as V_j ranges over all open cylinder sets such that $K_j \subset V_j$, the sets $V^c \cap \psi^{-1}(V_1 \times V_2)$ form the basis of a filter $\mathcal{F}$. Let $\mathcal{U}$ be an ultrafilter finer than $\mathcal{F}$ and let $\mathcal{U}_j$ be its image under ψ_j: $\mathcal{U}_j = \{U_j \subset W_j : \psi_j^{-1}(U_j) \in \mathcal{U}\}$. Then $\mathcal{U}_j$ is an ultrafilter, and $\bar{U}_j \cap K_j \neq \emptyset$ for the weak closure $\bar{U}_j$ of any U_j in $\mathcal{U}_j$; indeed, each weak neighborhood of K_j includes an open cylinder neighborhood of K_j. Since $\bar{U}_j \cap K_j$ is weakly compact, the ultrafilter $\mathcal{U}_j$ converges weakly. Hence, ψ being a homeomorphism, $\mathcal{U}$ itself converges weakly. Let w be its limit; then $w \in V^c$, since V^c is closed. Also, $\psi_j(w) \in K_j$, so $K \cap V^c$ is nonempty, contrary to assumption.

So $m(C) \geqq 1 - \varepsilon$ whenever C is a cylinder set including K. But then, by Prokhorov [16], m extends to a regular measure on $\mathcal{B}(W)$. □

If $|\cdot|_1$ and $|\cdot|_2$ are m-measurable seminorms which are not Mackey-continuous, $|\cdot|_1 \vee |\cdot|_2$ may fail to be m-measurable, as the following example shows. Let Y be the space $\mathcal{C}([0, 1])$ of continuous real functions on $[0, 1]$. Let X be the space of finite signed measure μ with *finite* support on $[0, 1]$ and $\mu(\{1/2\}) = 0$. X and Y are naturally in duality. For $f \in Y$ let

$$|f|_1 = \sup\{|f(t)|: 0 \leqq t \leqq 1/2\} ,$$

$$|f|_2 = \sup\{|f(t)|: 1/2 \leqq t \leqq 1\} .$$

The linear form $\mu \to \mu([0, 1/2])$ on X defines a cylinder set measure m on Y as follows. Given $\mu_1, \cdots, \mu_n \in X$ and Borel sets $B_1, \cdots, B_n \subset R$, let

$$m\left\{f: \int f d\mu_j \in B_j, j = 1, \cdots, n\right\}$$

$$\begin{aligned} &= 1 \qquad && \text{if } \mu_j([0, 1/2]) \in B_j,\ j = 1, \cdots, n ; \\ &= 0 && \text{otherwise} . \end{aligned}$$

Then $|\cdot|_1$ and $|\cdot|_2$ are m-measurable but $|\cdot|_1 \vee |\cdot|_2$ is not, as will now be proved. If $E \subset [0, 1/2)$ and $F \subset (1/2, 1]$ are finite sets, then $m\{f: f = 1 \text{ on } E$ and $f = 0$ on $F\} = 1$.

For $|\cdot|_1$, the G in Lemma 4 need only contain some function equal to 1 on $[0, 1/2]$. For $|\cdot|_2$, G need only contain a function equal to 0 on $[1/2, 1]$. Although the $|\cdot|_j$ are not Mackey-continuous, clearly all three conditions (a), (b), (c) in Theorem 2 hold for them.

Suppose now that the sup norm $|\cdot|_0 = |\cdot|_1 \vee |\cdot|_2$ were m-measurable. Let E_n and F_n be finite, $E_n \subset [0, 1/2)$, $F_n \subset (1/2, 1]$, where $E_n \cup F_n$ increases up to a dense set in $[0, 1]$. Let G be a finite-dimensional subspace of Y such that $m^*\{f: |f - G|_0 \leqq 1/3\} \geqq 2/3$. Then for all n, $\exists g_n \in G$: $|g_n - 1| \leqq 1/3$ on E_n and $|g_n| \leqq 1/3$ on F_n. Let $h_n = g_n/|g_n|_0$. Taking a subsequence, we can assume $|h_n - h|_0 \to 0$ for some h. Now

$$h(1/2) \geqq \liminf_{n\to\infty} \inf_{E_n} h_n \geqq 2 \limsup_{n\to\infty} \sup_{F_n} |h_n| \geqq 2\,|h(1/2)| .$$

Thus $h(1/2) = 0$. It follows that $h_n \to 0$ uniformly on E_n and F_n, and $h \equiv 0$, but $|h|_0 = 1$, a contradiction. Thus $|\cdot|_0$ is not measurable. In fact, as is easily seen, none of the three conditions in Theorem 2 holds for $|\cdot|_0$.

COROLLARY 2.1. *Let X, Y be spaces in duality, and let $|\cdot|_j$, $j = 1, 2, \cdots$ be Mackey-continuous, m-measurable seminorms on Y, m being a cylinder set measure on $\mathcal{C}(Y, X)$. Let W be the Fréchet space obtained from Y with respect to the topology defined by all the seminorms $|\cdot|_j$ by completion modulo the intersection of their null spaces. Then the cylinder set measure $m \circ \theta^{-1}$ on $\mathcal{C}(W, W')$ extends to a regular Borel measure.*

Proof. According to Lemma 5 one can assume without loss of generality that $|\cdot|_1 \leqq |\cdot|_2 \leqq \cdots$. Let W_j be the Banach space created by the j-th seminorm, and θ_j the canonical map: $Y \to W_j$. For $i \leq j$ let π_{ij} be the continuous linear map: $W_j \to W_i$ obtained by extending the map $\theta_j(y) \to \theta_i(y)$. Let π_i be the i-th coordinate function in $\prod_j W_j$, and let $Z = \{z \in \prod_j W_j : \pi_{ij}\pi_j(z) = \pi_i(z)$ whenever $i \leq j\}$. Z is closed in the product topology. There is an obvious map: $Y \to Z$, $\psi(y) = (\theta_1(y), \theta_2(y), \cdots)$. $\psi(y) = 0 \Leftrightarrow \theta(y) = 0$. A sequence $\theta(y_1), \theta(y_2), \cdots$ is Cauchy $\Leftrightarrow |y_i - y_j|_k \to 0$ as $i, j \to \infty$, for each fixed k; thus, $\Leftrightarrow \psi(y_i)$ is a Cauchy sequence in Z. Therefore the map $\theta(Y) \to \psi(y)$ extends to a linear homeomorphism from W into Z. Finally, the image of this map is all of Z. For suppose $z \in Z$. Choose $y_j \in Y$ such that $\|\pi_j(z) - \theta_j(y_j)\|_j < 1/j$. If $k \geq j$ then $\|\pi_j(z) - \theta_j(y_k)\|_j = \|\pi_{jk}\pi_k(z) - \pi_{jk}\theta_k(y_k)\|_j \leq \|\pi_k(z) - \theta_k(y_k)\|_k < 1/k$, so $\|\pi_j(z) - \theta_j(y_k)\|_j \to 0$ as $k \to \infty$, for each fixed j. Thus $\psi(y_k) \to z$. Thus, W has been identified with the projective limit of the system $\{W_j, \pi_{ij}\}$.

Now let μ_j be the regular extension of $m \circ \theta_j^{-1}$. For $i < j$ one has $\mu_i = \mu_j \circ \pi_{ij}^{-1}$. Indeed, the equality must hold for $\mathcal{C}(W_i, W_i')$, and this entails its holding on $\mathcal{B}(W_i)$, by Theorem 1. It follows then from the Kolmogorov consistency theorem in its projective limit version that there exists a regular Borel measure μ on W such that $\mu_i = \mu \circ \pi_i^{-1}$. See, for example, Bourbaki [4, § 4.3, Th. 2]. (To apply the version of the theorem given by Bochner [2, Th. 5.1.1] one can adjoin sets of measure 0 to make the maps π_{ij} onto.)

The assumption that the set of seminorms was countable could not have been dispensed with, as may be seen from the following example. Let $Y = R^A$, where A is an uncountable set, and let $X = R_A$, the subset of Y consisting of those functions: $A \to R$ which are nonzero in only finitely many places. The duality between X and Y is $\langle x, y \rangle = \sum_{\alpha \in A} x(\alpha)y(\alpha)$. The Mackey topology on Y is the same as $w(Y, X)$, and is just the product topology. For each α set $|y|_\alpha = |y(\alpha)|$; then $|\cdot|_\alpha$ is Mackey-continuous, and is m-measurable for any cylinder set measure m. The complete locally convex space gotten from Y by means of the seminorms $\{|\cdot|_\alpha, \alpha \in A\}$ may be identified with Y itself, θ being the identity map. And it is indeed the case that m extends to a countably additive measure μ on $\mathfrak{M}(Y, X)$. However, *regularity* may break down: the measure μ need not extend to a regular Borel measure on Y. For example, let μ be a product of probabilities on R without compact support.

By way of completing the picture:

COROLLARY 2.2. *Let (X, Y) be spaces in duality, θ continuous from the Mackey topology on Y into the locally convex Hausdorff space W, and having dense range. Let m be a cylinder set measure on $\mathcal{C}(Y, X)$ such that $m \circ \theta^{-1}$ extends to a regular Borel measure μ on W. Then for every continuous* *seminorm* $\|\cdot\|$ *on* W, *the seminorm* $|y| = \|\theta(y)\|$ *is* m-*measurable.*

Proof. Let W_1 be the completion of W for $\|\cdot\|$, and θ_1 the map: $W \to W_1$. Let $\theta_0 = \theta_1 \circ \theta$. Then θ_0 maps Y continuously into a dense subset of W_1, carrying $|\cdot|$ to the norm on W_1. Furthermore: $\mu \circ \theta_1^{-1} | \mathcal{B}(W_1)$ is a regular Borel measure, and extends $m \circ \theta_0^{-1}$, so by Theorem 2, $|\cdot|$ is m-measurable. □

4. Gross's notion of measurability, and GC sets

Now these ideas will be specialized to the canonical normal distribution n on the real Hilbert space H. Thus we take $X = Y = H$, the duality being given by the inner product, and n is the cylinder set measure whose finite-dimensional projections are all Gaussian, mean 0, variance 1. Alternately described: $\int e^{i(x,y)} dn(y) = \exp(-\|x\|^2/2)$. The following concept is due to Gross [13], although we call a seminorm "measurable by projections" when he calls it "measurable".

Definition. For any cylinder set measure m on H, the seminorm $|\cdot|$ will be called m-measurable by projections if for any $\varepsilon > 0$ there is some $G \in FD(H)$ such that if $F \in FD(H)$ and $F \perp G$ then $m((N_\varepsilon \cap F) + F^\perp) \geqq 1 - \varepsilon$, where $N_\varepsilon = \{x: |x| < \varepsilon\}$.

Obviously this definition implies what we have called m-measurability. Next, two definitions from Dudley [7]. Using the normal distribution n on H, we get a linear process $x \to (x, \cdot)^\sim$, $x \in H$.

Definition. The set $A \subset H$ is called a GB *set* if $\bigvee \{(x, \cdot)^\sim : x \in A\}$ is finite with probability 1 (where $\bigvee$ denotes the least upper bound in the lattice of random variables modulo sets of probability zero); equivalently, if the process $x \to (x, \cdot)^\sim$, $x \in A$, has a version with bounded path functions.

Definition. A is called a GC *set* if the process $x \to (x, \cdot)^\sim$, $x \in A$, has a version with *continuous* path functions.

Thus, a compact GC set is a GB set.

THEOREM 3. *Let* $|\cdot|$ *be a seminorm in Hilbert space* H, *and* n *the canonical normal distribution. Then the following statements are equivalent.*

(a) $|\cdot|$ *is continuous and* n-*measurable.*

(b) $\{y: |y| \leqq 1\}^0$ *is a compact* GC *set.*

(c) $|\cdot|$ *is* n-*measurable by projections.*

Proof. (a) $\Rightarrow$ (b): by Theorem 2, (a) implies that, if W is the Banach space obtained from H by means of $|\cdot|$, and θ is the map: $H \to W$, then $n \circ \theta^{-1}$ is the restriction to $\mathcal{C}(W, W')$ of a regular Borel measure μ on $\mathcal{B}(W)$. For each $w \in W$, the function $w' \to \langle w', w \rangle$ is bounded by $\|w\|$ on the unit sphere

S_1' of W'. Now, the linear process $w' \to (\theta^t(w'), \cdot)^\sim$ has the same joint distributions as $w' \to \langle w', \cdot \rangle$, regarded as random variables on $(W, \mathcal{B}(W), \mu)$. Thus $\bigvee \{(\theta^t(w'), \cdot)^\sim : w' \in S_1'\} < \infty$ with probability 1. But $\theta^t(S_1')$ is easily seen to be dense in $\{y: |y| \leqq 1\}^0$, so that if $x \in \{y: |y| \leqq 1\}^0$, $(x, \cdot)^\sim$ is a limit in L^2, and therefore in probability, of random variables of the form $(\theta^t(w'), \cdot)^\sim$, $w' \in S_1'$. Consequently, $\bigvee \{(x, \cdot)^\sim : x \in \{y: |y| \leqq 1\}^0\} < \infty$ with probability 1. So $\{y: |y| \leqq 1\}^0$ is a GB set. Since it is closed in H, it is compact, by Proposition 3.4 of Dudley [7].

Next, we observe that the map θ is a homeomorphism from S_1' in the weak* topology onto $\{y: |y| \leqq 1\}^0$. It suffices to show that $\theta^t \mid S_1'$ is continuous from S_1' in the weak* topology into H. Suppose w'_α converges weak* in S_1' to w'. Since $\theta^t(S_2')$ has compact closure in H, there exists a subnet $\theta^t(w'_{\alpha(\beta)})$ converging to some x. Then $(\theta^t(w'_{\alpha(\beta)}), y) = (w'_{\alpha(\beta)}, \theta(y)) \to (w', \theta(y)) = (\theta^t(w'), y)$, but also $\to (x, y)$. So $x = \theta^t(w')$, and $\theta^t(w'_{\alpha(\beta)}) \to \theta^t(w')$. Since this holds for all convergent subnets of $\theta^t(w'_\alpha)$, it follows that $\theta^t(w'_\alpha) \to \theta^t(w')$.

Finally: $w' \to \langle w', w \rangle$ is weak* continuous on S_1'; therefore $w' \to (\theta^t(w'), \cdot)^\sim$ has a version with weak* continuous path functions on S_1'. Since θ^t is a homeomorphism from S_1' in the weak* topology onto $\{y: |y| \leqq 1\}^0$, it follows that $x \to (x, \cdot)^\sim$, $x \in \{y: |y| \leqq 1\}^0$, has a version with continuous path functions, i.e. $\{y: |y| \leqq 1\}$ is a GC set.

(b) $\Rightarrow$ (c): this is part of Theorem 4.1 of Dudley [7].

(c) $\Rightarrow$ (a): As has already been remarked, it is obvious that n-measurability by projections implies n-measurability. It remains only to show continuity of $|\cdot|$. This is part of Corollary 5.4 in Gross [13], so we are done. However, as a matter of separate interest, we will provide a simple direct proof of the continuity of $|\cdot|$. The technique is a variant of that of Gross, but avoids many difficulties.

Let G be a finite-dimensional subspace of H such that $n(\{x: |P_F x| \geqq 1\}) < 1/4$ for any finite-dimensional $F \perp G$. $|\cdot|$ is bounded on the unit ball of G, by finite-dimensionality. For arbitrary y, $|y| \leqq |P_G y| + |P_{G^\perp} y|$. Thus, to show $|\cdot|$ is bounded on the unit ball of H, it suffices to do so for the unit ball of $G^\perp$.

Let $\|y\| = 1$, $y \perp G$, and let F be the subspace spanned by y, so $P_F x = (x, y)y$. Then $n(\{x: |P_F x| \geqq 1\}) < 1/4$. But the left hand side is just

$$(2\pi)^{-1/2} \int_{1/|y|}^{\infty} e^{-t^2/2} dt ,$$

so $|y|$ is bounded. □

Finally, one may ask: does *every* "reasonable" Gaussian measure of mean zero on a topological linear space arise as a countably additive extension of the image of the canonical normal distribution via a continuous linear map? The answer is yes, in many cases, as the following result shows:

THEOREM 4. *Let W be a locally convex space, W' its dual, and μ a mean zero Gaussian measure on the weak Borel sets of W, such that* $\sup\{\mu(K)$: *K convex and weakly compact*$\} = 1$. *Then there is a continuous map θ from a Hilbert space H into W such that, if n is the normal distribution on H, then $\mu \mid \mathcal{C}(W, W') = n \circ \theta^{-1}$. $\theta(H)$ will be dense in W if and only if no proper closed subspace of W has measure* 1.

Proof. Let H be the closure in $L^2(\mu)$ of the random variables $\langle w', \cdot\rangle^\sim$, $w' \in W'$, and let ψ: $W' \to H$ be defined by $\psi(w') = \langle w', \cdot\rangle^\sim$. If $w'_\alpha \to w'$ in the Mackey topology on W', i.e. uniformly on convex weakly compact sets in W, then $\langle w'_\alpha, \cdot\rangle^\sim \to \langle w', \cdot\rangle^\sim$ in probability, by assumption. Thus, since convergence in probability for Gaussian random variables implies mean square convergence, ψ is continuous from the Mackey topology on W' into H. Then ψ^t is continuous from H into the Mackey topology on W (see Schaefer [18, p. 130]). Since this topology is at least as fine as the original topology on W, ψ^t is continuous: $H \to W$. Let $\theta = \psi^t$; then $\psi = \theta^t$. It is trivial to check that $n \circ \theta^{-1} = \mu \mid \mathcal{C}(W, W')$. $\theta(H)$ is dense $\Leftrightarrow \psi$ is one-to-one $\Leftrightarrow \langle w', \cdot\rangle^\sim$ is the zero random variable only when $w' = 0 \Leftrightarrow \{w: \langle w', w\rangle = 0\}$ cannot have measure 1 for $w' \neq 0 \Leftrightarrow$ no proper closed subspace of W can have measure 1. □

It should be remarked that the above Hilbert space is not always separable.

A version of this theorem may be obtained for non-Gaussian measures on Fréchet spaces, as follows.

LEMMA 6. *Let $\xi_1, \xi_2, \cdots$ be non-negative random variables. Then $\exists$ positive real numbers $c_1, c_2, \cdots$ such that $\sum_{j=1}^\infty c_j\xi_j < \infty$ with probability* 1.

Proof. Choose K_j so that $Pr\{\xi_j > K_j\} < 1/2^j$. Then

$$Pr\{\xi_j > K_j \text{ for some } j > k\} \leqq \sum_{j=k+1}^\infty 1/2^j = 1/2^k .$$

Let $c_j = 1/2^j K_j$. Then:

$$Pr\{c_j\xi_j > 1/2^j \text{ for some } j > k\} \leqq 1/2^k ,$$

so

$$Pr\{\textstyle\sum_{j=k+1}^\infty c_j\xi_j \leqq 1/2^k\} \geqq 1 - 1/2^k ,$$

and $Pr\{\sum_{j=1}^\infty c_j\xi_j < \infty\} \geqq 1 - 1/2^k$ for all k, hence $=1$. □

Part of the idea of the following construction is hinted at in Veršik [20].

THEOREM 5. *Let μ be a regular Borel measure on the Fréchet space W.*

Then there exists a Hilbert space H, a cylinder set measure m on $\mathcal{C}(H, H)$ such that $x \to (x, \cdot)^{\sim}$ is continuous from H into random variables with respect to convergence in probability, and a continuous linear map $\theta: H \to W$ such that $m \circ \theta^{-1} = \mu \mid \mathcal{C}(W, W')$. If W is separable, θ may be taken to have dense range.

Proof. First, we construct H, θ, and m satisfying all the requirements except density of $\theta(H)$. Let $\|\cdot\|_1 \leqq \|\cdot\|_2 \leqq \cdots$ be a complete set of seminorms for W. Let $S_{j,a} = \{w: \|w\|_j \leqq a\} = aS_{j,1}$, and $S'_{j,a} = aS^0_{j,1} = (S_{j,a^{-1}})^0$. Then it is known that the topology of uniform convergence on compact subsets of W is the finest locally convex topology on W' which agrees with the weak* topology when restricted to each $S'_{j,a}$; (see Schaefer [18, IV. 6.3, p. 151]).

Choose $c_j > 0$ so that $\sum_{j=1}^{\infty} c_j \|w\|_j < \infty$ μ-almost everywhere by Lemma 6. Let

$$\Delta = \int e^{-\Sigma c_j \|w\|_j} d\mu(w) ,$$

and let

$$d\nu(w) = \Delta^{-1} e^{-\Sigma c_j \|w\|_j} d\mu(w) .$$

We claim that $\int |\langle w', w\rangle|^2 d\nu(w) < \infty$ for each $w' \in W'$. For there exist some k and $b > 0$ such that $|\langle w', w\rangle| \leqq b \|w\|_k$. Thus,

$$\int |\langle w', w\rangle|^2 d\nu(w) \leqq b^2\Delta^{-1} \int \|w\|_k^2 \, e^{-\Sigma_j c_j \|w\|_j} d\mu(w)$$

$$\leqq b^2\Delta^{-1} \int \|w\|_k^2 \, e^{-c_k \|w\|_k} d\mu(w) < \infty .$$

Let H be the closed linear subspace of $L^2(\nu)$ spanned by the equivalence classes of the functions $\langle w', \cdot\rangle$, $w' \in W'$. Let $\psi: W' \to H$ be defined by setting $\psi(w') =$ the equivalence class of $\langle w', \cdot\rangle$.

We now claim that ψ is weak* continuous on each $S'_{j,a}$. For let $w'_\alpha \in S'_{k,a}$ and $w'_\alpha \to w'$ in the weak* topology. Then for any compact set $K \subset W$, $\|\psi(w'_\alpha) - \psi(w')\|^2 = \int |\langle w'_\alpha - w', w\rangle|^2 d\nu(w) = \int_K |\langle w'_\alpha - w', w\rangle|^2 d\nu(w) + \int_{K^c} |\langle w'_\alpha - w', w\rangle|^2 d\nu(w) \leqq \sup\{|\langle w'_\alpha - w', w\rangle|^2: w \in K\} + 4a^2 \int_{K^c} \|w\|_k^2 \, d\nu(w)$. Since μ is regular and $\|\cdot\|_k^2 e^{-\Sigma_j c_j \|\cdot\|_j}$ is integrable, it follows that $\exists K$ compact such that $\int_{K^c} \|w\|_k^2 \, d\nu(w) < \varepsilon/a^2$. Then also $\exists \alpha_0$ such that $\alpha > \alpha_0 \Rightarrow |\langle w'_\alpha - w', w\rangle|^2 < \varepsilon/2$ for all $w \in K$, by the remarks in the first paragraph. So for $\alpha > \alpha_0$ one has $\|\psi(w'_\alpha) - \psi(w')\|^2 < \varepsilon$.

Thus ψ is continuous: $W' \to H$, when W' is given the topology of uniform convergence on compact sets (for if not, the sets $\psi^{-1}(S_a)$ could be used to get a finer locally convex topology agreeing with the weak* topology on all $S'_{k,a}$). A fortiori, ψ is continuous from the Mackey topology on W' (with respect to the dual pair $\langle W, W'\rangle$) to H. Then ψ^t exists, continuous, from H to the Mackey topology on W. Let $\theta = \psi^t$. Then θ is continuous: $H \to W$. It remains to construct m.

Consider the linear stochastic process on H gotten by assigning to the equivalence class $[f]$ in $L_2(\nu)$ of a function f the corresponding equivalence class in $L^0(\mu)$. This gives rise to a unique cylinder set measure m on H, characterized by

$$\int_H e^{i([f],y)}dm(y) = \int_W e^{if(w)}d\mu(w) .$$

The map: $L^2(\nu) \to L^0(\mu)$ so obtained is continuous from H into $L^0(\mu)$, with respect to the norm in H and convergence in probability in $L^0(\mu)$. Furthermore, for f of the form $\langle w', \cdot\rangle$, we have

$$\begin{aligned}\int_W e^{i\langle w',w\rangle}dm \circ \theta^{-1} &= \int_H e^{i\langle w',\theta(y)\rangle}dm(y) = \int_H e^{i(\theta^t w',y)}dm(y) \\ &= \int_W e^{i(\psi w',y)}dm(y) = \int_W e^{i\langle w',w\rangle}d\mu(w) ,\end{aligned}$$

so $m \circ \theta^{-1} = \mu \mid \mathcal{C}(W, W')$.

Now, assuming W to be separable, let (H_0, θ_0, m_0) be the (H, θ, m) constructed above. We shall define a Hilbert space H_1 and a continuous map $\theta_1: H_1 \to W$ with dense range. Then we define $H = H_0 \oplus H_1$, $\theta(x_0 \oplus x_1) = \theta_0(x_0) + \theta_1(x_1)$, and m on $\mathcal{C}(H, H)$ by $m = m_0 \circ i^{-1}$, where i is the injection: $H_0 \to H$.

The construction of H_1 and θ_1 is simple. Choose a sequence $w_1, w_2, \cdots$ in W with dense linear span, so normalized that $\| w_j \|_j \leqq 2^{-j}$. Let H_1 be the space l^2 of square-summable sequences $\{x_j\}$. Let $\theta_1(\{x_j\}) = \sum x_j w_j$. (The partial sums form a Cauchy sequence for $\| \cdot \|_k$ for all k, hence converge.) Now θ_1 is continuous from l^2 into $(W, \| \cdot \|_k)$ for all k and hence continuous. □

5. Two inequalities of L. Gross for normal measure

Here are two inequalities concerning the normal measure of convex centrally symmetric sets. Their original proofs in Gross [13, Lemmas 4.2 and 5.2] are rather difficult. These facts are basic to some results of Dudley [7] as well as Gross [13] which we used above in proving Theorem 3. It seems desirable, therefore, to present simple and direct proofs.

Let H be a finite dimensional Hilbert space, and C a centrally symmetric closed convex set in H. n_H will be the canonical normal measure on H.

(A) If T is a linear transformation in H with $\|T\| \leqq 1$, then $n_H(T(C)) \leqq n_H(C)$.
(B) If K is a subspace of H, then $n_H(C) \leqq n_H((C \cap K) + K^{\perp})$.

Proof of (A). Let H have dimension $k + 1$. For $k = 0$, the inequality is obvious, so assume $k \geqq 1$. Since n_H is orthogonally invariant, T may be assumed non-negative and self-adjoint, so that it has a diagonal matrix $\begin{pmatrix} t_0 & & 0 \\ & \ddots & \\ 0 & & t_k \end{pmatrix}$ for some orthonormal basis, with $0 \leqq t_j \leqq 1$. By continuity, one may assume $0 < t_j < 1$ for all j. Such a matrix may be written as a product of matrices of the form $\begin{pmatrix} s_0 & & 0 \\ & \ddots & \\ 0 & & s_k \end{pmatrix}$, where exactly one of the s_j is 1, and all the others are some constant t, $0 < t < 1$. Thus, it suffices to assume that $T = P + tP^{\perp}$ for some 1-dimensional projection P. Let L be the (one-dimensional) range of P, and $M = L^{\perp}$. For any set $S \subset H$, and $y \in M$, let

$$S_y = \{x \in L: x + y \in S\} .$$

Thus,

$$T(C)_y = \{x \in L: x + y \in T(C)\} = \{x \in L: x + t^{-1}y \in C\} = C_{t^{-1}y} .$$

So

$$n_H(T(C)) = \int_M n_L(T(C)_y) dn_M(y) = \int_M n_L(C_{t^{-1}y}) dn_M(y) ,$$

while $n_H(C) = \int_M n_L(C_y) dn_M(y)$. It will now be shown, for each $y \in M$, that $n_L(C_{t^{-1}y}) \leqq n_L(C_y)$. Obviously one may assume $y \neq 0$.

At this point we suggest that the reader draw a picture. Consider the plane N spanned by L and y. Let $D = C \cap N$; then D is a centrally symmetric convex set, and $C_{t^{-1}y}$ is the orthogonal projection in L of the intersection of D with the line $t^{-1}y + L$ in N; call this latter intersection I_t. Now: $I_{-t} = -I_t$. The convex hull of $I_t \cup I_{-t}$ is a parallelogram $\Pi \subset D$. The projection on L of $\Pi \cap (y + L)$ is an interval of the same length as $C_{t^{-1}y}$, but its midpoint is closer to 0. Consequently this projection has n_L-measure $\geqq$ that of $C_{t^{-1}y}$. But finally, $D \cap (y + L) \supset \Pi \cap (y + L)$, and the projection on L of $D \cap (y + L)$ is C_y. Thus $n_L(C_y) \geqq n_L(C_{t^{-1}y})$. □

Proof of (B). It follows from (A) that if T_k is defined for integers $k > 0$ by $T_k = kP_K^{\perp} + P_K$, then $n_H(T_k(C)) \geqq n_H(C)$, since C is obtained from the symmetric convex set $T_k(C)$ by applying T_k^{-1}, and $\|T_k^{-1}\| \leqq 1$. Now let $x = y + z$, $y \in K$ and $z \in K^{\perp}$. Suppose $y \notin C$. Then for some $\varepsilon > 0$, $S_\varepsilon(y)$ is disjoint from C. So $T_k(S_\varepsilon(y))$ is disjoint from $T_k(C)$. Now: $T_k^{-1}(y+z) = y + k^{-1}z \in S_\varepsilon(y)$ for $k^{-1} < \varepsilon/|z|$; so $x = y+z \notin T_k(C)$ when $k > \varepsilon^{-1}|z|$. Therefore $x \in \bigcap_k \bigcup_{j>k} T_j(C)$. So $\bigcap_k \bigcup_{j>k} T_j(C) \subset (C \cap K) + K^\perp$. Now, for each j, $n_H(T_j(C)) \geqq n_H(C)$, so $n_H(\bigcup_{i>k} T_j(C)) \geqq n_H(C)$ for each k, and

$$n_H((C \cap K) + K^\perp) \geqq n_H(\textstyle\bigcap_k \bigcup_{j>k} T_J(C)) \geqq n_H(C) . \qquad \square$$

The following is a related fact, proved by a method suggested by the work of Gross.

(C) Suppose L is one-dimensional, $M = L^\perp$, and for $x \in L$,

$$C_x = \{y \in M: x + y \in C\}, \quad |t| \leqq 1 .$$

Then $n_M(C_x) \geqq n_M(C_{tx})$.

Proof. Let B_r be the ball of radius r about 0 in M and let λ be Lebesgue measure in M. Whenever $x \in L$, $r > 0$, and $|t| \leqq 1$,

$$\frac{1}{2}(1+t)(C_x \cap B_r) + \frac{1}{2}(1-t)(C_{-x} \cap B_r) \subset B_r \cap C_{tx} ,$$

so $\lambda(B_r \cap C_{tx}) \geqq \lambda\{(1/2)(1+t)(C_x \cap B_r) + (1/2)(1-t)(C_x \cap B_r)\} \geqq \lambda(C_x \cap B_r)$ by the Brunn-Minkowski Theorem (Bonnesen and Fenchel [3, p. 88]). Since

$$n_M(A) = (2\pi)^{-m/2} \int_0^\infty \lambda(B_r \cap A) r e^{-r^2/2} dr ,$$

(where $m = \dim M$), for any measurable set $A \subset M$, we get the desired result. $\square$

Finally, here is an example related to (B).

(D) There exist a subspace K and a C such that $T(C)$ is never included in $(C \cap K) + K^\perp$ for any orthogonal transformation T.

Proof. Let $H = R^3$ and let K be the (x, y) plane. Let D be an equilateral triangle of side 1 in the plane $z = 1$, centered on the z axis. Let V be the set of 3 vertices of D. Let C be the symmetric convex hull of D.

Then $C \cap K$ is a regular hexagon, all of whose points are closer to the z axis than V is. Thus if T is orthogonal and $T(C) \subset (C \cap K) + K^\perp$, T must move all the vertices closer to the z axis. Let S be the sphere centered at 0 with V on its surface. Then $T(S) = S$. Call $S \cap \{z = 1\}$ the "Arctic circle". Then the points of $T(V)$ are either north of the Arctic circle or south of the Antarctic circle $S \cap \{z = -1\}$. They cannot be divided among both since $2 > 1$. So we can assume $T(V)$ is all strictly north of the Arctic circle. But this is impossible since T would take that circle into a circle of smaller radius. $\square$

MASSACHUSETTS INSTITUTE OF TECHNOLOGY
UNIVERSITY OF CALIFORNIA, BERKELEY
UNIVERSITY OF CALIFORNIA, BERKELEY

REFERENCES

[1] P. BILLINGSLEY, Convergence of probability measures, Wiley (1968).
[2] S. BOCHNER, Harmonic analysis and the theory of probability, University of California Press (1959).
[3] T. BONNESEN and W. FENCHEL, Theorie der Konvexen Körper, (Berlin, 1934).
[4] N. BOURBAKI, Eléments de Mathématiques, Livre VI (Intégration), Chapitre 9, Intégration sur les espaces topologiques séparés, Hermann, Paris (1970).
[5] S. CHEVET, "*p-ellipsoides de l^q, exposant d'entropie; mesures cylindriques gaussiennes*", C. R. Acad. Sci. Paris, **269** (1969), A658-A660.
[6] R. M. DUDLEY, *Random linear functionals*, Trans. Amer. Math. Soc. **136** (1969), 1-24.
[7] ———, *The sizes of compact subsets of Hilbert space and continuity of Gaussian processes*, J. Functional Analysis, **1** (1967), 290-330.
[8] ——— and J. FELDMAN, *On seminorms and probabilities*, preprint privately circulated (June 1969).
[9] J. FELDMAN, "*Some questions about measures in linear spaces*", in Proc. of the Conference on functional integration and quantum field theory, April 28-30, 1966, MIT, (Mimeographed notes).
[10] ———, *Making cylinder set measures countably additive by means of seminorms*, preprint privately circulated (May 1969).
[11] I. M. GELFAND and N. YA. VILENKIN, Applications of harmonic analysis, Generalized functions, Volume **4** (translated by A. Feinstein), Academic Press (1964).
[12] L. GROSS, Harmonic analysis on Hilbert space, Memoirs Amer. Math. Soc. **46** (1963).
[13] ———, *Measurable functions on Hilbert space*, Trans. Amer. Math. Soc. **105** (1962), 372-390.
[14] ———, "*Abstract Wiener spaces*", in Proc. 5th. Berkeley Symposium on Math. Stat. and Prob., University of California Press (1965).
[15] J. KUELBS, *Abstract Wiener spaces and applications to analysis*, Pac. J. Math. **31** (1969), 433-450.
[16] YU. V. PROKHOROV, "*The method of characteristic functionals*", in Proc. 4th Berkeley Symposium of Math. Stat. and Prob., University of California Press (1961), 403-419.
[17] H. SATO, *Gaussian measure on a Banach space and abstract Wiener space*, Nagoya Math. J. **36** (1969), 65-83.
[18] H. SCHAEFER, Topological Vector spaces, N. Y., MacMillan (1966).
[19] I. E. SEGAL, *Distributions in Hilbert space and canonical systems of operators*, Trans. Amer. Math. Soc. **88** (1958), 12-41.
[20] A. VERŠIK, *Duality in the theory of measure in linear spaces*, Soviet Math. Doklady, **7** (1966), 1210-1214.

(Received May 14, 1970)

Annals of Mathematics, **104** (1976), 391

Corrections to "On seminorms and probabilities, and abstract Wiener spaces"

By R. M. DUDLEY, J. FELDMAN and L. LECAM

As M. Ann Piech kindly informed us, the proposed new proof of the known inequality (A) in [2], p. 406, is incorrect. In dimension $\geqq 3$, not every diagonal matrix with eigenvalues in (0, 1) is a product of such matrices with one eigenvalue 1 and the rest all equal and less than 1. Instead, we can take one eigenvalue less than 1 and the rest equal to 1. Then we can use the Brunn-Minkowski inequality as in [2, (C) p. 407], and in other known proofs ([1], [3], [4]).

We are also grateful to Jim Kuelbs and others for pointing out to us that the Hilbert space mentioned in the middle of [2, p. 403], *is* always separable.

MASSACHUSETTS INSTITUTE OF TECHNOLOGY
UNIVERSITY OF CALIFORNIA, BERKELEY
UNIVERSITY OF CALIFORNIA, BERKELEY

REFERENCES

[1] T. W. ANDERSON, The integral of a symmetric unimodal function over a symmetric convex set and some probability inequalities, Proc. A.M.S. **6** (1955), 170-176.
[2] R. M. DUDLEY, J. FELDMAN and L. LECAM, On seminorms and probabilities, and abstract Wiener spaces, Ann. of Math. (2) **93** (1971), 390-408.
[3] C. FEFFERMAN, M. JODEIT and M. D. PERLMAN, A spherical surface measure inequality for convex sets, Proc. A.M.S. **33** (1972), 114-119.
[4] L. GROSS, Measurable functions on Hilbert space, Trans. A.M.S. **105** (1962), 372-390.

(Received August 3, 1976)

The Annals of Probability
1973, Vol. 1, No. 1, 66–103

SAMPLE FUNCTIONS OF THE GAUSSIAN PROCESS[1]

BY R. M. DUDLEY

Massachusetts Institute of Technology

This is a survey on sample function properties of Gaussian processes with the main emphasis on boundedness and continuity, including Hölder conditions locally and globally. Many other sample function properties are briefly treated.

The main new results continue the program of reducing general Gaussian processes to "the" standard isonormal linear process L on a Hilbert space H, then applying metric entropy methods. In this paper Hölder conditions, optimal up to multiplicative constants, are found for wide classes of Gaussian processes.

If H is L^2 of Lebesgue measure on R^k, L is called "white noise." It is proved that we can write $L = P(D)[x]$ in the distribution sense where x has continuous sample functions if $P(D)$ is an elliptic operator of degree $> k/2$. Also L has continuous sample functions when restricted to indicator functions of sets whose boundaries are more than $k - 1$ times differentiable in a suitable sense.

Another new result is that for the Lévy(-Baxter) theorem $\int_0^1 (dx_t)^2 = 1$ on Brownian motion, almost sure convergence holds for any sequence of partitions of mesh $o(1/\log n)$. If partitions into measurable sets other than intervals are allowed, the above is best possible: $\mathcal{O}(1/\log n)$ is insufficient.

TABLE OF CONTENTS

0. Introduction. This paper is an attempt to survey what is known about sample functions of Gaussian processes. Emphasis is given to continuity and boundedness properties, as in most of the literature, although other topics are treated. On many topics, new results are presented.

The principal idea is to study one Gaussian process, defined as follows. A sequence $\{X_n\}$ of random variables will be called *orthogaussian* iff they are independent with mean 0 and variance 1, $\mathcal{L}(X_j) \equiv N(0, 1)$. Now let H be a real,

Received February 4, 1972; revised June 12, 1972.

[1] This research was supported by National Science Foundation grant GP-29072.

AMS 1970 *subject classifications*. Primary, 60G15, Gaussian processes; 60G17, sample path properties; Secondary, 60G20, generalized stochastic processes.

Key words and phrases. Gaussian processes, sample functions, white noise, Hölder conditions, metric entropy.

E. Giné et al. (eds.), *Selected Works of R.M. Dudley*, Selected Works in Probability and Statistics,
DOI 10.1007/978-1-4419-5821-1_13, © Springer Science+Business Media, LLC 2010

infinite-dimensional Hilbert space. A Gaussian process L on H will be called *isonormal* iff L is a linear map from H into real Gaussian random variables with $EL(x) \equiv 0$ and $EL(x)L(y) \equiv (x, y)$ for all $x, y \in H$. (The term "isonormal" is due, I believe, to I. E. Segal, who originally called it the "normal distribution" (Segal (1954)).) Specifically if $\{\varphi_n\}$ is any orthonormal basis of H so that for $x \in H$, $x = \sum x_n \varphi_n$, let $L(x) = \sum x_n Y_n$ where the Y_n are orthogaussian.

As elaborated later on, most of the study of sample function continuity and boundedness of Gaussian processes reduces to the study of those sets $C \subset H$ on which the isonormal L has continuous or bounded sample functions, called GC-sets and GB-sets respectively. One measure of the size of a set C in H is the minimal number $N(C, \varepsilon)$ of sets of diameter $\leqq 2\varepsilon$ which cover it. GC-sets and GB-sets cannot quite be characterized in terms of $N(C, \varepsilon)$ (Dudley (1967) Proposition 6.10), but it still seems that such conditions are the most convenient general conditions now available. In Section 1 we give the weakest possible sufficient condition and the strongest possible necessary conditions on $N(C, \varepsilon)$ for the GC- and GB-properties. The best possible necessary condition for GB, namely $\limsup_{\varepsilon \downarrow 0} \varepsilon^2 \log N(C, \varepsilon) < \infty$, is new. The example proving it best possible contradicts one statement in an announcement by Sudakov (1971) while, on the other hand, the best possible necessary condition is proved by methods suggested in Sudakov's earlier note (1969), for which the inequality of Slepian ((1962) Lemma 1 page 468) is fundamental.

Section 2 treats moduli of continuity (Hölder conditions) on sample functions. If $\log N(C, \varepsilon)$ is a reasonably smooth function of ε asymptotically as $\varepsilon \downarrow 0$ (of course, it only changes by jumps!) we find uniform moduli of continuity which are optimal within a constant factor (Theorem 2.6 below). N. Kôno (1970) earlier found some results on moduli of continuity in terms of $N(C, \varepsilon)$.

Section 3 treats processes with stationary increments, which are accessible both by metric entropy methods (3.1) and by Fourier analytic methods (3.2). For these methods, "stationary increments" is fundamentally no worse than "stationary." This is the main point of Section 3, which contains no new hard results.

Given a measure space $(X, \mathscr{S}, \mu)$, the isonormal process L on $L^2(X, \mathscr{S}, \mu)$ is called *μ-noise*. Section 4 treats these processes. If μ is Lebesgue measure on R^k, we have what is called *white* noise W. Sub-section 4.1 shows that if $P(D)$ is an elliptic operator with constant coefficients of order m, then there is a process x_t, $t \in R^k$, with continuous sample functions and $P(D)[x_t] = W$ (in the distribution sense) iff $m > k/2$. In Sub-section 4.1 we restrict W to indicator functions χ_A of sets A whose boundaries are α times differentiable, obtaining continuous sample functions of the process $A \to W(\chi_A)$ if $\alpha > k - 1$ and not if $\alpha < k - 1$, conjecturally also not if $\alpha = k - 1$. Here α is not necessarily an integer. In this respect convex sets behave like sets with exactly twice differentiable boundaries. The proofs of these results depend on another paper (Dudley (1972b)). They answer questions raised by R. Pyke in Oberwolfach, March 1971.

Our final result on noise processes is an extension of the "Lévy–Baxter" theorem, on almost sure quadratic Brownian variation $\int_0^1 (dx_t)^2 = 1$, to partitions of measurable sets with mesh $o(1/\log n)$. In P. Lévy's original theorem the partitions were formed by successive refinement, allowing the mesh to approach 0 arbitrarily slowly; thus our results are independent.

Most of the rest of the paper surveys the literature. In Section 7 are two propositions showing that sample-continuity conditions for non-Gaussian processes are much different from Gaussian ones; sufficient conditions for the non-Gaussian case must be much stronger (7.1), while necessary conditions are weaker (7.2).

The bibliography at the end is no doubt uneven and incomplete. References to it are made with the date after the author's name. For a longer list, up to 1968, see J. Neveu (1968).

DEFINITION. A stochastic process $\{x_t, t \in T\}$ over a topological space T is *sample-continuous* if there is a version of the process with continuous sample functions, i.e., there is a countably additive probability measure on the space of continuous functions on T with the same joint distributions as x_t on finite subsets of T.

A real-valued process $\{x_t, t \in T\}$ will be called *sample-bounded* iff it has a version with bounded sample functions, i.e., there is a countably additive probability measure on the space of all bounded functions from T into X with the same joint distributions as x_t on finite subsets of T.

In this paper, we shall only consider real-valued Gaussian processes, although the results would carry over to complex processes easily. Gaussian measures can be defined on vector spaces, on locally compact Abelian groups (Urbanik (1960), Corwin (1970)) and on suitable homogeneous spaces.

The isonormal process L can be regarded as the only real Gaussian process in view of the theorem that Gaussian distributions are uniquely determined by their means and covariances. Thus let $\{x_t, t \in T\}$ be any real Gaussian process with mean $Ex_t = m_t$. Then $L(x_t - m_t) + m_t$ is another version of the same process, where we take H as the Hilbert space $L^2(\Omega, P)$. We can "forget" the specific joint distributions of $x_t - m_t$ over (Ω, P) and remember only the abstract, geometric Hilbert space structure of the function $t \to x_t - m_t \in H$. Then L will "remember" the joint distributions for us.

A set $C \subset H$ is called a GC-set iff L restricted to C is sample-continuous. C is called a GB-set iff L on C is sample-bounded. The following theorems from Dudley (1967) and Feldman (1971) show how these notions apply.

THEOREM 0.1. *Let $\{x_t, t \in T\}$ be a real Gaussian process over a metrizable space T with $Ex_t = m_t$. Then the following are equivalent.*

(a) *x_t is sample-continuous.*

(b) *$t \to L(x_t - m_t) + m_t$ is sample-continuous.*

(c) *$t \to m_t$ is continuous and $t \to L(x_t - m_t)$ is sample-continuous which implies $t \to x_t - m_t$ continuous $T \to H = L^2(\Omega, P)$.*

THEOREM 0.2. *Theorem* 0.1 *remains true if "continuous" is replaced by "bounded" throughout, and "metrizable space" replaced by "set." Also,* $L(x_t - m_t)$ *is sample-bounded if and only if* $\{x_t - m_t : t \in T\}$ *is a* GB-*set. A set B is* GB iff *its closed, convex, symmetric hull in H is* GB.

If B is GB then $\bar{L}(B) \equiv \sup\{|L(x)| : x \in A\}$, for any countable dense subset A of B, is a finite random variable which changes only with 0 probability if A is changed. Fernique (1970) and Landau and Shepp (1971) have shown that

$$\Pr(\bar{L}(B) > t) \leqq C \exp\{-t^2/2\alpha^2\}$$

for some $C < \infty$ and $\alpha < \infty$, specifically for any $\alpha > \sup\{\|x\| : x \in B\}$ (Marcus and Shepp (1971), Fernique (1971)). The latter's claim of an error in Landau–Shepp (1971) has been retracted. In any case Fernique's proofs are simpler. An error in an earlier preprint by Shepp alone was found by D. Cohn. Sudakov (1971) says his proof is based on a lemma in the preprint of Shepp. Thus Sudakov's theorem is open to doubt (see also the counter-example to a corollary, Remark after Theorem 1.1 below).

Turning now to GC-sets, we have the following theorem, due in present generality to J. Feldman (1971) with miscellaneous contributions by others (Jain and Kallianpur (1970); Dudley (1967) Theorem 4.6).

THEOREM 0.3. *Let T be a compact metric space and* $\{x_t, t \in T\}$ *a real Gaussian process with* $Ex_t \equiv 0$. *Let* $C \equiv \{x_t : t \in T\} \subset L^2(\Omega, P) = H$. *Assume* $t \to x_t$ *is continuous* $T \to H$. *Then the following are equivalent*:

(a) x_t *is sample-continuous.*

(b) C *is a* GC-*set.*

(c) *The closed, convex, symmetric hull of C is a* GC-*set.*

(d) *For every* $\varepsilon > 0$, $P(\bar{L}(C) < \varepsilon) > 0$.

(e) *For every orthornomal basis* $\{\varphi_n\}$ *of the linear span of C, the series* $\sum (x, \varphi_n)L(\varphi_n)$ *converges uniformly for* $x \in C$ *with probability* 1.

The Karhunen–Loève expansion of a Gaussian process $\{x_t, a \leqq t \leqq b\}$ is a special case of the orthogonal series in (e). That expansion involves the eigen-functions of the covariance kernel $K(s, t) = Ex_s x_t$, with respect to L^2 of Lebesgue measure. But for processes without stationarity properties, Lebesgue measure is not especially natural, and there is no good reason to single out the Karhunen–Loève expansion from other orthogonal expansions.

Dudley, Feldman and LeCam (1971) have shown, among other things, that the class of GC-sets is stable under vector sum (as is obvious for GB-sets, as well as homothetic invariance of both classes). It is also known that the compact GC-sets are precisely those compact GB-sets which are not "maximal" in a sense defined by compact operators (J. Feldman (1971)).

Fernique (1971) has given rather sharp sufficient conditions for the GB- and GC-properties, as follows. Let (K, μ) be a probability space. Let $G(K, \mu) = \{f : \exists \alpha > 0, \int_K \exp(\alpha f^2(x))\, d\mu(x) < \infty\}$. Then the μ-equivalence classes of functions in $G(K, \mu)$ form a Banach space for either of the norms

$$N(f) = \sup_{p \geq 1} (p!)^{-\frac{1}{2}p} \|f\|_{L^{2p}},$$
$$N'(f) = \inf \{\alpha > 0: \int \exp(f^2/\alpha^2) - 1 \, d\mu \leqq 1\}.$$

Here $N \leqq N' \leqq 2N$. Let $G_C(K, \mu)$ be the closure of the bounded functions in $G(K, \mu)$, a proper closed linear subspace in general. Then the dual space $G_C^*(K, \mu)$ consists of the μ-equivalence classes of functions f such that

$$\int |f|(\max(0, \log |f|))^{\frac{1}{2}} \, d\mu < \infty.$$

The following is a reformulation of Théorème 5 of Fernique (1971).

THEOREM (Fernique). *Suppose $C \subset H$ and B is a compact subset of H. Let μ be a finite measure on B. Let f be a real-valued function on $B \times C$ such that the map $x \to f(\cdot, x)$ is bounded from C into $G_C^*(B, \mu)$. Suppose that for all $x, y \in C$,*

$$(x, y) = \iint (s, t) f(s, x) f(s, y) \, d\mu(s) \, d\mu(t).$$

Then C is a GB*-set. Furthermore if $x \to f(\cdot, x)$ is continuous from C into G_C^* with weak-star topology, then C is a* GC*-set.*

If in the original Théorème 5 of Fernique (1971), one lets $K = [0, 1]$, $\Gamma(s, t) = 1$ for $s = t$ and 0 for $s \neq t$, then there are difficulties.

Fernique shows that his sufficient condition is also necessary in many cases, but to find f and B still requires ingenuity in different cases.

1. Continuity and boundedness; metric entropy. Let C be a subset of a metric space (S, d). Given $\varepsilon > 0$, let $N(C, \varepsilon)$ be the smallest n such that there exist sets $A_1, \cdots, A_n : C \subset \bigcup_{j=1}^{n} A_j$, and for each j, $d(x, y) \leqq 2\varepsilon$ for all $x, y \in A_j$. Let $H(C, \varepsilon) = \log N(C, \varepsilon)$. $H(C, \varepsilon)$ is called the *metric entropy* of C, following G. G. Lorentz (1966). Kolmogorov (1956), who invented this notion, called it "ε-entropy," as have most others who used it. On the other hand Posner, Rodemich and Rumsey (1969) use "ε-entropy" to mean infimum of information-theoretic entropy of partitions of C into sets of diameter $\leqq \varepsilon$, for a probability measure on C. Lorentz's term *metric entropy* seems well adapted as a name for the purely metric $H(C, \varepsilon)$ (there is no given or natural probability measure P on C here).

The *exponent of entropy* $r(C)$ is defined by

$$r(C) = \limsup_{\varepsilon \downarrow 0} \log H(C, \varepsilon)/\log(1/\varepsilon).$$

It is known that C is a GB-set in H if $r(C) < 2$ and not if $r(C) > 2$ (Sudakov (1969), Chevet (1970)). It is also known, however, (Dudley (1967) Proposition 6.10) that we may have $H(E, \varepsilon)/H(Oc, \varepsilon) \to 0$ as $\varepsilon \downarrow 0$ where Oc is a GC-set and GB-set while E is neither, $r(E) = r(Oc) = 2$. Thus inside $r(C) = 2$, there is an ambiguous range where $H(C, \varepsilon)$ does not determine whether C is a GB-set. V. N. Sudakov (1971) has recently announced a necessary and sufficient geometric condition on C for the GB-property, but this condition seems difficult to apply in practice, no proof is yet published, and the result is in doubt. Thus, despite the ambiguity at $r(C) = 2$, the metric entropy conditions seem still the most useful. Here is a summary of the best possible such conditions.

THEOREM 1.1. *Let C be a compact set in H. Then*

(a) *C is always a GC-set if $\int_0^1 H(C, x)^{\frac{1}{2}}\, dx < \infty$, in particular if for some $\delta > 0$ and all small enough ε, $H(C,\varepsilon) \leqq 1/\varepsilon^2|\log \varepsilon|^2[\log|\log \varepsilon|]^2 \cdots [\log \cdots \log|\log \varepsilon|]^{2+\delta}$.*

(b) *There are non-GB-sets C with*

$$H(C, \varepsilon) \leqq 1/\varepsilon^2|\log \varepsilon|^2[\log|\log \varepsilon|]^2 \cdots [\log \cdots \log|\log \varepsilon|]^2 .$$

(c) *C is never a GB-set if $\limsup_{\varepsilon \downarrow 0} \varepsilon^2 H(C, \varepsilon) = +\infty$, in particular if $H(C, \varepsilon) \geqq \varepsilon^{-2} \log \cdots \log|\log \varepsilon|$ for a sequence of values of $\varepsilon \downarrow 0$.*

(d) *There is a GB-set C with $\limsup_{\varepsilon \downarrow 0} \varepsilon^2 H(C, \varepsilon) > 0$.*

(e) *There are GC-sets C such that $\sup_{0<\varepsilon\leqq\delta} \varepsilon^2 H(C, \varepsilon) \downarrow 0$ arbitrarily slowly as $\delta \downarrow 0$.*

PROOF. For (a) note that $\int_0^1 H(C, x)^{\frac{1}{2}}\, dx < \infty$ iff $\sum_{n=1}^{\infty} H(C, 2^{-n})^{\frac{1}{2}}/2^n < \infty$. Then the result is stated in Dudley ((1967) Theorem 3.1). (A small error in the proof, noted by J. Neveu, is corrected in the proof of Theorem 2.1 below.) Sudakov ((1971) Theorem 5) has another approach to this fact. A related fact for processes $\{x_t, 0 \leqq t \leqq 1\}$ is due to Delporte (1964) with a neater formulation by Fernique (1964) and another recent proof by Garsia, Rodemich and Rumsey (1970).

For (b) we apply the examples of Fernique ((1964) Théorème 3, Remarque); for a more detailed discussion see Marcus and Shepp (1970), (1971).

For (c) we shall apply Slepian's inequality (1962) as in Sudakov's proof (1969) that $r(C) > 2$ implies C is not GB. I am grateful to S. Chevet, from whose presentation (1970b) the following lemma and proof are adapted, and to E. Giné who gave another exposition. Let Φ be the standard normal distribution function

$$\Phi(t) \equiv (2\pi)^{-\frac{1}{2}} \int_{-\infty}^{t} \exp(-x^2/2)\, dx \equiv 1 - F(t) .$$

LEMMA 1.2 (Sudakov–Chevet). *Let $a_1, \cdots, a_n \in H$, $1 \leqq M < \infty$, $0 < \varepsilon \leqq 1$, $\|a_j\| \leqq M$ for all j, and $\|a_i - a_j\| \geqq \varepsilon$ for $i \neq j$. Then*

$$F(1) \Pr\{L(a_j) \leqq 1, j = 1, \cdots, n\} < 2^{-n-1} + (2\pi)^{-\frac{1}{2}} \int_0^{\infty} \exp(-t^2/2)\Phi(Kt/\varepsilon)^n\, dt$$

where $K = (2(M^2 + 1))^{\frac{1}{2}}$.

PROOF. Let $\{e_j\}_{j=1}^{\infty}$ be an orthonormal basis of H such that $a_1, \cdots, a_n$ belong to the linear span of $e_1, \cdots, e_n$. Let H_{n+1} be the linear span of $e_1, \cdots, e_{n+1}$. Let G be the standard Gaussian measure on H_{n+1}, and $b_i = a_i - e_{n+1}$, $i = 1, \cdots, n$. Then

$$\begin{aligned} F(1)G\{z : (z, a_i) \leqq 1, i = 1, \cdots, n\} &= G\{z : (z, e_{n+1}) \geqq 1, (z, a_i) \leqq 1, i = 1, \cdots, n\} \\ &\leqq G\{z : (z, b_i) \leqq 0, i = 1, \cdots, n\} . \end{aligned}$$

Let $b_{ij} = (b_i, b_j)/\|b_i\| \|b_j\|$. Let θ be the angle between b_i and b_j, so that $b_{ij} = \cos\theta$. Then b_{ij} is largest for $i \neq j$ when $\|a_i\| = \|a_j\| = M$. Hence

$$\begin{aligned} b_{ij} &= 2\cos^2(\tfrac{1}{2}\theta) - 1 \\ &\leqq 2(M^2 + 1 - \varepsilon^2/4)/(M^2 + 1) - 1 \leqq 1 - \varepsilon^2/K^2 \leqq 1/(1 + \varepsilon^2/K^2) . \end{aligned}$$

Let $f_i = \varepsilon e_i/K - e_{n+1}$, $f_{ij} = (f_i, f_j)/\|f_i\| \|f_j\| = 1/(1 + \varepsilon^2/K^2)$ for $i \neq j$, $b_{ii} = f_{ii} = 1$. Thus Slepian's inequality ((1962) Lemma 1 page 468) can be applied to $b_i/\|b_i\|$ and $f_i/\|f_i\|$, giving

$$\begin{aligned} G\{z &: (z, b_i) \leqq 0, i = 1, \cdots, n\} \\ &\leqq G\{z : (z, f_i) \leqq 0, i = 1, \cdots, n\} \\ &= G\{z : \varepsilon(z, e_i)/K \leqq (z, e_{n+1}), i = 1, \cdots, n\} \\ &= (2\pi)^{-\frac{1}{2}} \textstyle\int_{-\infty}^{\infty} G\{z : (z, e_i) \leqq tK/\varepsilon, i = 1, \cdots, n\} \exp(-t^2/2)\, dt \\ &= (2\pi)^{-\frac{1}{2}} \textstyle\int_{-\infty}^{\infty} \Phi(Kt/\varepsilon)^n \exp(-t^2/2)\, dt \\ &= (2\pi)^{-\frac{1}{2}} \textstyle\int_0^{\infty} \exp(-t^2/2)[\Phi^n(Kt/\varepsilon) + \Phi^n(-Kt/\varepsilon)]\, dt \\ &\leqq 2^{-n-1} + (2\pi)^{-\frac{1}{2}} \textstyle\int_0^{\infty} \exp(-t^2/2)\Phi^n(Kt/\varepsilon)\, dt . \qquad \square \end{aligned}$$

To prove (c) from the lemma, we choose $\varepsilon_k \downarrow 0$ such that $\varepsilon_k^2 H(C, \varepsilon_k) \geqq k$, $k = 1, 2, \cdots$. In the lemma let $\varepsilon = \varepsilon_k$, $n = N(C, \varepsilon_k)$. Let $F(t) \equiv 1 - \Phi(t)$. Since $(1 - F)^n \leqq e^{-nF}$ it is enough to prove that $N(C, \varepsilon_k)F(s/\varepsilon_k) \to +\infty$ as $k \to \infty$ for each $s > 0$ (then we apply the dominated convergence theorem). We have

$$F(x) \geqq [\exp(-\tfrac{1}{2}x^2)]/6x \qquad \text{for } x \geqq 1 .$$

Thus it is enough to prove that

$$\lim_{k\to\infty} [H(C, \varepsilon_k) - |\log \varepsilon_k| - c\varepsilon_k^{-2}] = +\infty$$

for any $c > 0$, which is clear since $H(C, \varepsilon_k) \geqq k\varepsilon_k^{-2}$.

For (d) and (e),* let C_n be a cube of dimension n and side $2/n$ centered at 0. Let $n = n(k) = k^2$ and let the cubes $C_{n(k)}$ lie in orthogonal subspaces for $k = 1, 2, \cdots$.

Let $X_n \equiv \sup\{|L(x)| : x \in C_n\} = \sum_{j=1}^n |G_j|/n$ where G_j are orthogaussian. Then since $E|G_j| \equiv (2/\pi)^{\frac{1}{2}} < 1$, $EX_n < 1$ and $\sigma^2(X_n) < 1/n$. Thus by Chebyshev's inequality, $\Pr\{X_n \geqq 2\} < 1/n$. Hence $\sum_k \Pr(X_{n(k)} \geqq 2) < \infty$. If C is the closed convex hull of the $C_{n(k)}$, then C is a GB-set.

If $\varepsilon = \varepsilon_k \equiv (5n(k))^{-\frac{1}{2}}$ then

$$N(C, \varepsilon) \geqq N(C_{n(k)}, \varepsilon) \geqq e^{n(k)/8} = \exp(1/40\varepsilon^2)$$

by Lemma 3.6 of Dudley (1972b). Thus (d) is proved. This C is not a GC-set.

Let a_k be any sequence of positive numbers with $a_k \downarrow 0$. Then the convex hull of the sets $a_k C_{n(k)}$ is a GC-set, being a non-maximal GB-set (J. Feldman (1971)). Letting $a_k \downarrow 0$ slowly, we can make $\varepsilon^2 H(C, \varepsilon) \to 0$ as slowly as desired. $\square$

REMARK. Sudakov ((1971) Theorem 5) asserts that if $\int_0^1 \varepsilon^2\, dH(C, \varepsilon) = -\infty$ then C is not a GB-set. The set C constructed in the proof of part (d) above is a counter-example to Sudakov's claim.

---

* Note added in proof: J. Neveu suggests a simpler example: $C = \{\varphi_n(\log n)^{-\frac{1}{2}}, n \geqq 2\}$, φ_n orthonormal.

There is an $M < \infty$ such that $H(C, \varepsilon) \leqq M/\varepsilon^2$ for $0 < \varepsilon \leqq 1$, by part (c). Then $\int_0^1 \varepsilon^2 \, dH(C, \varepsilon) = -\infty$ by integration by parts, or as follows.

Let $\varepsilon(j) = 1/5^{\frac{1}{2}} j^j$ for $j = 1, 2, \cdots$, and $m = 1/5\varepsilon(j)^2$, an integer. Then

$$\begin{aligned} \int_{\varepsilon(j+1)}^{\varepsilon(j)} \varepsilon^2 \, dH(C, \varepsilon) &\leqq -\varepsilon(j+1)^2[H(C_m, \varepsilon(j+1)) - H(C, \varepsilon(j))] \\ &\leqq -\varepsilon(j+1)^2[40^{-1}\varepsilon(j+1)^{-2} - M\varepsilon(j)^{-2}] \\ &\leqq -1/50 \qquad \text{for } j \text{ large enough.} \end{aligned}$$

Hence $\int_0^1 \varepsilon^2 \, dH(C, \varepsilon) = -\infty$ while C is a GB-set.

APPLICATION. P. T. Strait (1966) considered Gaussian processes $\{x_t, t \in B\}$ where B is a rectangular solid of sides 2^{-n} in Hilbert space, showing that if

$$E|x_s - x_t|^2 \leqq K/|\log \|s - t\||^{4+\delta}$$

for some $\delta > 0$, $K < \infty$, then $\{x_t, t \in B\}$ is sample-continuous. Here $4 + \delta$ can be improved to $2 + \delta$ using Theorem 1.1 above, since

$$H(B, \varepsilon) = \mathcal{O}(|\log \varepsilon|^2), \qquad \varepsilon \downarrow 0.$$

Note that if C is a bounded set in a finite-dimensional Euclidean space then $H(C, \varepsilon) = \mathcal{O}(|\log \varepsilon|)$ so that the exponent $2 + \delta$ can be replaced by $1 + \delta$. Conversely for sets larger than B in the metric entropy sense, the bound on $E|x_s - x_t|^2$ would have to be correspondingly smaller to assure sample continuity.

2. Moduli of continuity.

DEFINITION. A function h from $[0, \infty)$ into $[0, \infty)$ is called a *modulus* (of continuity) iff both the following hold:

(a) h is continuous and $h(0) = 0$;
(b) $h(x) \leqq h(x + y) \leqq h(x) + h(y)$ for all $x, y \geqq 0$.

If is easily seen that h is a modulus iff there is a uniformly continuous function g with $h(t) = \sup\{|g(x + s) - g(x)| : x \in R, |s| \leqq t\}$ for all $t \geqq 0$. If x is restricted to lie in a bounded interval $I = [a, b]$, then there is a continuous g with

$$h(t) = \sup\{|g(x) - g(y)| : |x - y| \leqq t, x, y \in I\}$$

for $0 \leqq t \leqq b - a$; namely $g(x) = h(x - a)$.

2.1. *Uniform moduli.*

DEFINITION. Given a stochastic process $\{x_t; t \in T\}$ over a metric space (T, d), a modulus h will be called a (uniform) *sample modulus* for $\{x_t\}$ iff (for a suitable version of the process) for almost all ω there is a $K_\omega < \infty$ such that for all s, $t \in T$, $|x_s(\omega) - x_t(\omega)| \leqq K_\omega h(d(s, t))$.

Since $h \circ d$ is a metric, one can say that the process x is a.s. Lipschitzian for $h \circ d$.

Now suppose x_t is a Gaussian process with mean 0. Then h is a sample modulus for x_t iff $\{(x_s - x_t)/h(\|x_s - x_t\|) : s, t \in T\}$ is a GB-set in H, the Hilbert space

$L^2(\Omega, P)$ with usual covariance inner product. For the isonormal process L on a set $C \subset H$, a sample modulus will be called simply a *modulus*.

The following theorem, perphaps the main result of this paper, will be proved by the method used for Theorem 3.1 of Dudley (1967). I thank J. Neveu for pointing out an error in that proof requiring some changes to prove the stronger result here. Rather surprisingly, the difficulties concern the case of "small" sets C such as a sequence converging rapidly to a point. For connected or convex sets the earlier proof essentially suffices. A slightly weaker bound was proved by C. Preston (1972).

THEOREM 2.1. *Suppose* $C \subset H$. *Let* $f(h) = \int_0^h H(C, x)^{\frac{1}{2}}\, dx$. *Then* f *is a modulus for* L *on* C.

PROOF. If $f(1) = +\infty$ there is nothing to show, so we assume $f(1) < \infty$. Also we may assume C is infinite. Then $H(C, \varepsilon) \to +\infty$ as $\varepsilon \downarrow 0$. Let $H(C, x) \equiv H(x)$.

We define sequences $\delta_n \downarrow 0$, $\varepsilon_n \downarrow 0$ inductively as follows. Let $\varepsilon_1 = 1$. Given $\varepsilon_1, \cdots, \varepsilon_n$, let

$$\delta_n = 2 \inf \{\varepsilon : H(\varepsilon) \leqq 2H(\varepsilon_n)\} ,$$

$$\varepsilon_{n+1} = \min (\varepsilon_n/3, \delta_n) .$$

Then $\varepsilon_n \leqq 3(\varepsilon_n - \varepsilon_{n+1})/2$. Also if $\varepsilon_{n+1} = \delta_n$, then $\int_{\varepsilon_{n+1}}^{\varepsilon_n} H(x)^{\frac{1}{2}}\, dx \leqq 2H(\varepsilon_n)^{\frac{1}{2}}\varepsilon_n$, while otherwise $\varepsilon_{n+1} = \varepsilon_n/3$ and $\int_{\varepsilon_{n+1}}^{\varepsilon_n} H(x)^{\frac{1}{2}}\, dx \leqq 2\varepsilon_{n+1} H(\varepsilon_{n+1})^{\frac{1}{2}}$. Thus we have

$$\begin{aligned} \tfrac{2}{3} \textstyle\sum_{m=n}^{\infty} H(\varepsilon_m)^{\frac{1}{2}}\varepsilon_m &\leqq \textstyle\sum_{m=n}^{\infty} (\varepsilon_m - \varepsilon_{m+1})H(\varepsilon_m)^{\frac{1}{2}} \leqq f(\varepsilon_n) \\ &\leqq 4 \textstyle\sum_{m=n}^{\infty} \varepsilon_m H(\varepsilon_m)^{\frac{1}{2}} . \end{aligned}$$

So the convergence of the above integrals and sums is equivalent and they all converge.

Now for each n we can choose a set $A_n \subset C$ such that for any $x \in C$ there is a $y \in A_n$ with $\|x - y\| \leqq 2\delta_n$, and card $(A_n) \leqq \exp(2H(\varepsilon_n))$. Let $G_n = \{x - y : x, y \in A_{n-1} \cup A_n\}$. Then card $(G_n) \leqq 4 \exp(4H(\varepsilon_n))$. Let

$$P_n = \Pr \{\max \{|L(z)|/\|z\| : z \in G_n\} \geqq 3H(\varepsilon_n)^{\frac{1}{2}}\} .$$

Now we use the standard Gaussian tail estimate; for $T > 0$, $1 - \Phi(T) < \exp(-\frac{1}{2}T^2)/T$. Then for n large enough so that $3H(\varepsilon_n)^{\frac{1}{2}} \geqq 1$, we have

$$P_n \leqq 4 \exp\{4H(\varepsilon_n) - 9H(\varepsilon_n)/2\} \leqq 4 \exp\{-\tfrac{1}{2}H(\varepsilon_n)\} .$$

Since $H(\varepsilon_{n+2}) \geqq H(\delta_n/3) \geqq 2H(\varepsilon_n)$, $\sum P_n$ is dominated by a geometric series and hence converges. So for almost all ω there is an $n_0(\omega)$ such that for all $n \geqq n_0(\omega)$ we have

$$|L(z)| < 3\|z\| H(\varepsilon_n)^{\frac{1}{2}} \qquad \text{for all } z \in G_n .$$

Now for any $x \in C$ choose $A_n(x)$ with $\|x - A_n(x)\| \leqq 2\delta_n$. Then for almost all ω, $L(A_n(x))(\omega)$ is a Cauchy sequence for all $x \in C$. We choose a version of L such that whenever $n_0(\omega)$ is defined, $L(A_n(x))(\omega)$ converges to $L(x)(\omega)$ as $n \to \infty$ for all $x \in C$.

Now if $n \geqq n_0(\omega)$ and $\varepsilon_{n+1} < \|s - t\| \leqq \varepsilon_n$, $s, t \in C$, we have $\|A_n(s) - A_n(t)\| \leqq \|s - t\| + 4\delta_n$, and

$$\begin{aligned}|L(s) - L(t)|(\omega) &\leqq |L(A_n(s)) - L(A_n(t))|(\omega) + \textstyle\sum_{m=n}^{\infty} (|L(A_m(s)) - L(A_{m+1}(s))| \\ &\quad + |L(A_m(t)) - L(A_{m+1}(t))|)(\omega) \\ &\leqq (\|s - t\| + 4\delta_n)3H(\varepsilon_n)^{\frac{1}{2}} + \textstyle\sum_{m=n}^{\infty} 8\delta_m \cdot 3H(\varepsilon_{m+1})^{\frac{1}{2}} .\end{aligned}$$

Now note that $\delta_m \leqq 6\varepsilon_{m+1}$, so

$$\begin{aligned}|L(s) - L(t)|(\omega) &\leqq 75\|s - t\|H(\|s - t\|)^{\frac{1}{2}} + 144 \textstyle\sum_{m=n+1}^{\infty} \varepsilon_m H(\varepsilon_m)^{\frac{1}{2}} \\ &\leqq 291 f(\|s - t\|) .\end{aligned}$$

A modulus valid for small distances, on a totally bounded set, is also valid for all distances, possibly with a larger constant. □

EXAMPLE 2.2 (Brownian motion). For the usual Wiener process x_t, $0 \leqq t \leqq 1$, $\|x_s - x_t\| = |s - t|^{\frac{1}{2}}$. Thus $N(C, \varepsilon)$ is asymptotic to $1/4\varepsilon^2$ as $\varepsilon \downarrow 0$. Hence L on C has a modulus $f(x) = x|\log x|^{\frac{1}{2}}$. So x_t has a sample modulus $\varphi(h) = (h|\log h|)^{\frac{1}{2}}$. It is well known that, up to multiplicative constants which we are neglecting, φ is the best possible uniform sample modulus for Brownian motion (P. Lévy (1937), (1954)). Note that $y = o(f(y))$ as $y \downarrow 0$, and x_t is the convergent sum of independent functions of t with modulus $g(t) \equiv t$. Hence $\limsup_{|s-t| \downarrow 0} |x_s - x_t|/\varphi(\|x_s - x_t\|)$ is almost surely equal to a constant, which P. Lévy ((1954) Théorème 52.2 page 172) proved equal to $2^{\frac{1}{2}}$.

The following is also an easy consequence of Theorem 2.1.

COROLLARY 2.3. *Let $\{x_t, t \in K\}$ be a Gaussian process with mean 0 where (K, d) is a compact metric space. Let g be a modulus such that for all $s, t \in K$, $E(x_s - x_t)^2 \leqq g(d(s, t))^2$. Let $C = \{x_t : t \in K\}$ and assume that for some M, $\alpha < \infty$, $N(C, \varepsilon) \leqq M\varepsilon^{-\alpha}$ as $\varepsilon \downarrow 0$. Let $f(h) = |\log g(h)|^{\frac{1}{2}} g(h)$. Then f is a sample modulus for x_t.*

COROLLARY 2.4. *Suppose for some r, $0 < r < 2$, $\limsup_{\varepsilon \downarrow 0} \varepsilon^r H(C, \varepsilon) < \infty$. Then L on C has the modulus $f(x) = x^{1-\frac{1}{2}r}$.*

PROOF. For some $K < \infty$, $H(C, \varepsilon) \leqq K\varepsilon^{-r}$ for all small enough ε. Thus we have the modulus

$$f(x) = \textstyle\int_0^x t^{-\frac{1}{2}r}\, dt = x^{1-\frac{1}{2}r}/(1 - \frac{1}{2}r) . \qquad □$$

DEFINITION. We say f is a *weakly optimal* sample modulus of a process iff for any other modulus g of that process, $g \neq o(f)$, i.e.,

$$\liminf_{x \downarrow 0} f(x)/g(x) < \infty .$$

If the lim inf can be replaced by lim sup, we shall call f *strongly optimal*.

Even a strongly optimal modulus does not settle the finer question: for what moduli f is there for almost all ω a $\delta(\omega) > 0$ such that for $d(s, t) < \delta$, $|x_s(\omega) - x_t(\omega)| \leqq f(d(s, t))$. Such a modulus f is said to belong to the (uniform) *upper class*

$\mathscr{U}$ of the process x_t, while other moduli belong to the *lower class* $\mathscr{L}$. Usable, necessary and sufficient conditions for $f \in \mathscr{U}$ have been found for certain processes with stationary increments on a compact subset of R^k; see N. Kôno ((1970) Theorem 3).

DEFINITION. Given $0 < \delta < 1$, a function J is called *δ-slowly varying* as $x \downarrow 0$ iff $\lim_{x \downarrow 0} J(\delta x)/J(x) = 1$. Examples of δ-slowly varying functions include $J(x) = c|\log x|^\alpha(\log|\log x|)^\beta \cdots (\log \cdots \log|\log x|)^\zeta$ for any constants $c, \alpha, B, \cdots, \zeta$.

LEMMA 2.5. *If $H(x) = x^{-r}J(x)$ where $J \geqq 0$, $0 \leqq r < 2$, J is δ-slowly varying, $0 < \delta < 1$, and $H(x) \geqq H(y)$ for $0 < x \leqq y$, then for some $M < \infty$, $tH(t)^{\frac{1}{2}} \leqq \int_0^t H(x)^{\frac{1}{2}}\,dx \leqq MtH(t)^{\frac{1}{2}}$ for all small enough $t > 0$.*

PROOF. We have $tH(t)^{\frac{1}{2}} \leqq \int_0^t H^{\frac{1}{2}}$ by the monotonicity of H. Conversely, given $\varepsilon > 0$ such that $(1+\varepsilon)\delta^{1-\frac{1}{2}r} < 1$, we choose $\gamma > 0$ such that $J(\delta x) \leqq (1+\varepsilon)J(x)$ for $0 < x \leqq \gamma$. Then for $t \leqq \gamma$

$$\begin{aligned}\int_0^t H^{\frac{1}{2}} &= \sum_{n=0}^\infty \int_{\delta^{n+1}t}^{\delta^n t} H^{\frac{1}{2}} \\ &\leqq (1-\delta)t(1+\varepsilon)\delta^{-\frac{1}{2}r}H(t)^{\frac{1}{2}} \sum_{n=0}^\infty [(1+\varepsilon)\delta^{1-\frac{1}{2}r}]^n \leqq MtH(t)^{\frac{1}{2}}\end{aligned}$$

for some $M < \infty$. □

We abbreviate $H(C, x)$ to $H(x)$ or H below, so long as the set C intended is clear.

THEOREM. 2.6. *Let $C \subset H$. Assume that $r(C) > 0$, $\int_0^1 H^{\frac{1}{2}} < \infty$, and for some $M < \infty$,*

$$f(t) \equiv \int_0^t H^{\frac{1}{2}} \leqq MtH(t)^{\frac{1}{2}}$$

for t small enough. Then f is a weakly optimal modulus for L on C. If for some $\delta > 0$, $H(\varepsilon) \leqq (1-\delta)H(\delta\varepsilon)$ for ε small enough, then f is strongly optimal.

Before proving the above theorem, note that its hypotheses follow from those of Lemma 2.5 if $r > 0$.

DEFINITION. A set $C \subset H$ is called a GL-set iff the function $I(t) = t$ is a modulus for L on C.

PROPOSITION 2.7. *If $\limsup_{x \downarrow 0} H(C, x)/|\log x| = +\infty$, then C is not a GL-set.*

PROOF. I claim that $\limsup_{x \downarrow 0} N(C, \frac{1}{2}x)/N(C, x) = +\infty$. Otherwise, for some $M < \infty$, $N(C, 2^{-n}) \leqq M^n$ for all n, and for some k, $K < \infty$, $N(C, \varepsilon) \leqq K\varepsilon^{-k}$ for $0 < \varepsilon \leqq 1$, contrary to hypothesis. Given $\varepsilon > 0$, we choose a covering of C by a minimal number of sets $A_1, \cdots, A_{N(C,\varepsilon)}$, each of diameter $\leqq 2\varepsilon$. Let $x_1, \cdots, x_k$ be a maximal number of points such that $\|x_i - x_j\| > \varepsilon$ for $i \neq j$. Then $N(C, \varepsilon) \leqq k = k(\varepsilon)$ since the balls of radius ε and centers x_i cover C. One of the A_i contains at least $k(\varepsilon)/N(C, 2\varepsilon)$ of the x_j. Hence, for $n = 1, 2, \cdots$, there exist $\varepsilon_n > 0$ and a set $B_n \subset C$ with card $(B_n) \geqq n$ and $\varepsilon_n < \|x - y\| \leqq 4\varepsilon_n$ for $x \neq y \in B_n$. Thus the union of the sets $\{(x-y)/4\varepsilon_n : x, y \in B_n\}$ is not totally bounded and hence not GB. Thus, $C_N \equiv \{(x-y)/\|x-y\| : x, y \in C\}$ is not GB since if it were, the convex hull of $C_N \cup \{0\}$ would be GB. Thus C is not a GL-set. □

After this proof a little more will be said about GL-sets.

PROOF OF THEOREM 2.6. By Theorem 2.1, f is a modulus of L on C. Suppose there is another modulus $g = o(f)$. Then $g(t) = o(tH(t)^{\frac{1}{2}})$ as $t \downarrow 0$.

We have two cases:

Case I. $\limsup_{x \downarrow 0} x/g(x) > 0$, i.e., $\beta \equiv \liminf_{x \downarrow 0} g(x)/x < \infty$. Then, since g is a modulus, if $(n-1)x \leqq y \leqq nx$, n an integer, then $g(y) \leqq g(nx) \leqq ng(x)$, and $g(y)/y \leqq ng(x)/(n-1)x$. Letting $x \downarrow 0$, $n \to \infty$, gives $g(y)/y \leqq \beta$, so that the function $I(x) \equiv x$ is a modulus for L on C, contradicting Proposition 2.7 since $r(C) > 0$.

Case II. $\lim_{x \downarrow 0} x/g(x) = 0$. Then $x/g(x)$ extends to a continuous function on $[0, \infty)$.

If $\lim_{\varepsilon \downarrow 0} H(\varepsilon)/H(\frac{1}{2}\varepsilon) = 1$, then for $0 < \alpha < 1$ there is a $K < \infty$ such that $H(2^{-n-1}) \leqq K/(1-\alpha)^n$ for all n. Then for $1 > \varepsilon > 0$, $2^{-n-1} \leqq \varepsilon < 2^{-n}$ for some n, and

$$H(\varepsilon) \leqq K \exp\{|\log(1-\alpha)|\,|\log \varepsilon|/\log 2\},$$

a contradiction for $|\log(1-\alpha)|/\log 2 < r$. Thus for some $\alpha > 0$,

$$\liminf_{\varepsilon \downarrow 0} H(2\varepsilon)/H(\varepsilon) < 1 - \alpha .$$

Choose $\varepsilon_n \downarrow 0$ such that

$$H(2\varepsilon_n)/H(\varepsilon_n) < 1 - \alpha .$$

As in the proof of Proposition 2.7, choose sets $B_n \subset C$ such that $\varepsilon_n < \|x - y\| \leqq 4\varepsilon_n$ for $x \neq y \in B_n$ and

$$\begin{aligned} \operatorname{card}(B_n) \geqq N(C, \varepsilon_n)/N(C, 2\varepsilon_n) &= \exp\{H(C, \varepsilon_n) - H(C, 2\varepsilon_n)\} \\ &\geqq \exp\{\alpha H(C, \varepsilon_n)\} . \end{aligned}$$

Let $D_n = \{(x - y)/g(\varepsilon_n) : x, y \in B_n\}$. Then

$$N(D_n, \varepsilon_n/2g(\varepsilon_n)) \geqq \exp\{\alpha H(C, \varepsilon_n)\} .$$

For any $K < \infty$ we have for n large enough $H(C, \varepsilon_n) \geqq Kg(\varepsilon_n)^2/\alpha\varepsilon_n^2$. Letting $\kappa_n = \varepsilon_n/2g(\varepsilon_n)$ we have

$$H(D_n, \kappa_n) \geqq K\kappa_n^{-2} .$$

Thus by Theorem 1.1 (c), the union of all D_n is not a GB-set. Thus $C_g \equiv \{(x - y)/g(\|x - y\|) : x, y \in C\}$ is not a GB-set, since the convex hull of $4C_g$ includes each D_n. Thus g cannot be a modulus of L on C.

The above argument goes through as well for δ in place of $\frac{1}{2}$ if $0 < \delta < 1$. Then if $\limsup_{\varepsilon \downarrow 0} H(\varepsilon)/H(\delta\varepsilon) < 1$, we could choose $\varepsilon_n \downarrow 0$ to satisfy $g(\varepsilon_n)/\varepsilon_n H(C, \varepsilon_n)^{\frac{1}{2}} \downarrow 0$ and the same proof shows that f is a strongly optimal modulus. □

In one sense, Proposition 2.7 is best possible since any bounded open set C in a k-dimensional linear subspace is a GL-set with $\limsup_{\varepsilon\downarrow 0} H(C, \varepsilon)/|\log \varepsilon| = k$. On the other hand if $C = \{x_t : 0 \leqq t \leqq 1\}$ for the usual Wiener process x_t, then the lim sup is 2 while C is not GL (cf. Example 2.2 above). Thus, the behavior of $N(C, \varepsilon)$ does not determine whether C is GL. More relevant is $N(D, \varepsilon)$, where $D = \{(x - y)/\|x - y\| : x, y \in C\}$. If C is convex and infinite dimensional, then D is not totally bounded and not GB. But the GL-property (unlike the GB- or compact GC-property) is not preserved by taking the convex hull. Here are some infinite-dimensional GL-sets.

PROPOSITION 2.8. *Suppose* $\{x_t, 0 \leqq t \leqq 1\}$ *is a Gaussian process with mean* 0, *and* $x_t = \int_0^t y_s\, ds$, *where* y_s *is another such process and the integral is a Bochner integral in* H. *Assume that for some* $\alpha > 0$ *and* $M < \infty$, $\|y_s - y_t\| \leqq M/|\log|s - t||^{\frac{1}{2}+\alpha}$ *for* $0 \leqq s, t \leqq 1$. *Finally assume that for some* $\delta > 0$, $\|x_s - x_t\| \geqq \delta|s - t|$. *Let* $C = \{x_t : 0 \leqq t \leqq 1\}$. *Then* C *is a* GL-*set.*

PROOF. Since $t \to x_t$ is a bi-Lipschitz map it suffices to prove that $I(t) = t$ is a sample modulus for x_t on $[0, 1]$. The hypothesis implies that y_t is sample-continuous. Then we can write

$$x_t(\omega) = \int_0^t y_s(\omega)\, ds$$

and the result follows. □

Clearly, it is not enough for $t \to y_t$ to be continuous $T \to H$. The above proposition could be extended, replacing $[0, 1]$ by finite-dimensional sets and using Fréchet derivatives; the condition $\|x_s - x_t\| \geqq \delta|s - t|$ need only hold locally, and no doubt further improvements are possible.

2.2. *Processes on* R^k. Let $\{x_t, t \in K\}$ be a Gaussian process where K is a compact set in R^k. Let $C = \{x_t : t \in K\} \subset L^2(\Omega, P)$. For $\varepsilon, h > 0$ let

$$\Psi(h) = \sup\{E|x_s - x_t|^2 : |s - t| \leqq h\}^{\frac{1}{2}}, \qquad (2.9)$$
$$\eta(\varepsilon) = \sup\{h > 0 : \Psi(h) \leqq \varepsilon\}.$$

Then Ψ will be called the (uniform) QM *modulus* of $\{x_t, t \in K\}$. We assume x_t is continuous in quadratic mean (CQM), i.e., $\Psi(h) \downarrow 0$ as $h \downarrow 0$. Unless $\Psi \equiv 0$, $\eta(\varepsilon)$ is defined and non-decreasing for $\varepsilon > 0$ and small enough.

For some $M < \infty$ we have $N(K, \delta/2) \leqq M/\delta^k$ for $0 < \delta \leqq 1$. Hence $N(C, \Psi(\delta)) \leqq M/\delta^k$, $N(C, \varepsilon) \leqq M/\eta(\varepsilon)^k$. For $k = 1$ the following was shown by other methods in Garsia, Rodemich and Rumsey (1970), as far as β is concerned, and later extended to $k > 1$ by A. Garsia (1971).

THEOREM 2.10. *Let* $\{x_t, t \in K\}$ *be any* CQM *Gaussian process on a compact* $K \subset R^k$, *with* QM *modulus* Ψ, *mean* 0, *and* η *from* (2.9). *Let*

$$\alpha(h) = \int_0^{\Psi(h)} |\log \eta(x)|^{\frac{1}{2}}\, dx,$$
$$\beta(h) = \int_0^h |\log y|^{\frac{1}{2}}\, d\Psi(y),$$
$$\gamma(h) = \int_0^h \Psi(y)\, dy/y|\log y|^{\frac{1}{2}},$$
$$\kappa(h) = \Psi(h)|\log h|^{\frac{1}{2}}.$$

Then α, β *and* $\gamma + \kappa$ *are always sample moduli of* x_t; κ *is a modulus if* $\log \eta(x) \equiv x^{-r}J(x)$ *where* $0 \leqq r < 2$ *and for some* δ, $0 < \delta < 1$, J *is* δ*-slowly varying.*

PROOF. We represent $t \to x_t(\omega)$ by the composition $t \to x_t(\cdot) \to L(x_t(\cdot))(\omega)$, L isonormal. If f is a modulus for L on C, then $f \circ \Psi$ is a sample modulus for x_t. Thus by Theorem 2.1, α is a sample modulus for x_t.

To get β we substitute $x = \Psi(y)$. Note that $\eta(\Psi(y)) \geqq y$, so as $y \downarrow 0$, $|\log \eta(\Psi(y))|^{\frac{1}{2}} \leqq |\log y|^{\frac{1}{2}}$.

Then to get $\kappa + \gamma$ we use (Riemann–Stieltjes) integration by parts, yielding

$$\beta(h) \leqq \kappa(h) + \gamma(h) .$$

Lemma 2.5 proves the final statement. □

The modulus κ (with appropriate best possible constant multiples for the upper class, which we do not consider here) was first found for Brownian motion by P. Lévy ((1937), (1954) pages 168–172) and then for increasingly more general processes by Z. Ciesielski (1961), M. Marcus (1968 ff.) and N. Kôno (1970). The latter authors have also found other moduli under other conditions and further information not described here. The moduli in Theorem 2.10 are not always optimal (Marcus (1972)), but Theorem 2.6 says κ is optimal under the conditions on η stated at the end of Theorem 2.10 if $r > 0$.

M. Marcus ((1971) (4.4)) gives examples of Gaussian processes X, Y_1 and Y_2 such that

$$E(Y_1(t+h) - Y_1(t))^2$$
$$\leqq E(X(t+h) - X(t))^2 \leqq E(Y_2(t+h) - Y_2(t))^2$$

where Y_1 and Y_2 have the same weakly optimal sample modulus $[h|\log h|]^{\frac{1}{2}}$ and X has the weakly optimal modulus $[h \log |\log h|]^{\frac{1}{2}}$. Thus the modulus of the covariance map $t \to x_t \in H$ does not determine the optimal sample moduli. This is not surprising since we have seen that metric entropy does not determine the GL-, GB or GC-properties. See also Sub-section 3.1 below.

2.3. *Local moduli.* Now we consider $|x_s - x_t|$ as t approaches a fixed point s. A modulus f will be called a *local sample modulus* for x_t as s iff for almost all ω there is a $K_\omega < \infty$ such that $|x_s(\omega) - x_t(\omega)| \leqq K_\omega f(s - t|)$ for all t (in some neighborhood of s, hence for any compact set M, where K_ω also may depend on M).

Clearly any uniform sample modulus is also a local one everywhere. A modulus f is a local sample modulus of x_t at s iff $\{(x_t - x_s)/f(|s - t|) : t \in U\}$ is a GB-set for some open set $U \ni s$.

The proof of Theorem 2.6 above will show that the function $f(u) = uH(C, u)^{\frac{1}{2}}$ is an optimal local as well as uniform modulus for L on C, if in addition to the hypotheses of Theorem 2.6, we can choose a fixed point $x \in B_n$ for all large enough n (then the set D_n, also in that proof, can be formed using the fixed x). For $C = \{x_t : t \in U\}$ for example, where x_t has stationary increments (see Section

3) and U is open in R^k, local behavior at all points is the same so we can choose such an x.

In Theorem 2.6 we have $r(C) > 0$. For $r(C) = 0$ it is well known that we may have a local modulus smaller than the uniform one, with $|\log h|^{\frac{1}{2}}$ replaced by $(\log|\log h|)^{\frac{1}{2}}$. This is classical (due to Khinchin) for Brownian motion and has been extended to many other processes with stationary increments (see Section 3 below) by several workers including M. Marcus (1968 ff.), Sirao and H. Watanabe (1970), and N. Kôno (1970).

Sample behavior of x_t as $t \to \infty$ comes under the same rubric as behavior for $t \to 0$ for general processes, although not necessarily for processes with stationary increments.

For certain processes such that $E|x_s - x_t|^2$ behaves like $|s - t|^\alpha$, $0 < \alpha \leqq 2$, as $|s - t| \downarrow 0$, and $Ex_s x_t \to 0$ fast enough as $|s - t| \to \infty$, H. Watanabe (1970) found a necessary and sufficient condition on an increasing f so that $\lim \sup_{t \to \infty} x_t/f(t) \leqq 1$. (In his case the smallest such f are of the form $(Ex_t^2|\log t|)^{\frac{1}{2}}$ + terms of lower order of growth.) Such complete criteria have not yet been found for sets C in H, but it should be possible to get them for sets satisfying some good enough conditions. Such criteria cannot follow from Theorem 1.1 above although they might include conditions on metric entropy.

One can also consider local behavior along a given sequence $t_n \to s$ and ask how slowly t_n should converge to s for an optimal local modulus to be optimal also along that sequence. Here there are results for Brownian motion (Dudley (1972b)) with possibilities for generalization.

2.4. *Peano curves.* Consideration of stochastic processes with second moments as curves in Hilbert space goes back at least to Kolmogorov (1940). It is known that any suitably connected compact metric space is a continuous image of the unit interval. I shall give a specific construction for a compact, convex subset C of a Banach space, so that any compact convex GC-set or GB-set is represented as the range of a Gaussian process continuous in probability on $[0, 1]$, with some bound on the rate at which $E|x_s - x_t|^2 \to 0$ as $|s - t| \to 0$.

We construct a Peano curve C inductively. Choose $p \in C$ and let $f_0(t) = p$, $0 \leqq t \leqq 1$. At the nth stage we shall have a continuous function f_n from $[0, 1]$ into C. Here $[0, 1]$ is divided into k_n subintervals $I(n, 1), \cdots, I(n, k_n)$. These will be either "finished" or "unfinished" as defined below. On the closure of each interval $I(n, j)$, f_n will be linear. Let $A(n)$ be a set of minimal cardinality such that for each $x \in C$, $\|x - y\| \leqq 2^{1-n}$ for some $y \in A(n)$. Then card $A(n) \leqq N(C, 2^{-n})$. The values of f_n at the endpoints of the $I(n, j)$ will be precisely $\bigcup_{1 \leqq r \leqq n} A(r)$.

Given f_n and an interval $I(n, j) = [a, b]$, we let $f_{n+1} \equiv f_n$ on $I(n, j)$ if $I(n, j)$ is finished; it then becomes one of the finished intervals $I(n + 1, k)$. If $I(n, j)$ is unfinished, we let $f_{n+1}(a) = f_{n+1}((a + b)/2) = f_n(a)$, $f_{n+1}(b) = f_n(b)$ and call $[(a + b)/2, b]$ a finished interval $I(n + 1, r)$. At the first step, $[0, 1]$ is unfinished.

We divide $[a, (a+b)/2]$ into at most $N(C, 2^{-n-1})+1$ equal unfinished subintervals $I(n+1, s)$. At the endpoints of these intervals, except for a and $(a+b)/2$, the values of f_{n+1} will include all points x of A_{n+1} such that $\|x - f_n(a)\| \leqq 2^{-n}$. This completes the inductive construction. Clearly the f_n converge uniformly to a continuous function f from $[0, 1]$ onto C, with $\|f - f_n\| \leqq 2^{2-n}$.

The intervals $I(n, j)$ have length at least $1/2^n \prod_{s=1}^{n} [N(C, 2^{-s})+1] = \delta_n$. On each unfinished or newly finished interval, the value of f changes by at most 2^{4-n}. By induction, for s, t in a previously finished interval

$$|f(s) - f(t)|/|s-t| \leqq 2^{4-(n-1)}/\delta_{n-1} \leqq 2^{4-n}/\delta_n .$$

Thus whenever $|s-t| \leqq \delta_n$, we have $|f(s) - f(t)| \leqq 2^{5-n}$.

For example, if $H(C, \varepsilon)$ is asymptotic to ε^{-r} for some $r > 0$, then for some constant K, $\delta_n \geqq \exp(-K \cdot 2^{nr})$, and a modulus of continuity of f is $h(\delta) \equiv |\log \delta|^{-1/r}$, which is best possible in this case.

3. Stationary increments. A process x_t, $t \in R^k$, is said to have *stationary increments* iff the joint distributions of the random variables $\{x_{t_2} - x_{t_1}, \cdots, x_{t_{2n}} - x_{t_{2n-1}}\}$ are unchanged if some vector $t \in R^k$ is simultaneously added to all the t_j, for any $t_1, \cdots, t_{2n} \in R^k$.

The additive group R has many automorphisms which are highly pathological. Thus the assumption of stationarity (of increments) can be expected to yield convenient results only under further conditions.

We call a process $\{x_t\}$ *continuous in quadratic mean* (CQM) iff for all s,

$$\lim_{t \to s} E|x_s - x_t|^2 = 0 .$$

A Gaussian process is CQM iff it is continuous in probability, as is well known.

If a stationary Gaussian process $\{x_t(\omega)\}$ is measurable jointly in t and ω, then its characteristic function φ, with $Ex_0 x_h = \varphi(h)$, is measurable and hence continuous (Loève (1963) page 209) so that x_t is CQM. Since joint measurability is a rather minimal regularity assumption, we consider only CQM processes in this section. Then the variance $\sigma^2(h) = E|x_{t+h} - x_t|^2 \to 0$ as $h \to 0$.

3.1. *Moduli of continuity.* If x_t is any Gaussian process and $C = \{x_t\} \subset L^2(\Omega, P)$, then a sample modulus f for L on C and the QM modulus Ψ of x_t always give a sample modulus $f \circ \Psi$ for x_t (cf. Sub-section 2.2 above). In general, however, Ψ may reflect local rather than uniform behavior of $\{x_t\}$, so that $f \circ \Psi$ may well not be optimal even if f is optimal for L on C. For processes with stationary increments the QM modulus Ψ is the same at all t so that there seems a better chance for $f \circ \Psi$ to be optimal, especially if σ is non-decreasing for small h.

In fact, the modulus κ in Theorem 2.10, $\kappa(h) = \sigma(h)|\log h|^{\frac{1}{2}}$ both as a uniform and for $r(C) > 0$ as a local modulus, has been found for many processes with stationary increments by M. Marcus (1968 ff.), Sirao and Watanabe (1970), and N. Kôno (1970), all of whom got further, more precise results which will not be restated here.

Khinchin's local law

$$\limsup_{t\to\infty} x_t/\sigma(t)[2 \log \log t]^{\frac{1}{2}} = 1 \quad \text{a.s.}$$

has been extended from Brownian motion to certain other processes with stationary increments; see S. Orey (1971), who used estimates of J. Pickands III (1967). T. Sirao (1960) found the upper and lower classes for Lévy's Brownian motion on R^k with $E|x_s - x_t|^2 = |s - t|$, locally at 0 and ∞.

S. M. Berman (1964) proved that if X_n is a stationary Gaussian sequence with $EX_n = 0$, $EX_n^2 = 1$, and $\lim_{n\to\infty} (\log n)EX_1X_n = 0$, then $\max(X_1, \cdots, X_n) - (2 \log n)^{\frac{1}{2}} \to 0$ in probability. For orthogaussian X_j this is a theorem of Gnedenko. Berman (1962), in a closing remark, says that stationarity can be replaced by

$$\lim_{n\to\infty} nEX_iX_n = 0 \qquad \text{for all } i .$$

Cramér (1962) proved the analogous conclusion for real continuous-parameter stationary processes with a spectral density f of bounded variation with

$$\int_0^\infty \lambda^2(\log(1 + \lambda))^a f(\lambda)\, d\lambda < \infty$$

for some $a > 1$. Such processes have a.s. continuously differentiable sample functions. M. G. Shur (1965) improved "in probability" to "with probability one."

More refined properties of sample functions, such as the Hausdorff dimension of level sets and local times, have been treated first for Brownian motion (see Itô and McKean (1965)) and later for other processes with $E(X_t - X_0)^2 \sim C|t|^\alpha$, $0 < \alpha < 2$ (S. M. Berman (1970), S. Orey (1970)).

3.2. *Fourier analysis.* A measure $\mu \geqq 0$ on $R^k \sim \{0\}$ will be called a *Lévy-Khinchin* (LK) measure iff

$$\int_{|x|\leq 1} |x|^2\, d\mu(x) + \mu\{|x| \geqq 1\} < \infty ,$$

where $|x| = (x_1^2 + \cdots + x_k^2)^{\frac{1}{2}}$. (LK measures arise in the well-known formula for infinitely divisible characteristic functions.)

The following theorem was first stated by Kolmogorov (1940) for $k = 1$. In higher dimensions I do not know an explicit reference for it, although it is at least close to known results, e.g., A. M. Yaglom (1957), (1962).

THEOREM 3.1 (Kolmogorov–Yaglom *et al.*). *For any* CQM *complex-valued process* $\{x_t, t \in R^k\}$ *with mean* 0 *and stationary increments there is a unique* LK *measure* μ *on* $R^k \sim \{0\}$ *and a nonnegative definite real symmetric operator* A *on* R^k *such that for any* $s, t \in R$,

$$(3.2) \qquad E(x_s - x_0)(x_t - x_0)^- = \int (e^{i\lambda\cdot s} - 1)(e^{-i\lambda\cdot t} - 1)\, d\mu(\lambda) + As \cdot t .$$

If x_t *is real-valued, then* μ *is symmetric*: $\mu(B) \equiv \mu(-B)$. *Conversely given any such* μ *and* A *there is a complex Gaussian* CQM *process* x_t *with stationary increments such that* (3.2) *holds. If* μ *is symmetric then there is such a real Gaussian* x_t. *For* *any random variable y such that $x_t + y$ is a Gaussian process, $x_t + y$ will also satisfy* (3.2), *but no other Gaussian processes will satisfy* (3.2) *for a given μ and A.*

PROOF. Since $\{x_t\}$ is CQM, it has a jointly measurable version for which by Fubini's theorem $\int_K |x_t(\omega)|\, dt < \infty$ almost surely for any compact K. Thus x_t defines a random Schwartz distribution $X = [x_t]$ which has a gradient in the distribution sense grad $X \equiv (\partial X/\partial t_1, \cdots, \partial X/\partial t_k)$, a stationary (also called homogeneous) Gaussian generalized random field as treated by K. Itô (1956) and A. M. Yaglom (1957). (Note that $\int x_t\, \partial f/\partial t_j\, dt$ is a limit of linear combinations of increments of x_t.)

Let $\mathscr{D}$ denote the space of C^∞ complex-valued functions on R^k with compact support. We need the following known fact which I shall prove for completeness.

LEMMA. *If $f \in \mathscr{D}$ and $\int_{R^k} f\, dt = 0$ then for some $f_1, \cdots, f_k \in \mathscr{D}$, $f = \sum_{1 \leq j \leq k} \partial f_j/\partial t_j$.*

PROOF. We use induction on k. For $k = 1$ let $f_1(x) = \int_{-\infty}^x f(t)\, dt$. In general, fix a function $\alpha \in \mathscr{D}(R^1)$ with $\int_{-\infty}^\infty \alpha = 1$. Let $g = \int_{-\infty}^\infty f(t)\, dt_k$. Then by induction assumption $g = \sum \partial g_j/\partial t_j$ in $\mathscr{D}(R^{k-1})$. Now $f(t) - g(t_1, \cdots, t_{k-1})\alpha(t_k) = \partial f_k/\partial t_k$ for some $f_k \in \mathscr{D}$, and let $f_j(t) = g_j(t_1, \cdots, t_{k-1})\alpha(t_k)$ for $j < k$. □

Thus we know that X restricted to $\mathscr{D}_0 = \{\varphi \in \mathscr{D} : \int \varphi = 0\}$ is stationary. We apply A. M. Yaglom ((1957) Theorems 6, 6′) to obtain a unique measure $\mu \geqq 0$ on R^k, tempered at ∞, $\int_{|\lambda|<1} |\lambda|^2\, d\mu(\lambda) < \infty$, and a unique nonnegative Hermitian matrix J such that for all $\varphi, \Psi \in \mathscr{D}_0$,

$$EX(\varphi)X(\Psi)^- = \int (\mathscr{F}\varphi)(\lambda)(\mathscr{F}\Psi)(\lambda)^-\, d\mu(\lambda) + J \operatorname{grad} \mathscr{F}\varphi(0) \cdot \operatorname{grad} (\mathscr{F}\Psi)^-(0)\,, \tag{3.3}$$

where $\mathscr{F}$ denotes Fourier transform. (Itô (1956) Theorem 4.1 could also be applied to grad X to yield this result.)

Now let $\alpha_1 \in \mathscr{D}(R^1)$, $\int \alpha_1 = 1$, $\alpha_n(u) \equiv n\alpha_1(nu)$. Let $\beta_n \in \mathscr{D}(R^1)$, $\int \beta_n \equiv 1$, $|\beta_n(u)| \leqq 2e^{-2u}$, and $\lim_{n\to\infty} \beta_n(u) = e^{-2|u|}$ for all u.

X is a tempered random distribution since $\sup_{|t| \leq M} E|x_t|^2 = \mathscr{O}(M^2)$ as $M \to \infty$. Let $\Psi_n(t) = \prod_{1 \leq j \leq k} [\alpha_n(t_j) - \beta_n(t_j)]$. Then $E|X(\Psi_n)|^2$ remains bounded while $\mathscr{F}\Psi_n(\lambda) \to 1$ as $n, \lambda \to \infty$. Hence $\mu\{|\lambda| \geqq 1\} < \infty$.

Now let $\gamma_n(u_1, \cdots, u_k) = \prod_{1 \leq j \leq k} \alpha_n(u_j)$. In (3.3) let $\varphi(u) = \gamma_n(u - s) - \gamma_n(u)$, $\Psi(u) = \gamma_n(u - t) - \gamma_n(u)$ and let $n \to \infty$. Then grad $\mathscr{F}\varphi(0) \to is$, grad $\mathscr{F}\Psi(0) \to it$. Thus we obtain (3.2) by letting A be the real part of J.

If x_t is real, then the change $d\mu(\lambda) \to d\mu(-\lambda)$ does not change the covariance, so μ is symmetric.

For the converse, we need only apply existence theorems for Gaussian processes with given covariance (Doob (1953) Chapter II, Section 3), since the covariances are clearly nonnegative definite, and real if μ is symmetric. The distributions of increments $x_t - x_0$ are uniquely determined, so that all distributions of x_t are determined given x_0. Hence x_t is unique in law up to an additive random variable y. □

Now suppose x_t is a stationary CQM process on R^k. Then, as is well known, there is a finite measure $\mu \geqq 0$ on R^k called the *spectral measure* of x_t, such that $Ex_s x_t = \int e^{i\lambda(s-t)}\, d\mu(\lambda)$. Clearly x_t also has stationary increments. Its LK measure is the same μ.

Conversely suppose x_t is Gaussian with mean 0 and that its LK measure μ is finite. Let y_t be a stationary Gaussian process with μ as spectral measure. Let L be the isonormal process on $H = L^2(\mu)$, or μ-noise process. Then versions of x_t and y_t are

$$y_t = L(\lambda \to e^{it\cdot\lambda}), \qquad x_t = L(\lambda \to e^{it\cdot\lambda} - 1) = y_t - L(1).$$

Since the increments of x_t and y_t have the same distributions, x_t and y_t have the same continuity properties, moduli etc. In other words, for any set A, $\{x_t : t \in A\}$ is a GB- or GC-set iff $\{y_t : t \in A\}$ is one.

Now let x_t be any CQM Gaussian process with stationary increments, mean 0, and LK measure μ. Let y_t, z_t be other such processes with LK measures $\mu|_{|\lambda|\leq 1}$ and $\mu|_{|\lambda|>1}$ respectively.

Let y_t and z_t be independent. Then $y_t + z_t - y_0 - z_0$ is a version of $x_t - x_0$. grad $[y_t]$ is a stationary random field. Its components have spectral measures with compact support. Thus (cf. Belyaev (1959)), grad $[y_t]$ and hence $[y_t]$ are processes whose sample functions can be written as entire functions of exponential type, by the Paley–Wiener–Schwartz (1950), (1966a) theorem. Also z_t has continuity properties equivalent to those of the stationary process with spectral measure $\mu|_{|\lambda|>1}$. The following has been proved.

THEOREM 3.4. *Let $\mathscr{G}$ be a linear space of functions on R^k including all entire functions of exponential type* 1. *Let x_t be a Gaussian process* CQM *with stationary increments and* LK *measure μ. Let y_t be the stationary process with spectral measure $\mu|_{|\lambda|>1}$. If y_t has a version with sample functions in $\mathscr{G}$, then so does x_t.*

For stationary Gaussian processes x_t, $t \in R$, Fourier methods have been in general use. Nearly best possible conditions for sample continuity in terms of spectral measure were found by Hunt (1951). Periodic processes, e.g., with $x_0 \equiv x_{2\pi}$, have simpler Fourier series rather than Fourier transforms, yet the dependence of sample properties on spectral asymptotics seems to be very similar. So, suppose we have a random Fourier series

$$x_t = a_0 Y_0 + \textstyle\sum_{n=1}^{\infty} a_n(Y_n \cos nt + Z_n \sin nt) \tag{3.5}$$

where the Y_n and Z_n are all orthogaussian and $a_n > 0$. Let

$$s_n = [\textstyle\sum \{a_k^2 : 2^n < k \leqq 2^{n+1}\}]^{\frac{1}{2}}.$$

Kahane (1960) proved that $\sum s_n < \infty$ is a necessary condition for sample-continuity of x_t, while $s_n \leqq t_n$, t_n decreasing, and $\sum t_n < \infty$ is a sufficient condition. M. Nisio (1969) extended these results to nonperiodic stationary processes, replacing the discrete set $\{2^n + 1, \cdots, 2^{n+1}\}$ by the interval $(2^n, 2^{n+1}]$ in Fourier transform space. Her condition $E \sup_{t\in[0,1]} |X(t)| < \infty$ is equivalent to sample-boundedness by the Fernique–Landau–Shepp theorem and then to sample-continuity in view of stationarity (Belyaev (1961)). By Theorem 3.4 above, these results carry over to processes with stationary increments.

Marcus and Shepp ((1970) Section 5) showed that $\sum s_n < \infty$ is not sufficient for sample-continuity of (3.5) in general.

Lacunary random Fourier series, where in (3.5) $a_n = 0$ except for $n = n_j$, $n_{j+1}/n_j \geqq q > 1$, have been a useful source of examples for Gaussian processes (Fernique (1964), Marcus (1971)).

3.3. *Boundedness on R.* Yu. K. Belyaev (1958) proved that for every stationary Gaussian process $\{x_t, t \in R\}$ whose spectral measure μ is not purely atomic,

$$\sup \{|x_t| : t \in R\} = \infty \quad \text{almost surely.}$$

Suppose then that μ is purely atomic with masses μ_k at points λ_k, $k = 1, 2, \cdots$. Belyaev (1958) discovered that if $\sum \mu_k^{\frac{1}{2}} < \infty$ then the sample functions are bounded on R, while if $\sum \mu_k^{\frac{1}{2}} = \infty$ and the λ_k are all incommensurable then the sample functions are unbounded a.s. The reason is that for X_k orthogaussian, $\sum b_k|X_k|$ converges a.s. iff $\sum b_k < \infty$. If $\sum \mu_k^{\frac{1}{2}} < \infty$ then $\sum \mu_k^{\frac{1}{2}} X_k \exp(i\lambda_k t)$ a.s. converges uniformly on R. If $\sum \mu_k^{\frac{1}{2}} = +\infty$ and the λ_k are incommensurable, we can find t such that $\exp(i\lambda_k t) \simeq \operatorname{sgn} X_k$, $k = 1, \cdots, n$, $n \to \infty$.

For the λ_k all multiples of a fixed number, we have the random Fourier series of Kahane (1960) as discussed above; good necessary conditions and good sufficient conditions for boundedness are known but are not yet equivalent. If the λ_k are commensurable in a more complicated way, less seems to be known.

Let x_t be Gaussian continuous in probability with stationary increments and LK measure μ. If the sample functions of x_t are bounded on R, then for every $\varepsilon > 0$ so are the sample functions of the process with LK measure $\mu|_{|\lambda| \geqq \varepsilon}$, which is equivalent to sample-boundedness of the stationary process with spectral measure $\mu|_{|\lambda| \geqq \varepsilon}$. Hence μ is purely atomic, with masses μ_k at points λ_k. Again, $\sum \mu_k^{\frac{1}{2}} < \infty$ is sufficient for sample-boundedness on all of R, and necessary if the λ_k are incommensurable.

4. Noise processes. The isonormal linear process L will be called a *noise process* (for μ) in case the Hilbert space H is $L^2(\mu)$ for some measure μ. If μ is Lebesgue measure on a Euclidean space, the process is called *white noise*.

A linear process G on $L^2(\mu)$ will be called *μ-bounded* iff there is an $M < \infty$ such that $EG(f)^2 \leqq M \int |f|^2 \, d\mu$ for all $f \in L^2$. If G is Gaussian with mean 0 and μ-bounded, then we can write $G = L \circ A$ where A is a bounded linear operator. It is known that if L has continuous or bounded sample functions on a set $C \subset H$, then so does G (Dudley (1967) Proposition 4.1, Theorem 4.6; L. Gross (1962) Theorem 5).

A special case of interest is the "centered noise" L_c for μ where μ is a probability measure. Here

$$EL_c(f)^2 = \int |f|^2 \, d\mu - \left|\int f \, d\mu\right|^2 = \int \left|f - \int f \, d\mu\right|^2 d\mu \,.$$

In this case the operator A is projection onto the orthogonal complement of the constant functions. Since A differs from the identity only by the one-dimensional projection $f \to \int f\, d\mu$ onto constants, the asymptotic properties of L and L_c are essentially the same.

L_c arises as a limit (in some weak sense) of normalized empirical measures $n^{\frac{1}{2}}(\mu_n - \mu)$ where μ_n has mass $1/n$ at each of n points chosen independently with distribution μ. If we can prove continuity of L and hence of L_c on some classes of function or sets, then we can hope to prove some stronger central limit theorems.

If $\mu(A) < \infty$ then we define $L(A) = L(\chi_A)$, $L_c(A) = L_c(\chi_A)$.

4.1. *λ-bounded noise on R^k.* If μ is Lebesgue measure λ on R^k, then $L(A)$ has been studied mainly when A is a rectangle

$$A_t = [0, t_1] \times \cdots \times [0, t_k] .$$

Then $L(A_t)$ is a standard Wiener process for $k = 1$ and a generalized Brownian motion for $k > 1$ (Chentsov (1956)). The processes, of course, have sample functions continuous in t with probability 1. Here, at the suggestion of R. Pyke, more general sets will be considered. Let $Co(x) = Co(x_1, \cdots, x_m)$ be the convex hull of the points $x_1, \cdots, x_m \in R^k$.

THEOREM 4.1. *For a fixed k and $m > k$, let $g(x_1, \cdots, x_m, \omega) = G(Co(x), \omega)$, where G is a λ-bounded Gaussian process of mean 0. Then g has continuous sample functions on R^{mk} and, when restricted to x in a bounded set, has the sample modulus $h(u) = (u|\log u|)^{\frac{1}{2}}$.*

PROOF. See Theorem 2.1 and Corollary 2.3 above. Note that if $|x_j - y_j| \leqq \varepsilon$ for $j = 1, \cdots, m$, and x and y lie in a fixed bounded set B, then the Lebesgue measure of the symmetric difference, $\lambda(C(x) \triangle C(y))$, is $\mathcal{O}(\varepsilon)$ uniformly in x, $y \in B$ as $\varepsilon \downarrow 0$. The total surface area of the faces of $C(x)$ is bounded uniformly for $x \in B$. Now $E|G(C(x)) - G(C(y))|^2 = \mathcal{O}(\varepsilon)$ so the proof in Example 2.2 above applies to give the modulus, which incidentally is best possible. (Simple continuity was proved earlier, in Dudley (1965) Section 5.)

Now let $I(k, \alpha, M)$ be the class of all (indicator functions of) subsets of R^k with boundary functions having all derivatives of orders $\leqq \alpha$ bounded by M, as defined in Dudley (1972b).

THEOREM 4.2. *$I(k, \alpha, M)$ is a* GC-*set in $L^2(\lambda)$ if $\alpha > k - 1 \geqq 1$, and not a* GB-*set if $1 \leqq \alpha < k - 1$ or $0 < \alpha \leqq 1 < k - 1$ and $M > 0$.*

PROOF. This is a corollary of Theorems 1.1 above and 3.1 of Dudley (1972b). Note that the exponent of entropy of $I(k, \alpha, M)$ in H is twice its exponent of entropy in the d_λ metric, $d_\lambda(A, B) = \lambda(A \triangle B)$, since the metric induced from H is $d_\lambda^{\frac{1}{2}}$. □

REMARKS. If $\alpha < 1$ and $k = 2$ I conjecture that $I(2, \alpha, M)$ is not a GB-set for $M > 0$, based on the conjecture of equality in Dudley ((1972b) 3.4).

If $\alpha = k - 1$ I conjecture that $I(k, k - 1, M)$ is not a GB-set.

Let $C(U)$ be the class of all (indicator functions of) convex subsets of a given bounded open set $U \subset R^k$. An earlier investigation of λ-bounded processes on $C(U)$ was made by A. de Hoyos (1972), incorrectly.

THEOREM 4.3. *$C(U)$ is a* GB-*set in $L^2(\lambda)$ for $k = 1, 2$ and not* GB *for $k \geqq 4$.*

PROOF. This is a corollary of the Sudakov–Strassen theorem (1.1 above) and Dudley ((1972b) Theorem 4.1), with the same note as in the previous proof.

REMARK. For $k = 3$, I claim that $C(U)$ is not a GB-set. Although the theorems proved above do not apply directly, we can consider the spherical caps C_x at the end of the proof of Theorem 4.1 of Dudley (1972b). Then $\sum_{x \in A_\varepsilon} |L(C_x)|$ does not approach 0 as $\varepsilon \downarrow 0$, so $C(U)$ is not a GC-set. To prove it is not a GB-set, we could successively adjoin suitable unions of small caps (depending on ω) to a given convex set.

4.2. *Elliptic operators and white noise.* We know that the white noise process W on R^k can be written as $W = \partial^k[f]/\partial t_1 \cdots \partial t_k$ in the Schwartz distribution sense, where f is a process with continuous sample functions (Chentsov (1956), Dudley (1965)). It turns out that if we use elliptic operators a degree less than k will suffice.

For any polynomial P in k variables we let, as usual,

$$P(D) = P(-i\partial/\partial t_1, \cdots, -i\partial/\partial t_k) .$$

THEOREM 4.4. *Let $P(D)$ be an elliptic operator of degree m in k variables with constant coefficients. Let W be the white noise generalized random process on R^k. Then the following are equivalent:*

(i) *For almost all ω, the distribution solutions T of $P(D)T = W_\omega$, on any open set $U \subset R^k$, are continuous functions on U;*

(ii) *There is a process $T(t, \omega)$ with continuous sample functions on some open $U \subset R^k$ with $P(D)T = W$ in U;*

(iii) *Replace "continuous" by "L^2" in* (ii);

(iv) *$k < 2m$.*

PROOF. Let $\mathscr{D}(U)$ be the L. Schwartz space of test functions with support in U, $\mathscr{D} = \mathscr{D}(R^k)$ with usual topology and dual space $\mathscr{D}'$ of distributions (Schwartz (1966a)). By Minlos' theorem (Minlos (1959), Kolmogorov (1959)), W has some distributions W_ω as realizations. Then $P(D)T_\omega = W_\omega$ has distribution solutions T_ω (Hörmander (1964) Section 3.6). Thus

(i) implies (ii).

(ii) implies (iii) trivially.

(iii) $\Rightarrow$ (iv): we can assume $U = \{t : |t_j| < 4, j = 1, \cdots, k\}$. Let H_P be the inner product space of all $\varphi \in \mathscr{D}(U)$ with the norm

$$||\varphi||_P^2 = \int_U |P(-D)\varphi|^2 .$$

Then W restricted to H_P has a version with continuous sample functions for $||\cdot||_P$.

Hence, the identity from H_P into L^2 must be a Hilbert–Schmidt operator (Minlos (1959)).

Let $\eta \in \mathscr{D}(U)$ and $\eta = 1$ on $\pi U/4$. Let $\varphi_n(x) \equiv \eta(x)e^{in\cdot x}$, $n \in \mathbb{Z}^k$. Then $\|\varphi_n\|_P = \mathscr{O}(|n|^m)$ as $|n|^2 \equiv n_1^2 + \cdots + n_k^2 \to \infty$. Hence $\sum_{n\in Z^k, n\neq 0} |n|^{-2m} < \infty$, so $k < 2m$.

(iv) $\Rightarrow$ (i): All distribution solutions T of $P(D)T = 0$ on open sets are continuous (in fact, real analytic) (see Hörmander (1964)). Thus we need only get a continuous solution process T in the cube $V = \pi U/4$. Here we have

$$W = \sum_{n\in Z^k} X_n e^{in\cdot x}$$

where the X_n are orthogaussian. If φ is C^∞, all solutions S of $P(D)S = \varphi$ are C^∞. Also since P is elliptic, $P(n) = 0$ only for n in a finite set F. We assume $0 \in F$ even if $P(0) \neq 0$.

Now we need only prove that the Gaussian process

$$T_F(x, \omega) = \sum_{n\in Z^k \sim F} X_n(\omega)e^{in\cdot x}/P(n)$$

has continuous sample functions. For some $c > 0$, $|P(n)| \geqq c|n|^m$ for all $n \notin F$. Thus for any $x, y \in V$,

$$E|T_F(x) - T_F(y)|^2 \leqq c^{-2} \sum_{n\notin F} |e^{in\cdot x} - e^{in\cdot y}|^2/|n|^{2m} .$$

For some $C_1 < \infty$, $\sum_{|n|\geqq M} |n|^{-2m} \leqq C_1/M$ for all $M > 0$, since $k - 1 - 2m \leqq -2$. Also

$$\sum_{0<|n|<M} \frac{|e^{in\cdot x} - e^{in\cdot y}|^2}{|n|^{2m}} \leqq C_2 M|x - y|^2$$

for some $C_2 < \infty$. Thus $E|T_F(x) - T_F(y)|^2 \leqq C_3|x - y|$. Hence T_F has continuous sample functions. □

Theorem 4.4, as just proved, offers possibilities of extensions from R^k to other k-dimensional manifolds X with a μ-noise process, where μ is a measure which on local coordinate patches has a sufficiently smooth density with respect to Lebesgue measure. If X is a Riemannian manifold, there is an invariantly defined Laplace–Beltrami operator whose powers give elliptic operators of as high even order as desired. In Theorem 4.4 presumably constant coefficients could be replaced by sufficiently smooth, nonsingular coefficients.

4.3. *Lévy–Baxter theorems.* P. Lévy (1940) proved that if $x(t)$ is a standard Brownian motion, then with probability 1

$$\lim_{n\to\infty} \sum_{k=1}^{2^n} [x(k/2^n) - x((k-1)/2^n)]^2 = 1 .$$

This result was extended to some other Gaussian processes on [0, 1] by G. Baxter (1956), E. G. Gladyshev (1961), and V. G. Alekseev (1963), and to Lévy's Brownian motion with multidimensional time by S. M. Berman (1967) and P. T. Strait (1969). (R. Borges (1966) showed that the "generalization" by F. Kozin (1957) only applies to Brownian motion.)

Here we give another extension. Let $(X, \mathscr{S}, \mu)$ be any finite measure space:

$\mu \geqq 0$, $\mu(X) < \infty$. A *partition* of X will be a finite collection π of disjoint measurable sets whose union is X. The mesh of π is defined by

$$m(\pi) = \max \{\mu(A) : A \in \pi\} .$$

Let L be the μ-noise. We consider the sums $L(\pi)^2 = \sum_{A \in \pi} L(A)^2$. As $m(\pi) \to 0$, $L(\pi)^2 \to \mu(X)$ in law and hence in probability. P. Lévy ((1940) Section 4, Théorème 5) proved that $L(\pi_n)^2 \to \mu(X)$ almost surely if the π_n are nested, i.e., for all $A \in \pi_{n+1}$ there is a $B \in \pi_n$ with $A \subset B$. F. Kozin (1957) proved $L(\pi_n)^2 \to \mu(X)$ almost surely if $m(\pi_n) = o(n^{-2})$, for interval partitions. Most other authors have used $m(\pi_n) \leqq 2^{-n}$, considering more general Gaussian processes. Here for μ-noise we prove $m(\pi_n) = o(1/\log n)$ suffices for a.s. convergence, and that this is best possible.

THEOREM 4.5. *Let L be μ-noise for a finite measure space (X, μ). If $\{\pi_n\}$ is any sequence of partitions with $m(\pi_n) = o(1/\log n)$, then $L(\pi_n)^2 \to \mu(X)$ almost surely. If $(X, \mu) = ([0, 1], \lambda)$ there exist partitions π_n (not consisting of intervals) such that $m(\pi_n) \leqq (1/\log n)$ and $L(\pi_n)^2$ does not converge* a.s. *to* 1.

PROOF. Given a partition $\pi = \{A_1, \cdots, A_k\}$, let $a_j = \mu(A_j)$. We can assume $\mu(X) = 1$. Then $\sum a_j^2 \leqq m(\pi)$. We have $L(A_j)^2 = a_j X_j^2$ where the X_j are orthogaussian. Hence by a theorem of D. L. Hanson and F. T. Wright (1971), there are constants C_1 and C_2 such that for any $\varepsilon > 0$,

$$\Pr \{|\textstyle\sum_{A \in \pi} L(A)^2 - 1| \geqq \varepsilon\} \leqq 2 \exp\{-\min (C_1 \varepsilon/m(\pi), C_2 \varepsilon^2/m(\pi))\} .$$

Let $m(\pi_k) = \varepsilon_k/2 \log k$, where $\varepsilon_k \to 0$. Then

$$\Pr \{|\textstyle\sum_{A \in \pi_k} L(A)^2 - 1| \geqq \varepsilon_k\} = \mathcal{O}(k^{-2}) ,$$

so $L(\pi_n)^2 \to 1$.

Now for Lebesgue measure λ on $[0, 1]$ we choose λ-independent partitions π_n consisting of k_n sets of equal measure where k_n is the least integer $> \log n$, $n = 2, 3, \cdots$. Then $k_n L(\pi_n)^2$ has a χ^2 distribution with k_n degrees of freedom. Letting $k = k_n$ we have for any fixed $\varepsilon > 0$

$$\begin{aligned} \Pr \{L(\pi_n)^2 < 1 - \varepsilon\} &= \Pr \{kL(\pi_n)^2 < (1 - \varepsilon)k\} \\ &= 2^{(2-k)/2}\Gamma(k/2)^{-1} \textstyle\int_0^{[(1-\varepsilon)k]^{\frac{1}{2}}} r^{k-1} \exp(-r^2/2)\, dr \\ &\geqq 2^{(2-k)/2}\Gamma(k/2)^{-1}\{[(1 - \varepsilon)k]^{\frac{1}{2}} - 1\}^{k-1} e^{-(1-\varepsilon)k/2} , \end{aligned}$$

which by Stirling's formula is asymptotic as $k \to \infty$ to

$$\begin{aligned} &(e/(k - 2))^{(k-2)/2}(\pi k)^{-\frac{1}{2}}\{[(1 - \varepsilon)k]^{\frac{1}{2}} - 1\}^{k-1} e^{-(1-\varepsilon)k/2} \\ &\qquad \sim c([1 - \varepsilon]^{\frac{1}{2}} - k^{-\frac{1}{2}})^{k-1}/(1 - 2k^{-1})^{(k-2)/2} e^{\varepsilon k/2} \geqq \alpha(1 - 2\varepsilon)^k , \end{aligned}$$

for some constants c, α, and k large. Letting $\varepsilon = \frac{1}{4}$ and noting $k \sim \log n$ we have $\sum_n \alpha(1 - 2\varepsilon)^k = +\infty$.

Now let $\pi_n = \{A_1, \cdots, A_k\}$ and $\pi_m = \{B_1, \cdots, B_r\}$, $r = k_m$, be two of our independent partitions, $\lambda(A_j \cap B_i) = \lambda(A_j)\lambda(B_i) = 1/kr$. Let $Z = L([0, 1])$. Then

$E(L(A_i) - Z/k)(L(B_j) - Z/r) = 0$ for all i, j. Thus the set of random variables $\{L(A_i) - Z/k\}$ is independent of the set $\{L(B_j) - Z/r\}$. Let

$$Q(n) = \sum_i [L(A_i) - Z/k]^2 = L(\pi_n)^2 - Z^2/k .$$

Then $Q(n)$ is independent of $Q(m)$ for $m \neq n$. Since $\sum_n P(Q(n) < \frac{3}{4}) = +\infty$, we have a.s. $Q(n) < \frac{3}{4}$ infinitely often by Borel–Cantelli. Then since $Z^2/k \to 0$ a.s. as $n \to \infty$, we have a.s. $L(\pi_n)^2 < \frac{7}{8}$ infinitely often. □

Kozin (1957) attributes to Lévy (1940) the assertion that $L(\pi_n)^2 \to 0$ almost surely if card $(\pi_n) \to \infty$, $\mu = \lambda$ on $[0, 1]$ and π_n consist of intervals. No doubt Kozin meant to include the assumption $m(\pi_n) \to 0$. Even then, the mere assumption that π_n consists of intervals cannot correctly replace Lévy's assumption of nested partitions, for reasons indicated later by Lévy ((1965) page 192).

Whether $o(1/\log n)$ is best possible for interval partitions seems to be an open question. Also open, and perhaps not difficult, are questions of the best possible assumptions on speed of $m(\pi_n) \to 0$ for the various generalizations of Lévy's work.

5. Discontinuous Gaussian processes. As we shall see in Section 7, conditions for sample-continuity of Gaussian processes are relatively mild as compared with conditions for general non-Gaussian processes. Nevertheless it is of some interest to consider how the sample functions of Gaussian processes may behave even when they are not continuous.

It may be asked whether, instead of restricting L to suitable subsets of H, we can obtain a version of L with good properties defined on all of H. Certainly L on H is not sample-continuous. If we take any particular finite Borel measure μ on H, then L has a version which is jointly $\mu \times P$-measurable and is linear on H for each fixed ω. However, we cannot replace μ-measurability here by simultaneous measurability for all Borel measures μ, or absolute measurability, since every absolutely measurable linear form on H is continuous by a theorem of Douady (Schwartz (1966 b) Lemme 2).

By embedding the real line R in a compactification $\bar{R}$, we can obtain a regular Borel measure defining L on a compact Hausdorff space $\bar{R}^H$ (Kakutani (1943), E. Nelson (1959)) but, assuming the continuum hypothesis, this version of L is not jointly measurable, and such a large, non-metrizable compact has other bad properties (Dudley (1971 a), (1972 a)).

We have a probability measure P defining L on the algebraic dual space H^a of all linear forms on H, but this P has no extension regular for the weak topology $\sigma(H^a, H)$ (A. de Acosta (1971)).

Precisely because L is a universal model for all Gaussian processes, it is perhaps not surprising that L on the *entire* space H has the various pathological properties mentioned in the last few paragraphs.

Suppose $\{X_t, t \in T\}$ is a mean-zero Gaussian process, continuous in probability, on a separable metric space T. Then there exist orthogaussian random variables

Y_n and continuous functions f_n with $X_t = \sum Y_n f_n(t)$. Since finite partial sums are all continuous, such continuity properties as continuity at a fixed point, continuity everywhere, and even the existence of discontinuities of oscillation $\geqq \varepsilon$ for fixed $\varepsilon > 0$ on a fixed open set, all are "zero-one" properties. If X_t is not sample-continuous, then there exists a point t and an $\varepsilon > 0$ such that almost all sample functions oscillate by $\geqq \varepsilon$ in every neighborhood of t. Thus, unlike Markov processes with their isolated random jump discontinuities, Gaussian processes are discontinuous at fixed points. In fact, K. Itô and M. Nisio (1968) have proved given a Gaussian process x_t, $t \in [0, 1]$, there is a fixed function α such that almost every sample function $x_t(\omega)$ satisfies, for *all* t,

$$\limsup_{u,v \to t} |x_u - x_v| = \alpha(t) .$$

If a non-sample-continuous process on a separable metric T has probability laws invariant under a transitive group of homeomorphisms of T, then almost all sample functions oscillate by $\geqq \varepsilon > 0$ on every open set, since there is a countable base for the topology. Thus the sample functions are *everywhere* discontinuous. Further, the ε-oscillations pile up to produce infinite oscillations. On such matters see Yu. Belyaev (1961), S. M. Berman (1968), D. M. Eaves (1967), D. Cohn (1971), K. Itô and M. Nisio (1968), N. Jain and G. Kallianpur (1971).

Fernique (1971) proves that for any Gaussian process $\{x_t, t \in T\}$ with mean 0 and bounded measurable covariance, and any separable probability measure μ on T, the process has a version with almost all sample functions in the Banach space $G_C(\mu)$ (also defined in Section 1 above).

6. Infinitesimal σ-algebras and 0–1 laws. Let $\{x_t, t \in T\}$ be a mean-zero Gaussian process, continuous in probability, on a separable metric space (T, d). Then we can write $x_t(\omega) \equiv \sum f_n(t) X_n(\omega)$, f_n continuous, $\{X_n\}$ orthogaussian.

Suppose now that T is compact and $C \equiv \{x_t : t \in T\}$ is GC. A modulus f will be called a lim sup *modulus* for x_t iff for almost all ω,

$$\limsup \{x_t; f\} \equiv \limsup_{h \downarrow 0} \sup \{|x_s - x_t| : d(s, t) \leqq h\}/f(h) = 1 .$$

Clearly a lim sup modulus is a weakly optimal sample modulus as defined in Section 2 above. It is not clear whether it is strongly optimal.

Suppose g_n is a uniform modulus of continuity of f_n and assume that $\lim_{h \downarrow 0} g_n(h)/f(h) = 0$ for all n. This holds, for example, in case $T = C$, $t \to x_t$ is the identity, and $\lim_{h \downarrow 0} h/f(h) = 0$. Then $\limsup \{x_t, f\}$ is almost surely a constant (possibly infinite) by the zero-one law. If this constant is positive and finite, then some positive multiple of f is a lim sup modulus. For many processes, lim sup moduli are known, at least up to multiplicative constants. If f is a lim sup modulus, so is $f + o(f)$, but this apparently does not exhaust the class of lim sup moduli of the process. On the other hand it seems unclear whether lim sup moduli always exist in the cases we have been discussing:

QUESTION. Let C be a compact GC-set in H which is not a GL-set. Then is there always a lim sup modulus for L on C?

Corresponding questions can be asked for processes on [0, 1] and for local rather than uniform moduli, etc. For *non*-Gaussian processes it is known that local lim sup moduli need not exist: Gnedenko (1943), Rogozin (1968).

Let $\{x_t, 0 \leqq t \leqq 1\}$ be a Gaussian process with mean 0. Let $\mathscr{B}(\delta)$ be the smallest σ-algebra for which $\{x_t, 0 \leqq t \leqq \delta\}$ are all measurable, and let $\mathscr{B}(0^+) = \bigcap_{\delta>0} \mathscr{B}(\delta)$. If almost all sample functions of x_t have n derivatives at 0 then these derivatives are $\mathscr{B}(0^+)$-measurable functions which in general are nontrivial. Under assumptions of stationarity with spectral density behaving like an inverse power at ∞, Freidlin and Tutubalin (1962) showed that all $\mathscr{B}(0^+)$-measurable functions are measurable with respect to the derivatives which exist. Thus if x_t has non-differentiable sample functions then, under their conditions, every asymptotic property of the increments $x_s - x_t$ as $s \to t$ has probability 0 or 1. It seems reasonable to expect this zero-one law to hold under less restrictive conditions than those of Tutubalin and Freidlin. However, examples of a "gap" with $\mathscr{B}(0^+)$ not generated by derivatives are given by Levinson and McKean ((1964) pages 130–133) and Dym and McKean ((1970) page 1824). These papers also contain further relevant information on $\mathscr{B}(0^+)$.

At the other extreme, it may be asked which processes are entirely determined by their behavior in the neighborhood of one point. This question, for $T = L^p(\mu)$, has been considered by Bretagnolle and Dacunha-Castelle (1969). Such questions were also considered by P. Lévy (1948), (1965).

It is known rather generally that, given a Gaussian measure μ on a linear space X, the support of μ is a linear subspace of X (K. Itô (1970), G. Kallianpur (1970), (1971)). Here the support is defined as the smallest closed set whose complement has measure 0. There may also be some interest in considering non-closed supports. Let S be a nuclear Fréchet space with topology defined by a sequence of seminorms $|\cdot|_1 \leqq |\cdot|_2 \leqq \cdots$. For example, S may be L. Schwartz's test function space $\mathscr{S}$ or $\mathscr{D}(K)$, K compact. Let S' be the dual space of S, $S' = \bigcup S_n'$, where $S_n' = \{f \in S' : f \text{ continuous for } |\cdot|_n\}$. (Here S' is the inductive limit of the S_n', but in general this inductive limit is *not* strict, i.e., the embedding $S_n' \to S_{n+1}'$ is not a homeomorphism, contrary to an editorial insertion in my review (1969).) As noted by D. M. Eaves (1968), if μ is a Gaussian measure on S', then $\mu(S_n') = 1$ for some n. It happens often, for example if $S = \mathscr{S}$ or $\mathscr{D}(K)$, that S_n' is actually not closed and is dense in S', while $\mu(S_n') = 1$ is more interesting than $\mu(S') = 1$. For example, the smallest n such that $\mu(S_n') = 1$ may be the smallest n for which the sample generalized functions in S' are derivatives of order n of continuous functions, as in Section 4 above.

Further, many theorems assert that Gaussian measures live on subsets which are neither closed nor linear. For example, if $\{X_n\}$ are orthogaussian, $\lim_{n\to\infty} \sum_{j=1}^n X_j^2/n = 1$ a.s., and the set of sequences $\{X_j\}$ satisfying the condition is nonlinear. For an extension of this line of thought see, e.g., T. Hida and H. Nomoto (1964).

A zero-one law for Borel subgroups was proved by N. C. Jain (1971).

7. Non-Gaussian processes.

7.1. *Some counterexamples.* Let T be a bounded set in R^k. We know that if $\{x_t, t \in T\}$ is a Gaussian process with

$$E|x_s - x_t|^2 \leqq C/|\log|s - t||^{1+\delta} \qquad \text{for some } \delta > 0 \text{ and } C < \infty ,$$

then x_t is sample-continuous. The Gaussian property is strongly used.

H. Totoki (1962) proved that if x_t is any process on $T \subset R^k$ (Gaussian or not) such that for some α and $p > 0$, $E|x_s - x_t|^p \leqq C|t - s|^{k+\alpha}$, then x_t is sample-continuous. This condition is of course much stronger than the one stated above for Gaussian processes.

P. Bernard (1970) weakened Totoki's condition to

$$E|x_{t+h} - x_t|^p \leqq C|h|^k/|\log|h||^s , \qquad s > p + 1 .$$

Earlier, J. Delporte (1966) gave this sufficient condition for sample-continuity if $k = 2$ and $s > p$.

Now here are examples where $s = -1$ and we do not have sample-continuity, so that the power k in $|h|^k$ is best possible. We still do not know the best possible power of the logarithm, between -1 and p or $p + 1$.

PROPOSITION 7.1. *For $k = 1, 2, \cdots$, there is a stochastic process x_t, $t \in I^k$, where I^k is the unit cube in R^k, $I = [0, 1]$, such that*

$$E|x_s - x_t|^k \leqq |s - t|^k(1 + |\log|s - t||)$$

and (every version of) x_t has almost all sample functions unbounded.

PROOF. For each $n = 1, 2, \cdots$, we divide I^k into 2^{nk} equal, parallel cubes C_{nj}, $j = 1, \cdots, 2^{nk}$. Let f_{nj} be a function which is 1 at the center of C_{nj}, 0 outside C_{nj}, and linear on each line segment joining the center to the boundary of C_{nj}.

For each n let $j(n)$ be a random variable with $\Pr(j(n) = j) = 2^{-nk}$, $j = 1, \cdots, 2^{nk}$. Let the $j(n)$ be independent for different n.

Let

$$x_t(\omega) = \sum_{n=1}^{\infty} f_{n,j(n)(\omega)}(t) , \qquad t \in I^k .$$

The series converges in $L^1(I^k, \lambda)$, and for each ω it converges for almost all t. The cubes $C_{n,j(n)}$ have accumulation points t such that in every neighborhood of such t, x_t is unbounded. (Actually, such t are almost surely dense in I^k.)

For $k^{\frac{1}{2}}/2^{n+1} \leqq |s - t| \leqq k^{\frac{1}{2}}/2^n$, we have

$$\begin{aligned} E|x_s - x_t|^k &\leqq \textstyle\sum_{j=1}^{n} 2 \cdot (2^{(j+1)}|s - t|)^k/2^{jk} + \sum_{j>n} 2/2^{jk} ; \\ &\leqq 2^{k+1}n|s - t|^k + 2^{-nk-k+2} . \end{aligned}$$

Dividing the process by a suitable constant we have the result. □

Most proofs of sample continuity of Gaussian processes use just upper bounds on tail probabilities. Thus similar results do hold for non-Gaussian processes with tail bounds of Gaussian type, called "sub-Gaussian" processes; (see Kahane (1960), (1968), Kozačenko (1968), Hanson and Wright (1971)).

For processes stationary in quadratic mean, the assumptions on the spectrum for sample-continuity are again much stronger in general than they are for Gaussian processes; see, e.g., Kawata and Kubo (1970).

7.2. *Comparison of processes.* Marcus and Shepp ((1971) Lemma 1.5) proved that if X and Y are Gaussian processes with

$$E|Y(s) - Y(t)|^2 \leqq E|X(s) - X(t)|^2 , \qquad EX(t) = 0 ,$$

for $0 \leqq s \leqq t \leqq 1$, and X is sample-continuous, then so is Y. Their proof uses Slepian's inequality (1962). It may be amusing to note that the above result may fail for X non-Gaussian and Y Gaussian. Let H be the usual Hilbert space with an orthonormal basis $\{\varphi_n\}$. Let μ be a probability measure with mass 2^{-n-1} at $2^{n/2}\varphi_n$ and at $-2^{n/2}\varphi_n$ for $n = 1, 2, \cdots$. Let $N(f) = (f, x)$ where x has distribution μ. Then N has the same means and covariances as the isonormal process L: $EN(f) = 0$, $EN(f)N(g) = (f, g)$. Clearly N is sample-continuous on H. Hence:

PROPOSITION 7.2. *For any stochastic process $\{x_t, t \in T\}$ with $E|x_t|^2 < \infty$ and $E|x_s - x_t|^2 \to 0$ as $s \to t$, where T is any separable metric space, there is a process y_t with the same means and covariances as x_t and with continuous sample functions, namely $y_t = Ex_t + N(x_t - Ex_t)$.*

8. Miscellaneous topics on Gaussian sample functions. Several topics have not been treated at length in this survey, partly for lack of time and space. Here are brief remarks on a few of these topics.

8.1. *Multidimensional range.* Some questions arise for processes with multidimensional range which may be trivial for sample-continuous real-valued processes, such as the Hausdorff dimension or other measures of size of the range or trajectory, or of its intersection with a given set. D. Ray (1963) and S. J. Taylor (1964) proved sharp results for Brownian motion. See also Kahane (1968), Chapter 13 and S. Orey (1970). F. Spitzer (1958) considered asymptotic behavior of polar coordinates of plane-valued Brownian motion.

8.2. *Level crossings.* Let x_t be Gaussian CQM stationary with mean 0, covariance $Ex_s x_t = r(s - t)$, and spectral measure μ. Let $M(T, u)$ be the number of values of t with $x_t = u$ and $0 \leqq t \leqq T$. K. Itô (1964) proved that $EM(T, u) < \infty$ is equivalent to finiteness of the symmetric second derivative $r''(0)$ and hence to $\int \lambda^2 \, d\mu(\lambda) < \infty$. Thus such processes have first derivatives which are integrable to any finite power, but these derivatives are not necessarily sample-continuous. Itô proved rigorously a formula of S. O. Rice (1945) and V. Bunimovich (1951),

$$EM(T, u) = T\pi^{-1}(-r''(0)/r(0))^{\frac{1}{2}} \exp(-u^2/2r(0)) .$$

Others, e.g., Bulinskaya (1961), had proved the formula in increasing generality before Itô's final result.

If $EM(T, u) = +\infty$, can it happen that $M(T, u) < \infty$ almost surely? This question seems to be open.

As $u \to \infty$ for fixed T, $M(T, u)$ is asymptotically Poisson-distributed under suitable hypotheses (a recent reference is S. M. Berman (1971 b)).

On the other hand for fixed u, as $T \to \infty$ $M(T, u)$ is asymptotically normal (Malevich (1969)) under a list of conditions not reproduced here.

Yu. K. Belyaev and V. P. Nosko (1969) have several results on the asymptotic distribution of lengths of excursions above level u for stationary Gaussian (and other) processes. See also S. M. Berman (1971 a).

8.3. *Geometric properties in Hilbert space.* Let e_n be an orthonormal sequence in a Hilbert space H, $0 \leqq p < \infty$, and $\{a_n\}$ a decreasing sequence of positive real numbers. Let

$$B_p\{a_n\} = \{\sum x_n e_n : \sum |x_n/a_n|^p \leqq 1\},$$
$$B_\infty\{a_n\} = \{\sum x_n e_n : \sup |x_n/a_n| \leqq 1\}.$$

Then for $1 < p < \infty$, $B_p\{a_n\}$ is a GB-set iff $\sum |a_n Y_n|^q < \infty$ almost surely, where $1/p + 1/q = 1$ and Y_n are ortho-Gaussian. M. G. Sonis (1966) proved that this holds iff $\sum |a_n|^q < \infty$. For $p = q = 2$ this was classical.

Dudley (1967) proved $B_1\{a_n\}$ is a GB-set iff $a_n = \mathcal{O}(\log n)^{-\frac{1}{2}}$ and a GC-set iff $a_n = o(\log n)^{-\frac{1}{2}}$. Sonis (1966) had related, but more complicated conditions.

Let C be a closed, convex set in H. Let $V_n(C)$ be the supremum of n-dimensional Lebesgue measures of orthogonal projections of C into n-dimensional subspaces. Let

$$EV(C) = \limsup_{n\to\infty} (\log V_n)/n \log n .$$

My 1967 paper proved that $r(C) \geqq -2/(1 + 2EV(C))$ if $EV(C) < -\frac{1}{2}$ and conjectured that if $EV(C) < -1$, then $r(C) = -2/(1 + 2EV(C))$. It would follow that $r(C) < 2$, so that C is a GC-set. This conjecture remains open in general. My 1967 paper proved it for $B_p\{a_n\}$, $p = 1, 2, \infty$. S. Chevet (1969), (1970) proved if for $1 < p < 2$ and $2 < p < \infty$. For $1 < p \leqq \infty$, B_p is a GC-set iff it is a GB-set.

In my 1967 paper, Theorem 5.3 states that C is not a GB-set if

$$\sup_n [n^{-1} \log V_n(C) + \log n] = +\infty ,$$

in particular if $EV(C) > -1$. However, volumes, like $N(C, \varepsilon)$, cannot give a complete characterization of GC-sets or GB-sets (Dudley (1967) Proposition 6.10).

Sudakov (1969) announced that an ellipsoid $E = B_2\{a_n\}$ is a GB-set iff $\int_0^1 \varepsilon^2 \, dH(E, \varepsilon) > -\infty$. A proof of this can be based on Theorem 3 of B. S. Mityagin (1961) page 71. As we saw in Section 1 above, not every GB-set C satisfies $\int_0^1 \varepsilon^2 \, dH(C, \varepsilon) > -\infty$. There exist GB-sets, such as $B_1\{1/\log(n + 1)\}$, not included in any GB-ellipsoid.

8.4. *Differentiability.* A process has continuously differentiable sample functions iff it is the indefinite integral of a process with continuous sample functions.

For example, a Gaussian process with stationary increments and LK measure μ has m times continuously differentiable sample functions if for some $a > 1$,

$$\int_{|\lambda|\geq 1} \lambda^{2m}|\log \lambda|^a \, d\mu(\lambda) < \infty .$$

Yu. K. Belyaev (1959) proved that any stochastic process $\{x_t, 0 \leqq t \leqq 1\}$ with mean 0 and analytic covariance function $Ex_s x_t$ has a version with analytic sample functions. For Gaussian processes he proved that this sufficient condition is also necessary. For stationary processes with spectral measure μ he showed that sample functions x_t are analytic for $|t| \leqq r$ iff $\int_{-\infty}^{\infty} e^{r\lambda} \, d\mu(\lambda) < \infty$. Thus, for example, if μ has compact support or is itself Gaussian, then x_t has entire analytic sample functions. By Theorem 3.4 above these results on stationary processes carry over to processes with stationary increments, if we integrate over $|\lambda| > 1$.

8.5. *Examples of sample functions.* It may happen that almost all sample functions of a process are proved to have a certain property, yet it is difficult to construct any specific function with such a property. Thus, on the one hand, sample function theory can provide existence proofs simpler than constructive ones. On the other hand, construction of such functions is an interesting challenge; one such was met by K. Urbanik (1959).

8.6. *Local maxima.* Dvoretzky, Erdös and Kakutani (1961) proved that almost every Brownian path x_t has no points of increase, while it does have local maxima. In a side remark (page 105) inessential for their purposes, they state that "the set of points of maximum is, almost surely, of the power of the continuum in every open interval". But actually local maxima are a.s. countable, since strict local maxima form a countable union of discrete, hence countable sets, and for any rational $a < b < c$, we have a.s.

$$\max\{x_t : a < t < b\} \neq \max\{x_t : b < t < c\} ,$$

so a.s. all local maxima are strict. Also, a.s. the local maxima are dense. All this was noted by G. J. Foschini and R. K. Mueller (1970) with one lemma by L. Shepp.

Local behavior of Gaussian processes with smooth covariances near local maxima has been considered in several works by G. Lindgren (1971).

As to the possible values at local maxima, we have the following result, essentially due to D. Ylvisaker (1965), (1968).

THEOREM 8.1. *Let K be a compact metric space and $\{Y_t : t \in K\}$ any Gaussian process on K with continuous sample functions and such that for all $t \in K$,*

$$\sigma^2(t) \equiv E(Y_t - EY_t)^2 > 0 .$$

Then the distribution of $\max\{Y_t : t \in K\}$ *is absolutely continuous with respect to Lebesgue measure.*

PROOF. Ylvisaker (1965), (1968) proves that if in addition $\sigma^2(t) \equiv 1$ then the law of $\max\{Y_t: t \in K\}$ has a density f with respect to Lebesgue measure where

$$f(x) = \exp(-x^2/2)G(x)$$

and G is non-decreasing and finite, so that f is bounded on bounded sets. Then for a general Y_t, $x \in R$, and $0 < \varepsilon \leqq 1$, we have

$$\begin{aligned} \Pr\{x - \varepsilon \leqq \max Y_t \leqq x + \varepsilon\} \\ &= \Pr\{-\varepsilon \leqq \max(Y_t - x) \leqq \varepsilon\} \\ &\leqq \Pr\{-\varepsilon/\inf \sigma(t) \leqq \max(Y_t - x)/\sigma(t) \leqq \varepsilon/\inf \sigma(t)\}, \end{aligned}$$

noting that by sample continuity, σ is continuous so that $\inf \sigma > 0$. Hence by Ylvisaker's results there is some $K < \infty$, depending on Y_t and x but not on ε, such that the above probabilities are less than $K\varepsilon$. This suffices to prove the absolute continuity. □

Considering $-Y_t$, we get the same result for $|Y_t|$. If $\sigma^2(t) = 0$ for some t, let F be the compact set $\{t: \sigma^2(t) = 0\}$, and $\alpha = \max\{EY_t: t \in F\}$. Then $P(\max Y_t = \alpha)$ may be positive, or may be 0. It seems plausible that the law of $\max Y_t$ is absolutely continuous except for a possible atom at its essential infimum.

Under the hypotheses of Theorem 8.1, if g is a fixed continuous real function on K, there is probability 0 that there exists an open set U such that $Y_t = g(t)$ for some $t \in U$ and $Y_t \leqq g(t)$ for all $t \in U$ (Ylvisaker (1968)), since we can take a countable base for the open sets, and apply Theorem 8.1 to $Y_t - g(t)$. (This allows some simplification of the proof of Lemma 5A of Dudley (1971 b).)

Yuditskaya (1970) considers maxima of stationary isotropic processes on R^k over sets with some regularity conditions.

9. Relations to other major subjects.

9.1. *Equivalence and singularity.* Study of sample function properties of processes can be viewed as seeking sets in function space which contain almost all the sample functions yet which are as small as possible. Thus, in principle, sample function properties should help in proving equivalence or singularity of Gaussian measures. But, in practice I. E. Segal ((1958) Theorem 3) and J. Feldman (1959) solved the problem of equivalence and perpendicularity in terms of covariances. Their condition in terms of Hilbert–Schmidt operators usually seems easier to apply than sample function properties. In situations where their theorem does not apply (e.g., only one of the measures is Gaussian) sample function properties may be useful; see, e.g., my paper (1971 b).

9.2. *Prediction.* Again, in principle, knowledge of sample function behavior of processes should help to predict the future of a given process, given all or part of its past. But the existing prediction theory, both in the classical Kolmogorov–Wiener–Masani form with infinite past, and in the prediction of Gaussian processes from a finite segment of the past (Levinson and McKean (1964), Dym and McKean (1970)), concerns itself with subspaces and projections and hence with covariances rather than with sample functions. Of course, there are for example processes with analytic sample functions which can thus be perfectly predicted, but this intersection of prediction and sample function behavior seems relatively small and the theories seem to develop with little relation to each other.

9.3. *Diffusion.* Some interesting properties of Gaussian sample functions are connected with non-Gaussian diffusion processes such as Brownian motion with an absorbing or reflecting barrier, etc. There is an exposition by K. Itô and H. P. McKean, Jr. (1965).

REFERENCES

ACOSTA, A. DE (1971). On regular extensions of cylinder measures. Manuscript.

ALEKSEEV, V. G. (1963). New theorems on "almost sure" properties of realizations of Gaussian processes. *Litovsk. Mat. Sb.* **3** No. 2, 5–15.

BAXTER, G. (1956). A strong limit theorem for Gaussian processes. *Proc. Amer. Math. Soc.* **7** 522–527.

BELYAEV, YU. K. (1958). On the unboundedness of the sample functions of Gaussian processes. *Theor. Probability Appl.* **3** 327–329.

BELYAEV, YU. K. (1959). Analytic random processes. *Theor. Probability Appl.* **4** 402–409.

BELYAEV, YU. K. (1961). Continuity and Hölder conditions for sample functions of stationary Gaussian processes. *Proc. Fourth Berkeley Symp. Math. Statist. Prob.* **2**. Univ. of California Press, 23–33.

BELYAEV, YU. K. and NOSKO, V. P. (1969). Characteristics of excursions above a high level for a Gaussian process and its envelope. *Theor. Probability Appl.* **14** 296–309.

BERMAN, S. M. (1962). A law of large numbers for the maximum in a stationary Gaussian sequence. *Ann. Math. Statist.* **33** 93–97.

BERMAN, S. M. (1964). Limit theorems for the maximum term in stationary sequences. *Ann. Math. Statist.* **35** 502–516.

BERMAN, S. M. (1967). A version of the Lévy-Baxter theorem for the increments of Brownian motion of several parameters. *Proc. Amer. Math. Soc.* **18** 1051–1055.

BERMAN, S. M. (1968). Some continuity properties of Brownian motion with the time parameter in Hilbert space. *Trans. Amer. Math. Soc.* **131** 182–198.

BERMAN, S. M. (1969). Harmonic analysis of local times and sample functions of Gaussian processes. *Trans. Amer. Math. Soc.* **143** 269–281.

BERMAN, S. M. (1970). Gaussian processes with stationary increments: local times and sample function properties. *Ann. Math. Statist.* **41** 1260–1272.

BERMAN, S. M. (1971 a). Excursions above high levels for stationary Gaussian processes. *Pacific J. Math.* **36** 63–79.

BERMAN, S. M. (1971 b). Asymptotic independence of the numbers of high and low-level crossings of stationary Gaussian processes. *Ann. Math. Statist.* **42** 927–945.

BERNARD, P. (1970). Quelques propriétés des trajectoires des fonctions aléatoires stables sur R^k, *Ann. Inst. H. Poincaré Sect. B* **6** 131–151.

BORGES, R. (1966). A characterization of the normal distribution (a note on a paper by Kozin). *Z. Wahrscheinlichkeitstheorie und Verw. Gebiete* **5** 244–246.

BRETAGNOLLE, J. and DACUNHA-CASTELLE, D. (1969). Le déterminisme des fonctions laplaciennes sur certaines espaces de suites. *Ann. Inst. H. Poincaré Sect. B* **5** 1–12.

BULINSKAYA, E. V. (1961). On the mean number of crossings of a level by a stationary Gaussian process. *Theor. Probability Appl.* **6** 435–438.

BUNIMOVICH, V. I. (1951). Transgressions of intensities of fluctuation noises (in Russian). *Zhurnal Tekhnicheskoi Fiziki* **21** 625–628.

CHENTSOV, N. N. (1956). Wiener random fields depending on several parameters. *Dokl. Akad. Nauk SSSR* **106** 607–609.

CHENTSOV, N. N. (1957). Lévy Brownian motion for several parameters and generalized white noise. *Theor. Probability Appl.* **2** 265–266.

CHEVET, S. (1969). *p*-ellipsoides de l^q, exposant d'entropie, mesures cylindriques gaussiennes. *C. R. Acad. Sci. Paris Sec. A-B* **269** A658–A660.

CHEVET, S. (1970a). *p*-ellipsoides de l^q, mesures cylindriques gaussiennes. *Les Probabilités sur les Structures Algébriques* (Colloque, Clermont-Ferrand, 1969) (Paris, Centre National de la Recherche Scientifique), 55–73.

CHEVET, S. (1970b). Mesures de Radon sur R^n et mesures cylindriques. *Ann. Fac. Sci. Univ. Clermont* No. 43 (math., 6e fasc.) 91–158.

CIESIELSKI, Z. (1961). Hölder conditions for realizations of Gaussian processes. *Trans. Amer. Math. Soc.* **99** 403–413.

COHN, D. (1971). Manuscript.

CORWIN, L. (1970). Generalized Gaussian measure and a functional equation. *J. Functional Analysis* **5** 412–427.

CRAMÉR, H. (1962). On the maximum of a normal stationary stochastic process. *Bull. Amer. Math. Soc.* **68** 512–516.

CRAMÉR, H. and LEADBETTER, M. R. (1967). *Stationary and Related Stochastic Processes.* Wiley, New York.

CRAMÉR, H. (1966). On the intersections between the trajectories of a normal stationary stochastic process and a high level. *Ark. Mat.* **6** 337–349.

DELPORTE, L. (1964). Fonctions aléatoires presque sûrement continues sur un intervalle fermé. *Ann. Inst. H. Poincaré Sec. B* **1** 111–215.

DELPORTE, J. (1966). Fonctions aléatoires de deux variables à échantillons continus sur un domaine rectangulaire borné. *Z. Wahrscheinlichkeitstheorie und Verw. Gebiete* **6** 224–245.

DOOB, J. L. (1953). *Stochastic Processes.* Wiley, New York.

DUDLEY, R. M. (1965). Gaussian processes on several parameters. *Ann. Math. Statist.* **36** 771–788.

DUDLEY, R. M. (1967). The sizes of compact subsets of Hilbert space and continuity of Gaussian processes. *J. Functional Analysis* **1** 290–330.

DUDLEY, R. M. (1969) Review of D. M. Eaves (1968); *Math. Reviews* **38** No. 1731, page 324.

DUDLEY, R. M. (1971a). On measurability over product spaces. *Bull. Amer. Math. Soc.* **77** 271–274.

DUDLEY, R. M. (1971b). Non-linear equivalence transformations of Brownian motion. *Z. Wahrscheinlichkeitstheorie und Verw. Gebiete* **20** 249–258.

DUDLEY, R. M. (1972a). A counter-example on measurable processes. *Proc. Sixth Berkeley Symp. Math. Statist. Prob.* **2** 57–66. Univ. of California Press.

DUDLEY, R. M. (1972b). Metric entropy of some classes of sets with differentiable boundaries. To appear in *J. Approximation Theory.*

DUDLEY, R., FELDMAN, J. and LECAM, L. (1971). On seminorms and probabilities, and abstract Wiener spaces. *Ann. of Math.* **93** 390–408.

DVORETZKY, A., ERDÖS, P. and KAKUTANI, S. (1961). Nonincrease everywhere of the Brownian motion process. *Proc. Fourth Berkeley Symp. Math. Statist. Prob.* **2** 103–116. Univ. of California Press.

DYM, H. and MCKEAN, H. P., JR. (1970). Extrapolation and interpolation of stationary Gaussian processes. *Ann. Math. Statist.* **41** 1817–1844.

EAVES, D. M. (1967). Sample functions of Gaussian random homogeneous fields are either continuous or very irregular. *Ann. Math. Statist.* **38** 1579–1582.

EAVES, D. M. (1968). Random generalized functions of locally finite order. *Proc. Amer. Math. Soc.* **19** 1457–1463.

FELDMAN, J. (1971). Sets of boundedness and continuity for the canonical normal process. *Proc. Sixth Berkeley Symp. Math. Statist. Prob.* **2** 357–368. Univ. of California Press.

FELDMAN, J. (1959). Equivalence and perpendicularity of Gaussian processes. *Pacific J. Math.* **8** 699–708; Correction **9** 1295–1296.

FERNIQUE, X. (1964). Continuité des processus gaussiens. *C. R. Acad. Sci. Paris* **258** 6058–6060.

FERNIQUE, X. (1970). Intégrabilité des vecteurs gaussiens. *C. R. Acad. Sci. Paris Ser. A* **270** 1698–1699.

FERNIQUE, X. (1971). Régularité de processus gaussiens. *Invent. Math.* **12** 304–320.

FOSCHINI, G. J. and MUELLER, R. K. (1970). On Wiener process sample paths. *Trans. Amer. Math. Soc.* **149** 89–93.

FREIDLIN, M. I. and TUTUBALIN, V. N. (1962) On the structure of the infinitesimal σ-algebra of a Gaussian process. *Theor. Probability Appl.* **7** 196–199.

GARSIA, A., POSNER, E. and RODEMICH, E. (1968). Some properties of the measures on function spaces induced by Gaussian processes. *J. Math. Anal. Appl.* **21** 150–161.

GARSIA, A., RODEMICH, E. and RUMSEY, H., JR. (1970). A real variable lemma and the continuity of paths of some Gaussian processes. *Indiana Univ. Math. J.* **20** 565–578.

GARSIA, A. (1971). Continuity properties of Gaussian processes with multidimensional time parameter. *Proc. Sixth Berkeley Symp. Math. Statist. Prob.* **2** 369–374. Univ. of California Press.

GLADYSHEV, E. G. (1961). A new limit theorem for stochastic processes with Gaussian increments. *Theor. Probability Appl.* **6** 52–61.

GNEDENKO, B. V. (1943). On the growth of homogeneous random processes with independent single-type increments. *Dokl. Akad. Nauk SSSR* **40** 90–93.

GROSS, L. (1962). Measurable functions on Hilbert space. *Trans. Amer. Math. Soc.* **105** 372–390.

HANSON, D. L. and WRIGHT, F. T. (1971). A bound on tail probabilities for quadratic forms in independent random variables. *Ann. Math. Statist.* **42** 1079–1083.

HIDA, T. and NOMOTO, H. (1964). Gaussian measure on the projective limit space of spheres. *Proc. Japan Acad.* **40** 301–304.

HÖRMANDER, L. (1964). *Linear Partial Differential Operators*. Springer-Verlag, Berlin.

DE HOYOS, A. GUEVARA (1972). Continuity and convergence of some processes parametrized by the compact convex sets in R^s. *Z. Wahrscheinlichkeitstheorie und Verw. Gebiete* **23** 153–162.

HUNT, G. A. (1951). Random Fourier transforms. *Trans. Amer. Math. Soc.* **71** 38–69.

ITÔ, K. (1954). Stationary random distributions. *Mem. Coll. Sci. Kyoto Univ. Ser. A* (Math.) **28** 212–223.

ITÔ, K. (1956). Isotropic random current. *Proc. Third Berkeley Symp. Math. Statist. Prob.* **2** 125–132. Univ. of California Press.

ITÔ, K. (1964). The expected number of zeros of continuous stationary Gaussian processes. *J. Math. Kyoto Univ.* **3** 207–216.

ITÔ, K. (1970). The topological support of Gaussian measure on Hilbert space. *Nagoya Math. J.* **38** 181–184.

ITÔ, K. and MCKEAN, H. P., JR. (1965). *Diffusion Processes and Their Sample Paths*. Springer-Verlag, Berlin.

ITÔ, K. and NISIO, M. (1968). On the oscillation functions of Gaussian processes. *Math. Scand.* **22** 209–223.

JADRENKO, M. I. (1967). Local properties of sample functions of random fields. *Visnik Kiïv. Univ. Ser. Mat. Meh.* No. 9, 103–112. (English translation in *Selected Transl. Math. Statist. Prob.* **10** (1972) 233–245.)

JADRENKO, M. I. (1968). Continuity of sample functions of Gaussian random fields on Hilbert space. *Dopovidi Akad. Nauk Ukrain. RSR Ser. A* 734–736. (English translation in *Selected Transl. Math. Statist. Prob.* **10** (1972) 15–17.)

JAIN, N. C. (1971). A zero-one law for Gaussian processes. *Proc. Amer. Math. Soc.* **29** 585–587.

JAIN, N. C. and KALLIANPUR, G. (1970a). A note on uniform convergence of stochastic processes. *Ann. Math. Statist.* **41** 1360–1362.

JAIN, N. C. and KALLIANPUR, G. (1970b). Norm convergent expansions for Gaussian processes in Banach spaces. *Proc. Amer. Math. Soc.* **25** 890–895.

JAIN, N. C. and KALLIANPUR, G. (1971). Oscillation function of a multiparameter Gaussian process. Preprint.

KAC, M. and SLEPIAN, D. (1959). Large excursions of Gaussian processes. *Ann. Math. Statist.* **30** 1215–1228.

KAHANE, J.-P. (1960). Propriétés locales des fonctions à séries de Fourier aléatoires. *Studia Math.* **19** 1–25.

KAHANE, J.-P. (1968). *Some Random Series of Functions.* D. C. Heath, Lexington, Mass.

KAKUTANI, S. (1943). Notes on infinite product spaces, II. *Proc. Japan. Acad.* **19** 184–188.

KALLIANPUR, G. (1970). Zero-one laws for Gaussian processes. *Trans. Amer. Math. Soc.* **149** 199–211.

KALLIANPUR, G. and NADKARNI, M. (1971). Supports of Gaussian measures. *Proc. Sixth Berkeley Symp. Math. Statist. Prob.* **2** 375–388. Univ. of California Press.

KAWATA, T. and KUBO, I. (1970). Sample properties of weakly stationary processes. *Nagoya Math. J.* **39** 7–21.

KOLMOGOROV, A. N. (1940). Curves in Hilbert space which are invariant with respect to a one-parameter group of motions. *Dokl. Akad. Nauk SSSR* **26** 6–9; Wiener's spiral and other interesting curves in Hilbert space. *ibid.* 115–118.

KOLMOGOROV, A. N. (1956). On some asymptotic characteristics of totally bounded metric spaces (in Russian). *Dokl. Akad. Nauk SSSR* **108** 385–388.

KOLMOGOROV, A. N. (1959). A note to the papers of R. A. Minlos and V. Sazonov. *Theor. Probability Appl.* **4** 221–223.

KÔNO, N. (1970). On the modulus of continuity of sample functions of Gaussian processes. *J. Math. Kyoto Univ.* **10** 493–536.

KOZAČENKO, YU. V. (1968). Sufficient conditions for the continuity with probability one of sub-Gaussian processes (Ukrainian; Russian and English summaries). *Dopovīdī Akad. Nauk Ukrain. RSR Ser. A* 113–115.

KOZIN, F. (1957). A limit theorem for processes with stationary independent increments. *Proc. Amer. Math. Soc.* **8** 960–963.

LAMPERTI, J. (1965). On limit theorems for Gaussian processes. *Ann. Math. Statist.* **36** 304–310.

LANDAU, H. J. and SHEPP, L. A. (1971). On the supremum of a Gaussian process. *Sankhyā Ser. A* **32** 369–378.

LEVINSON, N. and MCKEAN, H. P., JR. (1964). Weighted trigonometrical approximation on R^1 with application to the germ field of a stationary Gaussian noise. *Acta Math.* **112** 99–143.

LÉVY, P. (1940). Le mouvement brownien plan. *Amer. J. Math.* **62** 487–550.

LÉVY, P. (1937, 1954). *Théorie de l'addition des variables aléatoires.* Gauthier-Villars, Paris.

LÉVY, P. (1948, 1965). *Processus stochastiques et mouvement brownien.* Gauthier-Villars, Paris.

LINDGREN, G. (1971). Extreme values of stationary normal processes. *Z. Wahrscheinlichkeitstheorie und Verw. Gebiete* **17** 39–47.

LOÈVE, M. (1960, 1963). *Probability Theory.* Van Nostrand, Princeton.

LORENTZ, G. G. (1966). Metric entropy and approximation. *Bull. Amer. Math. Soc.* **72** 903–937.

MALEVICH, T. L. (1969). Asymptotic normality of the number of crossings of level zero by a Gaussian process. *Theor. Probability Appl.* **14** 287–296.

MARCUS, M. B. (1968). Hölder conditions for Gaussian processes with stationary increments. *Trans. Amer. Math. Soc.* **134** 29–52.

MARCUS, M. B. (1970a). Hölder conditions for continuous Gaussian processes. *Osaka J. Math.* **7** 483–494.

MARCUS, M. B. (1970b). A bound for the distribution of the maximum of continuous Gaussian processes. *Ann. Math. Statist.* **41** 305–309.

MARCUS, M. B. (1972). Gaussian lacunary series and the modulus of continuity for Gaussian processes. Preprint.

MARCUS, M. B. and SHEPP, L. A. (1970). Continuity of Gaussian processes. *Trans. Amer. Math. Soc.* **151** 377–392.

MARCUS, M. B. and SHEPP, L. A. (1971). Sample behavior of Gaussian processes. *Proc. Sixth Berkeley Symp. Math. Statist. Prob.* **2** 423–442. Univ. of California Press.

MINLOS, R. A. (1959). Generalized random processes and their extension to measures. *Trudy Moskov. Mat. Obšč.* **8** 497–518. (English translation in *Selected Transl. Math. Statist. Prob.* **3** (1963) 291–314.)

MITYAGIN, B. S. (1961). Approximate dimension and bases in nuclear spaces. *Russian Math. Surveys* **16** No. 4 59–127.

NELSON, E. (1959). Regular probability measures on function space. *Ann. of Math.* **69** 630–643.

NEVEU, J. (1968). *Processus Aléatoires Gaussiens.* Les Presses de l'Université de Montréal.

NISIO, M. (1967). On the extreme values of Gaussian processes. *Osaka J. Math.* **4** 313–326.

NISIO, M. (1969). On the continuity of stationary Gaussian processes. *Nagoya Math. J.* **34** 89–104.

OREY, S. (1970). Gaussian sample functions and the Hausdorff dimension of level crossings. *Z. Wahrscheinlichkeitstheorie und Verw. Gebiete* **15** 249–256.

OREY, S. (1971). Growth rates of Gaussian processes with stationary increments. *Bull. Amer. Math. Soc.* **70** 609–612.

PICKANDS, J., III (1967). Maxima of stationary Gaussian processes. *Z. Wahrscheinlichkeitstheorie und Verw. Gebiete* **7** 190–223.

POSNER, E. C., RODEMICH, E. and RUMSEY, H., JR. (1969). Epsilon entropy of stochastic processes. *Ann. Math. Statist.* **40** 1272–1296.

PRESTON, C. (1972). Continuity properties of some Gaussian processes. *Ann. Math. Statist.* **43** 285–292.

RAY, D. (1963). Sojourn times and the exact Hausdorff measure of the sample path for planar Brownian motion. *Trans. Amer. Math. Soc.* **106** 436–444.

RICE, S. O. (1945). Mathematical analysis of random noise. *Bell System Tech. J.* **24** 46–156.

ROGOZIN, B. A. (1968). On the existence of exact upper sequences. *Theor. Probability Appl.* **13** 667–672.

SCHWARTZ, L. (1950, 1966 a). *Théorie des Distributions.* Hermann, Paris.

SCHWARTZ, L. (1966). Sur le théorème du graphe fermé. *C. R. Acad. Sci. Paris* **263** A602–A605.

SEGAL, I. E. (1954). Abstract probability spaces and a theorem of Kolmogoroff. *Amer. J. Math.* **76** 721–732.

SEGAL, I. E. (1958). Distributions in Hilbert space and canonical systems of operators. *Trans. Amer. Math. Soc.* **88** 12–41.

SHUR, M. G. (1965). On the maximum of a Gaussian stationary process. *Theor. Probability Appl.* **10** 354–357.

SIRAO, T. (1960). On the continuity of Brownian motion with a multidimensional parameter. *Nagoya Math. J.* **16** 135–136.

SIRAO, T. and WATANABE, H. (1970). On the upper and lower class for stationary Gaussian processes. *Trans. Amer. Math. Soc.* **147** 301–331.

SLEPIAN, D. (1962). The one-sided barrier problem for Gaussian noise. *Bell System Tech. J.* **41** 463–501.

SONIS, M. G. (1966). Certain measurable subspaces of the space of all sequences with a Gaussian measure. *Uspehi Mat. Nauk* **21** No. 5, 277–279.

SPITZER, F. (1958). Some theorems concerning 2-dimensional Brownian motion. *Trans. Amer. Math. Soc.* **87** 187–197.

STRAIT, P. T. (1966). Sample function regularity for Gaussian processes with the parameter in Hilbert space. *Pacific J. Math.* **19** 159–173.

STRAIT, P. T. (1969). On Berman's version of the Lévy-Baxter theorem. *Proc. Amer. Math. Soc.* **23** 91–93.

SUDAKOV, N. V. (1969). Gaussian and Cauchy measures and ε-entropy. *Soviet Math. Dokl.* **10** 310–313.

SUDAKOV, V. N. (1971). Gaussian random processes, and measures of solid angles in Hilbert space. *Dokl. Akad. Nauk SSSR* **197** 43–45, *Soviet Math. Dokl.* **12** 412–415.

TAYLOR, S. J. (1964). The exact Hausdorff measure of the sample path for planar Brownian motion. *Proc. Cambridge Philos. Soc.* **60** 253–258.

TOTOKI, H. (1962). A method of construction of measures on function spaces and its applications to stochastic processes. *Mem. Fac. Sci. Kyushu Univ. Ser. A* **15** 178–190.

URBANIK, K. (1959). An effective example of a Gaussian function. *Bull. Acad. Polon. Sci. Sér. Sci. Math. Astronom. Phys.* **7** 343–349.

URBANIK, K. (1960). Gaussian measures in locally compact abelian topological groups. *Studia Math.* **19** 77–88.

WATANABE, H. (1970). An asymptotic property of Gaussian processes. *Trans. Amer. Math. Soc.* **148** 233–248.

YAGLOM, A. M. (1957). Certain types of random fields in n-dimensional space, similar to stationary stochastic processes. *Theor. Probability Appl.* **2** 273–320.

YAGLOM, A. M. (1962). *An Introduction to the Theory of Stationary Random Functions* (translation by R. Silverman). Prentice-Hall, Englewood Cliffs.

YLVISAKER, N. D. (1965). The expected number of zeros of a stationary Gaussian process. *Ann. Math. Statist.* **36** 1043–1046.

YLVISAKER, N. D. (1968). A note on the absence of tangencies in Gaussian sample paths. *Ann. Math. Statist.* **39** 261–262.

YUDITSKAYA, P. I. (1970). The asymptotic behavior of the maximum of Gaussian random fields. *Dokl. Akad. Nauk SSSR* **194** 278–279.

DEPARTMENT OF MATHEMATICS
MASSACHUSETTS INSTITUTE OF TECHNOLOGY
CAMBRIDGE, MASSACHUSETTS 02139

The Annals of Probability
1979, Vol. 7, No. 2, 319-342

ON THE LOWER TAIL OF GAUSSIAN SEMINORMS

BY J. HOFFMANN-JØRGENSEN, L. A. SHEPP AND R. M. DUDLEY[1]
Aarhus Universitet, Bell Laboratories and Massachusetts Institute of Technology

Let E be an infinite-dimensional vector space carrying a Gaussian measure μ with mean 0 and a measurable norm q. Let $F(t) := \mu(q < t)$. By a result of Borell, F is logarithmically concave. But we show that F' may have infinitely many local maxima for norms $q = \sup_n |f_n|/a_n$ where f_n are independent standard normal variables. We also consider Hilbertian norms $q = (\Sigma b_n f_n^2)^{\frac{1}{2}}$ with $b_n > 0$, $\Sigma b_n < \infty$. Then as $t \downarrow 0$ we can have $F(t) \downarrow 0$ as rapidly as desired, or as slowly as any function which is $o(t^n)$ for all n. For $b_n = 1/n^2$ and in a few closely related cases, we find the exact asymptotic behavior of F at 0. For more general b_n we find inequalities bounding F between limits which are not too far apart.

1. Introduction. Let $\eta = (\eta_j)$ be a sequence of independent Gaussian, mean 0, variance 1, random variables in all of this paper. We shall then study the distribution of

$$q(\eta) \qquad \text{or} \quad q(\eta - a)$$

where $q : \mathbb{R}^\infty \to \overline{\mathbb{R}}_+ = [0, \infty]$ is a seminorm and $a \in \mathbb{R}^\infty$. In particular we shall study the behavior of $P(q(\eta) \leqslant t)$ as $t \to 0$.

In Section 3 we study *supremum norms*, that is, seminorms of the following form:

$$q(x) = \sup_n\{|x_n|/a_n\} \qquad \forall x = (x_n) \in \mathbb{R}^\infty \tag{1.1}$$

where (a_n) is a given sequence of positive numbers.

In Section 4 and Section 5 we study *Hilbertian norms*, that is, a seminorm of the following form:

$$q(x) = \{\Sigma_{n=1}^\infty \tau_n^2 x_n^2\}^{\frac{1}{2}} \qquad \forall x = (x_n) \in \mathbb{R}^\infty \tag{1.2}$$

where (τ_n) is a given sequence of positive numbers.

The setting above actually covers the following general case: Let E be a locally convex space and μ a Gaussian Radon probability on E, with mean 0; that is, μ is a Radon probability on E, whose finite dimensional marginals all are Gaussian and have mean 0. In particular if $x' \in E'$ (= the topological dual of E), then x' has a Gaussian distribution, when x' is considered as a random variable on $(E, \mathfrak{B}, \mu)$. Hence we have $E' \subseteq L^2(\mu)$, and so we may consider the L^2-closure of E', which we denote H'. Then H' is a Hilbert space and its dual H may be identified with a

Received September 26, 1977.
[1]Supported by the Danish National Research Foundation.
AMS 1970 *subject classifications.* Primary 60G15; secondary 60B99.
Key words and phrases. Gaussian processes, seminorms, measure of small balls, lower tail distribution.

E. Giné et al. (eds.), *Selected Works of R.M. Dudley*, Selected Works in Probability and Statistics,
DOI 10.1007/978-1-4419-5821-1_14, © Springer Science+Business Media, LLC 2010

subspace of E in the following manner:

$$H = \{x \in E | \varphi \to \varphi(x) \quad \text{is} \quad L^2\text{-continuous on} \quad H'\}$$
$$= \{x \in E | \exists K > 0 : |\langle x', x\rangle|^2 \leqslant K \textstyle\int_E \langle x', y\rangle^2 \mu(dy) \forall x' \in E'\}$$

H is the reproducing kernel Hilbert space (RKHS) of μ, and we define the Hilbert norm, $\|\cdot\|$, in H by

$$\|x\| = \sup\{|\langle x', x\rangle| : x' \in E', \textstyle\int_E \langle x', y\rangle^2 \mu(dy) \leqslant 1\}.$$

From [5] we have that $L^2(\mu)$ and H are separable, and we can find biorthonormal bases $\{f_j\} \subseteq E'$ and $\{e_j\} \subseteq H$ for H' and H, satisfying the following Karhunen-Loéve expansion:

(1.3) $f_1, f_2, \cdots$ are independent Gaussian, mean 0, variance 1, random variables on the probability space $(E, \mathfrak{B}, \mu)$.

$$\langle f_j, e_i\rangle = \delta_{ij} \tag{1.4}$$

$$x = \textstyle\sum_{j=1}^{\infty} \langle f_j, x\rangle e_j \quad \text{for} \quad \mu - \text{a.a.} \quad x \in E^{\infty} \tag{1.5}$$

$$H = \{x \in E | \textstyle\sum_{j=1}^{\infty} \langle f_j, x\rangle^2 < \infty\} \tag{1.6}$$

$$\|x\| = \{\textstyle\sum_{j=1}^{\infty} \langle f_j, x\rangle^2\}^{\frac{1}{2}}, \quad x \in H \tag{1.7}$$

So the study of a seminorm $r : E \to \overline{\mathbb{R}}_+$ reduces to the study of the seminorm

$$q(t) = r(\textstyle\sum_1^{\infty} t_j e_j)$$

on $\mathbb{R}^{\infty}$ (we put $q = \infty$ if the sum diverges).

Our original case is a special example taking $E = \mathbb{R}^{\infty}$ and μ equal to the infinite product of $N(0, 1)$. In this case we may take f_j to be the projection on the jth coordinate and e_i to be the ith unit vector, and we have

$$H = l^2 = \{x \in \mathbb{R}^{\infty} | \textstyle\sum_{n=1}^{\infty} |x_n|^2 < \infty\},$$
$$\|x\| = \{\textstyle\sum_{n=1}^{\infty} |x_n|^2\}^{\frac{1}{2}},$$

and the series in (1.5) converges for all $x \in \mathbb{R}^{\infty}$.

If q is a Borel measurable seminorm on $\mathbb{R}^{\infty}$ we define (μ is a given measure, and H and $\|\cdot\|$ are defined as above)

$$\|q\| = \sup\{q(x) | x \in H \, \|x\| \leqslant 1\}.$$

Then we have (Kallianpur [8], Borell [3], Marcus and Shepp [9])

THEOREM 1.1. *The two probabilities* $P(q(\eta) < \infty)$ *and* $P(q(\eta) = 0)$ *are* 0 *or* 1, *and* $q(\eta) < \infty$ a.s. *implies* $\|q\| < \infty$.

Moreover if $q(\eta) < \infty$ a.s. *then*

$$\lim_{t \to \infty} t^{-2} \log P(q(\eta) > t) = -\tfrac{1}{2}\|q\|^{-2}.$$

If $\|q\| = 0$, *then* $q(\eta) =$ *constant* a.s.

This theorem settles the behavior of the upper tail of the distribution function of q. Notice that q may be constant a.s. without being 0 a.s., e.g.

$$q(x) = \limsup_{n\to\infty}|x_n|(2\log n)^{-\frac{1}{2}}.$$

Then $q(\eta) = 1$ a.s., q is a seminorm and $\|q\| = 0$. If

$$q(x) = \limsup_{n\to\infty}|x_n|$$

then q is a seminorm with $\|q\| = 0$ and $q(\eta) = \infty$ a.s.

We shall mainly be concerned with seminorms of the form (1.1) or (1.2). These seminorms satisfy the following:

$$q(x) = \sup_n q(x_1, \cdots, x_n, 0, 0, \cdots) \qquad \forall x, \tag{1.8}$$

in contrast to the examples above. Note that (1.8) implies

(1.9) $q$is lower semicontinuous on $\mathbb{R}^\infty$, and so in particular Borel measurable.

$$q(x_1, \cdots, x_n, 0, 0, \cdots) \leqslant q(x_1, \cdots, x_m, 0, 0, \cdots) \qquad \forall m \geqslant n. \tag{1.10}$$

In [2] Borell introduces the class of 0-*convex measures*. A Radon probability μ on the locally convex space E is 0-convex if μ satisfies

$$\mu_*(\lambda A + (1-\lambda)B) \geqslant \mu(A)^\lambda \mu(B)^{1-\lambda} \tag{1.11}$$

for all $0 \leqslant \lambda \leqslant 1$ and all Borel sets A and B. Here μ_* denotes the inner measure generated by μ. Borell proves in [2] that μ is 0-convex if and only if all finite dimensional marginals are 0-convex and in [3] he proves that a probability μ on $\mathbb{R}^n$ is 0-convex, if and only if μ is concentrated on some affine subspace L of $\mathbb{R}^n$ with

$$\mu \ll \lambda_L \qquad \text{and} \quad \log\left(\frac{d\mu}{d\lambda_L}\right) \qquad \text{concave}$$

where λ_L is Lebesgue measure on L. In particular,

(1.12) any Gaussian measure is 0-convex.

Using this one easily establishes a conjecture of Marcus and Shepp (see [9], page 435) on the number of jumps of the distribution of q (see also Cirel'son [6]). Suppose that $q : \mathbb{R}^\infty \to \overline{\mathbb{R}}_+$ is a Borel measurable seminorm with $q(\eta) < \infty$ a.s. and put

$$F(t) = P(q(\eta) \leqslant t) \qquad t \geqslant 0,$$
$$C(q) = \inf\{t \geqslant 0 | F(t) > 0\}.$$

Then from (1.11) (with $A = \{q(x) \leqslant t\}$, $B = \{q(x) \leqslant s\}$) we find that

$$\log F(t) \qquad \text{is concave:} \quad \mathbb{R}_+ \to \overline{\mathbb{R}}_- = [-\infty, 0]. \tag{1.13}$$

Now a concave function is absolutely continuous on the interior of the set where it is finite. So we have

THEOREM 1.2. *If* $q(\eta) < \infty$ a.s., *then* F *admits right and left derivatives everywhere except possibly at* $t = C(q)$.

Moreover $F'(t)$ exists except possibly at countably many points t where F' has a jump downwards. Moreover $F''(t)$ exists Lebesgue-a.e. *and*

$$F(t) = p + \int_0^t F'(s)\,ds \qquad \text{for} \quad t \geqslant C(q)$$
$$= 0 \qquad \text{for} \quad t < C(q)$$

where $p = P(q(\eta) = C(q))$.

Hence, apart from a possible jump at $C(q)$, F is absolutely continuous. If $C(q) > 0$, then we shall see in Example 3.2 that $p = P(q(\eta) = C(q))$ may take any value in $[0, 1]$.

If $C(q) = 0$, then by Theorem 1.1 the jump p is either 0 or 1, the latter case occurs if and only if $q = 0$ a.s.

Finally let $B_q(a, r)$ denote the closed q-ball with center at a and radius r:

$$B_q(a, r) = \{x \in \mathbb{R}^\infty | q(x - a) \leqslant r\}.$$

2. The measure of a translated ball. Let q be a Borel measurable seminorm: $\mathbb{R}^\infty \to \overline{\mathbb{R}}_+$, then we put $\pi_n x = (x_1, \cdots, x_n, 0, 0, \cdots)$ and

$$q_n(x) = q(\pi_n x) \qquad \forall x \in \mathbb{R}^\infty,$$
$$q_n^*(x) = \sup\{|\textstyle\sum_{j=1}^n x_j y_j| : q_n(y) \leqslant 1\} \qquad \forall x \in \mathbb{R}^\infty.$$

Note that $q_n^*(x)$ is everywhere finite if and only if $q_n(x) = 0$ implies $\pi_n x = 0$. Obviously we have

$$|\textstyle\sum_{j=1}^n x_j y_j| \leqslant q_n(x) q_n^*(y) \qquad \text{if} \quad q_n(x) < \infty \tag{2.1}$$

(with the usual convention: $0 \cdot \infty = 0$). Note that if q satisfies (1.8) then by (1.10) we have

$$q_n(x) \uparrow q(x) \qquad \forall x. \tag{2.2}$$

THEOREM 2.1. *Let q be a Borel measurable seminorm on $\mathbb{R}^\infty$ with $q(\eta) < \infty$* a.s. *Then we have*

$$P(q(\eta - a) \leqslant t) \leqslant P(q(\eta) \leqslant t) \qquad \forall t \geqslant 0, \forall a \in \mathbb{R}^\infty. \tag{2.1.1}$$

Moreover if q satisfies (1.8) *and $a \in l^2$, then*

$$\exp\left(-\tfrac{1}{2}\|a\|^2\right) F(t) \leqslant F(t, a) \leqslant \exp\left(-\tfrac{1}{2}\|\pi_n a\|^2 + t q_n^*(a)\right) F(t) \tag{2.1.2}$$

for all $n \geqslant 1$ and all $t \geqslant 0$. Here $\|\cdot\|$ is the usual norm on l^2 and

$$F(t) = P(q(\eta) \leqslant t), \qquad F(t, a) = P(q(\eta - a) \leqslant t).$$

REMARK. (2.1.1) and (2.1.2) show that $F(t)$ and $F(t, a)$ are of the same order of magnitude as $t \to 0$ for $a \in l^2$. If $q_n(x) = 0$ implies $\pi_n x = 0$ then we have

$$F(t, a) \sim \exp\left(-\tfrac{1}{2}\|a\|^2\right) F(t) \qquad \text{as} \quad t \to 0$$

for $a \in l^2$.

PROOF. (2.1.1): Let $a \in \mathbb{R}^\infty$ and $t \geqslant 0$ be given, then we put $K = B_q(0, t)$ and $\alpha = F(t, a)$. Then K is convex closed and symmetric and

$$F(t, a) = \mu(K + a) = \alpha$$

where μ is the probability law of η on $\mathbb{R}^\infty$. Let

$$M = \{x \in \mathbb{R}^\infty | \mu(K + x) \geqslant \alpha\}.$$

Then M is symmetric, by symmetry of K and μ, and M is convex by (1.11). Moreover $a \in M$ hence $0 = \frac{1}{2}a + \frac{1}{2}(-a) \in M$. That is

$$F(t) = \mu(K) \geqslant \alpha = \mu(K + a) = F(t, a)$$

and (2.1.1) is proved.

(2.1.2): Let μ be the probability law of η, that is μ is the infinite product of $N(0, 1)$ and let μ_a be the probability law of $\eta - a$, that is μ shifted by a. If $a \in l^2$, then μ_a is absolutely continuous with respect to μ and

$$\mu_a(dx) = e^{-\frac{1}{2}\|a\|^2 - \langle x, a\rangle}\mu(dx)$$

where $\langle x, a\rangle = \Sigma_j x_j a_j$, whenever the sum converges, which it does μ-a.s. if $a \in l^2$. Hence

$$F(t, a) = \mu_a(q \leqslant t) = \int_{\{q \leqslant t\}} e^{-\frac{1}{2}\|a\|^2 - \langle x, a\rangle}\mu(dx)$$

and since μ and q are symmetric, the Cauchy-Schwarz inequality gives:

$$\begin{aligned} F(t) = \mu(q \leqslant t) &\leqslant \left\{\int_{\{q \leqslant t\}} e^{-\langle x, a\rangle}\mu(dx)\right\}^{\frac{1}{2}}\left\{\int_{\{q \leqslant t\}} e^{\langle x, a\rangle}\mu(dx)\right\}^{\frac{1}{2}} \\ &= \int_{\{q \leqslant t\}} e^{-\langle x, a\rangle}\mu(dx). \end{aligned}$$

So we have $F(t, a) \geqslant e^{-\frac{1}{2}\|a\|^2}F(t)$.

Let $a \in l^2$ and let $n \geqslant 1$ be given and fixed. Then we put $b = \pi_n a$ and $c = a - b$, and we find

$$\begin{aligned} F(t, a) &= \int_{\{q \leqslant t\}} e^{-\frac{1}{2}\|a\|^2 - \langle x, a\rangle}\mu(dx) \\ &= e^{-\frac{1}{2}\|b\|^2}\int_{\{q \leqslant t\}} e^{-\langle x, b\rangle - \langle x, c\rangle - \frac{1}{2}\|c\|^2}\mu(dx) \end{aligned}$$

and since $q_n(x) \leqslant q(x)$ we have (cf. (2.1))

$$|\langle x, b\rangle| = |\Sigma_1^n x_j a_j| \leqslant q_n(x)q_n^*(a) \leqslant t q_n^*(a)$$

for $x \in \{q \leqslant t\}$. Hence we find

$$\begin{aligned} F(t, a) &\leqslant \exp\left(-\tfrac{1}{2}\|b\|^2 + t q_n^*(a)\right)\int_{\{q \leqslant t\}} e^{-\frac{1}{2}\|c\|^2 - \langle x, c\rangle}\mu(dx) \\ &= \exp\left(-\tfrac{1}{2}\|b\|^2 + t q_n^*(a)\right)F(t, c) \\ &\leqslant \exp\left(-\tfrac{1}{2}\|b\|^2 + t q_n^*(a)\right)F(t) \end{aligned}$$

since $F(t, c) \leqslant F(t)$ by (2.1.1).

COROLLARY 2.2. *Let q be a Borel measurable seminorm with $q(\eta) < \infty$* a.s. *If $F = \{x \in \mathbb{R}^\infty | q(x) < \infty\}$ is q-separable, that is, if*

$$\forall \varepsilon > 0, \exists (x_j) \subseteq F \quad \text{so that} \quad F \subseteq \bigcup_{j=1}^{\infty} B_q(x_j, \varepsilon),$$

then $C(q) = 0$, that is, $F(t) > 0 \quad \forall t > 0$.

PROOF. If $C(q) > 0$, then by (2.1.1) we have

$$P(\eta \in B_q(x, \varepsilon)) = 0 \qquad \forall x \in \mathbb{R}^\infty \qquad \forall \varepsilon < C(q).$$

But then separability of F implies $P(\eta \in F) = 0$, which contradicts $q(\eta) < \infty$ a.s.

EXAMPLE 2.3. Let q be defined by

$$q(x) = \{\Sigma_{j=1}^{\infty} \sigma_j^2 x_j^2\}^{\frac{1}{2}}$$

where $\Sigma_j \sigma_j^2 < \infty$. Then $q(\eta) < \infty$ a.s. and

$$q_n^*(x) = \{\Sigma_{j=1}^{n} \sigma_j^{-2} x_j^2\}^{\frac{1}{2}}.$$

Taking $a_n = \sigma_n^{-1} e_n$ (e_n is the nth unit vector) gives

$$\|a_n\| = \|\pi_n a_n\| = \sigma_n^{-1}; \; q_n^*(a_n) = \sigma_n^{-2}.$$

So we have

$$\exp\left(-\tfrac{1}{2}\sigma_n^{-2}\right)F(t) \leqslant F(t, a_n) \leqslant \exp\left(-\left(\tfrac{1}{2} - t\right)\sigma_n^{-2}\right)F(t). \tag{2.3}$$

In particular we have

$$\Sigma_{n=1}^{\infty}\{-\log F(t, a_n)\}^{-1} < \infty \qquad \forall 0 \leqslant t < \tfrac{1}{2}. \tag{2.4}$$

Note that if $0 \leqslant t < (\frac{1}{2})^{\frac{1}{2}}$, then the balls $B_q(a_n, t)$, $n = 1, 2, \cdots$ are mutually disjoint and so

$$\Sigma_n F(t, a_n) < \infty \qquad \forall 0 \leqslant t < \left(\tfrac{1}{2}\right)^{\frac{1}{2}}, \tag{2.5}$$

which is much weaker than (2.4).

3. Sup-norms. We shall now consider seminorms of the form (1.1), so let $a_n > 0$ for all $n \geqslant 1$, and let

$$q(x) = \sup_n |x_n| / a_n. \tag{3.1}$$

Let Φ be the standard normal distribution function on $\mathbb{R}$, and put

$$R(t) = 2(1 - \Phi(t)) = \left(\frac{2}{\pi}\right)^{\frac{1}{2}} \int_t^{\infty} e^{-\frac{1}{2}x^2} dx.$$

Then we have the following elementary inequality:

$$\left(\frac{2}{\pi}\right)^{\frac{1}{2}}(1 + t)^{-1} e^{-\frac{1}{2}t^2} \leqslant R(t) \leqslant \frac{4}{3}\left(\frac{2}{\pi}\right)^{\frac{1}{2}}(1 + t)^{-1} e^{-\frac{1}{2}t^2} \qquad \forall t \geqslant 0 \tag{3.2}$$

and if $F(t) = P(q(\eta) \leqslant t)$, then

$$F(t) = \Pi_{j=1}^{\infty}(1 - R(ta_j)). \tag{3.3}$$

Hence $F(t) > 0$ if and only if $\Sigma_{j=1}^{\infty} R(ta_j) < \infty$, that is if and only if

$$\Sigma_{n=1}^{\infty} \frac{\exp\left(-\frac{1}{2}t^2a_n^2\right)}{1 + ta_n} < \infty.$$

Now we note that this sum converges for all $t > t_0$ if and only if

$$\Sigma_{n=1}^{\infty} \exp\left(-\tfrac{1}{2}t^2a_n^2\right) < \infty \qquad \forall t > t_0.$$

So if we define

$$C_0(a) = \inf\left\{t > 0 \mid \Sigma_{n=1}^{\infty} \exp\left(-\tfrac{1}{2}t^2a_n^2\right) < \infty\right\}, \tag{3.4}$$

then $C_0(a) = C(q)$, in other words:

$$C_0(a) = \inf\{t > 0 \mid F(t) > 0\}; \tag{3.5}$$

and from (3.2) and (3.3) we deduce:

$$F(t) \leqslant \exp\left\{-\left(\frac{2}{\pi}\right)^{\frac{1}{2}} \Sigma_{n=1}^{\infty} \frac{\exp\left(-\frac{1}{2}t^2a_n^2\right)}{1 + ta_n}\right\} \tag{3.6}$$

Now suppose that $a_n \geqslant a > 0 \quad \forall n \geqslant 1$, then by use of the inequality:

$$1 - x \geqslant \exp\left(\frac{x}{y}\log(1 - y)\right) \qquad \forall 0 \leqslant x \leqslant y \leqslant 1$$

we find (put $x = R(ta_n)$ and $y = R(ta)$):

$$F(t) \geqslant \exp\left\{-\psi(t)\Sigma_{n=1}^{\infty} \frac{4\exp\left(-\frac{1}{2}t^2a_n^2\right)}{3(1 + ta_n)}\right\} \tag{3.7}$$

where ψ is given by

$$\psi(t) = -\left(\frac{2}{\pi}\right)^{\frac{1}{2}} R(ta)^{-1} \log(1 - R(ta)). \tag{3.8}$$

It is easily checked that

$$\psi(t) \sim \left(\frac{2}{\pi}\right)^{\frac{1}{2}} \log\frac{1}{t} \qquad \text{as} \quad t \to 0. \tag{3.9}$$

So the estimates in (3.6) and (3.7) are fairly close together. Summarizing these observations we have proved:

THEOREM 3.1. *Let q be given by* (3.1), *and let C_0 be given by* (3.4). *Then we have*

$$q(\eta) < \infty \quad \text{a.s.} \quad \textit{if and only if} \quad C_0(a) < \infty. \tag{3.1.1}$$

$$C_0(a) = \inf\{t \mid F(t) > 0\}. \tag{3.1.2}$$

$$F(t) \leqslant \exp\left\{-\left(\frac{2}{\pi}\right)^{\frac{1}{2}} \Sigma_{n=1}^{\infty} \frac{\exp\left(-\frac{1}{2}t^2a_n^2\right)}{1 + a_nt}\right\}. \tag{3.1.3}$$

If $a_n \geq a > 0$ and ψ is given by (3.8), *then* $\psi(t) \sim -\left(\frac{2}{\pi}\right)^{\frac{1}{2}} \log t$ *as* $t \to 0$, *and*

$$(3.1.4) \qquad F(t) \geq \exp\left\{-\psi(t)\Sigma_{n=1}^{\infty} \frac{4\exp\left(-\frac{1}{2}t^2 a_n\right)}{3(1+a_n t)}\right\}$$

where F is the distribution function of $q(\eta)$.

EXAMPLE 3.2. Let $\alpha \geq 0$ and $\beta > 0$; then we consider the sequence $a_1 = a_2 = \beta$ and

$$(3.10) \qquad a_n = (2\beta^2(\log n + \alpha \log\log n))^{\frac{1}{2}} \qquad n \geq 3.$$

Then we have

$$\Sigma_{n=3}^{\infty}\exp\left(-\tfrac{1}{2}t^2 a_n^2\right) = \Sigma_{n=3}^{\infty} n^{-(\beta t)^2}(\log n)^{-\alpha(\beta t)^2}.$$

So we have $C_0(a) = 1/\beta$. Put $\gamma = 1/\beta$, then

$$(\log n)^{\frac{1}{2}} \leq 1 + \gamma a_n \leq \left(3 + 2\alpha^{\frac{1}{2}}\right)(\log n)^{\frac{1}{2}} \qquad \forall n \geq 3;$$

hence we find

$$\Sigma_{n=1}^{\infty} \frac{\exp\left(-\frac{1}{2}a_n^2\gamma^2\right)}{1+\gamma a_n} \leq e^{-\frac{1}{2}} + \Sigma_{n=3}^{\infty} n^{-1}(\log n)^{-\alpha-\frac{1}{2}}$$

$$\geq e^{-\frac{1}{2}} + \left(3 + 2\alpha^{\frac{1}{2}}\right)^{-1}\Sigma_{n=3}^{\infty} n^{-1}(\log n)^{-\alpha-\frac{1}{2}}.$$

If

$$s(\alpha) = \Sigma_{n=3}^{\infty} n^{-1}(\log n)^{-\alpha-\frac{1}{2}}$$

then (3.1.3) and (3.1.4) give

$$k\exp(-ms(\alpha)) \leq F(\gamma) \leq K\exp\left(-M\left(3 + 2\alpha^{\frac{1}{2}}\right)^{-1}s(\alpha)\right)$$

where k, m, K and M are positive finite constants not depending on α. So if $\alpha \leq \frac{1}{2}$ then $F(\gamma) = 0$, and if $\alpha > \frac{1}{2}$, then $F(\gamma) > 0$.

If $\alpha \downarrow \frac{1}{2}$, then $s(\alpha) \to \infty$ so $F(\gamma) \to 0$, and if $\alpha \to \infty$ then $s(\alpha) \to 0$ so $F(\gamma) \to 1$. Hence F can have a jump of any size $p \in [0, 1[$ at any point $\gamma \in]0, \infty[$.

However, from (3.3) it follows that $F(t) < 1$ for all $t > 0$. So F cannot have a jump of size 1, when q is a sup-norm. Also since q is a norm (i.e., $q(x) = 0$ implies $x = 0$), $C(q) > 0$, so F cannot have a jump at 0.

Now let

$$q_N(x) = \max_{1 \leq j \leq N}\{|x_j|/a_j\}$$

with a_n defined by (3.10). If $0 < b_2 < \gamma < b_1$ then

$$F(b_1) - F(b_2) \geq F(\gamma)$$

and since $q_N \to q$ and b_1 and b_2 are continuity points of F we can achieve the following lemma by taking α sufficiently large:

LEMMA 3.3. *Let* $0 < b_2 < b_1$ *and* $\varepsilon > 0$, *then there exist* $a_1, \cdots, a_N > 0$ *so that*

$$F_N(b_1) - F_N(b_2) > 1 - \varepsilon,$$

where F_N *is the distribution function of*

$$\max_{1 \leqslant j \leqslant N}\{|\eta_j|/a_j\}.$$

THEOREM 3.4. *Let* $\{b_j\}$ *be any strictly decreasing sequence of positive numbers. Then there exist sequences* $\{a_j\}$ *and* $\{m_j\}$ *so that*

$$b_{j+1} < m_j < b_j \qquad \forall j \geqslant 1, \tag{3.4.1}$$

$$F(b_j) - F(m_j) \geqslant 2(b_j - m_j) \qquad \forall j \geqslant 1, \tag{3.4.2}$$

$$F(m_j) \leqslant m_j - b_{j+1} \qquad \forall j \geqslant 1 \tag{3.4.3}$$

where F *is the distribution function of*

$$q(\eta) = \sup_n\{|\eta_n|/a_n\}.$$

In particular F *has a mode in each of the intervals*: $]b_{j+1}, b_{j-1}[$, $j = 2, 3, \cdots$, *in spite of the* log-*concavity of* F.

PROOF. Let $\mathfrak{F}$ denote the set of distribution functions of random variables of the form

$$Q = \max_{1 \leqslant j \leqslant N}\{|\eta_j|/a_j\}$$

with $N \geqslant 1$ and $a_1, \cdots, a_N$ positive. Then any infinite product of distributions from $\mathfrak{F}$ is the distribution function of $q(\eta)$ for some sup-norm q of the form (3.1).

The distribution F will be an infinite product

$$F(x) = \Pi_{j=1}^{\infty} F_j(x)$$

where $F_j \in \mathfrak{F}$. The F_j and m_j are defined inductively by:

$$F_j(b_j) - F_j(m_j) \geqslant 4(b_j - m_j)\Pi_{i=1}^{j-1} F_i(b_j)^{-1} \qquad \forall j \geqslant 1 \tag{i}$$

$$F_j(b_j) - F_j(m_j) \geqslant p_j \qquad \forall j \geqslant 1 \tag{ii}$$

$$F_j(b_j) - F_j(m_j) \geqslant 1 - (m_j - b_{j+1}) \qquad \forall j \geqslant 1 \tag{iii}$$

where (p_j) is any fixed sequence with $0 < p_j < 1$ and

$$\Pi_1^{\infty} p_j = \tfrac{1}{2}.$$

First m_1 is chosen so that $m_1 \in]b_2, b_1[$ and $4(b_1 - m_1) < 1$, then we choose $F_1 \in \mathfrak{F}$ by Lemma 3.3, such that

$$F_1(b_1) - F_1(m_1) > \max\{4(b_1 - m_1), p_1, 1 - (m_1 - b_2)\}.$$

Then (i)—(iii) are satisfied for $j = 1$.

If $F_1, \cdots, F_n$ and $m_1, \cdots, m_n$ are constructed, then we choose $m_{n+1} \in]b_{n+2}, b_{n+1}[$, so that

$$4(b_{n+1} - m_{n+1}) < \Pi_{i=1}^{n} F_i(b_{n+1})$$

(note that $F(t) > 0 \quad \forall t > 0 \quad \forall F \in \mathfrak{F}$). Then by Lemma 3.3 we can choose $F_{n+1} \in \mathfrak{F}$, so that (i)—(iii) holds for $j = n + 1$.

Now we note that (i)—(iii) imply

$$\text{(iv)} \qquad \Pi_{i=j+1}^{n} F_i(b_j) \geqslant \Pi_{i=j+1}^{n} F_i(b_i) \geqslant \Pi_{i=1}^{\infty} p_i = \tfrac{1}{2} \qquad \forall n \geqslant j+1,$$

$$\text{(v)} \qquad F_j(m_j) \leqslant 1 - \big(F_j(b_j) - F_j(m_j)\big) \leqslant m_j - b_{j+1}.$$

Now we put

$$F(x) = \Pi_{j=1}^{\infty} F_j(x), \qquad G_n(x) = \Pi_{j=1}^{n} F_j(x)$$

Then by (i) and (iv) we have for $j \leqslant n$:

$$\begin{aligned} G_n(b_j) - G_n(m_j) &\geqslant \big(G_j(b_j) - G_j(m_j)\big)\Pi_{i=j+1}^{n} F_i(b_j) \\ &\geqslant \tfrac{1}{2}\big(G_j(b_j) - G_j(m_j)\big) \\ &\geqslant \tfrac{1}{2} G_{j-1}(b_j)\big(F_j(b_j) - F_j(m_j)\big) \\ &\geqslant 2(b_j - m_j). \end{aligned}$$

So we see that F satisfies (3.4.2), and since $F \leqslant F_j$, it follows from (v) that F satisfies (3.4.3).

Since $F(b_j) > 0$ it follows from Theorem 1.2 that F is absolutely continuous on $]b, \infty[$, where $b = \lim_{n\to\infty} b_n$. Now (3.4.2) implies that $F'(x) \geqslant 2$ for some $x \in]m_j, b_j[$ and (3.4.3) implies that $F'(x) \leqslant 1$ for some $x \in]b_{j+1}, m_j[$. That is, F has at least one mode in each of the intervals $]b_{j+1}, b_{j-1}[$ for $j \geqslant 2$.

THEOREM 3.5. *Let $f : \mathbb{R}_+ \to \mathbb{R}_+$ be increasing, then there exist positive numbers $\{a_j\}$, so that*

$$0 < F(t) \leqslant f(t) \qquad \forall 0 < t \leqslant 1$$

where F is the distribution function of

$$Q = \sup_n\{|\eta_n|/a_n\}.$$

PROOF. Let $\{p_n\}$ be defined by

$$p_0 = f(\tfrac{1}{2}), \qquad p_n = f(2^{-n-1})/f(2^{-n}) \qquad \text{for} \quad n \geqslant 1$$

and let $\mathfrak{F}$ be defined as in the proof of Theorem 3.4, then we can find $F_n \in \mathfrak{F}$ so that

$$F_n(2^{-n+1}) - F_n(2^{-n}) \geqslant \max\{1 - 2^{-n}, 1 - p_n\}.$$

Let $F = \Pi_0^{\infty} F_n$, then F is the distribution of some $Q = q(\eta)$, where q is a sup-norm. Moreover if $2^{-n-1} \leqslant t \leqslant 2^{-n} (n \geqslant 0)$, then

$$\begin{aligned} F(t) &\leqslant F(2^{-n}) \leqslant \Pi_{j=0}^{n} F_j(2^{-n}) \leqslant \Pi_{j=0}^{n} F_j(2^{-j}) \\ &\leqslant \Pi_{j=0}^{n}\big(1 - \big(F_j(2^{-j+1}) - F_j(2^{-j})\big)\big) \leqslant \Pi_{j=0}^{n} p_j \\ &= f(2^{-n-1}) \leqslant f(t) \end{aligned}$$

and

$$F(t) \geqslant F(2^{-n-1}) \geqslant \Pi_{j=0}^{n+1} F_j(2^{-n-1}) \Pi_{j=n+2}^{\infty} F_j(2^{-j+1})$$
$$\geqslant \Pi_{j=0}^{n+1} F_j(2^{-n-1}) \Pi_{j=n+2}^{\infty}(1 - 2^{-j}) > 0$$

and the theorem is proved.

4. Hilbertian norms. We shall in this section study Hilbertian norms, that is, norms of the form (1.2). Before proceeding we shall assume that τ_n is given by $\tau_n = \tau(n)$, where $\tau : [1, \infty[\to \mathbb{R}_+$ satisfies

$$(4.1) \qquad \tau \quad \text{is decreasing,} \quad \tau(t) > 0 \quad \forall t \geqslant 1$$

$$(4.2) \qquad \tau(t) \leqslant t^{-\frac{1}{2}} \quad \forall t \geqslant 1$$

$$(4.3) \qquad \int_1^{\infty} \tau(t)^2 \, dt < \infty.$$

And we shall consider the norm

$$(4.4) \qquad q(x) = \left\{ \Sigma_{n=1}^{\infty} \tau(n)^2 x_n^2 \right\}^{\frac{1}{2}} \quad \forall x = (x_n) \in \mathbb{R}^{\infty}.$$

Note that $\Sigma \tau_n^2 < \infty$ by (4.1) and (4.3), so $Q = q(\eta)$ is finite a.s. Now let F_n and F^n be the two marginals:

$$F_n(t) = P\left(\Sigma_1^n \tau(j)^2 \eta_j^2 \leqslant t^2\right)$$
$$F^n(t) = P\left(\Sigma_{n+1}^{\infty} \tau(j)^2 \eta_j^2 \leqslant t^2\right).$$

If $B_n(t)$ denotes the euclidean ball of radius t centered at the origin, then we have

$$(4.5) \qquad F_n(t) = (2\pi)^{-n/2} \Pi_{j=1}^n \tau(j)^{-1} \int_{B_n(t)} \exp\left(-\tfrac{1}{2} \Sigma_{j=1}^n \tau(j)^{-2} x_j^2\right) dx$$

$$(4.6) \qquad F_n(s) F^n\left((t^2 - s^2)^{\frac{1}{2}}\right) \leqslant F(t) \leqslant F_n(t) \qquad \forall 0 \leqslant s \leqslant t,$$

since $Q^2 = Q_n^2 + R_n^2$ where

$$Q_n^2 = \Sigma_{j=1}^n \tau(j)^2 \eta_j^2, \qquad R_n^2 = \Sigma_{n+1}^{\infty} \tau(j)^2 \eta_j^2$$

and Q_n and R_n are independent.

THEOREM 4.1. *Let q be the seminorm given by* (4.4), *where τ satisfies* (4.1)—(4.3). *Let*

$$(4.1.1) \qquad \varphi(t) = t^{-\frac{1}{2}} \tau(t)^{-1} \quad \text{for} \quad t \geqslant 1.$$

If F is the distribution function of $Q = q(\eta)$, then there exists $A_1 > 0$, so that

$$(4.1.2) \qquad F(t) \leqslant A_1 \exp\{\int_1^x \log \varphi(y) \, dy + \log \varphi(x) + (x-1) \log t\}$$

for all $x \geqslant 1$ and all $t \in [0, 1]$.

REMARK. In applications of (4.1.2) one should try to minimize the right-hand side in x for t fixed. That is, take $x \geqslant 1$ to be a suitable solution to

$$\log \varphi(x) + \varphi'(x)/\varphi(x) + \log t = 0.$$

Ignoring the middle term one reasonable choice is $\varphi(x) = 1/t$ or $x = \varphi^{-1}(1/t)$.

PROOF. Let V_n be the volume of the n-dimensional unit ball. Then by Stirling's formula we have

$$V_n = \Gamma(\tfrac{1}{2})^n \Gamma(1 + \tfrac{1}{2}n)^{-1} = a_1(2\pi)^{\frac{1}{2}n} n^{-\frac{1}{2}(n+1)} e^{n/2} e^{-\theta/n} \tag{4.7}$$

where $a_1 = \pi^{-\frac{1}{2}} \doteq 0.56$ and $0 < \theta < \frac{1}{6}$. Hence by (4.5) and (4.6) we have

$$F(t) \leqslant a_1 \exp\{ -\Sigma_1^n \log \tau(j) - \tfrac{1}{2}(n+1)\log n + \tfrac{1}{2}n + n \log t\} \tag{4.8}$$

for all $n \geqslant 1$ and all $t \geqslant 0$. Since $f = -\log \tau$ is increasing we have

$$f(1) + \textstyle\int_1^n f(y)\, dy \leqslant \Sigma_1^n f(j) \leqslant f(n) + \int_1^n f(y)\, dy. \tag{4.9}$$

So we have for the exponent in (4.8):

$$\begin{aligned} &-\Sigma_1^n \log \tau(j) - \tfrac{1}{2}(n+1)\log n + \tfrac{1}{2}n + n \log t \\ &\leqslant -\textstyle\int_1^n \log \tau(y)\, dy - \log \tau(n) - \tfrac{1}{2}\int_1^n \log y\, dy + \tfrac{1}{2} - \tfrac{1}{2}\log n + n \log t \\ &= \textstyle\int_1^n \log \varphi(y)\, dy - \log \tau(n) - \tfrac{1}{2}\log n + n \log t + \tfrac{1}{2}. \end{aligned}$$

Now we note that $\varphi(y) \geqslant 1$ for $y \geqslant 1$ by (4.3). So if $n \leqslant x \leqslant n+1$ we have $x \leqslant 2n$ and we find

$$\begin{gathered} \textstyle\int_1^n \log \varphi(y)\, dy \leqslant \int_1^x \log \varphi(y)\, dy, \\ -\log \tau(n) - \tfrac{1}{2}\log n \leqslant -\log \tau(x) - \tfrac{1}{2}\log x + \tfrac{1}{2}\log 2 = \log \varphi(x) + \tfrac{1}{2}\log 2, \\ n \log t \leqslant (x-1)\log t \qquad \text{for} \quad 0 < t \leqslant 1. \end{gathered}$$

Inserting this in (4.8) gives

$$F(t) \leqslant A_1 \exp\{\textstyle\int_1^x \log \varphi(y)\, dy + \log \varphi(x) + (x-1)\log t\}$$

where

$$A_1 = a_1(2e)^{\frac{1}{2}} = (2e/\pi)^{\frac{1}{2}} \doteq 1.32. \tag{4.10}$$

THEOREM 4.2. *Let q be the seminorm given by* (4.4), *where τ satisfies* (4.1)–(4.3). *Suppose in addition that τ satisfies*

$$\log \tau(x) \qquad \textit{is convex} \tag{4.2.1}$$

$$\varphi(x) = x^{-\frac{1}{2}}\tau(x)^{-1} \quad \textit{increases to} \quad +\infty \quad \textit{on} \quad [1, \infty]. \tag{4.2.2}$$

If F is the distribution of $Q = q(\eta)$, then for some constant A_2 we have

$$F(t) \leqslant A_2 x^{-\frac{1}{2}}\tau(x)^{-\frac{1}{2}} e^{-H(x-1)} \qquad \textit{if} \quad x \geqslant 1 \qquad \textit{and} \quad \varphi(x) \leqslant t^{-1} \tag{4.2.3}$$

where H is defined by

$$H(x) = \int_1^x \frac{t\varphi'(t)}{\varphi(t)}\, dt \qquad \textit{for} \quad x \geqslant 1. \tag{4.2.4}$$

REMARK. Again in applications of (4.2.3) we have to choose an appropiate x. One possible choice is $\varphi(x) = 1/t$ or $x = \varphi^{-1}(1/t)$.

Proof. When $f = -\log \tau$ is increasing and concave one may improve (4.9) to

(4.11)

$$\tfrac{1}{2}(f(n+1) - f(2)) + f(1) + \int_1^n f(y)\,dy \leq \Sigma_1^n f(j) \leq \tfrac{1}{2}\big(f(1) + \tfrac{1}{2}f(n)\big) + \int_1^n f(y)\,dy$$

by estimating the integral over $[j-1, j]$ by the area of two trapezoids:

$$\tfrac{1}{2}(f(j) + f(j-1)) \leq \int_{j-1}^{j} f(y)\,dy \leq \tfrac{1}{2}(f(j) - f'(j) + f(j))$$

and noting that since f' is decreasing we have

$$\Sigma_{j=2}^n f'(j) \geq \int_2^{n+1} f'(x)\,dx = f(n+1) - f(2).$$

Using this we can estimate the exponent in (4.8) by

$$\begin{aligned} &-\Sigma_{j=1}^n \log \tau(j) - \tfrac{1}{2}(n+1)\log n + \tfrac{1}{2}n + n\log t \\ &\leq \int_1^n \log \tau(y)\,dy - \tfrac{1}{2}\{\log \tau(n) + \log \tau(1) + \int_1^n \log y\,dy + \log n - 1\} + n\log t \\ &= \int_1^n \log \varphi(y)\,dy - \tfrac{1}{2}\log(n\tau(n)) + n\log t + \tfrac{1}{2} - \tfrac{1}{2}\log \tau(1). \end{aligned}$$

By partial integration we find

$$\begin{aligned} \int_1^n \log \varphi(y)\,dy &= n\log \varphi(n) - \log \varphi(1) - \int_1^n \frac{t\varphi'(t)}{\varphi(t)}\,dt \\ &= n\log\varphi(n) + \log \tau(1) - H(n) \end{aligned}$$

since $\varphi(1) = \tau(1)^{-1}$. So we have

$$F(t) \leq a_2 \exp\big\{-H(n) + n\log(t\varphi(n)) - \tfrac{1}{2}\log(n\tau(n))\big\} \tag{4.12}$$

where

$$a_2 = a_1 \exp\big\{\tfrac{1}{2} + \tfrac{1}{2}\log \tau(1)\big\} \leq (e/\pi)^{\frac{1}{2}}$$

since $\log \tau(1) \leq 0$. Now suppose that $n \leq x \leq n + 1 (n \geq 1)$. Then $H(n) \geq H(x-1)$ since H is increasing by (4.2.2) and if $\varphi(x) \leq 1/t$, then

$$\log(t\varphi(n)) \leq \log(t\varphi(x)) \leq 0$$

and finally since $n \geq \tfrac{1}{2}x$

$$\begin{aligned} \log n + \log \tau(n) &\geq \log x + \log \tau(x) + \log(n/x) \\ &\geq \log(x\tau(x)) - \log 2. \end{aligned}$$

Inserting this in (4.12) gives

$$F(t) \leq A_2 x^{-\frac{1}{2}} \tau(x)^{-\frac{1}{2}} e^{-H(x-1)}$$

where

$$A_2 = a_2 2^{\frac{1}{2}} \leq (2e/\pi)^{\frac{1}{2}} \doteq 1.32, \tag{4.13}$$

proving the theorem.

Theorem 4.3. *Let q be given by* (4.4) *where τ satisfies* (4.1)–(4.3), *and let*

$$\psi(x) = \int_x^\infty \tau(t)^2\,dt \qquad \textit{for} \quad x \geq 1. \tag{4.3.1}$$

If F is the distribution of $Q = q(\eta)$, then for some $B_1 > 0$ we have
(4.3.2)

$$F(t) \geqslant B_1\left\{1 - \frac{\psi(x)}{t^2 - s^2}\right\}\exp\left\{\int_1^x \log \varphi(y)\, dy - \tfrac{1}{2}\log x + (x+1)\log s - \frac{s^2}{2\tau(x)^2}\right\}$$

for all $x \geqslant 1$ and all $0 \leqslant s \leqslant t \leqslant 1$. Here φ is defined as above in (4.1.1).

PROOF. From (4.5) and (4.7) we deduce the following lower bound of F_n:

$$F_n(t) \geqslant b_1 \exp\left\{-\Sigma_1^n \log \tau(j) - \tfrac{1}{2}(n+1)\log n + \tfrac{1}{2}n + n \log t - \frac{t^2}{2\tau(n)^2}\right\} \tag{4.14}$$

since we have

$$\tfrac{1}{2}\Sigma_{j=1}^n \tau(j)^{-2} x_j^2 \leqslant \frac{t^2}{2\tau(n)^2} \qquad \text{for all} \quad x \in B_n(t).$$

Here $b_1 = \pi^{-\frac{1}{2}} e^{-1/6}$.

Since $\log \tau(1) \leqslant 0$ we find from (4.9)

$$\begin{aligned}
&-\Sigma_1^n \log \tau(n) - \tfrac{1}{2}(n+1)\log n + \tfrac{1}{2}n + n \log t - \tfrac{1}{2}t^2\tau(n)^{-2} \\
&\geqslant -\int_1^n \log \tau(y)\, dy - \tfrac{1}{2}\int_1^n \log y\, dy + \tfrac{1}{2} - \tfrac{1}{2}\log n + n \log t - \tfrac{1}{2}t^2\tau(n)^{-2} \\
&= \int_1^n \log \varphi(y)\, dy - \tfrac{1}{2}\log n - \tfrac{1}{2}t^2\tau(n)^{-2} + \tfrac{1}{2} + n \log t \\
&\geqslant \int_1^x \log \varphi(y)\, dy - \tfrac{1}{2}\log x - \tfrac{1}{2}t^2\tau(x)^{-2} + \tfrac{1}{2} - \tfrac{1}{2}\log 2 + (x+1)\log t
\end{aligned}$$

for $n - 1 < x \leqslant n$, $n \geqslant 2$ and $0 \leqslant t \leqslant 1$. So we have

$$F_n(t) \geqslant b_2 \exp\left\{\int_1^x \log \varphi(y)\, dy - \tfrac{1}{2}\log x - \tfrac{1}{2}t^2\tau(x)^{-2} + (x+1)\log t\right\} \tag{4.15}$$

for $n \geqslant 2$, $0 \leqslant t \leqslant 1$ and $n - 1 < x \leqslant n$, where $b_2 = b_1(e/2)^{\frac{1}{2}}$.

Let $R_n^2 = \Sigma_{n+1}^\infty \eta_j^2 \tau(j)^2$. Then by Chebyshev's inequality we have

$$\begin{aligned}
F^n(u) &= 1 - P(R_n^2 > u^2) \geqslant 1 - u^{-2}ER_n^2 \\
&= 1 - u^{-2}\Sigma_{n+1}^\infty \tau(j)^2 \geqslant 1 - u^{-2}\int_x^\infty \tau(y)^2\, dy \\
&= 1 - u^{-2}\psi(x)
\end{aligned}$$

whenever $n - 1 < x \leqslant n$, $n \geqslant 2$. So by (4.6) and (4.15) we have

$$F(t) \geqslant B_1\left\{1 - \frac{\psi(x)}{t^2 - s^2}\right\}\exp\left\{\int_1^x \log \varphi(y)\, dy - \tfrac{1}{2}\log x + (x+1)\log s - \frac{s^2}{2\tau(x)^2}\right\}$$

for $0 \leqslant s \leqslant t \leqslant 1$ and $x \geqslant 1$, where B_1 and b_2 are given by the equation

$$B_1 = b_2 = (2\pi)^{-\frac{1}{2}} e^{\frac{1}{3}} = 0.56. \tag{4.16}$$

THEOREM 4.4. *Let q be given by (4.4) where τ satisfies (4.1)–(4.3) and (4.2.1)–(4.2.2). Let φ, ψ and H be given as above:*

$$\varphi(x) = x^{-\frac{1}{2}}\tau(x)^{-1} \qquad \text{for} \quad x \geqslant 1$$

$$\psi(x) = \textstyle\int_x^\infty \tau(y)^2\,dy \qquad \text{for} \quad x \geqslant 1$$

$$H(x) = \int_1^x \frac{t\varphi'(t)}{\varphi(t)}\,dt \qquad \text{for} \quad x \geqslant 1$$

If F is the distribution of $Q = q(\eta)$, then for some $B_2 > 0$, we have

$$F(t) \geqslant B_2 x^{-\frac{1}{2}}\tau(x)^{-\frac{1}{2}}\left\{1 - \frac{\psi(x)}{t^2 - s^2}\right\}\exp\left\{-H(x+1) - \frac{s^2}{2\tau(x)^2}\right\} \tag{4.4.1}$$

whenever $x \geqslant 1$ and $1/\varphi(x) \leqslant s \leqslant t \leqslant 1$.

PROOF. Using (4.11) we have for the exponent in (4.14):

$$\begin{aligned}
&-\textstyle\sum_1^n \log\tau(j) - \frac{1}{2}(n+1)\log n + \frac{1}{2}n + n\log t - \frac{1}{2}t^2\tau(n)^{-2}\\
&\quad\geqslant -\textstyle\int_1^n \log\tau(y)\,dy - \frac{1}{2}\log\tau(n+1) + \frac{1}{2}\log\tau(2) - \frac{1}{2}\int_1^n \log y\,dy\\
&\qquad + \tfrac{1}{2} - \tfrac{1}{2}\log n + n\log t - \tfrac{1}{2}t^2\tau(n)^{-2}\\
&\quad= \textstyle\int_1^n \log\varphi(y)\,dy - \frac{1}{2}\log(n\tau(n+1)) + n\log t - \frac{1}{2}t^2\tau(n)^{-2} + \frac{1}{2} + \frac{1}{2}\log\tau(2)\\
&\quad= -H(n) + n\log(t\varphi(n)) - \tfrac{1}{2}\log(n\tau(n+1)) - \tfrac{1}{2}t^2\tau(n)^{-2} + \alpha
\end{aligned}$$

where we have used the equality:

$$\textstyle\int_1^n \log\varphi(y)\,dy = -H(n) + n\log\varphi(n) + \log\tau(1)$$

and where $\alpha = \frac{1}{2} + \frac{1}{2}\log\tau(2) + \log\tau(1)$.

If $n - 1 \leqslant x \leqslant n$, $n \geqslant 2$ and $\varphi(x) \geqslant 1/t$, then

$$\begin{aligned}
-H(n) &\geqslant -H(x+1),\\
\log(t\varphi(n)) &\geqslant \log(t\varphi(x)) \geqslant 0,\\
-\tfrac{1}{2}\log(n\tau(n+1)) &= -\tfrac{1}{2}\log x + \tfrac{1}{2}\log(x/n) - \tfrac{1}{2}\log\tau(n+1)\\
&\geqslant -\tfrac{1}{2}\log(x\tau(x)) - \tfrac{1}{2}\log 2,
\end{aligned}$$

so we find as before

$$F(t) \geqslant B_2 x^{-\frac{1}{2}}\tau(x)^{-\frac{1}{2}}\left\{1 - \frac{\psi(x)}{t^2 - s^2}\right\}\exp\left\{-H(x+1) - \frac{s^2}{2\tau(x)^2}\right\}$$

where

$$B_2 = b_1\tau(1)^{-1}\left(\tfrac{1}{2}e\tau(2)\right)^{\frac{1}{2}}. \tag{4.17}$$

EXAMPLE 4.5. If $\tau(x) = x^{-\alpha}(\alpha > \frac{1}{2})$, then we have

$$F(t) \leqslant At^{\rho(1-\alpha)}\exp\left(-\left(\alpha - \tfrac{1}{2}\right)t^{-2\rho}\right) \tag{4.5.1}$$

$$F(t) \geqslant Bt^{\rho(3-\alpha)}\exp\left(-\alpha(1+\rho)^{\rho}t^{-2\rho}\right) \tag{4.5.2}$$

where $\rho = (2\alpha - 1)^{-1}$ and A and B are positive constants.

In this case the functions φ, ψ and H take the form:

$$\varphi(x) = x^{\alpha - \frac{1}{2}}, \qquad \psi(x) = \rho x^{1-2\alpha}, \qquad H(x) = \left(\alpha - \tfrac{1}{2}\right)(x-1).$$

Then $\varphi(t^{-2\rho}) = t^{-1}$, so putting $x = t^{-2\rho}$ in (4.2.3) gives (4.5.1).

Putting $x = s^{-2\rho}$ in (4.4.1) gives

$$F(t) \geqslant B_2 s^{\rho(1-\alpha)}\left\{1 - \frac{\rho s^2}{t^2 - s^2}\right\}\exp(-\alpha s^{-2\rho}) \tag{4.18}$$

for $0 \leqslant s \leqslant t \leqslant 1$. Now we take

$$s = t\left(\frac{1 - \frac{1}{2}t^{2\rho}}{\rho + 1}\right)^{\frac{1}{2}}.$$

Then we have

$$1 - \frac{\rho s^2}{t^2 - s^2} = \frac{t^2 - (\rho+1)s^2}{t^2 - s^2} = \frac{t^{2\rho}(1+\rho)}{2\rho + t^{2\rho}} \geqslant \tfrac{1}{2}t^{2\rho}$$

$$\begin{aligned} s^{-2\rho} &= (\rho+1)^{\rho}\left(1 - \tfrac{1}{2}t^{2\rho}\right)^{-\rho}t^{-2\rho} \\ &\leqslant (p+1)^{\rho}\left(1 + \tfrac{1}{2}t^{2\rho}\rho 2^{\rho+1}\right)t^{-2\rho} \\ &= (\rho+1)^{\rho}t^{-2\rho} + \quad \text{constant} \end{aligned}$$

where we in the last inequality used:

$$\left(1 - \tfrac{1}{2}t^{2\rho}\right)^{-\rho} = 1 + \tfrac{1}{2}t^{2\rho}\rho\xi^{-\rho-1} \leqslant 1 + \tfrac{1}{2}t^{2\rho}\rho 2^{\rho+1}$$

where $\frac{1}{2} \leqslant 1 - \frac{1}{2}t^{2\rho} \leqslant \xi \leqslant 1$. Inserting all this in (4.18) gives (4.5.2).

EXAMPLE 4.6. Let $\tau(x) = x^{-\frac{1}{2}}(1 + \log x)^{-1}$; then we have

$$F(t) \leqslant A\exp\left(-te^{t^{-1}-1}\right), \tag{4.6.1}$$

$$F(t) \geqslant B\exp\left(-\left(\tfrac{1}{2} + 3t^2\right)e^{t^{-2}+1}\right) \tag{4.6.2}$$

where A and B are positive constants.

In this case we have

$$\begin{aligned} \varphi(x) &= 1 + \log x, \\ \psi(x) &= (1 + \log x)^{-1}, \\ H(x) &= \int_1^x \frac{dt}{1 + \log t}, \end{aligned}$$

and since

$$\frac{d}{dt}\frac{t}{1+\log t} = \frac{\log t}{(1+\log t)^2} \leqslant \frac{1}{1+\log t} \qquad \forall t \geqslant 0,$$

$$\frac{d}{dt}\frac{t}{\log t - 1} = \frac{\log t - 2}{(\log t - 1)^2} \geqslant \frac{1}{1+\log t} \qquad \forall t \geqslant e^3,$$

we have

$$H(x) \geqslant \frac{x}{1+\log x} - 1$$

$$H(x) \leqslant H(e^3) + \frac{x}{\log x - 1} - \frac{e^3}{2} \qquad \forall x \geqslant e^3.$$

Let us choose $x = e^{1/t - 1}$ in (4.2.3); then we find

$$x^{-\frac{1}{2}}\tau(x)^{-\frac{1}{2}} = \exp\left(-1/(4t) + \tfrac{1}{4} - \tfrac{1}{2}\log t\right),$$

$$H(x-1) \geqslant \frac{x-1}{1+\log x} - 1 = te^{t^{-1}-1} - t - 1,$$

and since $t \leqslant 1/(4t) + \frac{1}{2}\log t$ for t sufficiently small (4.6.1) follows.

Let us then choose $x = e^{1/s} - 1$ in (4.4.1); then it is easily checked that $\varphi(x) \geqslant s^{-1}$, and we have

$$x^{-\frac{1}{2}}\tau(x)^{-\frac{1}{2}} = (e^{1/s} - 1)^{-\frac{1}{4}}\varphi(x)^{\frac{1}{2}} \geqslant e^{-1/(4s)},$$

$$\psi(x) = \varphi(x)^{-1} \leqslant s,$$

$$H(x+1) \leqslant K + se^{1/s}/(1-s)$$

$$\tfrac{1}{2}s^2\tau(x)^{-2} \leqslant \tfrac{1}{2}s^2\tau(x+1)^{-2} = \tfrac{1}{2}(1+s)^2 e^{1/s}.$$

Then we choose $s = t^2/(1+t^2)$, and we find

$$x^{-\frac{1}{2}}\tau(x)^{-\frac{1}{2}} \geqslant \exp\left(-\frac{1+t^2}{4t^2}\right),$$

$$1 - \frac{\psi(x)}{t^2 - s^2} \geqslant \frac{t^2 - s^2 - s}{t^2 - s^2} = \frac{t^4}{1 + t^4 + t^2} \geqslant \frac{t^4}{3} \qquad t \leqslant 1,$$

$$-H(x+1) \geqslant -K - t^2 e^{t^{-2}+1},$$

$$-\tfrac{1}{2}s^2\tau(x)^{-2} \geqslant -\tfrac{1}{2}(1+t^2)^2 e^{t^{-2}+1} = -\left(\tfrac{1}{2} + \tfrac{1}{2}t^4 + t^2\right)e^{t^{-2}+1}.$$

And since

$$-4\log t + \tfrac{1}{2}t^4 e^{t^{-2}+1} + \frac{1+t^2}{4t^2} \leqslant t^2 e^{t^{-2}+1}$$

for t sufficiently small, (4.6.2) follows by inserting the inequalities above in (4.4.1).

EXAMPLE 4.7. Let $\tau(x) = x^{-\frac{1}{2}}e^{-x}$; then we have

$$F(t) \leqslant At^{-\frac{3}{2}}\left(\log\frac{1}{t}\right)^{-\frac{1}{4}}\exp\left(-\tfrac{1}{2}\left(\log\frac{1}{t}\right)^2\right), \tag{4.7.1}$$

$$F(t) \geqslant Bt^{1+\log 2}\left(\log\frac{1}{t}\right)^{-\frac{1}{4}}\exp\left(-\tfrac{1}{2}\left(\log\frac{1}{t}\right)^2\right), \tag{4.7.2}$$

where A and B are positive constants.

In this case we have:

$$\varphi(x) = e^x, \qquad H(x) = \tfrac{1}{2}(x+1)(x-1),$$

$$\psi(x) = \int_x^\infty \frac{e^{-2t}}{t}\,dt \leqslant \frac{e^{-2x}}{2}.$$

Choosing $x = \log(1/t)$ in (4.2.3) gives

$$x^{-\frac{1}{2}}\tau(x)^{-\frac{1}{2}} = \left(\log\frac{1}{t}\right)^{-\frac{1}{4}}t^{-\frac{1}{2}},$$

$$-H(x-1) = -\tfrac{1}{2}x(x-2) = -\tfrac{1}{2}\left(\log\frac{1}{t}\right)^2 - \log t.$$

Now (4.7.1) follows by use of (4.2.3).

Choosing $x = \log(1/s)$ and $s = \frac{1}{2}t$ gives

$$x^{-\frac{1}{2}}\tau(x)^{-\frac{1}{2}} = \left(\log\frac{1}{t} + \log 2\right)^{-\frac{1}{4}}2^{\frac{1}{2}}t^{-\frac{1}{2}} \geqslant kt^{-\frac{1}{2}}\left(\log\frac{1}{t}\right)^{-\frac{1}{4}}$$

$$-H(x+1) = -\tfrac{1}{2}x(x+2) = -\tfrac{1}{2}\left(\log\frac{1}{t} + \log 2\right)^2 + \log t$$

$$= -\tfrac{1}{2}\left(\log\frac{1}{t}\right)^2 - \tfrac{1}{2}(\log 2)^2 + (1+\log 2)\log t$$

$$-\tfrac{1}{2}s^2\tau(x)^{-2} = -\tfrac{1}{2}s^2xe^{2x} = -\tfrac{1}{2}\log\frac{1}{t} - \tfrac{1}{2}\log 2$$

$$1 - \frac{\psi(x)}{t^2 - s^2} \geqslant 1 - \frac{e^{-2x}}{2(t^2-s^2)} = 1 - \frac{s^2}{2(t^2-s^2)} = \frac{5}{6}.$$

Now (4.7.2) follows from (4.4.1).

EXAMPLE 4.8. Let $f : \mathbb{R}_+ \to \mathbb{R}_+$ be increasing and with $\lim_{t\to 0} f(t) = 0$. Then there exists a function $\tau : [1, \infty[\to \mathbb{R}_+$ satisfying (4.1)–(4.3) and such that

$$F(t) \leqslant f(t) \qquad \forall 0 \leqslant t \leqslant 1/e \tag{4.8.1}$$

where F is the distribution of $Q = \{\Sigma_1^\infty \eta_j^2 \tau(j)^2\}$

Let $p(t) = [1/(et)]$ (here $[x]$ denote the integer part of x) and let $n_0 = 1$ and for $p \geqslant 1$:

$$n_p = 1 + \left[-2\log f\left(\frac{1}{e(p+1)}\right) - 2\log(e(p+1))\right].$$

We assume that $0 < f(t) \leqslant 1$ for $0 < t \leqslant 1$, $f(t)/t$ increases and $\lim_{t\to 0} f(t)/t = 0$, which is possible by substituting f by $tf(t)$. Let $\tau(1) = 1$, and

$$\tau(t) = p^{-1} n_p^{-\frac{1}{2}} \qquad \text{for} \quad n_{p-1} < t \leqslant n_p.$$

Then $n_p \leqslant n_{p+1}$ for all $p \geqslant 0$ and $n_p \to \infty$, since $f(x)/x$ increases and tends to 0 as $x \to 0$; (4.1)–(4.3) are easily checked, and

$$\varphi(t) = t^{-\frac{1}{2}} p (n_p)^{\frac{1}{2}} \quad \text{for} \quad n_{p-1} < t \leqslant n_p.$$

So

$$\begin{aligned}
\textstyle\int_{n_{p-1}}^{n_p} \log \varphi(t)dt &= (n_p - n_{p-1})\log\left(p(n_p)^{\frac{1}{2}}\right) - \tfrac{1}{2}(n_p(\log n_p - 1) - n_{p-1}(\log n_{p-1} - 1)) \\
&\leqslant (n_p - n_{p-1})\log\left(p(n_p)^{\frac{1}{2}}\right) - (n_p - n_{p-1})\left(\log(n_p)^{\frac{1}{2}} - \tfrac{1}{2}\right) \\
&= (n_p - n_{p-1})\left(\log p + \tfrac{1}{2}\right),
\end{aligned}$$

and

$$\textstyle\int_1^{n_p} \log \varphi(t)dt \leqslant (n_p - 1)\left(\log p + \tfrac{1}{2}\right).$$

Inserting this in the exponent of (4.1.2) with $x = n_p$ we get

$$F(t) \leqslant A_1 \exp\left(n_p\left(\log p + \tfrac{1}{2} + \log t\right) - \tfrac{1}{2} - \log t\right).$$

Taking $p = p(t)$ gives $p \leqslant (et)^{-1}$, so

$$\log p + \tfrac{1}{2} + \log t \leqslant -\tfrac{1}{2}$$

and $(1/e(p+1)) \leqslant t$ gives

$$-\tfrac{1}{2} n_p \leqslant \log f(t) + \log t$$

so

$$F(t) \leqslant e^{-\frac{1}{2}} A_1 f(t) \leqslant f(t)$$

since $e^{-\frac{1}{2}} A_1 = (2/\pi)^{\frac{1}{2}} \leqslant 1$ (cf. (4.10)).

EXAMPLE 4.9. Let $g :]0, 1] \to \mathbb{R}_+$ be an increasing function with $g(1) < 1$ and

$$g(t) = 0(t^n) \qquad \text{as} \quad t \to 0 \qquad \forall n > 1.$$

Then there exists a function $\tau : [1, \infty[\to \mathbb{R}_+$ satisfying (4.1)–(4, 3), and such that

$$F(t) \geqslant g(t) \qquad \forall t \in [0, 1] \tag{4.9.1}$$

where F is the distribution of $Q = \{\Sigma_1^\infty \eta_j^2 \tau(j)^2\}^{\frac{1}{2}}$.

There exist constants $A_n > 0$, so that

$$g(t) \leqslant A_n t^{n+3} \qquad \forall 0 \leqslant t \leqslant 1, \forall n \geqslant 1.$$

Hence if $B_n = \log A_n$ we have

$$\log g(t) \leqslant B_n + (n+3)\log t \qquad \forall t \in [0, 1] \forall n \geqslant 1.$$

Let $\tau_0^2 = 1$ and put

$$\alpha_n = e^{-B_n - 1} 2^{-n-7} (n+1)^{-\frac{1}{2}} \quad \forall n \geqslant 1.$$

Then we define τ_n^2 inductively by

$$\tau_n^2 = \min\{\alpha_n, 2^{-n}\tau_0^2, 2^{-n+1}\tau_1^2, \cdots, 2^{-1}\tau_{n-1}^2\}$$

for $n \geqslant 1$, and we put

$$\tau(t) = \tau_n \quad \text{for} \quad n < t \leqslant n+1 \quad \text{and} \quad n \geqslant 0.$$

Then τ satisfies (4.1)–(4.3), and we have

$$\psi(n) = \int_n^\infty \tau(t)^2 dt = \sum_{j=n}^\infty \tau_j^2 \leqslant \sum_{j=n}^\infty 2^{-(j-n)}\tau_n^2 = 2\tau_n^2.$$

Let $n \geqslant 1$ and $4\psi(n) \leqslant t^2 \leqslant 4\psi(n-1)$, then we shall apply (4.3.2) with $x = n$ and $s = \frac{1}{2}t$. The exponent in (4.3.2) gives ($s^2 \leqslant \psi(n-1) \leqslant 2\tau_{n-1}^2$)

$$\begin{aligned}
&\int_1^n \log \varphi(y)dy - \tfrac{1}{2}\log n + (n+1)\log s - \tfrac{1}{2}s^2\tau_{n-1}^{-2} \\
&\quad \geqslant -\tfrac{1}{2}\log n + (n+1)\log t - (n+1)\log 2 - 1 \\
&\quad = (B_{n-1} + (n+3)\log t) - 2\log t - B_{n-1} - \tfrac{1}{2}\log n - \log 2^{n+1} - 1 \\
&\quad \geqslant \log g(t) - \log t^2 + \log \alpha_{n-1} + 5\log 2 \\
&\quad \geqslant \log g(t) + 2\log 2
\end{aligned}$$

since $t^2 \leqslant 4\psi(n-1) \leqslant 8\tau_{n-1}^2 \leqslant 8\alpha_{n-1}$. The factor in (4.3.2) gives

$$1 - \frac{\psi(n)}{t^2 - s^2} = 1 - \frac{\psi(n)}{3s^2} \geqslant \frac{2}{3}$$

since $\psi(n) \leqslant s^2$. Now since $B_1 = 0, 56 \geqslant \frac{1}{2}$ we have

$$F(t) \geqslant B_1(8/3)g(t) \geqslant g(t)$$

for $t \in [0, 1]$ and $t \leqslant 2(\psi(0))^{\frac{1}{2}}$. However $\psi(0) \geqslant \tau_0^2 = 1$, so (4.9.1) holds.

5. Exact distributions. We shall now give some cases where the distribution of Q can be given in an exact form for certain Hilbertian norms q. Note that (3.3) gives the exact distribution for sup-norms. Let q and Q be given by

$$q(x) = \{\Sigma_{j=1}(x_{2j-1}^2 + x_{2j}^2)/(2\lambda_j)\}^{\frac{1}{2}}, \tag{5.1}$$

$$Q^2 = q(\eta)^2 = \Sigma_{j=1}^\infty(\eta_{2j-1}^2 + \eta_{2j}^2)/(2\lambda_j). \tag{5.2}$$

Let Q_n^2 denote the nth partial sum in (5.2). Since $(\eta_{2j-1}^2 + \eta_{2j}^2)/(2\lambda_j)$ is exponentially distributed with parameter λ_j, we have (see, e.g., [7], page 40)

$$P(Q_n \leqslant x) = 1 - \Sigma_{j=1}^n A_j^n e^{-\lambda_j x^2} \qquad \forall x \geqslant 0 \tag{5.3}$$

if $\lambda_i \neq \lambda_j \forall i \neq j$, and where A_j^n is defined by

$$A_j^n = \Pi_{k=1;\, k\neq j}^n (1 - \lambda_j/\lambda_k)^{-1} \quad \text{for} \quad j = 1, \cdots, n.$$

Now let

$$A_j = \Pi_{k\neq j}(1 - \lambda_j/\lambda_k)^{-1} = \Pi_{k\neq j}\frac{\lambda_k}{\lambda_k - \lambda_j} \tag{5.4}$$

and assume that (λ_j) satisfies:

(5.5) $$0 < \lambda_1 < \lambda_2 < \cdots,$$

(5.6) $$\Sigma_1^\infty \lambda_j^{-1} < \infty,$$

(5.7) $$\Sigma_1^\infty |A_j| e^{-\lambda_j x} < \infty \quad \forall x > 0.$$

Then we have for $j \leqslant n$

$$|A_j^n| = |A_j| \pi_{k=n+1}^\infty (1 - \lambda_j/\lambda_k) \leqslant |A_j|.$$

So by (5.3), (5.7) and the dominated convergence theorem we deduce:

(5.8) $$P(Q \leqslant x) = 1 - \Sigma_{j=1}^\infty A_j e^{-\lambda_j x^2} \qquad \forall x > 0.$$

And if we assume, in addition to (5.5)–(5.7), that we have

(5.9) $$\Sigma_{j=1}^\infty \lambda_j |A_j| e^{-\lambda_j x} < \infty \qquad \forall x > 0,$$

then the density, f, of Q is given by

(5.10) $$f(t) = 2t\Sigma_{j=1}^\infty \lambda_j A_j e^{-\lambda_j t^2}.$$

So under (5.5)–(5.7) the formulae (5.8) gives the distribution of Q defined by (5.2), and under (5.9) the density of Q is given by (5.10). Note that sign $A_j = (-1)^{j-1}$, so the series in (5.8) and (5.10) are alternating.

In order to use (5.8) and (5.10) we should be able to find A_j. One way is the following: suppose that we have given a product formula

$$\varphi(x) = \Pi_{k=1}^\infty (1 - \psi(x)/\lambda_k)$$

where φ and ψ are differentiable. Let x_j be a solution to $\psi(x_j) = \lambda_j$; then $\varphi(x_j) = 0$ and for $x \neq x_j$

$$\Pi_{k \neq j}(1 - \psi(x)/\lambda_k) = -\lambda_j \frac{\varphi(x) - \varphi(x_j)}{x - x_j} \left\{ \frac{\psi(x) - \psi(x_j)}{x - x_j} \right\}^{-1}.$$

Letting $x \to x_j$ gives

(5.11) $$A_j = \Pi_{k \neq j}(1 - \lambda_j/\lambda_k)^{-1} = -\frac{\psi'(x_j)}{\lambda_j \varphi'(x_j)}.$$

The series (5.8) and (5.10) will in general be divergent or at least slowly convergent at $x = 0$. But the Poisson summation formulae

(5.12) $$\Sigma_{n=-\infty}^\infty \cos(yn)\hat{f}(xn) = \frac{2\pi}{x} \Sigma_{k=-\infty}^\infty f\left(\frac{y + 2k\pi}{x}\right)$$

(f is an even density, $\hat{f}$ its Fourier transform) may in certain cases be used to transform the sums (5.8) and (5.10) into sums which are rapidly convergent for small x (see, e.g., [7], page 630 for the validity of (5.12)).

From the product formula (see [1], page 255)

$$\frac{\sin \pi x}{\pi x} = \Pi_{k=1}^\infty (1 - x^2 k^{-2}),$$

we find by (5.11):

$$\Pi_{k\neq j}(1 - k^2/j^2)^{-1} = 2(-1)^{j-1}.$$

So if $\lambda_j = j^2$ we have

$$P(Q \leqslant x) = 1 - \Sigma_{j=1}^{\infty} 2(-1)^{j-1} e^{-j^2x^2} = \Sigma_{j=-\infty}^{\infty}(-1)^j e^{-j^2x^2}. \tag{5.13}$$

If we put $f =$ the normal density in (5.12) we get

$$\Sigma_{n=-\infty}^{\infty}\cos(ny)e^{-\frac{1}{2}x^2n^2} = \frac{(2\pi)^{\frac{1}{2}}}{x}\Sigma_{k=-\infty}^{\infty}\exp\left(-\frac{(y+2k\pi)^2}{2x^2}\right). \tag{5.14}$$

Putting $y = \pi$ and $x = t2^{\frac{1}{2}}$ gives

$$\Sigma_{n=-\infty}^{\infty}(-1)^n e^{-t^2n^2} = \frac{2\pi^{\frac{1}{2}}}{t}\Sigma_{k=0}^{\infty}\exp\left(-\frac{(2k+1)^2\pi^2}{4t^2}\right).$$

The term for $k = 0$ is clearly dominant for small t, so we have

THEOREM 5.1. *Let*

$$Q = \left\{\Sigma_{j=1}^{\infty}\left(\eta_{2j-1}^2 + \eta_{2j}^2\right)/(2j^2)\right\}^{\frac{1}{2}}.$$

Then we have

$$P(Q \leqslant t) = \frac{2\pi^{\frac{1}{2}}}{t}\Sigma_{k=0}^{\infty}\exp\left(-\frac{(2k+1)^2\pi^2}{4t^2}\right) \qquad \forall t > 0, \tag{5.1.1}$$

$$P(Q \leqslant t) \sim \frac{2\pi^{\frac{1}{2}}}{t}\exp\left(-\frac{\pi^2}{4t^2}\right) \qquad \textit{as} \quad t \to 0. \tag{5.1.2}$$

From (5.11) and the product formulae (see [1], page 255):

$$\cos\left(\tfrac{1}{2}\pi x\right) = \Pi_{k=1}^{\infty}\left(1 - x^2/(2k-1)^2\right),$$

we find

$$\Pi_{k\neq j}\left(1 - (2j-1)^2/(2k-1)^2\right)^{-1} = \frac{4}{\pi}(-1)^{j-1}(2j-1)^{-1}.$$

So if $\lambda_j = (2j-1)^2$, then the density of Q is given by (cf. (5.10)):

$$\begin{aligned} f(t) &= \frac{8t}{\pi}\Sigma_{j=1}^{\infty}(2j-1)(-1)^{j-1}\exp(-(2j-1)^2t^2) \\ &= \frac{4t}{\pi}\Sigma_{j=-\infty}^{\infty}(2j-1)(-1)^{j-1}\exp(-(2j-1)^2t^2). \end{aligned}$$

Differentiating (5.14) with respect to y gives

$$\Sigma_{n=-\infty}^{\infty} n\sin(ny)e^{-\frac{1}{2}x^2n^2} = \frac{(2\pi)^{\frac{1}{2}}}{x}\Sigma_{k=-\infty}^{\infty}\frac{y+2k\pi}{x^2}\exp\left(-\frac{(y+2k\pi)^2}{2x^2}\right),$$

so putting $y = \pi/2$ and $x = t2^{\frac{1}{2}}$ gives

$$f(t) = \frac{4t}{\pi}\Sigma_{n=-\infty}^{\infty} n \sin\left(\tfrac{1}{2}n\pi\right)e^{-t^2n^2}$$

$$= \frac{4}{\pi^{\frac{1}{2}}}\Sigma_{k=-\infty}^{\infty}\frac{(4k+1)\pi}{4t^2}\exp\left(-\frac{(4k+1)^2\pi^2}{16t^2}\right)$$

$$= \frac{\pi^{\frac{1}{2}}}{t^2}\left\{\Sigma_{k=0}^{\infty}(4k+1)\exp\left(-\frac{(4k+1)^2\pi^2}{16t^2}\right) - \Sigma_{k=1}^{\infty}(4k-1)\exp\left(-\frac{(4k-1)^2\pi^2}{16t^2}\right)\right\}$$

$$= \frac{\pi^{\frac{1}{2}}}{t^2}\Sigma_{j=0}^{\infty}(2j+1)(-1)^j\exp\left(-\frac{(2j+1)^2\pi^2}{16t^2}\right).$$

For small t the term for $j = 0$ dominates the others. So we have

$$f(t) \sim \frac{\pi^{\frac{1}{2}}}{t^2}\exp\left(-\left(\frac{\pi}{4t}\right)^2\right) \qquad \text{as} \quad t \to 0.$$

Hence $F(t) = \int_0^t f(s)ds$, and satisfies

$$F(t) \sim \int_0^t \frac{\pi^{\frac{1}{2}}}{s^2}\exp\left(-\left(\frac{\pi}{4s}\right)^2\right)ds \qquad \text{as} \quad t \to 0$$

by l'Hospital's rule; but

$$\int_0^t \frac{\pi^{\frac{1}{2}}}{s^2}\exp\left(-\left(\frac{\pi}{4s}\right)^2\right)ds = 4\left(1 - \Phi\left(\frac{\pi}{t8^{\frac{1}{2}}}\right)\right)$$

and as $t \downarrow 0$,

$$1 - \Phi\left(\frac{\pi}{t8^{\frac{1}{2}}}\right) \sim (t\pi^{-1}8^{\frac{1}{2}})(2\pi)^{-\frac{1}{2}}\exp\left(-\frac{\pi^2}{16t^2}\right),$$

that is,

$$F(t) \sim 8t\pi^{-3/2}\exp\left(-\frac{\pi^2}{16t^2}\right).$$

And we have proved:

THEOREM 5.2. *Let*

$$Q = \left\{\Sigma_{j=1}^{\infty}\frac{\eta_{2j-1}^2 + \eta_{2j}^2}{2(2j-1)^2}\right\}^{\frac{1}{2}}$$

and let f denote the density of Q. Then we have

$$f(t) = \frac{\pi^{\frac{1}{2}}}{t^2}\Sigma_{j=0}^{\infty}(2j+1)(-1)^j\exp\left(-\frac{(2j+1)^2\pi^2}{16t^2}\right), \tag{5.2.1}$$

$$P(Q \leqslant t) \sim 8t\pi^{-3/2}\exp\left(-\frac{\pi^2}{16t^2}\right) \qquad \textit{as} \quad t \to 0. \tag{5.2.2}$$

THEOREM 5.3. *Let*

$$Q = \{\Sigma_{j=1}^{\infty} j^{-2}\eta_j^2\}^{\frac{1}{2}}.$$

Then we have

$$P(Q \leq t) \leq t^{-1}(2\pi)^{\frac{1}{2}}(1 + \varepsilon_1(t))\exp\left(-\frac{\pi^2}{8t^2}\right), \tag{5.3.1}$$

$$P(Q \leq t) \geq 4t\pi^{-3/2}2^{\frac{1}{2}}(1 + \varepsilon_2(t))\exp\left(-\frac{\pi^2}{8t^2}\right) \tag{5.3.2}$$

where $\varepsilon_j(t) \to_{t\to 0} 0$ *for* $j = 1, 2$.

PROOF. Let Q_1 and Q_2 be the random variables defined in Theorem 5.1 and Theorem 5.2 respectively. Then

$$Q_1/2^{\frac{1}{2}} \leq Q \leq 2^{\frac{1}{2}}Q_2,$$

so

$$P(Q \leq t) \leq P(Q_1 \leq 2^{\frac{1}{2}}t),$$

$$P(Q \leq t) \geq P(Q_2 \geq t/2^{\frac{1}{2}}),$$

and the theorem follows from (5.1.2) and (5.2.2).

Added in proof. We thank David Siegmund for calling our attention to a paper of T. W. Anderson and D. A. Darling (*Ann. Math. Statist.* **23** 191–212), where they give an exact series for the distribution of Q from Theorem 5.3, from which the exact behavior at $t = 0$ can be read off (Anderson and Darling, page 202).

REFERENCES

[1] BEHNCKE, H. and SOMMER, F. (1962). *Theorie der analytischen Funktionen einer komplexen Veränderlichen*, 2nd ed. Springer-Verlag, Berlin.

[2] BORELL, C. (1974). Convex measures on locally convex spaces. *Ark. Mat.* **120** 390–408.

[3] BORELL, C. (1975). The Brunn-Minkowski inequality in Gauss space. *Invent. Math.* **30** 207–216.

[4] BORELL, C. (1975). Convex set function in *d*-space. *Period Math. Hungar.* **6** 111–136.

[5] BORELL, C. (1976). Gaussian measures on locally convex spaces, *Math. Scan.* **38** 265–284.

[6] CIREL'SON, B. S. (1975). The density of the distribution of the maximum of a Gaussian process. *Theor. Probability Appl.* **20** 847–855.

[7] FELLER, W. (1970). *An Introduction to Probability Theory and its Application*, **2**, 2nd ed. Wiley, New York.

[8] KALLIANPUR, G. (1970). Zero-one laws for Gaussian processes. *Trans. Amer. Math. Soc.* **149** 199–211.

[9] MARCUS, M. B. and SHEPP, L. A. (1972). Sample behavior of Gaussian processes. *Proc. Sixth Berkeley Symp. Math. Stat. Prob.* **2** 423–441.

J. HOFFMANN-JØRGENSEN
MATEMATISK INSTITUT
AARHUS UNIVERSITET
NY MUNKEGADE
8000 AARHUS, DENMARK

L. A. SHEPP
BELL LABORATORIES
600 MOUNTAIN AVENUE
MURRAY HILL, NEW JERSEY 07974

R. M. DUDLEY
DEPARTMENT OF MATHEMATICS, ROOM 2-245
CAMBRIDGE, MASSACHUSETTS 02139

Part 4

Empirical Processes

Introduction

A fundamental question in Statistics is how well does the frequency of an event approach its probability when the number of repetitions of an experiment increases indefinitely, or how well does the average of the values of a function at the observed outcomes approach its expected value. 'How well' can be thought of in many ways, and one of them is: uniformly over what classes of sets or functions does convergence take place? Empirical process theory has its origin on this question. The first law of large numbers uniform over an infinite class of sets is the Glivenko-Cantelli theorem, from the thirties, that states that the convergence of the frequency to the probability, for real random variables, takes place uniformly over all the sets $(-\infty, x]$, $x \in \mathbf{R}$, almost surely. This was extended to much more general classes of events and functions by Blum in Ann. Math. Statist. 26 (1955) 527-529, and DeHardt in Ann. Math. Statist 42 (1971) 2050-2055 (metric entropy with inclusion, or bracketing), and finally Vapnik and Červonenkis, in seminal work of 1971, Theor. Probab. Appl. 16, 264-280 (and of 1981, Theor. Probab. Appl. 26, 532-553), gave non-random new combinatoric conditions, as well as random combinatoric and random entropy necessary and sufficient conditions, for the frequency to approach the probability uniformly over a class of sets (or a class of functions). Vapnik and Červonenkis announced first their results in the Doklady in 1968, and Dudley immediately noticed their significance in his review of their work for Math Reviews (MR0231431 (37 #6986)). Probably inspired by their work, Dudley (1978) considered the much more difficult question of uniformity over classes of sets and functions of the central limit theorem for the empirical measure. In the case of the empirical cumulative distribution function, where the class of sets is the family of half lines with finite right end point, the uniform law of large numbers is nothing but the Glivenko-Cantelli theorem, whereas its simplest analogue for the central limit theorem is Kolmogorov's theorem asserting the convergence in law of $\sqrt{n}\|F_n - F\|_\infty$, where F_n is the empirical distribution function of n i.i.d. random variables and F is their common distribution. Donsker in Ann. Math. Statist. 23 (1952) 277-281 (making precise a heuristic approach of Doob from 1949) considered the empirical distribution function as a stochastic process on $\mathbf{R}$ (or on $\mathcal{C} = \{(-\infty, x] : x \in \mathbf{R}\}$), $F_n(x) = P_n(-\infty, x] = \frac{1}{n}\sum_{i=1}^n \delta_{X_i}(-\infty, x]$, and showed that $\sqrt{n}(P_n - P)$ converges in law on the function space $l^\infty(\mathbf{R})$ to a Gaussian process (the P-Brownian bridge). Donsker's theorem contains Kolmogorov's as an easy consequence, and it is this theorem that Dudley took as a model for uniform central limit theorems.

In the first paper of this chapter, Dudley succeeds in proving very powerful central limit theorems for the empirical process, uniform over general classes of sets, that go far beyond the half lines or the half spaces, hence theorems that go far beyond Donsker's theorem. He considered two main kinds of conditions, namely in terms of metric entropy with inclusion (just as Blum-DeHardt for the LLN) and combinatorial conditions (that is, Vapnik-Červonenkis classes). Convergence in law in $l^\infty(\mathcal{C})$ is proved to be equivalent to an 'asymptotic equicontinuity' condition (reminiscent of Prohorov's criterion for uniform tightness in $C(S)$), and then the corresponding maximal inequalities are proved by 'chaining', in rough analogy to the way one proves sample continuity of Gaussian processes. For this, he must show that VC classes satisfy a uniform metric entropy condition, and this is his beautiful Lemma 7.13. He also shows that his general results apply to many important classes of sets like classes of sets with differentiable boundaries or the positivity sets of finite dimensional classes of functions. Moreover, he had to deal with serious measurability problems not always considered by previous authors. It is hard to overstate the importance and influence of this article as it not only contains some of the most frequently used results from empirical process theory but it has inspired many of its subsequent developments. In particular, the two main ways of measuring the complexity of classes of sets emphasized in the paper (bracketing entropy and VC-type combinatorial conditions) have been later extended by Dudley and others to the case of function classes, and they remain among the most important empirical processes tools in concrete applications. It is also difficult to overstate the importance of modern empirical process theory in Mathematical Statistics, from asymptotics, to non-parametrics, semiparametrics and learning theory. In many cases, the study of the asymptotic behavior of statistical

E. Giné et al. (eds.), *Selected Works of R.M. Dudley*, Selected Works in Probability and Statistics,
DOI 10.1007/978-1-4419-5821-1_15,

estimators can be reduced to limit properties or to inequalities for empirical processes, and the methods needed to solve such problems developed over the past 30 years often have their origin in Dudley's paper. Even the terminology introduced in this paper, such as VC-classes of sets, has been spreading far beyond the Probability literature and is now routinely used by computer scientists.

From the many contributions of Richard Dudley to empirical processes we have chosen three more very interesting articles, representative of his work. In the first, it is shown that relatively small classes of sets are too large for the central limit theorem, and the rates of convergence to zero of $\sup_{C\in\mathcal{C}} |(P_n-P)(C)|$ for classes $\mathcal{C}$ of sets of the form $\{x_d \le f(x_1,\ldots,x_{d-1})\} \subset \mathbf{R}^d$, where f and all its partial derivatives up to some order β are bounded and those of order β satisfy a Hölder condition, are determined up to a power of $\log n$, so that the powers of n in the bounds are sharp. The second article, by Dudley and Philipp, provides a way to bypass measurability problems inherent in the central limit theorem and the law of the iterated logarithm for empirical processes due to the facts that $l^\infty(\mathcal{C})$ (with the sup norm) is not separable if $\mathcal{C}$ is infinite, and that the empirical measure as a random element in this space is not Borel measurable; they do this by proving invariance principles for sums of independent not necessarily measurable random elements with values in not necessarily separable Banach spaces. The results of this article are quite useful and much stronger than just results on convergence in law. Moreover, this article provides a link between the general theory of empirical processes and deep strong approximation results for classical empirical processes in the real line. Finally, the third article considers classes of functions on which the uniform central limit theorem for the empirical process does hold for all probability measures P; this is very important in Statistics since usually the situation is that P, the common law of the data, is unknown. This article had definite influence on later work on central limit theorems for empirical processes holding uniformly over all probability distributions, which is related to a number of important statistical applications such as asymptotics of the bootstrap. This article also contains an important estimate of the metric entropy of the convex hull of a set of functions in terms of the metric entropy of the original, something that is of interest in geometry of Banach spaces and has been followed up by other authors.

The Annals of Probability
1978, Vol. 6, No. 6, 899–929

SPECIAL INVITED PAPER

CENTRAL LIMIT THEOREMS FOR EMPIRICAL MEASURES[1]

BY R. M. DUDLEY
Massachusetts Institute of Technology

Let $(X, \mathcal{Q}, P)$ be a probability space. Let $X_1, X_2, \cdots,$ be independent X-valued random variables with distribution P. Let $P_n := n^{-1}(\delta_{X_1} + \cdots + \delta_{X_n})$ be the empirical measure and let $\nu_n := n^{\frac{1}{2}}(P_n - P)$. Given a class $\mathcal{C} \subset \mathcal{Q}$, we study the convergence in law of ν_n, as a stochastic process indexed by $\mathcal{C}$, to a certain Gaussian process indexed by $\mathcal{C}$. If convergence holds with respect to the supremum norm $\sup_{C \in \mathcal{C}}|f(C)|$, in a suitable (usually nonseparable) function space, we call $\mathcal{C}$ a Donsker class. For measurability, X may be a complete separable metric space, $\mathcal{Q}$ = Borel sets, and $\mathcal{C}$ a suitable collection of closed sets or open sets. Then for the Donsker property it suffices that for some m, and every set $F \subset X$ with m elements, $\mathcal{C}$ does not cut all subsets of F (Vapnik-Červonenkis classes). Another sufficient condition is based on metric entropy with inclusion. If $\mathcal{C}$ is a sequence $\{C_m\}$ independent for P, then $\mathcal{C}$ is a Donsker class if and only if for some r, $\Sigma_m(P(C_m)(1 - P(C_m)))^r < \infty$.

1. Introduction. The statistics used in Kolmogorov-Smirnov tests are suprema of normalized empirical measures $n^{\frac{1}{2}}(P_n - P)$ or $(mn)^{\frac{1}{2}}(m + n)^{-\frac{1}{2}}(P_m - Q_n)$ over a class $\mathcal{C}$ of sets, namely the intervals $]-\infty, a]$, $a \in \mathbb{R}$. Donsker (1952) showed here that $n^{\frac{1}{2}}(P_n - P)$ converges in law, in the space $l^\infty(\mathcal{C})$ of all bounded functions on $\mathcal{C}$, to a Gaussian process. Later, Donsker's result was extended to the class of products of intervals parallel to the axes in $\mathbb{R}^k$ (Dudley (1966), (1967a)). Since $l^\infty(\mathcal{C})$ in the supremum norm is nonseparable, some measurability problems (overlooked by Donsker) had to be treated. Recently Révész (1976) proved an iterated logarithm law for a much more general class of sets

$$\bigcap_{1 \leqslant i \leqslant k}\{x : f_i(\{x_j : j \neq i\}) < x_i < g_i(\{x_j : j \neq i\})\}$$

where f_i and g_i have a fixed bound on their partial derivatives of orders $\leqslant k$, and P is the uniform measure on the unit cube. This paper will consider extensions of Donsker's theorem to suitable classes of sets in general probability spaces.

Section 2 will treat countable sequences of sets, with results in particular for independent or disjoint sequences. Sections 3 and 4 treat measurability questions, Section 4 on collections of closed or open sets. Section 5 introduces metric entropy with inclusions, and finds a sufficient condition applicable to bounded collections of convex sets in $\mathbb{R}^2$, or sets with more than $k - 1$ times differentiable boundaries in $\mathbb{R}^k$, if P has a bounded density with respect to Lebesgue measure. Section 6 shows how convergence in law of the one-sample measure $n^{\frac{1}{2}}(P_n - P)$ on a class $\mathcal{C}$

Received June 15, 1977; revised April 25, 1978.
[1]This research was partially supported by National Science Foundation Grant MCS76-07211 A01.
AMS 1970 *subject classifications.* Primary 60F05; Secondary 60B10, 60G17, 28A05, 28A40.
Key words and phrases. Central limit theorems, empirical measures, Donsker classes, Effros Borel structure, metric entropy with inclusion, two-sample case, Vapnik-Červonenkis classes.

E. Giné et al. (eds.), *Selected Works of R.M. Dudley*, Selected Works in Probability and Statistics, DOI 10.1007/978-1-4419-5821-1_16, © Springer Science+Business Media, LLC 2010

extends to the two-sample measure $(mn)^{\frac{1}{2}}(m+n)^{-\frac{1}{2}}(P_m - Q_n)$, where P_m and Q_n are independent empirical measures for P. Section 7 shows that for Vapnik-Červonenkis classes, satisfying suitable measurability conditions, Donsker's theorem holds for all P. Iterated logarithm laws uniformly on classes $\mathcal{C}$ will be treated in a separate paper by J. Kuelbs and the author.

Here are some definitions. Let (S, d) be a metric space (we have in mind a nonseparable space of bounded functions on $\mathcal{C}$ with supremum norm). Let $\mathfrak{B}_b :=$ $\mathfrak{B}_b(S, d)$ be the σ-algebra of subsets of S generated by all balls

$$B(x, \varepsilon) := \{y \in S : d(x, y) < \varepsilon\}, \qquad x \in S, \varepsilon > 0.$$

Then $\mathfrak{B}_b$ is a sub-σ-algebra of the σ-algebra $\mathfrak{B}$ of Borel sets, with $\mathfrak{B}_b = \mathfrak{B}$ for S separable. We have $\mathfrak{B}_b \subset \mathfrak{B}$ strictly if the smallest cardinality, γ, of a dense set in S is c, or 2^c, or $2^{2^c}, \cdots$, and hence for all nonseparable S we will treat (Dudley (1967a), proposition and following discussion). If $\gamma = \aleph_\omega$, possibly $\mathfrak{B}_b =$ $\mathfrak{B}$ (Talagrand, 1978).

I will call a probability measure a *law*, defined on $\mathfrak{B}_b$ unless otherwise specified. A sequence μ_n of laws will be said to *converge* to a law μ, $\mu_n \rightarrow_{\mathcal{C}} \mu$, if and only if $\int f\, d\mu_n \rightarrow \int f\, d\mu$ for every continuous, bounded, $\mathfrak{B}_b$-measurable real-valued f on S (Dudley (1966), (1967a)).

A net X_α of S-valued random variables on a probability space $(\Omega, \mathbb{S}, \Pr)$ is said to converge *almost uniformly* to Y if and only if for every $\varepsilon > 0$ there is a set $A \in \mathbb{S}$ with $\Pr(A) < \varepsilon$ and a β such that for all $\alpha \geqslant \beta$ and $\omega \notin A$, $d(X_\alpha, Y)(\omega) < \varepsilon$. This does not require that $\omega \rightarrow d(X_\alpha(\omega), Y(\omega))$ be measurable. Suppose however that for some separable subset $T \subset S$, $\Pr(Y \in T) = 1$. It is easily seen that the metric d is jointly measurable on $(S, \mathfrak{B}_b) \times (T, \mathfrak{B}_b)$, using only balls with centers in a countable dense subset of T. Then $d(X_\alpha, Y)$ is measurable. In this case, a sequence $X_n \rightarrow Y$ almost uniformly if and only if $X_n \rightarrow Y$ a.s., i.e., $d(X_n, Y) \rightarrow 0$ a.s. Wichura (1970) proved that given laws μ_α defined on σ-algebras $\mathcal{A}_\alpha$ with $\mathfrak{B}_b \subset \mathcal{A}_\alpha \subset \mathfrak{B}$, such that μ_0 has a separable support, then $\mu_\alpha \rightarrow_{\mathcal{C}} \mu_0$ if and only if there exists a probability space $(\Omega, \mathbb{S}, \Pr)$ with random variables X_α such that $X_\alpha^{-1}(B) \in \mathbb{S}$ for all $B \in \mathcal{A}_\alpha$, $\Pr(X_\alpha^{-1}(B)) = \mu_\alpha(B)$, and $d(X_\alpha, X_0) \rightarrow 0$ Pr-almost surely. The usefulness of Wichura's theorem will be seen, e.g., in Section 6.

Now let $(X, \mathcal{A}, P)$ be a probability space. Let $X_1, X_2, \cdots,$ be independent X-valued random variables with distribution P, defined on a countable product $(X^\infty, \mathcal{A}^\infty, P^\infty)$ of copies of $(X, \mathcal{A}, P)$. Let P_n be the random empirical measure

$$P_n := n^{-1}(\delta_{X_1} + \cdots + \delta_{X_n}),$$

where $\delta_x(A) := 1_A(x)$. Let ν_n be the normalized empirical measure $\nu_n := n^{\frac{1}{2}}(P_n - P)$.

Let W_P be the *P-noise* Gaussian process with parameter set $\mathcal{A}$, $EW_P(A) = 0$ and $EW_P(A)W_P(B) = P(A \cap B)$ for all $A, B \in \mathcal{A}$. Then W_P has independent values on disjoint sets. For each A let $G_P(A) := W_P(A) - P(A)W_P(X)$. Then G_P is a Gaussian process with parameter space $\mathcal{A}$ such that $EG_P(A) = 0$ and $EG_P(A)G_P(B) = P(A \cap B) - P(A)P(B)$ for all $A, B \in \mathcal{A}$.

The central limit theorem in finite-dimensional vector spaces tells us that, at least when restricted to a finite subclass of $\mathcal{A}$, ν_n converges in law to G_P.

Given a subclass $\mathcal{C} \subset \mathcal{A}$, let $l^\infty(\mathcal{C})$ denote the space of all bounded real-valued functions on $\mathcal{C}$, with supremum norm. Then $\nu_n(\cdot)(\omega) \in l^\infty(\mathcal{C})$ for all n and ω.

DEFINITIONS. Given a probability space $(X, \mathcal{A}, P)$ and $A, B \in \mathcal{A}$, let

$$d_P(A, B) := P(A \Delta B), \qquad \text{where} \quad A \Delta B := (A \setminus B) \cup (B \setminus A).$$

A class $\mathcal{C} \subset \mathcal{A}$ will be called a *$G_P UC$ class* if and only if G_P on $\mathcal{C}$ has a version such that for almost all ω, $C \to G_P(C)(\omega)$ is uniformly continuous on $\mathcal{C}$ for the pseudometric d_P.

Also, $\mathcal{C}$ will be called a $G_P B$ class if and only if G_P has a version which is a.s. bounded on $\mathcal{C}$. If $\mathcal{C}$ is both a $G_P B$ class and a $G_P UC$ class it will be called a $G_P BUC$ class.

Note that $\|1_A - 1_B\|_2 = d_P(A, B)^{\frac{1}{2}}$ where $\|\cdot\|_2$ is the norm in $L^2(X, \mathcal{A}, P)$. Thus these metrics define the same topology and uniform structure on $\mathcal{C}$.

If Y is a Gaussian variable independent of G_P, with $EY = 0$ and $EY^2 = 1$, and $W_P(A) := G_P(A) + P(A)Y$, $A \in \mathcal{A}$, then W_P is a (version of) P-noise.

We recall that if L is a linear map of a Hilbert space H to a space of Gaussian random variables with $EL(x) = 0$ and $EL(x)L(y) = (x, y)$ for all $x, y \in H$, then a set $C \subset H$ is called a *GC-set* (resp. *GB-set*) if and only if L restricted to $\mathcal{C}$ has a version with continuous (resp. bounded) sample functions (Dudley (1967b), (1973)). Let $I_{\mathcal{C}} := \{1_C : C \in \mathcal{C}\}$.

(1.0). PROPOSITION. *For any $\mathcal{C} \subset \mathcal{A}$, $\mathcal{C}$ is a $G_P BUC$ class if and only if the closure of $I_{\mathcal{C}}$ in $L^2(X, \mathcal{A}, P)$ is both a GB-set and a GC-set.*

PROOF. In view of the relations between G_P and W_P given above, where we can let $W_P(C) = L(1_C)$, G_P has a version uniformly continuous on a class $\mathcal{C} \subset \mathcal{A}$ if and only if W_P does, and likewise for sample continuity or boundedness. A *GB*-set must be totally bounded (Dudley (1967b), Proposition 3.4, page 295; (1973), Theorem 1.1(c), page 71). Functions on a totally bounded set $I_{\mathcal{C}}$ extend continuously to its compact closure if and only if they are uniformly continuous. The extension is still a version of the same process since it is continuous in probability. □

(1.1). PROPOSITION. *If $\mathcal{C}$ is countable, then in $l^\infty(\mathcal{C})$, $\mathcal{B}_b$ equals the smallest σ-algebra $\mathcal{B}_c$ for which all coordinate evaluations $f \to f(A)$, $A \in \mathcal{C}$, are measurable.*

PROOF. For any $f \in l^\infty(\mathcal{C})$ and $r > 0$, the closed ball

$$\begin{aligned} B^-(f, r) &:= \{g : \|g - f\|_\infty \leqslant r\} \\ &= \bigcap_{C \in \mathcal{C}} \{g : |g(C) - f(C)| \leqslant r\}. \end{aligned}$$

Thus $\mathcal{B}_b \subset \mathcal{B}_c$. Conversely, if $C \in \mathcal{C}$ and $x \in \mathbb{R}$, then

$$\{f : f(C) < x\} = \bigcup_n B(f_n, n)$$

where $f_n(C) := x - n$, $f_n(D) := 0$ for $D \neq C$, $D \in \mathcal{C}$, so $\mathcal{B}_c \subset \mathcal{B}_b$. □

For any $\mathcal{C}$, let $C_b(\mathcal{C}, d_P)$ be the space of all bounded real functions on $\mathcal{C}$ continuous for d_P. Let $D_0(\mathcal{C}, P)$ be the linear space of all functions $\phi + \psi$, where $\phi \in C_b(\mathcal{C}, d_P)$ and ψ is a finite linear combination of point masses, $\psi = \sum a_i \delta_{x(i)}$. Let $D(\mathcal{C}, P)$ be the closure of $D_0(\mathcal{C}, d_P)$ in $l^\infty(\mathcal{C})$ for the supremum norm.

The space $D(\mathcal{C}, P)$ can be considered as an extension of the usual space $D[0, 1]$ of functions on $[0, 1]$ continuous from the right with limits from the left, where $X = [0, 1]$, $\mathcal{C}$ is the class of all intervals $[0, c]$, $0 \leqslant c \leqslant 1$, and P is Lebesgue measure or any law on $[0, 1]$ with a strictly increasing distribution function.

However, as in this case, functions in D need not have a decomposition into a pure jump part and a continuous part: let $f = 0$ on $[1/(2n+1), 1/(2n)]$, $f(1/(2n)) = 1/n$, and let f be linear on $[1/(2n), 1/(2n-1)]$.

Since all our ν_n, and G_P for $G_P BUC$ classes, take values in $D_0(\mathcal{C}, P)$, it will be convenient for us to work in this incomplete space.

For a metric space (S, e) and $T \subset S$, the Borel sets in T are exactly the intersections with T of Borel sets in S. We have

$$\mathcal{B}_b(T, e) \subset \{A \cap T : A \in \mathcal{B}_b(S, e)\},$$

where the inclusion may be strict if T is nonseparable, e.g., if $e(x, y) = 1$ for $x \neq y$ in T.

DEFINITION. We say $\mathcal{C}$ is P-EM (*empirically measurable* for P) if and only if for all n, P_n is measurable from the measure-theoretic completion of $(X^\infty, \mathcal{Q}^\infty, P^\infty)$ to $(D_0(\mathcal{C}, P), \mathcal{B}_b)$.

A countable class $\mathcal{C} \subset \mathcal{Q}$ is always P-EM (by the easy direction of Proposition 1.1). More generally, if $\mathcal{C}$ has a countable subclass $\mathcal{D}$ such that for all $C \in \mathcal{C}$ there are $D(n) \in \mathcal{D}$ with $1_{D(n)}(x) \to 1_C(x)$ for all $x \in X$, then $\mathcal{C}$ is P-EM.

EXAMPLE. If P is Lebesgue measure on $[0, 1]$ and $\mathcal{C} = \{\{x\} : x \in E\}$ where E is a nonmeasurable set, then $\mathcal{C}$ is not P-EM, since $\sup_{C \in \mathcal{C}} |P_1(C)|$ is nonmeasurable. Here $\mathcal{C}$ is included in the P-EM class of all singletons.

Note that $\mathcal{B}_b$-measurability of P_n and ν_n are equivalent. For any P-EM class $\mathcal{C}$, let $\mathcal{L}(\nu_n)$ be the law (probability distribution) of ν_n on $(D_0(\mathcal{C}, P), \mathcal{B}_b)$.

DEFINITION. A P-EM class $\mathcal{C} \subset \mathcal{Q}$ will be called a *Donsker class* for P, or a *P-Donsker class*, if and only if it is a $G_P BUC$ class and we have convergence of laws $\mathcal{L}(\nu_n) \to \mathcal{L}(G_P)$ in $(D_0(\mathcal{C}, P), \mathcal{B}_b)$ for the supremum norm, where G_P is taken to have sample functions bounded and d_P-uniformly continuous on $\mathcal{C}$.

There are at least two definitions of convergence of laws, in spaces like $D_0(\mathcal{C}, P)$, different from ours. One the space $D[0, 1]$ of right-continuous functions on $[0, 1]$ with left limits, continuous at 1, Skorohod (1955) and Kolmogorov (1956) defined a complete separable metric topology, now called a "Skorohod topology," for which convergence to a continuous function is equivalent to uniform convergence. See also Billingsley (1968, Chapter 3). Replacing $[0, 1]$ by a cube $[0, 1]^q$, Neuhaus (1971) and Straf (1971) defined a Skorohod topology on a suitable function space $D[0, 1]^q$.

More generally, given a group G of $1-1$ transformations of a probability space X onto itself, with identity element e and a right-invariant metric d, Straf (1971) defines a metric for bounded real functions f, h on X by

$$\rho(f, h) := \inf_{g \in G}(d(e, g) + \sup_x |f(x) - h(g(x))|).$$

$D(X)$ is the ρ-closure of a suitable space of simple functions. Under some conditions, $(D(X), \rho)$ will be separable and topologically complete. But I do not know how to choose a suitable G in the generality of this paper.

Pyke and Shorack (1968) defined weak convergence for processes on $\mathbb{R}$ with bounded sample functions and laws μ_n by $\int f\, d\mu_n \to \int f\, d\mu_0$ for all bounded f which are continuous for the supremum norm and measurable for all μ_n. But the above example $\mathcal{C} = \{\{x\} : x \in E\}$, E nonmeasurable, indicates that there may not be enough such f here.

So, the definition requiring $\mathfrak{B}_b$-measurability will be used. Let $\Omega := X^\infty$ and $\Pr := P^\infty$. Then

$$\Pr^*(Y) := \inf\{\Pr(E) : E \supset Y\} \quad \text{for any set } Y \subset \Omega.$$

Given $\varepsilon > 0$ and $\delta > 0$, let

$$B_{\delta,\varepsilon} := \{f \in D_0(\mathcal{C}, P) : \text{for some } A, B \in \mathcal{C}, d_P(A, B) < \delta \text{ and } |f(A) - f(B)| > \varepsilon\}.$$

Here is a characterization of Donsker classes.

(1.2). THEOREM. *Given a probability space $(X, \mathcal{A}, P)$ and a P-EM class $\mathcal{C} \subset \mathcal{A}$, $\mathcal{C}$ is a Donsker class if and only if both*

(a) *$\mathcal{C}$ is totally bounded for d_P, and*
(b) *for any $\varepsilon > 0$ there is a $\delta > 0$ and an n_0 such that for $n \geqslant n_0$,*

$$\Pr^*\{\nu_n \in B_{\delta,\varepsilon}\} < \varepsilon.$$

PROOF. First assume (a) and (b). Then $\mathcal{C}$ has a countable d_P-dense set $\mathfrak{D}$. Applying the central limit theorem to finite subsets $\mathfrak{D}_m$ of $\mathfrak{D}$, using (b), and letting $\mathfrak{D}_m \uparrow \mathfrak{D}$, shows that $\mathfrak{D}$ is a $G_P UC$ class. Almost all sample functions of G_P then extend uniquely to $\mathcal{C}$ by uniform continuity. It follows that $\mathcal{C}$ is a $G_P UC$ class and a $G_P B$ class.

Let $s(f, g) := \sup_{C \in \mathcal{C}} |f(C) - g(C)|$. For $\gamma > 0$ and a set $K \subset D_0(\mathcal{C}, P)$ let $K^\gamma := \{g : s(f, g) < \gamma \text{ for some } f \in K\}$.

(1.3). LEMMA. 1.2(a) *and* (b) *and $\mathcal{C}$ P-EM imply that for any $\varepsilon > 0$ there is a compact set $K \subset C_b(\mathcal{C}, d_P)$ with metric s such that for any $\gamma > 0$, $\Pr\{\nu_n \in K^\gamma\} > 1 - \varepsilon$ for n large enough.*

PROOF. This is a variant of Dudley ((1966), Proposition 2). We may assume $0 < \varepsilon < 1$. By (b), take $\delta > 0$ such that $\Pr^*\{\nu_n \in B_{\delta,\varepsilon/2}\} < \varepsilon/4$ for $n \geqslant n_0 = n_0(\delta, \varepsilon)$. Let $\mathfrak{F}$ be a finite subset of $\mathcal{C}$ such that for all $C \in \mathcal{C}$, $d_P(A, C) < \delta$ for some $A \in \mathfrak{F}$, by (a). Let $\mathfrak{F}$ have k elements. Take M large enough so that

$(M-1)^{-2} < \varepsilon/k$. For each $A \in \mathcal{F}$,

$$\Pr\{|\nu_n(A)| > M-1\} < \varepsilon/(4k)$$

by Chebyshev's inequality. Thus

$$\Pr\{\sup_{A \in \mathcal{F}} |\nu_n(A)| > M-1\} < \varepsilon/4,$$

and

$$\Pr\{\sup_{C \in \mathcal{C}} |\nu_n(C)| > M\} < \varepsilon/2,$$

where the latter event is measurable by the P-EM assumption.

For $m = 1, 2, \cdots,$ and $\beta > 0$, let $\varepsilon(m) := \varepsilon/2^m$ and $A_{\beta, m} := B_{\beta, \varepsilon(m)}$. We choose a sequence $\{\beta(m)\}$ of positive numbers satisfying the following two conditions:

(I) $\beta(m+1) < \beta(m)/2$;
(II) For some sequence $\{n_0(m)\}$,

$$\Pr^*\{\nu_n \in A_{\beta(m), m}\} < \varepsilon(m) \quad \text{for all } n \geqslant n_0(m).$$

Let $\delta_m := \beta(m)\varepsilon/(2^{m+1}M)$. Then by (I), $\delta_{m+1} < \delta_m/4$. Let $A_m := A_{\beta(m), m}$. Now if $s(0, f) \leqslant M$, $f \notin A_m$, $C, D \in \mathcal{C}$, and $d_P(C, D) \geqslant \beta(j)$, then $|f(C) - f(D)| \leqslant 2M \leqslant \varepsilon d_P(C, D)/(2^j\delta_j)$, while if $d_P(C, D) < \beta(j)$, then $|f(C) - f(D)| \leqslant \varepsilon(j) = \varepsilon/2^j$. Thus for any $C, D \in \mathcal{C}$,

$$(*) \qquad |f(C) - f(D)| \leqslant \varepsilon 2^{-j} \max(1, d_P(C, D)/\delta_j).$$

Let F_m be the set of all $f \in D_0(\mathcal{C}, P)$ such that $s(0, f) \leqslant M$ and (*) holds for $j = 2, \cdots, m$ and all $C, D \in \mathcal{C}$. Then for $n \geqslant N = N(m)$ large enough, there is a measurable set $E_m \subset X^\infty$ such that $\Pr(E_m) > 1 - \varepsilon$ and for all $\omega \in E_m$, $\nu_n(\cdot)(\omega) \in F_m$.

Now let K be the set of all real functions g on $\mathcal{C}$ such that $s(0, g) \leqslant M$ and for all $j = 1, 2, \cdots, d_P(C, D) < \delta_j/2$ implies $|g(C) - g(D)| \leqslant 3\varepsilon/2^j$. Then $K \subset C_b(\mathcal{C}, d_P)$. Now $(\mathcal{C}, d_P)$ is totally bounded and K is a uniformly bounded, uniformly equicontinuous family of functions, complete for s. Thus by the Arzelà-Ascoli theorem (applied to the completion of $\mathcal{C}$ for d_P), (K, s) is compact.

Given a $\gamma > 0$, choose an integer $m > 1$ such that $\varepsilon/2^m < \gamma/2$. Let us show that $F_m \subset K^\gamma$. Take a maximal set $\mathcal{C}_m \subset \mathcal{C}$ such that $d_P(C, D) \geqslant \delta_m$ for all $C \neq D$ in $\mathcal{C}_m$. Then $\mathcal{C}_m$ is finite by (a). For all $C \in \mathcal{C}$, $d_P(C, D) < \delta_m$ for some $D \in \mathcal{C}_m$.

If $f \in F_m$ and $C, D \in \mathcal{C}_m$, then $|f(C) - f(D)| \leqslant \varepsilon d_P(C, D)/(2^m\delta_m)$, by (*). Then f on $\mathcal{C}_m$ can be extended to a function g on $\mathcal{C}$ with $|g(C) - g(D)| \leqslant \varepsilon 2^{-m} d_P(C, D)/\delta_m$ for all $C, D \in \mathcal{C}$ (McShane (1934)). Taking $\max(-M, \min(g, M))$, we can assume $s(0, g) \leqslant M$. Let us show that $g \in K$. For $i \geqslant m$, since

$$\varepsilon/(2^m\delta_m) = 2M/\beta(m) \leqslant 2M/\beta(i) = \varepsilon/(2^i\delta_i),$$

we have

$$|g(C) - g(D)| \leqslant \varepsilon/2^i \quad \text{for} \quad d_P(C, D) < \delta_i.$$

For $j < m$, given $C, D \in \mathcal{C}$ with $d_P(C, D) < \delta_j/2$, choose $C_m, D_m \in \mathcal{C}_m$ with $d_P(C, C_m) < \delta_m$ and $d_P(D, D_m) < \delta_m$. Then $d_P(C_m, D_m) < 2\delta_m + \delta_j/2 < \delta_j$, and

$$\begin{aligned}|g(C) - g(D)| \leqslant{}& |g(C) - g(C_m)| \\ &+ |f(C_m) - f(D_m)| + |g(D_m) - g(D)| \\ \leqslant{}& \varepsilon/2^m + \varepsilon/2^j + \varepsilon/2^m < 3\varepsilon/2^j,\end{aligned}$$

using (*) for the middle term. Thus $g \in K$. Now $s(f, g) < \gamma$ since for any $C \in \mathcal{C}$, there is $C_m \in \mathcal{C}_m$ with $d_P(C, C_m) < \delta_m$, and

$$\begin{aligned}|f(C) - g(C)| &\leqslant |g(C) - f(C_m)| + |g(C_m) - g(C)| \\ &\leqslant 2\varepsilon/2^m < \gamma.\end{aligned}$$

So $F_m \subset K^\gamma$ as desired. For $n \geqslant N(m)$, $\Pr^*(\nu_n \notin F_m) < \varepsilon$, so $\Pr(\nu_n \in K^\gamma) > 1 - \varepsilon$, noting that K^γ is a countable union of balls, hence $\mathcal{B}_b$ measurable. Thus Lemma 1.3 is proved.

Now, using Lemma 1.3, the sequence $\{\mathcal{L}(\nu_n)\}$ of laws on $(D_0(\mathcal{C}, P), \mathcal{B}_b)$ has a convergent subsequence, as in Dudley ((1966), Theorem 1; (1967a), Theorem). Any limit of such a subsequence is concentrated in a countable union of compact subsets K_n of $C_b(\mathcal{C}, d_P)$, with $\varepsilon = 1/n$ in Lemma 1.3. Now $\bigcup_n K_n$ is separable for s. By the finite-dimensional central limit theorem, these limits of subsequences must all equal $\mathcal{L}(G_P)$. Every subsequence of $\{\mathcal{L}(\nu_n)\}$ has a convergent subsubsequence. (In 1966 I carelessly called this property "precompact"; that should mean "having compact completion" for subsets of uniform spaces.) It follows that $\mathcal{L}(\nu_n) \to \mathcal{L}(G_P)$, i.e., $\mathcal{C}$ is a Donsker class.

Conversely if $\mathcal{C}$ is a Donsker class, hence a G_PB class, then (a) holds by Dudley ((1967b), Proposition 3.4). A version of G_P which has uniformly continuous, hence bounded sample functions on the totally bounded set $\mathcal{C}$ for d_P, gives a law $\mathcal{L}(G_P)$ with a separable support in $D_0(\mathcal{C}, P)$, concentrated in a countable union of compact subsets of $C_b(\mathcal{C}, d_P)$. Thus by Theorem 2 of Wichura (1970), we can assume here that $\nu_n \to G_P$ uniformly on $\mathcal{C}$, so that for any $\delta > 0$,

$$T_{n\delta} := \sup\{|\nu_n(A) - \nu_n(B)| : A, B \in \mathcal{C}, d_P(A, B) \leqslant \delta\}$$

converges almost uniformly (as defined above) for $n \to \infty$ to

$$T_\delta := \sup\{|G_P(A) - G_P(B)| : A, B \in \mathcal{C}, d_P(A, B) \leqslant \delta\}.$$

Here it is not claimed that $T_{n\delta}$ is a measurable random variable, but each T_δ is. For any $\varepsilon > 0$ there is a $\delta > 0$ such that

$$\Pr\{T_\delta > \varepsilon/2\} < \varepsilon/2, \qquad \text{and for some } n_0,$$
$$\Pr^*(|T_{n\delta} - T_\delta| > \varepsilon/2) < \varepsilon/2$$

for $n \geqslant n_0$. Thus $\Pr^*(T_{n\delta} > \varepsilon) \leqslant \varepsilon$, proving (b). □

2. Sequences of sets. Let $(X, \mathcal{A}, P)$ be a probability space.

(2.1). THEOREM. *Let* $\{A_m\}_{m=1}^{\infty}$ *be any sequence of measurable sets with* $P(A_m) = p_m$ *such that for some* $r < \infty$, $\Sigma_m p_m^r < \infty$. *Then* $\{A_m\}_{m \geqslant 1}$ *is a Donsker class.*

PROOF. We know that any countable collection $\{A_m\}$ is P-EM. Further $\mathcal{C} = \{A_m\}$ is totally bounded for d_P. Hence it remains only to verify condition (b) of Theorem 1.2.

By Theorem 3.1 in Dudley (1967b), $\mathcal{C}$ is a $G_P UC$-class. So for any $\varepsilon > 0$ there is a $\delta = \delta(\varepsilon) > 0$ such that

$$\Pr\left\{\sup\left\{|G_P(A_i) - G_P(A_j)| : d_P(A_i, A_j) < 3\delta\right\} > \varepsilon\right\} < \varepsilon. \tag{2.2}$$

It will be shown that

(2.3) There exist numbers μ and n_0 such that for all $n \geqslant n_0$,

$$\Pr\left\{\sup_{m \geqslant \mu} |\nu_n(A_m)| > \varepsilon\right\} < 2\varepsilon.$$

We define the binomial probabilities

$$b(k, n, p) := n! p^k q^{n-k} / k!\,(n-k)!, \quad \text{where} \quad q := 1 - p,$$

$$E(k, n, p) := \Sigma_{k \leqslant j \leqslant n} b(j, n, p),$$

$$B(k, n, p) := \Sigma_{0 \leqslant j \leqslant k} b(j, n, p)$$

where in B and E, k is not necessarily an integer.

To prove (2.3) it is enough to show that for any $\varepsilon > 0$ there are a μ and an n_0 such that for all $n \geqslant n_0$,

$$\Sigma_{m \geqslant \mu} E\left(np_m + \varepsilon n^{\frac{1}{2}}, n, p_m\right) < \varepsilon \quad \text{and} \tag{2.4}$$

$$\Sigma_{m \geqslant \mu} B\left(np_m - \varepsilon n^{\frac{1}{2}}, n, p_m\right) < \varepsilon. \tag{2.5}$$

We may assume $\frac{1}{2} \geqslant p_m \downarrow 0$ as $m \to \infty$.

To prove (2.5) we use the Chernoff-Okamoto inequality (Okamoto (1958), Lemmas 1, 2b′, or Hoeffding (1963), Theorem 1),

$$B(k, n, p) \leqslant \exp\left(-(np - k)^2 / 2npq\right),$$

if $p \leqslant \frac{1}{2}$ and $k \leqslant np$, which implies

$$B\left(np_m - \varepsilon n^{\frac{1}{2}}, n, p_m\right) \leqslant \exp\left(-\varepsilon^2 / 2p_m q_m\right).$$

Let $\delta = 1/r$. Then for some constant $K < \infty$, $p_m \leqslant K m^{-\delta}$ for all m. Thus since $\Sigma_m \exp(-m^\delta \varepsilon^2 / 2K) < \infty$, there is a μ large enough so that (2.5) holds for all n.

For series (2.4), Bernstein's inequality (Bennett (1962) or Hoeffding (1963)) gives

$$E\left(np + \varepsilon n^{\frac{1}{2}}, n, p\right) \leqslant \exp\left(-\varepsilon^2 / \left(2pq + \varepsilon n^{-\frac{1}{2}}\right)\right)$$

$$\leqslant \exp\left(-\varepsilon^2 / 6pq\right)$$

if $4pqn^{\frac{1}{2}} \geqslant \varepsilon > 0$ and hence, for $p \leqslant \frac{1}{2}$, if $2pn^{\frac{1}{2}} \geqslant \varepsilon$. Thus

$$\begin{aligned} S_1 &:= \Sigma_m\left\{ E\left(np_m + \varepsilon n^{\frac{1}{2}}, n, p_m\right) : 2p_m n^{\frac{1}{2}} \geqslant \varepsilon, m \geqslant \mu \right\} \\ &\leqslant \Sigma_m\left\{ \exp\left(-\varepsilon^2/6p_m\right) : 2p_m n^{\frac{1}{2}} \geqslant \varepsilon, m \geqslant \mu \right\} \\ &\leqslant \Sigma_m\left\{ \exp\left(-\varepsilon^2 m^\delta/6K\right) : m \geqslant \mu \right\} < \varepsilon/2 \end{aligned}$$

uniformly in n for μ large enough. Let

$$S_2 := \Sigma_m\left\{ E\left(np_m + \varepsilon n^{\frac{1}{2}}, n, p_m\right) : 2p_m n^{\frac{1}{2}} < \varepsilon \right\}.$$

The Chernoff-Okamoto inequality (Okamoto (1958), Lemma 1) gives

$$E(k, n, p) \leqslant (np/k)^k (nq/(n-k))^{n-k} \qquad \text{for} \quad k \geqslant np. \tag{2.6}$$

(2.7). LEMMA. *Whenever* $k \geqslant np$, $E(k, n, p) \leqslant (np/k)^k e^{k-np}$.

PROOF. By (2.2), we may prove $(nq/(n-k))^{n-k} \leqslant e^{k-\ }$. Let $x := nq/(n-k) \geqslant 1$. Then $x \leqslant e^{x-1}$, giving the result. □

Let $s(n) := n^{\frac{1}{2}}$. Then for $\varepsilon := (k - np)/s(n)$, $E(k, n, p) \leqslant (np/(np + \varepsilon n^{\frac{1}{2}}))^{np + \varepsilon s(n)} e^{\varepsilon s(n)}$, and $\varepsilon n^{\frac{1}{2}} - (np + \varepsilon n^{\frac{1}{2}})\ln(1 + \varepsilon/pn^{\frac{1}{2}}) = \varepsilon n^{\frac{1}{2}}(1 - (x+1)\ln(1 + x^{-1}))$ where $0 < x := n^{\frac{1}{2}}p/\varepsilon < \frac{1}{2}$. The function $f(x) := (1 + x)\ln(1 + x^{-1})$ is decreasing for all $x > 0$. Thus for $0 < x \leqslant \frac{1}{2}$, $f(x) \geqslant f(\frac{1}{2}) > 1.5$. Thus

$$\begin{aligned} \varepsilon n^{\frac{1}{2}}(1 - f(x)) &= \varepsilon n^{\frac{1}{2}}(1 - 2f(x)/3) - \varepsilon n^{\frac{1}{2}} f(x)/3 \\ &\leqslant -\varepsilon n^{\frac{1}{2}} f(x)/3. \end{aligned}$$

Hence

$$E(k, n, p) \leqslant \exp\left(-\left(\varepsilon n^{\frac{1}{2}} + np\right)\left(\ln\left(1 + \varepsilon/pn^{\frac{1}{2}}\right)\right)/3\right).$$

Thus

$$S_2 \leqslant \Sigma_m\left\{ \left(p_m n^{\frac{1}{2}}/\varepsilon\right)^{\varepsilon s(n)/3} : 2p_m n^{\frac{1}{2}} < \varepsilon \right\}.$$

Since $p_m \leqslant Km^{-\delta}$, we have $S_2 \leqslant S_3 + S_4$ where

$$\begin{aligned} S_3 &:= \Sigma_m\left\{ \left(p_m n^{\frac{1}{2}}/\varepsilon\right)^{\varepsilon s(n)/3} : 2Km^{-\delta} n^{\frac{1}{2}} < \varepsilon \right\} \\ &\leqslant \left(Kn^{\frac{1}{2}}/\varepsilon\right)^{\varepsilon s(n)/3} \Sigma_m\left\{ m^{-\delta\varepsilon s(n)/3} : m > G \right\} \end{aligned}$$

where $G := G_{nK\varepsilon\delta} := (2Kn^{\frac{1}{2}}/\varepsilon)^{1/\delta}$. Choosing n_1 large enough so that $\delta\varepsilon n_1^{\frac{1}{2}}/3 > 2$, we have for $n \geqslant n_1$

$$\begin{aligned} S_3 &\leqslant \left(Kn^{\frac{1}{2}}/\varepsilon\right)^{\varepsilon s(n)/3} \int_{G-1}^{\infty} x^{-\delta\varepsilon s(n)/3}\, dx \\ &\leqslant \left(Kn^{\frac{1}{2}}/\varepsilon\right)^{\varepsilon s(n)/3} \left(\delta\varepsilon n^{\frac{1}{2}}/3 - 1\right)^{-1} (G-1)^{1-\delta\varepsilon s(n)/3}. \end{aligned}$$

For fixed K, ε and δ, we find that the logarithm of the last expression is asymptotic to $-(\varepsilon n^{\frac{1}{2}}/3)\ln 2 \to -\infty$ as $n \to \infty$, so that $S_3 \to 0$. Thus $S_3 \leqslant \varepsilon/4$ for $n \geqslant n_2$ for some n_2.

Lastly,

$$S_4 := \Sigma_m\left\{\left(p_m n^{\frac{1}{2}}/\varepsilon\right)^{\varepsilon s(n)/3} : 2p_m n^{\frac{1}{2}} < \varepsilon \leqslant 2Km^{-\delta}n^{\frac{1}{2}}\right\}$$
$$\leqslant \left(2Kn^{\frac{1}{2}}/\varepsilon\right)^{1/\delta}(1/2)^{\varepsilon s(n)/3} \to 0$$

as $n \to \infty$, so $S_4 < \varepsilon/4$ for $n \geqslant n_3$ for some n_3. Hence for $n \geqslant \max(n_1, n_2, n_3)$, $S_2 < \varepsilon/2$ so (2.4) and hence (2.3) hold. We take μ large enough so that $p_m < \delta$ for all $m \geqslant \mu$. From the finite-dimensional central limit theorem and (2.2) we can find n_4 such that for $n \geqslant n_4$

$$\Pr\left\{\sup\left\{|\nu_n(A_i) - \nu_n(A_j)| : i, j \leqslant \mu \text{ and } d_P(A_i, A_j) < 3\delta\right\} > \varepsilon\right\} < \varepsilon.$$

This and (2.3) then imply condition (1.2b), completing the proof of Theorem 2.1.

(2.8). THEOREM. *If A_m are independent for P and $P(A_m) = p_m$, then $\{A_m\}$ is a Donsker class if and only if for some $r < \infty$, $\Sigma_m(p_m(1 - p_m))^r < \infty$.*

PROOF. Note that a subsequence of the A_m can converge for d_P if and only if their probabilities converge to 0 or 1. Hence, $\{A_m\}$ is Donsker if and only if the collection $\{A_m, X \setminus A_m\}$ of A_m and their complements is Donsker. Thus we may assume $p_m \leqslant \frac{1}{2}$.

Now, "if" follows from Theorem 2.1. Conversely, suppose $\Sigma_m p_m^n = +\infty$ for all n. Then for each n, $\Pr\{P_n(A_m) = 1$ for infinitely many $m\} = 1$ by the Borel-Cantelli lemma. Since $P_n(A_m) = 1$ implies $\nu_n(A_m) = n^{\frac{1}{2}}(1 - p_m) \to \infty$ as $n \to \infty$, $\{A_m\}$ is not a Donsker class. □

Also, any sequence of disjoint measurable sets is a Donsker class by Theorem 2.1 with $r = 1$.

Independent A_m with $P(A_m)\downarrow 0$ as $m \to \infty$ form a $G_P BUC$ class if $P(A_m)\log m \to 0$ as $m \to \infty$ (Dudley (1967b), Proposition 6.7, with $(\int 1_A^2\, dP) = P(A)$). This is weaker than the condition for $\{A_m\}$ to be Donsker, and shows that the binomial upper tail $E(k, n, p)$ for small p becomes substantially larger than the corresponding Gaussian tail.

3. Measurability. A *measurable space* is a pair $(X, \mathfrak{B})$ where X is a set and $\mathfrak{B}$ is a σ-algebra of subsets of X. Then $(X, \mathfrak{B})$ or $\mathfrak{B}$ is called *countably generated* iff there is some countable $\mathcal{C} \subset \mathfrak{B}$ such that $\mathfrak{B}$ is the smallest σ-algebra including $\mathcal{C}$.

Given a measurable space $(X, \mathfrak{B})$, a set $A \subset X$ will be called simply *measurable* if $A \in \mathfrak{B}$. If P is a law defined on $\mathfrak{B}$, A will be called *completion measurable* for P iff $d_P(A, B) = 0$ for some $B \in \mathfrak{B}$, i.e., $A \Delta B \subset C$ for some $C \in \mathfrak{B}$ with $P(C) = 0$. If A is completion measurable for all laws on $\mathfrak{B}$ (where B and C depend on P), it is called *universally measurable*.

If $(Y, \mathcal{E})$ is another measurable space, a function f from X into Y will be called (completion, resp. universally) measurable iff for all $E \in \mathcal{E}, f^{-1}(E)$ is (completion, resp. universally) measurable in X.

A *Polish* space is a topological space metrizable by a complete separable metric. A set A in a metric space is called a *Suslin* set iff there is a continuous function from a Polish space onto A. Note also that any Borel measurable function from a separable metric space into a metric space has separable range (Stone (1962), page 32, Theorem 16). A set in a metric space is Suslin iff it is the range of some Borel function on a Polish space. Thus any Borel set in a Polish space is Suslin.

All Suslin sets are universally measurable (e.g., Federer (1969), Theorem 2.2.12, page 69). Clearly a product of two Suslin sets with product topology is Suslin, a union of countably many Suslin sets is Suslin, and a continuous or Borel image of a Suslin set is Suslin. A countable intersection of Suslin sets is Suslin (e.g., Kuratowski (1966), pages 454, 478). As Suslin proved, a Suslin set whose complement is Suslin is a Borel set (e.g., Kuratowski (1966), pages 485–486), while there exist Suslin non-Borel sets (Kuratowski (1966), page 460).

A measurable space $(X, \mathcal{B})$ will be called *Suslin* iff there is a metric d on X for which (X, d) is Suslin and $\mathcal{B}$ is the σ-algebra of Borel sets.

DEFINITION. If $(X, \mathcal{B})$ and $(\mathcal{C}, \mathcal{S})$ are measurable spaces and $\mathcal{C} \subset \mathcal{B}$, we call $(X, \mathcal{B}; \mathcal{C}, \mathcal{S})$ a *chair*. The chair, or $(\mathcal{C}, \mathcal{S})$, will be called *admissible* iff the $\in$ relation, $\{\langle x, C\rangle : x \in C\}$, is a measurable subset of $X \times \mathcal{C}$ for the product σ-algebra of $\mathcal{B}$ and $\mathcal{S}$. We call $\mathcal{C}$ *admissible* iff there is some σ-algebra $\mathcal{S}$ for which $(\mathcal{C}, \mathcal{S})$ is admissible.

Note that if $(\mathcal{C}, \mathcal{S})$ is admissible, so is $(\mathcal{C}, \mathcal{U})$ for some countably generated $\mathcal{U} \subset \mathcal{S}$.

Suppose $\mathcal{B}$ has a countable set $\mathcal{A}$ of generators. Let $\mathcal{A}_0 := \mathcal{A}$. For each successor ordinal $\alpha + 1$, let $\mathcal{A}_{\alpha+1}$ be the collection of all complements and countable unions of sets in $\mathcal{A}_\alpha$. For each limit ordinal $\beta > 0$, let $\mathcal{A}_\beta := \bigcup_{\alpha<\beta} \mathcal{A}_\alpha$. The $\mathcal{A}_\alpha$ are called *Borel classes* (or *Banach classes*, cf. Aumann (1961)).

A collection $\mathcal{C} \subset \mathcal{B}$ will be said to be of ***bounded Borel class*** iff for some countable set $\mathcal{A}$ of generators of $\mathcal{B}$, and some countable ordinal α, $\mathcal{C} \subset \mathcal{A}_\alpha$. This notion does not depend on the choice of $\mathcal{A}$, although the specific ordinal α does.

We quote a theorem of Aumann ((1961), Theorem D); B. V. Rao ((1971), Theorem 3) gave a shorter proof.

(3.1). THEOREM (Aumann). *For any countably generated measurable space $(X, \mathcal{B})$ and $\mathcal{C} \subset \mathcal{B}$, $\mathcal{C}$ is admissible iff it is of bounded Borel class.*

Thus in any separable metric space, the collections of all open sets, closed sets, G_δ sets (countable intersections of open sets), F_σ sets (countable unions of closed sets), and countable sets are each admissible, etc. In Section 4, specific admissible σ-algebras will be put on collections of open or closed sets.

In [0, 1], for example, the collection of all Borel sets is not admissible.

We recall that $(\mathcal{C}, \mathcal{S})$ is called a *standard Borel space* iff there is a measurable isomorphism of it with a Polish space carrying its Borel σ-algebra. A metric space (X, d) is called *absolutely Borel* iff X is a Borel set in its completion for d. If X is separable, this property depends on d only through its topology (e.g., Parthasarathy (1967), page 22, Corollary 3.3). A separable metric space with Borel σ-algebra is a standard Borel space iff it is absolutely Borel (Parthasarathy (1967), pages 133–134).

For any probability space $(X, \mathfrak{B}, P)$ and $\mathcal{C} \subset \mathfrak{B}$, there is a smallest σ-algebra $\mathcal{S}_P$ of subsets of $\mathcal{C}$ for which P is a measurable function, generated by countably many sets $\{A : P(A) < r\}$, r rational. If $(\mathcal{C}, \mathcal{S})$ is admissible, let $\mathfrak{U}$ be the smallest σ-algebra including $\mathcal{S}$ and $\mathcal{S}_P$. Then $\mathfrak{U}$ is also admissible, and countably generated if $\mathcal{S}$ is; $\mathcal{S}$ always has a countably generated admissible sub-σ-algebra.

DEFINITION. A chair $(X, \mathfrak{B}; \mathcal{C}, \mathcal{S})$ is *ϵ-Suslin* iff it is admissible and both $(X, \mathfrak{B})$ and $(\mathcal{C}, \mathcal{S})$ are Suslin measurable spaces. Given a law P on $\mathfrak{B}$, the chair $(X, \mathfrak{B}; \mathcal{C}, \mathcal{S})$ is *$P\epsilon$-Suslin* iff it is ϵ-Suslin and all d_P-open subsets of $\mathcal{C}$ belong to $\mathcal{S}$.

If $(X, \mathfrak{B})$ is Suslin, it is countably generated, so $(\mathfrak{B}, d_P)$ and $(\mathcal{C}, d_P)$ are always separable.

DEFINITION. Given a probability space $(X, \mathfrak{B}, P)$ and $\mathcal{C} \subset \mathfrak{B}$, we call $\mathcal{C}$ *strongly P-EM* iff for any n, any real b_i, and independent random variables $X(i)$ in X with law $\mathcal{L}(X(i)) = P$, the map

$$\omega \to \Sigma_{1 \leqslant i \leqslant n} b_i \delta_{X(i)(\omega)}$$

is completion measurable into $(D_0(\mathcal{C}, P), \mathfrak{B}_b)$.

(3.2). PROPOSITION. *If $(X, \mathfrak{B}; \mathcal{C}, \mathcal{S})$ is $P\epsilon$-Suslin, then $\mathcal{C}$ is strongly P-EM.*

PROOF. For any $x \in X$, the function $C \to \delta_x(C)$ is $\mathcal{S}$-measurable on $\mathcal{C}$ since $(\mathcal{C}, \mathcal{S})$ is admissible. Thus for any $y(1), \ldots, y(k) \in X$ and real $b_1, \cdots, b_k$, the function $C \to \Sigma b_i \delta_{y(i)}(C)$ is $\mathcal{S}$-measurable on $\mathcal{C}$.

Let f be any d_P-continuous function on $\mathcal{C}$. Then for any real t, $\{C \in \mathcal{C} : f(C) > t\}$ is an open set for d_P, thus belongs to $\mathcal{S}$. Hence f is $\mathcal{S}$-measurable. In particular, P is $\mathcal{S}$-measurable. So $P + f + \Sigma b_i \delta_{y(i)}$ is $\mathcal{S}$-measurable.

Now it is enough to show that for independent $X(1), \cdots, X(n)$ with law P, real b_i, and any $\mathcal{S}$-measurable function g on $\mathcal{C}$,

$$\sup_{C \in \mathcal{C}} |\Sigma_{1 \leqslant i \leqslant n} b_i \delta_{X(i)(\cdot)}(C) - g(C)|$$

is measurable. We denote points of X^n by $x = \langle x(1), \cdots, x(n) \rangle$. For any $t \geqslant 0$ we define the set $E_t \subset X^n \times \mathcal{C}$ by

$$E_t := \{\langle x, C \rangle : |\Sigma_{1 \leqslant i \leqslant n} b_i \delta_{x(i)}(C) - g(C)| > t\}.$$

Then

$$E_t := \bigcup_F \{\langle x, C\rangle : x(i) \in C \quad \text{iff} \quad i \in F\}$$
$$\cap \{\langle x, C\rangle : |\Sigma_{i\in F} b_i - g(C)| > t\}$$

where the union runs over all 2^n subsets F of $\{1, \cdots, n\}$.

For each F, $\{\langle x, C\rangle : x(i) \in C$ iff $i \in F\}$ is product measurable since $(X, \mathfrak{B}; \mathcal{C}, \mathcal{S})$ is assumed admissible. Also, $\{C : |\Sigma_{i\in F} b_i - g(C)| > t\}$ is $\mathcal{S}$-measurable since g is an $\mathcal{S}$-measurable function. Thus E_t is jointly measurable.

Since $(X^n, \mathfrak{B}^n)$ and $(\mathcal{S}, \mathcal{C})$ are Suslin spaces, E_t is Suslin and its projection on X^n is Suslin, hence universally measurable. This projection equals

$$\{x : \sup_{C\in\mathcal{C}} |\Sigma_{1\leqslant i\leqslant n} b_i \delta_{x(i)}(C) - g(C)| > t\}. \qquad \square$$

The combination of admissibility and the Suslin property seems not so easy to satisfy. For example, let Co be the collection of all countable subsets of $I := [0, 1]$. Then Co is of bounded Borel class, hence admissible by 3.1. Also, Co has some Suslin measurable structures since its cardinal is c, but I do not know any admissible Suslin structure on Co.

For one thing, the structure generated by $\{C \in Co : x \in C\}$, for all $x \in I$, is not countably generated (Szpilrajn-Marczewski (1938)).

Or, take the space I^∞ of all sequences in I, with its standard Borel structure $\mathfrak{B}$. Take the equivalence relation $\equiv$: $\{x_n\} \equiv \{y_n\}$ iff $\{x_n\}$ and $\{y_n\}$ have the same range. Let $\mathfrak{B}/\equiv$ be the factor σ-algebra,

$$\mathfrak{B}/\equiv := \{A \in \mathfrak{B} : \quad \text{if} \quad x \equiv y \quad \text{then} \quad x \in A \quad \text{iff} \quad y \in A\}.$$

Then $\mathfrak{B}/\equiv$ is not countably generated (Freedman (1966), Lemma (5)).

For the present, our positive results are for families of open or closed sets (Section 4).

DEFINITION. Given sets X, Y and $E \subset X \times Y$, with projection $\pi_X E \subset X$ where $\pi_X(x, y) := x$, a *selector* for E is a function f from $\pi_X E$ into Y such that $\langle x, f(x)\rangle \in E$ for all $x \in \pi_X E$.

We state for later reference the following extension of a theorem of Lusin and Sierpiński, which is a consequence of Corollary 4.5 of Sion (1960):

(3.3). THEOREM. *Let X and Y be separable metric spaces and E a Borel set in $X \times Y$. Then there is a universally measurable selector f for E.*

4. Spaces of closed or open sets. Let (X, d) be any separable metric space. Then there is a metric e for the d topology of X such that (X, e) is totally bounded (e.g., Kelley (1955), page 125). If X is complete for some metric metrizing the same topology (not for e unless X is compact), X is called a *Polish* space. For any topological space X let $\mathfrak{B}(X)$ denote the Borel σ-algebra generated by the open sets.

For any metric space (X, d), nonempty $A \subset X$, and $\varepsilon > 0$, let

$$A^{\varepsilon} := \{y \in X : d(x, y) < \varepsilon \quad \text{for some} \quad x \in A\}, \qquad \text{and}$$
$${}_{\varepsilon}A := \{x \in X : y \in A \quad \text{whenever} \quad d(x, y) < \varepsilon\}.$$

The well-known *Hausdorff* pseudometric h is defined for A and B nonempty by

$$h(A, B) := h_d(A, B) := \inf\{\varepsilon > 0 : A \subset B^{\varepsilon} \text{ and } B \subset A^{\varepsilon}\}.$$

Then h_d is finite valued iff (X, d) is bounded. It is a metric on closed sets.

Let $\mathfrak{F}_0$ be the class of all nonempty closed subsets of a separable metric space X, with a totally bounded metric e. Take on $\mathfrak{F}_0$ the Hausdorff metric h_e defined by e. Then $(\mathfrak{F}_0, h_e)$ is a separable metric space; if X is Polish, then $(\mathfrak{F}_0, h_e)$ is Polish (Effros (1965)), although it is not complete unless X is compact. The σ-algebra $\mathfrak{B}(\mathfrak{F}_0)$ of Borel subsets of $\mathfrak{F}_0$ for h_e is called the *Effros* Borel structure on $\mathfrak{F}_0$. It does not depend on the totally bounded metric e except through its topology (Effros (1965)).

Let $\mathfrak{F}$ be the class $\mathfrak{F}_0 \cup \{\phi\}$ of all closed sets in X. If we make ϕ an isolated point of $\mathfrak{F}$, then $\mathfrak{F}$ is also Polish whenever X is. If $(\mathfrak{F}_0, \mathfrak{B}(\mathfrak{F}_0))$ is a Suslin or standard measurable space, so is $(\mathfrak{F}, \mathfrak{B}(\mathfrak{F}))$. We call $\mathfrak{B}(\mathfrak{F})$ the *Effros* Borel structure on $\mathfrak{F}$. It is also generated by all sets $\{F \in \mathfrak{F} : F \subset H\}$ for $H \in \mathfrak{F}$ (Christensen (1971), (1974)) since such collections are h-closed and a countable family of them separates elements of $\mathfrak{F}$. (Note however that for U open, $\{F \in \mathfrak{F} : F \subset U\}$ need not be Effros measurable or even Suslin, e.g., if U is the open unit ball in an infinite-dimensional Hilbert space: Christensen (1971), Theorem 8.) Here a *Suslin* subset of $\mathfrak{F}$ or $\mathfrak{F}_0$ will be the image of a Polish space by a map measurable for the Effros Borel structure of $\mathfrak{F}$ or $\mathfrak{F}_0$.

For any $\mathfrak{G} \subset \mathfrak{F}$ we have the naturally induced Borel structure $\mathfrak{B}(\mathfrak{G})$.

(4.1). PROPOSITION. *For any separable metric space (X, d) and any collection $\mathfrak{G} \subset \mathfrak{F}$ of closed sets, $(\mathfrak{G}, \mathfrak{B}(\mathfrak{G}))$ is admissible and for any law P on $\mathfrak{B}(X)$, P is $\mathfrak{B}(\mathfrak{G})$-measurable.*

PROOF. The set $\epsilon_{\mathfrak{F}} := \{\langle x, F\rangle : x \in F \in \mathfrak{F}\}$ is closed in $X \times \mathfrak{F}$, using the Hausdorff metric h_e on $\mathfrak{F}$. Since $(\mathfrak{F}, h_e)$ is separable, $\epsilon_{\mathfrak{F}}$ belongs to the product σ-algebra generated by rectangles $A \times B$ where $A \in \mathfrak{B}(X)$ and $B \in \mathfrak{B}(\mathfrak{F})$. Thus $(\mathfrak{G}, \mathfrak{B}(\mathfrak{G}))$ is admissible.

For any law P on X and $c > 0$, $\{F \in \mathfrak{F} : P(F) \geqslant c\}$ is closed for h_e. Thus P is upper semicontinuous on $\mathfrak{F}$ and hence Effros measurable on $\mathfrak{F}$ or any subset $\mathfrak{G}$. $\square$

(4.2). PROPOSITION. *For any separable metric (X, e) and law P on X, $(\mathfrak{F}, d_P)$ is separable and all open sets for d_P are Effros measurable.*

PROOF. Since $L^1(X, \mathfrak{B}(X), P)$ is separable, $(\mathfrak{F}, d_P)$ is separable, as is any subset $\mathfrak{G} \subset \mathfrak{F}$. Thus, any open set for d_P is a countable union of open balls. For any fixed closed set F, the function $C \to P(C \setminus F)$ is upper semicontinuous and hence Effros measurable. Also, $C \to P(F \setminus C)$ is lower semicontinuous and Effros measurable. Thus, $C \to P(C \Delta F) = d_P(C, F)$ is Effros measurable. Thus d_P-open balls, and d_P-open sets, are all Effros measurable. □

(4.3). PROPOSITION. *If X is Polish and $\mathcal{G}$ is any Suslin subset of $\mathcal{F}$ (for h_e), with Effros Borel structure $\mathcal{B}(\mathcal{G})$, then $(X, \mathcal{B}(X); \mathcal{G}, \mathcal{B}(\mathcal{G}))$ is $P\epsilon$-Suslin for any law P on $\mathcal{B}(X)$.*

PROOF. This follows from Propositions 4.1 and 4.2.

Let $\mathcal{U}$ be the class of all open sets in X. On $\mathcal{U} \setminus \{X\}$ we have the metric $h_{(e)}$ defined as the Hausdorff metric of complements:

$$h_{(e)}(U, V) := h_e(X \setminus U, X \setminus V).$$

In the notation defined at the beginning of this section, we have ${}_\varepsilon V \subset U$ iff $(X \setminus U) \subset (X \setminus V)^\varepsilon$. Thus

$$h_{(e)}(U, V) = \inf\{\varepsilon > 0 : {}_\varepsilon U \subset V \quad \text{and} \quad {}_\varepsilon V \subset U\}.$$

Now $h_{(e)}$ has various properties which follow from those of h_e on $\mathcal{F}_0$. For a totally bounded metric e on X, we will call the σ-algebra $\mathcal{B}(\mathcal{U} \setminus \{X\})$ of Borel sets for $h_{(e)}$ in $\mathcal{U} \setminus \{X\}$ the *Effros* Borel structure, and likewise for the Borel structure induced on $\mathcal{U}$ by making X an isolated point. Then we have an induced measurable structure on any subset $\mathcal{V} \subset \mathcal{U}$. We conclude from Proposition 4.3:

(4.4). PROPOSITION. *If X is a Polish space, and $\mathcal{V}$ is any Suslin subset of $\mathcal{U}$ for $h_{(e)}$, with Effros Borel structure $\mathcal{B}(\mathcal{V})$, then $(X, \mathcal{B}(X); \mathcal{V}, \mathcal{B}(\mathcal{V}))$ is $P\epsilon$-Suslin for any law P on $\mathcal{B}(X)$.*

Let X be a set and G a collection of real-valued functions on X. Let $\operatorname{pos}(G)$ be the collection of all sets

$$\operatorname{pos}(g) := \{x \in X : g(x) > 0\}, g \in G.$$

Recall that by Lindelöf's theorem any locally compact separable metric space is a countable union of compact sets.

(4.5). PROPOSITION. *Let (X, e) be a locally compact, separable metric space and G a collection of continuous real functions on X. Suppose we are given a σ-algebra $\mathcal{S}$ of subsets of G such that for each $x \in X$, $g \to g(x)$ is $\mathcal{S}$-measurable. Then the map $g \to \operatorname{pos}(g)$ is measurable from $(G, \mathcal{S})$ to $\mathcal{U}$ with Effros Borel structure.*

PROOF. It suffices to show that for any open $V \subset X$, $\{g \in G : V \subset \operatorname{pos}(g)\} \in \mathcal{S}$. Let $V = \bigcup_{n \geqslant 1} K_n$ with K_n compact. For each n let A_n be a countable dense set in K_n. For any $\varepsilon > 0$, let

$$G(n, \varepsilon) := \{g \in G : g(x) \geqslant \varepsilon \quad \text{for all} \quad x \in A_n\} \in \mathcal{S}.$$

Now $V \subset \operatorname{pos}(g)$ iff $g \in \bigcap_{n \geqslant 1} \bigcup_{m \geqslant 1} G(n, 1/m)$. □

Under the conditions of Proposition 4.5, if $(G, \mathcal{S})$ is a Suslin measurable space, then $\operatorname{pos}(G)$ is a Suslin set, and by Proposition 4.4, $(X, \mathcal{B}(X); \operatorname{pos}(G), \mathcal{B}(\operatorname{pos}(G)))$ is $P\epsilon$-Suslin.

5. Metric entropy and inclusion.

DEFINITION. Let $(X, \mathcal{A}, P)$ be a probability space and $\mathcal{C} \subset \mathcal{A}$. For each $\varepsilon > 0$ let $N_I(\varepsilon) := N_I(\varepsilon, \mathcal{C}, P)$ be the smallest n such that for some $A_1, \cdots, A_n \in \mathcal{A}$ (not necessarily in $\mathcal{C}$), for every $A \in \mathcal{C}$ there exist i, j with $A_i \subset A \subset A_j$ and $P(A_j \setminus A_i) < \varepsilon$.

Let $N(\varepsilon) := N(\varepsilon, \mathcal{C}, P)$ be the smallest n such that $\mathcal{C} = \bigcup_{1 \leqslant j \leqslant n} \mathcal{C}_j$ for some sets $\mathcal{C}_j$ with $\sup\{d_P(A, B) : A, B \in \mathcal{C}_j\} \leqslant 2\varepsilon$ for each j. Log $N(\varepsilon)$ is called a *metric entropy* and $\log N_I(\varepsilon)$ will be called a *metric entropy with inclusion*.

Dehardt (1971) proved in effect that if $N_I(\varepsilon, \mathcal{C}, P) < \infty$ for all $\varepsilon > 0$, and $\mathcal{C}$ is P-EM, then

$$\Pr\{\lim_{n\to\infty} \sup_{A \in \mathcal{C}} |(P_n - P)(A)| = 0\} = 1.$$

We recall that $\mathcal{C}$ is a $G_P BUC$ class if $\int_0^1 (\log N(x^2))^{\frac{1}{2}}\, dx < \infty$ (Dudley (1967b); (1973), page 71). (Note that the L^2 norm of 1_A is $P(A)^{\frac{1}{2}}$, hence the x^2.) If $\mathcal{C}$ is the collection of all finite subsets of [0, 1] and P is Lebesgue measure, $\mathcal{C}$ is not a Donsker class although $d_P(A, B) = 0$ for all $A, B \in \mathcal{C}$. So hypotheses on $N(\varepsilon)$ will not imply the Donsker property. The following sufficient condition on $N_I(\varepsilon)$ is the same as the above condition on $N(\varepsilon)$ for the $G_P BUC$ property. Note however that if $P(A_m) = m^{-\frac{1}{2}}$ with A_m independent for P, then $\mathcal{C} := \{A_m\}_{m \geqslant 1}$ is Donsker by Theorem 2.1, while $N_I(\frac{1}{2}, \mathcal{C}, P) = +\infty$.

(5.1). THEOREM. *If $\mathcal{C}$ is P-EM and $\int_0^1 (\log N_I(x^2))^{\frac{1}{2}}\, dx < \infty$ then $\mathcal{C}$ is a Donsker class.*

PROOF. Since $\mathcal{C}$ is totally bounded for d_P, it is enough to verify Theorem 1.2b.

Let $0 < \varepsilon < 1$. Since $N_I(x)\uparrow$ as $x \downarrow 0$, the hypothesis implies $x \log N_I(x) \to 0$. So there is a $\gamma > 0$ such that

$$N_I(x) \leqslant \exp(\varepsilon^2/(600x)), \qquad 0 < x \leqslant \gamma. \tag{5.2}$$

Take $\alpha > 0$ small enough so that

$$\exp(-\varepsilon^2/(1800\alpha)) < \varepsilon/4. \tag{5.3}$$

The hypothesis on N_I is equivalent to

$$\int_0^1 (\log N_I(y))^{\frac{1}{2}} y^{-\frac{1}{2}}\, dy < \infty$$

and to

$$\Sigma_{i \geqslant 1} (2^{-i} \log N_I(2^{-i}))^{\frac{1}{2}} < \infty.$$

Take u large enough so that

$$\Sigma_{i \geqslant u} (2^{-i} \log N_I(2^{-i}))^{\frac{1}{2}} < \varepsilon/64, \tag{5.4}$$

and such that

$$\Sigma_{i \geqslant 0} \exp(-2^{i+u}\varepsilon^2/(9000(i+1)^4)) < \varepsilon/32. \tag{5.5}$$

Let $\delta_0 := 2^{-r}$ for $r \geqslant u$ and r large enough so that $\delta_0 \leqslant \min(\alpha, \gamma)$. For $k = 0, 1, 2, \cdots$, let $\delta_k := \delta(k) := \delta_0/2^k = 1/2^{k+r}$. Let $m(k) := N_I(\delta_k, \mathcal{C}, P)$ and $b_k := (2^{-k} \log m(k))^{\frac{1}{2}}$.

Take sets $A_{k1}, \cdots, A_{km(k)}$ as in the definition of $N_I(\delta_k, \mathcal{C}, P)$, so that for each $C \in \mathcal{C}$ and $k = 0, 1, 2, \cdots$, there are $r(k) := r(k, C)$ and $s(k) := s(k, C)$ with $A_{kr(k)} \subset C \subset A_{ks(k)}$ and $P(A_{ks(k)} \setminus A_{kr(k)}) < \delta_k$. Let $B_k := B_k(C) := A_{ks(k)} \setminus A_{k+1, s(k+1)}$ and $D_k := D_k(C) := A_{k+1, s(k+1)} \setminus A_{ks(k)}$. Then $P(B_k) < \delta_k$ and $P(D_k) < \delta_{k+1} < \delta_k$.

Let $n_0 := n_0(\varepsilon) := \varepsilon^2/(256\delta_0^2)$. (Note that $\delta_0 \leqslant \alpha < \varepsilon^2/1800$, so that $n_0 > 12{,}000/\varepsilon^2 \to \infty$ as $\varepsilon \downarrow 0$.) For each $n > n_0$ there is a unique $k = k(n)$ such that

$$\tfrac{1}{2} < 8\delta_k n^{\frac{1}{2}}/\varepsilon \leqslant 1. \tag{5.6}$$

Then for each n, $k = k(n)$, $\delta = \delta_k$, each $C \in \mathcal{C}$, $r = r(k, C)$, and $s = s(k, C)$, we have

$$\nu_n(A_{kr}) - \varepsilon/8 \leqslant \nu_n(A_{kr}) - \delta n^{\frac{1}{2}} \leqslant \nu_n(C) \leqslant \nu_n(A_{ks}) + \varepsilon/8. \tag{5.7}$$

We have

$$\begin{aligned} &|\nu_n(A_{ks(k)}) - \nu_n(A_{0s(0)})| \\ &\leqslant \Sigma_{0 \leqslant i < k} |\nu_n(A_{is(i)}) - \nu_n(A_{i+1, s(i+1)})| \\ &\leqslant \Sigma_{0 \leqslant i < k} |\nu_n(B_i)| + |\nu_n(D_i)|. \end{aligned} \tag{5.8}$$

Let β_i be the collection of all sets $B = A_{is} \setminus A_{i+1,t}$ or $A_{i+1,t} \setminus A_{is}$ with $P(B) < \delta_i$. Then for each $C \in \mathcal{C}$, $B_i(C)$ and $D_i(C) \in \beta_i$. The number of sets in β_i is bounded by

$$\operatorname{card}(\beta_i) \leqslant 2m(i)m(i+1). \tag{5.9}$$

We have $\log m(i) = 2^i b_i^2$. Let

$$d_i := \max\left((i+1)^{-2}\varepsilon/32,\ 4b_{i+1}2^{-r/2}\right).$$

Then by (5.4),

$$\Sigma_{i \geqslant 0} d_i < \varepsilon/8. \tag{5.10}$$

For each $i < k = k(n)$, by (5.6), $n^{\frac{1}{2}}\delta_i > n^{\frac{1}{2}}\delta_k > \varepsilon/16$. Thus by (5.10), $d_i \leqslant 2n^{\frac{1}{2}}\delta_i$. Bernstein's inequality (Bennett (1962)) gives for each $B \in \beta_i$

$$\begin{aligned} P_{inB} &:= \Pr\{|\nu_n(B)| > d_i\} \\ &\leqslant 2\exp\left(-d_i^2/\left(2pq + d_i n^{-\frac{1}{2}}\right)\right), \end{aligned}$$

where $1 - q := p := P(B) < \delta_i$.

Thus $P_{inB} \leqslant 2\exp(-d_i^2/(4\delta_i))$. Let $M_i := 4m(i)m(i+1) \leqslant 4m(i+1)^2 = 4\exp(2^{i+1}b_{i+1}^2)$. Then using (5.9) we have

$$\begin{aligned} P_{in} &:= \Pr\{|\nu_n(B)| > d_i \quad \text{for some} \quad B \in \beta_i\} \\ &\leqslant M_i \exp(-d_i^2/(4\delta_i)) = M_i \exp(-2^i d_i^2/(4\delta_0)) \\ &\leqslant 4\exp(2^i(2b_{i+1}^2 - d_i^2/(4\delta_0))). \end{aligned}$$

Now by definition of d_i, $2b_{i+1}^2 \leqslant d_i^2/(8\delta_0)$, and

$$P_{in} \leqslant 4\exp(-2^i d_i^2/(8\delta_0)) \leqslant 4\exp\left(-2^{i+r}\varepsilon^2/\left(8(32)^2(i+1)^4\right)\right).$$

Thus by (5.5),

$$\Sigma_{0 \leqslant i \leqslant k} P_{in} < \varepsilon/8. \tag{5.11}$$

Now with $k := k(n)$, $\delta := \delta_k$, let

$$V_n := \sup\{|\nu_n(A_{kr}) - \nu_n(A_{ks})|: A_{kr} \subset A_{ks}, P(A_{ks} \setminus A_{kr}) < \delta, r, s = 1, \cdots, m(k)\}.$$

Let $Q_n := \Pr(V_n > \epsilon/8)$. Then by Bernstein's inequality, and (5.6),

$$\begin{aligned} Q_n &\leqslant m(k)^2 \exp\left(-\varepsilon^2 64^{-1}/\left(2\delta_k + \varepsilon 8^{-1} n^{-\frac{1}{2}}\right)\right) \\ &\leqslant m(k)^2 \exp(-\varepsilon^2/(128\delta_k + 128\delta_k)) \\ &\leqslant \exp(2^k(2b_k^2 - \varepsilon^2/(256\delta_0))). \end{aligned}$$

Now for $j := k + r$,

$$\begin{aligned} 2b_k^2 &= 2^{1-k} \log N_I(2^{-j}) \\ &= 2^{r+1}(2^{-j} \log N_I(2^{-j})) \\ &\leqslant 2^r\varepsilon^2/300 \quad \text{by (5.2)}. \end{aligned}$$

Thus

$$\begin{aligned} Q_n &\leqslant \exp(-2^{k+r}\varepsilon^2/1800) \\ &\leqslant \exp(-2^r\varepsilon^2/1800) < \varepsilon/4 \end{aligned}$$

by (5.3) and choice of r.

If $V_n < \varepsilon/8$ then by (5.7), $|\nu_n(C) - \nu_n(A_{k, s(k, C)}| \leqslant \varepsilon/4$ for all $C \in \mathcal{C}$. Then by (5.8), (5.10) and (5.11),

$$\Pr^*\{\sup_{C \in \mathcal{C}} |\nu_n(C) - \nu_n(A_{0s(0, C)})| > \varepsilon/2\} < \varepsilon/2.$$

We also have

$$\begin{aligned} P_0 &:= \Pr\{\sup\{|\nu_n(A_{0i}) - \nu_n(A_{0j})| : P(A_{0i}\Delta A_{0j}) < 3\delta_0\} > \varepsilon/4\} \\ &\leqslant m(0)^2 \exp\left(-\varepsilon^2 16^{-1}/\left(6\delta_0 + \varepsilon 4^{-1} n^{-\frac{1}{2}}\right)\right). \end{aligned}$$

For $n \geqslant n_0$, as in (5.6), $n^{-\frac{1}{2}}\varepsilon/4 \leqslant 4\delta_0$, so

$$P_0 \leqslant m(0)^2 \exp(-\varepsilon^2/(160\delta_0)).$$

Now by (5.2) and choice of δ_0, we have $m(0)^2 \leqslant \exp(2\varepsilon^2/(600\delta_0))$, so

$$P_0 \leqslant \exp(-\varepsilon^2/(250\delta_0)) < \varepsilon/4$$

by (5.3) since $\delta_0 \leqslant \alpha$. Thus, for $n \geqslant n_0$,

$$\Pr^*\{\sup\{|\nu_n(C) - \nu_n(D)| : C, D \in \mathcal{C}, P(C\Delta D) < \delta_0\} > \varepsilon\} < \varepsilon,$$

proving (1.2b) in this case. □

NOTE. If $\mathcal{C}$ is, e.g., the collection of all half-planes in $\mathbb{R}^2$, the sets A_{kj} cannot be chosen in $\mathcal{C}$. In this case, for suitable P, the A_{kj} can be chosen as intersections of two half-planes.

We recall some definitions from Dudley (1974). Given a continuous function f from the sphere $S^{k-1} := \{x \in \mathbb{R}^k : |x| = 1\}$ into $\mathbb{R}^k$, let $I(f)$ denote the open set of all $y \in \mathbb{R}^k \setminus \text{range}\,(f)$ such that in $\mathbb{R}^k \setminus \{y\}$, f is not homotopic to a constant map. For $\alpha > 0$ and $M > 0$ let $G(k, \alpha, M)$ be the set of all f which with all their partial derivatives of orders $\leqslant \alpha$ are bounded in norm by M (as in Dudley (1974), page 229).

Let $J(f) := I(f) \cup \text{range}\ (f)$. Let $I(k, \alpha, M) := \{I(f) : f \in G(k, \alpha, M)\}$, $J(k, \alpha, M) := \{J(f) : f \in G(k, \alpha, M)\}$. (To correct equation 3.2 in Dudley (1974), put $J(k, \alpha, M)$ in place of $I(k, \alpha, M)$.)

If $|f(\theta) - g(\theta)| < \varepsilon$ for all $\theta \in S^{k-1}$, then ${}_\varepsilon I(f) \subset I(g) \subset J(g) \subset J(f)^\varepsilon$ by Lemma 2.1 of Dudley (1974). Also, $J(f)^\varepsilon \setminus {}_\varepsilon I(f) \subset (\text{range}\, f)^\varepsilon$.

I understand from R. Pyke that Sun (1976) has found a theorem along the following lines. At this writing I have not seen his precise statements or proofs.

(5.12). THEOREM (Sun). *For any law P on $\mathbb{R}^k$, $k \geqslant 2$, which has a bounded density with respect to Lebesgue measure λ, any $M < +\infty$ and $\alpha > k - 1$, $I(k, \alpha, M)$ and $J(k, \alpha, M)$ are Donsker classes for P.*

PROOF. Since $\alpha > k - 1 \geqslant 1$, the proof of Theorem 3.1 of Dudley (1974) shows that for some $N = N(k, \alpha, M) < +\infty$, $\lambda(J(f)^\varepsilon \setminus {}_\varepsilon I(f)) \leqslant N\varepsilon$ for all $f \in G(k, \alpha, M)$. Let $T := N\ \text{ess.sup}(dP/d\lambda)$.

If $f_1, \cdots, f_r \in G(k, \alpha, M)$ are such that for each $f \in G(k, \alpha, M)$, $\sup\{|f - f_j|(\theta) : \theta \in S^{k-1}\} \leqslant \varepsilon/T$ for some j, let $A_j := {}_{\varepsilon/T}I(f_j)$, $A_{j+r} := J(f_j)^{\varepsilon/T}$, $j = 1, \cdots, r$, to obtain $N_I(\varepsilon, \mathcal{C}, P) \leqslant 2r$, for $\mathcal{C} = I(k, \alpha, M)$ or $J(k, \alpha, M)$.

Thus by Theorem 3 of Clements (1963) as in the proof of Theorem 3.1 of Dudley (1974) the hypothesis on N_I in Theorem 5.1 above holds.

Let d be the usual Euclidean distance on $\mathbb{R}^k$. For the distance $s(f, g) := \sup_\theta d(f(\theta), g(\theta))$, $G(k, \alpha, M)$ is a compact set of functions, for any $\alpha > 0$, by the Arzelà-Ascoli theorem. The map $g \to I(g)$ is continuous from $G(k, \alpha, M)$ onto $I(k, \alpha, M)$ with the metric $h_{(d)}$ as defined after 4.3 above, since if $s(g_m, g) \to 0$ and $\varepsilon > 0$, then for m large, $I(g_m) \supset {}_\varepsilon I(g)$ and ${}_\varepsilon I(g_m) \subset I(g)$ as in Dudley ((1974), Lemma 2.1, proof of Theorem 3.1). (Note however that $g \to I(g)$ is not continuous for the Hausdorff metric h_d.)

So, $I(k, \alpha, M)$ is compact for $h_{(d)}$ and hence absolutely Borel. Note that all sets in $I(k, \alpha, M)$ are included in a fixed compact set, so that there is no need to use a totally bounded metric on all of $\mathbb{R}^k$ to obtain the Effros Borel structure. Then by

Propositions 4.4 and 3.2 above, $I(k, \alpha, M)$ is strongly P-EM. Then by 5.1 it is a Donsker class.

Likewise, $g \to J(g)$ is continuous from $G(k, \alpha, M)$ onto $J(k, \alpha, M)$ with the Hausdorff metric h_d, so $J(k, \alpha, M)$ is compact and absolutely Borel. By Propositions 4.3 and 3.2, $J(k, \alpha, M)$ is strongly P-EM and by 5.1 a Donsker class. □

In the unit cube of $\mathbb{R}^k$ let $R(k, \alpha, M)$ be the class of sets defined by Révész (1976), of the form

$$\{x : f_i(\{x_j\}_{j \neq i}) < x_i < g_i(\{x_j\}_{j \neq i}), i = 1, \cdots, k\},$$

where f_i and g_i, each defined on the unit cube of $\mathbb{R}^{k-1}$, have partial derivatives of all orders $\leqslant \alpha$ bounded by M in absolute value. Aside from differences in the values of M, the classes $\bigcup_M R(k, \alpha, M) := R(k, \alpha)$ and $\bigcup_M I(k, \alpha, M) := I(k, \alpha)$ are different: for $k = 2$, $I(2, 1)$ contains a set bounded by a "figure 8" which is not in $R(2, 1)$. I do not know whether $R(k, \alpha) \subset I(k, \alpha)$. For P with bounded density, one can show as in (5.2) that $R(k, \alpha, M)$ is a Donsker class if $\alpha > k - 1$.

For any set $U \subset \mathbb{R}^k$ let $cnv(U)$ denote the class of all open convex subsets of U. The following result, in the case of the uniform Lebesgue probability on the square I^2, is due to Bolthausen (1976).

(5.13). THEOREM (Bolthausen). *For any law P on $\mathbb{R}^2$ having a bounded density with respect to Lebesgue measure, and any bounded convex open U, $cnv(U)$ is a Donsker class for P.*

PROOF. We apply Theorem 4.1 of Dudley (1974) and its proof, and Theorem 5.1 above and its proof, as in Theorem 5.12. To see that $cnv(U)$ is P-EM we apply 4.4 as in the proof of Theorem 5.12, noting that $cnv(U)$ is compact for $h_{(e)}$. Thus 5.13 is proved.

In $\mathbb{R}^3$, a collection $cnv(U)$ is not $G_P BUC$ (Dudley (1973), page 87, Remark), and a fortiori not Donsker.

6. The two-sample case. We will call $\mathcal{C}$ *2-sample* P-EM iff for any independent empirical measures P_m and Q_n for P, $P_m - Q_n$ is completion measurable into $(D_0(\mathcal{C}, P), \mathcal{B}_b)$. Note that if $\mathcal{C}$ is strongly P-EM, it is 2-sample P-EM (let $b_1 = \cdots = b_m = 1/m, b_{m+1} = \cdots = b_{m+n} = -1/n$). Thus Propositions 3.2, 4.3 and 4.4 give conditions which imply that Suslin classes of closed or open sets are 2-sample P-EM.

(6.1). THEOREM. *Let $(X, \mathcal{A}, P)$ be a probability space and $\mathcal{C}$ a Donsker class, $\mathcal{C} \subset \mathcal{A}$, with $\mathcal{C}$ 2-sample P-EM. Let $\{P_m\}$ and $\{Q_n\}$ be independent empirical measures for P. Then as m and $n \to \infty$,*

$$\mathcal{L}\left((mn)^{\frac{1}{2}}(m+n)^{-\frac{1}{2}}(P_m - Q_n)\right) \to \mathcal{L}(G_P) \quad \text{in} \quad D_0(\mathcal{C}, P).$$

PROOF. Since $\mathcal{C}$ is a Donsker class, it is $G_P BUC$, and $\mathcal{L}(G_P)(T) = 1$ for some separable set $T \subset D_0(\mathcal{C}, P)$. According to Wichura ((1970), Theorem 2), there is a probability space $(\Omega, \mathcal{S}, \mu)$ with random variables A_m and B_n such that the sequence $\{A_m\}$ is independent of $\{B_n\}$, A_m and B_n are measurable into $(D_0(\mathcal{C}, P), \mathfrak{B}_b)$, $\mathcal{L}(A_n) = \mathcal{L}(B_n) = \mathcal{L}(n^{\frac{1}{2}}(P_n - P))$ for all n, $A_m \to H$ a.s. as $m \to \infty$ and $B_n \to J$ a.s. as $n \to \infty$ in $(D_0(\mathcal{C}, P), \|\cdot\|_\infty)$ where $\mathcal{L}(H) = \mathcal{L}(J) = \mathcal{L}(G_P)$, and H and J are independent.

As noted in the discussion of almost uniform convergence in the introduction above, since $\mu(H \in T) = 1$, $\sup_{C \in \mathcal{C}} |A_m - H|(C)$ is actually a measurable random variable, converging to 0 a.s. as $m \to \infty$, and likewise for $|B_n - J|$. Let

$$D_{mn} := (m+n)^{-\frac{1}{2}}\left(n^{\frac{1}{2}}A_m - m^{\frac{1}{2}}B_n\right).$$

Then

$$\mathcal{L}\left((mn)^{\frac{1}{2}}(m+n)^{-\frac{1}{2}}(P_m - Q_n)\right)$$
$$= \mathcal{L}\left((mn)^{\frac{1}{2}}(m+n)^{-\frac{1}{2}}\left(m^{-\frac{1}{2}}A_m + P - n^{-\frac{1}{2}}B_n - P\right)\right) = \mathcal{L}(D_{mn}).$$

Let

$$E_{mn} := (m+n)^{-\frac{1}{2}}\left(n^{\frac{1}{2}}H - m^{\frac{1}{2}}J\right), \quad \text{and}$$
$$F_{mn} := (m+n)^{-\frac{1}{2}}\left(n^{\frac{1}{2}}(A_m - H) - m^{\frac{1}{2}}(B_n - J)\right).$$

Then $D_{mn} = E_{mn} + F_{mn}$, $\mathcal{L}(E_{mn}) = \mathcal{L}(G_P)$ for all m and n, and $F_{mn} \to 0$, uniformly on $\mathcal{C}$ almost surely, and almost uniformly in $l^\infty(\mathcal{C})$, as m and $n \to \infty$. (I do not claim that F_{mn} is measurable.) Now Theorem 6.1 will be a consequence of the following, letting $\alpha = \langle m, n\rangle$, $Y_\alpha = D_{mn}$, and $Z_\alpha = E_{mn}$.

(6.2). LEMMA. *Let (S, d) be any metric space and Y_α, Z_α nets of random variables, measurable into $(S, \mathfrak{B}_b)$, with $d(Y_\alpha, Z_\alpha) \to 0$ almost uniformly (we do not assume $d(Y_\alpha, Z_\alpha)$ is measurable) and $\mathcal{L}(Z_\alpha) \to \mu$ as $\alpha \to \infty$, where $\mu(T) = 1$ for some separable $T \subset S$. Then $\mathcal{L}(Y_\alpha) \to \mu$.*

PROOF. Let f be continuous and $\mathfrak{B}_b$-measurable on S with $\sup|f| \leqslant 1$. Given $\varepsilon > 0$, for each $x \in T$ take $\delta_x := \delta(x) > 0$ such that whenever $d(x, u) < 3\delta_x$, we have $|f(x) - f(u)| < \varepsilon$. By Lindelöf's theorem, the open cover $\{B(x, \delta_x)\}_{x \in T}$ of T has a countable subcover. So there is a finite set $K \subset T$ such that $\mu(U(K)) > 1 - \varepsilon$ where $U(K) := \bigcup_{x \in K} B(x, \delta_x)$. Let $g(u) := \max(0, \min(1, 2 - \min_{x \in K} d(x, u)/\delta_x))$. Then $1_{U(K)} \leqslant g \leqslant 1_{V(K)}$ where $V(K) := \bigcup_{x \in K} B(x, 2\delta_x)$, and g is continuous and $\mathfrak{B}_b$ measurable. Thus $\lim_\alpha Eg(Z_\alpha) = \int g\, d\mu > 1 - \varepsilon$, so for some β,

$$\Pr(Z_\alpha \in V(K)) \geqslant Eg(Z_\alpha) > 1 - \varepsilon$$

for $\alpha \geqslant \beta$. Let $\delta := \min_{x \in K} \delta_x$. For some $\gamma \geqslant \beta$, we have $|Ef(Z_\alpha) - \int f\, d\mu| < \varepsilon$ and $\Pr^*(d(Y_\alpha, Z_\alpha) \geqslant \delta) < \varepsilon$ for $\alpha \geqslant \gamma$. Then, except on some event W with $\Pr(W) < 2\varepsilon$, we have $Z_\alpha \in V(K)$ and $d(Y_\alpha, Z_\alpha) < \delta$, so that for some $x \in K$, $d(Z_\alpha, x) <$

$2\delta_x$, $d(Y_\alpha, x) < 3\delta_x$, and $|f(Y_\alpha) - f(Z_\alpha)| < 2\varepsilon$. Thus $|Ef(Y_\alpha) - Ef(Z_\alpha)| < 6\varepsilon$, so $|Ef(Y_\alpha) - \int f\, d\mu| < 7\varepsilon$. $\square$

If $X = \mathbb{R}^1$ and $\mathcal{C}$ is the class of intervals, $\sup_{A \in \mathcal{C}} G_P(A)$ has the same law for all continuous P. Likewise, $\mathcal{L}(\sup_{A \in \mathcal{C}} |G_P(A)|)$ is the same for all P without atoms. Then, results like Theorem 6.1 apply to testing whether two unknown continuous distribution functions, from which finite samples have been taken, are equal. If $X = \mathbb{R}^k$ for $k > 1$, such laws depend on P, but perhaps they are the same for more restricted classes of laws P.

Limit theorems in the literature for the two-sample case have often been stated under restrictive conditions such as m/n converging to a positive constant. Theorem 6.1 shows that no such restriction is necessary. The proof will give rates of convergence in the two-sample case if one has them (in a suitable form) in the one-sample case.

7. Universal Donsker classes and Vapnik-Červonenkis classes. Given a set X, a collection $\mathcal{C}$ of subsets of X will be called a *universal Donsker class* (UDC) iff it is a Donsker class for every probability measure on the σ-algebra generated by $\mathcal{C}$.

EXAMPLES. If $X = \mathbb{R}^k$, the class of all rectangles $\Pi_{1 \leqslant j \leqslant k}]a_j, b_j]$ is a UDC (Donsker (1952), for $k = 1$; Dudley (1966), for $k > 1$).

DEFINITIONS. Given a class $\mathcal{C}$ of subsets of a set X and a finite set $F \subset X$, let $\Delta^{\mathcal{C}}(F)$ be the number of different sets $C \cap F$ for $C \in \mathcal{C}$. For $n = 1, 2, \cdots$, let $m^{\mathcal{C}}(n) := \max\{\Delta^{\mathcal{C}}(F) : F \text{ has } n \text{ elements}\}$. Let

$$V(\mathcal{C}) := \inf\{n : m^{\mathcal{C}}(n) < 2^n\}$$
$$= +\infty \quad \text{if} \quad m^{\mathcal{C}}(n) = 2^n \ \forall n.$$

Vapnik and Červonenkis (1971) introduced $\Delta^{\mathcal{C}}$, $m^{\mathcal{C}}$ and $V(\mathcal{C})$. If $m^{\mathcal{C}}(n) < 2^n$ for some n, i.e., if $V(\mathcal{C}) < +\infty$, we will call $\mathcal{C}$ a *Vapnik-Červonenkis class* (VCC).

Here is the main result of this section:

(7.1). THEOREM. *If $\mathcal{C}$ is a* VCC *and for some σ-algebras $\mathcal{Q} \supset \mathcal{C}$ in X and $\mathcal{S}$ in $\mathcal{C}$, $(X, \mathcal{Q}; \mathcal{C}, \mathcal{S})$ is $P\epsilon$-Suslin, then $\mathcal{C}$ is a Donsker class for P.*

Before proving Theorem 7.1 we will go through some other facts. First, recalling Proposition 4.5, here is one way to generate VCC's. It is related to results in Cover (1965).

(7.2). THEOREM. *Let G be any m-dimensional real vector space of real functions on a set X. Then $V(\mathrm{pos}(G)) = m + 1$ (or, if $\mathrm{card}(X) = m$, $\mathrm{pos}(G)$ contains all subsets of X).*

PROOF. First, suppose $A \subset X$ and $\mathrm{card}(A) = m + 1$. Take the map $r : G \to \mathbb{R}^A$ which restricts functions in G to the set A. Then r, as a linear map of an m-dimensional real vector space into one of higher dimension, cannot be onto. For the usual inner product $(\cdot\ ,\ \cdot)$ on $\mathbb{R}^A$, let $v \neq 0$ be a vector orthogonal to $r(G)$. We may assume $A_+ := \{x \in A : v(x) > 0\}$ is nonempty, since otherwise we can take $-v$ (thanks to M. Artin for this remark). If $A_+ \in \text{pos}(G)$, take $f \in G$ with $\{f > 0\} = A_+$. Then $(r(f), v) > 0$, a contradiction. So $A_+ \notin \text{pos}(G)$ and $V(\text{pos}(G)) \leqslant m + 1$.

On the other hand, $\dim(G) = m$ implies that for some subset A of X with $\text{card}(A) = m$, $r(G) = \mathbb{R}^A$, so all subsets of A are of the form $B \cap A$, $B \in \text{pos}(G)$, and $m < V(\text{pos}(G)) = m + 1$. □

One example of a finite-dimensional G on $X = \mathbb{R}^k$ is the collection of all polynomials of degree $\leqslant d$ for any fixed $d < \infty$.

Vapnik and Červonenkis (1971) proved inequalities relating $m^{\mathcal{C}}(n)$ to its values for the special case of half-spaces of $\mathbb{R}^k$, as follows. Let

$$(7.3) \qquad {}_N C_{\leqslant k} := \Sigma_{j=0}^k \binom{N}{j}, \qquad \text{where} \quad \binom{N}{j} := 0 \quad \text{for} \quad j > N.$$

Recalling the notation ${}_N C_k := \binom{N}{k}$ for the number of k-element subsets of an N-element set, ${}_N C_{\leqslant k}$ is the number of subsets with *at most* k elements.

By *r-flat* I will mean a linear variety of $\mathbb{R}^k$ of dimension r, i.e., a set of the form $\{x \in \mathbb{R}^k : A(x - v) = 0\}$ where $v \in \mathbb{R}$ and A is a $k \times k$ matrix of rank $k - r$. A $(k-1)$-flat in $\mathbb{R}^k$ will be called a *hyperplane*.

Let $\mathcal{H}$ be the collection of all half-spaces $\{x : (v, x) > c\}$ for $x, v \in \mathbb{R}^k$, $v \neq 0$, and $c \in \mathbb{R}$. Let $\mathcal{H}(0)$ be the subcollection of half-spaces bounded by hyperplanes through 0 (i.e., with $c = 0$). Let $\eta_k(N)$ be the maximum number of open regions into which $\mathbb{R}^k$ is decomposed by N hyperplanes $H_1, \cdots, H_N$ of H_N. Then the maximum is attained for $H_1, \cdots, H_N$ in "general position," i.e., if

$$H_j = \{x : (x, v_j) = c_j\}, \qquad j = 1, \cdots, N,$$

then any k or fewer of the v_j are linearly independent. Schläfli (1901) shows that

$$(7.4) \qquad \eta_k(N) = {}_N C_{\leqslant k}.$$

Steiner (1826) had proved this for $k \leqslant 3$. If F is a set of N points in $\mathbb{R}^k$, then

$$(7.5) \qquad \Delta^{\mathcal{H}}(F) \leqslant 2\Sigma_{r=0}^k \binom{N-1}{r} = 2_{N-1}C_{\leqslant k},$$

and equality is attained if the points of F are in general position, i.e., no $k + 1$ of them are in any hyperplane (Cover (1965), page 330; Harding (1967); Watson (1969)). One also has

$$(7.6) \qquad \Delta^{\mathcal{H}(0)}(F) \leqslant 2_{N-1}C_{\leqslant k-1}$$

with equality iff every nonempty subset of F with at most k elements is linearly independent (Schläfli (1901), page 211; Cover (1965), noting varying definitions of "general position" on page 230; Harding (1967)). More generally, if for a fixed j-flat A, $\mathcal{H}(j)$ is the set of all half-spaces bounded by hyperplanes including A, then

$$(7.7) \qquad \Delta^{\mathcal{H}(j)}(F) \leqslant 2_{N-1}C_{\leqslant k-j-1}$$

(Harding (1967)).

Without using (7.4)–(7.7), but directly from the definition (7.3) and the recurrence relation

(7.8) $${}_{N}C_{\leqslant k} = {}_{N-1}C_{\leqslant k} + {}_{N-1}C_{\leqslant k-1},$$

Vapnik and Červonenkis ((1971), Lemma 1) prove:

(7.9). THEOREM (Vapnik-Červonenkis). *If X is any set, $\mathcal{C}$ any collection of subsets of X, and $V(\mathcal{C}) \leqslant v$, then $m^{\mathcal{C}}(n) < {}_{N}C_{\leqslant v}$ for all $n \geqslant v$.*

They note that ${}_{n}C_{\leqslant k} \leqslant n^k + 1$. (Their 1974 book, pages 214–219, shows that $m^{\mathcal{C}}(n) \leqslant {}_{n}C_{\leqslant V(\mathcal{C})-1}$. Note: in the 1971 paper and the 1974 book, pages 97 and 214, are three disagreeing definitions of "$\Phi(k, n)$.") They prove that for $n > k \geqslant 1$, ${}_{n}C_{\leqslant k} \leqslant 1.5n^k/k!$. Hence

(7.10) $$\text{for} \quad n > v := V(\mathcal{C}) \geqslant 1,$$
$$m^{\mathcal{C}}(n) \leqslant 1.5n^{v-1}/(v-1)! < n^v.$$

For $n < v$, $m^{\mathcal{C}}(n) = 2^n \leqslant 2^v \leqslant n^v$. If $v = 0$, $\mathcal{C}$ is empty. Thus (without using (7.10)) we have:

(7.11) For any collection $\mathcal{C}$ of sets, $m^{\mathcal{C}}(n) \leqslant n^{V(\mathcal{C})}$ for all $n \geqslant 2$, and $m^{\mathcal{C}}(n) \leqslant n^{V(\mathcal{C})} + 1$ for all $n \geqslant 0$.

Now for any sets $A_1, \cdots, A_m$, let $\mathcal{A}(A_1, \cdots, A_m)$ denote the algebra of subsets of X generated by $A_1, \cdots, A_m$.

(7.12). PROPOSITION. *For any* VCC *$\mathcal{C}$ and any $k < +\infty$,*

$$\mathcal{A}_k(\mathcal{C}) := \bigcup \{\mathcal{A}(A_1, \cdots, A_k) : A_1, \cdots, A_k \in \mathcal{C}\} \quad \textit{is a VCC.}$$

PROOF. By induction, we may assume $k = 2$. Let $\mathcal{D} := \{A \cap B : A, B \in \mathcal{C}\}$. Then $m^{\mathcal{D}}(n) \leqslant m^{\mathcal{C}}(n)^2 \leqslant (n^{V(\mathcal{C})} + 1)^2 < 2^n$ for n large, so $\mathcal{D}$ is a VCC.

We may assume $\phi \in \mathcal{C}$ and $X \in \mathcal{C}$. If $\mathcal{S} := \{A \setminus B : A, B \in \mathcal{C}\}$ then $\mathcal{S}$ is a VCC as above. A finite union of VCC's is likewise a VCC. Now every set in $\mathcal{A}(A, B)$ is a union of some of the four atoms $A \cap B$, $A \setminus B$, $B \setminus A$, and $(X \setminus A) \setminus B$. Unions of at most four sets can be treated also as above, completing the proof.

(7.13). LEMMA. *If $(X, \mathcal{A}, P)$ is a probability space, $\mathcal{C} \subset \mathcal{A}$, $\mathcal{C}$ is a* VCC *and $v := V(\mathcal{C})$, there is a constant $K = K(v)$ (not depending on P) such that for $0 < \varepsilon \leqslant \frac{1}{2}$,*

$$N(\varepsilon, \mathcal{C}, P) \leqslant K\varepsilon^{-v}|\ln \varepsilon|^v.$$

PROOF. Suppose $A_1, \cdots, A_m \in \mathcal{C}$, and $P(A_i \Delta A_j) \geqslant \varepsilon$ for $i \neq j$. We may assume $m \geqslant 2$. If $n \geqslant 2$ is so large that $m(m-1)(1-\varepsilon)^n < 2$, then $\Pr\{P_n(A_i \Delta A_j) > 0$ for all $i \neq j\} > 0$. In that case, $m \leqslant m^{\mathcal{C}}(n) \leqslant n^v$ by (7.11). If we take the smallest n for which $m^2(1-\varepsilon)^n < 2$, then $m^2(1-\varepsilon)^{n-1} \geqslant 2$ so $n - 1 \leqslant (2 \ln m - \ln 2)/|\ln(1-\varepsilon)|$, $n \leqslant (2 \ln m)/\varepsilon$, and $m \leqslant (2 \ln m)^v \varepsilon^{-v}$.

For some $m_0 = m_0(v) < +\infty$, $(2 \ln m)^v \leqslant m^{1/(v+1)}$ for $m \geqslant m_0$, and then $m \leqslant \varepsilon^{-v-1}$, so $\ln m \leqslant (v+1)|\ln \varepsilon|$. Hence

$$m \leqslant K(v)\varepsilon^{-v}|\ln \varepsilon|^v \quad \text{for} \quad 0 < \varepsilon \leqslant \tfrac{1}{2}$$

if $K(v) = \max(m_0, 2^{v+1}(v+1)^v)$. Thus, choosing at most m points $\geqslant \varepsilon$ apart, the balls of radius ε with these centers cover $\mathcal{C}$, proving Lemma 7.13.

Now to prove Theorem 7.1, given a probability space $(X, \mathcal{A}, P)$ and VCC $\mathcal{C} \subset \mathcal{A}$, with $(X, \mathcal{A}; \mathcal{C}, \mathcal{S})$ $P\epsilon$-Suslin for some $\mathcal{S}$, $\mathcal{C}$ is totally bounded for d_P by Lemma 7.13, and P-EM by Proposition 3.2, so it will suffice to verify 1.2b.

For any $\delta > 0$, let $\mathcal{C}(\delta) := \{A \setminus B : A, B \in \mathcal{C}, P(A \setminus B) \leqslant \delta\}$, $\mathcal{D}(\delta) := \{\langle A, B\rangle \in \mathcal{C} \times \mathcal{C} : A \setminus B \in \mathcal{C}(\delta)\}$. Note $\mathcal{C}(\delta) \subset \mathcal{C}(1)$.

Let $d_P^{(2)}$ be the product pseudometric on $\mathcal{C} \times \mathcal{C}$, $d_P^{(2)}(\langle A, B\rangle, \langle C, D\rangle) := d_P(A, C) + d_P(B, D)$. Then $\mathcal{D}(\delta)$ is closed in $\mathcal{C} \times \mathcal{C}$ for $d_P^{(2)}$. By (7.13), $(\mathcal{C}, d_P)$ is separable, so $\mathcal{D}(\delta)$ is measurable for the product σ-algebra $\mathcal{S} \times \mathcal{S}$.

We can define $\nu_n(\cdot)(\omega)$ for ω in a product X^∞ of copies of $(X, \mathcal{A}, P)$,

$$\omega = \langle x_1, x_2, \cdots, \rangle, \qquad x(j)(\omega) := x_j.$$

For any $r > 0$, let

$$E_{\delta r} := \{\langle \omega, A, B\rangle : |\nu_n(A \setminus B)(\omega)| > r, \langle A, B\rangle \in \mathcal{D}(\delta)\}.$$

Then as in the proof of Proposition 3.2, $E_{\delta r}$ is jointly measurable in the Suslin measurable space $(X^\infty \times \mathcal{C} \times \mathcal{C}, \mathcal{A}^\infty \times \mathcal{S} \times \mathcal{S})$. The projection of $E_{\delta r}$ into X^∞ is measurable for the completion of P^∞. Thus $\sup_{C \in \mathcal{C}(\delta)}|\nu_n(C)|$ is a measurable random variable. By 1.2b, it will suffice to prove that for any $\varepsilon > 0$ there is a $\delta > 0$ and an n_0 such that

$$\Pr\{\sup_{C \in \mathcal{C}(\delta)}|\nu_n(C)| > \varepsilon\} < \varepsilon \qquad \text{for} \quad n \geqslant n_0.$$

By Proposition 7.12, $\mathcal{C}(1)$ is a VCC. By Lemma 7.13, take a w with $V(\mathcal{C}(1)) := v < w < +\infty$ and $N < \infty$ such that $N(\gamma/2, \mathcal{C}(1), P) \leqslant N\gamma^{-w}$ for $0 < \gamma < 1$. Given a δ, $0 < \delta \leqslant 1$, to be chosen later, for each $j = 0, 1, 2, \cdots,$ take sets $A_{j1}, \cdots, A_{jm(j)} \in \mathcal{C}(\delta)$ such that for each $A \in \mathcal{C}(\delta)$, $P(A \Delta A_{ji}) \leqslant \delta/2^j$ for some i. Then we can take $m(j) \leqslant N2^{jw}/\delta^w$, $m(0) = 1$, and $A_{01} = \phi$.

For each integer $j \geqslant 1$ and $1 \leqslant i \leqslant m(j)$, take a $k = k(j, i)$ such that

$$P(A_{ji}\, \Delta A_{j-1,k}) \leqslant 2\delta/2^j.$$

Let $\mathcal{C}_j$ be the collection of all sets $A \setminus B$ or $B \setminus A$, $A = A_{ji}$, $B = A_{j-1,k(j,i)}$. There are at most $2m(j)$ such sets, all with probability $\leqslant 2\delta/2^j$. Take $0 < \varepsilon < 1$. Then $P_j := \Pr\{|\nu_n(A)| \geqslant \varepsilon/j^2$ for some $A \in \mathcal{C}_j\} \leqslant 2m(j)\sup\{\Pr\{|\nu_n(A)| \geqslant \varepsilon/j^2 : P(A) \leqslant 2\delta/2^j\}\}$. Given $p \leqslant 2\delta/2^j$, we have by Bernstein's inequality (Bennett (1962), or Hoeffding (1963))

$$E_{jnp} := E\big(np + \varepsilon n^{\frac{1}{2}}/j^2, n, p\big) \leqslant \exp\Big(-\varepsilon^2/\big(2j^4pq + j^2\varepsilon n^{-\frac{1}{2}}\big)\Big)$$

$$\leqslant \exp\Big(-\varepsilon^2 n^{\frac{1}{2}}/\big(4j^4 n^{\frac{1}{2}}\delta 2^{-j} + j^2\varepsilon\big)\Big).$$

We will treat this by cases according to which term in the denominator is larger.

In case $4j^2n^{\frac{1}{2}}\delta > 2^j\varepsilon$, we say $j \in J(n, \varepsilon, \delta)$. Then $E_{jnp} \leqslant \exp(-\varepsilon^2 2^j/8j^4\delta)$. If $j \notin J(n, \varepsilon, \delta)$, then $E_{jnp} \leqslant \exp(-\varepsilon n^{\frac{1}{2}}/2j^2)$. Setting $B_{jnp} := B(np - \varepsilon n^{\frac{1}{2}}/j^2, n, p)$ we have the same upper bounds for B_{jnp} just shown for E_{jnp}. Let $j(n) := [n^{\frac{1}{8}}]$. Then

$$\Sigma_{1 \leqslant j \leqslant j(n)} P_j \leqslant S_1 + S_2 \qquad \text{where}$$

$$S_1 := 2\Sigma_{j \leqslant j(n), j \in J(n, \varepsilon, \delta)} N 2^{jw}\delta^{-w} \exp(-\varepsilon^2 2^j/8j^4\delta),$$

$$S_2 := 2\Sigma_{j \leqslant j(n), j \notin J(n, \varepsilon, \delta)} N 2^{jw}\delta^{-w} \exp(-\varepsilon n^{\frac{1}{2}}/2j^2).$$

Now S_1 is a partial sum of a convergent infinite series, whose value approaches 0 as $\delta \to 0$ for fixed ε and w. Thus for some $\delta_1(\varepsilon) > 0$, $S_1 < \varepsilon/3$ for $0 < \delta \leqslant \delta_1(\varepsilon)$. *We now choose* $\delta = \delta_1(\varepsilon)$. Next, $S_2 \leqslant 2n^{\frac{1}{8}}N2^{j(n)w}\delta^{-w} \exp(-\varepsilon n^{\frac{1}{4}}/2) \to 0$ as $n \to \infty$, so for some $n_0(\delta)$, $S_2 \leqslant \varepsilon/3$ for $n \geqslant n_0(\delta)$.

Let $D_{ni} := A_{j(n), i}$. Since $A_{01} = \phi$, we have

$$|\nu_n(D_{ni})| \leqslant \Sigma_{r=1}^{j(n)} |\nu_n(C_r)| + |\nu_n(D_r)|$$

for some C_r and $D_r \in \mathcal{C}_r$. Then since $\Sigma_{j \geqslant 1} j^{-2} = \pi^2/6 < 2$, we have

$$\text{(7.14)} \qquad \Pr\{\max_i |\nu_n(D_{ni})| \geqslant 4\varepsilon\} < 2\varepsilon/3, \qquad n \geqslant n_0(\delta).$$

Let

$$\mathfrak{D}_{n\delta i} := \{\langle A, B\rangle \in \mathfrak{D}(\delta) : P((A \setminus B)\,\Delta D_{ni}) \leqslant \delta/2^{j(n)}\}.$$

The functions $\langle A, B\rangle \to P((A \setminus B) \setminus D_{ni})$, $\langle A, B\rangle \to P(D_{ni} \setminus (A \setminus B))$, and hence $\langle A, B\rangle \to P((A \setminus B)\,\Delta D_{ni})$, are all continuous on the separable pseudometric space $\mathcal{C} \times \mathcal{C}$ for $d_P^{(2)}$, thus $\mathcal{S} \times \mathcal{S}$ measurable, so $\mathfrak{D}_{n\delta i} \in \mathcal{S} \times \mathcal{S}$. For each j and i, the map

$$\langle \omega, A, B\rangle \to \delta_{x(j)(\omega)}((A \setminus B) \setminus D_{ni})$$

is measurable ($\mathcal{Q}^\infty \times \mathcal{S} \times \mathcal{S}$) by the ϵ-Suslin assumption. Thus the function

$$\langle \omega, A, B\rangle \to \nu_n(\omega)((A \setminus B) \setminus D_{ni})$$

is jointly measurable. Likewise, so is

$$\langle \omega, A, B\rangle \to \nu_n(\omega)(D_{ni} \setminus (A \setminus B)).$$

As in the proof of Proposition 3.2, since $\mathfrak{D}_{n\delta i} \in \mathcal{S} \times \mathcal{S}$,

$$\omega \to \sup\{\nu_n(\omega)(D_{ni} \setminus (A \setminus B)) : \langle A, B\rangle \in \mathfrak{D}_{n\delta i}\}$$

is completion measurable for P^∞, and likewise if sup is replaced by inf or $\sup\{|\cdots| : \cdots\}$, or $D_{ni} \setminus (A \setminus B)$ by $(A \setminus B) \setminus D_{ni}$. For each $E \in \mathcal{C}(\delta)$ there is some i such that $d_P(E, D_{ni}) \leqslant \delta/2^{j(n)}$. Let $\mathfrak{B}_{n\delta} := \mathfrak{B}(n, \delta)$ be the collection of all sets $F = E \setminus D_{ni}$ or $F = D_{ni} \setminus E$ where $E \in \mathcal{C}(\delta)$ and $P(F) \leqslant \delta/2^{j(n)}$. Now $\omega \to \sup\{|\nu_n(F)| : F \in \mathfrak{B}_{n\delta}\}$ is P^∞-completion measurable. Let

$$Q_{n\varepsilon} := \Pr\{\sup\{|\nu_n(B)| : B \in \mathfrak{B}_{n\delta}\} \geqslant \varepsilon\}.$$

Then by (7.14),

(7.15) $$\Pr\{\sup_{A\in\mathcal{C}(\delta)}|\nu_n(A)| > 6\varepsilon\} \leqslant 2\varepsilon/3 + Q_{n\varepsilon}.$$

Let

$$C_{n\varepsilon\delta i} := \{\langle\omega, A, B\rangle \in X^\infty \times \mathfrak{D}_{n\delta i} : \nu_n((A \setminus B)\setminus D_{ni}) \geqslant \varepsilon\}.$$

Then $C_{n\varepsilon\delta i}$ is measurable ($\mathcal{Q}^\infty \times \mathcal{S} \times \mathcal{S}$). By Theorem 3.3, let $\omega \to \langle C_{i1}, C_{i2}\rangle(\omega)$ be a universally measurable selector for $C_{n\varepsilon\delta i}$, mapping a universally measurable subset of X^∞ into $\mathfrak{D}_{n\delta i}$. Let $C_i(\omega) := (C_{i1}(\omega)\setminus C_{i2}(\omega))\setminus D_{ni}$. Then $\omega \to P(C_i(\omega))$ is universally measurable, and so is $\omega \to \nu_n(C_i(\omega))$.

Uhlmann ((1966), Satz 6) and Jogdeo and Samuels (1968) showed that the median of a binomial distribution is within 1 of its mean. Thus for any fixed set $A \in \mathcal{Q}$, $\Pr(\nu_n(A) \leqslant n^{-\frac{1}{2}}) \geqslant \frac{1}{2}$. Let ν_n and ν_n' be two independent copies of the normalized empirical measure $n^{\frac{1}{2}}(P_n - P)$, with $\nu_n(\cdot)(\omega)$ and $\nu_n'(\cdot)(\omega')$ defined for $\langle\omega, \omega'\rangle$ in a product space $X^\infty \times X^\infty$ with product probability. We write "$\exists C_i(\omega)$" iff $C_i(\omega)$ is defined, i.e., iff $\exists\langle A, B\rangle : \langle\omega, A, B\rangle \in C_{n\varepsilon\delta i}$, so that $\nu_n(C_i(\omega)) \geqslant \varepsilon$. Then for each i, $\{\omega : \exists C_i(\omega)\}$ is a universally measurable event. If $\exists i\exists C_i(\omega)$, let $i(\omega)$ be the least such i, and write "$\exists i(\omega)$." Then since ν_n' is independent of ω, we have

(7.16) $$\Pr\{\exists i(\omega) \text{ and } \nu_n'(C_{i(\omega)}(\omega)) \leqslant n^{-\frac{1}{2}}\} \geqslant \Pr\{\exists i(\omega)\}/2.$$

Likewise, if we replace $(A \setminus B)\setminus D_{ni}$ by $D_{nu}\setminus(A\setminus B)$ in the definition of $C_{n\varepsilon\delta i}$, we get another measurable set $D_{n\varepsilon\delta u}$ with a selector $\langle D_{u1}, D_{u2}\rangle(\omega)$. Let $D_u(\omega) := D_{nu} \setminus (D_{u1}\setminus D_{u2})(\omega)$. Let $d(\omega)$ be the least u, if one exists, such that $\exists D_u(\omega)$. Then $\Pr\{\exists d(\omega) \text{ and } \nu_n'(D_{d(\omega)}(\omega)) \leqslant n^{-\frac{1}{2}}\} \geqslant \Pr\{\exists d(\omega)\}/2$.

Replacing "$\geqslant \varepsilon$" by "$\leqslant -\varepsilon$" in the definition of $C_{n\varepsilon\delta i}$, we get a measurable set $E_{n\varepsilon\delta i}$ with a selector $\langle E_{i1}, E_{i2}\rangle$, $E_i(\omega) := (E_{i1}\setminus E_{i2})\setminus D_{ni}$, and let $e(\omega) := \min\{i : \exists E_i(\omega)\}$. Doing the same for $D_{n\varepsilon\delta u}$ we get $\langle F_{u1}, F_{u2}\rangle$, $F_u(\omega)$, and $f(\omega) := \min\{u : \exists F_u(\omega)\}$. Then

(7.17) $$\Pr\{\exists e(\omega) \text{ and } \nu_n'(E_{e(\omega)}(\omega)) \geqslant -n^{-\frac{1}{2}}\} \geqslant \Pr\{\exists e(\omega)\}/2,$$

and likewise for $f(\omega)$.

Let $X(1), \cdots, X(2n)$ be independent and identically distributed with law P, $\nu_n := n^{\frac{1}{2}}(P_n - P)$, $\nu_n' = n^{\frac{1}{2}}(P_n' - P)$, where $P_n' := n^{-1}\Sigma_{n<j\leqslant 2n}\delta_{X(j)}$. As in Vapnik and Červonenkis (1971), conditional on $P_{2n} = P_n + P_n'$, the distribution of $\langle X(1), \cdots, X(2n)\rangle$ is obtained by averaging over all permutations of the integers $1, 2, \cdots, 2n$. For fixed P_{2n} and a set A with $r_n(A) := 2nP_{2n}(A)$,

$$\Pr\{nP_n(A) \geqslant s, nP_n'(A) \leqslant t | P_{2n}\}$$

is the hypergeometric probability of choosing at least s white balls and at most t black balls in a random sample of $r := r_n(A)$ balls without replacement from an urn containing n white and n black balls. For fixed s and t, this event is equivalent to drawing at least s white balls if $r \leqslant s + t$, or at most t black balls if $r \geqslant s + t$. Thus the probability is maximized when $r = s + t$, and then it is $H := H(t, r, n, 2n) := \Sigma_{j \leqslant t}\binom{n}{r-j}\binom{n}{j}/\binom{2n}{r}$.

On the event in (7.16), $s \geqslant np + \varepsilon n^{\frac{1}{2}}$ and $t \leqslant np + 1$ where $p := P(A)$, so $s - t \geqslant \varepsilon n^{\frac{1}{2}} - 1 > 2$ for n large enough. Hence, by an inequality of Uhlmann (1966), hypergeometric tails are smaller than corresponding binomial tails; specifically $H \leqslant E(s, n, (s + t)/2n)$. Then by Lemma 2.7 above, $H \leqslant ((s + t)/2s)^s \exp((s - t)/2)$. Now

$$s \cdot \ln((s + t)/2s) = s \cdot \ln(1 - (s - t)/2s) \leqslant -(s - t)/2 - (s - t)^2/8s,$$

so $H \leqslant \exp(-(s - t)^2/8s)$. The same inequality holds for $s \geqslant np - 1$ and $t \leqslant np - \varepsilon n^{\frac{1}{2}}$, as in (7.17), if n is large.

By (7.11), $\Delta^{\mathcal{C}}\{X(1), \cdots, X(2n)\} \leqslant (2n)^v$, $\Delta^{\mathcal{B}(n,\delta)}\{X(1), \cdots, X(2n)\} \leqslant 2(2n)^{2v}$, and using (7.16) and (7.17),

$$\begin{aligned} Q_{n\varepsilon} &= E(\Pr\{\sup\{|\nu_n(B)| : B \in \mathcal{B}_{n\delta}\} \geqslant \varepsilon | P_{2n}\}) \\ &\leqslant \sup(\Pr\{\sup\{|\nu_n(B)| : B \in \mathcal{B}_{n\delta}\} \geqslant \varepsilon | P_{2n}\}) \\ &\leqslant 8(2n)^{2v} \sup(H) \leqslant 8(2n)^{2v} \sup\big(\exp(-(s - t)^2/(8s))\big) \end{aligned}$$

where the supremum is over values of $p \leqslant \delta/2^{j(n)}$, $j(n) = [n^{\frac{1}{8}}]$, and either $s \geqslant np + \varepsilon n^{\frac{1}{2}}$ and $t \leqslant np + 1$, or $s \geqslant np - 1$ and $t \leqslant np - \varepsilon n^{\frac{1}{2}}$. The function $s \to (s - t)^2/s$ is increasing for $t < s$, so for $n > 2\varepsilon^{-2}$ we have

$$\begin{aligned} Q_{n\varepsilon} &\leqslant 8(2n)^{2v} \exp\Big(-\big(\varepsilon n^{\frac{1}{2}} - 1\big)^2/8\big(np + \varepsilon n^{\frac{1}{2}}\big)\Big) \\ &\leqslant 8(2n)^{2v} \exp\Big(-\varepsilon^2 n/32\big(n\delta/2^{j(n)} + \varepsilon n^{\frac{1}{2}}\big)\Big) \to 0 \end{aligned}$$

as $n \to \infty$ for fixed $\varepsilon > 0$, $\delta > 0$. So for some $n_1 = n_1(\delta, \varepsilon)$, $Q_{n\varepsilon} \leqslant \varepsilon/3$ for $n \geqslant n_1$. Hence by (7.15), for $\delta = \delta_1(\varepsilon)$ and $n \geqslant \max(n_0, n_1)(\delta, \varepsilon)$, $\Pr\{|\nu_n(A)| \geqslant 6\varepsilon$ for some $A \in \mathcal{C}(\delta)\} < \varepsilon$, proving Theorem 7.1.

(7.18). COROLLARY. *If X is a Polish space, $\mathcal{C}$ is a* VCC *in X, and $\mathcal{C}$ is a Suslin measurable collection of closed sets, or of open sets, for the Effros Borel structure, then $\mathcal{C}$ is a universal Donsker class.*

PROOF. We apply Theorem 7.1, Proposition 3.2, and 4.3 or 4.4. □

Note also that the VCC's of open sets given by Theorem 7.2 (and/or their closed complements) are Suslin by Proposition 4.5 if X is locally compact and separable, and G consists of continuous functions.

If P is Lebesgue measure on [0, 1] and $\mathcal{C} = \{\{x\} : x \in E\}$ where E is non-measurable, then $\mathcal{C}$ is a VCC but not P-EM. In this case, $\mathcal{C}$ can be enlarged to the set of all singletons, a VCC with good measurability properties. We may ask whether any VCC can be enlarged in such a way. The following shows that closure for the Hausdorff metric may lose the VCC property.

(7.19). PROPOSITION. *In* $[0, 1]$ *there is a* VCC $\mathcal{C}$, *consisting of finite sets, whose closure* $\bar{\mathcal{C}}$ *for the Hausdorff metric contains all (closed) sets in* $[0, 1]$.

PROOF. We enumerate the primes: $p_1 = 2, p_2 = 3, p_3 = 5, \ldots$. Let $\mathcal{C}$ consist of all finite sets $A = \{x_1, \cdots, x_k\}$ with $x_j \in [0, 1]$ such that for each $j = 1, \cdots, k-1$, $x_{j+1} = p_{2j-1}^{m(j)} x_j / p_{2j}^{n(j)}$ for some integers $m(j)$ and $n(j)$. Then for any set E with 3 elements, if $E \subset A \in \mathcal{C}$, then $E = \{x_t, x_u, x_v\}$ for some x_j as above. By unique factorization, whenever $x_t \in B \in \mathcal{C}$ and $x_v \in B$, then $x_u \in B$ so $\{x_t, x_v\} \neq B \cap E$. Hence $\mathcal{C}$ is a VCC.

For each j, the set of rationals of the form p_{2j-1}^m / p_{2j}^n is dense in $]0, \infty[$ (since their logarithms are dense in $\mathbb{R}$: for any irrational ξ, the multiples $m\xi$ are dense mod 1). Thus any finite set $\{y_1, \cdots, y_k\} \subset [0, 1]$ can be approximated as well as desired by some $\{x_1, \cdots, x_k\} \in \mathcal{C}$. $\square$

For classes $\mathcal{C}$ with $V(\mathcal{C}) = +\infty$, it may still happen that $\Delta^{\mathcal{C}}(X_1, \cdots, X_n) < 2^n$ with high probability. Then, even if $\mathcal{C}$ is not a P-Donsker class, one may use $\Delta^{\mathcal{C}}$ in proving that $\sup_{C \in \mathcal{C}} |P_n - P|(C) \to 0$, as in the original work of Vapnik and Červonenkis (1971). Steele (1977) has substantial further results along these lines.

Acknowledgment. Many thanks to a referee for finding an error in an earlier version of Theorem 5.1.

REFERENCES

AUMANN, R. J. (1961). Borel structures for function spaces. *Illinois J. Math.* **5** 614–630.

BENNETT, G. (1962). Probability inequalities for the sum of independent random variables. *J. Amer. Statist. Assoc.* **57** 33–45.

BILLINGSLEY, P. (1968). *Convergence of Probability Measures*. Wiley, New York.

BOLTHAUSEN, E. (1976). On weak convergence of an empirical process indexed by the closed convex subsets of $\mathbf{R}^2$. (Preprint).

CHERNOFF, H. (1952). A measure of asymptotic efficiency for tests based on the sum of observations. *Ann. Math. Statist.* **23** 493–507.

CHRISTENSEN, J. P. R. (1971). On some properties of Effros Borel structure on spaces of closed subsets. *Math. Ann.* **195** 17–23.

CHRISTENSEN, J. P. R. (1974). *Topology and Borel Structure*. North-Holland, Amsterdam; American Elsevier, New York.

CLEMENTS, G. F. (1963). Entropies of several sets of real valued functions. *Pacific J. Math.* **13** 1085–1095.

COVER, T. M. (1965). Geometric and statistical properties of systems of linear inequalities with applications to pattern recognition. *IEEE Trans. Elec. Comp.* **EC-14** 326–334.

DEHARDT, J. (1971). Generalizations of the Glivenko-Cantelli theorem. *Ann. Math. Statist.* **42** 2050–2055.

DONSKER, M. D. (1952). Justification and extension of Doob's heuristic approach to the Kolmogorov-Smirnov theorems. *Ann. Math. Statist.* **23** 277–281.

DUDLEY, R. M. (1966). Weak convergence of probabilities on nonseparable metric spaces and empirical measures on Euclidean spaces. *Illinois J. Math.* **10** 109–126.

DUDLEY, R. M. (1967a). Measures on non-separable metric spaces. *Illinois J. Math.* **11** 449–453.

DUDLEY, R. M. (1967b). The sizes of compact subsets of Hilbert space and continuity of Gaussian processes. *J. Functional Analysis* **1** 290–330.

DUDLEY, R. M. (1973). Sample functions of the Gaussian process. *Ann. Probability* **1** 66–103.

DUDLEY, R. M. (1974). Metric entropy of some classes of sets with differentiable boundaries. *J. Approximation Theory* **10** 227–236.

EFFROS, E. G. (1965). Convergence of closed subsets in a topological space. *Proc. Amer. Math. Soc.* **16** 929–931.

FEDERER, H. (1969). *Geometric Measure Theory*. Springer, Berlin.

FREEDMAN, D. (1966). On two equivalence relations between measures. *Ann. Math. Statist.* **37** 686–689.

HARDING, E. F. (1967). The number of partitions of a set of N points in k dimensions induced by hyperplanes. *Proc. Edinburgh Math. Soc. (Ser. II)* **15** 285–289.

HAUSDORFF, F. (1937). *Set Theory*, 3rd ed. Transl. by J. Aumann et al. (New York, Chelsea, 1962).

HOEFFDING, W. (1963). Probability inequalities for sums of bounded random variables. *J. Amer. Statist. Assoc.* **58** 13–30.

JOGDEO, K. and SAMUELS, S. M. (1968). Monotone convergence of binomial probabilities and a generalization of Ramanujan's equation. *Ann. Math. Statist.* **39** 1191–1195.

KELLEY, J. L. (1955). *General Topology*. Van Nostrand, Princeton.

KOLMOGOROV, A. N. (1956). On Skorohod convergence. *Theor. Probability Appl.* **1** 215–222. *(Teor. Verojatnost. i Primenen.* **1** 239–247, in Russian.)

KOLMOGOROV, A. N. and TIHOMIROV, V. M. (1959). ε-entropy and ε-capacity of sets in functional spaces. *Uspehi Mat. Nauk* **14**, #2 (86), 3–86 (in Russian); (1961) *Amer. Math. Soc. Transl.* (Ser. 2), **17** 277–364.

KURATOWSKI, K. (1966). *Topology*, **1**. Academic Press, New York.

MCSHANE, E. J. (1934). Extension of range of functions. *Bull. Amer. Math. Soc.* **40** 837–842.

NEUHAUS, G. (1971). On weak convergence of stochastic processes with multidimensional time parameter. *Ann. Math. Statist.* **42** 1285–1295.

OKAMOTO, MASASHI (1958). Some inequalities relating to the partial sum of binomial probabilities. *Ann. Inst. Statist. Math.* **10** 29–35.

PARTHASARATHY, K. R. (1967). *Probability Measures on Metric Spaces*. Academic Press, New York.

PHILIPP, W. (1973). Empirical distribution functions and uniform distribution mod 1. In *Diophantine Approximation and its Applications* (C. F. Osgood, ed.) 211–234. Academic Press, New York.

PYKE, R. and SHORACK, G. (1968). Weak convergence of a two-sample empirical process and a new approach to Chernoff-Savage theorems. *Ann. Math. Statist.* **39** 755–771.

RAO, B. V. (1971). Borel structures for function spaces. *Colloq. Math.* **23** 33–38.

RÉVÉSZ, P. (1976). Three theorems of multivariate empirical process. *Lecture Notes in Math.* **566** 106–126.

SCHLÄFLI, LUDWIG (1901, posth.). *Theorie der vielfachen Kontinuität*, in *Gesammelte Math. Abhandlungen I* (Basel, Birkhäuser, 1950).

SION, M. (1960). On uniformization of sets in topological spaces. *Trans. Amer. Math. Soc.* **96** 237–245.

SKOROHOD, A. V. (1955). On passage to the limit from sequences of sums of independent random variables to a homogeneous random process with independent increments. *Dokl. Akad. Nauk. SSSR* **104** 364–367 (in Russian).

STEELE, J. M. (1978). Empirical discrepancies and subadditive processes. *Ann. Probability* **6** 118–127.

STEINER, J. (1826). Einige Gesetze über die Theilung der Ebene und des Raumes. *J. Reine Angew. Math.* **1** 349–364.

STONE, A. H. (1962). Non-separable Borel sets. *Rozprawy Mat.* **28** (41 pp.).

STRAF, M. L. (1972). Weak convergence of stochastic processes with several parameters. *Proc. Sixth Berkeley Symp. Math. Statist. Prob.* **2** 187–221. Univ. of California Press.

SUN, TZE-GONG (1976). Ph.D. dissertation, Dept. of Mathematics, Univ. of Washington, Seattle.

SZPILRAJN, E. (1938). Ensembles indépendants et mesures non séparables. *C. R. Acad. Sci. Paris* **207** 768–770.

TALAGRAND, M. (1978). Les boules peuvent elles engendrer la tribu borélienne d'un espace métrisable non séparable? (Preprint).

UHLMANN, W. (1966). Vergleich der hypergeometrischen mit der Binomial-Verteilung. *Metrika* **10** 145–158.

VAPNIK, V. N. and ČERVONENKIS, A. YA. (1971). On the uniform convergence of relative frequencies of events to their probabilities. *Theor. Probability Appl.* **16** 264–280. (*Teor. Verojatnost. i Primenen.* **16** 264–279, in Russian.)

VAPNIK, V. N. and ČERVONENKIS, A. YA. (1974). *Theory of Pattern Recognition* (in Russian). Nauka, Moscow.

WATSON, D. (1969). On partitions of n points. *Proc. Edinburgh Math. Soc.* **16** 263–264.

WICHURA, MICHAEL J. (1970). On the construction of almost uniformly convergent random variables with given weakly convergent image laws. *Ann. Math. Statist.* **41** 284–291.

DEPARTMENT OF MATHEMATICS
ROOM 2-245, M.I.T.
CAMBRIDGE, MASSACHUSETTS 02139

The Annals of Probability
1979, Vol. 7, No. 5, 909–911

CORRECTIONS TO "CENTRAL LIMIT THEOREMS FOR EMPIRICAL MEASURES"

BY R. M. DUDLEY

Massachusetts Institute of Technology

In [1], page 926, line 15 is *not* $\leqslant 8(2n)^{2v}\sup(H)$. Let us correct this and some other errors and obscurities.

In the proof of (2.7), "(2.2)" should be "(2.6)."

In the last three lines of page 905, replace Pr by Wichura's probability measure, say $\Pr_W$. Take a sequence $\{h_m\}$ dense in the set of all uniformly continuous $h \notin B_{\delta, \varepsilon/2}$ on $(\mathcal{C}, d_p)$. Then

$$M := \bigcup_m B(h_m, \varepsilon/4) \in \mathcal{B}_b \quad \text{in} \quad D_0(\mathcal{C}, P),$$

$\Pr(v_n \notin M) < \varepsilon$ for n large enough (using P–EM), and $M \cap B_{\delta,\varepsilon} = \varnothing$, proving (b).

In (5.4), $\varepsilon/64$ should be $\varepsilon/96$. Two lines after (5.9), page 915, $4b_{i+1}$ should be $6b_{i+1}$; line 4 from below, 2^{i+1} should be 2^{i+2}; last line, $2^i(2$ should be $2^i(4$; page 916, first line, 2 should be 4; on pages 916–917, $n \geqslant n_0$ should be $n > n_0$; on page 917, line 2, ε should be 3ε (twice). On page 921, 4th line after (7.3), $v \in \mathbb{R}^k$. Five lines further, delete "of H_N." On page 922, (7.9), N should be n; line 9, replace "$n > k \geqslant 1$" by "$n \geqslant k + 2 \geqslant 2$"; in the line after (7.10), replace "$n < v$" by "$2 \leqslant n \leqslant v$."

On page 923, to clarify the choice of δ, in the proof of Theorem 7.1 after the third display, replace "Given a δ, $0 < \delta \leqslant 1$, to be chosen later," by: "For $0 < \delta \leqslant 1$, $\mathcal{C}(\delta) \subset \mathcal{C}(1)$. Thus

$$N(\delta/2, \mathcal{C}(\delta), P) \leqslant N(\delta/2, \mathcal{C}(1), P) \leqslant N\delta^{-w}$$

for $0 < \delta < 1$. Then choose a $\delta := \delta_1(\varepsilon) > 0$ such that for $0 < \gamma \leqslant \delta_1(\varepsilon)$,

$$2N\Sigma_{j=1}^{\infty} 2^{jw}\gamma^{-w}\exp(-\varepsilon^2 2^j/(8j^4\gamma)) < \varepsilon/3,$$

which is possible since: for each $\gamma > 0$, the series converges; the jth term converges to 0 as $\gamma \downarrow 0$, monotonically for $\gamma < \varepsilon^2 2^j/(8j^4 w) > 1$ for j large enough, so we have dominated convergence."

Then on page 924, after the first display, replace the first three sentences "Now . . . $\delta = \delta_1(\varepsilon)$." by "Then $S_1 < \varepsilon/3$ by choice of δ." On page 925, line 7 up, $(P_n + P_n')/2$.

In lines 5 and 4 up, replace "For fixed . . . $P_{2n}\}$" by: "Let $\mathcal{S}_{2n}$ be the σ-algebra of events $\{\langle X_1, \ldots, X_{2n}\rangle \in V\}$ where $V \in \mathcal{Q}^{2n}$ and V is preserved by all $(2n)!$

Received April 4, 1979. Revised in proof, July 10, 1979.

permutations of coordinates. Then if a measurable random variable is a function of P_{2n}, it is measurable for $\mathcal{S}_{2n}$. For any measurable set A let $r_n(A) := 2nP_{2n}(A)$. Then $\Pr\{nP_n(A) \geq s, nP_n'(A) \leq t | \mathcal{S}_{2n}\}$" $\cdots$.

On page 926, replace line 4, "On ... so" by "If $s \geq \varepsilon n^{1/2}$ and $t = 1$, then". Replace lines 12–21, "By (7.11), $\cdots \to 0$" by: "We have $m^{\mathfrak{B}(n,\delta)}(2n) \leq m^{\mathcal{C}(1)}(2n)^2 \leq (2n)^{2v}$ by (7.11), and $Q_{n\varepsilon} \leq \Sigma_{i=1}^4 P_{(i)}$, where if J is the event on the left in (7.16),

$$P_{(1)} := \Pr(\exists i(\omega)) \leq 2\Pr(J) = 2E\Pr(J|\mathcal{S}_{2n})$$

$$\leq 2E\Pr\{\exists W \in \mathfrak{B}_{n\delta}: v_n(W) \geq \varepsilon \text{ and } v_n'(W) \leq n^{-1/2}|\mathcal{S}_{2n}\}.$$

For n large, $\sup\{nP(W): W \in \mathfrak{B}_{n\delta}\} < 1$, so

$$P_{(1)} \leq 2E\Pr\{\exists W \in \mathfrak{B}_{n\delta}: n^{1/2}P_n(W) \geq \varepsilon \text{ and } nP_n'(W) \leq 1|\mathcal{S}_{2n}\}.$$

Let the 4^n subsets of $\{1, 2, \cdots, 2^n\}$ be $Y(1), \cdots, Y(4^n)$. Now $(X^{2n}, \mathcal{Q}^{2n})$, being Suslin, is isomorphic as a measurable space to a subset of a countable product of lines, on which there is a lexicographical linear ordering $\preccurlyeq$ with measurable graph in X^{4n}. Then we can arrange $X_1, \cdots, X_{2n}$ in order: $X_{(1)} \preccurlyeq X_{(2)} \preccurlyeq \cdots \preccurlyeq X_{(2n)}$. The $X_{(i)}$ are all $\mathcal{S}_{2n}$-measurable random variables. Thus for each $j = 1, \cdots, 4^n$, and measurable $W \in \mathcal{Q}$, let $I(j, W) := \{\omega: X_{(i)} \in W \text{ if and only if } i \in Y(j)\}$. Let $X^{(2n)}(\omega) := \langle X_{(1)}, \cdots, X_{(2n)}\rangle$. Then $X^{(2n)}$ is measurable from $\langle X^\infty, \mathcal{S}_{2n}\rangle$ into $\langle X^{2n}, \mathcal{Q}^{2n}\rangle$. For each i and j, the set $I(i, j) :=$

$$\{\langle A, B, X^{(2n)}(\omega)\rangle: \langle A, B\rangle \in \mathfrak{D}_{n\delta i}, \omega \in I(j, (A \setminus B) \setminus D_{ni})\}$$

is $\mathcal{S} \times \mathcal{S} \times \mathcal{Q}^{2n}$ measurable in $\mathcal{C} \times \mathcal{C} \times X^{2n}$. Let

$$I(j) := \{\omega: \exists i, A, B, \langle A, B, X^{(2n)}(\omega)\rangle \in I(i, j)\}.$$

Then using (3.3), $I(j)$ is measurable for the Pr-completion of $\mathcal{S}_{2n}$. Let $A(j) := \{X_{(i)}: i \in Y(j)\}$. For any measurable set $A \in \mathcal{Q}$ let $H(A, n)$ denote the event $(n^{1/2}P_n(A) \geq \varepsilon \text{ and } nP_n'(A) \leq 1\}$. Then

$$\begin{aligned} P_{(1)} &\leq 2E(E(\Sigma_{1 \leq j \leq 4^n} 1_{I(j)} 1_{H(A(j), n)}|\mathcal{S}_{2n})) \\ &= 2E(\Sigma_{1 \leq j \leq 4^n} 1_{I(j)} E(1_{H(A(j), n)}|\mathcal{S}_{2n})) \\ &\leq 2E(\Sigma_{1 \leq j \leq 4^n} 1_{I(j)} \sup_{A \in \mathcal{Q}} \Pr(H(A, n)|\mathcal{S}_{2n})) \\ &\leq 2(2n)^{2v} E \sup_A \Pr(H(A, n)|\mathcal{S}_{2n}). \end{aligned}$$

The last conditional probability is of the hypergeometric form H treated above, with $s \geq \varepsilon n^{1/2}$ and $t = 1$. For $P_{(2)}$, $P_{(3)}$ and $P_{(4)}$, we replace $i(\cdot)$ by d, e and f respectively, and make the other appropriate changes. For n large, $P_{(3)} = P_{(4)} = 0$, and

$$Q_{n\varepsilon} \leq 4(2n)^{2v} \exp(-(\varepsilon n^{1/2} - 1)^2/(8\varepsilon n^{1/2})) \to 0" \ldots.$$

Many thanks to Peter Gaenssler for the corrections to Theorem 5.1, to J. Berruyer and R. Carmona for noting a gap in the proof of Theorem 1.2, and to Roy Erickson and Joel Zinn for useful discussions of the proof of Theorem 7.1.

REFERENCES

[1] DUDLEY, R. M. (1978). Central limit theorems for empirical measures. *Ann. Probability* **6** 899–929.

DEPARTMENT OF MATHEMATICS
ROOM 2-245
MASSACHUSETTS INSTITUTE OF TECHNOLOGY
CAMBRIDGE, MASSACHUSETTS 02139

Z. Wahrscheinlichkeitstheorie verw. Gebiete
61, 355–368 (1982)

Zeitschrift für
Wahrscheinlichkeitstheorie
und verwandte Gebiete

Empirical and Poisson Processes on Classes of Sets or Functions Too Large for Central Limit Theorems*

R. M. Dudley

Massachusetts Institute of Technology, Dept. of Mathematics, Room 2-245, Cambridge, Mass. 02139, USA

Summary. Let P be the uniform probability law on the unit cube I^d in d dimensions, and P_n the corresponding empirical measure. For various classes $\mathscr{C}$ of sets $A \subset I^d$, upper and lower bounds are found for the probable size of $\sup\{|P_n - P)(A)|: A \in \mathscr{C}\}$. If $\mathscr{C}$ is the collection of lower layers in I^2, or of convex sets in I^3, an asymptotic lower bound is

$$((\log n)/n)^{1/2} (\log \log n)^{-\delta - 1/2} \quad \text{for any } \delta > 0.$$

Thus the law of the iterated logarithm fails for these classes.

If $\alpha > 0$, β is the greatest integer $< \alpha$, and $0 < K < \infty$, let $\mathscr{C}$ be the class of all sets $\{x_d \leqq f(x_1, \ldots, x_{d-1})\}$ where f has all its partial derivatives of orders $\leqq \beta$ bounded by K and those of order β satisfy a uniform Hölder condition $|D^p(f(x) - f(y))| \leqq K|x-y|^{\alpha-\beta}$. For $0 < \alpha < d-1$ one gets a universal lower bound $\delta n^{-\alpha/(d-1+\alpha)}$, for a constant $\delta = \delta(d, \alpha) > 0$. When $\alpha = d-1$ the same lower bound is obtained as for the lower layers in I^2 or convex sets in I^3. For $0 < \alpha \leqq d-1$ there is also an upper bound equal to a power of $\log n$ times the lower bound, so the powers of n are sharp.

1. Introduction

First let us define a collection of sets in d-dimensional Euclidean space $\mathbb{R}^d$ whose boundaries are given by functions with derivatives up through some order β bounded and satisfying a Hölder condition of order $0 < \alpha - \beta \leqq 1$. Specifically, let $\alpha > 0$ and $K > 0$. Let β be the greatest integer $< \alpha$. Let

$$D^p = \partial^{[p]} / \partial x_1^{p_1} \ldots \partial x_d^{p_d}, \qquad [p] = p_1 + \ldots + p_d,$$

for p_i integers $\geqq 0$, $p = (p_1, \ldots, p_d)$. For a function f on $\mathbb{R}^d$ such that $D^p f$ is continuous whenever $[p] \leqq \beta$, let

* This research was partially supported by National Science Foundation Grant MCS-79-04474

0044-3719/82/0061/0355/$02.80

$$\|f\|_\alpha = \max_{[p]\leq\beta} \sup\{|D^p f(x)|: x\in \mathbb{R}^d\} + \max_{[p]=\beta} \sup_{x\neq y}\{|D^p f(x)-D^p f(y)|/|x-y|^{\alpha-\beta}\}$$

where $|u|=(u_1^2+\ldots+u_k^2)^{1/2}$, $u\in\mathbb{R}^k$.

Let I^d be the unit cube $\{x\in\mathbb{R}^d\colon 0\leq x_j\leq 1,\ j=1,\ldots,d\}$. Let $x_{(d)}=(x_1,\ldots,x_{d-1})$, $x\in\mathbb{R}^d$. Then let $\mathscr{C}(\alpha,d,K)$ be the collection of all sets of the form

$$\{x\in I^d\colon 0\leq x_d\leq f(x_{(d)})\}$$

for all f on $\mathbb{R}^{d-1}$ with $\|f\|_\alpha\leq K$.

As the most difficult result in this paper (Theorem 3) gives an asymptotic *lower* bound for empirical and Poisson processes over classes $\mathscr{C}(d-1,d,K)$, the relatively small class $\mathscr{C}(\alpha,d,K)$ just defined will suffice here. Larger classes with the same degree of boundary smoothness, but allowing unions and intersections of a fixed (for each class) number of such sets have been defined and studied previously ([5, 6, 14], Révész (1976a-b)). Corresponding asymptotic results will also hold for them.

Throughout, let P be the uniform Lebesgue measure on I^d. Let $X(1)$, $X(2),\ldots,$ be independent and identically distributed with law P, $P_n=n^{-1}\sum_{j=1}^{n}\delta_{X(j)}$, and $\nu_n=n^{1/2}(P_n-P)$. Let $\mathscr{A}$ be the σ-algebra of measurable sets completed for P.

The main results of the paper will be stated in this section and proved in later sections.

For any collection $\mathscr{C}$ of measurable sets and finite signed measure μ let $\|\mu\|_{\mathscr{C}}=\sup_{A\in\mathscr{C}}|\mu(A)|$. If $\mathscr{C}=\mathscr{C}(\alpha,d,K)$ let

$$\|\mu\|_{\alpha,d,K}=\|\mu\|_{\mathscr{C}(\alpha,d,K)}.$$

Theorem 1. *If $0<\alpha<d-1$ then for any $K>0$ there is a $\delta>0$ such that for all possible values of P_n, we have*

$$\|P_n-P\|_{\alpha,d,K}\geq\delta n^{-\alpha/(d-1+\alpha)}.$$

Remarks. Theorem 1 is related to a result of N.S. Bakhvalov (1959) and will be proved, by a technique similar to his, in Sect. 2.

Specifically, Bakhvalov (1959, Teorema 1, p. 6) implies that for each $d=1,2,\ldots,$ and $\alpha>0$, there is a $\gamma=\gamma(d,\alpha)>0$ such that for all possible values of P_n, we have

$$\sup\{\textstyle\int f d(P_n-P)\colon \|f\|_\alpha\leq 1\}\geq\gamma n^{-\alpha/d}. \tag{1.1}$$

This gives information for empirical measures mainly when $\alpha\leq d/2$, since for $\alpha>d/2$ the supremum even over $\pm f$ for a single $f\neq 0$ tends to be of order $n^{-1/2}$.

In (1.1) one can, replacing γ by $\gamma/2$ if necessary, restrict the supremum to those f with $\int f dP=0$, implying the results of Kaufman and Philipp (1978, Sect.

4) with some improvements: $\delta=1$, $\varepsilon=0$, and no independence, lacunarity or other probabilistic assumption is needed.

The next result applies to a general probability space $(X, \mathscr{A}, Q)$. Here $v_n = n^{1/2}(Q_n - Q)$. For a collection $\mathscr{C} \subset \mathscr{A}$ and $\varepsilon > 0$ let [6]

$$N_I(\varepsilon, \mathscr{C}, Q) = \inf\{m\colon \exists A_1, \ldots, A_m \in \mathscr{A}\colon \forall A \in \mathscr{C}\ \exists i,j\colon A_i \subset A \subset A_j \text{ and } Q(A_j \setminus A_i) < \varepsilon\}.$$

Let $\Pr^*(A) = \inf\{\Pr(U)\colon U \supset A\}$.

Theorem 2. *If $\mathscr{C} \subset \mathscr{A}$ and for some constants ζ, $1 \leqq \zeta < \infty$, and $K < \infty$, $N_I(\varepsilon, \mathscr{C}, Q) \leqq \exp(K\varepsilon^{-\zeta})$ for $0 < \varepsilon \leqq 1$, then*

$$\Pr^*\{\|v_n\|_{\mathscr{C}} > n^\theta (\log n)^\eta\} \to 0$$

as $n \to \infty$, where $\theta = (\zeta - 1)/(2\zeta + 2)$ and for any $\eta > 2/(\zeta + 1)$.

Remarks. The classes $\mathscr{C}(\alpha, d, K)$ will satisfy the hypothesis of Theorem 2 for $\zeta = (d-1)/\alpha$ (Kolmogorov and Tikhomirov, 1959, Sect. 5, Theorems XIII–XV; [5, (3.2), as corrected, 1979, with its proof]). Then in Theorem 2, we get $\theta = \frac{1}{2} - \frac{\alpha}{d-1+\alpha}$. This θ cannot be reduced, by Theorem 1. Conversely, the exponent $\alpha/(d-1+\alpha)$ in Theorem 1 cannot be improved. It remains to find the best exponent η for $\log n$ in Theorem 2, a problem left open here.

The condition $\zeta \geqq 1$ in Theorem 2 is necessary, as we clearly cannot have $\theta < 0$ even for $\mathscr{C}$ consisting of a single set A with $0 < Q(A) < 1$.

Theorem 2 will also be proved in Sect. 2.

On I^d, for each $\lambda > 0$ let X_λ be the Poisson point process with intensity measure λP. That is, for each measurable set $A \subset I^d$, $X_\lambda(A)$ has a Poisson distribution with parameter $\lambda P(A)$, and for disjoint measurable sets A_i, the $X_\lambda(A_i)$ are independent. Let Y_λ be the centered process

$$Y_\lambda(A, \omega) = X_\lambda(A, \omega) - \lambda P(A),$$

which has mean 0 and still has independent values on disjoint sets. In Theorems 3 and 4 P is again uniform on I^d; $\ln = \log$.

Theorem 3. *If $0 < \alpha = d-1$ then for any $K > 0$ and $\delta > 0$ there is a $c > 0$ such that as $\lambda \to +\infty$*

$$\Pr\{\|Y_\lambda\|_{\alpha,d,K} > c(\lambda \ln \lambda)^{1/2} (\ln \ln \lambda)^{-0.5-\delta}\} \to 1$$

and as $n \to \infty$,

$$\Pr\{n \|P_n - P\|_{\alpha,d,K} > c(n \ln n)^{1/2} (\ln \ln n)^{-0.5-\delta}\} \to 1.$$

Theorem 3 will be proved in Sects. 3–4. Kaufman (1980) proved that $\|v_n\|_{d-1,d,K}$ is unbounded in probability.

If $\alpha > d-1$, then $\|v_n\|_{\alpha,d,K}$ is bounded in probability (and further, central limit theorems and laws of the iterated logarithm hold: Révész, 1976b; Sun and Pyke, 1982; [6, 14]).

If $\alpha < d-1$, then it was known (without Theorem 1 above) that the central limit theorem must fail since the limiting Gaussian process is almost surely unbounded over $\mathscr{C}(\alpha, d, K)$ [4, Theorem 4.2]. For $\alpha = d-1$, the power η of $\log n$, between 1/2 (Theorem 3) and 1 (Theorem 2) remains to be settled.

A *lower layer* in $\mathbb{R}^2$ is a set B such that if $(x, y) \in B$, $u \leqq x$ and $v \leqq y$, then $(u, v) \in B$.

Steele (1978; Sect. 7), Wright (1981), and others they cite, prove laws of large numbers (Glivenko-Cantelli theorems) uniformly over the lower layers, for suitable P, as had R. Ranga Rao (1962) for convex sets, in all dimensions.

For the central limit theorem or law of the iterated logarithm the critical dimension is 2 for the lower layers and 3 for the convex sets. For these classes the central limit theorem fails since $\|v_n\|_{\mathscr{C}}$ is unbounded in probability [7]. The next result, apparently for the first time, allows us to conclude that $\|v_n\|_{\mathscr{C}}/(\log\log n)^{1/2}$ is also almost surely unbounded for $\mathscr{C}$ = lower layers in $\mathbb{R}^2$ or convex sets in $\mathbb{R}^3$ (for previous results see e.g. Stute, 1977).

Theorem 4. *The conclusion of Theorem 3 also holds if $d=2$ and $\mathscr{C}(1, 2, K)$ is replaced by the collection of all lower layers, or by the set of all lower layers in $\mathscr{C}(1, 2, K)$. For $d=3$, the conclusion of Theorem 3 holds if $\mathscr{C}(2, 3, K)$ is replaced by the collection of all convex sets in $\mathbb{R}^3$.*

Theorem 4 will be proved in Sect. 5.

Acknowledgment. Thanks to Walter Philipp for several conversations, specifically for the case $\zeta = 1$ in Theorem 2, where he has proved a more refined result, and for pointing out W. Schmidt's work. Thanks also to Rae Shortt for a helpful comment.

2. Proof of Theorems 1 and 2

First, to prove Theorem 1, let f be a C^∞ function on $\mathbb{R}^{d-1}$, 0 outside I^{d-1}, with $f > 0$ in the interior of I^{d-1}. For any ε with $0 < \varepsilon \leqq 1$, set $f_\varepsilon(x) = \varepsilon^\alpha f(x/\varepsilon)$. Then

$$\|f_\varepsilon\|_\alpha \leqq \|f\|_\alpha. \tag{2.1}$$

Given n, let $j = j_n = [(2nc)^{1/(\alpha+d-1)}]$ where $[x]$ is the greatest integer $\leqq x$ and $c = K \int f\,dx/(2\|f\|_\alpha)$. Then $1/(2n) \leqq cj^{1-d-\alpha}$. We have

$$\sup_{n \geqq 1} 2ncj_n^{1-d-\alpha} = M < \infty.$$

Let $\theta_n = j_n^{\alpha+d-1}/(2nc)$. Then $1/M \leqq \theta_n \leqq 1$.

Let $h = K\theta_n f/(2\|f\|_\alpha)$. We decompose I^{d-1} into j^{d-1} disjoint cubes C_i, $i = 1, \ldots, j^{d-1}$, of side $1/j$. Let $c_i \in I^{d-1}$ be the vertex of C_i closest to 0. Let

$$A_i = \{x \in I^d \colon 0 < x_d \leqq j^{-\alpha} h(j(x_{(d)} - c_i))\}.$$

Then the sets A_i are disjoint. Since by (2.1)

$$K\theta_n \|f_{1/j}\|_\alpha/(2\|f\|_\alpha) \leqq K/2,$$

the union of any set of the A_i, together with $\{x \in I^d: x_d = 0\}$ forms a set in $\mathscr{C}(\alpha, d, K)$. We have for each i

$$P(A_i) = j^{1-d-\alpha} \int h\,dx = j^{1-d-\alpha}\,\theta_n c = 1/(2n),$$

and $P_n(A_i) = 0$ or $P_n(A_i) \geqq 1/n$. Either at least half the A_i have $P_n(A_i) = 0$, or at least half have $P_n(A_i) \geqq 1/n$. In either case,

$$\|P_n - P\|_{\alpha, d, K} \geqq j^{d-1}/(4n) \geqq cj^{-\alpha}/(2M) \geq \delta n^{-\alpha/(\alpha+d-1)}$$

for some $\delta = \delta(\alpha, d, K) > 0$, proving Theorem 1.

Proof of Theorem 2. The proof is similar to that of [6, Theorem 5.1]. Given n, let

$$k(n) = [(\tfrac{1}{2} - \theta) \cdot \log_2 n - \eta \log_2 \log n] \sim (\tfrac{1}{2} - \theta) \cdot \log_2 n$$

as $n \to \infty$ ($\log_2 = \log$ to base 2).

For $k = 1, 2, \ldots$, let $N(k) = N_I(2^{-k}, \mathscr{C}, Q)$. By its definition choose sets $A_{ki} \in \mathscr{A}$, $i = 1, \ldots, N(k)$, such that for all $A \in \mathscr{C}$ there are i and j with $A_{ki} \subset A \subset A_{kj}$ and $Q(A_{kj} \setminus A_{ki}) \leqq 2^{-k}$. Let $A_{01} = \emptyset$ (empty set) and $A_{02} = X$. For each $A \in \mathscr{C}$ and $k = 0, 1, \ldots$, choose such an $i = i(k, A)$ and $j = j(k, A)$. Then for $k \geqq 1$,

$$Q(A_{k, i(k,A)} \,\Delta\, A_{k-1, i(k-1, A)}) \leqq 2^{2-k},$$

where Δ denotes symmetric difference.

Let $\mathscr{B}(k)$ be the collection of all sets $B = A_{ki} \setminus A_{k-1,j}$ or $A_{k-1,j} \setminus A_{ki}$ or $A_{kj} \setminus A_{ki}$ such that $Q(B) \leqq 2^{2-k}$ $(k \geqq 1)$. Then $\operatorname{card}(\mathscr{B}(k)) \leqq 2N(k-1)\,N(k) + N(k)^2 \leqq 3\exp(2K\,2^{k\zeta})$.

For each $B \in \mathscr{B}(k)$ we have by Bernstein's inequality (Bennett, 1962) for any $t > 0$

$$\Pr\{|v_n(B)| > t\} \leqq \exp(-t^2/(2^{3-k} + tn^{-1/2})). \tag{2.2}$$

Set $t = t_{n,k} = n^\theta (\log n)^\eta c k^{-1-\delta}$ for a δ such that $0 < \delta < 1$ and $\eta > (2+2\delta)/(1+\zeta)$, as is possible since $\eta > 2/(1+\zeta)$, and where $c = \delta/(1+\delta)$. Then for $1 \leqq k \leqq k(n)$,

$$2^{3-k} \geqq 8n^{\theta - 1/2} (\log n)^\eta \geqq t_{n,k}\, n^{-1/2}.$$

Then (2.2) gives

$$\Pr\{|v_n(B)| > t_{n,k}\} \leqq \exp(-t_{n,k}^2/2^{4-k}).$$

Hence

$$\begin{aligned} p_{nk} &= \Pr\{\sup_{B \in \mathscr{B}(k)} |v_n(B)| > t_{n,k}\} \\ &\leqq 3\exp(2K\,2^{k\zeta} - 2^{k-4}\,t_{n,k}^2) \\ &= 3\exp(2K\,2^{k\zeta} - 2^{k-4}\,c^2 n^{2\theta} (\log n)^{2\eta} k^{-2-2\delta}). \end{aligned}$$

We have $2\theta/(\tfrac{1}{2} - \theta) = \zeta - 1 \geqq 0$. For $k \leqq k(n)$ we have $n^{0.5-\theta} \geqq 2^k (\log n)^\eta$, so $n^{2\theta} \geqq 2^{k(\zeta-1)} (\log n)^{\eta(\zeta-1)}$, and

$$p_{nk} \leqq 3\exp(2K\,2^{k\zeta} - 2^{k\zeta-4} c^2 (\log n)^{\eta(\zeta+1)} k^{-2-2\delta}).$$

Now $k \leqq \log n$ since $\frac{1}{2} - \theta < \log 2$, so

$$p_{nk} \leqq 3 \exp(2^{k\zeta}(2K - c^2 2^{-4}(\log n)^\gamma)$$

where $\gamma = \eta(\zeta+1) - 2 - 2\delta > 0$ by choice of δ. Thus for n large enough so that $(\log n)^\gamma > 32 K/c^2$,

$$\sum_{k=1}^{k(n)} p_{nk} \leqq 3(\log n) \exp(4K - c^2 2^{-3}(\log n)^\gamma) \to 0$$

as $n \to \infty$.

Let $\mathscr{E}_n$ be the event that $|v_n(B)| \leqq t_{n,k}$ for all $B \in \mathscr{B}(k)$, $k = 1, \ldots, k(n)$. Then $\Pr(\mathscr{E}_n) \to 1$ as $n \to \infty$. On $\mathscr{E}_n$, for each $A \in \mathscr{C}$ and $i = i(k(n), A)$, $j = j(k(n), A)$,

$$|v_n(A_{k(n),i})| \leqq 2 \sum_{k=1}^{k(n)} t_{n,k} = 2cn^\theta (\log n)^\eta \sum_{k \geqq 1} k^{-1-\delta}$$
$$\leqq 2n^\theta (\log n)^\eta,$$

and $|v_n(A_{k(n),j} \setminus A_{k(n),i})| \leqq n^\theta (\log n)^\eta$. Now

$$n^{1/2} Q(A_{k(n),j} \setminus A_{k(n),i}) \leqq n^{1/2}/2^{k(n)} < 2n^\theta (\log n)^\eta.$$

Thus $n^{1/2} Q_n(A_{k(n),j} \setminus A_{k(n),i}) \leqq 3n^\theta (\log n)^\eta$, so $|v_n(A \setminus A_{k(n),i})| \leqq 3n^\theta (\log n)^\eta$. Thus on $\mathscr{E}_n$, $|v_n(A)| \leqq 5n^\theta (\log n)^\eta$ for our arbitrary $A \in \mathscr{C}$. Using a smaller η we can drop the 5, proving Theorem 2.

If $\alpha = d - 1$, the method of proof of Theorem 1 shows that for larger n there are smaller and smaller sets on which v_n is not small. But Theorem 3 gives better information.

The proof of (1.1) above by Bakhvalov (1959) is like the proof of Theorem 1 here, replacing $d-1$ by d, and letting $j = [(2n)^{1/d}] + 1$. On the $j^d \gtrsim 2n$ little cubes C_i let $g = 0$ on those with $P_n(C_i) > 0$. On all other C_i let $g(x) = -f_{1/j}(x - c_i)$. Then $\|g\|_\alpha \leqq \|f\|_\alpha$ and using g gives (1.1). Bakhvalov notes, in turn, that Kolmogorov had used a similar method to prove a lower bound for the metric entropy of classes $\{f\colon \|f\|_\alpha \leqq K\}$ in the supremum norm [13]. Later, W. Schmidt (1975) applied such a method to classes of convex sets.

3. Lemmas: Poissonization and Random Sets

First, let us relate empirical and Poisson processes ("Poissonization"). Consider the following property of a function f defined for large enough $x > 0$:

(*) for any $\varepsilon > 0$ there is a $\delta > 0$ such that whenever

$$x > 1/\delta \quad \text{and} \quad \left|1 - \frac{y}{x}\right| < \delta \quad \text{then} \quad \left|1 - \frac{f(y)}{f(x)}\right| < \varepsilon.$$

If f is continuous, and slowly varying (Karamata), i.e. for all $k > 0$, $f(kx)/f(x) \to 1$ as $x \to +\infty$; or if f is regularly varying, i.e. for some real r, $f(x) \equiv x^r L(x)$ where L is slowly varying, then (*) holds (see e.g. Feller, Vol. II,

VIII.8, Lemma 2). The following Lemma will treat more general situations than are needed in this paper. For any probability space $(S, \mathscr{A}, P)$ we can define the Poisson processes X_λ and Y_λ, as in Sect. 1 for the uniform P, and the empirical processes P_n and ν_n.

(3.1) **Lemma.** *Let $(S, \mathscr{A}, P)$ be a probability space and $\mathscr{C} \subset \mathscr{A}$, where we assume that for each n and constant t, $\sup_{A \in \mathscr{C}} (P_n - tP)(A)$ is measurable. Let f be a function satisfyng (*) such that $f(\lambda) \to +\infty$ and $f(\lambda)/\lambda \to 0$ as $\lambda \to +\infty$. Suppose that for some constant $c > 0$ and some κ with $0 \leqq \kappa \leqq 1$, we have*

$$\liminf_{\lambda \to \infty} \Pr\{\sup_{A \in \mathscr{C}} Y_\lambda(A) \geqq c f(\lambda)\, \lambda^{1/2}\} = \kappa.$$

Then for $0 < K < c$ we have

$$\liminf_{n \to \infty} \Pr\{\sup_{A \in \mathscr{C}} \nu_n(A)/f(n) \geqq K\} \geqq \kappa.$$

Proof. Let $n(\lambda)$ be a Poisson variable with parameter λ, independent of the $X(i)$. Then $n(\lambda) P_{n(\lambda)}$ has the properties of X_λ, as is well known (Kac, 1949; Csörgő and Révész, 1981, pp. 250–251). Also, $n(\lambda)$ can be defined from X_λ or Y_λ by $n(\lambda) = X_\lambda(S) = (Y_\lambda + \lambda P)(S)$. We can then write

$$(3.2) \qquad \begin{aligned} Y_\lambda &= n(\lambda)\, P_{n(\lambda)} - \lambda P = n(\lambda)(P_{n(\lambda)} - P) + (n(\lambda) - \lambda)\, P, \\ Y_\lambda/\lambda^{1/2} &= (n(\lambda)/\lambda)^{1/2}\, \nu_{n(\lambda)} + (n(\lambda) - \lambda)\, \lambda^{-1/2} P, \end{aligned}$$

where if $n(\lambda) = 0$, we replace $P_{n(\lambda)}$ and $\nu_{n(\lambda)}$ by 0. As $\lambda \to \infty$, $\Pr(n(\lambda) > 0) \to 1$, $n(\lambda)/\lambda^{1/2} \to 1$ in probability, and $(n(\lambda) - \lambda)\lambda^{-1/2}$ is bounded in probability.

From this construction we see that $\sup_{A \in \mathscr{C}} Y_\lambda(A)$ is also measurable.

If the Lemma fails, then there is a $\theta < \kappa$ and an infinite sequence of values $m = m_k \to +\infty$ such that

$$\Pr(\sup_{A \in \mathscr{C}} \nu_m(A) \geqq K f(m)) \leqq \theta.$$

Take $0 < \varepsilon < 1/3$ such that $K(1 + 7\varepsilon) < c$. Then take a $\delta > 0$ such that (*) holds for f. We may assume $\delta < 1/2$ and $(1 + \delta)(1 + 5\varepsilon) < 1 + 6\varepsilon$. We may also assume that for all k, $m = m_k \geqq 2/\delta$ and $1 + 2\varepsilon < K\varepsilon f(m)^{1/2}$.

Let $\delta_m = (f(m)/m)^{1/2}$. Then since $f(m)/m \to 0$ we may assume that $\delta_m < \delta/2$ for all $m = m_k$. For any $m = m_k$, if $(1 - \delta_m) m \leqq n \leqq m$ and $A \in \mathscr{A}$, then $m P_m(A) \geqq n P_n(A)$, so $m(P_m - P)(A) \geqq n(P_n - P)(A) - m\delta_m$, and

$$\nu_m(A) \geqq (n/m)^{1/2} \nu_n(A) - f(m)^{1/2} \geqq (1 + \delta)^{-1} \nu_n(A) - f(m)^{1/2},$$

or $\nu_n(A) \leqq (1 + \delta)(\nu_m(A) + f(m)^{1/2})$. Now

$$\left|1 - \frac{m}{n}\right| = \frac{m}{n} - 1 < 2\delta_m < \delta \quad \text{implies} \quad \left|\frac{f(m)}{f(n)} - 1\right| < \varepsilon$$

and $1/f(n) < (1 + \varepsilon)/f(m)$, so if $\nu_n(A) \geqq 0$ then

$$\nu_n(A)/f(n) \leqq (1 + 2\varepsilon)(\nu_m(A) f(m)^{-1} + f(m)^{-1/2}).$$

So since $(1+2\varepsilon)f(m)^{-1/2}<K\varepsilon$,

$$\Pr\{\sup_{A\in\mathscr{C}} v_n(A)\geqq Kf(n)(1+3\varepsilon)\}\leqq\theta.$$

For each $m=m_k$, set $\lambda=(1-\delta_m/2)m$. Then as $k\to\infty$, since $f(m_k)\to\infty$ we will have

$$\Pr((1-\delta_m)m\leqq n(\lambda)\leqq m)\to 1.$$

Then for any γ with $\theta<\gamma<\kappa$ and k large enough, since the $X(i)$ are independent of $n(\lambda)$,

$$\Pr\{\sup_{A\in\mathscr{C}} v_{n(\lambda)}(A)\geqq Kf(n(\lambda))(1+3\varepsilon)\}<\gamma.$$

By $(*)$, for $(1-\delta_m)m\leqq n\leqq m$ we have $\left|1-\frac{f(n)}{f(\lambda)}\right|<\varepsilon$, so that for k large enough we may assume

$$\Pr\{\sup_{A\in\mathscr{C}} v_{n(\lambda)}(A)\geqq Kf(\lambda)(1+5\varepsilon)\}<\gamma.$$

By (3.2), we then have for k large enough

$$\Pr\{\sup_{A\in\mathscr{C}} Y_\lambda(A)\geqq K\lambda^{1/2}f(\lambda)(1+7\varepsilon)\}<(\gamma+\kappa)/2<\kappa,$$

a contradiction, proving Lemma 3.1.

Note. Pyke (1968) has related estimates with and without Poissonization.

For any real x let $x^+=\max(x,0)$.

(3.3) **Lemma.** *There is a constant $c>0$ such that whenever z has a Poisson law with parameter $m\geqq 1$ then*

$$E(z-m)^+\geqq cm^{1/2}.$$

Proof. Let j be the greatest integer $\leqq m$. Then by a telescoping sum and Stirling's formula with an error bound (e.g. Feller, Vol. I, Sect. II.9, p. 54),

$$\begin{aligned}E(z-m)^+ &= \sum_{k>m} e^{-m}m^k(k-m)/k!\\ &\geqq me^{-m}m^j/((j/e)^j(2\pi j)^{1/2}e^{1/(12j)})\\ &\geqq (m^{j+1}/j^{j+1/2})e^{-13/12}(2\pi)^{-1/2}\geqq cm^{1/2}\ \text{(with } c=0.135).\quad \text{Q.E.D.}\end{aligned}$$

Now the Poisson process X_λ has the property that for any two disjoint measurable sets A, $B\in\mathscr{A}$, given the σ-algebra generated by X_λ on B and its measurable subsets, or any sub-σ-algebra $\mathscr{G}$ of that σ-algebra, the conditional distribution of $X_\lambda(A)$ given $\mathscr{G}$ is Poisson with parameter $\lambda P(A)$. This property will be extended to suitable random sets $A=C_\omega$ and $B=L_\omega$ where $P(A)$ is $\mathscr{G}$-measurable.

More generally, let $(X,\mathscr{A})$ be a measurable space. Let $\mathscr{A}_f\subset\mathscr{A}$ be such that for any $A,B\in\mathscr{A}_f$ and $C\in\mathscr{A}$, $A\cup B\in\mathscr{A}_f$ and $A\cap C\in\mathscr{A}_f$. (For example if $(X,\mathscr{A},\mu)$ is a σ-finite measure space, we can take $\mathscr{A}_f=\{A\in\mathscr{A}: \mu(A)<\infty\}$.) Let

$Y\colon \langle A, \omega\rangle \to Y(A)(\omega)$ be a real-valued stochastic process indexed by $A \in \mathscr{A}_f$, with $\omega \in \Omega$ for a probability space $(\Omega, \mathscr{B}, \Pr)$, such that for any disjoint $A_1, \ldots, A_n \in \mathscr{A}_f$, $Y(A_j)$ are independent, $1 \leqq j \leqq m$, and $Y(A_1 \cup A_2) = Y(A_1) + Y(A_2)$. Then we say Y has *independent pieces.* Clearly each Y_λ is such a process.

For each $C \in \mathscr{A}$ let $\mathscr{B}_C$ be the smallest σ-algebra for which all $Y(A)(\cdot)$ are measurable for $A \subset C$, $A \in \mathscr{A}_f$.

Let G be a function from Ω into $\mathscr{A}$. Then (by analogy with stopping times) we call $G(\cdot)$ a *stopping set* iff for all $C \in \mathscr{A}$, $\{\omega\colon G(\omega) \subset C\} \in \mathscr{B}_C$.

Given a stopping set $G(\cdot)$, let B_G denote the σ-algebra of all sets $B \in \mathscr{B}$ such that for all $C \in \mathscr{A}$, $B \cap \{G \subset C\} \in \mathscr{B}_C$.

(3.4) **Lemma.** *Suppose Y has independent pieces and we have random sets $G_j(\omega) = G(j)(\omega)$, $j = 0, \ldots, m$, and $A(\omega)$ in $\mathscr{A}_f$ such that:*

i) *$G_0(\omega)$ is a fixed set $G_0 \in \mathscr{A}_f$;*

ii) *for all ω, $G_0 \subset G_1(\omega) \subset \ldots \subset G_m(\omega)$, and $G_m(\omega) \cap A(\omega) = \emptyset$;*

iii) *each $G_j(\omega)$, and $A(\omega)$, has only countably many possible values $G(j, i) = G_{ji}$ and $C_i = C(i)$ respectively;*

(iv) *for all i and for $1 \leqq j \leqq m$, $\{G_j(\cdot) = G_{ji}\} \in \mathscr{B}_{G(j-1)}$ and $\{A(\cdot) = C_i\} \in \mathscr{B}_{G(m)}$. Then the $G_j(\cdot)$ are all stopping sets, and the conditional probability*

$$\Pr\{Y(A)(\cdot)) \leqq t \mid \mathscr{B}_{G(m)}\} = \sum_i 1_{\{A(\omega) = C(i)\}} \Pr\{Y(C_i) \leqq t\}$$

almost surely, for each $t \in \mathbb{R}$.

Proof. First it will be shown by induction on j that $G_j(\cdot)$ are stopping sets. Clearly G_0 is. For the induction step, given $C \in \mathscr{A}$ and $j \geqq 1$,

$$\{G_j \subset C\} = \bigcup_i \{\{G_j = G_{ji}\}\colon G_{ji} \subset C\}.$$

For each i, $\{G_j = G_{ji}\} \in \mathscr{B}_{G(j-1)}$. If $G_{ji} \subset C$, then by ii),

$$\{G_j = G_{ji}\} = \{G_j = G_{ji}\} \cap \{G_{j-1} \subset C\} \in \mathscr{B}_C$$

by definition of $\mathscr{B}_{G(j-1)}$ and the induction hypothesis. Thus $\{G_j \subset C\}$, as a countable union of sets in $\mathscr{B}_C$, is in $\mathscr{B}_C$, so G_j is a stopping set.

If $A(\omega) = C_i$ and $G_m(\omega) = G_{mj}$ for some ω, then $C_i \cap G_{mj} = \emptyset$ by ii), so $Y(C_i)$ is independent of $\mathscr{B}_{G_{mj}}$. Let $B_i = \{A(\cdot) = C_i\} \in \mathscr{B}_{G(m)}$ by iv). For each j, $\{G_m = G_{mj}\} \in \mathscr{B}_{G(m-1)}$, by iv), so by ii),

$$\{G_m = G_{mj}\} = \{G_m = G_{mj}\} \cap \{G_{m-1} \subset G_{mj}\} \in \mathscr{B}_{G_{mj}}.$$

For any $B \in \mathscr{B}_{G(m)}$, $B \cap \{G_m = G_{mj}\} = B \cap \{G_m \subset G_{mj}\} \cap \{G_m = G_{mj}\} \in \mathscr{B}_{G_{mj}}$. So a.s.

$$\begin{aligned}
\Pr(Y(A) \leqq t \mid \mathscr{B}_{G_m}) &= \sum_{i,j} \Pr(Y(A) \leqq t \mid \mathscr{B}_{G_m}) 1_{B_i} 1_{\{G_m = G_{mj}\}} \\
&= \sum_{i,j} \Pr(Y(C_i) \leqq t \mid \mathscr{B}_{G_{mj}}) 1_{B_i} 1_{\{G_m = G_{mj}\}} \\
&= \sum_i \Pr(Y(C_i) \leqq t) 1_{B_i}. \quad \text{Q.E.D.}
\end{aligned}$$

Note. Evstigneev (1977, Theorem 1) proves a strong Markov property for suitable random fields, indexed by closed subsets of a Euclidean space, which might also be used, with some work, in place of Lemma 3.4 above.

4. Proof of Theorem 3

$\|P_n - P\|_{\alpha,d,K}$ is a measurable random variable [6, proof of (5.12)]. Writing $X_\lambda = n(\lambda) P_{n(\lambda)}$ as in the proof of Lemma (3.1) we see that $\| Y_\lambda \|_{\alpha,d,K}$ is also measurable.

Theorem 3 for Y_λ implies it for $P_n - P$, taking a smaller value of c and using (3.1). Let us prove it for Y_λ.

By assumption $d \geqq 2$. Let $J = [0, 1[$, so

$$J^{d-1} = \{x \in \mathbb{R}^{d-1}: 0 \leqq x_j < 1, j = 1, \ldots, d-1\}.$$

Let $x_{(d)} = \langle x_1, \ldots, x_{d-1} \rangle$, $x \in \mathbb{R}^d$.

Let f be a C^∞ function on $\mathbb{R}^{d-1}$ with $f(x) = 0$ outside the unit cube J^{d-1} and $f(x) > 0$ for x in the interior of J^{d-1}. We choose f so that

$$\sup_x \sup_{[p] \leqq d-1} |D^p f(x)| \leqq 1$$

and such that at all points x of the sub-cube

$$\{x: 1/3 \leqq x_j \leqq 2/3, j = 1, \ldots, d-1\}, \quad f(x) = \sup\{f(t): t \in I^{d-1}\} = \gamma,$$

say, $\gamma < 1$.

A sequence of sets and functions, some of them random, will be defined recursively as follows. As the jth stage, $j = 1, 2, \ldots$, J^{d-1} is decomposed into $3^{j(d-1)}$ disjoint sub-cubes A_{ji} of side 3^{-j} for $i = 1, \ldots, 3^{j(d-1)}$, where each A_{ji} is also a Cartesian product of left closed, right open intervals.

Let B_{ji} be a cube of side 3^{-j-1}, concentric with and parallel to A_{ji}. Let x_{ji} be the point of A_{ji} closest to 0 and y_{ji} the point of B_{ji} closest to 0. For $\delta > 0$ and $j = 1, 2, \ldots$, let

$$c_j = Cj^{-1} (\log(j+1))^{-1-2\delta}$$

where the contant $C > 0$ is chosen so that $\sum_{j \geqq 1} c_j \leqq 1$.

For $x \in \mathbb{R}^{d-1}$ let

$$f_{ji}(x) = c_j 3^{-j(d-1)} f(3^j(x - x_{ji})),$$
$$g_{ji}(x) = c_j 3^{-(j+1)(d-1)} f(3^{j+1}(x - y_{ji})).$$

Then the supports of f_{ji} and g_{ji} are the closures of A_{ji} and B_{ji} respectively. Note that on B_{ji}, $f_{ji}/g_{ji} \geqq 3^{d-1} > 1$.

Let $S_0 = 1/2$. We will define recursively a sequence of random variables $s_{ji}(\omega) = \pm 1$ and let

$$S_k = S_0 + \sum_{j=1}^{k} \sum_{i=1}^{3^{j(d-1)}} s_{ji}(\omega) f_{ji}. \tag{4.0}$$

Then since $\sum_{j \geqq 1} c_j 3^{-j(d-1)} \leqq 1/3$, $d \geqq 2$, we have $0 < S_k < 1$ for all k.

Given S_{j-1}, $j \geqq 1$, let $C_{ji} = C_{ji}(\omega) = \{x \in J^d \colon |x_d - S_{j-1}(x_{(d)})(\omega)| < g_{ji}(x_{(d)})\}$. Let $s_{ji}(\omega) = +1$ if $Y_\lambda(C_{ji}(\omega)) > 0$, otherwise $s_{ji}(\omega) = -1$. This completes the recursive definition of the s_{ji} and so of the S_k.

Recursively, one sees that each C_{ji} has only finitely many possible values, each on a measurable event, so the s_{ji} and S_k are all measurable.

Since the interiors of the A_{ji} are disjoint, we have for any $k \geqq 1$

$$\sup_{[p] \leqq d-1} \sup_x |D^p S_k(x)| \leqq \sum_{j=1}^{k} \sup_{[p] \leqq d-1} |D^p f_{j1}(x)| \leqq \sum_{j=1}^{k} c_j < 1.$$

The volume of $C_{ji}(\omega)$ is always

$$P(C_{ji}(\omega)) = 2\int g_{ji}\, dx = 2\mu c_j / 9^{(j+1)(d-1)} \tag{4.1}$$

where $0 < \mu = \int f dx < 1$.

Now let us show that $C_{ji}(\omega)$ for different i, j are always disjoint. For $1 \leqq j \leqq k$ and any ω, i and $x \in B_{ji}$ we have

$$\begin{aligned} |S_k(x) - S_{j-1}(x)|(\omega) &\geqq \gamma c_j 3^{-j(d-1)} - \sum_{r>j} \gamma c_r 3^{-r(d-1)} \\ &\geqq \gamma c_j 3^{-j(d-1)}(1 - \tfrac{1}{2}) \\ &\geqq \gamma c_j 3^{-j(d-1)}/2 \geqq \sup_y (g_{ji} + \sup_r g_{k+1,r})(y). \end{aligned}$$

Thus if $s_{ji}(\omega) = +1$, then for any r and any $x \in B_{ji}$,

$$S_k(x)(\omega) - g_{k+1,r}(x) \geqq S_{j-1}(x)(\omega) + g_{ji}(x),$$

so $C_{ji}(\omega)$ is disjoint from $C_{k+1,r}(\omega)$. They are likewise disjoint if $s_{ji}(\omega) = -1$, interchanging $+$ and $-$, $\geqq$ and $\leqq$.

For the same j and different i, the $C_{ji}(\omega)$ are disjoint since they project into disjoint B_{ji}.

Given $\lambda > 0$ let $r = r(\lambda)$ be the largest j, if one exists, such that $2\lambda\mu c_j \geqq 9^{(j+1)(d-1)}$. Then as $\lambda \to +\infty$, $r(\lambda) \sim (\log \lambda)/((d-1)\log 9)$.

Let $G_0(\omega) = \emptyset$. For $m = 1, 2, \ldots$, let

$$G(m)(\omega) = G_m(\omega) = \bigcup \{C_{ji}(\omega) \colon j \leqq m\}.$$

Let $H_m(\omega) = \{x \colon 0 \leqq x_d \leqq S_m(x_{(d)})(\omega)\}$, so that for all ω, $H_m \in \mathscr{C}(d-1, d, 2d)$. Let $A_m(\omega) = H_m(\omega) \setminus G_m(\omega)$. Then from the disjointness proof,

$$H_m(\omega) \cap G_m(\omega) = \bigcup \{C_{ji}(\omega) \colon j \leqq m,\ s_{ji} = +1\}.$$

For each m, each of these sets has finitely many possible values, each on a measurable event. For each m, we apply Lemma 3.4 to these G_j, with $A = A_m$. Then each G_j is $\mathscr{B}_{G(j-1)}$ measurable, $A_m(\cdot)$ is $\mathscr{B}_{G(m)}$ measurable, and the other hypotheses of Lemma (3.4) clearly hold. Thus, conditional on $\mathscr{B}_{G(m)}$, $X_\lambda(A)$ is Poisson with parameter $\lambda P(A_m(\omega))$. Also,

$$Y_\lambda((H_m \cap G_m)(\omega)) = \sum_{j=1}^{m} \sum_{i=1}^{3^{j(d-1)}} Y_\lambda(C_{ji})^+. \tag{4.2}$$

Now $P(C_{ji})$ does not depend on ω nor i, and $C_{ji}(\cdot)$ is $\mathscr{B}_{G(j-1)}$ measurable. Thus by Lemma (3.4), applied to $A=C_{mi}$, replacing G_m by $G_m \setminus C_{mi}$, the $Y_\lambda(C_{ji})$ for different i or j are jointly independent, and each has the law of $Y_\lambda(C)$ for a fixed set C with $P(C)=2\int g_{j1}\,dx$.

Taking $m=r(\lambda)$, (4.2) is a sum of independent nonnegative parts of centered Poisson variables with parameters $\lambda P(C_{ji}) \geqq 1$, by (4.1). Thus by Lemma (3.3), for a constant $c>0$,

$$\begin{aligned}
\lambda^{-1/2} E Y_\lambda((H_m \cap G_m)(\omega)) &\geqq c \sum_{j=1}^{r(\lambda)} \sum_{i=1}^{3^{j(d-1)}} P(C_{ji})^{1/2} \\
&= c \sum_{j=1}^{r(\lambda)} 3^{j(d-1)} (2\mu c_j / 9^{(j+1)(d-1)})^{1/2} \qquad \text{by (4.1)} \\
&= c\, 3^{1-d} (2\mu)^{1/2} \sum_{j=1}^{r(\lambda)} (C j^{-1} (\log(j+1))^{-1-2\delta})^{1/2} \qquad (\text{def. of } c_j) \\
&= c (2\mu C)^{1/2}\, 3^{1-d} \sum_{j=1}^{r(\lambda)} j^{-1/2} (\log(j+1))^{-0.5-\delta} \\
&\geqq a_d (\log(r(\lambda)+1))^{-0.5-\delta} \sum_{j=1}^{r(\lambda)} j^{-1/2} \\
&\geqq 2 a_d (r(\lambda)^{1/2} - 1)(\log(r(\lambda)+1))^{-0.5-\delta}
\end{aligned}$$

for some constant $a_d>0$. For λ large the above is

$$\geqq 3 b_d (\log \lambda)^{1/2} (\log\log \lambda)^{-0.5-\delta}$$

for some $b_d>0$.

By independence of the $Y_\lambda(C_{ji})$, the variance of $Y_\lambda((H_{r(\lambda)} \cap G_{r(\lambda)})(\omega)$ is less than

$$\sum_{j=1}^{r(\lambda)} \sum_{i=1}^{3^{j(d-1)}} \lambda P(C_{ji}) = \lambda \sum_{j=1}^{r(\lambda)} 3^{j(d-1)} 2\mu C j^{-1} (\log(j+1))^{-1-2\delta} q^{-(j+1)(d-1)} < \lambda.$$

Thus by Chebyshev's inequality,

$$\Pr\{\lambda^{-1/2} Y_\lambda(H_{r(\lambda)} \cap G_{r(\lambda)})(\omega) \geqq 2 b_d (\log \lambda)^{1/2} (\log\log \lambda)^{-0.5-\delta}\} \to 1$$

as $\lambda \to +\infty$.

By Lemma (3.4), the conditional distribution of $Y_\lambda(A_{r(\lambda)}(\omega))(\omega)$ given $\mathscr{B}_{G(r(\lambda))}$ is that of $Y_\lambda(D)$ for $P(D)=P(A_{r(\lambda)}(\omega))$, where $EY_\lambda(D)^2 \leqq \lambda$ for all D. Thus $EY_\lambda (D_{r(\lambda)})^2 \leqq \lambda$, $Y_\lambda(A_{r(\lambda)})/\lambda^{1/2}$ is bounded in probability, and

$$\Pr\{Y_\lambda(H_{r(\lambda))}) \geqq b_d (\lambda \log \lambda)^{1/2} (\log\log \lambda)^{-0.5-\delta}\} \to 1$$

as $\lambda \to +\infty$. This proves Theorem 3 (for Y_λ) if $K \geqq 2d$. For smaller values of $K>0$ we can just multiply the constant C (in c_j, f_{ji} and g_{ji}) by $K/(2d)$, completing the proof of Theorem 3.

5. Convex Sets and Lower Layers

We have measurability of the relevant norms as in [6, (4.3), (4.4), (5.13)]. Theorem 4 will be proved first for the convex sets in case $d=3$. In the proof of Theorem 3 in the last section, let us take $d=3$ and define S_k, $k \geqq 1$, by (4.0) with a *new* definition of S_0:

$$S_0 = \tfrac{3}{4} - (x_1 - \tfrac{1}{2})^2 - (x_2 - \tfrac{1}{2})^2.$$

Then since $\sum_{j \geqq 1} c_j/9^j < 1/4$, $0 < S_k < 1$ for all k. The rest of the proof remains the same, to prove Theorem 4 for the convex sets, except for the consideration of second derivatives, as follows. We now get

$$\sup_{[p] \leqq 2} \sup_x |D^p(S_k - S_0)(x)| \leqq 1.$$

Let $H(f)$ denote the Hessian matrix $H_{ij} = \partial^2 f/\partial x_i \partial x_j$ for the function f. Then $H(S_0) \equiv \begin{pmatrix} -2 & 0 \\ 0 & -2 \end{pmatrix}$ and $H(S_k - S_0)$ is symmetric with all its entries in $[-1, 1]$. Hence $-H(S_k)$ is everywhere nonnegative definite. Since S_k is C^∞, it is concave (Roberts and Varberg, 1973, pp. 100, 103) so that the set

$$H_{r(\lambda)}(\omega) = \{x \colon 0 \leqq x_d \leqq S_{r(\lambda)}(x_{(d)})(\omega)\}$$

is now convex. Thus Theorem 4 is proved for the convex sets in $\mathbb{R}^3$.

To prove Theorem 4 for lower layers, we take $d=2$ in the proof in Sect. 4 and make the following changes. Choose C now so that $\sum_j c_j < 1/2$. Let $S_0 = \frac{3}{4} - \frac{x}{2}$. Then $0 < S_k < 1$ for all k, since $\sum_{j \geqq 1} c_j 3^{-j} < 1/6$. Also

$$S_k'(x) \leqq -\tfrac{1}{2} + \sum_{j \geqq 1} c_j < 0$$

for all k and x. Thus each H_k is a lower layer. The rest of the proof works as before, proving Theorem 4.

References

1. Bakhvalov [Bahvalov], N.S.: On approximate calculation of multiple integrals (in Russian). Vestnik Mosk. Univ. Ser. Mat. Mekh. Astron. Fiz. Khim. **1959**, no. 4, 3–18 (1959)
2. Bennett, G.: Probability inequalities for sums of independent random variables. J. Amer. Statist. Assoc. **57**, 33–45 (1962)
3. Csörgő, M., Révész, P.: Strong Approximations in Probability and Statistics. Akadémiai Kiadó, Budapest (1981)
4. Dudley, R.M.: Sample functions of the Gaussian process. Ann. Probab. **1**, 66–103 (1973)
5. Dudley, R.M.: Metric entropy of some classes of sets with differentiable boundaries. J. Approximation Theory **10**, 227–236 (1974); Correction, ibid. **26**, 192–193 (1979)
6. Dudley, R.M.: Central limit theorems for empirical measures. Ann. Probab. **6**, 899–929 (1978); Correction, ibid. **7**, 909–911 (1979)

7. Dudley, R.M.: Lower layers in $\mathbb{R}^2$ and convex sets in $\mathbb{R}^3$ are not *GB* classes. Lecture Notes in Math. no. **709**, 97–102. Berlin-Heidelberg-New York: Springer 1979
8. Evstigneev, I.V.: "Markov times" for random fields. Theor. Probability Appls. **22**, 563–569 =Teor. Verojatnost. i Primenen. **22**, 575–581 (1977)
9. Feller, W.: An Introduction to Probability Theory and its Applications, vol. I, 3d ed.; vol. II, 2d ed. New York: Wiley 1968 and 1971
10. Kac, M.: On deviations between theoretical and empirical distributions. Proc. Nat. Acad. Sci. USA **35**, 252–257 (1949)
11. Kaufman, R., Walter Philipp: A uniform law of the iterated logarithm for classes of functions. Ann. Probab. **6**, 930–952 (1978)
12. Kaufman, Robert: Smooth functions and Gaussian processes. Approximation Theory III, ed. E.W. Cheney, p. 561–564. N.Y.: Academic Press 1980
13. Kolmogorov, A.N., Tikhomirov, V.M.: ε-entropy and ε-capacity of sets in functional spaces. Amer. Math. Soc. Transl. (Ser. 2) **17**, 277–364 (1961)=Uspekhi Mat. Nauk. **14**, vyp. 2(86), 3–86 (1959)
14. Kuelbs, J., Dudley, R.M.: Log log laws for empirical measures. Ann. Probab. **8**, 405–418 (1980)
15. Pyke, R.: The weak convergence of the empirical process with random sample size. Proc. Cambridge Philos. Soc. **64**, 155–160 (1968)
16. Rao, R. Ranga: Relations between weak and uniform convergence of measures with applications. Ann. Math. Statist. **33**, 659–680 (1962)
17. Révész, P.: On strong approximation of the multidimensional empirical process. Ann. Probab. **4**, 729–743 (1976)
18. Révész, P.: Three theorems of multivariate empirical process. Empirical Distributions and Processes, ed. P. Gaenssler and P. Révész. Lecture Notes in Math. **566**, 106–126. Berlin-Heidelberg-New York: Springer 1976
19. Roberts, A.W., Varberg, D.E.: Convex Functions. New York: Academic Press 1973
20. Schmidt, W.: Irregularities of distribution IX. Acta Arith. **27**, 385–396 (1975)
21. Steele, J. Michael: Empirical discrepancies and subadditive processes. Ann. Probab. **6**, 118–127 (1978)
22. Stute, W.: Convergence rates for the isotrope discrepancy. Ann. Probab. **5**, 707–723 (1977)
23. Sudakov, V.N.: Gaussian and Cauchy measures and ε-entropy. Soviet Math. Doklady **10**, 310–313 (1969)
24. Sun, Tze-Gong, Pyke, R.: Weak convergence of empirical processes. Technical Report, Dept. Statist. Univ. Washington, Seattle (1982)
25. Wright, F.T.: The empirical discrepancy over lower layers and a related law of large numbers. Ann. Probab. **9**, 323–329 (1981)

Received April 29, 1982; in final form June 15, 1982

Z. Wahrscheinlichkeitstheorie verw. Gebiete
62, 509–552 (1983)

Zeitschrift für
Wahrscheinlichkeitstheorie
und verwandte Gebiete

Invariance Principles for Sums of Banach Space Valued Random Elements and Empirical Processes*

R.M. Dudley[1] and Walter Philipp[2]

[1] Room 2-245, M.I.T., Cambridge, MA 02139, USA
[2] Department of Mathematics, University of Illinois, Urbana, IL 61801, USA

Summary. Almost sure and probability invariance principles are established for sums of independent not necessarily measurable random elements with values in a not necessarily separable Banach space. It is then shown that empirical processes readily fit into this general framework. Thus we bypass the problems of measurability and topology characteristic for the previous theory of weak convergence of empirical processes.

Contents

1. Introduction

Let F be a distribution function with mean zero and variance one. Then on some probability space there exist a sequence $\{X_j, j \geqq 1\}$ of independent random variables with common distribution function F and a sequence $\{Y_j, j \geqq 1\}$ of independent standard normal random variables such that

$$n^{-1/2} \max_{k \leqq n} |\sum_{j \leqq k} X_j - Y_j| \to 0 \quad \text{in probability} \tag{1.1}$$

* Both authors were partially supported by NSF grants. This work was done while the second author was visiting the M.I.T. Mathematics Department

E. Giné et al. (eds.), *Selected Works of R.M. Dudley*, Selected Works in Probability and Statistics,
DOI 10.1007/978-1-4419-5821-1_18, © Springer Science+Business Media, LLC 2010

(Freedman, 1971, (130), p. 83; Major, 1976). This result improves upon and is conceptually much simpler than Donsker's functional central limit theorem, commonly stated in terms of laws on $C[0, 1]$ (see e.g. Billingsley, 1968, p. 68).

There is an additional benefit to be gained when we formulate the corresponding result on empirical distribution functions in the spirit of (1.1). Let $\{x_j, j \geqq 1\}$ be a sequence of independent random variables each having a uniform distribution over $[0, 1]$. The empirical distribution function of a sample of size n is defined as

$$F_n(s) = n^{-1} \sum_{j \leqq n} 1\{x_j \leqq s\}, \qquad 0 \leqq s \leqq 1.$$

Let

$$X_j(s) = 1\{x_j \leqq s\} - s, \qquad 0 \leqq s \leqq 1. \tag{1.2}$$

Then $\{X_j, j \geqq 1\}$ is a sequence of independent identically distributed $(D[0, 1], \|\cdot\|)$-valued random variables with mean zero and uniformly bounded by 1 and $s \to n(F_n(s) - s)$ is the n-th partial sum of this sequence. (Here $\|\cdot\|$ denotes the supremum norm.) Komlós, Major and Tusnády (1975, Theorem 4) proved that: on some probability space there exist a sequence $\{x_j, j \geqq 1\}$ of independent random variables each having uniform distribution over $[0, 1]$ and a sequence $\{Y_j, j \geqq 1\}$ of independent $C[0, 1]$-valued random variables, each with the distribution of the standard Brownian bridge, such that for the X_j defined by (1.2)

$$\|\sum_{j \leqq n} X_j - Y_j\| = \mathcal{O}(\log n)^2 \text{ a.s.} \tag{1.3}$$

This improved on the original theorem of Kiefer (1972), the first of this type, which gave an error term $\mathcal{O}(n^{\varepsilon + 1/3})$, $\varepsilon > 0$. Such a result not only improves (it implies for instance the Smirnov-Chung law of the iterated logarithm) and is conceptually simpler than Donsker's theorem on empirical distribution functions, but also avoids the problems of measurability and topology caused by the fact that $(D[0, 1], \|\cdot\|)$ is not separable (see e.g. Billingsley, 1968, p. 153).

In this paper we shall use this idea to reformulate and strengthen the results of [12, 14, 15, 43] on empirical processes while removing the measurability conditions in most of them. We do this by proving invariance principles for sums of not necessarily measurable random elements with values in a not necessarily separable Banach space and by showing that empirical processes fit easily into this setup.

Before we state our main results we need to introduce some notation. Let $(A, \mathscr{A}, P)$ be a probability space. To provide an adequate setting for our results we consider $(A^\infty, \mathscr{A}^\infty, P^\infty)$, the countable product of copies of $(A, \mathscr{A}, P)$, with coordinates $\{x_j\}_{j \geqq 1}$ and define $(\Omega, \Sigma, \Pr)$ as the product of $(A^\infty, \mathscr{A}^\infty, P^\infty)$ and a copy of the unit interval with Lebesgue measure. This last makes Ω rich enough.

Let $(S, \|\cdot\|)$ be a Banach space and let h be a mapping: $X \to S$, not necessarily measurable. We call $X_j = h(x_j)$, $j \geqq 1$, a sequence of independent *identically formed* random elements. We also define

$$\Pr^*(E) = \inf\{\Pr(B) \colon B \supset E,\ B \in \Sigma\}, \qquad E \subset \Omega.$$

For not necessarily measurable real-valued functions g_n on Ω we say $g_n \to 0$ *in probability* iff $\lim_{n\to\infty} \Pr^*\{|g_n| > \varepsilon\} = 0$ for every $\varepsilon > 0$. We say $g_n \to 0$ *in* L^p iff there is a sequence $\{f_n, n \geqq 1\}$ of measurable functions $f_n \geqq |g_n|$ with $f_n \to 0$ in $L^p(\Omega, \Sigma, \Pr)$.

Theorem 1.1. *Let $\{X_j, j \geqq 1\}$ be a sequence of independent identically formed S-valued random elements $X_j = h(x_j)$, $j \geqq 1$. Suppose that for each $m \geqq 1$ there is a mapping $\Lambda_m \colon S \to S$ with the following properties:*

(1.4) *The linear span $L_m S$ of $\Lambda_m S$ is finite-dimensional, $m \geqq 1$.*

(1.5) *For each $m \geqq 1$ there is an $n_0 = n_0(m)$ such that for all $n \geqq n_0$*

$$\Pr^*\{n^{-1/2} \|\sum_{j \leqq n} X_j - \Lambda_m X_j\| \geqq 1/m\} \leqq 1/m.$$

(1.6) *For each $m \geqq 1$ the mapping $\Lambda_m \circ h$ is a measurable function from $(A, \mathscr{A})$ into $L_m S$.*

$$E(\Lambda_m X_1) = 0, \qquad E\|\Lambda_m X_1\|^2 < \infty, \qquad m \geqq 1. \tag{1.7}$$

Let T be the completion of the linear span of $\bigcup_{m \geqq 1} \Lambda_m S$, so T is a separable Banach space. Then there exists a sequence $\{Y_j, j \geqq 1\}$ of independent identically distributed T-valued Gaussian random variables defined on $(\Omega, \Sigma, \Pr)$ *such that*

$$EY_1 = 0 \tag{1.8}$$

$$E(s(Y_1)\, t(Y_1)) = \lim_{m\to\infty} E\{s(\Lambda_m X_1)\, t(\Lambda_m X_1)\}, \qquad s, t \in T', \tag{1.9}$$

where T' is the dual space of T, and as $n \to \infty$

$$n^{-1/2} \max_{k \leqq n} \|\sum_{j \leqq k} X_j - Y_j\| \to 0 \qquad \text{in Pr. and in } L^p \text{ for any } p < 2. \tag{1.10}$$

Moreover, if there is a function $F \geqq \|X_1\|$ with $EF^2 < \infty$ then the convergence in (1.10) *is also in L^2.*

Remarks. Condition (1.5) can be viewed as a tightness condition, close to the "flat concentration" condition of A. de Acosta (1970, p. 279) in separable Banach spaces, cf. also Pisier (1975, Théorème 3.1).

In our current applications (Theorem 1.3 and Sects. 6, 7 below) the maps Λ_m will be linear. We allow them to be non-linear in view of some difficulties of linear approximation shown to exist by Enflo (1973). Kuelbs (1976, Lemma 2.1) defines linear Λ_m in case S is separable and $E\|X_1\|^2 < \infty$. In any case non-linearity of Λ_m makes little difference in our proofs.

In § 8 below we show that if S is separable in Theorem 1.1, then X_j must actually be measurable for (1.10) to hold (although for applications to empirical processes we do need non-measurable X_j in non-separable spaces S). This suggests that the finite-dimensional measurability assumption (1.6) is not too restrictive.

We also note that (1.10) and an argument given by Pisier (1975, p. III.10) imply the existence of a function $F \geqq \|X_1\|$ with $EF^p < \infty$ for any $p<2$ (proved first, as Pisier states, by Jain, 1976a, Theorem 5.7, at least for symmetric variables) but not for $p=2$ (Jain, 1976b).

We define (as have others) the function L by setting $Lx=\log(x \vee e)$ and $L_2 = LL = L \circ L$. Pisier (1975) showed that in a separable Banach space, the central limit theorem and $E\|X_1\|^2 < \infty$ imply a compact law of the iterated logarithm. Heinkel (1979) proved a refinement, replacing $E\|X_1\|^2$ by $E\|X_1\|^2/L_2\|X_1\|$, partly based on methods of Kuelbs and Zinn (1979) who also (independently, as Heinkel notes) proved the refinement (Goodman, Kuelbs and Zinn, 1981). Our next result further improves these facts in different directions, dropping separability and weakening measurability assumptions and strengthening the conclusion to an almost sure invariance principle.

Theorem 1.2. *If in addition to the hypotheses of Theorem 1.1 there exists a measurable function $F \geqq \|X_1\|$ with $E(F^2/LLF) < \infty$, then the sequence $\{Y_j, j \geqq 1\}$ can be chosen such that (instead of (1.10)) there are measurable functions U_n with*

$$\|\sum_{j \leqq n} X_j - Y_j\| \leqq U_n = o((nLLn)^{1/2}) \text{ a.s.}, \quad n \to \infty. \tag{1.11}$$

Since $\{Y_j, j \geqq 1\}$ is a sequence of independent identically distributed Gaussian random variables with values in a separable Banach space, $\{Y_j, j \geqq 1\}$ satisfies a compact as well as a functional law of the iterated logarithm. Hence it is an immediate consequence of Theorem 1.2 that $\{X_j, j \geqq 1\}$ also satisfies the compact and the functional law of the iterated logarithm. (For details and precise statements of these results in separable spaces see [44, (1.4), (1.5), (1.19), (1.20)].)

The significance of Theorems 1.1 and 1.2 primarily lies in their applications to empirical processes. The empirical measure P_n is defined as

$$P_n(B) = n^{-1} \sum_{j \leqq n} 1\{x_j \in B\}, \quad B \in \mathscr{A} \tag{1.12}$$

and the normalized empirical measure ν_n as

$$\nu_n = n^{1/2}(P_n - P). \tag{1.13}$$

Theorem 1.3. *Let $\mathscr{F} \subset \mathscr{L}^2(A, \mathscr{A}, P)$ be a class of functions such that*

(1.14) *$\mathscr{F}$ is totally bounded in $\mathscr{L}^2$.*

(1.15) *For any $\varepsilon > 0$ there is a $\delta > 0$ and n_0 such that for all $n \geqq n_0$*

$$\Pr{}^*[\sup\{|\textstyle\int (f-g)\, d\nu_n| : f, g \in \mathscr{F}, \int (f-g)^2\, dP < \delta^2\} > \varepsilon] < \varepsilon.$$

Then there exists a sequence $\{Y_j, j \geqq 1\}$ of independent identically distributed Gaussian processes, defined on the probability space Ω, indexed by $f \in \mathscr{F}$ and with sample functions of Y_1 almost surely uniformly continuous on $\mathscr{F}$ in the $\mathscr{L}^2$ norm such that

$$EY_1(f) = 0 \quad \textit{for all } f \in \mathscr{F}, \tag{1.16}$$

(1.17) $$EY_1(f)\,Y_1(g)=\int fg\,dP-\int f\,dP\cdot\int g\,dP \quad \textit{for all } f,g\in\mathscr{F}$$

and as $n\to\infty$

(1.18) $$n^{-1/2}\max_{k\leqq n}\sup_{f\in\mathscr{F}}\Big|\sum_{j\leqq k}f(x_j)-\int f\,dP-Y_j(f)\Big|\to 0$$

in Pr. *as well as in* L^p *for any* $p<2$.

If in addition, there is a function $F\geqq|f|$ *for all* $f\in\mathscr{F}$ *with* $\int F^2\,dP<\infty$ *then we also have* L^2 *convergence in* (1.18).

If only $\int F^2/LLF\,dP<\infty$ *then the* Y_j *can be chosen such that, instead of* (1.18), *we have with probability* 1 *for some measurable* U_n

(1.19) $$\sup_{f\in\mathscr{F}}\Big|\sum_{j\leqq n}f(x_j)-\int f\,dP-Y_j(f)\Big|\leqq U_n=o((n\,LLn)^{1/2}).$$

Remark. If $\mathscr{F}=\{1_C\colon C\in\mathscr{C}\}$ for a collection $\mathscr{C}$ of sets, then under a measurability condition on $\mathscr{C}$ Theorem 1.2 of [12] states that the central limit theorem holds for empirical measures with respect to uniform convergence on $\mathscr{C}$ if and only if both (1.14) and (1.15) hold. The corresponding result for a class of functions is [14, Theorem 1.3].

Thus Theorem 1.3 applies to all classes $\mathscr{F}$ or $\mathscr{C}$ previously proved to be "Donsker classes" [12, 14–16]. Those papers defined spaces $D_i(\mathscr{F}, P)$ of bounded functions on $\mathscr{F}$, analogous to the Skorohod space $D[0,1]$, but depending on P. On these spaces D_i special σ-algebras were defined, e.g. generated by coordinate evaluations and balls for the sup norm. The present formulation allows us to work more simply in the space of all bounded functions on $\mathscr{F}$, with no special σ-algebra. We also pass (Theorem 1.2) from a central limit theorem to a law of the iterated logarithm under a sharp moment condition without further measurability assumptions such as those in [43], cf. also Kolčinski (1981b).

Theorem 1.3 follows easily from Theorems 1.1 and 1.2 and Lemma 1.4 below which takes care of the uniform continuity: let $m\geqq 1$ and $\varepsilon=1/m$. Choose δ and n_0 according to (1.15). Let $e_P(f,g)=(\int(f-g)^2\,dP)^{1/2}$. Since by (1.14) $\mathscr{F}$ is totally bounded for e_P there exist $f_k=f_{km}\in\mathscr{F}$, $1\leqq k\leqq N(\delta)$, say, such that for each $f\in\mathscr{F}$ there is a k with

(1.20) $$e_P(f,f_k)<\delta,\ k=k(f)\ \text{minimal.}$$

Hence by (1.15) in view of (1.12) and (1.13), for all $n\geqq n_0$

(1.21) $$\Pr{}^*\Big\{n^{-1/2}\sup_{f\in\mathscr{F}}\Big|\sum_{j\leqq n}(f-f_k)(x_j)-\int(f-f_k)\,dP\Big|>1/m\Big\}<1/m,$$

where f_k are chosen according to (1.20).

Now let S be the space of all bounded real-valued functions on $\mathscr{F}$. For $\psi\in S$ let $\|\psi\|=\sup\{|\psi(f)|\colon f\in\mathscr{F}\}$. Then $(S,\|\cdot\|)$ is a Banach space (non-separable for $\mathscr{F}$ infinite). We define the mapping $h\colon A\to S$ by setting $h(x)(f)=f(x)-\int f\,dP$ for each $x\in A$, and the mapping $\Lambda_m\colon S\to S$ by setting $\Lambda_m\psi(f)=\psi(f_k)$ with f_k from (1.20), for each $\psi\in S$. Letting $X_j=h(x_j)$ (as usual), we have

(1.22) $$(\Lambda_m X_j)(f)=(\Lambda_m h(x_j))(f)=f_k(x_j)-\int f_k\,dP,\ f\in\mathscr{F}.$$

Obviously, $\dim L_m S = N(\delta) < \infty$. Then (1.5) follows from (1.20) and (1.21) since we can assume without loss of generality that $\delta(\varepsilon)$ decreases as ε decreases. Next, (1.6) and (1.7) clearly hold. Now $(T, \|\cdot\|)$ is the closed linear span in S of the ranges of Λ_m, a separable Banach space. Theorem 1.1 implies that there exist independent identically distributed Gaussian variables $Y_j \in T$ satisfying (1.8), (1.9) and (1.10).

We next state a Lemma to be proved in Sect. 3:

Lemma 1.4. *If* (1.14) *and* (1.15) *hold and* Y_j *are* i.i.d. *variables in* T *satisfying* (1.8), (1.9) *and* (1.10) *then there is a Borel set* $W \subset T$, *consisting of functions uniformly continuous on* $\mathscr{I}$ *for* e_P, *such that* $Y_j \in W$ a.s.

Using this Lemma we now obtain Y_j as desired to satisfy (1.16), (1.17) and (1.18). Theorem 1.2 (to be proved in Sect. 5) then implies (1.19), proving Theorem 1.3.

To establish an invariance principle in the form (1.18) for empirical processes (or (1.19), if e.g. $\mathscr{I}$ is uniformly bounded) it is enough to prove (1.14) and (1.15). These two conditions have been established in [12, 14–16, 37, 56b] in various cases. Here we mention only a few of these results and refer the reader to Sect. 7 for a more complete treatment of the subject.

We consider the special case $\mathscr{I} = \{1_C : C \in \mathscr{C}\}$ for a collection $\mathscr{C} \subset \mathscr{A}$ of sets. For each $\delta > 0$ we define $N_I(\delta) = N_I(\delta, \mathscr{C}, P)$ to be the smallest number d of sets $A_1, \ldots, A_d \in \mathscr{A}$ with the following property. For each $C \in \mathscr{C}$ there exist A_r and A_s, $1 \leqq r, s \leqq d$ such that $A_r \subset C \subset A_s$ and $P(A_s \setminus A_r) < \delta$. Recall that $\log N_I(\delta)$ is called a metric entropy with inclusion [12]. Inspection of the proof of Theorem 5.1 of [12] shows that, as Roy Erickson and Joel Zinn pointed out to us,

$$\int_0^1 (\log N_I(x^2))^{1/2} \, dx < \infty \tag{1.23}$$

without any further measurability assumptions implies both (1.14) and (1.15). Hence we have the following result.

Theorem 1.5. *Let* $\mathscr{C} \subset \mathscr{A}$ *be a class of sets for which the entropy condition* (1.23) *holds. Then there exists a sequence* $\{Y_j, j \geqq 1\}$ *of independent identically distributed Gaussian processes, defined on the same probability space, indexed by* $C \in \mathscr{C}$ *and with sample functions of* Y_1 *almost surely uniformly continuous on* $\mathscr{C}$ *in the* d_P-*pseudometric which is defined by*

$$d_P(C, D) = P(C \triangle D), \quad C, D \in \mathscr{A}, \tag{1.24}$$

where $\{Y_j, j \geqq 1\}$ *has the following properties:*

$$E Y_1(C) = 0 \quad \textit{for all } C \in \mathscr{C}, \tag{1.25}$$

$$E\{Y_1(C)\, Y_1(D)\} = P(C \cap D) - P(C)\, P(D) \tag{1.26}$$

for all $C, D \in \mathscr{C}$, *and as* $n \to \infty$

$$n^{-1/2} \max_{k \leqq n} \sup_{C \in \mathscr{C}} \left| \sum_{j \leqq k} 1\{x_j \in C\} - P(C) - Y_j(C) \right| \to 0 \tag{1.27}$$

in probability as well as in L^2. *Or,* Y_j *can be chosen to satisfy, instead of* (1.27), *as* $n \to \infty$

$$(1.28) \qquad \sup_{C \in \mathscr{C}} |\sum_{j \leq n} 1\{x_j \in C\} - P(C) - Y_j(C)| = o((nL_2 n)^{1/2}) \quad \text{a.s.}$$

I.S. Borisov (1981) has shown that the sufficient condition (1.23) on N_I cannot be weakened, being necessary in case $\mathscr{C}$ is the collection of all subsets of a countable set X, where it is equivalent to $\sum_{x \in X} P\{x\}^{1/2} < \infty$ (cf. also [16]).

A collection $\mathscr{C} \subset \mathscr{A}$ is called a Vapnik-Červonenkis class if for some $n < \infty$, no set F with n elements has all its subsets of the form $C \cap F$, $C \in \mathscr{C}$. The Vapnik-Červonenkis number $V(\mathscr{C})$ denotes the smallest such n.

Note. $V(\mathscr{C})$ does not depend on P.

For Vapnik-Červonenkis classes $\mathscr{C}$ satisfying some measurability conditions - which to a certain extent are also necessary [16] - it is shown [12, Theorem 7.1, Correction], [56a], that (1.14) and (1.15) hold for $\mathscr{F} = \{1_B : B \in \mathscr{C}\}$, and hence all the conclusions of Theorem 1.5 will hold. Conversely, if Y_1 exists as in Theorem 1.5 for all P on $\mathscr{A}$, then $V(\mathscr{C}) < \infty$ [16].

Bounds on the growth of $\dim L_m S$ as $m \to \infty$ in (1.4) will give, in Sect. 6 below, improvements of the error term in Theorem 1.2. Applying these results in Sect. 7 to empirical processes for classes of sets we obtain sharper error terms in (1.28), first if $\log N_I(x) \leq c x^{-\tau}$ for constants $c < \infty$ and $0 \leq \tau < 1$. For Vapnik-Červonenkis classes of sets we can improve the error term in (1.28) to $\mathcal{O}(n^{1/2-\lambda})$, for any $\lambda < 1/(2700\, V(\mathscr{C}))$.

Let us recall that a) for any k-dimensional real vector space $\mathscr{V}$ of functions on X, with $k < \infty$, the collection $\mathscr{C}$ of all sets of the form $\{x \in X : f(x) > 0\}$, $f \in \mathscr{V}$, is a Vapnik-Červonenkis class with $V(\mathscr{C}) = k+1$ [12, Theorem 7.2]; b) for any $\mathscr{C}$ with $V(\mathscr{C}) < \infty$ and $m < \infty$, the collection of all sets formed from elements of $\mathscr{C}$ by at most m Boolean operations (unions, intersections, and set differences) is also a Vapnik-Červonenkis class [12, Proposition 7.12]; c) if $V(\mathscr{C}_i) < \infty$, $i = 1, \ldots, m < \infty$, then $V(\bigcup_{i \leq m} \mathscr{C}_i) < \infty$. Combining a), b) and c) one obtains a good supply of Vapnik-Červonenkis classes; for example, polyhedra in $\mathbb{R}^k$ with at most m faces. The sets $\{y : y_j \leq x_j,\ 1 \leq j \leq k\}$, $x \in \mathbb{R}^k$, used in defining empirical distribution functions, form a still more special Vapnik-Červonenkis class, with $V(\mathscr{C}) = k+1$ [63, Proposition 2.3].

In Theorem 7.5 we apply Theorem 1.3 to weighted empirical distribution functions, thereby improving the compact law of the iterated logarithm, due to Goodman, Kuelbs and Zinn (1981), to an almost sure invariance principle.

Throughout, for any functions $f, g, f \ll g$ means the same as $f = \mathcal{O}(g)$, i.e. f/g is bounded, as $n \to \infty$ or under whatever is the designated condition.

2. Independent Random Elements

We need to generalize several lemmas on sums of independent random variables to random elements, not necessarily measurable, but independent in an extended sense. Let $(\Omega, \mathcal{S}, P)$ be a probability space. Let $\mathcal{L}^0(\Omega, \mathcal{S}, P)$ be the set of $\mathcal{S}$-measurable functions $f: \Omega \to [-\infty, \infty]$. Let $L^0(\Omega, \mathcal{S}, P)$ denote the set of equivalence classes of functions in $\mathcal{L}^0(\Omega, \mathcal{S}, P)$ for the relation of equality P-almost surely. In $L^0(\Omega, \mathcal{S}, P)$ (which, with values $\pm\infty$ allowed, is *not* a vector space) we define a metric by

$$d(f, g) = \inf\{\varepsilon > 0: P(|\tan^{-1} f - \tan^{-1} g| > \varepsilon) < \varepsilon\}.$$

Then (L^0, d) is a separable metric space.

For given $\mathcal{J} \subset \mathcal{L}^0$ let

$$j = \text{ess.}\inf \mathcal{J}$$

iff $j \leqq h$ almost surely for all $h \in \mathcal{J}$ and whenever $g \leqq h$ almost surely for all $h \in \mathcal{J}$ then $g \leqq j$ almost surely. For all $\mathcal{J} \subset \mathcal{L}^0$ the ess. inf $\mathcal{J}$ exists and is uniquely determined with probability 1. Indeed, we can choose a sequence $\{j_n, n \geqq 1\}$ in $\mathcal{J}$ with $\int \tan^{-1} j_n dP \downarrow \inf\{\int \tan^{-1} j dP: j \in \mathcal{J}\}$. Then $\min_{k \leqq n} j_k \downarrow j = \text{ess.}\inf \mathcal{J} \in \mathcal{L}^0(\Omega, \mathcal{S}, P)$ (cf. e.g. Vulikh, 1967, pp. 79–79). The next two lemmas are straightforward. The notion of "measurable cover function" f^* was defined by Eames and May (1967), by a different method (for bounded functions) which turns out to be equivalent. See also May (1973).

Lemma 2.1. *For each function* $f: \Omega \to [-\infty, \infty]$

$$f^* = \text{ess.}\inf\{j \in \mathcal{L}^0(\Omega, \mathcal{S}, P): j \geqq f\}$$

exists and is $\mathcal{S}$*-measurable. Moreover, we can take* $f^* \geqq f$ *everywhere. Further, for all* $g: \Omega \to [-\infty, \infty]$,

$$(f+g)^* \leqq f^* + g^* \text{ a.s. } \textit{if both sums are defined} \text{ a.s.};$$
$$(f-g)^* \geqq f^* - g^* \text{ a.s. } \textit{if both differences are defined} \text{ a.s.}$$

Remark. If $f > -\infty$ and $g > -\infty$ everywhere then also $f^* > -\infty$ and $g^* > -\infty$ everywhere so $f+g$ and $f^* + g^*$ are everywhere defined.

Lemma 2.2. *Let* $(S, \|\cdot\|)$ *be a vector space with a seminorm* $\|\cdot\|$. *Then for all* $X, Y: \Omega \to S$,

$$\|X+Y\|^* \leqq (\|X\| + \|Y\|)^* \leqq \|X\|^* + \|Y\|^* \text{ a.s.},$$
$$\|cX\|^* = |c|\, \|X\|^* \text{ a.s. } \textit{for each } c \in \mathbb{R}.$$

Remarks. For many calculations Lemma 2.2 allows us to treat $\|X\|^*$ much as $\|X\|$. Can $\|\cdot\|^*$ be made into an actual seminorm on a space of S-valued functions? Let $(\Omega, \mathcal{S}, P)$ be any complete probability space and $(S, \|\cdot\|)$ any seminormed vector space. Let $\mathcal{J}^\infty(S, P)$ be the set of all bounded functions f from Ω into S. Then using a lifting of the space $\mathcal{L}^\infty(\Omega, \mathcal{S}, P)$ of real-valued functions (A. and C. Ionescu Tulcea, 1961) we can define $\|f\|^*$ so that for all $\omega, f \to \|f\|^*(\omega)$ is actually a seminorm on $\mathcal{J}^\infty(S, P)$. For unbounded functions, however, unless P is atomic one does not have a lifting of $L^p(\Omega, \mathcal{S}, P)$, $1 \leqq p < \infty$ (A. and C. Ionescu Tulcea, 1962, Theorem 7).

Lemma 2.3. *Let $(A_j, \mathscr{A}_j, P_j)$ be any probability spaces. Let $f_j\colon A_j \to [0, \infty]$ be any functions, $j=1, \ldots, n$. Then on the Cartesian product $\prod_{j=1}^{n} (A_j, \mathscr{A}_j, P_j)$ with coordinate functions x_j,*

$$\left(\prod_{j=1}^{n} f_j(x_j)\right)^* = \prod_{j=1}^{n} f_j^*(x_j) \text{ a.s.} \tag{2.0}$$

where we set $0 \cdot \infty = 0$. If $n=2$ and $f_1 \equiv 1$ then the same holds for any $f_2\colon A_2 \to [-\infty, \infty]$.

Proof. Clearly $\left(\prod_{j=1}^{n} f_j\right)^* \leqq \prod_{j=1}^{n} f_j^*$. For the converse, by induction, we can take $n=2$. Suppose g is measurable on $A_1 \times A_2$ and for $\varepsilon \geqq 0$ let

$$C(\varepsilon) := \{(x, y)\colon g(x, y) + \varepsilon < f_1^*(x) f_2^*(y)\}.$$

Suppose $(P_1 \times P_2)(C(0)) > 0$. Then for some $\varepsilon > 0$, $(P_1 \times P_2)(C(\varepsilon)) > 0$. Fix such an ε. For $m = 1, 2, \ldots$, let $B_m = \{y\colon m < f_2^*(y) < +\infty\}$. Then for some m, $(P_1 \times P_2)(C(\varepsilon) \setminus (A_1 \times B_m)) > 0$. Fix such an m and let $D := C(\varepsilon) \setminus (A_1 \times B_m)$, $D_x := \{y\colon \langle x, y \rangle \in D\}$, and $H := \{x\colon P_2(D_x) > 0\}$. Suppose $f_1(x) f_2(y) \leqq g(x, y)$ everywhere. Let $x \in H$. If $f_1(x) = +\infty$, then $f_2 \geqq 0$ and for P_2-almost all $y \in D_x$, $f_1(x) f_2(y) < f_1^*(x) f_2^*(y)$, so $f_2(y) = 0 = f_2^*(y)$, a contradiction. If $0 < f_1(x) < \infty$, then for P_2-almost all $y \in D_x$, $f_2^*(y) \leqq g(x, y)/f_1(x)$, so

$$f_2^*(y) < (f_1^*(x) f_2^*(y) - \varepsilon)/f_1(x).$$

Then $f_2^*(y) < +\infty$, so $f_2^*(y) \leqq m$. If $f_2^*(y) \leqq 0$, we get a contradiction since $f_1^*(x) \geqq f_1(x) > 0$. So for any such y, $0 < f_2^*(y) \leqq m$ and

$$f_1(x) < f_1^*(x) - \varepsilon/m.$$

If $f_1 \equiv 1$ this is a contradiction and finishes the proof in that case. In the case $f_j \geqq 0$, $j = 1, 2$, we have

$$f_1(x) \leqq \max(0, f_1^*(x) - \varepsilon/m)$$

for all $x \in H$. If $f_1^* > 0$ on some subset J of H with $P_1(J) > 0$ this allows f_1^* to be chosen smaller, a contradiction. So $f_1 = f_1^* = 0$ a.e. on H, but then $0 \leqq g < 0$ on D, again a contradiction. Q.E.D.

Lemma 2.4. *Let*

$$(\Omega, \mathscr{S}, P) = (\Omega_1 \times \Omega_2 \times \Omega_3, \mathscr{S}_1 \times \mathscr{S}_2 \times \mathscr{S}_3, P_1 \times P_2 \times P_3)$$

and denote the projections $\pi_i\colon \Omega \to \Omega_i$ $(i = 1, 2, 3)$. Then for any bounded non-negative function f

$$E\{f^*(\omega_1, \omega_3) \mid (\pi_1, \pi_2)^{-1}(\mathscr{S}_1 \times \mathscr{S}_2)\} = E\{f^*(\omega_1, \omega_3) \mid \pi_1^{-1}(\mathscr{S}_1)\} \text{ a.s. } (P).$$

Proof. By Lemma 2.3 (with $f_2(\omega_2)=1$) $f^*(\omega_1,\omega_3)$ equals P-almost surely a measurable function not depending on ω_2 thus is independent of $\pi_2^{-1}(\mathcal{S}_2)$. □

Recall the definitions of convergence in probability or in L^p for non-measurable functions just above Theorem 1.1.

Lemma 2.5. *Let $X:\Omega\to\mathbb{R}$. Then, for all $t\in\mathbb{R}$ and $\varepsilon>0$*

$$P^*(X\geqq t)\leqq P(X^*\geqq t)\leqq P^*(X\geqq t-\varepsilon).$$

For any $X_n:\Omega\to\mathbb{R}, X_n\to 0$ in probability or in L^p if and only if $|X_n|^*\to 0$ in probability or in L^p, respectively.

Proof. Since $\{X\geqq t\}\subset\{X^*\geqq t\}$ which is measurable the first inequality follows. To prove the second inequality consider the sets $C_j=\{\omega: X\geqq j\varepsilon\}$, $j\in\mathbb{Z}$. Let $D_j\supset C_j$ be a measurable cover of C_j, i.e. $P^*(C_j)=P(D_j)$. Without loss of generality we can assume that the sequence $\{D_j\}$ is non-increasing. We have $\bigcup_j D_j=\bigcup_j C_j=\Omega$ since $X(\omega)>-\infty$ for all ω. Let

$$\begin{aligned} Y(\omega)&:=(j+1)\,\varepsilon \quad \text{on } D_j\setminus D_{j+1} \quad j\in\mathbb{Z}\\ &:=+\infty \quad \text{on } \bigcap_j D_j. \end{aligned}$$

We claim that for all ω

$$X^*(\omega)\leqq Y(\omega). \tag{2.1}$$

Here Y is measurable. Where $Y(\omega)=+\infty$ the result is clear. Otherwise, $\omega\in D_j\setminus D_{j+1}$ for some j and so $\omega\notin C_{j+1}$. Thus $X(\omega)<(j+1)\,\varepsilon=Y(\omega)$; this proves (2.1).

Given $t\in\mathbb{R}$ there is a unique $j\in\mathbb{Z}$ such that $j\varepsilon\leqq t<(j+1)\,\varepsilon$. Then

$$P(X^*\geqq t)\leqq P(X^*\geqq j\,\varepsilon)\leqq P(Y\geqq j\,\varepsilon).$$

But $\{Y\geqq j\,\varepsilon\}=D_{j-1}$. Thus

$$P(D_{j-1})=P^*(C_{j-1})=P^*(X\geqq (j-1)\,\varepsilon)\leqq P^*(X\geqq t-2\,\varepsilon).$$

The second sentence of the lemma follows directly. □

As in Sect. 1 let $(A^\infty,\mathcal{A}^\infty,P^\infty)$ be the product of countably many copies of $(A,\mathcal{A},P)$. Then the sequence $\{x_j, j\geqq 1\}$, where x_j is the j-th coordinate of $x\in A^\infty$, is a sequence of independent random variables with common distribution P. We extend the concept of independence by calling $\{X_j, j\geqq 1\}$ a sequence of independent elements, if we can write each $X_j=h_j(x_j)$ for some function h_j with domain X. If h_j is a measurable function from $(A,\mathcal{A})$ into some measurable space this implies the usual notion of independence.

The next lemma is an extension of Lemma 2.1 of Kuelbs (1977). Notice that neither the existence (nor the vanishing) of EX_j is needed.

Lemma 2.6. *Let $\{X_j, 1 \leqq j \leqq n\}$ be an independent sequence where the random elements $X_j = h_j(x_j)$ take values in some vector space $(S, \|\cdot\|)$. Let $S_n := \sum_{j \leqq n} X_j$ and $\tau_n \geqq \sum_{j \leqq n} E\|X_j\|^{*2}$. Suppose that*

$$\|X_j\| \leqq M, \quad 1 \leqq j \leqq n. \tag{2.2}$$

Then for $0 \leqq \gamma \leqq 1/(2M)$

$$E\exp(\gamma\|S_n\|^*) \leqq \exp(3\gamma^2\tau_n + \gamma E\|S_n\|^*). \tag{2.3}$$

It follows that for any $K > 0$

$$\Pr\{\|S_n\|^* \geqq K\} \leqq \exp(3\gamma^2\tau_n - \gamma(K - E\|S_n\|^*)), \quad 0 \leqq \gamma \leqq 1/(2M). \tag{2.4}$$

Remarks. If $K < E\|S_n\|^*$ then the infimum of the right side of (2.4) is attained at $\gamma = 0$. If $0 \leqq K - E\|S_n\|^* \leqq 3\tau_n/M$ then the infimum is attained at $\gamma = (K - E\|S_n\|^*)/(6\tau_n)$ and equals $\exp(-(K - E\|S_n\|^*)^2/(12\tau_n))$. If $K - E\|S_n\|^* > 3\tau_n/M$ then the infimum for $0 \leqq \gamma \leqq 1/(2M)$ is at $1/(2M)$ and equals

$$\exp(3\tau_n/(4M^2) - (K - E\|S_n\|^*)/(2M)) \leqq \exp(-(K - E\|S_n\|^*)/(4M)).$$

Proof. We first observe that (2.2) is equivalent to $\|X_j\|^* \leqq M$, $1 \leqq j \leqq n$. Throughout the proof of Lemma 2.1 of Kuelbs (1977) we replace $\|\cdot\|$ by $\|\cdot\|^*$ and note that $E_{k+1}\|Y_k\|^* = E_k\|Y_k\|^*$ by our Lemma 2.4 with $\omega_1 = (X_1, \dots, X_{k-1})$, $\omega_2 = X_k$ and $\omega_3 = (X_{k+1}, \dots, X_n)$. (Trivial correction: in the first equation of (2.9) in [41] replace "$-$" by "$+$".) After obtaining [41], (2.4) with $\|\cdot\|^*$ in place of $\|\cdot\|$ we set $\gamma = \varepsilon/(2b_n)$. Note that given $\gamma > 0$ and $M > 0$ we can choose ε, b_n and $c > 0$ satisfying the three conditions $\gamma = \varepsilon/(2b_n)$, $M \leqq b_n c$ and $\varepsilon c \leqq 1$, if and only if $\gamma \leqq 1/(2M)$. This gives our (2.3). Since for any $K \geqq 0$,

$$\Pr\{\|S_n\|^* \geqq K\} \leqq \exp(-\gamma K)\, E\exp(\gamma\|S_n\|^*)$$

relation (2.4) follows. □

We also need an extension of Ottaviani's inequality.

Lemma 2.7. *Let $\{X_j, 1 \leqq j \leqq n\}$ be an independent sequence where the random elements X_j assume values in a vector space $(S, \|\cdot\|)$. Write $S_n = \sum_{j \leqq n} X_j$ and suppose that*

$$\max_{j \leqq n} \Pr(\|S_n - S_j\|^* > \alpha) = c < 1.$$

Then

$$\Pr(\max_{j \leqq n} \|S_j\|^* > 2\alpha) \leqq (1 - c)^{-1} \cdot \Pr(\|S_n\|^* > \alpha).$$

Proof. In Lemma 3.21 of Breiman (1968, p. 45) and its proof we replace $|\cdot|$ by $\|\cdot\|^*$ throughout using Lemmas 2.1 and 2.2, specifically $\|S_j\|^* \leqq \|S_n\|^* + \|S_n - S_j\|^*$. In the step where the independence is to be used we argue as follows.

Let $\omega_1=(x_{j+1},\dots,x_n)$ and $\omega_2=(x_1,\dots,x_j)$. Then $F(\omega_1,\omega_2)=S_n-S_j$ only depends on ω_1. By Lemma 2.3 $\|S_n-S_j\|^*$ only depends on ω_1 and thus is independent of the event $\{j^*=j\}$ in the usual sense. The remainder of the proof requires no changes. □

The following lemma is based on an argument of Kahane (1968, p. 16).

Lemma 2.8. *Let* $\{X_j, 1\leqq j\leqq n\}$ *be an independent sequence where the random elements* X_j *assume values in a vector space* $(S, \|\cdot\|)$. *Write* $S_n=\sum_{j\leqq n} X_j$ *and* $S_0=0$. *Suppose that for some* $K>0$

$$\Pr(\|S_j-S_i\|^*\geqq K)\leqq \tfrac{1}{2} \qquad 0\leqq i<j\leqq n. \tag{2.5}$$

Then for any $t>K$ *and* $s\geqq 0$

$$\Pr\{\|S_n\|^*\geqq 4t+s\}\leqq 4(\Pr\{\|S_n\|^*\geqq t\})^2+\Pr\{\max_{m\leqq n}\|X_m\|^*\geqq s\}.$$

Proof. Let $T(\omega)=\min\{j\geqq 1: \|S_j(\omega)\|^*\geqq 2t\}$. Then by Lemma 2.1

$$\begin{aligned}\Pr\{\|S_n\|^*\geqq 4t+s\} &= \sum_{1\leqq m\leqq n}\Pr\{T=m, \|S_n\|^*\geqq 4t+s\}\\ &\leqq \sum_{1\leqq m\leqq n}\Pr\{T=m, \|S_n\|^*\geqq 4t+s, \|X_m\|^*<s\}\\ &\quad +\Pr\{\max_{m\leqq n}\|X_m\|^*\geqq s\}.\end{aligned}\tag{2.6}$$

By Lemma 2.2, $\|S_n\|^*\leqq\|S_{m-1}\|^*+\|X_m\|^*+\|S_n-S_m\|^*$, so the sum on the right side of (2.6) does not exceed

$$\begin{aligned}&\sum_{1\leqq m\leqq n}\Pr\{T=m, \|S_n-S_m\|^*\geqq 2t\}\\ &\quad=\sum_{1\leqq m\leqq n}\Pr\{T=m\}\cdot\Pr\{\|S_n-S_m\|^*\geqq 2t\}\\ &\quad\leqq \Pr\{\max_{0\leqq m<n}\|S_n-S_m\|^*\geqq 2t\}\cdot\sum_{1\leqq m\leqq n}\Pr\{T=m\}\\ &\quad\leqq 2\Pr\{\|S_n\|^*\geqq t\}\cdot\Pr\{\max_{m\leqq n}\|S_m\|^*\geqq 2t\}\\ &\quad\leqq 4(\Pr\{\|S_n\|^*\geqq t\})^2\end{aligned}$$

using Lemma 2.7 twice, the first time reversing the order of summation. Also note that the independence is used in the same way as in the proof of Lemma 2.7. The lemma follows now from (2.6).

A random element X will be called *symmetric* if $X=h(x)-h(x')$ where $h: A\to S$ and x and x' are independent A-valued random variables on Ω with the same law. The proofs of the next two lemmas follow those in the given references, with $\|\cdot\|^*$ replacing $\|\cdot\|$ and other changes just as in the proofs of Lemma 2.7 and 2.8 above.

Lemma 2.9 (P. Lévy inequality). *Let $\{X_j, 1 \leqq j \leqq n\}$ be a sequence of independent symmetric random elements with values in $(S, \|\cdot\|)$ and partial sums S_k. Then*

$$\Pr(\max_{j \leqq n} \|S_j\|^* > \alpha) \leqq 2 \Pr(\|S_n\|^* > \alpha).$$

Proof. See Kahane (1968, p. 12).

Lemma 2.10. *Let $\{X_j, 1 \leqq j \leqq n\}$ be as above. Then for all $s, t > 0$*

$$\Pr\{\|S_n\|^* \geqq 2t + s\} \leqq 4(\Pr\{\|S_n\|^* \geqq t\})^2 + \Pr\{\max_{m \leqq n} \|X_m\|^* \geqq s\}.$$

Proof. See Kahane [35, p. 16], [26, p. 164], or [31, Lemma 3.4].

Lemma 2.11. *Let S and T be Polish spaces and Q a law on $S \times T$, with marginal μ on S. Let $(\Omega, \mathscr{B}, P)$ be a probability space and X be a random variable on Ω with values in S and law $\mathscr{L}(X) = \mu$. Assume that there is a random variable U on Ω, independent of X, with values in a separable metric space R and law $\mathscr{L}(U)$ on R having no atoms. Then there exists a random variable Y on Ω with values in T and $\mathscr{L}(\langle X, Y \rangle) = Q$.*

Proof. First, we may assume R is complete, hence Polish. Now, any uncountable Polish space is Borel-isomorphic with $[0, 1]$ (e.g. Parthasarathy, 1967, p. 14). Every Polish space is Borel-isomorphic with some compact subset of $[0, 1]$. Thus there is no loss of generality in assuming that $S = T = R = [0, 1]$ with the usual topology, metric and Borel structure.

Next, we take a disintegration of Q on $[0, 1] \times [0, 1]$ (e.g. N. Bourbaki, 1959, Chap. 6, pp. 58–59). Namely, there is a map $s \to Q_s$ from S into the set of all laws on T and such that $\int Q_s \, d\mu(s) = Q$, i.e. for any bounded, Borel measurable function f on $[0, 1] \times [0, 1]$ we have

$$\int f(s, t) \, dQ(s, t) = \int_0^1 \int_0^1 f(s, t) \, dQ_s(t) \, d\mu(s)$$

where all these integrals are defined (possibly for the completion of μ).

For each s let F_s be the distribution function of Q_s and for $0 \leqq t \leqq 1$ let $F_s^{-1}(t) = \inf\{z : F_s(z) \geqq t\}$. We may assume that U has uniform distribution over $[0, 1]$, since if H is the distribution function of U, which has no atoms, then $H(U)$ is uniformly distributed over $[0, 1]$. Now for each t, the map $s \to F_s^{-1}(t)$ is measurable since $F_s^{-1}(t) \leqq \alpha$ iff $F_s(\alpha) \geqq t$, and since the map $s \to F_s(\alpha) = Q_s([0, \alpha])$ is measurable by a property of the disintegration. Since $F_s^{-1}(\cdot)$ is non-decreasing and left-continuous we have

$$F_s^{-1}(t) = \lim_{n \to \infty} \sum_{j=0}^{n} F_s^{-1}(j/n) \, 1\{j/n \leqq t < (j+1) n\}$$

which is jointly measurable in (s, t). Hence

$$Y(\omega) := F_{X(\omega)}^{-1}(U(\omega))$$

is a random variable. Moreover, for any bounded Borel function g on $[0, 1] \times [0, 1]$ we have using Fubini's theorem and the fact that $\lambda \circ (F_s^{-1})^{-1} = Q_s$ (here

λ denotes Lebesgue measure)

$$\begin{aligned}\int g\,dQ &= \int_0^1\int_0^1 g(s,t)\,dQ_s(t)\,d\mu(s)\\ &= \int_0^1\int_0^1 g(s,F_s^{-1}(t))\,dt\,d\mu(s)\\ &= \int_0^1\int_0^1 g(s,F_s^{-1}(t))\,d(\mu\times\lambda)(s,t)\\ &= Eg(X,F_X^{-1}(U)) = Eg(X,Y)\end{aligned}$$

since X and U are independent and thus $\mathscr{L}(\langle X, U\rangle)=\mu\times\lambda$. Consequently $\mathscr{L}(\langle X, Y\rangle)=Q$. □

Note Added in Proof

Lemma 2.11 also follows from the proof of Theorem 1 of Skorohod, Theory Prob. Appl. **21**, 628–632 (1976).

We thank E. Berger for this remark.

For two laws μ and ν on a separable metric space (S,ρ) recall the Prohorov distance defined by

$$\pi(\mu,\nu):=\inf\{\varepsilon>0\colon \mu(A)\leqq\nu(A^{\varepsilon})+\varepsilon \text{ for all closed } A\subset S\}$$

where

$$A^{\varepsilon}=\{x\in S\colon \inf_{y\in A}\rho(x,y)<\varepsilon\}.$$

The following result is a special case of an extension by Dehling (1982, Prop. 5.1 and Lemma 5.1) of a theorem of Yurinskii (1977, Theorem 1). Where Yurinskii assumed third moments and a Euclidean (Hilbert) norm, Dehling by truncation used $(2+\delta)$th moments, $0<\delta\leqq 1$, and Banach norms (via Lindenstrauss and Tzafriri, 1977, p. 17) as follows:

Lemma 2.12. *Let $\{\xi_j, j\geqq 1\}$ be independent, identically distributed random variables in a d-dimensional Banach space $(B,\|\cdot\|)$, $d<\infty$, with $E\xi_1=0$ and $E\|\xi_1\|^{2+\delta}<\infty$ for a δ with $0<\delta\leqq 1$. Then for the Gaussian law μ with mean 0 and the covariance of ξ_1, and $T_n:=\sum_{j\leqq n}\xi_j$, we have*

$$\pi(\mathscr{L}(n^{-1/2}T_n),\mu)\leqq C d^{4/3} n^{-\delta/9}(E\|\xi_1\|^{2+\delta}+1)^{1/4}$$

where C is an absolute constant.

Remarks. If in the proof of Dehling (1982, Prop. 5.1) we use the truncation $Y_i=\xi_i\cdot 1\{\|\xi_i\|\leqq n^{3/(2\delta+6)}\}$ then we can replace $n^{-\delta/9}$ by $n^{\varepsilon-\delta/(2\delta+6)}$ for any $\varepsilon>0$ or, apparently, for $\varepsilon=0$ (Senatov, 1980). Some reduction in the exponent 4/3 is also easy. But in our applications in Theorems 6.1 and 6.2 below this would lead to no major improvement.

We will use the following fact, which is essentially [4, Lemma A1], and which is also a special case of a generalized Vorob'ev theorem (Shortt, 1982, Theorem 2.6; Vorob'ev, 1962).

Lemma 2.13. *Let X, Y and Z be Polish spaces. Suppose μ is a law on $X\times Y$ and ν a law on $Y\times Z$ such that μ and ν have the same marginal on Y. Then there is a law on $X\times Y\times Z$ with marginals μ on $X\times Y$ and ν on $Y\times Z$.*

3. Proof of Theorem 1.1 and Lemma 1.4

To prove Theorem 1.1 we first show the existence of the desired Gaussian limit distribution. Let $k, m, r \geqq 1$ and let Λ_k, Λ_m, Λ_r be the corresponding maps as given in Theorem 1.1. We consider the sequence of independent identically distributed random vectors $\{(\Lambda_k X_j, \Lambda_m X_j, \Lambda_r X_j), j \geqq 1\}$. Let $0<\varepsilon<1/2$. Applying (1.5) twice we obtain for fixed $k, m \geqq 6/\varepsilon$ and all $n \geqq n_0(k) \vee n_0(m)$

$$\Pr\{n^{-1/2} \|\sum_{j \leqq n} (\Lambda_k X_j - \Lambda_m X_j)\| > \varepsilon/2\} < \varepsilon/2. \tag{3.1}$$

We write

$$U_{nkmr} = n^{-1/2} \sum_{j \leqq n} (\Lambda_k X_j, \Lambda_m X_j, \Lambda_r X_j). \tag{3.2}$$

On the finite-dimensional space $L_k S \times L_m S \times L_r S$ we have the sum norm $\|(u, v, w)\| = \|u\| + \|v\| + \|w\|$. By the central limit theorem in this space there is a mean zero Gaussian measure μ_{kmr} such that the Prohorov distance π for the sum norm satisfies

$$\pi(\mathscr{L}(U_{nkmr}), \mu_{kmr}) < \varepsilon/2, \qquad n \geqq n_1(\varepsilon, k, m, r). \tag{3.3}$$

We denote the marginals of the Gaussian law μ_{kmr} by μ_{km}, μ_{kr}, μ_{mr}, μ_k, μ_m and μ_r correspondingly. (Letting $n \to \infty$ in (3.2) shows that the notation is consistent, e.g. μ_6 and $\mu_{4,7}$ are well-defined.) Now the measures μ_k, μ_{km} and μ_{kmr} can be regarded as Borel probability measures on the separable Banach space T, $T \times T$ or $T \times T \times T$ respectively (rather than just on finite-dimensional subspaces) for $\|\cdot\|$ on T and corresponding norms on $T \times T$, $T \times T \times T$. Then (3.1) implies

$$\mu_{mr}\{(v, w) \in T \times T: \|v - w\| > \varepsilon\} < \varepsilon, \qquad m, r \geqq 6/\varepsilon. \tag{3.4}$$

On $T \times T$ we take the norm $\|(u, v)\| = \|u\| + \|v\|$. We rewrite (3.4) in the form

$$\mu_{kmr}\{(u, v, w): \|(u, v) - (u, w)\| > \varepsilon\} < \varepsilon, \qquad m, r \geqq 6/\varepsilon, \; k \geqq 1,$$

and obtain for the corresponding Prohorov distance π on $T \times T$

$$\pi(\mu_{km}, \mu_{kr}) \leqq \varepsilon, \qquad m, r \geqq 6/\varepsilon, \; k \geqq 1. \tag{3.5}$$

Consequently, for each $k \geqq 1$, $\{\mu_{km}, m \geqq 1\}$ is a Cauchy sequence for the Prohorov metric. Since $T \times T$ is complete we conclude by Theorem 1.11 of Prohorov (1956) that for each $k \geqq 1$ there is a law $\mu_{k\infty}$ on $T \times T$ such that

$$\mu_{km} \to \mu_{k\infty} \qquad \text{as } m \to \infty. \tag{3.6}$$

Thus by (3.4)

$$\mu_{k\infty}\{(u, v): \|u - v\| > \varepsilon\} \leqq \varepsilon, \qquad k \geqq 6/\varepsilon \tag{3.7}$$

and so for some law μ_∞ on T,

$$\mu_k \to \mu_\infty \qquad \text{as } k \to \infty. \tag{3.8}$$

Since μ_{km} has marginals μ_k and μ_m we conclude from (3.6) and (3.8) that $\mu_{k\infty}$ is Gaussian with marginals μ_k and μ_∞.

We now start with the construction of the Gaussian variables Y_j. For $k \geqq 1$ fixed for the time being, let $\{Z_{kj}, Z_j), j \geqq 1\}$ be a sequence of independent identically distributed random vectors on some probability space Ω' with values in $T \times T$ and

$$(3.9) \qquad \mathscr{L}(Z_{kj}, Z_j) = \mu_{k\infty}, \quad j \geqq 1.$$

Since $\mu_{k\infty}$ is centered Gaussian we conclude from (3.7) and (3.9) that for all $n \geqq 1$

$$(3.10) \qquad \Pr\{n^{-1/2} \| \sum_{j \leqq n} (Z_{kj} - Z_j) \| > \varepsilon\} \leqq \varepsilon, \quad k \geqq 6/\varepsilon.$$

By Lévy's inequality (Lemma 2.9 above) we obtain for $n \geqq 1$

$$(3.11) \qquad \Pr\{n^{-1/2} \max_{m \leqq n} \| \sum_{j \leqq m} (Z_{kj} - Z_j) \| > \varepsilon\} \leqq 2\varepsilon.$$

We now let $k > 6/\varepsilon$. By (3.3) $\{\Lambda_k X_j, j \geqq 1\}$ satisfies the central limit theorem with limit μ_k. Next by [53, with correction] there exists on some probability space Ω'' a sequence $\{V_{kj}, j \geqq 1\}$ of independent random variables having the same distribution as $\{\Lambda_k X_j, j \geqq 1\}$ and a sequence $\{W_{kj}, j \geqq 1\}$ of independent identically distributed random variables with common distribution μ_k such that

$$(3.12) \qquad n^{-1/2} \max_{m \leqq n} \| \sum_{j \leqq m} (V_{kj} - W_{kj}) \| \to 0 \quad \text{in prob.}$$

By Lemma 2.13 we may assume that $W_{kj} = Z_{kj}$, $\Omega' = \Omega''$. This together with (3.11) implies that for some $n_2(\varepsilon, k) \geqq n_0(k)$

$$(3.13) \qquad \Pr\{n^{-1/2} \max_{m \leqq n} \| \sum_{j \leqq m} (V_{kj} - Z_j) \| > 3\varepsilon\} < 3\varepsilon, \quad n \geqq n_2(\varepsilon, k).$$

In view of (3.13) and (1.5) it might appear that we are done because

$$(3.14) \qquad \mathscr{L}(\{V_{kj}, j \geqq 1\}) = \mathscr{L}(\{\Lambda_k X_j, j \geqq 1\}).$$

But there are two more hurdles to overcome. First, the sequence $\{Z_j, j \geqq 1\}$ depends on k. We use an idea of Major, also applied in [53] in the same context to construct a universal sequence $\{Z_j, j \geqq 1\}$. We choose $\varepsilon := \varepsilon_p := 2^{-p-3}$, $p = 1, 2, \ldots$ and accordingly $k = k(p) = 2^{p+6} > 6/\varepsilon_p + 1$. In view of (3.13) we obtain for each $p = 1, 2, \ldots$ two sequences $\{V_j^{(p)}, j \geqq 1\}$ and $\{Z_j^{(p)}, j \geqq 1\}$ with the following properties:

$$(3.15) \qquad \begin{gathered} V_j^{(p)} = V_{k(p)j}, \quad j \geqq 1, \\ \mathscr{L}(\{Z_j^{(p)}, j \geqq 1\}) = \mathscr{L}(\{Z_j, j \geqq 1\}) \quad (\text{i.i.d. } \mu_\infty) \end{gathered}$$

and for some $n_3(p)$, which we choose to satisfy $n_3(p) \geqq n_2(2^{-p-6}, k(p))$, and all $n \geqq n_3$

$$(3.16) \qquad \Pr\{n^{-1/2} \max_{m \leqq n} \| \sum_{j \leqq m} V_j^{(p)} - Z_j^{(p)} \| > 2^{-p}\} < 2^{-p}.$$

Moreover, we can assume without loss of generality that the V-sequences are independent of one another and that the Z-sequences are independent of one another. Put $r(p)=\sum_{q\leq p} n_3(q)$. We define

$$V_j=V_j^{(p)} \quad \text{and} \quad Z'_j=Z_j^{(p)} \quad \text{if } r(p)<j\leqq r(p+1). \tag{3.17}$$

Then $\{V_j, j\geqq 1\}$ and $\{Z'_j, j\geqq 1\}$ are sequences of independent random variables. Moreover, it will be shown that for each $\varepsilon>0$ there is an $n_4(\varepsilon)$ such that

$$\Pr\{n^{-1/2}\max_{m\leqq n}\|\sum_{j\leqq m}(V_j-Z'_j)\|>4\varepsilon\}<4\varepsilon, \qquad n\geqq n_4. \tag{3.18}$$

Indeed, let s be such that $2^{-s}<\varepsilon$ and let $N_0=N_0(\varepsilon)$ be so large that for all $n\geqq N_0$

$$\Pr\{n^{-1/2}\max_{m\leqq r(s)}\|\sum_{j\leqq m}V_j\|>\varepsilon\}<\varepsilon$$

and

$$\Pr\{n^{-1/2}\max_{m\leqq r(s)}\|\sum_{j\leqq m}Z'_j\|>\varepsilon\}<\varepsilon.$$

Let $n\geqq\max(N_0, n_3(s))=n_4(\varepsilon)$ and keep it fixed. We choose M such that $r(M)<n\leqq r(M+1)$. Then $n\geqq n_3(p)$ for $p\leqq M$ and by (3.16), (3.17) and stationarity

$$\begin{aligned}\max_{m\leqq n}\|\sum_{j\leqq m}(V_j-Z'_j)\| &\leqq \max_{m\leqq r(s)}\|\sum_{j\leqq m}V_j\|+\max_{m\leqq r(s)}\|\sum_{j\leqq m}Z'_j\| \\ &\quad+\sum_{p=s}^{M-1}\max_{r(p)<m\leqq r(p+1)}\left\|\sum_{j=r(p)+1}^{m}(V_j-Z'_j)\right\| \\ &\quad+\max_{r(M)<m\leqq n}\left\|\sum_{j=r(M)+1}^{m}(V_j-Z'_j)\right\| \\ &\leqq 2\varepsilon n^{1/2}+\sum_{p=s}^{M}2^{-p}\cdot n^{1/2}<4\varepsilon n^{1/2}\end{aligned}$$

except on a set of probability $<4\varepsilon$. This proves (3.18). Since the sequence $\{Z'_j, j\geqq 1\}$ does not depend on ε we have passed the first hurdle.

The sequences $\{X_j, j\geqq 1\}$ and $\{Z'_j, j\geqq 1\}$ are defined on probability spaces, not necessarily the same. Lemma 2.13, successfully employed earlier in the proof, cannot be applied at this point since the X_j are not necessarily measurable. We apply Lemma 2.11 instead. For $j\geqq 1$ define $p:=p(j)$ by $r(p)<j\leqq r(p+1)$ and let $\rho(j):=2^{p(j)+6}$. By (3.14), (3.15) and (3.17)

$$\mathscr{L}(\{A_{\rho(j)}X_j, j\geqq 1\})=\mathscr{L}(\{V_j, j\geqq 1\}).$$

Hence by Lemma 2.11 with

$$Q=\mathscr{L}(\{V_j, j\geqq 1\}, \{Z'_j, j\geqq 1\}), \qquad X=\{A_{\rho(j)}X_j, j\geqq 1\}$$

defined on the appropriate Banach spaces and U a random variable uniformly distributed over $[0,1]$ we obtain a random variable $\{Y_j, j\geqq 1\}$, say, defined on the original probability space Ω such that $Q=\mathscr{L}(\{A_{\rho(j)}X_j, j\geqq 1\}, \{Y_j, j\geqq 1\})$.

Thus by (3.18)

$$n^{-1/2} \max_{m \leqq n} \| \sum_{j \leqq m} (\Lambda_{\rho(j)} X_j - Y_j) \| \to 0 \quad \text{in Pr. } (n \to \infty). \tag{3.19}$$

For the proof of (1.10) it remains to show that

$$n^{-1/2} \max_{m \leqq n} \| \sum_{j \leqq m} (X_j - \Lambda_{\rho(j)} X_j) \| \to 0 \quad \text{in Pr. } (n \to \infty). \tag{3.20}$$

This follows in the same way as (3.18). Since $n_3(p) \geqq n_2(\varepsilon_p, k(p)) \geqq n_0(2^{p+6})$ for all $p \geqq 1$, we have by (1.5)

$$\Pr^* \{ n^{-1/2} \| \sum_{j \leqq n} (X_j - \Lambda_{k(p)} X_j) \| \geqq 2^{-p-6} \} \leqq 2^{-p-6}, \qquad n \geqq n_3(p).$$

Hence by Lemmas 2.5 and 2.7

$$\Pr^* \{ n^{-1/2} \max_{k \leqq n} \| \sum_{j \leqq k} (X_j - \Lambda_{k(p)} X_j) \| > 2^{-p} \} < 2^{-p}, \quad n \geqq n_3(p).$$

In the proof of (3.18) we replace V_j and Z'_j by X_j and $\Lambda_{\rho(j)} X_j$, respectively and obtain for given $\varepsilon > 0$ and for some $n_5(\varepsilon)$ and all $n \geqq n_5(\varepsilon)$

$$\max_{k \leqq n} \| \sum_{j \leqq k} (X_j - \Lambda_{\rho(j)} X_j) \| \leqq 4 \varepsilon n^{1/2} \tag{3.21}$$

except on a set of Pr*-measure $< 4\varepsilon$. This proves (3.20) and thus convergence in pr. in (1.10).

Next we show (1.8) and (1.9). Recall that $\{\Lambda_k X_j, j \geqq 1\}$ satisfies the central limit theorem with limit measure μ_k. Thus μ_k is a mean zero Gaussian measure. Moreover, for each $s \in T'$

$$E\{s^2(\Lambda_k X_j)\} = E\{s^2(Z_{k1})\} \to E\{s^2(Z_1)\} = E\{s^2(Y_1)\}$$

by (3.8) and since $s(Z_{k1})$ and $s(Z_1)$ are Gaussian. But $E\{s(Y_1)\, t(Y_1)\}$ is determined by $E\{s^2(Y_1)\}$.

We now prove the statements about L^p-convergence in (1.10). As in Pisier (1975, Proposition 2.1 and Remarque 2.1), we have

$$\sup_{n \geqq 1} E \| n^{-1/2} S_n \|^{*p} < \infty \qquad \text{for each } p < 2. \tag{3.22}$$

We follow the next to last paragraph of Sect. 3 on p. 80 of [53] (replacing λ by $\lambda/2$ on the right in the display) and obtain L^p-convergence for each $p < 2$.

For $p = 2$ we observe that by (1.5) with $m = 10^4$ and Chebyshev's inequality applied to $\sum_{j \leqq n} \Lambda_m X_j$

$$\Pr^* \{ \| S_n \| \geqq \alpha n^{1/2} \} < 10^{-3} \qquad \text{for all } n \geqq n_0 \tag{3.23}$$

for some $\alpha > 0$. To prove L^2-convergence in (1.10) we replace (3.22) above by the following lemma.

Lemma 3.1. *Let $\{X_j, j \geqq 1\}$ be a sequence of independent identically formed random elements $X_j = h(x_j)$ with values in a seminormed vector space S and $E\|X_1\|^{*2} < \infty$. Suppose that (3.23) holds for some α and n_0. Then there is an $N = N(\alpha, n_0, \mathscr{L}(\|X_1\|^*))$ such that for all $n \geqq N$*

$$E\|S_n\|^{*2} \leqq 500\,\alpha^2 n.$$

Proof. By stationarity and (3.23), for $0 \leqq i \leqq j \leqq n$,

$$\Pr\{\|S_j - S_i\|^* > 2\alpha n^{1/2}\} \leqq \Pr^*\{\|S_j - S_i\| > \alpha n^{1/2}\} < 10^{-3} \tag{3.24}$$

if $j - i \geqq n_0$. But if $j - i < n_0$ then by Markov's inequality

$$\Pr\{\|S_j - S_i\|^* > 2\alpha n^{1/2}\} \leqq (j-i)\,E\|X_1\|^*/(2\alpha n^{1/2}) < 10^{-3}$$

if $n \geqq 10^6 n_0^2 E\|X_1\|^{*2}/\alpha^2$. We apply Lemma 2.8 with $K = 2\alpha n^{1/2}$ and obtain for all $t \geqq 4\alpha^2 n$

$$\begin{aligned}\Pr\{\|S_n\|^* \geqq 10t^{1/2}\} &\leqq 4\Pr\{\|S_n\|^* \geqq 2t^{1/2}\}^2 + n\Pr\{\|X_1\|^* \geqq 2t^{1/2}\}\\ &\leqq 4 \cdot 10^{-3}\Pr\{\|S_n\|^* \geqq 2t^{1/2}\} + n\Pr\{\|X_1\|^* \geqq 2t^{1/2}\}\end{aligned}$$

using (3.24). Thus with $u = 4\alpha^2 n$

$$\begin{aligned}10^{-2}E\|S_n\|^{*2} &= \int_0^\infty \Pr\{\|S_n\|^* \geqq 10t^{1/2}\}\,dt\\ &\leqq u + 4 \cdot 10^{-3}\int_u^\infty \Pr\{\|S_n\|^* \geqq 2t^{1/2}\}\,dt\\ &\quad + n\int_u^\infty \Pr\{\|X_1\|^* \geqq 2t^{1/2}\}\,dt\\ &\leqq 4\alpha^2 n + 4 \cdot 10^{-3} \cdot \tfrac{1}{4} \cdot E\|S_n\|^{*2} + o(n)\end{aligned}$$

since $E\|X_1\|^{*2} < \infty$. Hence for some N and $n \geqq N$

$$E\|S_n\|^{*2} \leqq n(4\alpha^2 + o(1))(10^{-2} - 10^{-3})^{-1} < 500\,\alpha^2 n. \quad \square \tag{3.25}$$

The proof of Theorem 1.1 is complete.

Proof of Lemma 1.4. Let $G_n := (Y_1 + \ldots + Y_n)/n^{1/2}$. Then $\mathscr{L}(G_n) \equiv \mathscr{L}(Y_1)$ on T and (1.10) implies that $\|G_n - v_n\| \to 0$ in probability. Given $\varepsilon > 0$, take $\delta = \delta(\varepsilon) > 0$ and n_0 from (1.15). Take $n \geqq n_0$ large enough so that

$$\Pr^*\{\|G_n - v_n\| > \varepsilon\} < \varepsilon. \tag{3.26}$$

For $\psi \in S$ let

$$p_\delta(\psi) = \sup\{|\psi(f) - \psi(g)| : f, g \in \mathscr{F}, e_P(f, g) < \delta\}.$$

Then p_δ is a seminorm on S with $p_\delta(\psi) \leqq 2\|\psi\|$ for all $\psi \in S$. Relation (1.15) gives $\Pr^*\{p_\delta(v_n) > \varepsilon\} < \varepsilon$, $n \geqq n_0$. Thus we have by (3.26)

$$\Pr^*\{p_\delta(G_n) > 3\varepsilon\} < 2\varepsilon.$$

But p_δ is continuous, hence measurable on the separable space T where the laws $\mathscr{L}(G_n)=\mathscr{L}(Y_1)$, so $\Pr\{p_\delta(Y_1)>3\varepsilon\}<2\varepsilon$. Let $\gamma(k):=\delta(2^{-k})$, and

$$W_k:=\{\psi\in S\colon p_{\gamma(k)}(\psi)\leqq 3/2^k\}.$$

Then $\Pr\{Y_1\notin W_k\}<2^{1-k}$. Let $W=\bigcup_{j\geqq 1}\bigcap_{k\geqq j} W_k$. Then W is a Borel set in T, consisting of functions uniformly continuous on $\mathscr{J}$, and $\Pr\{Y_1\in W\}=1$ by the Borel-Cantelli Lemma. □

4. A Bounded Law of the Iterated Logarithm

Let $a_n:=a(n):=(2n\,L_2\,n)^{1/2}$ and $S_n:=\sum_{j\leqq n} X_j$.

Theorem 4.1. *Let $\{X_j, j\geqq 1\}$ be a sequence of independent identically formed random elements $X_j=h(x_j)$ with values in a seminormed vector space $(S, \|\cdot\|)$. Suppose that for some $\kappa>0$ and n_0 we have*

$$\Pr{}^*\{\|S_n\|>\kappa n^{1/2}\}<10^{-3} \quad \textit{for all } n\geqq n_0 \tag{4.1}$$

and that

$$E\{(\|X_1\|^2/L_2\|X_1\|)^*\}<\infty. \tag{4.2}$$

Then

$$\limsup_{n\to\infty} \|S_n\|^*/a_n\leqq 2^{28}\,\kappa \quad a.s.$$

The proof is a minor modification of the proof of Theorem 4.1 of Goodman, Kuelbs and Zinn (1981). We first observe that we can assume without loss of generality $\kappa=1$. To symmetrize the random elements X_j we choose a sequence $\{x_j', j\geqq 1\}$ of independent identically distributed A-valued random variables independent of the sequence $\{x_j, j\geqq 1\}$ with the same law. (Specifically, let x_j' also be coordinates in a product of copies of A.) Let $X_j'=h(x_j')$. As in [23] let

$$I(n):=\{2^n+1,\ldots,2^{n+1}\},\quad \alpha(t):=t/L_2\,t,$$
$$\beta(t):=t\,L_2\,t,\quad \alpha_n:=\beta^{-1}(2^n),\quad \beta_n:=\alpha^{-1}(2^n)$$

and for $j\in I(n)$ let

$$\begin{aligned} U_j&:=X_j\,1\{\|X_j\|^{*2}<\alpha_n\},\\ V_j&:=X_j\,1\{\alpha_n\leqq\|X_j\|^{*2}\leqq\beta_n\},\\ W_j&:=X_j\,1\{\beta_n<\|X_j\|^{*2}\}\end{aligned} \tag{4.3}$$

and define U_j', V_j', W_j' accordingly replacing X_j by X_j'. We then set

$$u_j=U_j-U_j',\quad v_j=V_j-V_j'\quad\text{and}\quad w_j=W_j-W_j'. \tag{4.4}$$

Then u_j, v_j, w_j are symmetric and $u_j+v_j+w_j=X_j-X_j'$.

Lemma 4.1. *We have*

$$\sum_{n\geqq 1} \Pr\{\|\sum_{j\in I(n)} w_j\|^* \geqq 2a(2^n)\} < \infty.$$

Proof. The series in question does not exceed $\sum_{j\geqq 1} \Pr^*\{w_j \neq 0\} < \infty$ in view of (4.2).

Lemma 4.2. *We have*

$$\sum_{n\geqq 1} \Pr\{\|\sum_{j\in I(n)} v_j\|^* \geqq a(2^n)\} < \infty.$$

Proof. As in [23] we set $Z_j = 2^n v_j/a(2^n)$ for $j\in I(n)$. We are to show

$$\sum_{n\geqq 1} \Pr\{\|\sum_{j\in I(n)} Z_j\|^* \geqq 2^n\} < \infty. \tag{4.5}$$

This can be done exactly as in the proof of Lemma 4.2 of [23] by verifying

$$j^{-1}\|Z_j\|^* \to 0 \quad \text{a.s.} \tag{4.6}$$

$$\sum_{n\geqq 1} \Lambda^2(n) < \infty, \qquad \Lambda(n) := \sum_{j\in I(n)} 4^{-n} E\|Z_j\|^{*2} \tag{4.7}$$

and

$$\lim_{k\to\infty} E\|k^{-1}\sum_{j\leqq k} Z_j\|^* = 0. \tag{4.8}$$

Now (4.6) is easy [23, (4.11)] and (4.7) can be proved in the same way as [23, (4.12)], with the following changes. We replace $\|\cdot\|$ by $\|\cdot\|^*$ everywhere. Convergence of any random variables $U_n \to 0$, in probability or a.s., is replaced by $\|U_n\|^* \to 0$ in the same sense (cf. Lemma 2.5 above). After [23, (4.14)] replace the next "$\Lambda(n)=$" by "$\Lambda(n)/4 \leqq$". Then in (4.15), (4.16) and (4.19), divide $\Lambda(n)^2$ by 16.

Correct [23, (4.16)] by replacing $T := (Lm - LM)^+/\log 2$ by $T+1$. The additional term multiplied by $+1$ has finite expectation by (4.2) so all is well. Above [23, (4.18)], choose $c>1$. Correct [23, (4.20)] by changing the second "$=$" to "$\leqq$" and in its last line replace "$>$" by "$\geqq$". Note that $\gamma'(t) \leqq L_3 t/L_2 t$, $L_3 t > 1$, stated before (4.20), is actually used just after (4.21).

For the proof of [23, (4.21)] we use the proof of Proposition 2.1 and Remarque 2.1 of Pisier (1975), applying our Lemmas 2.5, 2.9 and 2.2. In the next display after (4.21), replace $\lambda(n)$ (a typo) by $\Lambda(n)/4$, obtaining our (4.7) as desired.

To prove our (4.8) we have to estimate $E\|\sum_{j\leqq k} v_j\|^*$. (For the details see [23, p. 729].) Instead of using Lemma 2.3 of [23] we use once more the proof of Pisier (1975), Proposition 2.1 and Remarque 2.1 to obtain for $N \geqq n_0$

$$\Pr\{\|S_N - S'_N\|^* \geqq 12 N^{1/2} n\} \leqq 1/n^2 \qquad \text{for all } n \geqq 1.$$

By (4.4) and stationarity we obtain for all $m \geqq 0$ and $N \geqq n_0$

$$E\left\|\sum_{j=m+1}^{m+N} u_j+v_j+w_j\right\|^* < 9N^{1/2}$$

and thus by symmetry

$$E\left\|\sum_{j=m+1}^{m+N} u_j\right\|^* \leqq 9N^{1/2}, \quad E\left\|\sum_{j=m+1}^{m+N} v_j\right\|^* \leqq 9N^{1/2}. \tag{4.9}$$

We are now ready to apply the proof of Theorem 1 of Kuelbs and Zinn (1979). Fix δ, $0<\delta<1$. For $n \geqq 1$ we set

$$\begin{aligned} \xi_n &= \sum_{j\in I(n)} Z_j 1\{\|Z_j\|^* \leqq \Lambda(n)^{1/4}\, 2^{n+1}\} \\ \eta_n &= \sum_{j\in I(n)} Z_j 1\{\|Z_j\|^* \geqq 2^{n-1}\delta\} \\ \zeta_n &= \sum_{j\in I(n)} Z_j 1\{\Lambda(n)^{1/4}\cdot 2^{n+1} < \|Z_j\|^* < 2^{n-1}\delta\}. \end{aligned} \tag{4.10}$$

We bypass the question whether or not ξ_n, η_n and ζ_n are symmetric by symmetrizing them. Let $\tilde{\xi}_n=\xi_n-\xi_n'$, $\tilde{\eta}_n=\eta_n-\eta_n'$ and $\tilde{\zeta}_n=\zeta_n-\zeta_n'$ where we define ξ_n', η_n' and ζ_n' as in (4.10) replacing Z_j by an independent copy Z_j'. Then by (4.8) and (4.10)

$$E\,\|\tilde{\xi}_n+\tilde{\eta}_n+\tilde{\zeta}_n\|^* \leqq 2\Big(E\,\Big\|\sum_{j\leqq 2^{n+1}} Z_j\Big\|^* + E\,\Big\|\sum_{j\leqq 2^n} Z_j\Big\|^*\Big)=o(2^n).$$

Thus by symmetry

$$E\,\|\tilde{\xi}_n\|^*=o(2^n).$$

As in the proof of Theorem 1 of [42] we obtain

$$\Pr\{\|\tilde{\xi}_n+\tilde{\eta}_n+\tilde{\zeta}_n\|^* \geqq 2^{n-1}\} \leqq A_n$$

for some A_n with $\sum A_n<\infty$. By a standard desymmetrization argument and (4.10) we obtain Lemma 4.2.

Lemma 4.3. *We have*

$$\sum_{n\geqq 1} \Pr\Big\{\Big\|\sum_{j\in I(n)} u_j\Big\|^* \geqq 2^{24}\, a(2^n)\Big\} < \infty.$$

Proof. It is enough to show

$$\sum_{k\geqq 1} \Pr\Big\{\Big\|\sum_{j\leqq 2^k} u_j\Big\|^* \geqq 2^{22}\, a(2^k)\Big\} < \infty. \tag{4.11}$$

For the proof of (4.11) we apply Proposition 4.3 of Pisier (1975) and its proof to the sequence $\{u_j, j\geqq 1\}$. By (4.9) and (4.3) the hypotheses are satisfied with $\alpha=9$. We also note that $H=2^{12}$ in [55, Lemma 4.1].

We now can finish the proof of Theorem 4.1. By Lemmas 4.1, 4.2, 4.3 and (4.3) we obtain

$$\sum_{n\geqq 1} \Pr\{\| \sum_{j\in I(n)} X_j - X'_j\|^* \geqq 2^{25}\, a(2^n)\} < \infty$$

and by stationarity and Lemma 2.9

$$\sum_{n\geqq 1} \Pr\{\max_{k\leqq 2^n} \|S_k - S'_k\|^* \geqq 2^{25}\, a(2^n)\} < \infty.$$

By (4.1) and Lemma 2.7 and a standard desymmetrization argument we finally obtain

$$\sum_{n\geqq 1} \Pr\{\max_{k\leqq 2^n} \|S_k\|^* \geqq 2^{26}\, a(2^n)\} < \infty.$$

Theorem 4.1 follows now from the Borel-Cantelli lemma and the triangle inequality.

5. Proof of Theorem 1.2

We use the notation of Sect. 3. We apply Theorem 4.1 to the sequence $\{Z_{kj}-Z_j, j\geqq 1\}$ of independent identically distributed T-valued random variables. In view of (3.10), for each $k\geqq 10^4$, setting $\kappa := 6/k$ we obtain

$$\limsup_{n\to\infty} a_n^{-1} \|\sum_{j\leqq n} Z_{kj} - Z_j\| \leqq 2^{31}/k \quad \text{a.s.} \tag{5.1}$$

By (3.3) the sequence $\{\Lambda_k X_j, j\geqq 1\}$ satisfies the central limit theorem with limit measure μ_k. Next we infer that there is a sequence $\{V_{kj}, j\geqq 1\}$ of independent identically distributed random variables having the same distribution as $\{\Lambda_k X_j, j\geqq 1\}$ and a sequence $\{W_{kj}, j\geqq 1\}$ of independent identically distributed random variables with common distribution μ_k such that as $n\to\infty$

$$\|\sum_{j\leqq n} V_{kj} - W_{kj}\| = o(a_n) \quad \text{a.s.} \tag{5.2}$$

Existence of such V_{kj}, W_{kj} follows from Corollary 1 of [52]. We take the opportunity to make a few corrections and remarks on [52]. In checking the proof of Corollary 1 [52], the reader can omit the beginning of Sect. 3 through the end of Subsection 3.1 since we can set $\prod_N = \text{identity}$, $\xi_\nu = x_\nu$.

In [52, (3.22)] set $H_k = (t_k, t_{k+1}]$. On p. 179, line 4, p. 172 refers to Hartman and Wintner (1941). Both ξ_{n_k} in (3.2.16) and $\xi^*_{n_k}$ in (3.2.17) should be $\xi^*_{t_{k+1}}$, and in the next display $\mu^3_{n_k}$ should be $\mu^3_{t_{k+1}}$. In (3.2.20) 1/2 should be 1/8 and in the next display "(2" should be "2(". In the proof of Lemma 3.7 [52], last line of display, the exponent on α should be $-1/4$. On p. 285, line 9 replace "Kolmogorov's existence theorem" by "Lemma A1 of" [4]. On the right side of (3.3.6) replace α by 5α. At the beginning of the proof of Lemma 3.8 [52] insert: "We may assume without loss of generality $E\|x_1\|^2 < 1/4$. Redefine $\varepsilon(\nu)$ and $\lambda(\nu)$ in terms of $\|x_\nu\|$ rather than $|\xi_\nu|$." Below (3.4.2) add a factor 2 in the definition of c.

As a matter of fact Lemma 3.8 [52] is redundant as we shall demonstrate below when we prove (5.5). This completes our remarks on [52].

By the argument following relation (3.12) using Lemma 2.13 we can assume $W_{kj} \equiv Z_{kj}$ (on some probability space). Then (5.1) and (5.2) imply

$$\limsup_{n\to\infty} a_n^{-1} \| \sum_{j\leq n} V_{kj} - Z_j \| \leqq 2^{31}/k \quad \text{a.s.} \tag{5.3}$$

Again the sequence $\{Z_j, j \geqq 1\}$ may depend on k. We again use Major's idea to construct a universal sequence $\{Y_j, j \geqq 1\}$. We choose $k = k(p) = 2^{p+6}$ and obtain for each $p = 1, 2, \ldots$ two sequences of independent identically distributed random variables, say $\{V_j^{(p)}, j \geqq 1\}$, $\{Z_j^{(p)}, j \geqq 1\}$ with the following properties:

$$V_j^{(p)} = V_{k(p)j}, j \geqq 1, \qquad \mathscr{L}(\{Z_j^{(p)}, j \geqq 1\} = \mathscr{L}(\{Z_j, j \geqq 1\})$$

and

$$\limsup_{n\to\infty} \| \sum_{j\leq n} V_j^{(p)} - Z_j^{(p)} \| / a_n \leqq 2^{-p+25} \quad \text{a.s.} \tag{5.4}$$

Moreover, we can assume without loss of generality that the V-sequences, considered as T^∞-valued random variables, are independent and that the same is true for the Z-sequences. Using (5.4) we apply monotone convergence in order to obtain a sequence $\{s(p), p \geqq 1\}$ of integers with $s(p) \uparrow \infty$ such that

$$\Pr\{ \sup_{n \geqq s(p)} \| \sum_{j\leq n} V_j^{(p)} - Z_j^{(p)} \| / a_n \geqq 2^{-p+26} \} \leqq 2^{-p}$$

and thus by the Borel-Cantelli lemma

$$\sup_{n \geqq s(p)} \| \sum_{j\leq n} V_j^{(p)} - Z_j^{(p)} \| / a_n \ll 2^{-p} \quad \text{a.s.} \tag{5.5}$$

Next we apply Theorem 4.1 to the sequences $\{X_j - \Lambda_{k(p)} X_j, j \geqq 1\}$ and obtain for $p \geqq 10$

$$\limsup_{n\to\infty} \| \sum_{j\leq n} X_j - \Lambda_{k(p)} X_j \|^* / a_n \leqq 2^{-p+25} \quad \text{a.s.} \tag{5.6}$$

and thus by the above argument, for a possibly larger sequence $s(p) \uparrow \infty$

$$\sup_{n \geqq s(p)} \| \sum_{j\leq n} X_j - \Lambda_{k(p)} X_j \|^* / a_n \ll 2^{-p} \text{ a.s.} \tag{5.7}$$

We take $s(p)$ large enough to satisfy both (5.5) and (5.7). We note that for each $p \geqq 1$ the sequences $\{\Lambda_{k(p)} X_j, j \geqq 1\}$ and $\{V_j^{(p)}, j \geqq 1\}$ have the same laws. Thus by Lemma 2.11 we can assume without loss of generality that $V_j^{(p)} = \Lambda_{k(p)} X_j$, for each p, and that $\{Z_j^{(p)}, j \geqq 1\}$ having the desired joint distribution with $\{V_j^{(p)}, j \geqq 1\}$ are all defined on the original probability space.

Without loss of generality we can take $s(q) \geqq n_3(q)$ as in (3.16). Let $r(p) := \sum_{q \leqq p} s(q)$, so that $r(p) \geqq s(p)$. Thus (5.5) and (5.7) remain valid if $s(p)$ is replaced by $r(p)$. Then define $\rho(j)$ and Y_j as in the proof of (3.19). By the remarks in the preceding paragraph

$$\Lambda_{\rho(j)} X_j = V_j^{(p)}, \quad Y_j = Z_j^{(p)} \quad \text{if } r(p) < j \leqq r(p+1). \tag{5.8}$$

By (3.19) and (3.20) there exists a subsequence $\{p(t), t \geqq 0\}$ such that

$$\Big\| \sum_{j \leqq r(p(t))} X_j - Y_j \Big\|^* = o(r(p(t))^{1/2}) \quad \text{a.s.} \tag{5.9}$$

We finally show that the sequence $\{Y_j, j \geqq 1\}$ has the desired properties. Let n be given and find t such that $r(p(t)) < n \leqq r(p(t+1))$. Next find h such that $r(p(t)+h) < n \leqq r(p(t)+h+1)$. Then by (5.8)

$$\begin{aligned}
\Big\| \sum_{j \leqq n} X_j - Y_j \Big\|^* \leqq & \Big\| \sum_{j \leqq r(p(t))} X_j - Y_j \Big\|^* \\
& + \sum_{p(t) \leqq p < p(t)+h} \Big\| \sum_{r(p) < j \leqq r(p+1)} X_j - \Lambda_{k(p)} X_j \Big\|^* \\
& + \sum_{p(t) \leqq p < p(t)+h} \Big\| \sum_{r(p) < j \leqq r(p+1)} \Lambda_{k(p)} X_j - Y_j \Big\| \\
& + \sup_{n \geqq m > r(p(t)+h)} \Big\| \sum_{j \leqq m} X_j - \Lambda_{k(p(t)+h)} X_j \Big\|^* \\
& + \sup_{n \geqq m > r(p(t)+h)} \Big\| \sum_{j \leqq m} \Lambda_{k(p(t)+h)} X_j - Z_j^{(p(t)+h)} \Big\| \\
& + \Big\| \sum_{j \leqq r(p(t)+h)} X_j - \Lambda_{k(p(t)+h)} X_j \Big\|^* \\
& + \Big\| \sum_{j \leqq r(p(t)+h)} \Lambda_{k(p(t)+h)} X_j - Z_j^{(p(t)+h)} \Big\| \\
\leqq & \Big\| \sum_{j \leqq r(p(t))} X_j - Y_j \Big\|^* \\
& + 2 \sum_{p(t) \leqq p \leqq p(t)+h} \Big\| \sum_{j \leqq r(p)} X_j - \Lambda_{k(p)} X_j \Big\|^* \\
& + 2 \sum_{p(t) \leqq p \leqq p(t)+h} \Big\| \sum_{j \leqq r(p)} Z_j^{(p)} - \Lambda_{k(p)} X_j \Big\| \\
& + \sup_{n \geqq m \geqq r(p(t)+h)} \Big\| \sum_{j \leqq m} X_j - \Lambda_{k(p(t)+h)} X_j \Big\|^* \\
& + \sup_{n \geqq m \geqq r(p(t)+h)} \Big\| \sum_{j \leqq m} Z_j^{(p(t)+h)} - \Lambda_{k(p(t)+h)} X_j \Big\|.
\end{aligned}$$

We divide the inequality by a_n and take the limes superior as $n \to \infty$. By (5.9) the first term on the right side of the last inequality tends to 0. The lim sup of the last two terms is $\ll \limsup 2^{-p(t)-h}$ by (5.7) and (5.5). Similarly the lim sup of the second and third term is

$$\ll \limsup \sum_{p(t) \leqq p \leqq p(t)+h} 2^{-p} \ll \limsup 2^{-p(t)} = 0.$$

Thus

$$\Big\| \sum_{j \leqq n} X_j - Y_j \Big\|^* = o(a_n) \quad \text{a.s.}$$

6. Improvement of the Error Term

In this section we improve the error term in Theorem 1.2 under two additional sets of hypotheses, thereby giving a partial extension of Theorem 1.1 of [47]. This theorem can be completely extended to conform with the underlying theme of the present paper. But we refrain from carrying out this program since we do not have reasonable applications for this more general theorem.

Theorem 6.1. *Let* $\{X_j, j \geqq 1\}$ *be a sequence of independent identically formed random elements with values in* $(S, \|\cdot\|)$. *Suppose that* $E\|X_1\|^{*2+\delta} < \infty$ *for some* $0 < \delta \leqq 1$. *Let* $\{\Lambda_m, m \geqq 1\}$ *be a sequence of mappings as described in Theorem 1.1 having the following additional properties.*

(6.1) *The* Λ_m *are linear maps with* $\sup_{m \geqq 1} \|\Lambda_m\| < \infty$.

If in condition (1.5)

$$n_0(m) \leqq C_1 m^D, \quad m \geqq 1 \tag{6.2}$$

and if

$$\dim \Lambda_m S \leqq C_2 \exp(C_3 m^\beta), \quad m \geqq 1 \tag{6.3}$$

for some constants $C_i \geqq 1$ $(i=1,2,3)$, $D > 2$ *and* $\beta > 0$ *then the error term in* (1.11) *can be improved to* $\mathcal{O}(n^{1/2} (\log n)^{-\theta})$ *for any* $\theta < 1/(2\beta)$.

If, instead of (6.3)

$$\dim \Lambda_m S \leqq C_4 m^\gamma \tag{6.4}$$

for some C_4, $\gamma \geqq 1$ *then the error term in* (1.11) *can be improved to* $\mathcal{O}(n^{1/2-\lambda})$ *where* $\lambda = \kappa^2/(600\gamma)$, $\kappa = \min(\delta, 4/(D-2))$.

The proof of Theorem 6.1 follows a by now well-established method [3], [44, Theorem 2], [54], [47, Theorem 1.1], [2], [10, Theorem 2], etc. In all of these papers explicit bounds on the probabilities of errors easily could have been established, by collecting the relevant probability bounds before the Borel-Cantelli lemma is applied. In the case of our Theorem 6.1 the corresponding result is as follows.

Theorem 6.2. *Assume that the hypotheses of Theorem 6.1 hold. If* (6.3) *holds then one can choose the sequence* $\{Y_j, j \geqq 1\}$ *such that for any* $\theta < 1/(2\beta)$ *and any constant* $B < \infty$

$$\Pr\{\max_{m \leqq n} \|\sum_{j \leqq m} X_j - Y_j\|^* \geqq n^{1/2} (\log n)^{-\theta}\} \ll (\log n)^{-B}.$$

If however, condition (6.4) *holds then with* $\lambda = \kappa^2/(600\gamma)$ *the sequence* $\{Y_j, j \geqq 1\}$ *can be chosen so that for some constant* $H < \infty$

$$\Pr\{\max_{m \leqq n} \|\sum_{j \leqq m} X_j - Y_j\|^* \geqq H n^{1/2-\lambda}\} \ll n^{-\kappa/28}.$$

Theorem 6.1 is an immediate consequence of Theorem 6.2. If (6.3) holds we let $n=2^{k+1}$ and applying the Borel-Cantelli lemma we obtain

$$\max_{2^k<m\leqq 2^{k+1}} \|\sum_{j\leqq m} X_j-Y_j\| \ll 2^{k/2}k^{-\theta} \quad \text{a.s.}$$

since we can choose $B=2$. Similarly, if (6.4) holds we let $n=k^\rho$ with $\rho=[56/\kappa]$, instead.

Hence it remains to prove Theorem 6.2. Since $\kappa\leqq\delta$, $EF^{2+\kappa}<\infty$. So we may replace δ by κ and assume in the proofs that $\kappa=\delta$, i.e. that $D\leqq 2+4/\delta$ in (6.2). Then, let $\alpha:=6\delta/7$. The hypotheses made so far all hold with α in place of δ: $EF^{2+\alpha}<\infty$ and $D\leqq 2+4/\alpha$. In the conclusion under (6.4), $\lambda:=\delta^2/(600\gamma)<\alpha^2/(400\gamma)$. In Theorem 6.2 we can and will take $B\geqq 1$.

We can assume without loss of generality

$$10^6\, C_1^2(1+\sup_{m\geqq 1}\|\Lambda_m\|)\,\|F\|_{2+\delta}\leqq\alpha. \tag{6.5}$$

Let

$$V_{jm}:=V_j(m):=X_j-\Lambda_m X_j, \qquad m,j\geqq 1$$

and

$$\begin{aligned} V'_{jm}&:=V_{jm}\,1\{\|X_j\|^*\leqq j^{1/(2+\alpha)}\}:=V'_j(m)\\ V''_{jm}&:=V_{jm}-V'_{jm}=V_{jm}\,1\{\|X_j\|^*>j^{1/(2+\alpha)}\}. \end{aligned} \tag{6.6}$$

Lemma 6.1. *Let $m\geqq 10^4$. Then for all $n\geqq C_1 m^{2+4/\alpha}$ and $s\geqq 0$*

$$\Pr\left\{\left\|\sum_{j=s+1}^{s+n} V'_{jm}\right\|^*\geqq 2\,n^{1/2}/m\right\}\leqq 10^{-3}.$$

Proof. We have by Lemma 2.5, (6.5) and (6.6) for all $n\geqq m^{2+4/\alpha}$ and $s\geqq 0$

$$\begin{aligned} &\Pr\left\{\left\|\sum_{j=s+1}^{s+n} V''_{jm}\right\|^*\geqq n^{1/2}/m\right\}\\ &\leqq \Pr\left\{\sum_{j=s+1}^{s+n}\|V''_{jm}\|^*\geqq n^{1/2}/m\right\}\\ &\leqq n^{-1/2}\,m\sum_{j=s+1}^{s+n} E\{\|V_{jm}\|^*\,1\{\|X_j\|^*\geqq j^{1/(2+\alpha)}\}\}\\ &\leqq n^{-1/2}\,m(1+\|\Lambda_m\|)^{2+\alpha}\cdot E\,\|X_1\|^{*2+\alpha}\cdot\sum_{j=s+1}^{s+n} j^{-(1+\alpha)/(2+\alpha)}\\ &<10^{-5}\,m n^{-\alpha/(4+2\alpha)}\leqq 10^{-5}. \end{aligned}$$

The lemma follows now from (6.6), (1.5) and the stationarity of V_{jm}.

Lemma 6.2. *Let $m\geqq 10^4$. Then for all $n\geqq m^{14+16/\alpha}$ and $s\geqq 0$*

$$E\left\|\sum_{j=s+1}^{s+n} V'_{jm}\right\|^{*2}\leqq 500\,n/m^2.$$

Proof. Write $S'_n = \sum_{j=1}^{n} V'_{j+s,m}$. Then by Lemma 6.1 we have for all $0 \leqq i < j \leqq n$

$$\Pr\{\|S'_j - S'_i\|^* \geqq 2\, n^{1/2}/m\} \leqq 10^{-3} \tag{6.7}$$

if $j - i \geqq C_1 m^{2+4/\alpha}$. But if $j - i < C_1 m^{2+4/\alpha}$ we have by (6.5), (6.6) and Markov's inequality for $n \geqq m^{6+8/\alpha}$

$$\Pr\{\|S'_j - S'_i\|^* \geqq n^{1/2}/m\} \leqq m\, n^{-1/2}\, C_1\, m^{2+4/\alpha}(1 + \|\Lambda_m\|) \cdot E\,\|X_1\|^* \leqq 10^{-3}.$$

Thus (6.7) holds for all $n \geqq m^{6+8/\alpha}$. We now follow the proof of Lemma 3.1 with S_n replaced by S'_n and $\alpha = 1/m$. Then by the proof of (3.25) we have for all $n \geqq m^{14+16/\alpha}$ $(= m^2 \cdot m^{2(6+8/\alpha)} \geqq m^2 \cdot 10^6\, m^{2(6+8/\alpha)}\, E\,\|X_1\|^{*2}$ by (6.5) and since $\delta \leqq 1)$

$$E\,\|S'_n\|^{*2} \leqq n\left(4/m^2 + \int_u^\infty \Pr\{\|X_1\|^* \geqq 2\, t^{1/2}\}\, dt\right)(10^{-2} - 10^{-3})^{-1}. \tag{6.8}$$

The result follows now at once since the integral in (6.8) does not exceed $E\,\|X_1\|^{*2+\alpha}\, u^{-\alpha/2}/\alpha \leqq m^{-2}$.

From now on we assume that (6.3) holds. (The case (6.4) will be treated at the end of this section.) We put $\rho = 1/(1+\beta) < 1$ and take any ζ with $0 < \zeta < \rho$. Then $\beta\,\zeta < 1 - \rho$. Let

$$m_0 := 1, \quad m_k := m(k) := k^\zeta, \qquad k \geqq 1, \tag{6.9}$$

$$t_0 := 0, \quad t_k := t(k) := [\exp(k^{1-\rho})], \qquad k \geqq 1 \tag{6.10}$$

and

$$n_k := n(k) := t_{k+1} - t_k \sim \text{const} \cdot t_k\, k^{-\rho}. \tag{6.11}$$

Let $H_k := H(k) := \{j\colon\ t_k < j \leqq t_{k+1}\} = \{t_k + 1, \ldots, t_{k+1}\}$.

Proposition 6.1. *For any $B < \infty$ there is a $C = C_{6.1}(B)$ such that as $k \to \infty$*

$$\Pr\{\|\sum_{j \in H(k)} V_j(m_k)\|^* \geqq C(n_k\, L_2\, t_k)^{1/2}/m_k\} \ll k^{-B}.$$

Proof. Let $r_k := r(k) := m_k^{14+16/\alpha}$ and define new random elements

$$U_h := U_h(k) := \sum_{j=1}^{r(k)} V'_{t(k)+r(k)(h-1)+j}(m_k),$$

$$h = 1, 2, \ldots, l_k := l(k) := [n_k/r_k],$$

$$U_{l(k)+1} := \sum_{j=t(k)+l(k)r(k)+1}^{t(k+1)} V'_j(m_k).$$

Then we get using (6.6) and (6.9)

$$\|U_h\| < r_k\, t_{k+1}^{1/(2+\alpha)} := M_k, \qquad 1 \leqq h \leqq l_k + 1. \tag{6.12}$$

For k large, $m_k \geqq 10^4$ and by Lemma 6.2,

$$E\,\|U_h\|^{*2} \leqq 500\, r_k\, m_k^{-2}, \qquad 1 \leqq h \leqq l_k,$$

and

$$E\,\Big\|\sum_{h \leqq l(k)} U_h\Big\|^* \leqq 500\, n_k^{1/2}/m_k.$$

We apply now Lemma 2.6 and the Remarks after it to the sequence $\{U_h, h \leqq l_k\}$, with $M := M_k$, $n := l_k$, $\tau := \tau_n := 500 r_k\, l_k\, m_k^{-2}$ and $K := K_k := \xi(n_k L_2 t_k)^{1/2}/m_k$ for a constant $\xi > (6000\,B/(1-\rho))^{1/2}$.

Then for k large, $E\,\|S_n\|^* < K_k < 3\,\tau_n/M_k$, so

$$\begin{aligned}\Pr\Big\{\Big\|\sum_{h \leqq l(k)} U_h\Big\|^* \geqq K_k\Big\} &\leqq \exp(-(K - E\,\|S_n\|^*)^2/(12\,\tau_n)) \\ &\leqq \exp(-\xi^2 L_2\, t_k/6000) \ll \exp(-B \log k) = k^{-B}.\end{aligned}$$

Since $M_k = o(K_k)$ as $k \to \infty$, the latter bound still holds for $\sum_{h \leqq l(k)+1} U_h$, because of (6.12).

Next,

$$\begin{aligned}(6.13)\qquad \Pr\{V_j'(m_k) &\neq V_j(m_k) \quad \text{for some } j \in H(k)\} \\ &\leqq \sum_{j > t(k)} \Pr\{\|X_1\|^* > j^{1/(2+\alpha)}\} \\ &\leqq E\,\|X_1\|^{*2+\alpha}\, 1\{\|X_1\|^* > t_k^{1/(2+\alpha)}\} \\ &\leqq E\,\|X_1\|^{*2+\delta}\, t_k^{(\alpha-\delta)/(2+\alpha)} \leqq t_k^{(\alpha-\delta)/3} = t_k^{-\delta/21} \ll k^{-B}.\end{aligned}$$

This completes the proof.

Next we set

$$\begin{aligned}(6.14)\qquad X_j' &:= X_j\, 1\{\|X_j\|^* \leqq j^{1/(2+\alpha)}\}, \\ X_j'' &:= X_j - X_j' = X_j\, 1\{\|X_j\|^* > j^{1/(2+\alpha)}\}.\end{aligned}$$

Lemma 6.3. *There is a constant C_5 such that for all $s \geqq 0$ and $n \geqq 1$*

$$E\left\|\sum_{j=s+1}^{s+n} X_j'\right\|^* \leqq C_5\, n^{1/2}.$$

Proof. From (6.6) and Lemma 6.2 with $\mu = 10^4$ we obtain

$$\begin{aligned}E\left\|\sum_{j=s+1}^{s+n} X_j'\right\|^* &\leqq E\left\|\sum_{j=s+1}^{s+n} V_j'(\mu)\right\|^* + E\left\|\sum_{j=s+1}^{s+n} \Lambda_\mu X_j\right\|^* \\ &\quad + \sum_{j=s+1}^{s+n} E\{\|\Lambda_\mu X_j\|\, 1\{\|X_j\|^* > j^{1/(2+\alpha)}\}\} \\ &\ll n^{1/2} + \sum_{j=s+1}^{s+n} \|\Lambda_\mu\|\, E\,\|X_j\|^{*2+\alpha} \cdot j^{-(1+\alpha)/(2+\alpha)} \ll n^{1/2}.\end{aligned}$$

Note that we also have used the fact that $\{\Lambda_\mu X_j, j \geqq 1\}$ is a sequence of independent identically distributed random vectors with mean zero and finite second moment (1.7).

Proposition 6.2. *For any $B<\infty$ there is a $C=C_{6.2}(B)$ large enough so that as $k\to\infty$*

$$\Pr\{\max_{n\in H(k)}\|\sum_{t(k)<j\leqq n} X_j\|^*\geqq C(n_k L_2 t_k)^{1/2}\}\ll k^{-B}.$$

Proof. Using (6.13) it suffices to prove this for X'_j in place of X_j. Let $S'_n:=\sum_{j\leqq n} X'_j$. By (6.11) and Markov's inequality we have for all k

$$\max_{n\in H(k)}\Pr\{\|S'_{t(k+1)}-S'_n\|^*>2\,C_5\,n_k^{1/2}\}\leqq 1/2.$$

Hence by Lemma 2.7 we have for all k, and $\xi>2\max(C_5, 12\,B/(1-\rho))^{1/2}$

$$\begin{aligned}(6.16)\qquad \Pr\{\max_{n\in H(k)}\|S'_n-S'_{t(k)}\|^*\geqq 2\,\xi(n_k L_2 t_k)^{1/2}\}\\ \leqq 2\Pr\{\|S'_{t(k+1)}-S'_{t(k)}\|^*\geqq\xi(n_k L_2 t_k)^{1/2}\}.\end{aligned}$$

To estimate this last probability we apply Lemma 2.6 to the sequence $\{X'_j, j\in H_k\}$ with $K=\xi(n_k L_2 t_k)^{1/2}$, $n=\tau_n=n_k$ and $M=t_{k+1}^{1/(2+\alpha)}$ and for k large obtain the bound

$\exp(-\xi^2 L_2 t_k/48)\ll k^{-B}$. Q.E.D.

Let $0<\varepsilon<1/4$. For $m\geqq 6/\varepsilon$ let $\{(Z_{mj}, Z_j), j\geqq 1\}$ be any sequence of independent random variables with $\mathscr{L}(Z_{mj}, Z_j)=\mu_{m\infty}, j\geqq 1$ (as defined in (3.6)).

Proposition 6.3. *For any $B<\infty$ there is a $C=C_{6.3}(B)$ large enough so that as $k\to\infty$*

$$\Pr\{\|\sum_{j\in H(k)} Z_{m(k)j}-Z_j\|\geqq C(n_k L_2 t_k)^{1/2}/m_k\}\ll k^{-B}.$$

Proof. By (3.10) we have for each $m\geqq 6/\varepsilon$

$$\Pr\{n^{-1/2}\|\sum_{j\leqq n} Z_{mn}-Z_j\|\geqq\varepsilon\}\leqq\varepsilon<1/4.$$

Hence by the Fernique-Landau-Shepp inequality [19, p. 1699]

$$\Pr\{\sum_{j\leqq n}\|Z_{mj}-Z_j\|\geqq\eta\,\varepsilon\,n^{1/2}\}\leqq\exp(-\eta^2/24)$$

for any $\eta\geqq 1$. Putting $n=n_k$, $\varepsilon=6/m_k$ and $\eta=(24\,B(1-\rho)^{-1}L_2 t_k)^{1/2}$ we obtain the result.

Next, let $F_k:=\mathscr{L}(n_k^{-1/2}\sum_{j\in H(k)}\Lambda_{m(k)}X_j)$ and

$$G_k:=\mathscr{L}(n_k^{-1/2}\sum_{j\in H(k)} Z_{m(k)j})=\mathscr{L}(Z_{m(k)1})=\mu_{m(k)}.$$

Then by Lemma 2.12, (6.3), (6.9), (6.10) and (6.11) we obtain for the Prohorov distance

(6.17) $$\pi(F_k, G_k) \ll n_k^{-\alpha/9} \exp(\tfrac{4}{3} C_2 m_k^\beta) \ll n_k^{-\alpha/10},$$

i.e. for some constant C_6,

$$\pi(F_k, G_k) \leqq C_6 n_k^{-\alpha/10}, \qquad k \geqq 0.$$

Let $\mathscr{S}_k := L_{m(k)} S$. Then by a classic theorem of Strassen (1965) there is a law J_k on $\mathscr{S}_k \times \mathscr{S}_k$ with marginals F_k and G_k on $\mathscr{S}_k$ such that

$$J_k \{\langle x, y\rangle : \|x - y\| > C_6 n_k^{-\alpha/10}\} \leqq C_6 n_k^{-\alpha/10}, \qquad k \geqq 0.$$

Thus by Lemma 2.13 applied to $X = \mathscr{S}_k^{n(k)}$, $Y = Z = \mathscr{S}_k$, there are independent $\mathscr{S}_k$-valued random variables X_{kj}, $j = 1, \ldots, n_k$, all with law $\mathscr{L}(X_{kj}) = \mathscr{L}(\Lambda_{m(k)} X_1)$, and a random variable Γ_k with law G_k such that

$$\Pr\left\{\left\|\Gamma_k - n_k^{-1/2} \sum_{j=1}^{n(k)} X_{kj}\right\| > C_6 n_k^{-\alpha/10}\right\} \leqq C_6 n_k^{-\alpha/10}.$$

Now we apply Lemma 2.13 to $X = \mathscr{S}_k^{n(k)}$, $Y = \mathscr{S}_k$, and $Z = (\mathscr{S}_k \times T)^{n(k)}$ to obtain, for each $k \geqq 0$, independent random variables $(\Gamma_{kj}, W_{kj}) \in \mathscr{S}_k \times T$, $j = 1, \ldots, n_k$, each with law $\mu_{m(k)\infty}$ (as in (3.6)), so that each Γ_{kj} has the Gaussian law $G_k = \mu_{m(k))}$ and such that

(6.18) $$\Pr\left\{n_k^{-1/2} \left\|\sum_{j=1}^{n(k)} \Gamma_{kj} - X_{kj}\right\| > C_6 n_k^{-\alpha/10}\right\} \leqq C_6 n_k^{-\alpha/10}.$$

(The above two applications of Lemma 2.13 could be replaced by one application of the generalized Vorob'ev theorem: Shortt, 1982, Theorem 2.6.)

Now we take the countable product of the laws

$$\mathscr{L}\{\langle X_{kj}, \Gamma_{kj}, W_{kj}\rangle : j = 1, \ldots, n_k\}_{k \geqq 0}$$

on the Polish space $\prod_{k \geqq 0} (\mathscr{S}_k \times \mathscr{S}_k \times T)^{n(k)}$ to obtain a joint law for all these random variables; they are independent for different values of k.

The law of $\{X_{kj}\}_{k \geqq 0,\, j = 1, \ldots, n(k)}$ equals that of

$$\{\Lambda_{m(k)} X_{j + t(k)}\}_{k \geqq 0,\, j = 1, \ldots, n(k)}.$$

So by Lemma 2.11 we may in fact take

(6.19) $$X_{kj} \equiv \Lambda_{m(k)} X_{j + t(k)}$$

and define (Γ_{kj}, W_{kj}) on the original probability space Ω. Then we define

(6.20) $$Y_j := W_{k, j - t(k)} \quad \text{for } t_k < j \leqq t_{k+1}, \qquad k \geqq 0.$$

Thus Y_j are independent and identically distributed with law μ_∞ on T as desired. Then as $M \to \infty$ we let $\mu := [M^{1/2}]$ and estimate

$$\begin{aligned}
&\max_{t(M)<n\leqq t(M+1)} \|\sum_{j\leqq n} X_j - Y_j\|^* \\
&\quad \leqq \max_{n\in H(m)} \|\sum_{t(M)<j\leqq n} X_j\|^* + \max_{n\in H(m)} \|\sum_{t(M)<j\leqq n} Y_j\| \\
&\quad + \sum_{\mu\leqq k<M} \|\sum_{r\leqq n(k)} \Gamma_{kr} - W_{kr}\| + \sum_{\mu\leqq k<M} \|\sum_{j\in H(k)} X_j - \Lambda_{m(k)} X_j\|^* \\
&\quad + \sum_{\mu\leqq k<M} \|\sum_{r\leqq n(k)} X_{kr} - \Gamma_{kr}\| + \sum_{j\leqq t(\mu)} \|X_j\|^* + \|Y_j\| \\
&\quad := A_1 + A_2 + A_3 + A_4 + A_5 + A_6,
\end{aligned}$$

say, using (6.19) and (6.20), with $1 \leqq r = j - t_k$. We will prove that for $1 \leqq B < \infty$, $\theta < 1/(2\beta)$, and $i = 1, 2, \ldots, 6$, as $M \to \infty$

$$\Pr\{A_i \geqq t_M^{1/2}(Lt_M)^{-\theta}\} \ll M^{-B}. \tag{6.21}$$

For $i=1$, Proposition 6.2 gives for $C \leqq C_{6.2}(B)$

$$\Pr\{A_1 \geqq C(n_M L_2 t_M)^{1/2}\} \ll M^{-B}.$$

Now for $\theta < \varphi < \rho/2(1-\rho) = 1/(2\beta)$,

$$(n_M L_2 t_M)^{1/2} \ll (t_M M^{-\rho} LM)^{1/2} \ll t_M^{1/2}(Lt_M)^{-\varphi}. \tag{6.22}$$

This implies (6.21) for $i=1$.

Since Y_j are Gaussian, the Landau-Shepp-Fernique inequality ([19], [20], or [48]) implies $E\|Y_1\|^3 < \infty$. Thus, replacing X_j by Y_j in Prop. 6.2 and its proof, we obtain (6.21) for $i=2$.

For $i=3$, Proposition 6.3 gives, for $C \geqq C_{6.3}(2B+1)$,

$$\Pr\{\|\sum_{r\leqq n(k)} \Gamma_{kr} - W_{kr}\| \geqq C(n_k L_2 t_k)^{1/2}/m_k\} \ll k^{-2B-1}. \tag{6.23}$$

We have

$$\begin{aligned}
\sum_{k<M} (n_k L_2 t_k)^{1/2}/m_k &\ll \sum_{M/2<k\leqq M} \exp(k^{1-\rho}/2)\, k^{-\zeta-\rho/2} Lk \\
&\ll \exp(M^{1-\rho}/2)\, M^{-\zeta+\rho/2} LM \\
&\ll t_M^{1/2}(Lt_M)^{(-\zeta+\rho/2)/(1-\rho)} L_2 t_M.
\end{aligned}$$

For $\theta < \varphi < 1/(2\beta) = \rho/2(1-\rho)$, and ζ close enough to ρ, we have $(-\zeta+\rho/2)/(1-\rho) < -\varphi < -\theta$. Then

$$\sum_{k<M} (n_k L_2 t_k)^{1/2}/m_k \ll t_M^{1/2}(Lt_M)^{-\varphi}, \tag{6.24}$$

and

$$\sum_{k\geqq\mu} k^{-2B-1} \leqq (\mu-1)^{-2B} \ll M^{-B}. \tag{6.25}$$

Thus (6.21) holds for $i=3$.

For $i=4$, Proposition 6.1 gives, for $C \geqq C_{6.1}(2B+1)$,

$$\Pr\{\|\sum_{j\in H(k)} X_j - \Lambda_{m(k)} X_j\|^* \geqq C(n_k L_2 t_k)^{1/2}/m_k\} \ll k^{-2B-1}.$$

Then by (6.24) and (6.25), we have (6.21) also for $i=4$.

For A_5 we use (6.18) and note that

$$\sum_{k<M} n_k^{(5-\alpha)/10} \ll M n_M^{(5-\alpha)/10} \ll n_M^{(10-\alpha)/20}$$
$$\ll t_M^{1/2} (\log t_M)^{-\varphi}, \quad \varphi < 1/(2\beta),$$

while

$$\sum_{k\geqq\mu} n_k^{-\alpha/10} \ll \sum_{k\geqq\mu} k^{-2B-1} \ll \mu^{-2B} \ll M^{-B}.$$

So (6.21) holds for $i=5$. For $i=6$,

$$\Pr\{\sum_{j\leqq t(\mu)} \|X_j\|^* + \|Y_j\| \geqq n_M^{1/3}\} \leqq \sum_{j\leqq t(\mu)} E(\|X_j\|^* + \|Y_j\|)/n_M^{1/3}$$
$$\ll t(\mu)\, n_M^{-1/3} \ll M^{-B},$$

while $n_M^{1/3} \ll t_M^{1/2}(Lt_M)^{-\varphi}$, $\theta < \varphi < 1/(2\beta)$, so (6.21) holds for $i=6$.

As $M\to\infty$, $t_M < n \leqq t_{M+1}$ implies $n \sim t_M$ and $\log n \sim M^{1-\rho}$, so $M^{-B} \ll (\log n)^{-B}$. Again letting $\theta \uparrow 1/(2\beta)$ gives Theorem 6.2 in the (6.3) case.

If instead of (6.3), condition (6.4) holds, then *instead of* (6.9), (6.10) and (6.11) we define, for $\rho := 189\gamma/\alpha^2$,

(6.26) $$m_k := k, \quad n_k := [k^\rho] \quad \text{and} \quad t_k := \sum_{j<k} n_j.$$

Then $t_k \sim k^{\rho+1}/(\rho+1)$ as $k\to\infty$. Other definitions in terms of these remain the same. In place of Props. 6.1 and 6.2 we now will have

Proposition 6.4. *There is a $C = C_{6.4}$ large enough so that as $k\to\infty$*

$$\Pr\{\|\sum_{j\in H(k)} V_j(m_k)\|^* \geqq C(n_k Lt_k)^{1/2}\} \ll k^{-10\gamma/\alpha}$$

and

$$\Pr\{\max_{n\in H(k)} \|\sum_{t(k)<j\leqq n} X_j\|^* \geqq C(n_k Lt_k)^{1/2}\} \ll k^{-10\gamma/\alpha}.$$

Proof. Wherever $L_2 t_k$ appeared in the (6.3) case we put Lt_k in the (6.4) case. Our choice of ρ in (6.26) implies

$$14 + 16/\alpha + (\rho+1)/(2+\alpha) < \rho/2 - 1,$$

so following the proof of Proposition 6.1 we have $M_k = o(K_k)$. If we take $\xi > \xi_1(B)$ large enough, $B = 10\gamma/\alpha$, we still get a bound k^{-B} from Lemma 2.6. In the last terms of (6.13) we now get $t_k^{-\delta/21} \ll k^{-10\gamma/\alpha}$. The proof of Proposition 6.2 adapts likewise. Q.E.D.

Next, in Proposition 6.3 we (as always) replace $L_2 t_k$ by Lt_k. In its proof we need only replace η by $5(BLt_k)^{1/2}$.

Then Lemma 2.12, (6.4) and (6.26) give, in place of (6.17),

(6.27) $$\pi(F_k, G_k) \ll n_k^{-\alpha/9} k^{4\gamma/3} \ll k^{-R}$$

where $R := (\alpha\rho - 2\gamma)/9 > 1$, so $\pi(F_k, G_k) \leqq C_7 k^{-R}$ for some constant C_7. The latter bound replaces $C_6 n_k^{-\alpha/10}$ everywhere; specifically, in place of (6.18) we have

(6.28) $$\Pr\left\{n_k^{-1/2} \left\| \sum_{j=1}^{n(k)} \Gamma_{kj} - X_{kj} \right\| > C_7 k^{-R}\right\} \ll k^{-R}.$$

The proof then runs unchanged until (6.21) except that we redefine $\mu := [M^{4/9}]$. In place of (6.21) it will be shown that for some constants D_i,

(6.29) $$\Pr\{A_i \geqq D_i t_M^{-\lambda+1/2}\} \ll M^{-8\gamma/\alpha}.$$

We first note that

(6.30) $$\rho/2 < (\rho+1)(\tfrac{1}{2} - \lambda).$$

Then letting $D_1 = C_{6.4}$ and applying Proposition 6.4 (latter half) gives (6.29) for $i=1$. As in the (6.3) case, the proof of Propositions 6.2 and 6.4 adapts to Y_j in place of X_j to give D_2 for which (6.29) holds, $i=2$. For $i=3$,

(6.31) $$\sum_{k<M} (n_k Lt_k)^{1/2}/m_k \ll \sum_{k<M} k^{-1+\rho/2} Lk \ll M^{\rho/2} LM \ll t_M^{-\lambda+1/2}$$

by (6.30) again. Since $\lambda < 1/6$, using Proposition 6.3 as adapted (L in place of L_2) gives a D_3 large enough such that, noting

(6.32) $$\sum_{k \geqq \mu} k^{-3B-1} \ll \mu^{-3B} \ll M^{-B},$$

we obtain (6.29) for $i=3$. The case $i=4$ follows likewise, using Prop. 6.4 (first half).

For $i=5$ we use (6.28). Then

$$\sum_{k<M} n_k^{1/2} k^{-R} \ll \sum_{k<M} k^{-R+\rho/2} \ll M^{1-R+\rho/2} \ll t_M^{-\lambda+1/2}$$

since $1+\rho\lambda < R$. Also,

$$\sum_{k \geqq \mu} k^{-R} \ll \mu^{1-R} \ll M^{-8\gamma/\alpha}$$

giving (6.29) for $i=5$. For $i=6$, Markov's inequality gives for $D_6 := 1$

$$\Pr\{\sum_{j\leqq t(\mu)}\|X_j\|^*+\|Y_j\|\geqq t_M^{-\lambda+1/2}\}$$
$$\leqq \sum_{j\leqq t(\mu)} E(\|X_j\|^*+\|Y_j\|)\, t_M^{\lambda-1/2}$$
$$\ll t_\mu t_M^{\lambda-1/2}\ll M^{(\rho+1)(\lambda-1/18)}\ll M^{-10\gamma/\alpha}$$

and (6.29) follows for $i=6$. Now $t_M<n\leqq t_{M+1}$ implies

$$M^{-8\gamma/\alpha}\ll t_M^{-8\gamma/(\alpha(\rho+1))}\ll n^{-\delta/28}.$$

Setting $H:=\sum_{i\leqq 6} D_i$ we obtain Theorem 6.2 in the (6.4) case. Q.E.D.

Remarks. In the above proof, the large size of ρ in the (6.4) case was primarily used in the proof of Proposition 6.4 (first display). Once ρ is large, then to satisfy (6.30), λ must be small. Since we suspect that the result is far from best possible we did not seek the largest possible λ using our methods.

7. Application to Empirical Processes

In this section Theorems 6.1 and 6.2 will be applied to empirical processes, giving speeds of convergence in Theorem 1.3 uniformly over suitable families $\mathscr{I}$ of functions. Our rates of convergence, proved in some generality, are relatively slow, but are sufficient to imply, for example, some "upper and lower class" results (cf. Corollary 4 of [44]). On the other hand for special classes of functions on, or sets in, Euclidean spaces, defined by differentiability conditions, and under some more or less severe restrictions on the underlying probability laws, much faster rates have been obtained (Révész, 1976; Ibero, 1979a, b).

For a collection $\mathscr{C}$ of sets we take $\mathscr{I}=\{1_B\colon B\in\mathscr{C}\}$. Let $(A,\mathscr{A},P)$ be a probability space. First, under hypotheses on $\log N_I$, the metric entropy with inclusion, stronger than (1.23), here are rates of convergence in Theorem 1.5.

Theorem 7.1. *Let $\mathscr{C}$ be a collection of measurable sets with* $\log N_I(x,\mathscr{C},P)\leqq cx^{-\tau}$, $x>0$, *for some constants* $c>0$ *and* $0\leqq\tau<1$. *Then for any* $H<\infty$ *and* $\theta<(1-\tau)/(4\tau)$, *we can choose* Y_j *in Theorem* 1.3 *to improve* (1.28) *to*

$$\text{(7.1)}\quad \Pr{}^*\{n^{-1/2}\max_{k\leqq n}\sup_{B\in\mathscr{C}}|\sum_{j\leqq k}1_B(x_j)-P(B)-Y_j(1_B)|>(\log n)^{-\theta}\}\ll(\log n)^{-H},$$

and for some measurable U_n, almost surely

$$\text{(7.2)}\qquad \sup_{B\in\mathscr{C}}|\sum_{j\leqq n}1_B(x_j)-P(B)-Y_j(1_B)|\leqq U_n=o(n^{1/2}(\log n)^{-\theta}).$$

Proof. Given $m\geqq 1$, let $\varepsilon=1/(3m)$. In the proof of Theorem 5.1 of [12], to satisfy (5.2) and (5.3) there we can set $\alpha=\gamma=c_1\varepsilon^{2/(1-\tau)}$ for a small enough $c_1=c_1(\tau,c)>0$. We have (5.4) as corrected, i.e.

$$\sum_{i\geqq u}(2^{-i}\log N_I(2^{-i}))^{1/2}<\varepsilon/96,$$

if $2^{-u} < c_2 \varepsilon^{2/(1-\tau)}$ for some $c_2 = c_2(\tau, c)$. There is a constant $K = K(\tau)$ such that

$$\exp(-K\varepsilon^{-2\tau/(1-\tau)}) < \varepsilon/200, \quad 0 < \varepsilon \leqq 1.$$

Then there is a $c_3(\tau) \leqq \min(c_1(\tau), c_2(\tau))$ small enough so that for all $j \geqq 0$,

$$2^j > c_3(\tau) K(\tau) 9000(j+1)^5.$$

Then both (5.4) and (5.5) of [12] hold if $2^{-u} < c_3(\tau) \varepsilon^{2/(1-\tau)}$. Thus we can take $\delta_0 := 2^{-u}$ for the least such u. Then

$$n_0(\varepsilon) > \varepsilon^{-(2+2\tau)/(1-\tau)} c_3(\tau)^{-2}/256 \geqq n_0(\varepsilon)/4.$$

We then obtain (1.15) above with its ε replaced by $1/m$ and $\delta^2 = \delta_0 \geqq c_3(\tau)(3m)^{-2/(1-\tau)}/2$, according to [12, p. 917, line 2] with its ε corrected to $3\varepsilon = 1/m$. Then in the proof of Theorem 1.3 above we can let

$$\dim \Lambda_m S = N_I(\delta_0) \leqq \exp(c\delta_0^{-\tau}) \leqq \exp(c_4(\tau) m^{2\tau/(1-\tau)})$$

for some $c_4(\tau) > 0$, and $n_0(m) \leqq c_5(\tau) m^{(2+2\tau)/(1-\tau)}$ for some $c_5(\tau) > 0$.

So condition (6.3) holds with $\beta = 2\tau/(1-\tau)$ and (6.2) holds with $D = 2(1+\tau)/(1-\tau)$. Thus Theorem 6.2 applies with θ as stated. Q.E.D.

Corollary 7.2. *Let P on $\mathbb{R}^2$ have a bounded density with respect to the Lebesgue measure and let $\mathscr{C} = \mathscr{C}(U)$ be the collection of all convex subsets of a bounded open set $U \subset \mathbb{R}^2$. Then for any $\theta < 1/4$ and $H < \infty$ we can choose Y_j in Theorem* 1.5 *such that* (7.1) *and* (7.2) *hold.*

Proof. We can apply Theorem 7.1 with $\tau = 1/2$ according to Bronštein (1976). □

In $\mathbb{R}^3$, Theorem 1.5 does not hold for the convex sets [13].

Let $J(k, \alpha, M)$ be the collection of compact subsets of $\mathbb{R}^k$ with boundaries defined by functions with all partial derivatives of orders $\leqq \alpha$ bounded by M, as defined in [11, with Correction] or [12, p. 917].

Corollary 7.3. *For P with bounded support and bounded density with respect to Lebesgue measure on $\mathbb{R}^k$, $k \geqq 1$, if $\alpha > k-1$ then for any $r > (k-1)/\alpha$, $\theta := (1-r)/(4r)$ and $H < \infty$,* (7.1) *and* (7.2) *hold for $\mathscr{C} = J(k, \alpha, M)$, $M < \infty$.*

Proof. We may assume $r < 1$. Then Theorem 7.1 applies to any τ such that $(k-1)/\alpha < \tau < r < 1$ [11] and the rest follows. □

Remark. Révész (1976) obtained in Corollary 7.3, if $k = \alpha = 2$, for regions with 1-1 boundary curves, if P is the uniform measure on the unit square or has a sufficiently regular density, (7.2) with $\mathcal{O}(n^{1/2}(\log n)^{-\theta})$ replaced by $\mathcal{O}(n^{12/25})$, a much better result in that case. We suppose that for such P, Révész' method would also yield $\mathcal{O}(n^{12/25})$ in (7.2) for the convex sets as in Corollary 7.2. But we do not see how to extend Révész' construction to the generality of Theorem 7.1.

Next, for Vapnik-Červonenkis classes of sets (mentioned late in Sect. 1 above) we do need [16] some measurability condition, such as the following [15].

Given a probability space $(A, \mathscr{A}, P)$ and $\mathscr{C} \subset \mathscr{A}$, we say $\mathscr{C}$ is *image $P\in$-Suslin* iff there is a measurable space $(W, \mathscr{S})$ and a map G of W onto $\mathscr{C}$ such that:

1) $\{\langle x, w\rangle : x \in G(w)\}$ is $\mathscr{A} \times \mathscr{S}$-measurable;

2) for the pseudo-metric $d_P(B, C) = P(B \triangle C)$ and any d_P-open set $U \subset \mathscr{C}$, $G^{-1}(U) \in \mathscr{S}$;

3) $(W, \mathscr{S})$ is Suslin, i.e. for some Polish space V there is a map H of V onto W with $H^{-1}(B)$ Borel measurable for all $B \in \mathscr{S}$. Also, $(A, \mathscr{A})$ is Suslin.

Theorem 7.4. *Suppose $(A, \mathscr{A}, P)$ is a probability space, $\mathscr{C} \subset \mathscr{A}$, $V(\mathscr{C}) < \infty$ and $\mathscr{C}$ is image $P\in$-Suslin. Then for any $\lambda < 1/(2700\, V(\mathscr{C}))$*

$$\Pr{}^* \{n^{-1/2} \max_{k \leqq n} \sup_{B \in \mathscr{C}} |\sum_{j \leqq k} 1_B(x_j) - P(B) - Y_j(1_B)| > n^{-\lambda}\} \ll n^{-1/50} \tag{7.3}$$

and for some measurable U_n, almost surely

$$\sup_{B \in \mathscr{C}} |\sum_{j \leqq n} 1_B(x_j) - P(B) - Y_j(1_B)| \leqq U_n = \mathcal{O}(n^{1/2-\lambda}). \tag{7.4}$$

Proof. Examination of the proof of [12, Theorem 7.1, Correction] (clarified in [15]) shows that given $\varepsilon > 0$ we can take the δ there to satisfy, for a small enough constant $c > 0$,

$$\delta = c\varepsilon^2 / |\ln \varepsilon| \tag{7.5}$$

(specifically see [12, Correction, p. 909, last display]). By the last two lines of the proof of [12, Theorem 7.1], we have for $n \geqq n_0(\delta)$ large enough

$$\Pr\{\sup\{|\nu_n(B \setminus C)| : B, C \in \mathscr{C}, P(B \setminus C) < \delta\} \geqq 6\varepsilon\} < \varepsilon.$$

Then in the proof of Theorem 1.3 above, replacing δ^2 in (1.15) by δ and taking $m = 6/\varepsilon$, we apply [12, Lemma 7.13] to obtain a linear mapping Λ_m onto a subspace of dimension $\ll \delta^{-w}$ for any $w > V(\mathscr{C})$. Thus by (7.5) we have $\dim \Lambda_m S \ll \varepsilon^{-2w}$ for each $w > V(\mathscr{C})$. So in (6.4) we can take any $\gamma > 2V(\mathscr{C})$. In the hypothesis of Theorem 6.1 we have $\delta = 1$.

In the proof of [12, Theorem 7.1] we must take $n \geqq \max(n_0(\delta), n_1(\varepsilon))$ where for ε small it suffices to take $n_0(\delta) \geqq \delta^{-r}$ for any $r > 8$ and $n_1(\varepsilon) \geqq \varepsilon^{-r}$ for $r > 2$, so (6.2) holds for any $D > 8$. Thus the conclusions of Theorems 6.1 and 6.2 in the (6.4) case hold for any $\kappa < 2/3$, so for $\lambda < 1/(2700\, V(\mathscr{C}))$ and $\kappa/28 < 1/42$. Q.E.D.

Remarks. Let $(A, \mathscr{A}, P)$ be the unit interval with Lebesgue measure and let $\mathscr{C}_k$ be the collection of all unions of at most k intervals. Then $V(\mathscr{C}_k) = 2k+1$. Yet from Komlós, Major and Tusnády (1975, Theorem 4) one can get (7.4) with $\mathcal{O}(n^{.5-\lambda})$ replaced by $\mathcal{O}(\log n)^2$, for any P. So this rate holds for nontrivial classes with arbitrarily large $V(\mathscr{C})$. It is not known, in fact, whether it or even the rate $\mathcal{O}(\log n)$ may hold for arbitrary Vapnik-Červonenkis classes. Révész (1976, Lemma 12) considers the class $P(\kappa, 1)$ of polygons with at most κ sides where P is uniform on the unit square in $\mathbb{R}^2$ or has a regular enough density, and obtains (7.4) with $(n^{.5-\lambda})$ replaced by $\mathcal{O}(n^{3/8} \log^2 n)$, which although much better than our rate, depends on the specific assumptions made on P and is much slower than the Komlós-Major-Tusnády rate.

To obtain empirical distribution functions one just takes the collection $\mathscr{C}$ of all sets $B_x \subset \mathbb{R}^k$ where

$$B_x = \{y: y_j \leqq x_j, j=1, \ldots, k\}, \qquad x \in \mathbb{R}^k.$$

Here $V(\mathscr{C}) = k+1$ [63, Proposition 2.3], so that Theorem 7.4 improves somewhat on Theorem 1 of Philipp and Pinzur (1980), again for arbitrary P (for P e.g. uniform on the unit cube, Révész [58] obtains a much smaller error term even over a much larger collection $\mathscr{C}$ of sets).

Given a set $\mathscr{F} \subset \mathscr{L}^2(A, \mathscr{A}, P)$ of functions, and $\varepsilon > 0$, let $N_I(\varepsilon, \mathscr{F}, P)$ be the smallest m such that for some $f_1, \ldots, f_m \in \mathscr{L}^2(A, \mathscr{A}, P)$, for all $f \in \mathscr{F}$ there are $i, j \leqq m$ with $f_i \leqq f \leqq f_j$ and $\int f_j - f_i \, dP < \varepsilon$. Then $\log N_I(\varepsilon, \mathscr{F}, P)$ has been called a *metric entropy with bracketing* [14].

Suppose that for some $F \in \mathscr{L}^2$, $|f| \leqq F$ for all $f \in \mathscr{F}$. In [14] it is shown that if $F \in \mathscr{L}^p$ for some $p > 2$ and $\log N_I(\varepsilon, \mathscr{F}, P) = \mathcal{O}(\varepsilon^{-r})$ for some $r < 1 - 2/p$ as $\varepsilon \downarrow 0$, then (1.14) and (1.15) hold. For this, or any such future result under weaker conditions on p and r, one can again obtain invariance principles (1.18) or (1.19) without needing further measurability conditions on the class $\mathscr{F}$. For such classes, the proof in [14] gives dimensions too large for our Theorems 6.1 and 6.2 to apply.

Pollard [56b] shows that if $\mathscr{C} \subset \mathscr{A}$ is a countable Vapnik-Červonenkis class, $F \in \mathscr{L}^2(A, \mathscr{A}, P)$, and $\mathscr{F} = \{F 1_B: B \in \mathscr{C}\}$, then (1.14) and (1.15) hold for $\mathscr{F}$. Thus Theorem 1.3 applies to give either (1.18) or (1.19).

We end this section with an application to weighted empirical distribution functions, which improves somewhat on one direction of results of O'Reilly (1974, Theorem 2), James (1975), and Goodman, Kuelbs and Zinn (1981, Theorem 6.1).

Theorem 7.5. *Let $\{W_j, j \geqq 1\}$ be a sequence of independent random variables with uniform distribution on $[0, 1]$. Define*

$$X_j = \omega(s)(1\{W_j \leqq s\} - s), \qquad 0 < s < 1$$
$$= 0 \qquad \textit{else},$$

where ω is a real function with the following properties:

(i) *ω is continuous and positive on $(0, 1)$*

(ii) *for some $\gamma > 0$, we have ω is nonincreasing (nondecreasing) on $(0, \gamma]$ ($[1-\gamma, 1)$ respectively)*

(iii) *For all $\varepsilon > 0$ and $i = 1, 2$*

$$\int_0^1 s^{-1} \exp(-\varepsilon/k_i(s)) \, ds < \infty$$

where $k_1(s) = s\omega^2(s)$ and $k_2(s) = s\omega^2(1-s)$.

(iv) *We have*

$$\int_0^1 \omega^2(s)/L_2\left(\frac{1}{s(1-s)}\right) ds < \infty.$$

Then there exists a sequence $\{Y_j, j \geqq 1\}$ *of independent identically distributed* $C[0,1]$*-valued Gaussian random variables, indexed by* $s \in [0,1]$, *and with sample functions almost surely continuous (in* s*) such that*

(i)′ $EY_1(s)=0,\ 0 \leqq s \leqq 1,$

(ii)′ $E\{Y_1(s)\,Y_1(t)\}=\omega(s)\,\omega(t)\,s(1-t),\ 0 \leqq s \leqq t \leqq 1,$

(iii)′ $n^{-1/2}\,E\{\max_{k\leqq n}\ \sup_{0\leqq s\leqq 1} |\sum_{j\leqq k} X_j(s) - Y_j(s)|\} \to 0$

or, instead of (iii)′, *for some measurable* V_n

(iv)′ $\sup_{0\leqq s\leqq 1} |\sum_{j\leqq n} X_j(s) - Y_j(s)| \leqq V_n = o((nLLn)^{1/2})$ *a.s.*

Proof. Under conditions (i), (ii) and (iii), O'Reilly (1974, Theorem 2) proves that the law of $Z_n := (\sum_{1\leqq j\leqq n} X_j)/n^{1/2}$ converges to that of Y_1 in the Polish space $D[0,1]$ with Skorohod topology. By Skorohod's theorem [61] there exist U_n with $\mathscr{L}(U_n)=\mathscr{L}(Z_n)$, $n \geqq 1$, $\mathscr{L}(U_0)=\mathscr{L}(Y_1)$ and $U_n \to U_0$ a.s. for the Skorohod metric. Since the limit process has continuous sample functions, also $U_n \to U_0$ for the sup norm. Now $\langle f, g\rangle \to \sup|f-g|$ is jointly measurable on $D[0,1] \times D[0,1]$ since we can restrict the supremum to rational arguments. So $\sup|U_n - U_0| \to 0$ a.s. and in probability. Since $\mathscr{L}(U_0)$ is tight on a Polish space $C[0,1]$, for any $m \geqq 1$ there is a map Λ_m of $D[0,1]$ into $C[0,1]$ with finite-dimensional range (consisting of piecewise linear functions with given vertices) such that

$$\Pr\{\sup|U_0 - \Lambda_m U_0| > 1/(3m)\} < 1/(3m).$$

For some n_0 we have for $n \geqq n_0$

$$\Pr\{\sup|U_n - U_0| > 1/(3m)\} < 1/(3m).$$

Since the Λ_m can be defined by interpolation and are linear we have $\sup|\Lambda_m f| \leqq \sup|f|$ for all $f \in D[0,1]$, so

$$\Pr\{\sup|\Lambda_m U_n - \Lambda_m U_0| > 1/(3m)\} < 1/(3m).$$

Combining we have

$$\Pr\{\sup|\Lambda_m U_n - U_n| > 1/m\} < 1/m,$$

and likewise for Z_n, i.e. we have the tightness condition (1.5). The other hypotheses in Theorem 1.1 are easily checked, so we obtain (iii)′. Now $E\|X_1\|^2/L_2\|X_1\| < \infty$ is equivalent to (iv), assuming (i), (ii) and (iii): Goodman, Kuelbs and Zinn (1981, Lemma 6.2) give a proof, on which we have the following comments.

In their statement of Theorem 2 of O'Reilly (1974), specifically in (6.3), $k_2(t)$ should be $t\omega^2(1-t)$ (or else in the k_2 case, t^{-1} should be replaced by $(1-t)^{-1}$). By (i), $\sup_{\gamma\leqq s\leqq 1-\gamma} \omega(s) < \infty$. Next, $\sup_{0<s<\gamma} s\omega(s) < \infty$ by (iii) (cf. [23, Lemma 6.1]), and likewise $\sup_{1-\gamma\leqq s<1} (1-s)\,\omega(s) < \infty$. Thus integrability conditions for $\|X_1\|$ depend only on what happens for $W_1 < \gamma$ or $W_1 > 1 - \gamma$; by symmetry, we need only consider $W_1 < \gamma$. Then $\sup\{X_1(s)\colon W_1 \leqq s < \gamma\}$ is attained at $s = W_1$ and equals $\omega(W_1)(1 - W_1)$. Since the suprema over other intervals are uniformly bounded as above and we can assume $W_1 < \gamma \leqq 1/3$ here, $E\|X_1\|^2/L_2\|X_1\| < \infty$ is equivalent to

$$E(\omega^2(W_1)/L_2\,\omega(W_1)) < \infty. \tag{7.6}$$

Since $\omega(t) \ll 1/t$ as $t \downarrow 0$, (7.6) implies (iv). Conversely if (iv) holds, let $x_n = x(n) = \sup\{x \leqq \gamma\colon \omega(x) \geqq n\}$. If $\omega(\cdot)$ is bounded for $x \leqq \gamma$, then (7.6) holds, so we can assume that the x_n are all defined. Then $x_n \downarrow 0$. Now we have

$$\sum_n n^2 \int_{x(n+1)}^{x(n)} dx/L_2(1/x) < \infty \tag{7.7}$$

and (7.6) is equivalent to

$$\sum_n n^2(x_n - x_{n+1})/L_2 n < \infty. \tag{7.8}$$

To prove (7.8), first note that the sum of those terms such that $x_n \leqq 2n^{-4}$ clearly converges. For the terms with $x_{n+1} \geqq x_n/2 > n^{-4}$, note that for $x_{n+1} \leqq x$, $1/L_2(1/x) \geqq 1/L_2(n^4) \sim 1/L_2 n$ as $n \to \infty$, so the sum of such terms in (7.8) also converges. Terms in (7.8) with $x_{n+1} < x_n/2 > n^{-4}$ are at most doubled if we replace x_{n+1} by $x_n/2$, and then their sum converges as in the last case. So (7.8) and hence (7.6) are proved. Thus we can apply Theorem 1.2, completing the proof of Theorem 7.5.

8. Necessity of Measurability in Separable Normed Spaces

It is well known that, for example, the law of an empirical distribution function in $D[0, 1]$ need not be defined on the Borel σ-algebra for the (non-separable) supremum norm. Thus results such as our Theorem 1.1 really need to allow non-measurability of the variables X_j. On the other hand we assumed in (1.6) that suitable finite-dimensional variables are measurable. We now clarify the roles of such assumptions by showing that in any separable Banach space, a weak central limit theorem (or *a fortiori*, invariance principles such as ours) can only hold for measurable variables. The passage from one-dimensional to separable Banach spaces will be easy. First we have a one-dimensional result.

We have the usual inner measure $\Pr_*(B) := \sup\{\Pr(C)\colon C \subset B\}$ and set $f_* := -((-f)^*) = \operatorname{ess.sup}\{g\colon g \leqq f,\ g \text{ measurable}\}$.

Theorem 8.1. *Let $X_n = h(x_n)$, $n = 1, 2, \ldots$, where x_n are i.i.d. random variables with values in some measurable space A and h is any real-valued function (not assumed measurable) on A. Let $S_n := X_1 + \ldots + X_n$. Suppose $\mathscr{L}(S_n/n^{1/2}) \to N(0, 1)$ in the sense that for all t,*

$$\lim_{n\to\infty} \Pr^*(S_n/n^{1/2} \leqq t) = \lim_{n\to\infty} \Pr_*(S_n/n^{1/2} \leqq t) = N(0, 1)(]-\infty, t]).$$

Then h is measurable for the completion of $\mathscr{L}(x_1)$, so that X_i are measurable, $EX_i=0$ and $EX_i^2=1$.

Proof. It is classical that if h is measurable then $EX_i=0$ and $EX_i^2=1$ (Gnedenko and Kolmogorov, 1968, §35, Theorem 4, p. 181). Suppose h is non-measurable for $\mathscr{L}(x_1)$, so that X_n are non-measurable, and consider $X_{n*} \leqq X_n \leqq X_n^*$.

Let B be the measurable set on which $X_1^*=+\infty$. If $P(B)>0$ then P restricted to subsets of B must be non-atomic. We always have $X_1^*>-\infty$ everywhere. For some numbers $M_n \uparrow +\infty$ we have $P(X_1^* \leqq -M_n) \leqq n^{-3}$. Then $P(\min_{j \leqq n} X_j^* \leqq -M_n) \leqq n^{-2}$. Then by the Borel-Cantelli lemma, almost surely for n large enough $X_j^* \geqq -M_n$ for all $j \leqq n$, so $\sum_{1 \leqq j \leqq n} \{X_j^*: X_j^*<0\} \geqq -nM_n$.

On B define a measurable, finite valued $Y_1 \geqq 0$ such that for n large enough $P(Y_1 \geqq nM_n+2n) \geqq n^{-1/2}$. Let $Y_1 = X_1^*-1$ outside B. Define Y_n from X_n for all n just as Y_1 from X_1. Then

$$P(\max_{j \leqq n} Y_j \geqq nM_n+2n) \geqq 1-(1-n^{-1/2})^n \to 1$$

rapidly as $n \to \infty$, so almost surely for n large enough, there is a $j \leqq n$ with $Y_j \geqq nM_n+2n$ and thus $\sum_{1 \leqq j \leqq n} Y_j \geqq n$. Since $Y_j < X_j^*$ for all j, $P^*(X_j \geqq Y_j,\ j=1,\ldots,n)=1$ by independence and Lemma 2.3. Then $P^*(S_n/n^{1/2} \geqq n^{1/2}) \to 1$ as $n \to \infty$, a contradition. Thus $P(B)=0$ and X_1^* is finite valued a.s.

Let $B_j := B(j) := \{X_j \geqq X_j^* - 2^{-j}\}$. Then $P^*(B_j)=1$, $j=1,2,\ldots$. Let $C_n = \bigcap_{j=1}^n B_j$. We apply Lemma 2.3 to $P_j := \mathscr{L}(x_j)$ and $f_j := 1_{B(j)}$, giving $P^*(C_n)=1$. On C_n, $S_n \leqq X_1^*+\ldots+X_n^* \leqq S_n+1$. Thus $\mathscr{L}((X_1^*+\ldots+X_n^*)/n^{1/2}) \to N(0,1)$. Hence $EX_1^*=0$. Likewise $EX_{1*}=0$. Thus $X_{1*}=X_1=X_1^*$ a.s., i.e. X_1 is completion measurable. Q.E.D.

Corollary 8.2. *Let S be any real vector space and $X_n=h(x_n)$ where x_n are independent, identically distributed random variables with values in some measurable space $(A, \mathscr{A})$ and h is any function from A into S (not assumed measurable). Let F be a collection of linear functionals: $S \to \mathbb{R}$. Suppose that for each $f \in F$, the central limit theorem as in Theorem 8.1 holds for the $f(h(x_j))$, with some limit law $N(0, \sigma_f^2)$, $\sigma_f^2>0$. Then each $f \circ h$, $f \in F$, is measurable for the completion of $\mathscr{L}(x_1)$ on $\mathscr{A}$.*

Corollary 8.3. *If S is a separable normed space, the hypotheses of Theorem 1.1 imply that the X_j are completion measurable for the Borel σ-algebra on S.*

Proof. Let $F=S'$, the dual Banach space. The conclusion of Theorem 1.1 implies the hypotheses of Corollary 8.2. Let $\mathscr{I}$ be the σ-algebra of all subsets B of S such that $h^{-1}(B)$ is measurable for the completion of $\mathscr{L}(x_1)$ on A. Then all elements of S' are $\mathscr{I}$ measurable. Since S is separable, $\mathscr{I}$ must contain all Borel sets. □

Corollary 8.4. *If the hypotheses of Theorem 1.1 hold and H is a bounded linear operator from S into a separable Banach space, then the $H(X_i)$ must be measurable for the completion of $\mathscr{L}(x_1)$.*

References

1. Acosta, Alejandro de: Existence and convergence of probability measures in Banach spaces. Trans. Amer. Math. Soc. **152**, 273–298 (1970)
2. Berkes, István, Morrow, G.: Strong invariance principles for mixing random fields. Z. Wahrscheinlichkeitstheorie verw. Geb. **57**, 15–37 (1981)
3. Berkes, I., Philipp, Walter: An almost sure invariance principle for the empirical distribution function of mixing random variables. Z. Wahrscheinlichkeitstheorie verw. Geb. **41**, 115–137 (1977)
4. Berkes, I., Philipp, Walter: Approximation theorems for independent and weakly dependent random vectors. Ann. Probability **7**, 29–54 (1979)
5. Billingsley, Patrick: Convergence of Probability Measures. New York: Wiley 1968
6. Borisov, I.S.: Some limit theorems for empirical distributions. Abstracts of Reports. Third Vilnius Conf. Probability. Theory Math. Statist. **I**, 71–72 (1981)
7. Bourbaki, N.: Eléments de math. Première partie, livre VI, Intégration. Paris: Hermann 1959
8. Breiman, Leo: Probability. Reading Mass.: Addison-Wesley 1968
9. Bronštein, E.M.: ε-entropy of convex sets and functions. Siberian Math. J. **17**, 393–398 (transl. from Sibirsk. Mat. Ž. **17**, 508–514) (1976)
10. Dehling, Herold: Limit theorems for sums of weakly dependent Banach space valued random variables. Z. Wahrscheinlichkeitstheorie verw. Geb., to appear (1983)
11. Dudley, R.M.: Metric entropy of some classes of sets with differentiable boundaries. J. Approximation Th. **10**, 227–236 (1974); Correction, ibid. **26**, 192–193 (1979)
12. Dudley, R.M.: Central limit theorems for empirical measures. Ann. Probability **6**, 899–929 (1978); Correction, ibid. **7**, 909–911 (1979)
13. Dudley, R.M.: Lower layers in $\mathbb{R}^2$ and convex sets in $\mathbb{R}^3$ are not GB classes. Probability in Banach Spaces II (Proc. Conf. Oberwolfach, 1978). Lecture Notes in Math. **709**, 97–102. Berlin-Heidelberg-New York: Springer 1979
14. Dudley, R.M.: Donsker classes of functions, Statistics and Related Topics (Proc. Symp. Ottawa, 1980), pages 341–352. New York Amsterdam: North-Holland 1981
15. Dudley, R.M.: Vapnik-Červonenkis Donsker classes of functions. Aspects statistiques et aspects physiques des processus gaussiens (Proc. Colloque C.N.R.S. St.-Flour, 1980), C.N.R.S., Paris, 251–269 (1981b)
16. Durst, Mark, Dudley, R.: Empirical processes, Vapnik-Chervonenkis classes and Poisson processes. Probability and Mathematical Statistics (Wrocław) **1**, 109–115 (1981)
17. Eames, W., May, L.E.: Measurable cover functions. Canad. Math. Bull. **10**, 519–523 (1967)
18. Enflo, Per: A counterexample to the approximation problem in Banach spaces. Acta Math. **130**, 309–317 (1973)
19. Fernique, Xavier: Intégrabilité des vecteurs gaussiens. C.R. Acad. Sci. Paris Sér. A **270**, A1698–1699 (1970)
20. Fernique, X.: Régularité de processus gaussiens. Invent. Math. **12**, 304–320 (1971)
21. Freedman, D.: Brownian motion and diffusion. San Francisco; Holden-Day 1971
22. Gnedenko, B.V., Kolmogorov, A.N.: Limit distributions for sums of independent random variables (transl. K.L. Chung). Reading, Mass.: Addison-Wesley 1949, 1954, 1968
23. Goodman, Victor, Kuelbs, James, Zinn, Joel: Some results on the LIL in Banach space with applications to weighted empirical processes. Ann. Probability **9**, 713–752 (1981)
24. Hartman, Philip, Wintner, Aurel: On the law of the iterated logarithm. Amer. J. Math. **63**, 169–176 (1941)
25. Heinkel, B.: Relation entre théorème central-limite et loi du logarithme itéré dans les espaces de Banach. Z. Wahrscheinlichkeitstheorie verw. Geb. **49**, 211–220 (1979)
26. Hoffmann-Jørgensen, J.: Sums of independent Banach space valued random variables. Studia Math. **52**, 159–186 (1974)
27. Ibero, M.: Approximation forte du processus empirique fonctionnel multidimensionnel. Bull. Sci. Math. **103**, 409–422 (1979a)
28. Ibero, M.: Intégrale stochastique par rapport au processus empirique multidimensionnel. C.R. Acad. Sci. Paris Sér. A **288**, A165–168 (1979b)
29. Ionescu Tulcea, A., and C.: On the lifting property I. J. Math. Anal. Appl. **3**, 537–546 (1961)
30. Ionescu Tulcea, A. and C.: On the lifting property II: Representation of linear operators on spaces L_{E^r}, $1 \leqq r < \infty$. J. Math. Mech. (Indiana) **11**, 773–795 (1962)

31. Jain, N.C., Marcus, M.B.: Integrability of infinite sums of independent vector-valued random variables. Trans. Amer. Math. Soc. **212**, 1–36 (1975)
32. Jain, N.C.: Central limit theorem in a Banach space, Probability in Banach Spaces (Proc. Conf. Oberwolfach, 1975). Lecture Notes in Math. **526**, 113–130. Berlin-Heidelberg-New York: Springer 1976
33. Jain, N.C.: An example concerning CLT and LIL in Banach space. Ann. Probability **4**, 690–694 (1976b)
34. James, B.R.: A functional law of the iterated logarithm for weighted empirical distributions. Ann. Probability **3**, 762–772 (1975)
35. Kahane, J.-P.: Some random series of functions. Lexington, Mass.: D.C. Heath 1968
36. Kiefer, J.: Skorohod embedding of multivariate rv's, and the sample df. Z. Wahrscheinlichkeitstheorie verw. Geb. **24**, 1–35 (1972)
37. Kolčinski, V.I.: On the central limit theorem for empirical measures. Teor. Verojatnost. i. Mat. Statist. (Kiev) **1981**, no. 24, 63–75 (in Russian). Theory Probability and Math. Statist. no. 24, 71–82 (1981a)
38. Kolčinski, V.I.: On the law of the iterated logarithm in Strassen's form for empirical measures. Teor. Verojatnost. i. Mat. Statist. (Kiev) **1981**, no. 25, 40–47. Theory Probability and Math. Statist. no. 25, 43–50 (1981b)
39. Komlós, J., Major, P., Tusnády, G.: An approximation of partial sums of independent $RV'-s$, and the sample DF. I. Z. Wahrscheinlichkeitstheorie verw. Geb. **32**, 111–131 (1975)
40. Kuelbs, J.: A strong convergence theorem for Banach space valued random variables. Ann. Probability **4**, 744–771 (1976)
41. Kuelbs, J.: Kolmogorov's law of the iterated logarithm for Banach space valued random variables. Illinois J. Math. **21**, 784–800 (1977)
42. Kuelbs, J., Zinn, J.: Some stability results for vector valued random variables. Ann. Probability **7**, 75–84 (1979)
43. Kuelbs, J., Dudley, R.M.: Log log laws for empirical measures. Ann. Probability **8**, 405–418 (1980)
44. Kuelbs, J., Philipp, Walter: Almost sure invariance principles for partial sums of mixing B-valued random variables. Ann. Probability **8**, 1003–1036 (1980)
45. Lindenstrauss, J., Tzafriri, L.: Classical Banach Spaces I. Berlin-Heidelberg-New York: Springer 1977
46. Major, Péter: Approximation of partial sums of i.i.d. rv's when the summands have only two moments. Z. Wahrscheinlichkeitstheorie verw. Geb. **35**, 221–229 (1976)
47. Marcus, Michael, B., Philipp, Walter: Almost sure invariance principles for sums of B-valued random variables with applications to random Fourier series and the empirical characteristic process. Trans. Amer. Math. Soc. **269**, 67–90 (1982)
48. Marcus, M.B., Shepp, L.A.: Sample behavior of Gaussian processes. Proc. 6th Berkeley Sympos. Math. Statist. Probab. **2**, 423–442. Univ. Calif. Press (1971)
49. May, L.E.: Separation of functions. Canad. Math. Bull. **16**, 245–250 (1973)
50. O'Reilly, Neville E.: On the convergence of empirical processes in sup-norm metrics. Ann. Probability **2**, 642–651 (1974)
51. Parthasarathy, K.R.: Probability measures on metric spaces. New York: Academic Press 1967
52. Philipp, Walter: Almost sure invariance principles for sums of B-valued random variables, Probability in Banach Spaces II (Proc. Conf. Oberwolfach, 1978). Lecture Notes in Math. **709**, 171–193. Berlin-Heidelberg-New York: Springer 1979
53. Philipp, Walter: Weak and L^p-invariance principles for sums of B-valued random variables. Ann. Probability **8**, 68–82 (1980); Correction, ibid. (to appear)
54. Philipp, Walter, Pinzur, Laurence: Almost sure approximation theorems for the multivariate empirical process. Z. Wahrscheinlichkeitstheorie verw. Geb. **54**, 1–13 (1980)
55. Pisier, G.: Le théorème de la limite centrale et la loi du logarithme itéré dans les espaces de Banach. Séminaire Maurey-Schwartz 1975–76, Exposé IV, École Polytechnique, Paris (1975)
56a. Pollard, D.: Limit theorems for empirical processes. Z. Wahrscheinlichkeitstheorie verw. Geb. **57**, 181–195 (1981)
56b. Pollard, D.: A central limit theorem for empirical processes. [To appear in J. Australian Math. Soc. Ser. A]
57. Prohorov, Yu.V.: Convergence of random processes and limit theorems in probability theory. Theor. Probability Appl. **1**, 157–214 (1956)

58. Révész, Pál: On strong approximation of the multidimensional empirical process. Ann. Probability **4**, 729–743 (1976)
59. Senatov, V.V.: Some estimates of the rate of convergence in the multidimensional central limit theorem. Soviet Math. Doklady **22**, 462–466 = Dokl. Akad. Nauk SSSR **254** no. 4, 809–812 (1980)
60. Shortt, R.M.: Existence of laws with given marginals and specified support. Ph.D. thesis, M.I.T. (1982)
61. Skorohod, A.V.: Limit theorems for stochastic processes. Theor. Probability Appl. **1**, 261–290 (1956)
62. Strassen, V.: The existence of probability measures with given marginals. Ann. Math. Statist. **36**, 423–439 (1965)
63. Vorob'ev, N.N.: Consistent families of measures and their extensions. Theor. Probability Appl. **7**, 147–163 (1962)
64. Vulikh, B.Z.: Introduction to the theory of partially ordered spaces (transl. by L.F. Boron). Gröningen: Walters-Noordhoff 1967
65. Wenocur, R.S., Dudley, R.M.: Some special Vapnik-Chervonenkis classes. Discrete Math. **33**, 313–318 (1981)
66. Yurinskii, V.V.: On the error of the Gaussian approximation for convolutions. Theor. Probability Appl. **22**, 236–247 (1977)

Received March 24, 1982

Added in Proof

For a better proof of Theorem 8.1 and other improvements see R.M. Dudley, Ecole d'été de probabilités de St.-Flour, 1982, Theorem 3.3.1, etc., Lecture Notes in Math. (to appear).

The Annals of Probability
1987, Vol. 15, No. 4, 1306–1326

UNIVERSAL DONSKER CLASSES AND METRIC ENTROPY[1]

By R. M. Dudley

Massachusetts Institute of Technology

Let $(X, \mathscr{A})$ be a measurable space and $\mathscr{F}$ a class of measurable functions on X. $\mathscr{F}$ is called a universal Donsker class if for every probability measure P on $\mathscr{A}$, the centered and normalized empirical measures $n^{1/2}(P_n - P)$ converge in law, with respect to uniform convergence over $\mathscr{F}$, to the limiting "Brownian bridge" process G_P. Then up to additive constants, $\mathscr{F}$ must be uniformly bounded. Several nonequivalent conditions are shown to imply the universal Donsker property. Some are connected with the Vapnik–Červonenkis combinatorial condition on classes of sets, others with metric entropy. The implications between the various conditions are considered. Bounds are given for the metric entropy of convex hulls in Hilbert space.

0. Introduction. When $\mathscr{F}$ is a universal Donsker class, then for independent, identically distributed (i.i.d.) observations $X_1, \dots, X_n$ with an unknown law P, for any f_i in $\mathscr{F}$, $i = 1, \dots, m$, $n^{-1/2}\{f_i(X_1) + \cdots + f_i(X_n)\}_{1 \le i \le m}$ is asymptotically normal with mean vector $n^{1/2}\{\int f_i(x)\, dP(x)\}_{1 \le i \le m}$ and covariance matrix $\int f_i f_j\, dP - \int f_i\, dP \int f_j\, dP$, uniformly for $f_i \in \mathscr{F}$. Then, for certain statistics formed from the $f_i(X_k)$, even where f_i may be chosen depending on the X_k, there will be asymptotic distributions as $n \to \infty$. For example, for χ^2 statistics, where f_i are indicators of disjoint intervals, depending suitably on $X_1, \dots, X_n$, whose union is the real line, X^2 quadratic forms have limiting distributions [Roy (1956) and Watson (1958)] which may, however, not be χ^2 distributions and may depend on P [Chernoff and Lehmann (1954)]. Universal Donsker classes of sets are, up to mild measurability conditions, just classes satisfying the Vapnik–Červonenkis combinatorial conditions defined later in this section [Durst and Dudley (1981) and Dudley (1984) Chapter 11]. The use of such classes allows a variety of extensions of the Roy–Watson results to general (multidimensional) sample spaces [Pollard (1979) and Moore and Stubblebine (1981)]. Vapnik and Červonenkis (1974) indicated applications of their families of sets to classification (pattern recognition) problems. More recently, the classes have been applied to tree-structured classification [Breiman, Friedman, Olshen and Stone (1984), Chapter 12].

The use of functions more general than indicators gives additional potentially useful freedom in constructing statistics. For example, there may be advantages to procedures based on spaces of smooth functions, which contain no nontrivial indicators. Le Cam, Mahan and Singh (1983) give a rather general extension of the Chernoff–Lehmann approach to "quadratic forms, or related objects."

Or, if $\mathscr{F}$ is an (infinite-dimensional) ellipsoid, the square of the supremum of an empirical measure over $\mathscr{F}$ is a sum of squared integrals, approximable by

---

Received August 1985; revised August 1986.

[1]This research was partially supported by National Science Foundation Grant DMS-8506638.

AMS 1980 *subject classifications.* Primary 60F17, 60F05; secondary 60G17, 60G20.

Key words and phrases. Central limit theorems, Vapnik–Červonenkis classes.

E. Giné et al. (eds.), *Selected Works of R.M. Dudley*, Selected Works in Probability and Statistics, DOI 10.1007/978-1-4419-5821-1_19,

finite sums, thus easier to compute than suprema over most families of functions. Ellipsoids seem difficult to obtain as symmetric convex hulls of classes of indicators.

The present paper will give no specific statistical procedures, but rather a general approach to sufficient conditions for the universal Donsker property.

First, some terminology and previous results will be recalled. Most appeared in Dudley (1984). Let $(X, \mathscr{A}, P)$ be a probability space. Let $P(f) := \int f\,dP$ for each integrable function f. Let G_P be a Gaussian process indexed by $\mathscr{L}^2(X, \mathscr{A}, P)$ with mean 0 and covariance $EG_P(f)G_P(g) = P(fg) - P(f)P(g)$ for all f and g in $\mathscr{L}^2$. Let $\rho_P(f, g) := (E((G_P(f) - G_P(g))^2))^{1/2}$. Such a process will be called *coherent* if each sample function $G_P(\cdot)(\omega)$ is bounded and uniformly continuous on $\mathscr{F}$ with respect to ρ_P. A class $\mathscr{F} \subset \mathscr{L}^2$ will be called *pregaussian* or *P-pregaussian* (formerly called G_PBUC) iff a coherent G_P process exists on $\mathscr{F}$. Let $(X^\infty, \mathscr{A}^\infty, P^\infty)$ be the countable product of copies of $(X, \mathscr{A}, P)$, with coordinates $x(1), x(2), \dots$. Let $P_n := n^{-1}(\delta_{x(1)} + \cdots + \delta_{x(n)})$, the nth empirical measure for P. Let $\nu_n := n^{1/2}(P_n - P)$. Let $(\Omega, \mathscr{S}, \Pr)$ be the product of $(X^\infty, \mathscr{A}^\infty, P^\infty)$ and the unit interval with Lebesgue measure. For any real-valued function G on $\mathscr{F}$ let $\|G\|_{\mathscr{F}} := \sup\{|G(f)|: f \in \mathscr{F}\}$. Call $\mathscr{F}$ a *functional Donsker class* for P iff on Ω, there exist independent coherent G_P processes $Y_1, Y_2, \dots$, such that for each $\varepsilon > 0$,

$$\lim_{n\to\infty} \Pr{}^*\Big\{n^{-1/2} \max_{k \le n} \|k(P_k - P) - Y_1 - \cdots - Y_k\|_{\mathscr{F}} > \varepsilon\Big\} = 0.$$

An alternate definition, due to Hoffmann-Jørgensen (1984), is that $\mathscr{F}$ be P-pregaussian and for every bounded $\|\cdot\|_{\mathscr{F}}$-continuous real function h on the set of all bounded functions on $\mathscr{F}$,

$$\int^* h(\nu_n)\, d\Pr \to \int h(G_P)\, d\Pr;$$

this definition is equivalent to the previous one according to Hoffmann-Jørgensen (1984), Talagrand (1987) and/or Dudley (1985), Theorem 5.2. Thus, if, in addition, $h(\nu_n)$ is Pr-measurable, it converges in law to $h(G_P)$. We have uniform convergence of $Eh(\nu_n)$ to $Eh(G_p)$ for h in any uniformly bounded, $\|\cdot\|_{\mathscr{F}}$-equicontinuous family of functions h for which $h(\nu_n)$ are measurable, as follows from an extended Wichura theorem [Dudley (1985), Theorem 4.1].

A collection $\mathscr{C}$ of sets is said to *shatter* a set F iff every subset of F is of the form $B \cap F$ for some $B \in \mathscr{C}$. The supremum of cardinalities of finite sets shattered by $\mathscr{C}$ will be called $S(\mathscr{C})$. A collection $\mathscr{C}$ is called a *Vapnik–Červonenkis* (VC) *class* iff $S(\mathscr{C}) < \infty$.

A measurable space $(S, \mathscr{B})$ will be called *standard* iff there exists a metric d on S for which (S, d) is a complete separable metric space and $\mathscr{B}$ is the Borel σ-algebra. A measurable space $(Y, \mathscr{U})$ will be called *Suslin* iff $\mathscr{U}$ is countably generated and separates points of Y and there is a standard space $(S, \mathscr{B})$ and a measurable function from S onto Y. Given a measurable space $(X, \mathscr{A})$, a collection $\mathscr{F}$ of measurable functions on X will be called *image admissible Suslin via* $(Y, \mathscr{U}, T)$ iff $(X, \mathscr{A})$ and $(Y, \mathscr{U})$ are Suslin and T is a function from Y

onto $\mathscr{F}$ such that the function $\langle x, y\rangle \mapsto T(y)(x)$ is jointly measurable. Here, equivalently, $(Y, \mathscr{U})$ can be taken to be standard.

For a measurable space $(X, \mathscr{A})$, a collection $\mathscr{F}$ of measurable real-valued functions on X will be called a *universal Donsker class* iff it is a functional Donsker class for every probability measure (law) P on $\mathscr{A}$. If $\mathscr{F}$ is a collection of indicators of sets, $\mathscr{F} = \{1_B\colon B \in \mathscr{C}\}$, and a universal Donsker class, or even pregaussian for all P, then $\mathscr{C}$ must be a Vapnik–Červonenkis class [Durst and Dudley (1981) and Dudley (1984), Theorem 11.4.1]. Conversely, an image admissible Suslin VC class is a universal Donsker class [Dudley (1978), Section 7, and (1984), Theorems 11.1.2 and 11.3.1]. The measurability hypotheses cannot be completely removed [Durst and Dudley (1981) and Dudley (1984), Theorem 11.4.2]. The question remains open as far as I know: What (measurability) conditions on a family of sets, together with the VC property, are equivalent to the universal Donsker property?

For classes of functions a quantitative condition characterizing the universal Donsker property up to measurability is not known (to me). Several nonequivalent extensions of the VC property to families of functions will be shown to be sufficient (under suitable measurability), but not necessary.

The remaining sections of the paper are: 1. Statements of conditions. 2. The easier implications. 3. Small non-VC hull classes and dual density. 4. Stability properties. 5. Metric entropy of convex hulls in Hilbert space. 6. Sharpness of the conditions. 7. Notes on weighted processes.

1. Statements of conditions. Here is a first result. For any real function f let $\operatorname{diam}(f) := \sup f - \inf f$.

(1.1) PROPOSITION. *For any universal Donsker class* $\mathscr{F}$,

$$\sup_{f \in \mathscr{F}} \operatorname{diam}(f) < \infty.$$

PROOF. If not, take $x_n := x(n) \in X$, $y_n := y(n) \in X$ and $f_n \in \mathscr{F}$ with $f_n(x_n) - f_n(y_n) > 2^n$ for all n. Let

$$P := \sum_{n=1}^{\infty} \left(\delta_{x(n)} + \delta_{y(n)}\right)/2^{n+1},$$

a law on all subsets of X and thus on $\mathscr{A}$. Then for P,

$$E(f_n - Ef_n)^2 \geq 2^{-n-1} \inf_{c \in \mathbb{R}} \left\{ \left(f_n(x_n) - c\right)^2 + \left(f_n(y_n) - c\right)^2 \right\}$$

$$= 2^{-n-2}\left(f_n(x_n) - f_n(y_n)\right)^2 > 2^{n-2}, \qquad n = 1, 2, \dots.$$

Thus, for the ρ_P (standard deviation) metric, $\mathscr{F}$ is unbounded, so not a Donsker class for P [Dudley (1984), Theorem 4.1.1] and not a universal Donsker class. □

Say a function h on $\mathscr{F}$ *ignores additive constants* if $h(f) = h(f + c)$ whenever $f \in \mathscr{F}$, c is a constant and $f + c \in \mathscr{F}$. Let Y be a coherent G_P

process. Then if $f \in \mathscr{F}$ and $f + c \in \mathscr{F}$ for a constant c, $\rho_P(f, f + c) = 0$ so $Y(f) \equiv Y(f + c)$. Then Y is consistently extended to the set $\mathscr{F} + \mathbb{R}$ of all functions $f + c$, $f \in \mathscr{F}$, $c \in \mathbb{R}$, setting $Y(f + c) \equiv Y(f)$. This extension is coherent, so $\mathscr{F} + \mathbb{R}$ is pregaussian.

Now let $\mathscr{F}$ be a functional Donsker class for P. Note that each $k(P_k - P)$ is defined and ignores additive constants on $\mathscr{F} + \mathbb{R}$. Extending each Y_j in the definition to $\mathscr{F} + \mathbb{R}$ as above, we have $\|\alpha\|_{\mathscr{F}+\mathbb{R}} \equiv \|\alpha\|_{\mathscr{F}}$ for each $\alpha = k(P_k - P) - Y_1 - \cdots - Y_k$. Thus, $\mathscr{F} + \mathbb{R}$ is a functional Donsker class for P. Hence, if $\mathscr{F}$ is a universal Donsker class, so is $\mathscr{F} + \mathbb{R}$.

Any subset of a functional Donsker class for P is also, thus any subset of a universal Donsker class is a universal Donsker class. For an arbitrary real function $c(\cdot)$ on $\mathscr{F}$, $\mathscr{F}$ is a universal Donsker class iff $\{f - c(f): f \in \mathscr{F}\}$ is. Let $b\mathscr{F} := \{bf: f \in \mathscr{F}\}$ for a constant $b \neq 0$. Let $U_j(bf) := bY_j(f)$, $f \in \mathscr{F}$, for Y_j as in the definition of a functional Donsker class. Then each U_j is a coherent G_P process on $b\mathscr{F}$, so that replacing $\mathscr{F}$ by $b\mathscr{F}$ and Y_j by U_j we see that $b\mathscr{F}$ is a functional Donsker class iff $\mathscr{F}$ is, and thus a universal Donsker class iff $\mathscr{F}$ is. So in finding sufficient conditions for a class $\mathscr{F}$ to be a universal Donsker class, it will be enough to consider uniformly bounded classes of functions [letting $c(f) := \inf f$], where we can also assume $0 \leq f \leq 1$ for all $f \in \mathscr{F}$.

Now several conditions to be shown sufficient for the universal Donsker property will be defined.

Definitions. Let $\mathscr{F}$ be a class of functions on X with $f(x) \geq 0$ for all $f \in \mathscr{F}$ and $x \in X$. Let $\mathscr{C}$ be a class of subsets of X.

For each $f \in \mathscr{F}$ and $t \in \mathbb{R}$, the set $\{x \in X: f(x) > t\}$ will be called a *major set* of f and of $\mathscr{F}$. I call $\mathscr{F}$ a *major class* for $\mathscr{C}$ iff all the major sets of $\mathscr{F}$ are in $\mathscr{C}$. If $\mathscr{C}$ is a VC class, then I call $\mathscr{F}$ a *VC major class* (for $\mathscr{C}$).

The *subgraph* of the function $f \geq 0$ will be defined as

$$\operatorname{sub}(f) := \{\langle x, t\rangle \in X \times \mathbb{R} : 0 \leq t \leq f(x)\}.$$

For a class $\mathscr{D}$ of subsets of $X \times \mathbb{R}$, say $\mathscr{F}$ is a *subgraph class* for $\mathscr{D}$ iff $\operatorname{sub}(f) \in \mathscr{D}$ for all $f \in \mathscr{F}$. If $\mathscr{D}$ is a VC class, then I call $\mathscr{F}$ a *VC subgraph class* (for $\mathscr{D}$). [VC subgraph classes have previously been called "VC graph" classes [Alexander (1984, 1987)] or "Polynomial classes" [Pollard (1984), pages 17 and 34, and (1985), Section 6].]

Given a class $\mathscr{F}$ of functions and $0 < M < \infty$, let $H(\mathscr{F}, M)$ be M times the symmetric convex hull of $\mathscr{F}$, that is, the class of all functions g such that for some $k < \infty$, $f_j \in \mathscr{F}$ and real t_j, $j = 1, \dots, k$, $\sum_{j=1}^k |t_j| \leq M$ and $g(x) = \sum_{j=1}^k t_j f_j(x)$ for all x. Then let $\overline{H}_s(\mathscr{F}, M)$ [sequential closure of $H(\mathscr{F}, M)$] be the smallest class $\mathscr{G}$ of functions including $H(\mathscr{F}, M)$ such that for any $g_n \in \mathscr{G}$ with $g_n(x) \to g(x)$ as $n \to \infty$ for all x, we have $g \in \mathscr{G}$. Say that $\mathscr{F}$ is a *hull class* for $\mathscr{C}$ (a class of sets identified with its class of indicator functions) iff $\mathscr{F} \subset \overline{H}_s(\mathscr{C}, M)$ for some $M < \infty$. Then $\mathscr{F}$ will be called a *VC hull class* if $S(\mathscr{C}) < \infty$. I call $\mathscr{F}$ a *VC subgraph hull class* if $\mathscr{F} \subset \overline{H}_s(\mathscr{G}, M)$ for some $M < \infty$ and VC subgraph class $\mathscr{G}$. [The "secondary VC graph difference classes," for which Alexander

(1985b), Theorem 2.2, proves extended Kiefer inequalities, are VC subgraph hull classes.]

For any metric space (S, d) and $\varepsilon > 0$ let $D(\varepsilon, S) := D(\varepsilon, S, d) := \sup\{m\colon$ for some $x_1, \dots, x_m \in S$, $d(x_i, x_j) > \varepsilon$ for $1 \le i < j \le m\}$. Given a law Q on $\mathscr{A}$, $1 \le p < \infty$ and $\varepsilon > 0$ let $D^{(p)}(\varepsilon, \mathscr{F}, Q) := D(\varepsilon, \mathscr{F}, e_{Q,p})$, where $e_{Q,p}(f, g) := (\int |f - g|^p \, dQ)^{1/p}$. Let $D^{(p)}(\varepsilon, \mathscr{F}) := \sup_Q D^{(p)}(\varepsilon, \mathscr{F}, Q)$, where the supremum is over all laws Q concentrated in finite sets. [These definitions are due to Kolčinskiĭ (1981) and Pollard (1982).] Say that $\mathscr{F}$ satisfies *Pollard's entropy condition* iff

$$\int_0^1 \left(\log D^{(2)}(\varepsilon, \mathscr{F})\right)^{1/2} d\varepsilon < \infty. \tag{1.2}$$

Let $D^{(\infty)}(\varepsilon, \mathscr{F}) := D(\varepsilon, \mathscr{F}, d_\infty)$, where $d_\infty(f, g) := \sup_x |(f - g)(x)|$.

Then clearly $D^{(2)}(\varepsilon, \mathscr{F}) \le D^{(\infty)}(\varepsilon, \mathscr{F})$.

Note that for any uniformly bounded class $\mathscr{F}$ of measurable functions, $M < \infty$ and any law Q on $\mathscr{A}$, $H(\mathscr{F}, M)$ is dense in $\overline{H}_s(\mathscr{F}, M)$ for $e_{Q,p}$, $1 \le p < \infty$, as will be used for $p = 2$.

Some other conditions to be considered are as follows:

(1.3) For some $r < \infty$, $\quad D^{(2)}(\varepsilon, \mathscr{F}) = O(\varepsilon^{-r})$ as $\varepsilon \downarrow 0$.

(1.4) $\mathscr{F}$ is a sequence $\{f_j\}_{j \ge 1}$ of measurable functions on X with $\operatorname{diam}(f_j) = o((\log j)^{-1/2})$ as $j \to \infty$.

$(1.4)^{\text{co}}$ $\quad \mathscr{F} \subset \overline{H}_s(\{f_j\}, M)$, where $M < \infty$ and $\{f_j\}$ satisfies (1.4).

Condition (1.2) with sup instead of L^2 norm becomes

$$\int_0^1 \left(\log D^{(\infty)}(x, \mathscr{F})\right)^{1/2} dx < \infty. \tag{1.5}$$

Most of the rest of the paper is devoted to showing what implications hold between the various conditions just defined, all of which imply the universal Donsker property. Figure 1 illustrates the implications, where h) requires some measurability conditions. Some implications which are obvious (given others) are not indicated. Most of the cases are handled by Theorem 2.1 below; "a)," "b)," etc. in Figure 1 indicate parts of Theorem 2.1. The condition

$$\log D^{(2)}(\varepsilon, \mathscr{F}) = O(\varepsilon^{-2}), \quad \text{as } \varepsilon \downarrow 0 \tag{1.6}$$

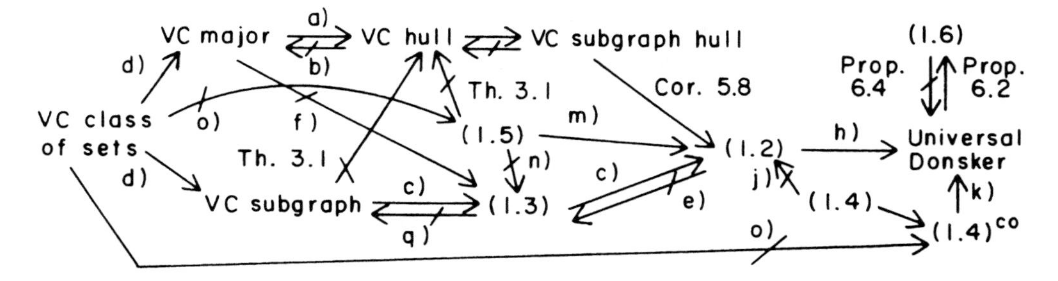

FIG. 1.

is shown in Section 6 to be a sharp (Proposition 6.3) necessary condition (Proposition 6.2) for the universal Donsker property but not sufficient (Proposition 6.4). These facts and Theorem 2.1h) show that $D^{(2)}$ comes close to characterizing the universal Donsker property, but does not. [On (1.4) see Theorem 2.1o).] Some of the stronger conditions are often easier to check and apply. Also, they may imply stronger inequalities or speeds of convergence, which may vary notably even between VC classes of sets, according to Beck (1985). A statistician has some freedom of choice in the class of functions.

2. The easier implications.

This section will prove most of the implications and nonimplications in Figure 1 relatively easily.

(2.1) THEOREM. *Let $\mathscr{F}$ be a class of measurable functions f on X with $0 \leq f(x) \leq 1$ for all x, and $\mathscr{C}$ a class of measurable subsets of X.*

a) *If all the major sets of $\mathscr{F}$ are in $\mathscr{C}$, then $\mathscr{F}$ is also a hull class for $\mathscr{C}$. Thus, every VC major class $\mathscr{F}$ is also a VC hull class.*

b) *There exist VC hull classes $\mathscr{F}$ which are not VC major classes [i.e., $\{\{x\colon f(x) > t\}\colon f \in \mathscr{F},\ t \in \mathbb{R}\}$ is not a VC class].*

c) *(Pollard) For every VC subgraph class $\mathscr{F}$,* (1.3) *and* (1.2) *hold.*

d) *For any VC class $\mathscr{C}$ of sets, its set of indicator functions is both a VC subgraph and a VC major class.*

e) *There exist classes $\mathscr{F}$ of functions satisfying* (1.2) *and not* (1.3).

f) *There exist VC major (thus VC hull) classes which do not satisfy* (1.3), *thus are not VC subgraph classes.*

g) *There are VC subgraph classes [thus satisfying* (1.3) *and* (1.2)] *which are not VC major.*

h) *(Pollard) If $\mathscr{F}$ satisfies* (1.2) *and is image admissible Suslin, it is a universal Donsker class.*

i) *If $\mathscr{F}$ is a functional Donsker class for P, then so is $\overline{H}_s(\mathscr{F}, M)$ for any $M < \infty$.*

j) *There exist sequences $\{f_j\}$ satisfying* (1.4) *but not Pollard's entropy condition* (1.2).

k) *(Paulauskas and Heinkel) If $\mathscr{F}$ satisfies* (1.4) *or* $(1.4)^{co}$, *then $\mathscr{F}$ is a universal Donsker class.*

m) *Condition* (1.5) *implies* (1.2).

n) (1.5) *does not imply* (1.3), *thus does not imply that $\mathscr{F}$ is a VC subgraph class.*

o) *An infinite collection of indicators of sets is never included in $\overline{H}_s(\mathscr{F}, M)$ for any $M < \infty$ and class $\mathscr{F}$ satisfying* (1.5) *or sequence $\mathscr{F} = \{f_j\}$ satisfying* (1.4).

p) (1.5) *does not imply that $\mathscr{F}$ is a VC major class.*

q) (1.3) *does not imply that $\mathscr{F}$ is a VC subgraph class.*

REMARK. As noted, if $\mathscr{F}$ is the class of indicators of sets in $\mathscr{C}$, then up to measurability conditions, the universal Donsker property of $\mathscr{F}$ is equivalent to the Vapnik–Červonenkis property of $\mathscr{C}$, and hence to the intermediate conditions in Figure 1, namely the VC major, VC hull and VC subgraph (hull) properties of $\mathscr{F}$ as well as (1.2) and (1.3) [that (1.2) implies (1.3) for classes of sets may be surprising]. Some of these equivalences can be seen directly (without measurability assumptions). Conditions (1.4) and (1.5) do not join in the equivalence.

PROOF. For the examples in parts e), j) and q), $A(n)$ will always be independent sets with $P(A(n)) = 1/2$, $n = 1, 2, \dots$, for some law P, specifically Lebesgue measure on $[0, 1]$.

a) Let $\mathscr{F}$ be a major class for $\mathscr{C}$. Given $f \in \mathscr{F}$ let

$$f_n := \frac{1}{n} \sum_{j=1}^{n} 1_{\{f > j/n\}} = \sum_{j=0}^{n-1} \frac{j}{n} 1_{\{j/n < f \le (j+1)/n\}}.$$

Then $f_n \to f$ as $n \to \infty$ (even uniformly), so $\mathscr{F}$ is a hull class for $\mathscr{C}$ [this argument was used previously in Dudley (1981), Theorem 1.9].

b) Let $X = \mathbb{R}^2$ with usual Borel σ-algebra. Let $\mathscr{C}$ be the collection of all open lower left quadrants

$$\{\langle x, y\rangle : x < a,\ y < b\}, \qquad a \in \mathbb{R},\ b \in \mathbb{R}.$$

Then $\mathscr{C}$ is a VC class [$S(\mathscr{C}) = 2$, see Wenocur and Dudley (1981), Proposition 2.3, and Dudley (1984), Corollary 9.2.15]. Let

$$\mathscr{F} := \left\{ \sum_{n=1}^{\infty} 1_{C(n)}/2^n : C(n) \in \mathscr{C} \right\}.$$

Then $\mathscr{F}$ is clearly a hull class for $\mathscr{C}$ (this did not depend on the particular choice of $\mathscr{C}$). The sets $\{f > 0\}$, $f \in \mathscr{F}$, are exactly the countable unions of sets in $\mathscr{C}$. These are all the open "lower layers" in the plane [e.g., Dudley (1984), Section 7.2] and do not form a VC class. For example, their intersections with the line $x + y = 1$ shatter any finite subset of the line.

c) See Pollard (1984), page 27, Lemma 25, with $F \equiv 1$, and page 34, Lemma 36.

d) $\mathscr{C}$ is a VC class in X iff $\{B \times [0, 1]: B \in \mathscr{C}\}$ is a VC class in $X \times \mathbb{R}$: "if" is straightforward, and "only if" is a special case of Assouad (1983), Proposition 2.5b, and Dudley (1984), Theorem 9.2.6, for $\times$. The class of major sets for $\{1_A : A \in \mathscr{C}\}$ is $\mathscr{C} \cup \{\varnothing, X\}$, a VC class iff $\mathscr{C}$ is.

e) and n) Let $Lx := \max(1, \log x)$ and $f_n := 1_{A(n)}/Ln$. Then the sequence $\mathscr{F} := \{f_n\}_{n \ge 1}$ satisfies (1.4), (1.5) and (1.2). For each subset F of $\{1, \dots, n\}$ choose a point

$$x(F) \in \bigcap_{j \in F} A(j) \cap \bigcap_{j \notin F,\ j \le n} X \setminus A(j).$$

Set $Q := 2^{-n} \Sigma_F \delta_{x(F)}$. Then $A(j)$ for $j = 1, \dots, n$ are independent for Q with probability 1/2. We have for $i < j \le n$,

$$\int (f_i - f_j)^2 \, dQ = \int f_i^2 + f_j^2 \, dQ - \frac{1}{2LiLj}$$
$$= \frac{1}{2}\left(\frac{1}{(Li)^2} + \frac{1}{(Lj)^2} - \frac{1}{LiLj}\right) \ge \frac{1}{2(Lj)^2} \ge \frac{1}{2(Ln)^2}.$$

So $D^{(2)}(\varepsilon, \mathscr{F}, Q) \ge \exp(1/2^{1/2}\varepsilon) - 1$ for $0 < \varepsilon < 1$ and (1.3) fails, proving n) and e).

f) Let $\mathscr{F}$ be the set of all nondecreasing right-continuous functions f on $\mathbb{R}$ with $0 \le f \le 1$. Then $\mathscr{F}$ is a VC major class, since half-lines $[x, \infty[$ or $]x, \infty[$, $-\infty \le x \le \infty$, form a VC class $\mathscr{C}$ with $S(\mathscr{C}) = 1$. To show that (1.3) fails for $\mathscr{F}$, first note that for any measurable f and g and law Q, the L^2 distance $e_Q(f, g) := e_{Q,2}(f, g) \ge \int |f - g| \, dQ$, so $D^{(2)}(\varepsilon, \mathscr{F}) \ge D^{(1)}(\varepsilon, \mathscr{F})$, $\varepsilon > .0$.

The next steps use some facts about lower layers [Dudley (1984), Section 7.2]. For Lebesgue measure P on $[0, 1]$,

$$D^{(1)}(\varepsilon, \mathscr{F}, P) = D(\varepsilon, \mathscr{L}\mathscr{L}_{2,1}, d_\lambda), \qquad 0 < \varepsilon < 1,$$

where $\mathscr{L}\mathscr{L}_{2,1}$ is the set of all lower layers in the unit square I^2 in $\mathbb{R}^2$ and d_λ is the Lebesgue measure of the symmetric difference of sets in I^2. Thus, for some $c > 0$, $D^{(1)}(\varepsilon, \mathscr{F}, P) \ge e^{c/\varepsilon}$ as $\varepsilon \downarrow 0$. For each ε with $0 < \varepsilon < 1$, there is a law Q with finite support and $D^{(1)}(\varepsilon, \mathscr{F}) \ge D^{(1)}(\varepsilon, \mathscr{F}, Q) \ge e^{c/\varepsilon} - 1$, so (1.3) fails, proving f).

g) and p) These will follow from Theorem 3.1, but here are short proofs. Let $f_n := n^{-1} + n^{-2} 1_{B(n)}$, $n = 1, 2, \ldots$, where $B(n)$ is any sequence of measurable sets. Then $f_n \downarrow 0$, so the subgraphs of f_n also decrease. Being linearly ordered by inclusion, they form a VC class $\mathscr{C}$ with $S(\mathscr{C}) = 1$ [Wenocur and Dudley (1981), Corollary 2.2, and Dudley (1984), Theorem 9.2.4). So $\mathscr{F} := \{f_n\}_{n \ge 1}$ is a VC subgraph class. We have $\{f_n > 1/n\} = B(n)$, and the $B(n)$ need not form a VC class, so $\mathscr{F}$ need not be a VC major class, proving g). Now $\mathscr{F}$ has $D^{(\infty)}(\varepsilon, \mathscr{F}) \le 2 + 1/\varepsilon$ for $0 < \varepsilon < 1$, so $\mathscr{F}$ satisfies (1.5), proving p).

h) This essentially follows from a theorem of Pollard (1982), proved with the specific "image admissible Suslin" measurability condition in Dudley (1984), Section 11.3, with $F \equiv 1$.

i) For any subset $\mathscr{F}$ of a real vector space V and real function Y on $\mathscr{F}$, call Y *prelinear* iff whenever $\sum_{i=1}^m a_i f_i = 0$ for real a_i and f_i in $\mathscr{F}$, then $\sum_{i=1}^m a_i Y(f_i) = 0$.

(2.2) LEMMA. *For any coherent G_P process Y on some $\mathscr{F}$ there is such a process Z on the ρ_P-closed convex symmetric hull $\mathscr{K}$ of $\mathscr{F}$ with all sample functions of Z prelinear on $\mathscr{K}$ and a.s. $Z \equiv Y$ on $\mathscr{F}$. Thus, almost all sample functions of Y are prelinear on $\mathscr{F}$.*

PROOF. See Dudley (1985), Theorem 5.1, in whose proof the three-series theorem should and need only be applied when $f \in \mathscr{F}$. □

Let $\mathscr{F}$ be a uniformly bounded functional P-Donsker class and $\mathscr{H} :=$ $\{h$: for some $f_n \in \mathscr{F}$, $f_n(x) \to h(x)$ for all $x\}$. Each $k(P_k - P)$ is continuous for bounded pointwise convergence, and each coherent Y_j in the definition of functional P-Donsker class extends naturally to $\mathscr{H}$. For each $\alpha := k(P_k - P) - Y_1 - \cdots - Y_k$, we have $\|\alpha\|_{\mathscr{H}} \equiv \|\alpha\|_{\mathscr{F}}$. Thus, $\mathscr{H}$ is a functional P-Donsker class. Hence, so is $\mathscr{G} := \overline{H}_s(\mathscr{F}, M)$, we have $\|\alpha\|_{\mathscr{G}} = M\|\alpha\|_{\mathscr{F}}$, and i) follows.

j) An example will be based on one in the proof of Dudley (1967), Proposition 6.10. Let $a_n := \alpha_n(Ln)^{-1/2}$ where $\alpha_n \downarrow 0$ slowly; specifically, $\alpha_n := (LLn)^{-1/2}$. Let $f_n := a_n 1_{A(n)}$. Then for each n, taking Q as in the proof of e), for $1 \leq i < j \leq n$ we have

$$\int (f_i - f_j)^2 \, dQ = \tfrac{1}{2}\Big(\alpha_i^2(Li)^{-1} + \alpha_j^2(Lj)^{-1} - \alpha_i\alpha_j(LiLj)^{-1/2}\Big)$$

$$\geq \tfrac{1}{2}\alpha_j^2(Lj)^{-1} > \varepsilon^2$$

for $j \leq n$ if $\alpha_n^2 > 2\varepsilon^2 Ln$, or, equivalently, $LnLLn < 1/(2\varepsilon^2)$. Let $n(\varepsilon)$ be the largest n for which this holds. Then as $\varepsilon \downarrow 0$, $LLn(\varepsilon) \sim 2L(1/\varepsilon)$, so $Ln(\varepsilon) \sim (4\varepsilon^2 L(1/\varepsilon))^{-1}$, and $\int_0^1 (Ln(\varepsilon))^{1/2} \, d\varepsilon = +\infty$. So $\{f_j\}_{j \geq 1}$ satisfies (1.4) but not (1.2), proving j).

k) By part i), we may assume (1.4), then apply a central limit theorem in the Banach space c_0 stated by Paulauskas (1980), Corollary. Heinkel (1983), Theorem 1, gave a relatively long proof. Here is a direct proof.

By an inequality of Hoeffding (1963), (2.3) of Theorem 2, for any measurable f with $0 \leq f \leq 1$, $n = 1, 2, \ldots$, and any law P,

$$\Pr\{\nu_n(f) \geq y\} \leq \exp(-2y^2), \text{ for each } y \geq 0.$$

Considering $1 - f$, we have

$$\Pr\{|\nu_n(f)| \geq y\} \leq 2\exp(-2y^2).$$

Let $a_j := \operatorname{diam}(f_j) = \alpha_j(Lj)^{-1/2}$, so that $\alpha_j \to 0$ as $j \to \infty$. We may assume that $0 \leq f_j \leq a_j > 0$. Then for any $\varepsilon > 0$,

$$\Pr\{|\nu_n(f_j)| > \varepsilon\} = \Pr\{|\nu_n(f_j/a_j)| > \varepsilon/a_j\} \leq 2\exp\big(-2\varepsilon^2/a_j^2\big)$$

$$= 2\exp\big(-(2\varepsilon^2 Lj)/\alpha_j^2\big)$$

$$= 2j^{-2\varepsilon^2/\alpha(j)^2}, \text{ where } \alpha(j) := \alpha_j,\ j \geq 3.$$

Since $\varepsilon^2/\alpha_j^2 \geq 1$ for j large,

$$\sum_j \Pr\{|\nu_n(f_j)| > \varepsilon\} \text{ converges, uniformly in } n.$$

If $f_i = f_j$ a.s. for P, then $\nu_n(f_i) = \nu_n(f_j)$ a.s. for P^∞. Thus, $\{f_j\}_{j \geq 1}$ form a functional Donsker class [by the proof of Theorem 5.2.1 of Dudley (1984)], proving k).

m) follows easily from the definitions.

Recall that n) was proved with e) and p) with g).

For o), recall that the convex hull of a totally bounded set is totally bounded [by Mazur's theorem, e.g., Dunford and Schwartz (1958), page 416].

q) Take $\mathscr{F} := \{1_{A(n)}/n\}_{n \geq 1}$. Then for $0 < \varepsilon < 1$, $D^{(2)}\varepsilon, \mathscr{F}) \leq D^{(\infty)}(\varepsilon, \mathscr{F}) \leq 1 + 1/\varepsilon$, so (1.3) holds. For $m = 1, 2, \ldots,$ the sets $A(1), \ldots, A(2^m)$ shatter some set B with m elements [Assouad (1983), 2.12 and 2.13b, and Dudley (1984), Theorem 9.3.2, which lacks the Assouad reference]. Then the class $\mathscr{D}$ of all subgraphs of functions in $\mathscr{F}$ shatters $\{\langle x, 2^{-m-1}\rangle: x \in B\}$, so $S(\mathscr{D}) = +\infty$, proving q) and Theorem 2.1. □

3. Small non-VC hull classes and dual density. This section will show that there are sequences converging to 0 in supremum norm like any power of n which are not VC hull classes. The proof applies the notion of dual density [Assouad (1983)].

(3.1) THEOREM. *There exist sequences $\mathscr{F} := \{b_n 1_{A(n)}\}_{n \geq 1}$, specifically with $b_n = 1/n^v$ for any positive integer v, which satisfy* (1.3), (1.4) *and* (1.5), *and there exist VC subgraph classes $\mathscr{G}$, such that $\mathscr{F}$ and $\mathscr{G}$ are not VC hull classes.*

PROOF. Two lemmas will be proved.

(3.2) LEMMA. *Let $A_1, \ldots, A_n$ be jointly independent events in X with $P(A_j) = 1/2$, $j = 1, \ldots, n$. Let $B_1, \ldots, B_n$ be any events and set $D := \bigcup_{1 \leq j \leq n} A_j \vartriangle B_j$. Then the algebra $\mathscr{B}$ generated by $B_1, \ldots, B_n$ has at least $2^n(1 - P(D))$ atoms.*

PROOF. For any $F \subset \{1, \ldots, n\}$, let

$$A_F := \bigcap_{j \in F} A_j \cap \bigcap_{j \notin F,\, j \leq n} X \setminus A_j,$$

and likewise define B_F. The atoms of $\mathscr{B}$ are those B_F which are nonempty. For each F, $P(A_F) = 1/2^n$. If a point of A_F is not in D, it is also in B_F, which is then nonempty. Since D can include at most $2^n P(D)$ of the 2^n events A_F, the lemma follows. □

(3.3) LEMMA. *Suppose $A_j := A(j)$ are independent events with $P(A_j) = 1/2$ for all j, and $\mathscr{C}$ is a class of events such that for some $K < \infty$ and $u < \infty$, for each j there is an event D_j such that $P(A_j \vartriangle D_j) < \eta_j$, where D_j is in an algebra generated by at most Kj^u elements of $\mathscr{C}$ and $\sum_j \eta_j < 1$. Then $\mathscr{C}$ is not a VC class.*

PROOF. By Lemma 3.2, for each $m = 1, 2, \ldots,$ the algebra $\mathscr{D}$ generated by $D_1, \ldots, D_m$ has at least $2^m \alpha$ atoms, where $\alpha := 1 - \sum_j \eta_j > 0$. On the other hand, $\mathscr{D}$ is generated by at most

$$\sum_{1 \leq j \leq m} Kj^u \leq K(m+1)^{u+1}/(u+1)$$

sets in $\mathscr{C}$. Now, for a VC class $\mathscr{C}$, according to Assouad (1983), 1.3d and 2.13e, there is some $t < \infty$ and a $C < \infty$ such that the number of atoms of the algebra generated by k elements of $\mathscr{C}$ is at most Ck^t. Then $2^m \alpha$ is bounded above by a polynomial in m [of degree $(u+1)t$], a contradiction. □

Now to prove Theorem 3.1, let $A(m)$ and P be as in the proof of Theorem 2.1. Suppose the class $\mathscr{F} := \{1_{A(m)}/m^v\}_{m \geq 1}$ were a VC hull class for some $\mathscr{C}$. Clearly, $D^{(\infty)}(\varepsilon, \mathscr{F}) \leq m$ for the least m such that $1/m^v \leq \varepsilon$, so $m < \varepsilon^{-1/v} + 1$. Thus, (1.4) and (1.5) both hold for $\mathscr{F}$ as stated.

Suppose $P(A) = 1/2$ and that $a + b1_A \in \overline{H}_s(\mathscr{C}, M)$ for some $a \geq 0$, $b > 0$ and $0 < M < \infty$. Replacing a by a/M and b by b/M, we can assume that $M = 1$. If $S(\mathscr{C}) = r < \infty$, then for any $w > r$ there is a constant $C < \infty$ such that $D(\varepsilon, \mathscr{C}, d_P) < C\varepsilon^{-w}$ for $0 < \varepsilon < 1$ [Assouad (1983), Proposition 4.3, and Dudley (1984), Theorem 9.3.1], where $d_P(B, C) := P(B \vartriangle C)$. Take $w = r + 1$. Given $0 < \beta < 1$, take $C(j) \in \mathscr{C}$ and t_j with $\Sigma_j |t_j| \leq 1$ such that $P(|a + b1_A - \Sigma_j t_j 1_{C(j)}|) < \beta$. Choose $D_i := D(i) \in \mathscr{C}$, $i = 1, \dots, m$, such that for each j, $P(C_j \vartriangle D_i) \leq \beta$ for some $i := i(j)$, with $m \leq C\beta^{-r-1}$. Then $P(|a + b1_A - f|) < 2\beta$, where $f := \Sigma_j t_j 1_{D(i(j))}$. Let $B := \{f > a + b/2\}$. Then B is in the algebra generated by $D_1, \dots, D_m$, and $P(B \vartriangle A) < 4\beta/b$. We apply this to $A = A(k)$ for each k, with $a = a(k) = 0$, $b = b(k) = 1/k^v$ and $\beta = \beta_k := 1/(8k^{2+v})$. We obtain sets $B := B(k)$ and $m := m(k) \leq Nk^u$ for some $N < \infty$ and $u := (r + 1)(2 + v) < \infty$. To apply Lemma 3.3, let $\eta_k := 1/(2k^2)$, giving a contradiction if $\mathscr{C}$ is a VC class. We obtain a VC subgraph class $\mathscr{G}$ as in the proof of Theorem 2.1g), letting $a := a(k) := 1/k$ and $b := b(k) = 1/k^2$. So Theorem 3.1 is proved. □

Notes. The counterexamples in the proofs of Theorem 2.1b), e), f), g), j), n), o), p), q) and Theorem 3.1, all are, or can be taken to be, image admissible Suslin.

It is not settled here whether classes satisfying (1.5) are necessarily VC subgraph hull classes.

4. Stability properties. Some of the properties treated above are preserved by some operations. If $\mathscr{F}$ is a major class for $\mathscr{C}$, then so is $\{cf: f \in \mathscr{F}, c > 0\}$, clearly. The property of being a subgraph class for some $\mathscr{D}$ is not necessarily preserved by taking constant multiples, but it is for many of the classes $\mathscr{D}$ in applications.

(4.1) Proposition [Assouad (1983), Proposition 2.15]. *Let $\mathscr{C}$ be a VC class and let $\mathscr{E}$ be its closure for pointwise convergence of indicator functions. Then $\mathscr{E}$ is also a VC class, with $S(\mathscr{E}) = S(\mathscr{C})$.*

Proof (for completeness). Let F be a finite set, $A \in \mathscr{E}$, and let $A(\alpha)$ be a net of sets in $\mathscr{C}$ with $1_{A(\alpha)}(x) \to 1_A(x)$ for all x. Then for some α, $A(\alpha) \cap F = A \cap F$, so $\mathscr{C}$ and $\mathscr{E}$ give the same intersections with finite sets and shattered finite sets. □

Sometimes, to preserve norms for empirical measures and/or for measurability, it may be useful to consider subsets of the closure, such as the sequential closure. Given $\mathscr{C}$, its *monotone derived class* will be the class $\mathscr{D}$ of all sets $D \subset X$ such that for some $C_n \in \mathscr{C}$, either $C_n \uparrow D$ or $C_n \downarrow D$. If $\mathscr{F}$ is a VC major class for $\mathscr{C}$, then the sets $\{x \in X: f(x) \geq t\}$, $f \in \mathscr{F}$, $t \geq 0$, are in $\mathscr{D}$, so they can also be assumed to be in $\mathscr{C}$ without increasing $S(\mathscr{C})$.

(4.2) PROPOSITION. *If $\mathscr{F}$ is a major class for $\mathscr{C}$, $\mathscr{H}$ is the set of all nondecreasing functions from $\mathbb{R}$ into $[0,1]$, and*

$$\mathscr{G} := \{h \circ f\colon f \in \mathscr{F},\ h \in \mathscr{H}\},$$

then $\mathscr{G}$ is a major class for the monotone derived class $\mathscr{D}$ of $\mathscr{C}$. Thus, if $\mathscr{F}$ is a VC major class, so is $\mathscr{G}$.

PROOF. Given $h \in \mathscr{H}$ and $t \geq 0$, let $s := \inf\{y\colon h(y) > t\}$. Then $h^{-1}(]t,\infty[) =]s,\infty[$ or possibly $[s,\infty[$ if h has a jump just to the left of s, $h(s^-) < h(s)$. So for $f \in \mathscr{F}$,

$$(h \circ f)^{-1}(]t,\infty[) = f^{-1}(]s,\infty[) \in \mathscr{C} \subset \mathscr{D}$$

or

$$f^{-1}([s,\infty[) \in \mathscr{D}. \qquad \square$$

For any VC class $\mathscr{C}$, if we set

$$\mathscr{E} := \mathscr{C} \cup \{X \setminus A\colon A \in \mathscr{C}\}, \tag{4.3}$$

then $\mathscr{E}$ is also a VC class [e.g., by Vapnik and Červonenkis (1971), Theorem 1]. Thus, in Proposition 4.2 if $\mathscr{H}$ is replaced by the set of all monotone functions from $\mathbb{R}$ into $[0,1]$, $\mathscr{F}$ is a major class for the monotone derived class of $\mathscr{E}$ and thus still a VC major class. Here is a further extension:

(4.4) PROPOSITION. *Given $0 < M < \infty$ let V_M be the set of all functions from $\mathbb{R}$ into itself with total variation $\leq M$. Let $\mathscr{F}$ be a major class for $\mathscr{C}$, a functional Donsker class for P. Then $\mathscr{J} := \{h \circ f\colon h \in V_M,\ f \in \mathscr{F}\}$ is a functional Donsker class for P.*

PROOF. By subtracting constants which do not matter, as noted after Proposition 1.1, we may consider just functions h in V_M with $h(x) \to 0$ as $x \to -\infty$. For each such h, we have $h = u - v$ for some nondecreasing functions u and v, each with total variation $\leq M$, also going to 0 at $-\infty$. Let $\mathscr{G}$ be the class of all functions $u \circ f$, where u is nondecreasing, $f \in \mathscr{F}$ and $0 \leq u \leq 1$. Then by Proposition 4.2, $\mathscr{G}$ is a major class for $\mathscr{D}$, the monotone derived class of $\mathscr{C}$. By Theorem 2.1a) and its proof, $\mathscr{G}$ is a hull class for $\mathscr{D}$, with $\mathscr{G} \subset \overline{H}_s(\mathscr{D},1)$. It follows that $\mathscr{J} \subset \overline{H}_s(\mathscr{D},2M) \subset \overline{H}_s(\mathscr{C},2M)$. Thus, Theorem 2.1i) gives the result. $\square$

If $\mathscr{F}$ is a finite-dimensional real vector space of measurable functions on X, then either it contains the constants or the vector space W spanned by $\mathscr{F}$ and constants has dimension one larger. The class of sets

$$\{\{x\colon f(x) > t\}\colon f \in \mathscr{F},\ t \in \mathbb{R}\} = \{\{x\colon g(x) > 0\}\colon g \in W\}$$

is a VC class $\mathscr{C}$ with index $S(\mathscr{C})$ equal to the dimension of W [Dudley (1978), Theorem 7.2].

Let $\mathscr{F}$ be a VC major class which is image admissible Suslin via $(Y, \mathscr{U}, T)$. For example, $\mathscr{F}$ may be any finite-dimensional real vector space of measurable functions on X. Let $\mathscr{C} := \{\{x: f(x) > t\}: f \in \mathscr{F}, t \in \mathbb{R}\}$. Then $\mathscr{C}$ is a VC class and is image admissible Suslin via $(Y \times \mathbb{R}, \mathscr{U} \times \mathscr{B}, T')$, where $\mathscr{B}$ is the Borel σ-algebra and $T'(y, t)(x) := 1_{]t,\infty[}(T(y)(x))$. Then for V_M in Proposition 4.4, the given class of compositions is a universal Donsker class. On some possible applications and related results see Pollard (1985), Section 6.

Next, classes of products of functions will be considered.

(4.5) PROPOSITION. *Let $\mathscr{F}$ and $\mathscr{G}$ be classes of real functions on X and $\mathscr{C}$ and $\mathscr{D}$ classes of subsets of X. Let*

$$\mathscr{C} \underline{\cap} \mathscr{D} := \{A \cap B: A \in \mathscr{C}, B \in \mathscr{D}\}.$$

If for some K and M, $\mathscr{F} \subset \overline{H}_s(\mathscr{C}, K)$ and $\mathscr{G} \subset \overline{H}_s(\mathscr{D}, M)$, then $\{fg: f \in \mathscr{F}, g \in \mathscr{G}\} \subset \overline{H}_s(\mathscr{C} \underline{\cap} \mathscr{D}, KM)$.

PROOF. Suppose $\Sigma_i|s_i| \le K$, $\Sigma_j|t_j| \le M$, $A(i) \in \mathscr{C}$ and $B(j) \in \mathscr{D}$. Then $(\Sigma_i s_i 1_{A(i)} \Sigma_j t_j 1_{B(j)}) \equiv \Sigma_{i,j} s_i t_j 1_{A(i) \cap B(j)}$ and $\Sigma_{i,j}|s_i t_j| \le KM$. Taking limits of sequences then gives the result. □

Recall that if $\mathscr{C}$ and $\mathscr{D}$ are VC classes, then so is $\mathscr{C} \underline{\cap} \mathscr{D}$ [Assouad (1983), Proposition 2.5b, and Dudley (1984), Theorem 9.2.6]. Thus, if $\mathscr{F}$ and $\mathscr{G}$ are VC hull classes, so is their set of products fg.

(4.6) PROPOSITION. *There exist VC major classes $\mathscr{F}$ and $\mathscr{G}$ such that $\{\langle x, y\rangle \mapsto f(x)g(y): f \in \mathscr{F}, g \in \mathscr{G}\}$ is not a VC major class. Both $\mathscr{F}$ and $\mathscr{G}$ can be taken as the set of increasing functions on $\mathbb{R}$.*

PROOF. Let g be a fixed positive, continuous, strictly increasing function on $\mathbb{R}$, say $g(y) := e^y$. Let h be any continuous, strictly decreasing function. Let $f(x) := 1/g(h(x))$. Then f is positive, continuous and strictly increasing. We have $f(x)g(y) = 1$ on the graph $y = h(x)$. Thus, $f(x)g(y) < 1$ for $y < h(x)$ and $f(x)g(y) > 1$ iff $y > h(x)$. But the class of all sets where $y > h(x)$ for h continuous and strictly decreasing is not a VC class; it shatters all finite subsets of the line $x + y = 0$, for example. □

Thus, the VC hull property, with more stability, and which follows from the VC major property, may be more useful. Inequalities for suprema over VC subgraph hull classes, as good as those for VC classes $\mathscr{C}$ of sets [Alexander (1984)] extend immediately to VC subgraph hull classes, as near the end of the proof of Theorem 2.1i). In this sense VC subgraph hull classes are "just as good" as VC classes of sets despite the failure of several converse implications in Figure 1.

5. Metric entropy of convex hulls in Hilbert space. Let H be a real Hilbert space. For any subset B of H let $\mathrm{co}(B)$ be its convex hull.

(5.1) THEOREM. *Suppose $\|x\| \le 1$ for all $x \in B$ and for some $K < \infty$ and $\lambda < \infty$, $D(\varepsilon, B) \le K\varepsilon^{-\lambda}$ for $0 < \varepsilon \le 1$. Let $s := 2\lambda/(2 + \lambda)$. Then for any $t > s$*

there are constants C_1 and C_2, depending only on K, λ and t, such that

$$D(\varepsilon, \mathrm{co}(B)) \leq C_1 \exp\left(C_2 \varepsilon^{-t}\right), \qquad 0 < \varepsilon \leq 1.$$

PROOF. We may assume $K \geq 1$. Choose any $x_1 \in B$. Given $B(n) := \{x_1, \ldots, x_{n-1}\}$, $n \geq 2$, let $d(x, B(n)) := \inf_{y \in B(n)} \|x - y\|$ and $\delta_n := \sup_{x \in B} d(x, B(n))$. If $\delta_n > 0$ choose $x_n \in B$ with $d(x_n, B(n)) > \delta_n/2$. (The estimates to be proved in general will also hold in case B is finite, so that $\delta_n = 0$ for some n.) Then for all n, $K(2/\delta_n)^\lambda \geq D(\delta_n/2, B) \geq n$, so $\delta_n \leq M n^{-1/\lambda}$ for all n where $M := 2K^{1/\lambda}$.

Let $0 < \varepsilon \leq 1$. Let $N := N(\varepsilon)$ be the next integer larger than $(4M/\varepsilon)^\lambda$. Then $\delta_N < \varepsilon/4$. Let $G := B(N)$. For each $x \in B$ there is an $i = i(x) \leq N$ with $\|x - x_i\| \leq \delta_N$. For any convex combination $z = \sum_{x \in B} z_x x$, where $z_x \geq 0$ and $\sum_{x \in B} z_x = 1$, let $z_N := \sum_{x \in B} z_x x_{i(x)}$. Then $\|z - z_N\| \leq \delta_N < \varepsilon/4$, so

$$D(\varepsilon, \mathrm{co}(B)) \leq D(\varepsilon/2, \mathrm{co}(G)). \tag{5.2}$$

To bound the latter, let $m := m(\varepsilon) := [\varepsilon^{-s}]$ (integer part). Then for each i with $m < i \leq N$, there is a $j := j(i) \leq m$ such that

$$\|x_i - x_j\| \leq \delta_{m+1} \leq M \varepsilon^{s/\lambda}. \tag{5.3}$$

Let $\Lambda_m := \{\{\lambda_j\}_{1 \leq j \leq m}: \lambda_j \geq 0, \sum_{1 \leq j \leq m} \lambda_j = 1\}$. On $\mathbb{R}^m$ we have the l_p metrics

$$\rho_p(\{x_j\}, \{y_j\}) := \left(\sum_{1 \leq j \leq m} |x_j - y_j|^p\right)^{1/p}.$$

Let $\gamma := \varepsilon/6$ and $\delta := \gamma/(2m^{1/2})$. The δ-neighborhood of Λ_m for ρ_2 is included in a ball of radius $1 + \delta < 13/12$. We have $D(2\delta, \Lambda_m, \rho_2)$ centers of disjoint balls of radius δ included in the neighborhood. Comparing volumes of balls gives

$$\begin{aligned} D(\gamma, \Lambda_m, \rho_1) &\leq D(2\delta, \Lambda_m, \rho_2) \leq 13^m m^{m/2} \varepsilon^{-m} \\ &\leq \exp(m\{L(1/\varepsilon) + (Lm)/2 + \log(13)\}) \\ &\leq \exp(C_3 \varepsilon^{-s} L(1/\varepsilon)) \leq \exp\left(C_4 \varepsilon^{-t}\right), \qquad 0 < \varepsilon \leq 1, \end{aligned}$$

for some constants C_3, C_4.

For each $j = 1, \ldots, m$, let $A(j)$ consist of x_j and the set of all x_i, $i = m + 1, \ldots, N$, such that $j(i) = j$. Take a maximal set $S = S(\varepsilon) \subset \Lambda_m$ with $\rho_1(x, y) > \gamma$ for any $x \neq y$ in S. For a given $\lambda = (\lambda_1, \ldots, \lambda_m) \in S$, let

$$F_\lambda := \left\{x \in \mathrm{co}(G): x = \sum_{y \in G} \mu_y y, \text{ where } \mu_y \geq 0 \text{ and } \sum_{y \in A(j)} \mu_y = \lambda_j, \text{ for all } j = 1, \ldots, m\right\}.$$

For any $x \in \mathrm{co}(G)$, and μ_y as in the definition of F_λ, there is a $\lambda \in S$ and $z \in F_\lambda$ with $\|z - x\| \leq \gamma$. Thus,

$$\begin{aligned} D(\varepsilon, \mathrm{co}(B)) &\leq D(\varepsilon/2, \mathrm{co}(G)) \leq D\left(\gamma, \bigcup_{\lambda \in S} F_\lambda\right) \\ &\leq \mathrm{card}(S) \max_\lambda D(\gamma, F_\lambda) \leq \exp\left(C_4 \varepsilon^{-t}\right) \max_\lambda D(\gamma, F_\lambda). \end{aligned} \tag{5.4}$$

To estimate the latter factor, let $\lambda \in S$. We may assume $\lambda_j > 0$ for all j. For any $x \in F_\lambda$, let $x^{(j)} := \sum_{y \in A(j)} \mu_y y$. Let Y_j be a random variable with values in $A(j)$ and $P(Y_j = y) = \mu_y/\lambda_j$ for each $y \in A(j)$. Then $EY_j = x^{(j)}/\lambda_j := z_j$. Take $Y_1, \ldots, Y_m$ to be independent and let $Y := \sum_{1 \le j \le m} \lambda_j Y_j$. Then $EY = x$ and

$$E\|Y - x\|^2 = E\left\| \sum_{1 \le j \le m} \lambda_j (Y_j - z_j) \right\|^2 = \sum_{1 \le j \le m} \lambda_j^2 E\|Y_j - z_j\|^2,$$

since $Y_j - z_j$ are independent and have mean 0.

Now the diameter of $A(j)$ is at most $2M\varepsilon^{s/\lambda}$ by (5.3), and z_j is a convex combination of elements of $A(j)$. Thus,

$$E\|Y_j - z_j\|^2 \le 4M^2 \varepsilon^{2s/\lambda}$$

and for any set $F \subset \{1, \ldots, m\}$,

$$E\left\| \sum_{j \in F} \lambda_j (Y_j - z_j) \right\|^2 \le \sum_{j \in F} \lambda_j^2 4M^2 \varepsilon^{2s/\lambda} \le 4M^2 \Big(\max_{j \in F} \lambda_j \Big) \varepsilon^{2s/\lambda}$$

and

$$E\left\| \sum_{j \in F} \lambda_j (Y_j - z_j) \right\| \le 2M \Big(\max_{j \in F} \lambda_j \Big)^{1/2} \varepsilon^{s/\lambda}.$$

The following argument is based on an idea of Maurey, see the proofs of Pisier (1981), Lemma 2, and Carl (1982), Lemma 1. Let $Y_{j1}, Y_{j2}, \ldots, Y_{jk}$ be independent with the distribution of Y_j, with Y_{ji} also independent for different j. Then

$$E\left\| \sum_{j \in F} \lambda_j \left(\left(\frac{1}{k} \sum_{i=1}^{k} Y_{ji} \right) - z_j \right) \right\| \le 2M \Big(\max_{j \in F} \lambda_j \Big)^{1/2} \varepsilon^{s/\lambda} / k^{1/2}.$$

Thus, there exist $y_{ji} \in A(j)$, $i = 1, \ldots, k$, such that

$$\left\| \sum_{j \in F} \lambda_j \left(\left(\frac{1}{k} \sum_{i=1}^{k} y_{ji} \right) - z_j \right) \right\| \le 2M \Big(\max_{j \in F} \lambda_j \Big)^{1/2} \varepsilon^{s/\lambda} / k^{1/2}. \tag{5.5}$$

Take $v > 0$ such that $s + v < t$. Let $F(0) := \{j \le m \colon \lambda_j \ge \varepsilon^v\}$. Let $k(0)$ be the smallest integer k such that

$$k \ge 6400 M^2 \varepsilon^{-2 + 2s/\lambda} = 6400 M^2 \varepsilon^{-s}.$$

Then the expressions in (5.5) are at most $\varepsilon/40$.

Let r be the smallest positive integer such that $\varepsilon^v/4^r \le (\varepsilon^{1-s/\lambda}/(80M))^2$. For $u = 1, 2, \ldots, r$, let

$$F(u) := \left\{ j \le m \colon \varepsilon^v/4^u \le \lambda_j < \varepsilon^v/4^{u-1} \right\},$$

and let $k = k(u)$ be the smallest integer k such that $2^{2-u} M \varepsilon^{s/\lambda}/k^{1/2} < \varepsilon/(40r)$, i.e., $k > 100 M^2 4^{4-u} r^2 \varepsilon^{-2+2s/\lambda}$. Thus, for some constant C_5, $k(u) \le 1 + C_5 4^{-u} \varepsilon^{-s} (L(1/\varepsilon))^2$. The y_{ji} for $F = F(u)$ will be called $y_{ji}^{(u)}$ (they depend on x).

Let $F(r+1) := \{j \le m: \lambda_j < \varepsilon^v/4^r\}$. Let $k(r+1) = 1$. For $F = F(r+1)$ and $k = k(r+1)$, (5.5) is bounded above by $\varepsilon/40$. Let $y_{j1}^{(r+1)} := x_j$, a single choice for each j.

Combining terms for $u = 0, 1, \ldots, r+1$, we see that each $x \in F_\lambda$ can be approximated within $3\varepsilon/40 < \varepsilon/12$ by a convex combination determined uniquely by the $k(u)$-tuples $(y_{j1}^{(u)}, \ldots, y_{jk(u)}^{(u)})$, $j \in F(u)$, $u = 0, 1, \ldots, r+1$. Each $A(j)$ has at most N elements, so that for given $u \le r$ and $j \le m$, there are at most $N^{k(u)}$ ways of choosing the $y_{ji}^{(u)}$. Now $\operatorname{card}(F(u)) \le 4^u/\varepsilon^v$, so the number of ways to choose the $y_{ji}^{(u)}$ for given $u \le r$ is at most $\exp\{(\log N)(4^u\varepsilon^{-v} + C_5\varepsilon^{-s-v}L(1/\varepsilon)^2)\}$. Thus, the total number of ways to choose all the $y_{ji}^{(u)}$ gives $D(\varepsilon/6, F_\lambda) \le \exp(C_6\{\varepsilon^{-s-v}L(1/\varepsilon)^4 + L(1/\varepsilon)4^r/\varepsilon^v\})$ for some C_6. By definition of r, $4^r/\varepsilon^v \le C_7\varepsilon^{2s/\lambda-2} = C_7\varepsilon^{-s}$ for some C_7. Thus, $D(\varepsilon/6, F_\lambda) \le \exp(C_8\varepsilon^{-t})$ for some C_8. Combining with (5.4) completes the proof of Theorem 5.1. □

(5.6) REMARK. The exponent $s = 2\lambda/(2+\lambda)$ in Theorem 5.1 is sharp, as shown by Dudley (1967), Proposition 6.12, where $B = \{\pm n^{-1/\lambda}e_n\}_{n \ge 1}$ and $\{e_n\}$ is an orthonormal basis. More generally, Carl (1982) treats Banach spaces of type 2 in place of Hilbert space.

(5.7) COROLLARY. *If $\mathscr{G}$ is a uniformly bounded class of measurable functions such that for some $K < \infty$ and $0 < \lambda < \infty$, $D^{(2)}(\varepsilon, \mathscr{G}) \le K\varepsilon^{-\lambda}$ for $0 < \varepsilon \le 1$, then for any $t > s = 2\lambda/(2+\lambda)$, and for the constants C_1 and C_2 of Theorem* 5.1,

$$D^{(2)}\big(\varepsilon, \overline{H}_s(\mathscr{G}, 1)\big) \le C_1\exp\big(C_2\varepsilon^{-t}\big), \quad \text{for } 0 < \varepsilon \le 1.$$

(5.8) COROLLARY. *If $\mathscr{F}$ is a VC subgraph hull class, then $\mathscr{F}$ satisfies Pollard's entropy condition* (1.2).

PROOF. Let $\mathscr{F} \subset \overline{H}_s(\mathscr{G}, M)$ for $M < \infty$, where $\mathscr{G}$ is a VC subgraph class. Theorem 2.1c) gives condition (1.3) for $\mathscr{G}$, so Corollary 5.7 applies, with $s < t < 2$, and (1.2) for $\mathscr{F}$ follows. □

(5.9) EXAMPLE. Let $\mathscr{C}$ be the collection of all intervals $]a, b]$, $0 \le a \le b \le 1$. Let G be the space of functions f satisfying $|f(x)| \le 1/2$ and $|f(x) - f(y)| \le |x - y|$ for all $x, y \in]0, 1]$, $f(x) = 0$ elsewhere. Functions in G have total variation at most 2 (at most 1 on the open interval $]0, 1[$ and $1/2$ at each end). As in the proof of Proposition 4.4, each such $f = g - h$, where g and h are each nondecreasing and 0 for $x \le 0$ and $df = dg - dh$ is the Jordan decomposition of the signed measure df. Thus, g and h have equal total variations ≤ 1 and $G \subset \overline{H}_s(\mathscr{C}, 2)$. Next, $D^{(2)}(\varepsilon, G) \ge \exp(c/\varepsilon)$ as $\varepsilon \downarrow 0$ for some constant $c > 0$, by considering Lebesgue measure P on $[0, 1]$ (or finite approximations to it), noting that the $L^2(P)$ norm is larger than the $L^1(P)$ norm, and applying Dudley (1984), Theorem 7.1.10. Here since $S(\mathscr{C}) = 2$, the exponent λ can be any number larger than 2 [Assouad (1983), Proposition 4.3], so for $s = 2\lambda/(2+\lambda)$, $\lambda \downarrow 2$, t can be any number larger than 1. So the exponent s of Theorem 5.1 is also sharp for this example.

6. Sharpness of the conditions. How sharp are the various sufficient conditions for the universal Donsker property in Figure 1? First note that all but (1.4), $(1.4)^{co}$ and (1.5) hold for every universal Donsker class of indicators of sets, since such classes of sets are VC [Durst and Dudley (1981) and exposition in Dudley (1984), Theorem 11.4.1]. So we have sharpness when restricted to classes of indicators. Thus, we have:

(6.1) REMARK. Let $\mathscr{C}$ be a class of measurable sets. If either every uniformly bounded major class for $\mathscr{C}$, or the convex hull of the indicators of sets in $\mathscr{C}$, is a universal Donsker class, then $\mathscr{C}$ is a VC class.

In this sense "VC major (with enough measurability) implies universal Donsker" is sharp in that "VC" cannot be replaced by any weaker condition on families of sets. On the other hand, "VC major" is one of the strongest conditions in Figure 1. It turns out that $D^{(2)}$ entropy conditions, much like and because of the L^2 entropy conditions for Gaussian processes, come close to characterizing, but do not characterize, the universal Donsker property. Note that by Pollard's theorem (2.1h) above), for any $\delta > 0$,

$$\log D^{(2)}(\varepsilon, \mathscr{F}) = O(1/\varepsilon^{2-\delta}), \quad \text{as } \varepsilon \downarrow 0$$

is (with measurability) sufficient for $\mathscr{F}$ to be a universal Donsker class.

(6.2) PROPOSITION. *$\log D^{(2)}(\varepsilon, \mathscr{F}) = O(\varepsilon^{-2})$ as $\varepsilon \downarrow 0$ is necessary for $\mathscr{F}$ to be a universal Donsker class.*

PROOF. If not, there exists a universal Donsker class $\mathscr{F}$ and $\varepsilon_k \downarrow 0$ such that $\log D^{(2)}(\varepsilon_k, \mathscr{F}) > k^3/\varepsilon_k^2$ for all $k = 1, 2, \dots$. Take probability laws P_k such that

$$\log D^{(2)}(\varepsilon_k, \mathscr{F}, P_k) > k^3/\varepsilon_k^2, \qquad k = 2, 3, \dots,$$

and let P be a law with $P \geq \sum_{k \geq 2} P_k/k^2$. Then for any measurable f and g,

$$\left(\int (f-g)^2\, dP\right)^{1/2} \geq \left(\int (f-g)^2\, dP_k\right)^{1/2} \Big/ k.$$

Let $\delta_k := \varepsilon_k/k$. Then

$$\log D^{(2)}(\delta_k, \mathscr{F}, P) \geq \log D^{(2)}(\varepsilon_k, \mathscr{F}, P_k) > k^3/\varepsilon_k^2 = k/\delta_k^2.$$

Thus, $\mathscr{F}$ is not pregaussian for P [Dudley (1973), Theorem 1.1 (c)], so not a universal Donsker class. □

An example will show that Proposition 6.2 is sharp. Let $A(j)$ be disjoint, nonempty measurable sets, $j = 1, 2, \dots$. For $x = \{x_j\}_{j \geq 1}$ let $\|x\|_2 := (\sum_j x_j^2)^{1/2}$. Let

$$\mathscr{E} := \mathscr{E}\left(\{1_{A(j)}\}_{j \geq 1}\right) := \left\{\sum_j x_j 1_{A(j)} : \|x\|_2 \leq 1\right\},$$

the ellipsoid with center 0 and semiaxes $1_{A(j)}$.

(6.3) PROPOSITION. *$\mathscr{E}$ is a universal Donsker class and*

$$\liminf_{\delta \downarrow 0} \delta^2 \log D(\delta, \mathscr{E}) > 0.$$

PROOF. Let P be any probability law with $p_j := P(A(j))$, $j = 1, 2, \ldots$. We may assume that $p_j > 0$ for all j, since the union of all those $A(j)$ with $p_j = 0$ is a fixed set of probability 0 on which G_P and ν_n can be taken to be identically 0 a.s.

Let $\varepsilon > 0$. For any k and n, let

$$\|\nu_n\|_{2,k} := \left(\sum_{j \geq k} \nu_n(A(j))^2\right)^{1/2}.$$

Then for all n,

$$E\|\nu_n\|_{2,k}^2 \leq \sum_{j \geq k} p_j \to 0 \quad \text{as } k \to \infty.$$

Take $k = k(\varepsilon)$ large enough so that

$$\Pr\{\|\nu_n\|_{2,k} > \varepsilon/3\} < \varepsilon/2.$$

If $\|\nu_n\|_{2,k} \leq \varepsilon/3$, $\|x\|_2 \leq 1$ and $\|y\|_2 \leq 1$, then

$$\left|\nu_n\left(\sum_{j \geq k} (x_j - y_j) 1_{A(j)}\right)\right| \leq 2\varepsilon/3.$$

Also, $\Pr\{\|\nu_n\|_{2,1} > 2/\varepsilon\} < \varepsilon/2$.

Let $\delta := \delta(\varepsilon) := (\min_{j<k} p_j)\varepsilon^2/6$. For each $x = \{x_j\}_{j \geq 1}$ let $f_x := \sum_j x_j 1_{A(j)}$. Then if $f_x, f_y \in \mathscr{E}$ and $(\int (f_x - f_y)^2 \, dP)^{1/2} < \delta$, $(\sum_{j<k}(x_j - y_j)^2)^{1/2} < \varepsilon^2/6$, so

$$\left|\sum_{j<k} \nu_n((x_j - y_j) 1_{A(j)})\right| \leq \|\nu_n\|_{2,1}\varepsilon^2/6 < \varepsilon/3,$$

if $\|\nu_n\|_{2,1} \leq 2/\varepsilon$. Then except on one event with probability ε, we have $|\nu_n(f_x - f_y)| < \varepsilon$ for all such x and y.

Clearly, $\mathscr{E}$ is totally bounded in $L^2(P)$. Thus, $\mathscr{E}$ is a P-Donsker class [Dudley (1984), Theorem 4.1.10] and since P was arbitrary, a universal Donsker class.

Now let $0 < \delta < 1$. To bound $D^{(2)}(\delta, \mathscr{E})$ from below, let $m = [1/(4\delta^2)]$ (integer part). Let $P = P^{(m)}$ be a law with $P(A(j)) = 1/m$ for $j = 1, \ldots, m$. Then in $L^2(P)$, $\mathscr{E}$ is an m-dimensional ball of radius $m^{-1/2}$. The balls of radius 2δ with centers at a maximal set of points δ apart in $\mathscr{E}$ cover the ball of radius $m^{-1/2} + \delta$. Thus,

$$D^{(2)}(\delta, \mathscr{E}, P) \geq (m^{-1/2} + \delta)^m/(2\delta)^m = \left(\frac{1}{2}\left(\frac{1}{\delta m^{1/2}} + 1\right)\right)^m$$

$$\geq \left(\tfrac{3}{2}\right)^m \geq \left(\tfrac{2}{3}\right)\exp\left(\log\left(\tfrac{3}{2}\right)/4\delta^2\right),$$

and the result follows. □

(6.4) PROPOSITION. *There is a uniformly bounded class $\mathscr{F}$ of measurable functions, which is not a universal Donsker class, such that*

$$\log D^{(2)}(\varepsilon, \mathscr{F}) \leq 3/(\varepsilon^2 L(1/\varepsilon)), \quad \textit{as } \varepsilon \downarrow 0.$$

PROOF. Let $B(j)$, $j = 1, 2, \ldots$, be disjoint, nonempty measurable sets, $a_j := 1/(jLj)^{1/2}$ and $\mathscr{F} := \{\sum_{j \geq 1} x_j 1_{B(j)}\colon\ x_j = \pm a_j \text{ for all } j\}$. Take c such that

$\sum_j p_j = 1$ where $p_j = c(a_j/LLj)^2$. Then $\sum_j a_j p_j^{1/2} = +\infty$. Take a probability measure P with $P(B(j)) = p_j$ for all j. Then

$$E \sup_{f \in \mathscr{F}} |G_P(f)| = E \sum_j a_j |G_P(1_{B(j)})|$$

$$= \sum_j a_j (2/\pi)^{1/2} \big(p_j(1-p_j)\big)^{1/2} \geq \alpha \sum_j a_j p_j^{1/2} = +\infty,$$

where $\alpha = (1-p_1)^{1/2}/2 > 0$. Thus, by the Landau–Shepp–Fernique theorem [Fernique (1970)], $\mathscr{F}$ is not pregaussian for P, hence not a universal Donsker class.

For any probability law Q, $r < \infty$, and $x_j = \pm a_j$,

$$\left(\int \left(\sum_{j \geq r} x_j 1_{B(j)}\right)^2 dQ\right)^{1/2} = \left(\sum_{j \geq r} a_j^2 Q(B(j))\right)^{1/2} \leq a_r.$$

Given $\varepsilon > 0$, take the smallest $r = r(\varepsilon)$ for which $a_r < \varepsilon/2$. Then $D^{(2)}(\varepsilon, \mathscr{F}) \leq 2^r$ so $\log D^{(2)}(\varepsilon, \mathscr{F}) \leq r \log 2$. As $\varepsilon \downarrow 0$, we have $a_r \sim \varepsilon/2$, $\log(1/\varepsilon) \sim \log(2/\varepsilon) \sim -\log a_r \sim \log(r)/2$ and $\varepsilon/2 \sim 1/(r(\varepsilon)(2L(1/\varepsilon)))^{1/2}$, so $r(\varepsilon) \sim 2/(\varepsilon^2 L(1/\varepsilon))$. □

Propositions 6.3 and 6.4 together show that there is no characterization of the universal Donsker property in terms of $D^{(2)}$. Pollard's integral condition (1.2) is not sharp in the example of Proposition 6.4. [*Added in proof*: M. Talagrand has shown that (1.2) is sharp.] On this point see also Giné and Zinn (1984), Section 5.

7. Notes on weighted processes. In Dudley (1985), for Theorem 6.3 I gave credit to "O'Reilly, Chibisov et al." The "et al." should include Csörgő, Csörgő and Horváth (1986) and Csörgő, Csörgő, Horváth and Mason (1986), who proved parts e), f) and g) of that theorem, also removed the continuity assumption from the weight function, and gave an example like Example 7.2 of my paper. So I claim no priority (rather than "somewhat... perhaps") for these matters. I am very grateful to Miklos and Sandor Csörgő for sending me their and co-authors' reports (in 1983) and for pointing out their specific results. See also Alexander (1985a, b, 1987).

Acknowledgment. I thank the referee for useful comments.

REFERENCES

ALEXANDER, K. S. (1984). Probability inequalities for empirical processes and a law of the iterated logarithm. *Ann. Probab.* **12** 1041–1067.

ALEXANDER, K. S. (1985a). The central limit theorem for weighted empirical processes indexed by sets. *J. Multivariate Anal.* To appear.

ALEXANDER, K. S. (1985b). Rates of growth for weighted empirical processes. In *Proc. of the Conference Berkeley in Honor of Jerzy Neyman and Jack Kiefer* (L. M. Le Cam and R. A. Olshen, eds.) **2** 475–493. Wadsworth, Monterey, Calif.

ALEXANDER, K. S. (1987). The central limit theorem for empirical processes on Vapnik–Červonenkis classes. *Ann. Probab.* **15** 178–203.

ASSOUAD, P. (1983). Densité et dimension. *Ann. Inst. Fourier (Grenoble)* **33** (3) 233–282.

BECK, J. (1985). Lower bounds on the approximation of the multivariate empirical process. *Z. Wahrsch. verw. Gebiete* **70** 289–306.

BREIMAN, L., FRIEDMAN, J. H., OLSHEN, R. A. and STONE, C. J. (1984). *Classification and Regression Trees.* Wadsworth, Belmont, Calif.

CARL, B. (1982). On a characterization of operators from l_q into a Banach space of type p with some applications to eigenvalue problems. *J. Funct. Anal.* **48** 394–407.

CHERNOFF, H. and LEHMANN, E. L. (1954). The use of maximum likelihood estimates in χ^2 tests for goodness of fit. *Ann. Math. Statist.* **25** 579–586.

CSÖRGŐ, M., CSÖRGŐ, S., and HORVÁTH, L. (1986). *An Asymptotic Theory for Empirical Reliability and Concentration Processes. Lecture Notes in Statist.* **33**. Springer, New York.

CSÖRGŐ, M., CSÖRGŐ, S., HORVÁTH, L. and MASON, D. (1986). Weighted empirical and quantile processes. *Ann. Probab.* **14** 31–85.

DUDLEY, R. M. (1967). The sizes of compact subsets of Hilbert space and continuity of Gaussian processes. *J. Funct. Anal.* **1** 290–330.

DUDLEY, R. M. (1973). Sample functions of the Gaussian process. *Ann. Probab.* **1** 66–103.

DUDLEY, R. M. (1978). Central limit theorems for empirical measures. *Ann. Probab.* **6** 899–929. Correction. **7** (1979) 909–911.

DUDLEY, R. M. (1981). Donsker classes of functions. In *Statistics and Related Topics* (Proc. Symp. Ottowa, 1980) 341–352. North-Holland, Amsterdam.

DUDLEY, R. M. (1984). A course on empirical processes. *École d'Été de Probabilités de Saint Flour XII–*1982. *Lecture Notes in Math.* **1097** 1–142. Springer, New York.

DUDLEY, R. M. (1985). An extended Wichura theorem, definitions of Donsker class, and weighted empirical distributions. *Lecture Notes in Math.* **1153** 141–178. Springer, New York.

DUNFORD, N. and SCHWARTZ, J. T. (1958). *Linear Operators, Part I: General Theory.* Wiley, New York.

DURST, M. and DUDLEY, R. M. (1981). Empirical processes, Vapnik–Červonenkis classes and Poisson processes. *Probab. Math. Statist.* **1** 109–115.

FERNIQUE, X. (1970). Intégrabilité des vecteurs gaussiens. *C. R. Acad. Sci. Paris Sér. A* **270** 1698–1699.

GINÉ, E. and ZINN, J. (1984). Some limit theorems for empirical processes. *Ann. Probab.* **12** 929–989.

HEINKEL, B. (1983). Majorizing measures and limit theorems for c_0-valued random variables. In *Probability in Banach Spaces IV* (A. Beck and K. Jacobs, eds.). *Lecture Notes in Math.* **990** 136–149. Springer, New York.

HOEFFDING, W. (1963). Probability inequalities for sums of bounded random variables. *J. Amer. Statist. Assoc.* **58** 13–30.

HOFFMANN-JØRGENSEN, J. (1984). Personal communication.

KOLČINSKIĬ, V. I. (1981). On the central limit theorem for empirical measures. *Theory Probab. Math. Statist.* **24** 71–82.

LE CAM, L., MAHAN, C. and SINGH, A. (1983). An extension of a theorem of H. Chernoff and E. L. Lehmann. In *Recent Advances in Statistics: Papers in Honor of Herman Chernoff* (M. H. Rizvi, J. S. Rustagi and D. Siegmund, eds.) 303–337. Academic, New York.

MOORE, D. S. and STUBBLEBINE, J. B. (1981). Chi-square tests for multivariate normality with application to common stock prices. *Comm. Statist. A—Theory Methods* **10** 713–738.

PAULAUSKAS, V. (1980). On the central limit theorem in the Banach space c_0. *Dokl. Akad. Nauk SSSR* **254** 286–288.

PISIER, G. (1981). Remarques sur un résultat non publié de B. Maurey. *Séminaire d'Analyse Fonctionelle 1980–1981* V.1–V.12. Ecole Polytechnique, Centre de Mathématiques, Palaiseau.

POLLARD, D. (1979). General chi-square goodness-of-fit tests with data-dependent cells. *Z. Wahrsch. verw. Gebiete* **50** 317–331.

POLLARD, D. (1982). A central limit theorem for empirical processes. *J. Austral. Math. Soc. Ser. A* **33** 235–248.

POLLARD, D. (1984). *Convergence of Stochastic Processes.* Springer, New York.

POLLARD, D. (1985). New ways to prove central limit theorems. Preprint.

ROY, A. R. (1956). On χ^2 statistics with variable intervals. Technical Report 1, Dept. of Statistics, Stanford Univ.

TALAGRAND, M. (1987). The Glivenko–Cantelli problem. *Ann. Probab.* **15** 837–870.

VAPNIK, V. N. and ČERVONENKIS, A. YA. (1971). On the uniform convergence of relative frequencies of events to their probabilities. *Theory Probab. Appl.* **16** 264–280.

VAPNIK, V. N. and ČERVONENKIS, A. YA. (1974). *Teoriya Raspoznavaniya Obrazov: Statisticheskie Problemy Obucheniya* (*Theory of Pattern Recognition: Statistical Problems of Learning*). (In Russian.) Nauka, Moscow. German ed.: *Theorie der Zeichenerkennung*, by W. N. Wapnik and A. J. Tscherwonenkis, translated by K. G. Stöckel and B. Schneider (S. Unger and K. Fritzsch, eds.). Akademie-Verlag, Berlin, 1979 (*Elektronisches Rechnen und Regeln*, Sonderband **28**).

WATSON, G. S. (1958). On chi-square goodness-of-fit tests for continuous distributions. *J. Roy. Statist. Soc. Ser. B* **20** 44–61.

WENOCUR, R. S. and DUDLEY, R. M. (1981). Some special Vapnik–Červonenkis classes. *Discrete Math.* **33** 313–318.

DEPARTMENT OF MATHEMATICS
ROOM 2-245
MASSACHUSETTS INSTITUTE OF TECHNOLOGY
CAMBRIDGE, MASSACHUSETTS 02139

Part 5

Nonlinear Operators and p–Variation

Introduction

The classical theory of statistical functionals aims at developing estimation methods when the unknown parameter to be estimated is a function of a probability distribution. Consider a random variable X with distribution function F_θ, where $F_\theta \in S := \{ F_\theta : \theta \in \Theta \subset \mathbf{R}^d \}$. Then in many cases θ can be looked at as a value $\theta = T(F)$ of a mapping T defined on S. A natural estimator of θ, based on observations $X_1, \ldots, X_n$, is $T(F_n)$, where F_n is the empirical distribution function. The mapping T from S to $\mathbf{R}^d$, or to a subset Θ of $\mathbf{R}^d$, is then called a statistical functional. More generally, the range Θ of a statistical functional could be a function space (a nonparametric case).

The theory of statistical functionals was initiated by R. von Mises around 1936. He adopted differentiation of functionals as originated in the works of V. Volterra and R. Gateaux as a natural tool for an analysis of statistical functionals. The main result of the theory was the asymptotic normality of $\sqrt{n}[T(F_n) - T(F)]$ proved under suitable hypotheses. Further progress in this direction was achieved by investigating several classical estimators using differentiation of statistical functionals. The usual domain of definition of statistical functionals was the space $D[0,1]$ of right-continuous functions with left limits endowed with the supremum norm. Compact differentiability was adopted since Reeds in 1976 showed that it holds in the supremum norm while Fréchet differentiability fails in many important cases.

Further progress in the theory of statistical functionals was initiated by R. M. Dudley who suggested replacing the supremum norm on the domain of statistical functionals by p-variation norms $\|\cdot\|_{[p]}$, $1 \le p < \infty$. The main results in this direction are presented in the three papers of this chapter. In the first paper, Dudley proved that $\sqrt{n}\|F_n - F\|_{[p]}$ is bounded in probability uniformly in n if $p > 2$. (The reverse implication also holds as shown in the third paper.) To prove this, a bound of the ε-entropy $\log N(\varepsilon, \mathcal{F}_r, \rho_{2,Q})$ of a ball $\mathcal{F}_r$ of the Banach space of functions of bounded r-variation with $0 < r < 2$ uniformly over the class of all probability measures Q on $[0,1]$ is established. This bound is sufficient to show that the class of functions $\mathcal{F}_r$ is a universal Donsker class. This together with an ingenious duality argument based on L. C. Young's integrability theorem (1936) are sufficient tools to prove the stated boundedness in probability of the sequence $\sqrt{n}\|F_n - F\|_{[p]}$ provided $2 < p < \infty$. Y.-C. Huang strengthened this result in his Ph. D. thesis (1994) by showing that from the construction of Komlós, Major and Tusnády (1975) one obtains a sequence of Brownian bridges which approximates empirical processes $\sqrt{n}(F_n - F)$ in the p-variation norm with an optimal rate of convergence to zero.

Also in the first paper of this chapter, a significant observation is that, due to L. C. Young's integrability theorem, Fréchet differentiability of the extended Riemann-Stieltjes integral as a bilinear operator holds with respect to the p-variation norm and the r-variation seminorm when $p^{-1} + r^{-1} > 1$. Application of this fact is not so straightforward because the uniform boundedness in probability of the empirical process holds in the p-variation norm only with $p > 2$, while the preceding differentiability condition $p^{-1} + p^{-1} > 1$ (when $r = p$) holds only if $p < 2$. To overcome the difficulty Dudley calculated an order of growth in n of $\sqrt{n}\|F_n - F\|_{[p]}$ when $1 \le p \le 2$ (Theorem 2 in the third paper of this chapter). He showed then how it helps by proving asymptotic normality with remainder bounds for the Mann-Whitney statistic. The order of growth of the p-variation of the empirical process was improved when $1 \le p < 2$ by J. Qian in her Ph. D. thesis (1997).

A nonlinear operator T on a domain U with norm $\|\cdot\|$ is Fréchet differentiable at $F \in U$ if there exists a bounded linear operator $T'(F)$ such that the remainder $T(F+h) - T(F) - T'(F)(h)$ is $o(\|h\|)$ as $\|h\| \to 0$. The observation that the exact order $O(\|h\|^\gamma)$ for some $\gamma > 1$ can also be useful is the main point of the second paper of this chapter. Besides this, Dudley also proved Fréchet differentiability in p-variation norm of the important composition operator $(F, G) \to F \circ G$ and inverse operator $H \to H^{-1}$.

On several occasions Dudley emphasized basic reasons why p-variation norms should be used in place of the supremum norm when considering the differentiability of some statistical functionals. First, Fréchet derivatives often exist for the p-variation norm while they do not for the supremum norm. Second, uniformly over all possible norms on spaces containing empirical distribution functions, the p-variation

E. Giné et al. (eds.), *Selected Works of R.M. Dudley*, Selected Works in Probability and Statistics,
DOI 10.1007/978-1-4419-5821-1_20,

norm gives remainder bounds in the differentiation which are of the smallest possible order in a range of cases. Third, the p-variation norm has the good property of being invariant under all strictly increasing, continuous transformations of **R** onto itself, just as the supremum norm does. Fourth, as he proved, the central limit theorem for the empirical process in the p-variation norms holds for $p>2$.

The Annals of Probability
1992, Vol. 20, No. 4, 1968–1982

FRÉCHET DIFFERENTIABILITY, p-VARIATION AND UNIFORM DONSKER CLASSES[1]

BY R. M. DUDLEY

Massachusetts Institute of Technology

Differentiability of functionals of the empirical distribution function is extended. The supremum norm is replaced by p-variation seminorms, which are the pth roots of suprema of sums of pth powers of absolute increments of a function over nonoverlapping intervals. Fréchet derivatives often exist for such norms when they do not for the supremum norm. For $1 < q < 2$, classes of functions uniformly bounded in q-variation are universal and uniform Donsker classes: The central limit theorem for empirical measures holds with respect to uniform convergence over such a class, also uniformly over all probability laws on the line. The integral $\int F\,dG$ was defined by L. C. Young if F and G are of bounded p- and q-variation respectively, where $p^{-1} + q^{-1} > 1$. Thus the normalized empirical distribution function $n^{1/2}(F_n - F)$ is with high probability in sets of uniformly bounded p-variation for any $p > 2$, uniformly in n.

1. Introduction. Classically, a statistical functional is defined on a space of distribution functions F on the real line with the supremum norm. The values may be real or themselves functions such as the quantile function F^{-1}. Nonlinear functionals are studied via their derivatives. This paper and a related one [Dudley (1991a)] will show that the differentiability properties originally proved by Reeds (1976), also treated in Fernholz (1983), can be improved substantially.

Differentiability of functionals is defined in general as follows. Let $(X, \|\cdot\|)$ and $(Y, |\cdot|)$ be two Banach spaces. Let U be an open set in X and u a point of U. A function T defined on U with values in Y is said to be *Fréchet differentiable* at u if there is a bounded linear operator A from X into Y such that

$$|T(x) - T(u) - A(x-u)| = o(\|x-u\|)$$

as $x \to u$. Equivalently, for every bounded set B in X,

$$\sup\{|T(u+tv) - T(u) - tA(v)|: v \in B\}/t \to 0 \quad \text{as } t \to 0,$$

where t is a real variable. If $\mathscr{C}$ is a collection of subsets of X, T will be called *$\mathscr{C}$-differentiable* at u if "every bounded set" is replaced by "each set in $\mathscr{C}$." Then if $\mathscr{C}$ is the class of all compact sets for a topology, T is said to be *compactly* or *Hadamard differentiable* at u for the given topology, which in past work has usually been, but need not be, that of a norm.

Received February 1991; revised October 1991.
[1]Research partially supported by an NSF grant.
AMS 1980 *subject classifications*. Primary 60F17, 62G30; secondary 26A42, 26A45.
Key words and phrases. Wilcoxon statistics, Riemann–Stieltjes integral, L. C. Young integral.

E. Giné et al. (eds.), *Selected Works of R.M. Dudley*, Selected Works in Probability and Statistics, DOI 10.1007/978-1-4419-5821-1_21, © Springer Science+Business Media, LLC 2010

Reeds (1976) and Fernholz (1983) showed that once the domain and range of functionals are suitably specified, in several cases functionals of statistical interest are compactly but not Fréchet differentiable at some points. To my knowledge, most if not all such examples were for the sup norm $\|\cdot\|_\infty$ in $\mathbb{R}$, where $\|F\|_\infty := \sup_t |F(t)|$. Although Fréchet differentiability holds in some other useful cases [e.g., Fernholz (1983), Proposition 6.1.3, page 72, Corollary 6.1.4, page 75, and Hampel, Ronchetti, Rousseeuw and Stahel (1986), page 54], many statisticians have been convinced that in general it is too strong [e.g., Huber (1981), page 37, Gill (1989), page 101, and Serfling (1980), page 220]. But Serfling (1980), page 218, mentions the "choice of norm": A larger norm increases the chances for Fréchet differentiability, and will work well if probability limit theorems still hold for the larger norm.

This paper expands on the "choice of norm." It will be shown that in p-variation norms for suitable values of p, defined in Sections 2 and 3, uniform central limit theorems still hold. In this paper, the result will be applied to the bilinear functional $(F, G) \mapsto \int F\,dG$ (Section 4). Another paper [Dudley (1991a)] will treat the inverse operator $F \mapsto F^{-1}$ and composition $(F, G) \mapsto F \circ G$.

In a nonseparable normed space, specifically $D[0,1]$ with $\|\cdot\|_\infty$, compact sets are very small compared to open, bounded sets. If F is a continuous distribution function and K is a compact subset of $D[0,1]$ for the sup norm, the probability that the empirical distribution function $F_n \in K$, or $n^{1/2}(F_n - F) \in K$, is 0! So there are technical problems in applying norm-compact differentiability. The problems have been treated by modifying empirical distribution functions to be continuous [e.g., Fernholz (1983), pages 32–42]. Or, the definition of compact differentiation can (also for very general sample spaces and norms) be modified to hold "tangentially to a subspace" [Gill (1989), page 102, and Pons and Turckheim (1989)]. Compact differentiation can also be applied to almost surely convergent realizations. At any rate, no such modification is needed when we have $\mathscr{C}$-differentiability for a class $\mathscr{C}$ of sets of functions such that $n^{1/2}(F_n - F)$ satisfies a tightness condition for sets in $\mathscr{C}$, as will be shown in Proposition 3.7 and Corollary 3.8 for classes of bounded p-variation for $p > 2$.

For a probability measure P and observations $X_1, X_2, \ldots,$ i.i.d. with law P, we form the empirical measures $P_n := n^{-1}(\delta_{X_1} + \cdots + \delta_{X_n})$ and the empirical process $\nu_n := n^{1/2}(P_n - P)$. If $\mathscr{F}$ is a class of measurable functions and ν is a signed measure, let $\|\nu\|_{\mathscr{F}} := \sup\{|\int f\,d\nu| : f \in \mathscr{F}\}$. Here a class of measurable sets can be identified with the class of indicator functions of the sets. The "sup norm" or "Kolmogorov norm" of the empirical process is the supremum of its absolute value over the family $\mathscr{H}$ of all sets (half-lines) $]-\infty, x]$, so $\|\cdot\|_\infty \equiv \|\cdot\|_{\mathscr{H}}$. $\mathscr{H}$ is taken into itself by all nondecreasing transformations of $\mathbb{R}$ and is taken one to one and onto itself by all strictly increasing, continuous transformations of $\mathbb{R}$ onto itself. It is known that a central limit theorem holds in the sup norm for all probability laws P on $\mathbb{R}$, at a rate which is uniform in P. On extensions to general sample spaces and classes $\mathscr{F}$, see Dudley (1987) for the "universal Donsker" property and Giné and Zinn (1991) for the uniformity in P.

The sup norm is defined by a relatively small and quite tractable class $\mathscr{H}$ of sets, while it also encompasses larger classes of functions: Let BV1 be the class of all functions f from $\mathbb{R}$ into itself of total variation at most 1, with $f(x) \to 0$ as $x \to +\infty$. Let DE1 be the class of nonincreasing functions in BV1. Then

$$\|\cdot\|_\infty = \|\cdot\|_{\mathrm{DE1}} \le \|\cdot\|_{\mathrm{BV1}} \le 2\|\cdot\|_{\mathrm{DE1}}$$

[e.g., Dudley (1987), proof of Theorem 2.1i], although BV1 and DE1 are in many ways much larger than $\mathscr{H}$. So it is not surprising that the sup norm has long been considered as the main norm for empirical processes and differentiability on $\mathbb{R}$. But p-variation norms, besides their other advantages, are also preserved by any continuous increasing function from $\mathbb{R}$ onto itself.

I suggested [Dudley (1989b, 1990)] that statistical functionals might be Fréchet differentiable for norms $\|\cdot\|_{\mathscr{F}}$ for universal Donsker classes $\mathscr{F}$ other than $\mathscr{H}$. It will be seen that classes of functions of uniformly bounded q-variation for $1 < q < 2$ are universal (and uniform) Donsker classes $\mathscr{F}$ on $\mathbb{R}$. Via the duality theory in Section 3, this is equivalent to $n^{1/2}(F_n - F)$ being with high probability in sets of uniformly bounded p-variation, $p > 2$.

Compact differentiability for $\|\cdot\|_\infty$ implies Fréchet differentiability for $\|\cdot\|_{\mathscr{F}}$ directly if all bounded sets for $\|\cdot\|_{\mathscr{F}}$ are compact for $\|\cdot\|_\infty$. But matters are not quite so easy, since no such $\mathscr{F}$ is a universal Donsker class.

PROPOSITION 1.1. *There is no universal Donsker class $\mathscr{F}$ of Borel-measurable functions on $\mathbb{R}$ such that all bounded sets of finite signed measures for $\|\cdot\|_{\mathscr{F}}$ are compact or even separable for $\|\cdot\|_\infty$.*

PROOF. Any universal Donsker class $\mathscr{F}$ is uniformly bounded up to additive constants [Dudley (1987), Proposition 1.1]. Thus the set B of all differences $\delta_x - \delta_y$ for x and y in $\mathbb{R}$ is bounded for $\|\cdot\|_{\mathscr{F}}$. But if $x \ne z$, $\|\delta_x - \delta_y - (\delta_z - \delta_y)\|_{\mathscr{H}} = 1$, so B is not separable for the supremum norm $\|\cdot\|_\infty = \|\cdot\|_{\mathscr{H}}$. $\square$

This paper does not treat inverse or implicit function theorems and M-estimators, on which I hope to complete a separate paper [Dudley (1991b)].

The present paper is organized as follows: Section 2 reviews p-variation and proves a uniform central limit theorem (Donsker property) for classes of functions of bounded p-variation, $p < 2$. Section 3 treats Young's duality theory of r-variation spaces W_r, W_s via $(F,G) \mapsto \int F\,dG$, $F \in W_r$, $G \in W_s$, $1/r + 1/s > 1$. Section 4 treats differentiability of the functional $(F,G) \mapsto \int F\,dG$ and compares it with the results of Gill (1989).

2. Functions of bounded *p*-variation and Donsker classes. A function f from an interval $J \subset \mathbb{R}$ into $\mathbb{R}$ has *p-variation* defined by

$$v(f,p) := v(f,p,J) := \sup\left\{ \sum_{i=1}^{n} |f(x_i) - f(x_{i-1})|^p : x_0 < x_1 < \cdots < x_n \in J, \right.$$
$$\left. x_0 \in J,\ n = 1, 2, \dots \right\}.$$

For $p = 1$, this is the usual total variation. Apparently the notion of p-variation was first defined by Wiener (1924), who treated mainly the case $p = 2$, "quadratic variation." For $p \neq 2$, the first major work, including the duality theory (Section 3), was done in the late 1930s by L. C. Young, partly with E. R. Love; see Young (1936) and Love and Young (1937). More recently, Bruneau (1974) treats p-variation, mainly in a different direction ("fine" variation). There are recent applications in probability theory [e.g., Bertoin (1989), Xu (1986, 1988) and Pisier and Xu (1987, 1988)]. Apparently p-variation for empirical processes was not previously addressed explicitly, but see the discussion after Theorem 2.2 below. The following will be proved.

THEOREM 2.1. *For any (bounded or unbounded) interval $J \subset \mathbb{R}$, any p with $0 < p < 2$ and any $M < \infty$, $\mathscr{F}_{p,M} := \{f\colon J \mapsto \mathbb{R},\ v(f,p) \le M\}$ is a universal Donsker class.*

PROOF. Recall the notion of Kolchinskii–Pollard entropy $\log D^{(2)}$ for the L^2 norm over finite sets, where $\int_0^1 (\log D^{(2)}(\varepsilon, \mathscr{F}))^{1/2}\, d\varepsilon < \infty$ and some measurability conditions imply that $\mathscr{F}$ is a universal Donsker class by a theorem of Pollard (1982), stated with more general measurability conditions in Dudley (1987), Theorem 2.1h.

We can assume without loss of generality that $M = 1$.

For any $f \in \mathscr{F} := \mathscr{F}_{p,1}$, we have $\operatorname{diam} f := \sup f - \inf f$ satisfying $(\operatorname{diam} f)^p \le 1$, so $\operatorname{diam} f \le 1$. So $\mathscr{F}$ is uniformly bounded up to additive constants and in proving the universal Donsker property we can replace the functions f in $\mathscr{F}$ by the functions $f - \inf f$, so that we can assume $0 \le f \le 1$, $f \in \mathscr{F}$ [cf. Dudley (1987)].

We can assume $p \ge 1$ and $J = \mathbb{R}$. Given $f \in \mathscr{F}$, let $h(x) := v(f, p,]-\infty, x])$, the total p-variation of f up to x. Then h is a nondecreasing function. For any $x < y$, $|f(y) - f(x)|^p \le h(y) - h(x)$, so $|f(y) - f(x)| \le (h(y) - h(x))^{1/p}$. Thus f is a function of h, $f(x) \equiv g(h(x))$, for a function g which satisfies a Hölder condition $|g(u) - g(v)| \le |u - v|^{1/p}$ for all u and v in the range of h, as was known [Love and Young (1937), page 244, and Bruneau (1974), page 3]. Since $0 < 1/p \le 1$, $e(u,v) := |u - v|^{1/p}$ is a metric, and the ranges of g and h are included in $[0,1]$, so we can assume g is defined and satisfies the same Hölder condition on all of $[0,1]$ into $[0,1]$ by the Kirszbraun–McShane extension theorem [e.g., Dudley (1989a), Theorem 6.1.1].

Let G be the set of all functions from $[0,1]$ into itself satisfying the given Hölder condition. Then there are constants C_1 and C_2 such that for $0 < \varepsilon \le 1$, there is a set $G_\varepsilon \subset G$, dense within ε for the supremum norm and containing at most $C_1 \exp(C_2 \varepsilon^{-p})$ functions [Kolmogorov and Tihomirov (1959), Theorem XIII].

Let H be the set of nondecreasing functions from $\mathbb{R}$ into $[0,1]$. Then H is included in the set of sequential pointwise limits of convex combinations of members of the set H_1 of all indicators of half-lines, $1_{[a,\infty[}$ [e.g., Dudley (1987), proof of Theorem 2.1(a), and since an open half-line is a countable union of closed half-lines). For any probability measure Q on $\mathbb{R}$ (such as one with finite support), for $0 < \varepsilon \le 1$ there is a set of at most $2/\varepsilon^2$ members of H_1, dense within ε^2 in H_1 in the $L^1(Q)$ norm and so dense within ε in H_1 in the $L^2(Q)$ norm. Thus, for any $t > 1$, there are constants C_3 and C_4 such that for $0 < \varepsilon \le 1$ there is a set $H_\varepsilon \subset H$, dense in H within ε^p in the $L^2(Q)$ norm and containing at most $C_3 \exp(C_4 \varepsilon^{-pt})$ functions [Dudley (1987), Theorem 5.1]. Choose t so that $pt < 2$.

Now take any $f \in \mathscr{F}$. Then $f \equiv g \circ h$ for some $g \in G$ and $h \in H$. Given $0 < \varepsilon \le 1$, take $\gamma \in G_\varepsilon$ with $\sup|g - \gamma| \le \varepsilon$ and $\theta \in H_\varepsilon$ with $\|h - \theta\|_2 \le \varepsilon^p$ for Q. Then

$$\begin{aligned}\| f - (\gamma \circ \theta) \|_2 &\le \|g \circ h - \gamma \circ h\|_2 + \|\gamma \circ h - \gamma \circ \theta\|_2 \\ &\le \varepsilon + \|\gamma \circ h - \gamma \circ \theta\|_2.\end{aligned}$$

Next,

$$\begin{aligned}\int (\gamma \circ h - \gamma \circ \theta)^2 \, dQ(x) &\le \int |(h - \theta)(x)|^{2/p} \, dQ(x) \\ &\le \left(\int |(h - \theta)(x)|^2 \, dQ(x) \right)^{1/p} \le \varepsilon^{2p/p} = \varepsilon^2.\end{aligned}$$

So $\| f - (\gamma \circ \theta)\|_2 \le 2\varepsilon$. So we have found a set $\mathscr{F}_\varepsilon$ dense within 2ε in $\mathscr{F}$ for the $L^2(Q)$ norm, where for some constants C and D, the number of functions in $\mathscr{F}_\varepsilon$ is at most $C \exp(D\varepsilon^{-pt})$. This implies Pollard's entropy condition. So it will be enough to show that the image admissible Suslin measurability condition holds.

Any function f of bounded p-variation is in the set $E(\mathbb{R})$ of real functions on $\mathbb{R}$ having right limits on $[-\infty, \infty[$ and left limits on $]-\infty, \infty]$. Let $D(\mathbb{R})$ be the set of all right-continuous functions in $E(\mathbb{R})$. Let $c_0(\mathbb{R})$ be the set of all functions h from $\mathbb{R}$ into $\mathbb{R}$ such that for each $\varepsilon > 0$, $\{x: |h(x)| > \varepsilon\}$ is finite. Then h is 0 except on a countable set.

Given $f \in E(\mathbb{R})$, let $g(x) := f(x+) := \lim\{f(y): y \downarrow x\}$ for all x. Then since $f \in E(\mathbb{R})$, we have $g \in D(\mathbb{R})$, and f is continuous except at most on a countable set C, so $g(x) = f(x)$ for $x \notin C$. Let $h := f - g$. For any $\varepsilon > 0$, f can only have finitely many jumps of heights larger than ε, so $h \in c_0(\mathbb{R})$.

Now $h \in c_0(\mathbb{R})$ if and only if there is a sequence $\{x_i\}$ of distinct real numbers and a sequence $y_i \to 0$ such that $h(x) = \sum_i y_i 1\{x = x_i\}$. The set of such sequences $\{x_i\}$ is a countable intersection of open sets $\{x_i \ne x_j\}$, $i \ne j$, in the product topology, so it is Polish [e.g., Dudley (1989a), Theorem 2.5.4]. The set of sequences $\{y_i\}$, $y_i \to 0$, is a Banach space with supremum norm. Then evaluation of h is jointly measurable, so $c_0(\mathbb{R})$ is image admissible Suslin.

Let $\mathscr{S}(T)$ be the smallest σ-algebra on a set of real functions with domain including T such that evaluation at each point of T is measurable. For $f \in D[0, 1]$, let $f_n(x) := f(k/n)$ for $(k-1)/n \le x < k/n$, $k = 1, \dots, n$, $f_n(1) := f(1)$. Then $f_n(x) \to f(x)$ for all x, and $(x, f) \mapsto f_n(x)$ is jointly measurable for the Borel σ-algebra in $x \in \mathbb{R}$ and the σ-algebra $\mathscr{S}(T)$, where T is the set of rational numbers. Since the given σ-algebras are Borel σ-algebras of Polish spaces, the same holds for $(x, f) \mapsto f(x)$, so $D(\mathbb{R})$ is image admissible

Suslin, and so by addition $E(\mathbb{R})$ is image admissible Suslin. In this class, the set of functions of p-variation bounded by 1 is the image of a Borel set, since we can consider the p-variation on finite sets of rationals or points x_i for h in $c_0(\mathbb{R})$. □

Now it will be shown that $\mathscr{F}_{p,M}$ is not only a universal but a *uniform* Donsker class, in the sense that the central limit theorem for empirical measures for uniform convergence over $\mathscr{F}_{p,M}$ also holds uniformly in P. Giné and Zinn (1991) gave a general, natural definition of the uniform Donsker property in terms of dual-bounded-Lipschitz metrics, and found a striking Gaussian characterization. Here, a different property will be defined, closer to the traditional uniformity over distribution functions on $\mathbb{R}$.

DEFINITION. Let $\mathscr{F}$ be a class of real-valued measurable functions on a measurable space $(X, \mathscr{A})$. Then $\mathscr{F}$ is a *dominated uniform Donsker class* if there is a law λ on $(X, \mathscr{A})$ for which $\mathscr{F}$ is a Donsker class, and for every law P on $(X, \mathscr{A})$ there is a measurable function f_P from X into itself such that the image measure $\lambda \circ f_P^{-1} = P$, and such that the class $\mathscr{F} \circ f_p$ of all compositions $f \circ f_p$, $f \in \mathscr{F}$, is included in $\mathscr{F}$.

When the condition just defined holds, other definitions of uniform Donsker class could be verified, but the details will just be sketched here. If λ_n are empirical measures for λ, then $\lambda_n \circ f_P^{-1}$ have all the properties of the empirical measures P_n, so $n^{1/2}(P_n - P)$ can be written as $n^{1/2}(\lambda_n - \lambda) \circ f_P^{-1}$. For a class $\mathscr{F}$ of functions, we then have by the image measure theorem:

$$\begin{aligned}\|n^{1/2}(P_n - P)\|_{\mathscr{F}} &= n^{1/2} \sup\left\{\left|\int f d(\lambda_n - \lambda) \circ f_P^{-1}\right| : f \in \mathscr{F}\right\} \\ &= n^{1/2} \sup\left\{\left|\int f \circ f_P \, d(\lambda_n - \lambda)\right| : f \in \mathscr{F}\right\} \\ &= \|n^{1/2}(\lambda_n - \lambda)\|_{\mathscr{F} \circ f_P}.\end{aligned}$$

Moreover, for the limiting Gaussian processes G_P and G_λ, for any $\mathscr{F} \subset \mathscr{L}^2(P)$, $f \mapsto G_\lambda(f \circ f_P)$ has the same distributions as G_P on $\mathscr{F}$ since it is Gaussian with mean 0 and the same variances and covariances. It then follows, for example by existence of almost surely convergent realizations [Dudley (1985)] that the convergence in limit theorems for P_n for any P is at least as fast as for λ_n.

For the classes in Theorem 2.1 we have the following theorem.

THEOREM 2.2. *Each $\mathscr{F}_{p,M}$ on $\mathbb{R}$ for $0 < p < 2$ and $M < \infty$ is a dominated uniform Donsker class.*

PROOF. Let λ be Lebesgue measure on $[0, 1]$. Let P be any probability measure on $\mathbb{R}$ with distribution function F. For $0 < y < 1$, let $F^{-1}(y) :=$

$\inf\{x\colon F(x) \geq y\}$. Take f_P as the function F^{-1}. Then the image measure $\lambda \circ (F^{-1})^{-1} = P$ as desired. If $f \in \mathscr{F}_{p,M}$, then for any nondecreasing function G, such as $G = F^{-1}$, we have $f \circ G \in \mathscr{F}_{p,M}$ since for any $y_1 < \cdots < y_n$, letting $x_i := G(y_i)$ gives

$$\sum_{i=2}^{n} |f(G(y_i)) - f(G(y_{i-1}))|^p = \sum_{i=2}^{n} |f(x_i) - f(x_{i-1})|^p,$$

which (with Theorem 2.1 for λ) completes the proof. □

Gilles Pisier, in a letter dated September 3, 1991, has kindly pointed out the following: The work of Pisier and Xu (1987) and Xu (1986, 1988) easily implies at least a fact close to Theorem 2.2, which I will call Theorem 2.2′, where convergence in distribution of ν_n is replaced by boundedness in L^2: $\sup_P \sup_n E\|\nu_n\|^2_{\mathscr{F}} < \infty$ for $\mathscr{F} = \mathscr{F}_{p,M}$.

3. Duality inequalities for *r*-variation spaces. For $1 \leq r < \infty$ and $-\infty \leq a < b \leq +\infty$, let $W_r := W_r[a, b]$ be the class of all real-valued functions f on $[a, b]$ with finite r-variation $v(f, r) = v(f, r, [a, b]) < \infty$. For any $f \in W_r$, we have the seminorm $\|f\|_{(r)} := v(f, r)^{1/r}$, which is 0 only for constants. If f has finite r-variation on an open interval $]a, b[$, then it has a limit as $x \to a$ or b respectively, even if $a = -\infty$ or $b = +\infty$, in which case $f(-\infty)$ or $f(+\infty)$ is defined as the corresponding limit.

For $f \in W_r[a, b]$, let $\|f\|_{[r]} := \|f\|_\infty + \|f\|_{(r)}$, where $\|f\|_\infty := \sup_x|f(x)|$. Then $\|\cdot\|_{[r]}$ is a norm on W_r. It is not hard to show that W_r is complete for $\|\cdot\|_{[r]}$ and so W_r is a Banach space.

To define integrals $\int F\,dG$, first consider the Riemann–Stieltjes integral defined as follows: Given an interval $[a, b]$, a *partition* will be a finite sequence $x_0 = a < x_1 < \cdots < x_n = b$. A *Riemann–Stieltjes sum* for F, G and the given partition will be any sum $\sum_{i=1}^{n} F(t_i)(G(x_i) - G(x_{i-1}))$, where $x_{i-1} \leq t_i \leq x_i$ for $i = 1, \ldots, n$. Then the *Riemann–Stieltjes integral* $\int_a^b F\,dG$ exists and equals c if for every $\varepsilon > 0$ there exists a partition $\pi = \{x_0, x_1, \ldots, x_n\}$ as above such that for all partitions τ including π and all Riemann–Stieltjes sums S based on τ, $|S - c| < \varepsilon$. [This is one of two definitions most often given in the literature, and is sometimes called the Moore–Pollard definition. The other definition requires convergence of sums S to c as the *mesh* $\max_i(x_i - x_{i-1}) \to 0$.] When the integral exists, we say $F \in \mathscr{R}(G)$. Hildebrandt (1938) is a survey on Riemann–Stieltjes integration, with references to earlier sources for most results.

Riemann–Stieltjes integrals have been considered mainly when one of F and G is continuous and the other is of bounded variation. The theory of p-variation provides a class of cases, to be given below, where neither F nor G is of bounded variation.

The integral $\int F\,dG$ will not be defined as a Riemann–Stieltjes integral, even if both F and G are of bounded variation, if they are discontinuous on the same side of the same point [Hildebrandt (1938), 4.13, "σ" case; for one example, see Rudin (1976), page 138, Example 3]. So the integral has to be defined otherwise. For a function h and point x, let $h(x-) = \lim_{y \uparrow x} h(y)$ and $h(x+) = \lim_{y \downarrow x} h(y)$. If $h \in W_r[a,b]$ for any $r < \infty$, then $h(x+)$ exists for $a \le x < b$ and $h(x-)$ exists for $a < x \le b$. The following is known [Hildebrandt (1938), Theorem 5.32].

LEMMA 3.1. *The Riemann–Stieltjes integral* $\int_a^b F\,dG$ *exists if* G *is of bounded variation and right continuous* $[G(x+) = G(x),\ a \le x < b]$ *while* F *is left continuous* $[F(x) = F(x-),\ a < x \le b]$ *and has right limits* $F(x+)$ *for* $a \le x < b$.

The *Young's* or (Y_1) *integral*

$$\begin{aligned}(Y_1)\int_a^b F\,dG := \int_a^b F(x+)\,dG(x-) &+ \sum_{a<x<b} (F(x) - F(x+))(G(x+) - G(x-)) \\ &+ (F(a) - F(a+))(G(a+) - G(a)) \\ &+ F(b)(G(b) - G(b-))\end{aligned}$$

will be so defined if $\int_a^b F(x+)\,dG(x-)$ exists as a Riemann–Stieltjes integral, and in the sum, the summands are 0 except for at most countably many values of x and the series is absolutely convergent. Here in the integral, $G(a-)$ is replaced by $G(a)$ [and $F(b+)$ by $F(b)$, which does not matter since $x \mapsto G(x-)$ is left continuous at b]. Note that if F is continuous from the right, $F(x) \equiv F(x+)$, the (Y_1) integral becomes

$$\int_a^b F(x)\,dG(x-) + F(b)(G(b) - G(b-)).$$

W. H. Young (1914) contributed to defining extended Riemann–Stieltjes integrals. L. C. Young (1936), pages 263–265, shows that $(Y_1)\int_a^b F\,dG$ exists if $F \in W_p[a,b]$ and $G \in W_q[a,b]$, where $p^{-1} + q^{-1} > 1$. [Hildebrandt (1938) distinguishes different integrals by putting symbols such as N, Y and/or σ before the integral sign.

An alternate definition of integral is obtained by applying the definition of the (Y_1) integral to the interval in the reverse order, taking $-(Y_1)\int_{-b}^{-a} F(-x)\,dG(-x)$. The resulting integral, which will also be called a Young integral, is here defined as

$$\begin{aligned}(Y_2)\int_a^b F\,dG := \int_a^b F(x-)\,dG(x+) &+ \sum_{a<x<b} (F(x) - F(x-))(G(x+) - G(x-)) \\ &+ F(a)(G(a+) - G(a)) \\ &+ (F(b) - F(b-))(G(b) - G(b-)),\end{aligned}$$

which will, as with the (Y_1) integral, be said to exist if the integral on the right exists as a Riemann–Stieltjes integral and the sum is absolutely convergent, and here $G(b+)$ is replaced by $G(b)$ in the integral [and $F(a-)$ by $F(a)$, which does not matter since $x \mapsto G(x+)$ is right continuous at a]. The inversion $x \mapsto -x$ shows that the (Y_2) integral likewise exists for $F \in W_p$ and $G \in W_q$ with $p^{-1} + q^{-1} > 1$. The two integrals actually agree when defined, as follows.

THEOREM 3.2. *Suppose that F and G are two bounded real functions on $[a, b]$ having right limits on $[a, b[$ and left limits on $]a, b]$. Then if both of the integrals $(Y_1)\int_a^b F\,dG$ and $(Y_2)\int_a^b F\,dG$ exist, they are equal.*

PROOF. Since the Riemann–Stieltjes integral $I_1 := \int_a^b F(x+)\,dG(x-)$ exists, for any $\varepsilon > 0$ there is a partition τ given by $a = x_0 < x_1 < \cdots < x_n = b$ of $[a, b]$ such that any Riemann–Stieltjes sum S for I_1 based on a partition which is a refinement of τ differs from the integral by at most ε. Here $S = \sum_{i=1}^n F(y_i+)[G(x_i-) - G(x_{i-1}-)]$ for any y_i in the interval $[x_{i-1}, x_i]$, replacing $a-$ by a and $b+$ by b. The same holds for $I_2 := \int_a^b F(x-)\,dG(x+)$, so taking a common refinement of the two partitions, we can assume τ is such that it holds for both Riemann–Stieltjes integrals. Also, τ can be chosen to contain enough of the points of discontinuity of F and G such that the sums in the definitions of the (Y_1) and (Y_2) integrals over all other points are at most ε each.

For the given partition τ, consider refinements formed by adjoining points u_i at which G is continuous with $x_{i-1} < u_i < x_i$ for $i = 1, \ldots, n$. [Since $G(x-)$ and $G(x+)$ exist for $a < x < b$ by assumption, G is continuous except at most on a countable set.] We can then form Riemann–Stieltjes sums for I_1 by evaluating F at $x_{i-1}+$ for $[x_{i-1}, u_i]$ and at x_i- (a limit of points $x+$ as $x \uparrow x_i$) for $[u_i, x_i]$. In forming sums for I_2, we can evaluate F at the same places, interchanging the reasons for the lower and higher of the two intervals. Then $I_1 - I_2$ differs by at most 2ε from

$$\sum_{i=1}^{n} F(x_i-)[G(x_i-) - G(x_i+)] + F(x_{i-1}+)[G(x_{i-1}+) - G(x_{i-1}-)]$$

$$= R(a, b) + \sum_{i=1}^{n-1} [G(x_i+) - G(x_i-)][F(x_i+) - F(x_i-)],$$

where

$$R(a, b) := F(a+)[G(a+) - G(a)] - F(b-)[G(b) - G(b-).$$

So if we subtract the (Y_2) from the (Y_1) integral, all the terms in the approximating sums cancel so the integrals differ at most by 4ε. Let $\varepsilon \downarrow 0$ to complete the proof. $\square$

When G has bounded variation, Young's integrals are equal to Lebesgue–Stieltjes integrals, as has been known at least under some conditions [e.g., Hildebrandt (1938), page 275; Young (1936), page 266].

LEMMA 3.3. *Let F be a function on $[a,b]$ which has left and right limits at all points. Let ν be a finite signed measure on $]a,b]$ with $G(x) := \nu([a,x])$, $a \le x \le b$. Then $\int_a^b F\,d\nu = (Y_2)\int_a^b F\,dG$.*

PROOF. Since G is right continuous, $G(x+) \equiv G(x)$, and $G(x) - G(x-) = \nu(\{x\})$, which is 0 by assumption if $x = a$. We can write $G = G_c + G_{\rm at}$, where G_c is continuous and $G_{\rm at}$ is the distribution function of the purely atomic part $\nu_{\rm at}$ of ν, $G_{\rm at}(x) \equiv \nu_{\rm at}(]a,x]) \equiv \sum_{a<y\le x}\nu(\{y\})$. Then the sum in the definition of (Y_2) integral (with the $x = a, b$ terms) is $\int_a^b F(x) - F(x-)\,d\nu_{\rm at}(x)$.

The next fact follows directly from Billingsley (1968), page 110, Lemma 1, interchanging the sides on which intervals are closed and functions are continuous.

LEMMA 3.4. *A function F from a closed interval $[a,b]$ into $\mathbb{R}$ is left continuous with right limits if and only if F is a uniform limit of step functions which are finite sums of the form $\sum_i c_i 1_{]a_i,b_i]}$ for some real c_i and left open, right closed intervals $]a_i,b_i]$.*

Each step function as in Lemma 3.4 is left continuous with right limits and so in $\mathscr{R}(G)$ with $\int 1_{]c,d]}\,dG \equiv G(d) - G(c)$, $a \le c < d \le b$. Also, clearly, a uniform limit of functions in $\mathscr{R}(G)$ is in $\mathscr{R}(G)$ and the integrals converge [e.g., Hildebrandt (1938), 5.41]. So under the conditions of Lemma 3.3, since $x \mapsto F(x-)$ is left continuous with right limits, we have $\int_a^b F(x-)\,d\nu(x) = (Y_2)\int_a^b F(x-)\,dG(x)$. For any $\varepsilon > 0$, $|F(x) - F(x-)| > \varepsilon$ for at most finitely many values of x, so $F(x) - F(x-)$ is a uniform limit of functions with finite support, and its integrals for dG and $d\nu$ equal those for $dG_{\rm at}$ and $d\nu_{\rm at}$ respectively. Lemma 3.3 then follows. □

If $G(a-)$ is defined, let

$$(Y_2)\int_{a-}^b F\,dG := (Y_2)\int_a^b F\,dG + F(a)(G(a) - G(a-)).$$

For this integral, Lemma 3.3 extends to the case where ν has an atom at a, with $G(a-) = 0$ and $G(a) = \nu(\{a\})$.

THEOREM 3.5 (L. C. Young). *Suppose on an interval $[a,b]$ we have $f \in W_p$ and $g \in W_q$, where $s := 1/p + 1/q > 1$. Then $(Y_i)\int_a^b f\,dg$ is well defined for $i = 1, 2$, and for any ξ in $[a,b]$,*

$$\left|(Y_i)\int_a^b (f(x) - f(\xi))\,dg(x)\right| \le (1 + \zeta(s))\|f\|_{(p)}\|g\|_{(q)}, \qquad i = 1,2,$$

where ζ is the Riemann zeta function $\zeta(s) := \sum_{n\ge 1} n^{-s}$. Also,

$$\left|(Y_i)\int_a^b f(x)\,dg(x)\right| \le (1 + \zeta(s))\|f\|_{[p]}\|g\|_{(q)}, \qquad i = 1,2.$$

PROOF. The first inequality is given in Young (1936), pages 264–266, for the (Y_1) integral, and follows for the (Y_2) integral by the transformation $x \mapsto -x$. The second inequality then holds, since for any ξ,

$$\left|\int_a^b f(\xi)\,dg(x)\right| \le |f(\xi)|(\sup - \inf)(g) \le \|f\|_\infty \|g\|_{(q)}. \qquad \square$$

As $s \downarrow 1$, the constant $1 + \zeta(s)$ goes to $+\infty$. Young (1936), Section 7, shows that this is inevitable: For $s = 1$, $1 + \zeta(1)$ cannot be replaced in Theorem 3.5 by any finite constant. But there is a converse inequality, as follows. For $\delta > 0$, let $v(f, r, [a,b], \delta)$ be the supremum of all sums $\Sigma_{i=1}^n |f(x_i) - f(x_{i-1})|^r$, where $a \le x_0 < x_1 < \cdots < x_n \le b$, and where $x_i - x_{i-1} \le \delta$ for all $i = 1, \dots, n$. For $1 < r < \infty$, let $C_r^* := C_r^*[a,b]$ be the set of all functions f in W_r such that $v(f, r, [a,b], \delta) \to 0$ as $\delta \downarrow 0$. Clearly, each function in C_r^* is continuous. [Young (1936), page 261, defines the class $W_r^*[a,b]$ of functions in W_r such that $\lim_{\delta \downarrow 0} v(f, r, [a,b], \delta)$ equals the r-variation coming from the jumps of f. Then C_r^* is the set of continuous functions in W_r^*.] A dual norm will be defined with respect to C_r^*: For any linear functional h on C_r^*, let

$$\|h\|'_{[r]} := \sup\{|h(f)|: f \in C_r^*, \|f\|_{[r]} \le 1\}.$$

THEOREM 3.6 (Love and Young). *Let $1 < s < \infty$ and let $L \in (C_s^*)'$. Then there exists a function $g \in W_r$, where $r^{-1} + s^{-1} = 1$ such that $L(f) = \int_a^b f\,dg$, the Riemann–Stieltjes integral existing for all $f \in C_s^*$, and where $\|g\|_{(r)} \le 2^{2+1/r}\|L\|'_{[s]}$.*

PROOF. Love and Young (1937), page 248, give this fact except for using the norm $\|f\|_{[p,a]} := |f(a)| + \|f\|_{(p)}$ in place of $\|\cdot\|_{[p]}$ and having a factor of $2^{1+1/r}$. The conclusion follows since $|f(a)| \le \|f\|_\infty \le \|f\|_{[p,a]}$, so $\|f\|_{[p,a]} \le \|f\|_{[p]} \le 2\|f\|_{[p,a]}$ for all f. $\square$

Now given any finite signed measure ν on the Borel sets of $[a,b]$, let G be the distribution function $G(x) := \nu([a,x])$. Then $\int f\,d\nu = \int f\,dG$, where the Riemann–Stieltjes integral on the right exists for all continuous f [by Lemma 3.1, or, e.g., Kolmogorov and Fomin (1970), page 368]. Since G is of bounded variation, we also have $\|G\|_{(t)} < \infty$ for $1 \le t < \infty$.

Suppose $\int f\,d\nu = \int f\,dg$ for all $f \in C_s^*$ where $g \in W_r$, $r^{-1} + s^{-1} = 1$. For $a < c \le b$, consider the functions $f_n(x) := \max(0, \min(1, n(c - x)))$ on $[a,b]$. Then $f_n \uparrow f := 1_{[a,c[}$ and f_n are uniformly of bounded variation. Let $c(n) := c - 1/n$ and $d(n) := c - 1/n^2$. Then as $n \to \infty$,

$$\int_a^b f_n\,dg = \left(\int_a^{c(n)} + \int_{c(n)}^{d(n)} + \int_{d(n)}^{c}\right) f_n\,dg \to g(c-) - g(a),$$

because the first integral on the right equals $g(c_n) - g(a)$ which converges to the given limit, the third integral goes to 0 since $0 \le f_n \le 1/n$ on $[d(n), c]$, and the second integral goes to 0 since the r-variation of g on the half-open interval $[c(n), c[$ goes to 0 while the 1-variation of f_n remains bounded by 1, so Theorem 3.5 applies. It follows that $g(c-) - g(a)$ is uniquely determined by ν. Likewise, taking $h_n \downarrow 1_{[a,c]}$, so is $g(c+) - g(a)$, and clearly $g(b) - g(a) = \int_a^b 1\, dg$.

By adding a constant to g, we can take $g(a) = 0$. Then on the set of functions $g \in W_r[a,b]$ with the same $g(x-)$ for $a < x \leq b$, the same $g(x+)$ for $a \leq x < b$, the same $g(b)$, and the same $\int f\,dg$ for all $f \in C_s^*$, $\|g\|_{(r)}$ is minimized when g is right continuous at x for $a < x < b$, because for such a g and x, in any r-variation sum, if x appears, it can be replaced by $x + \delta$ for $\delta \downarrow 0$ such that g is continuous at $x + \delta$, so that the supremum of such sums is the same as the supremum of sums in which x does not appear. A different value of $g(x)$ could only make the r-variation larger.

Also considering f_n in C_s^* with $f_n \downarrow 1_{\{a\}}$ and $\|f_n\|_{[s]} \equiv 2$, we get $|\nu(\{a\})| = |\lim_{n\to\infty} \int f_n\, d\nu| \leq 2\|\nu\|'_{[s]}$. So we obtain the following result.

PROPOSITION 3.7. *If ν is a finite signed measure on $[a,b]$, $G(x) \equiv \nu([a,x])$, $1 < s < \infty$ and $s^{-1} + r^{-1} = 1$, then*

$$\|G\|_{(r)} \leq 2^{2+1/r}\|\nu\|'_{[s]} \quad \textit{and} \quad \|G\|_{[r]} \leq (2^{3+1/r} + 2)\|\nu\|'_{[s]}.$$

Note that conversely, by Theorem 3.5, for any $\varepsilon > 0$,

$$\|\nu\|'_{[s]} \leq \left(1 + \zeta\left(\frac{1}{s} + \frac{1}{r-\varepsilon}\right)\right)\|G\|_{(r-\varepsilon)}.$$

COROLLARY 3.8. *For any r with $2 < r < \infty$ and $\varepsilon > 0$, there is an $M < \infty$ such that for any distribution function F on $\mathbb{R}$ and its empirical distribution functions F_n, $\Pr\{\|n^{1/2}(F_n - F)\|_{[r]} > M\} < \varepsilon$ for all n.*

PROOF. $\|F_n - F\|_{[r]}$ is a measurable random variable since in both $\|F_n - F\|_{(r)}$ and $\|F_n - F\|_\infty$ one can restrict to the rational numbers. Let P be the law with distribution function F and P_n its empirical measures. Apply Proposition 3.7 to $\nu := \nu_n := n^{1/2}(P_n - P)$, $a := -\infty$, $b := +\infty$ and $s := r/(r-1)$. Then $1 < s < 2$, and we can treat $\|\nu_n\|'_{[s]}$ in place of $\|n^{1/2}(F_n - F)\|_{[r]}$. The unit ball of C_s^*, being a subset of that of W_s, is a dominated uniform Donsker class by Theorem 2.2, and the result follows. □

COROLLARY 3.9. *Let $T(\cdot)$ be a functional on some open set U in W_r, $r > 2$, such that T is Fréchet differentiable with derivative $L(\cdot)$ at some distribution function $F \in U$ with respect to $\|\cdot\|_{[r]}$. Then for the empirical distribution functions F_n,*

$$T(F_n) = T(F) + L(F_n - F) + o_p(n^{-1/2}) \quad \textit{as } n \to \infty,$$

where the $o_p(n^{-1/2})$ is uniform in F.

PROOF. By definition of Fréchet differentiability, the remainder is $o(\|F_n - F\|_{[r]})$, which is uniformly $o_p(n^{-1/2})$ by Corollary 3.8. □

Some functionals with the differentiability stated in Corollary 3.9 are the inverse operator $F \mapsto F^{-1}$ and the composition operator $(F, G) \mapsto F \circ G$ with respect to F, while G varies in L^p, under suitable conditions [Dudley (1991a)].

Corollary 3.8 also follows directly from:

(a) the results of Pisier and Xu, discussed after Theorem 2.2 above, and
(b) again, the Love and Young (1937) duality.

There is still another proof of Corollary 3.8, by martingales [Dudley (1991c)].

4. Bilinear operations and Young integrals. Let X, Y and Z be three Banach spaces and let B be a function from $X \times Y$ into Z which is *bilinear*, so that $B(\cdot, y)$ is linear on X for each $y \in Y$ and $B(x, \cdot)$ is linear on Y for each $x \in X$. Then B provides its own partial derivatives with respect to each variable, since

$$B(x+u, y) - B(x, y) \equiv B(u, y) \quad \text{and} \quad B(x, y+v) - B(x, y) \equiv B(x, v).$$

The remainder in differentiating with respect to both variables is

$$B(x+u, y+v) - B(x, y) - B(u, y) - B(x, v) \equiv B(u, v).$$

So B is jointly Fréchet differentiable from $X \times Y$ into Z if $\|B(u, v)\| = o(\|u\| + \|v\|)$ as $\|u\| \to 0$ and $\|v\| \to 0$. If B is a *bounded* bilinear form in the sense that for some $K < \infty$, $\|B(u, v)\| \le K\|u\|\,\|v\|$ for all $u \in X$ and $v \in Y$, then B is jointly Fréchet differentiable. If X, Y and Z are Banach spaces, boundedness is equivalent to joint continuity, and even to separate continuity, where $B(x, y)$ is continuous in x for each fixed y and in y for each fixed x [e.g., Schaefer (1966), page 88].

Functionals of the form $B(F, G) := \int F\, dG$ are bilinear. Such functionals are basic in the duality theory of functions of bounded p-variation as in Section 3. On the other hand, for empirical distribution functions F_m and G_n, $\int F_m\, dG_n$ is (up to normalization) a two-sample Wilcoxon statistic. Gill (1989), Lemma 3, page 110 shows that B is compactly differentiable for supremum norms at F_0, G_0 on $D[-\infty, \infty] \times E_1$ if E_1 is the set of functions in $D[-\infty, \infty]$ of total variation at most C for some $C < \infty$, and F_0 has bounded variation. Gill (1989), Lemma 1, page 105, then gives an extension, compactly differentiable for sup norms, of B to $D[-\infty, \infty] \times D[-\infty, \infty]$, at (F_0, G_0) of bounded variation. But this extension simply deletes the remainder term $B(f, g)$ in

$$B(F_0 + f, G_0 + g) \equiv B(F_0, G_0) + B(F_0, g) + B(f, G_0) + B(f, g)$$

if $G_0 + g \notin E_1$, in other words, if $G_0 + g$ has total variation greater than C. The resulting function is no longer bilinear, and is discontinuous even along lines. As Gill notes, for any probability distribution functions G and $G + g$ such as $G + g = G_n$, if $C \ge 1$, then G and $G + g$ are both in E_1, while if $C \ge 2$, then also $g \in E_1$, so the extension makes no difference.

Young's p-variation duality theory (Section 3) provides an alternative approach.

THEOREM 4.1. *The integrals* $(F, G) \mapsto (Y_i)\int F\,dG$, $i = 1, 2$, *are Fréchet differentiable, on* $W_r \times W_s$ *for* $\|\cdot\|_{[r]}, \|\cdot\|_{(s)}$ *whenever* $r^{-1} + s^{-1} > 1$.

PROOF. By a theorem of Young (the latter part of Theorem 3.5 above), the integrals are bounded bilinear forms for the given norm and seminorm, so they are Fréchet differentiable. □

Corollary 3.8 and Theorem 4.1 cannot be applied directly if F and G are both replaced by empirical processes, since we cannot take both $r > 2$ and $s > 2$. On the other hand, it may sometimes be useful that empirical processes do have sample functions in r-variation spaces also for $1 \le r \le 2$, without the uniformity of Corollary 3.8.

In this case [unlike the composition and inverse operators treated by Reeds (1976), Fernholz (1983) and Dudley (1991a)], the derivative need not be taken at F_0 or G_0 having stronger smoothness properties; neither of them has to be of bounded variation. While the $\|\cdot\|_{[r]}$ norm for f is stronger than the sup norm, recall that compact sets for the sup norm are small while bounded sets for $\|\cdot\|_{[r]}$ can be quite large (nonseparable) for the sup norm as well as for the $\|\cdot\|_{[r]}$ norm itself.

Acknowledgments. I thank Richard Gill, Evarist Giné, Jose Gonzalez-Barrios, Gilles Pisier, Aad van der Vaart, Jon Wellner and the referee for helpful comments on this paper and the subject generally.

REFERENCES

BERTOIN, J. (1989). Sur une intégrale pour les processus à α-variation bornée. *Ann. Probab.* **17** 1521–1535.

BILLINGSLEY, P. (1968). *Convergence of Probability Measures.* Wiley, New York.

BRUNEAU, M. (1974). *Variation totale d'une fonction. Lecture Notes in Math.* **413**. Springer, New York.

DUDLEY, R. M. (1985). An extended Wichura theorem, definitions of Donsker class, and weighted empirical distributions. *Probability in Banach Spaces V. Lecture Notes in Math.* **1153** 141–178. Springer, New York.

DUDLEY, R. M. (1987). Universal Donsker classes and metric entropy. *Ann. Probab.* **15** 1306–1326.

DUDLEY, R. M. (1989a). *Real Analysis and Probability.* Wadsworth and Brooks/Cole, Pacific Grove, Calif.

DUDLEY, R. M. (1989b). Comment on "Asymptotics via empirical processes" by D. Pollard. *Statist. Sci.* **4** 354.

DUDLEY, R. M. (1990). Nonlinear functionals of empirical measures and the bootstrap. In *Probability in Banach Spaces VII* (E. Eberlein, J. Kuelbs and M. B. Marcus, eds.). *Progress in Probability* **21** 63–82. Birkhäuser, Boston.

DUDLEY, R. M. (1991a). Differentiability of the composition and inverse operators for regulated and a.e. continuous functions. Preprint.

DUDLEY, R. M. (1991b). Implicit functions of empirical measures and robust statistics. Unpublished manuscript.

DUDLEY, R. M. (1991c). The order of the remainder in derivatives of composition and inverse operators for Φ-variation norms. Preprint.

FERNHOLZ, L. T. (1983). *von Mises Calculus for Statistical Functionals. Lecture Notes in Statist.* **19**. Springer, New York.

GILL, R. D. (1989). Non- and semi-parametric maximum likelihood estimators and the von Mises method. I. *Scand. J. Statist.* **16** 97–128.

GINÉ, E. and ZINN, J. (1991). Gaussian characterization of uniform Donsker classes of functions. *Ann. Probab.* **19** 758–782.

HAMPEL, F. R., RONCHETTI, E. M., ROUSSEEUW, P. J. and STAHEL, W. A. (1986). *Robust Statistics: The Approach Based on Influence Functions.* Wiley, New York.

HILDEBRANDT, T. H. (1938). Definitions of Stieltjes integrals of the Riemann type. *Amer. Math. Monthly* **45** 265–278.

HUBER, P. J. (1981). *Robust Statistics.* Wiley, New York.

KOLMOGOROV, A. N. and FOMIN, S. V. (1970). *Introductory Real Analysis.* Translated and edited by R. A. Silverman. Prentice-Hall, Englewood Cliffs, NJ.

KOLMOGOROV, A. N. and TIHOMIROV, V. M. (1961). ε-entropy and ε-capacity of sets in functional spaces. *Amer. Math. Soc. Transl.* (*Ser.* 2) **17** 277–364. [*Uspehi Mat. Nauk* **14** no. 2 (1959) 3–86.]

LOVE, E. R. and YOUNG, L. C. (1937). Sur une classe de fonctionelles linéaires. *Fund. Math.* **28** 243–257.

PISIER, G. and XU, Q. (1987). Random series in the real interpolation spaces between the spaces v_p. In *Geometrical Aspects of Functional Analysis* (J. Lindenstrauss and V. Milman, eds.). *Lecture Notes in Math.* **1267** 185–209. Springer, New York.

PISIER, G. and XU, Q. (1988). The strong p-variation of martingales and orthogonal series. *Probab. Theory Related Fields* **77** 497–514.

POLLARD, D. B. (1982). A central limit theorem for empirical processes. *J. Austral. Math. Soc. Ser. A* **33** 235–248.

POLLARD, D. (1989). Asymptotics via empirical processes (with discussion). *Statist. Sci.* **4** 341–366.

PONS, O. and DE TURCKHEIM, E. (1989). Méthode de von Mises, Hadamard-différentiabilité et bootstrap dans un modèle non paramétrique sur un espace métrique. *C. R. Acad. Sci. Paris Sér. I* **308** 369–372.

REEDS, J. A., III (1976). On the definition of von Mises functionals. Ph.D. dissertation, Dept. Statistics, Harvard Univ.

RUDIN, W. (1976). *Principles of Mathematical Analysis*, 3d ed. McGraw-Hill, New York.

SCHAEFER, H. H. (1966). *Topological Vector Spaces.* MacMillan, New York.

SERFLING, R. J. (1980). *Approximation Theorems of Mathematical Statistics.* Wiley, New York.

WIENER, N. (1924). The quadratic variation of a function and its Fourier coefficients. *J. Math. Phys.* **3** 72–94.

XU, Q. (1986). Espaces d'interpolation réels entre les espaces V_p: propriétés géométriques et applications probabilistes. *Séminaire d'Analyse Fonctionelle 1985–86–87* 77–123. Univ. Paris VI–VII; also in Xu (1988), II, with "Remarques sur l'exposé..." (3 pages).

XU, Q. (1988). Thèse de doctorat, math., Univ. Paris VI.

YOUNG, L. C. (1936). An inequality of the Hölder type, connected with Stieltjes integration. *Acta Math.* **67** 251–282.

YOUNG, W. H. (1914). On integration with respect to a function of bounded variation. *Proc. London Math. Soc.* (2) **13** 109–150.

DEPARTMENT OF MATHEMATICS
MASSACHUSETTS INSTITUTE OF TECHNOLOGY
CAMBRIDGE, MASSACHUSETTS 02139

The Annals of Statistics
1994, Vol. 22, No. 1, 1–20

THE ORDER OF THE REMAINDER IN DERIVATIVES OF COMPOSITION AND INVERSE OPERATORS FOR p-VARIATION NORMS[1]

BY R. M. DUDLEY

Massachusetts Institute of Technology

Many statisticians have adopted compact differentiability since Reeds showed in 1976 that it holds (while Fréchet differentiability fails) in the supremum (sup) norm on the real line for the inverse operator and for the composition operator $(F, G) \mapsto F \circ G$ with respect to F. However, these operators are Fréchet differentiable with respect to p-variation norms, which for $p > 2$ share the good probabilistic properties of the sup norm, uniformly over all distributions on the line.

The remainders in these differentiations are of order $\|\cdot\|^{\gamma}$ for $\gamma > 1$. In a range of cases p-variation norms give the largest possible values of γ on spaces containing empirical distribution functions, for both the inverse and composition operators. Compact differentiability in the sup norm cannot provide such remainder bounds since, over some compact sets, differentiability holds arbitrarily slowly.

1. Introduction. The theory of differentiable statistical functionals began with work of von Mises [e.g., von Mises (1936, 1947) and Filippova (1961)]. A nonlinear functional T is defined, for example, on distribution functions. Von Mises differentiated T at a distribution function F along lines. For T to have a (Gâteaux) derivative at F means that, in the direction of a function h,

$$T(F + th) = T(F) + tT'(F)(h) + o(|t|) \quad \text{as } t \to 0. \tag{1.1}$$

Here $T'(F)(\cdot)$ is a bounded linear operator on functions h, for example, of the form

$$T'(F)(h) = \int g\,dh \quad \text{for some function } g \text{ (depending on } F\text{).} \tag{1.2}$$

It was well known that such differentiability is most useful if the $o(|t|)$ is uniform for h in some sets. The object was to analyze the behavior of $T(F_n)$ for empirical distribution functions F_n, and the F_n do not approach F along lines as $n \to \infty$. If the $o(|t|)$ is uniform for h in bounded sets for some norm $\|\cdot\|$, or equivalently,

$$T(F + h) - T(F) - T'(F)(h) = o(\|h\|) \quad \text{as } \|h\| \to 0, \tag{1.3}$$

then T is *Fréchet* differentiable at F for $\|\cdot\|$. For distribution functions on

Received October 1991; revised April 1993.

[1]Research partially supported by NSF grants and a Guggenheim fellowship.

AMS 1991 *subject classifications*. Primary 62G30, 58C20, 26A45; secondary 60F17.

Key words and phrases. Fréchet derivative, compact derivative, Hadamard derivative, Gâteaux derivative, Bahadur–Kiefer theorems, Orlicz variation.

E. Giné et al. (eds.), *Selected Works of R.M. Dudley*, Selected Works in Probability and Statistics, DOI 10.1007/978-1-4419-5821-1_22, © Springer Science+Business Media, LLC 2010

the line, the first natural choice for the norm was the *supremum* (sup) norm $\|h\|_\infty := \sup_x |h(x)|$.

In robust statistics [Huber (1981), Sections 1.5 and 2.5] the derivative, or specifically the g in (1.2), has been called the influence function [Hampel, Ronchetti, Rousseeuw and Stahel (1986)]. Differentiability is also an issue in semiparametric statistics [e.g., Gill (1989) and Bickel, Klaassen, Ritov and Wellner (1993)].

In his landmark thesis Reeds [(1976), Section 6.4] considered two functionals (or *operators*, since their values are also functions): the *composition operator* $(F, G) \mapsto F \circ G$, where $(F \circ G)(x) \equiv F(G(x))$, and the *inverse operator*, $H \mapsto H^{-1}$ where $H^{-1}(y) := \inf\{x: H(x) \geq y\}$. Here F and H are in a space $D[0,1]$ or $D(\mathbb{R})$ of right-continuous functions with left limits. The inverse operator takes a distribution function into a quantile function. Then, applying the composition operator with H^{-1} as G, we get $F \circ H^{-1}$, the quantile–quantile function.

In this introduction, let us consider differentiation at the uniform $U[0,1]$ distribution function $F(x) := x$, $0 \leq x \leq 1$. For a fixed y, say $y = \frac{1}{2}$, the quantile functional $H \mapsto H^{-1}(y)$ is not differentiable even at $U[0,1]$ along lines in $D[0,1]$ (as reviewed in Proposition 2.7), although it is for $C[0,1]$ (see Section 3). Even on $C[0,1]$, the inverse operator is differentiable with the sup norm on its range for functions in the range restricted to $[a,b]$, $0 < a < b < 1$, but not on all of $[0,1]$. Reeds then naturally considered the L^p norms, $\|h\|_p := (\int_0^1 |h(y)|^p dy)^{1/p}$, $1 \leq p < \infty$, on the range of the inverse operator. Then, to get the quantile–quantile function, L^p norms of G are taken on the domain of the composition operator.

Easy examples [Proposition 2.8(b), (c)] show that under these conditions the inverse operator and the composition operator with respect to F are not Fréchet differentiable for the sup norm on their domains. Among Reeds' major results are that in these cases the $o(|t|)$ does hold uniformly over *compact* sets in the sup norm. Although Reeds' (1976) thesis was unpublished, his work is the main ingredient of the survey by Fernholz (1983).

Following the work of Reeds and Fernholz, while Fréchet differentiability is used when it holds, statisticians have worked with compact differentiability as apparently the best kind that held in enough generality. However, if the sup norm is replaced by p-variation norms $\|\cdot\|_{[p]}$ (defined in Section 2), Fréchet differentiability holds for both the composition and inverse operators [Dudley (1991)]. Moreover, the norms $\|\cdot\|_{[p]}$ share the excellent properties of the sup norm of being invariant under all strictly monotone continuous transformations of the line onto itself, and they satisfy (for some values of $p, p < 2$ or $p > 2$ depending on which side of duality one is on) uniform central limit theorems (Donsker properties) [Dudley (1992a)].

The present paper treats the size of the remainder in the Fréchet differentiability and finds bounds for our two operators in (1.3), where $o(\|h\|_{[p]})$ is replaced by $O(\|h\|_{[p]}^\gamma)$ for some $\gamma > 1$. For the inverse operator we get (for $F = U[0,1]$, so $F^{-1} = F$)

$$\|(F+g)^{-1} - F^{-1} + g\|_p = O(\|g\|_{[p]}^\gamma), \quad \text{for } \gamma = 1 + 1/p, \tag{1.4}$$

as a special case of Corollary 2.4. Theorem 2.2 gives a corresponding remainder bound for joint Fréchet differentiability of the composition operator. The powers γ are shown (in Proposition 2.5 and Theorem 2.2) to be optimal, not only for p-variation norms but for an arbitrary norm on distribution functions. Prior to Dudley (1991) and the present paper, the composition and inverse operators under the given conditions had not been shown to be Fréchet differentiable for any norm, to my knowledge. Note that for twice-differentiable functionals one expects an exponent $\gamma = 2$, so our operators are not that smooth if $p > 1$.

Actually, the remainder in differentiation of the inverse operator had been studied earlier and in precise detail in the main case of statistical interest, where $g = F_n - F$ (F_n empirical), by Bahadur (1966) and Kiefer (1967, 1970). Reeds (1976) and Fernholz (1983) did not cite this work of Bahadur and Kiefer. Take any $p > 2$ in (1.4), giving any $\gamma < \frac{3}{2}$. From $\|F_n - F\|_{[p]} = O_p(n^{-1/2})$ [Dudley (1992a), Corollary 3.8] we get in (1.4) a remainder

$$\left\|F_n^{-1} - F^{-1} + F_n - F\right\|_p = O_p\left(n^{-\gamma/2}\right), \tag{1.5}$$

where $\gamma = 1 + p^{-1}$. Letting $p \downarrow 2$ gives $-\gamma/2 \downarrow -\frac{3}{4}$; or, as $p \uparrow 2$, results of Dudley (1992b) also give $-\frac{3}{4}$ as limiting exponent. Here $-\frac{3}{4}$ is the best possible, as Kiefer proved. See also Theorem 2.6 and Section 5.

The usual space containing distribution functions on [0, 1]—and the one Reeds and Fernholz treat—is the space $D[0, 1]$ of right-continuous functions with left limits. However, $D[0, 1]$ with supremum norm is nonseparable (has no countable dense subset). In such a space compact differentiability has some substantial drawbacks:

1. *Compact sets—in a nonseparable normed space—are extremely small.* The notion that compact differentiability is good when Fréchet differentiability fails may have come from separable spaces, where indeed the compact sets are about as large as one can expect, short of bounded sets. However, for any sup-norm-compact set K in $D[0,1]$ and for any nonatomic distribution F, the probability that the empirical distribution function F_n is in K is 0. How inconvenient this is may be indicated by the variety of ingenious devices statisticians have applied to getting around it, including smoothing, "tangential" differentiability and use of almost surely convergent realizations.
2. Unlike the Fréchet case, compact differentiability does not give helpful uniform error bounds like (1.4). To the contrary, the rate of $o(|t|)$ in (1.1) depends on the compact set and can be arbitrarily slow, as Proposition 2.1 shows. In other words, in contrast to drawback 1, *some compact sets are inconveniently large.* One can think of them as only extending out in a separable and hence very small set of directions in the nonseparable space, but, in those directions, sticking out far enough to slow down the $o(|t|)$ just short of making it fail. When applied to empirical distribution functions (as it can be but, by drawback 1, only via some modification), compact differentiability would only give a remainder $o_p(n^{-1/2})$ even when a better rate holds.

On the other hand, in separable spaces compact differentiability may be more useful although drawback 2 will still hold there.

There are many statistical functionals or operators for which compact differentiability was the first to be proved. For some of these, Fréchet differentiability (even allowing different norms) has not yet been proved. An important example of such an operator is the product integral [Gill and Johansen (1990)].

One of Reeds' main points was to separate analytical from probabilistic parts of the work, although the asymptotic results obtained may not be the best possible. The p-variation norms evidently come closer to giving optimal results, although there are limitations; (1.4) is sharp as it stands, but when applied to empirical distribution functions in (1.5) it is not: (1.5) holds (as Kiefer proved) with $p = \infty$ on the left and $O_p(n^{-3/4}(\log n)^{1/2})$ on the right.

The composition and inverse operators would seem to be of interest not only in statistics but in analysis more generally. However, analysts seem not to have developed the theory needed. They have treated the composition operator almost exclusively for fixed F, as a special case of the so-called Nemitsky operator, surveyed by Appell and Zabrejko (1990). Also, analysts seem to have given little if any attention to differentiability of the inverse operator in cases where H is not 1–1, but empirical distribution functions are never 1–1.

The main results will be stated in Section 2 and proved in Section 4. Section 3 compares different kinds of differentiability as applied to empirical distribution functions, and Section 5 treats Orlicz variation and martingales.

2. Statements of results. There are a number of possible definitions of differentiability. Let X be a vector space with a norm $\|\cdot\|$ and let T be a functional defined on an open set U in X. (Here we have in mind for X a function space containing distribution functions on $\mathbb{R}$.) Let T take values in another vector space Y with a norm $|\cdot|$. Note that values of the composition and inverse operators are again functions.

T is said to be Gâteaux differentiable at a point $x_0 \in U$ if there is a bounded linear operator $T'(x_0)(\cdot)$ from X into Y such that, for each $v \in X$,

$$|T(x_0 + tv) - T(x_0) - tT'(x_0)(v)| = o(|t|) \quad \text{as } t \to 0. \tag{2.1}$$

The chain rule for composition fails for Gâteaux differentiability. Filippova and von Mises considered Gâteaux differentiability, with further probabilistic hypotheses.

Let $\mathcal{C}$ be a class of bounded subsets of X containing all the finite sets. Then T will be called *$\mathcal{C}$-differentiable* at x_0 if (2.1) holds uniformly for $v \in A$ for all $A \in \mathcal{C}$ [Sebastião e Silva (1956)]. If $\mathcal{C}$ is the family of all bounded sets in X, or equivalently if $\mathcal{C}$ consists of the one set $B_1 := \{x \in X: \|x\| \leq 1\}$, then T is said to be *Fréchet* differentiable at x_0. Then we have more simply

$$|T(x_0 + u) - T(x_0) - T'(x_0)(u)| = o(\|u\|) \quad \text{as } \|u\| \to 0. \tag{2.2}$$

This is the main kind of differentiability in use among analysts. It is the strongest possible kind of $\mathcal{C}$-differentiability, since sets in $\mathcal{C}$ are bounded. (If $\mathcal{C}$

contains an unbounded set, e.g., in a one-dimensional space X, $\mathcal{C}$-differentiability would imply that T is linear at least in a one-sided neighborhood of x_0.) There seems to be little doubt that Fréchet's is the most useful form of differentiability, when it holds.

Alternatively, if $\mathcal{C}$ is the class of all compact sets in X, we get *compact* or *Hadamard* differentiability, emphasized by statisticians since the work of Reeds (1976) and Fernholz (1983). In the Introduction and in detailed results below are shown the advantages of Fréchet over compact differentiability. However, the Reeds–Fernholz theory can also be extended in a very different direction. One can enlarge the class of compact sets for the sup norm to much larger classes over which $\mathcal{C}$-differentiability still holds: the class of compact sets for the Skorohod topology or the still much larger class of "uniformly Riemann" sets [Dudley (1991), Theorems 4.5 and 5.1]. In other words, referring to drawbacks 1 and 2 of (sup-)norm-compact differentiability in the Introduction, if one is willing to accept the arbitrarily slow remainder rates (drawback 2), it is far from necessary to accept also drawback 1. The Skorohod-compact sets do carry the normalized empirical distribution functions $n^{1/2}(F_n - F)$ with high probability, uniformly in n. The slow rates for compact (and not Fréchet) differentiability are given as follows:

PROPOSITION 2.1. *Let $(X, \|\cdot\|)$ be a normed linear space and $x \in X$. Let T be a function from a neighborhood of x into a normed space $(Y, |\cdot|)$, Gâteaux but not Fréchet differentiable at x. Then, for any sequence $\varepsilon_n \downarrow 0$ (however slowly), there exist a sequence $v_n \to 0$ in X and numbers t_n such that $0 < t_n \leq 1/n$ and*

$$\left|T(x + t_n v_n) - T(x) - t_n T'(x)(v_n)\right| \geq \varepsilon_n t_n.$$

(Proofs are given in Section 4.) Compact differentiability for a given norm is presumably of interest mainly when Fréchet differentiability fails. Then, however, since the set $\{0\} \cup \{v_n\}_{n\geq 1}$ from Proposition 2.1 is compact, the remainder in equation (2.1) is $o(|t|)$ only very slowly on some compact sets.

In this paper, Fréchet differentiability will be considered with respect to ψ-variation norms $\|f\|_{[\psi]} := \|f\|_\infty + \|f\|_{(\psi)}$, where $\|f\|_\infty := \sup_x |f(x)|$, $\|f\|_{(\psi)} := \inf\{K > 0: \; v_\psi(f/K) \leq 1\}$ and

$$v_\psi(f) := \sup\left\{\sum_i \psi(|f(x_i) - f(x_{i-1})|): x_0 < x_1 < \cdots\right\},$$

and where ψ is assumed to be a convex, strictly increasing function on $[0, \infty[$, with $\psi(0) = 0$, $\lim_{x\downarrow 0} \psi(x)/x = 0$ and $\lim_{x\uparrow\infty} \psi(x)/x = +\infty$. Such a ψ will be called an *Orlicz function*. Here $v_\psi(f)$ is called the *ψ-variation* of f. $\|\cdot\|_{[\psi]}$ is indeed a norm: subadditivity is proved, for example, in Musielak and Orlicz [(1959), Section 3.03], and the other properties of a norm are clear.

The best known cases are where $\psi(x) = x^p, 0 \leq x < \infty, 1 \leq p < \infty$. Here x^p is an Orlicz function for $1 < p < \infty$, while $\|f\|_{(1)}$ is the usual total variation of f. Then ψ-variation is called p-variation and written $v_p(f)$, and we set $\|f\|_{(p)} := \|f\|_{(\psi)} = v_p(f)^{1/p}$, $\|f\|_{[p]} := \|f\|_{[\psi]}$.

The notions of Orlicz-variation and p-variation are not very familiar. I did not know of them in advance, but found that they seemed to arise naturally in the research being reported here and in Dudley (1991), (1992a) and (1992b). One can arrive at them perhaps as follows. For the composition operator, one has a remainder term $f(G+g)-f(G)$ which one would like to be $o(|f|+|g|)$ in some sense. In showing differentiability one might think f itself should be differentiable, but a moment's thought suggests that one only needs some continuity property of f so that $f(G+g)-f(G)$ is smaller than f itself. Having a bounded derivative is equivalent to satisfying a Lipschitz condition $|f(x)-f(y)| \le C|x-y|$. For present purposes it will be enough if f satisfies a weaker *Hölder condition* of order α, $|f(x)-f(y)| \le C|x-y|^\alpha$ for some $\alpha, 0<\alpha<1, C<\infty$. (This is one way of defining differentiability of fractional order α.)

Next, if we want a function space suitable for nonparametric statistics, we would like a space, with a norm, invariant under all increasing homeomorphisms—in other words strictly increasing, continuous functions from the line onto itself—just as the space $D(\mathbb{R})$ of bounded, right-continuous functions with left limits and supremum norm is invariant. Also, to deal with empirical distribution functions and the like, we would like functions that are not necessarily continuous. We can satisfy both these desires by taking the set of all functions $f \circ g$ where f is α-Hölder and g is nondecreasing and bounded (not necessarily continuous). The set of such $f \circ g$ is exactly the set of functions of bounded p-variation for $p = 1/\alpha$ [e.g., Dudley (1992a), proof of Theorem 2.1].

The test of a function space and norm, like any other mathematical objects, is how well they serve their purposes, and the results given here and in Dudley (1991) and (1992a) seem to show that p-variation spaces and norms serve well for differentiability of some interesting functionals of empirical distribution functions.

One may then ask why one needs the more complicated-looking Orlicz variation beyond just p-variation. There are two reasons. One is that Orlicz functions allow one to refine and interpolate between the p-variation norms; for example, x^p can be multiplied by logarithmic factors. The best possible Orlicz function may give somewhat better results than any particular x^p, as will be seen in Section 5. Second, one can take Orlicz functions ψ decreasing to 0 as $x \downarrow 0$ faster than any power. Classes of bounded ψ-variation can then be larger than classes of bounded p-variation. It turns out that each set in a very large class of sets called *uniformly regulated* sets, among which are all compact sets for the Skorohod topology (hence also for the supremum norm) in $D[0,1]$, is of bounded ψ-variation for some ψ [Dudley (1991), Theorems 2.2 and 2.6], which is not the case for p-variation.

Returning to the present development, an Orlicz function ψ is said to satisfy the Δ_2 condition if, for some constant $K<\infty, \psi(2x) \le K\psi(x)$ for all $x>0$. For the space of measurable functions f on a finite measure space such that $\int \psi(|f|)\, d\mu < \infty$ (Orlicz class), the Δ_2 condition is only important for large x, say $x \ge x_0 > 0$ [see, e.g., Krasnosel'skii and Rutickii (1961), page 23]. For ψ-variation spaces, the condition is only needed for small x, say $0 < x \le x_0$ [Musielak and Orlicz (1959)], but here, for simplicity, it is assumed for all

$x > 0$. Clearly, the functions x^p satisfy $\Delta_2, 1 < p < \infty$.

If an Orlicz function ψ satisfies Δ_2, then by convexity we have a "quasi-subadditivity" property: for any $x, y > 0$,

$$\psi(x+y) = \psi((2x+2y)/2) \leq (\psi(2x)+\psi(2y))/2 \leq K(\psi(x)+\psi(y))/2.$$

Fréchet differentiability holds for the operators $(f,g) \mapsto (F+f)\circ(G+g)$ and $f \mapsto (F+f)^{-1}$ with respect to the $\|\cdot\|_{[\psi]}$-norm of f under suitable conditions on F, G and g [Dudley (1991), Theorems 4.5 and 5.1]. Here, in some cases, the $o(\cdot)$ condition will be improved to $O(\|f\|_{[p]}^{\eta})$ for some $\eta > 1$ depending on p and other hypotheses. For the inverse operator there are bounds for the remainder in terms of p- and ψ-variation (Theorems 2.3 and 2.6 and Section 5).

Let $V_p(\mathbb{R})$ be the set of all real-valued functions f on $\mathbb{R}$ for which $\|f\|_{(p)} < \infty$. In the differentiation of the composition operator, one of the two remainder terms [Dudley (1991), Section 5] is $f(G+g) - f(G)$. If f is Lipschitz, with $\|f\|_L = K$, for example, if f has a bounded derivative with $\|f'\|_\infty = K$, and if g is bounded, then we have

$$\|f(G+g)-f(G)\|_\infty \leq K\|g\|_\infty \leq (\|f\|_L^2 + \|g\|_\infty^2)/2.$$

I have found in the literature outside of statistics two papers on joint differentiability of composition of nonlinear functions, with respect to norms: Gray (1975) and Brokate and Colonius (1990). Both papers took g bounded and f with bounded derivative, at least locally, with f and g both possibly Banach-valued. Reeds (1976) and Fernholz (1983) treat the much deeper case where f need not be Lipschitz and g varies in an L^p space and so may be unbounded. As far as I know, only statisticians have worked on these cases.

Let μ be a finite measure. In a composition $f\circ g$ where f is bounded, the size of large values of g is not important. It will turn out that for g the pseudometric

$$\rho_p(g_1,g_2) := \left(\int (\min(1,|g_1-g_2|))^p \, d\mu\right)^{1/p}$$

works the same as the usual $L^p(\mu)$ distance for $1 \leq p < \infty$. The function $m(y) := \min(y,1)$ for $y > 0$ is nondecreasing, concave and so subadditive. By the Minkowski inequality [e.g., Dudley (1989), Theorem 5.1.5], ρ_p is indeed a pseudometric. It is easily seen to metrize convergence in measure.

The other remainder term, $F(G+g)-F(G)-F'(G)g$, does not depend on f, so the following will give an explicit bound for the dependence of the remainder on f, and some optimality properties of the bound:

THEOREM 2.2. *Let $\beta > 0$. Let G be an increasing function on $[0,1]$ with $G(y) - G(x) \geq \beta(y-x)$, for $0 \leq x \leq y \leq 1$. Let $1 \leq p < \infty$ and $1 \leq s < \infty$. Then there is a constant $C < \infty$ such that, for $f \in V_p(\mathbb{R})$ and $g \in L^s[0,1]$,*

$$\|f(G+g)-f(G)\|_p \leq C\|g\|_s^{s/(p(1+s))}\|f\|_{[p]} \leq C(\|g\|_s^{\eta} + \|f\|_{[p]}^{\eta}),$$

where $\eta := 1 + s/(p(1+s)) > 1$. *Conversely, if* $\|\cdot\|$ *is any norm on a space* V *of functions containing the function* $h(x) = 1_{\{x>0\}}, U(x) = x, 0 \le x \le 1$, *and there are constants* C, α *and* γ *such that*

$$\|f \circ (U+g) - f \circ U\|_p \le C\|g\|_s^\alpha \|f\|^\gamma,$$

for $\|f\| \le 1$, *then* $\gamma \le 1$ *and* $\alpha \le s/(p(1+s))$. *In both parts of this theorem,* $\|\cdot\|_p$ *can be replaced by* $\rho_p(0,\cdot)$ *and/or* $\|g\|_s$ *by* $\rho_s(0,g)$.

Next, consider the inverse operator. Let G be an increasing, continuous function from $[0,1]$ onto itself, having a derivative everywhere on the open interval $]0,1[$ which extends to a continuous, strictly positive function G' on $[0,1]$. Such a G will be called an increasing *diffeomorphism* of $[0,1]$. Let $g \in R[0,1]$, the space of all bounded real functions on $[0,1]$, continuous almost everywhere for Lebesgue measure [Dudley (1991)]. Then it is known [e.g., Dudley (1991), Theorem 4.5] that the derivative of the inverse operator $g \mapsto (G+g)^{-1}$ at $g = 0$ is the linear operator $g \mapsto -(g \circ G^{-1})/(G' \circ G^{-1})$, bounded for the sup norm on g since by assumption $\inf_{[0,1]} G' > 0$. The remainder in the differentiation is

$$R_g := (G+g)^{-1} - G^{-1} + (g \circ G^{-1})/(G' \circ G^{-1}) \quad \text{on } [0,1].$$

Let us recall a definition of modulus of continuity. For a function f and $\delta > 0$, let $\omega(\delta, f) := \sup\{|f(x) - f(y)|: |x-y| \le \delta\}$. We then have the following.

THEOREM 2.3. *For any increasing diffeomorphism* G *of* $[0,1]$, *so that* $0 < \beta := \inf G' \le \eta := \sup G' < \infty$, *and for any Orlicz function* ψ *satisfying* Δ_2, *there are constants* $C_1, C_2 < \infty$ *such that, for* $\gamma := \|g\|_\infty$,

$$\int_0^1 \psi(|R_g(y)|)\, dy \le C_1 \left\{\gamma \left[v_\psi\left(\frac{g}{\beta}\right) + \psi(C_2\gamma)\right] + \psi\left(2\gamma\beta^{-2}\omega\left(\frac{10\gamma}{\beta}, G'\right)\right)\right\}.$$

For $\psi(u) \equiv u^p, 1 \le p < \infty$, *we have*

$$\|R_g\|_p \le C\left\{\gamma^{1/p}\|g\|_{[p]} + \frac{\gamma}{\beta^2}\omega\left(\frac{10\gamma}{\beta}, G'\right)\right\}.$$

If G satisfies a stronger smoothness condition, specifically a Hölder condition on G', then, taking $\psi(x) = x^p$, Theorem 2.3 yields the following.

COROLLARY 2.4. *If in addition to the hypotheses of Theorem 2.3 we have* $|G'(x) - G'(y)| \le D(y-x)^{1/p}$ *for* $0 \le x \le y \le 1$, *where* D *is a constant, then there is a constant* $K < \infty$ *such that*

$$\|R_g\|_p \le K\|g\|_{[p]}^{1+1/p}, \quad \text{for } g \in V_p[0,1],\ \|g\|_{[p]} \le 1.$$

Now, an example will show that for a given value of p and any norm $\|\cdot\|$, $\|R_g\|_p \le K_p\|g\|^\alpha$ implies $\alpha \le 1+1/p$, since a remainder in Gâteaux differentiability along one line is of order $t^{1+1/p}$. Thus, by Corollary 2.4, $\|\cdot\| = \|\cdot\|_{[p]}$ achieves the largest possible power of the norm for the remainder in this case.

PROPOSITION 2.5. *Let $f = 1_{[a,b]}, 0 < a < b < 1$. Then for $0 < t < \min(b-a, 1-b)$ and $U(x) \equiv x, 0 \le x \le 1$,*

$$\left\|(U+tf)^{-1} - U + tf\right\|_p = C_p t^{1+1/p},$$

for some constant C_p. Thus, for any normed space $(Y, |\cdot|)$ of functions on $[0,1]$ containing U and $1_{[a,b]}$, if the inverse operator is Fréchet differentiable at U from $(Y,|\cdot|)$ to $L^p[0,1]$ with $\|(U+g)^{-1} - U + g\|_p = O(|g|^\alpha)$ as $|g| \to 0$ in Y, then $\alpha \le 1 + 1/p$.

For a probability distribution function F and its empirical distribution functions F_n, $n^{1/2}\|F_n - F\|_{[p]}$ is bounded in probability uniformly in n and F for $p > 2$ [Dudley (1992a), Corollary 3.8]; another proof is given in Section 5. "Bahadur–Kiefer theorems" treat the size of the remainder R_g in Theorem 2.3 when $G \equiv F$ and $g \equiv g_n \equiv F_n - F$ [Bahadur (1966) and Kiefer (1967, 1970)]. Most of the results treat the case where F is the uniform distribution function U on $[0,1]$. Here, $G' \equiv 1$ on $[0,1]$, so $\omega(\cdot, G') \equiv 0$. So Corollary 2.4 gives $\|R_{g_n}\|_2 \le \|R_{g_n}\|_p = O_p(n^{(-1/2)(1+1/p)})$. Letting $p \downarrow 2$, we get $\|R_{g_n}\|_2 = O_p(n^{-t})$ for any $t < \frac{3}{4}$. A more precise result, not claimed as new, gives the size of the remainder for the L^2 norm in probability:

THEOREM 2.6. *For the L^2 norm of the Bahadur–Kiefer remainder we have $\|R_{g_n}\|_2 = O_p(n^{-3/4})$.*

Kiefer [1967, (1.6)] showed that, for each y, $n^{3/4}R_{g_n}(y)$ has a specific nontrivial limiting distribution (depending on y). It follows that $O_p(n^{-3/4})$ cannot be improved. For the supremum norm, Kiefer (1970) found a limiting distribution for $n^{3/4}\|R_{g_n}\|_\infty/(\log n)^{1/2}$ and proved for $G = U$ that

$$\limsup_{n\to\infty} \|R_{g_n}\|_\infty n^{3/4}(\log n)^{-1/2}(\log\log n)^{-1/4} = 2^{-1/4}$$

almost surely. Shorack (1982) gave another proof. Let $\alpha_n := n^{1/2}(G_n - G)$. Kiefer (1970) proved that $n^{3/4}\|R_{g_n}\|_\infty/(\|\alpha_n\|_\infty \log n)^{1/2} \to 1$ in probability. Shorack (1982) in one direction, and Deheuvels and Mason (1990) in the other, showed that this convergence is almost sure. Thus the asymptotic magnitude of $\|R_{g_n}\|_\infty$ follows from that of $\|\alpha_n\|_\infty$. Kiefer (1970) also proved his results for G with second derivative G'' bounded above as well as G' bounded below by $\beta > 0$, replacing R_{g_n} by $(G' \circ G^{-1})R_{g_n}$. See also Csörgő and Réveś́ [(1978), Theorem 4, page 891].

Now let $\|\cdot\|$ be any norm on a function space containing the centered empirical distribution functions $F_n - F$ and such that $n^{1/2}\|F_n - F\|$ is measurable and does not converge to 0 in probability. (This seems a rather mild assumption on $\|\cdot\|$.) Then if $\|F_n - F\| = O_p(n^{-\beta})$, we have $\beta \leq \frac{1}{2}$. This and the second half of Proposition 2.5 show that for $p > 2$ (unlike $p < 2$) a separation of analytic and probabilistic methods does not given optimal results: For $2 < p < \infty$, Kiefer's correct order of magnitude (in probability),

$$\|U_n^{-1} - U + (U_n - U)\|_p = O_p(n^{-3/4}),$$

does not follow by combining Fréchet differentiability of the inverse operator in the norm $\|\cdot\|$ (even with the best possible power bound for the remainder) with an estimate $\|U_n - U\| = O_p(n^{-\beta})$.

Next, it will be seen that the inverse operator, even at such a smooth distribution F as the uniform, is not differentiable in a neighborhood of F (even in the Gâteaux sense). So the inverse operator is not C^1 and has no second derivative. The last statement (at least) in the following proposition was essentially known [Fernholz (1983), page 66, gives a related example], but it recalls a notable fact.

PROPOSITION 2.7. *Let $U(x) = x$, $0 \leq x \leq 1$, and $f(x) := 1_{[0,1/2]}(x)$. Then, for any fixed $t > 0$, $g \mapsto (U + tf + g)^{-1}$ is not Lipschitz for the L^p norm on the range, $1 < p < \infty$, even along the same direction $g = uf$ as $u \downarrow 0$, and so is not Gâteaux differentiable in that direction. For $p = \infty$ the operator is not continuous. For $p = 1$ it is Lipschitz along the given line but still not Gâteaux differentiable. Allowing also $t < 0$ and for fixed $y = \frac{1}{2}$, $g \mapsto (U + g)^{-1}(y)$ is not Gâteaux differentiable along the line $g = tf$, $t \to 0$.*

So, by the way, despite the remainder bound, the inverse operator on $D[0,1]$ at U does not satisfy (in any norm) the definition of "smooth" given in Wong and Severini [(1991), page 610].

Next are easy examples showing non-Fréchet differentiability of the composition and inverse operators with respect to the sup norm, variously on domains and/or ranges. Such examples must have been known although I do not have references for them.

PROPOSITION 2.8. *At $F = G = U[0,1]$, the following hold.*

(a) *The composition operator is not jointly Gâteaux differentiable on $D[0,1]$ for the* sup *norm on the range: for some $f, g \in D[0,1]$,*

$$\|tf \circ (G + sg) - tf \circ G\|_\infty \neq o(|t| + |s|) \quad \text{as } s, t \to 0.$$

(b) *The composition operator is not jointly differentiable into L^2 for the* sup *norm on the domain*

$$\|f \circ (G + g) - f \circ G\|_2 \neq o(\|f\|_\infty + \|g\|_\infty) \quad \text{as } \|f\|_\infty + \|g\|_\infty \to 0.$$

(c) *The inverse operator is not Fréchet differentiable at G for the* sup *norm on the domain and the L^2 (or* sup*) norm on the range*

$$\|(G + g)^{-1} - G^{-1} + g\|_2 \neq o(\|g\|_\infty) \quad \text{as } \|g\|_\infty \to 0.$$

3. Kinds of differentiability. This section will list several forms of differentiability. The question will be which forms are better adapted to proving asymptotic normality of suitable functionals of the empirical d.f. F_n, say on $[0, 1]$. Let us consider differentiation at a sufficiently regular distribution such as the uniform distribution $U[0, 1]$. Three kinds of Fréchet differentiability are as follows:

(FF) for functionals with finite-dimensional values, Fréchet differentiability for functions between finite-dimensional spaces;
(FP) Fréchet differentiability for norms $\|\cdot\|$ stronger than the sup norm, such as p-variation norms, on subspaces of $D[0, 1]$ such that $\|F_n - F\|$ is finite, or better still $O_p\left(n^{-1/2}\right)$;
(FS) Fréchet differentiability for the sup norm on $D[0, 1]$.

Here (FS), when it holds, always implies (FP), but neither necessarily implies nor is implied by (FF), since no relation is assumed between the norm on the finite-dimensional space and any norm on distribution functions.

(FF) has the advantage, when it holds, that in finite dimensions we can often hope for more than one derivative, giving multiterm Taylor expansions, while the nonlinear functionals of interest on infinite-dimensional spaces tend to have no more than one derivative, as seen in Section 2.

Three kinds of differentiability of compact type are as follows:

(CC) compact differentiability in $C[0, 1]$ for the sup norm, applied to empirical distribution functions by replacing them, for example, by piecewise-linear rather than piecewise-constant functions;
(CD) compact differentiability in $D[0, 1]$ for the sup norm, applied by way of almost surely convergent realizations;
(CR) $\mathcal{C}$-differentiability, where $\mathcal{C}$ may be the class of sets compact for the Skorohod topology, or the still larger uniformly regulated or uniformly Riemann sets as defined in Dudley (1991).

Here (CR) implies (CD) implies (CC) and neither converse holds. All the compact forms share the disadvantage of slow convergence of remainders to 0, as shown in Proposition 2.1, when Fréchet differentiability fails. When all three hold, (CR) has the advantage that normalized empirical distribution functions $n^{1/2}(F_n - F)$ belong, with high probability uniformly in n, to sets of the classes mentioned in (CR), without the need for a.s. convergent realizations as in (CD) or smoothing as in (CC).

The forms of differentiability will be compared for some functionals of interest.

3.1. *Specific quantiles and the sup norm on the range.* The functional $F \mapsto F^{-1}(y)$ for a fixed y, as seen in the last statement of Proposition 2.7, is not even Gâteaux differentiable on $D[0, 1]$, so (CD) and (FP) and the stronger forms (FS) and (CR) must all fail. (FF) applies as follows: If F and G are any two distribution functions with $F(x_\alpha) = G(u_\alpha) = \alpha$, $F'(x_\alpha) > 0$ and $G'(u_\alpha) > 0$, then asymptotic normality of the sample α quantile of F follows from that for

G by the one-dimensional delta-method. One interesting choice of G is standard exponential, when the order statistics are partial sums of independent exponential variables with different scale parameters, as in a classic paper of Rényi (1953). Rényi's representation helps to treat the joint distribution of several order statistics. The more standard choice is to take G as uniform $U[0,1]$. Then the distribution function at x of the rth order statistic for sample size n and a general distribution function F is the beta$(r, n-r+1)$ distribution function at $F(x)$ [e.g., Kendall and Stuart (1977), Section 14.2, page 347]. The simple normal approximation to the beta can be much improved by a Cornish–Fisher expansion [e.g., Pratt (1968), page 1467, Molenaar (1970), page 72, or Holt (1986)]. The expansion can then be combined with a Taylor expansion of F around x_α.

The compact differentiability (CC) also applies to individual quantiles as shown by Reeds [(1976), page 127] and Esty, Gillette, Hamilton and Taylor (1985). Gill [(1989), page 107] proves compact differentiability in D, tangentially to C, a form intermediate between (CC) and (CD) which applies to empirical distributions. The compact forms seem to give less precise information than (FF).

The supremum norm on the range of the operator $F \mapsto F^{-1}$ can be treated in the same way if one restricts to an interval $[a, b]$, $0 < a < b < 1$ [e.g., Esty, Gillette, Hamilton and Taylor (1985) and Gill (1989)]. It has also been treated on $(0, 1)$ via classical (stochastic) methods [e.g., Csörgő and Révész (1978)].

3.2. *The inverse operator, and composition operator with respect to f, for L^p norms on the range.* Reeds showed that here (CD) holds while (FS) fails. We now have that (FP) holds with the error bounds given in Section 2. Also, all the given forms of (CR) hold for these operators [Dudley (1991)]. The main theme of the present paper is that (FP) is preferable because of the remainder bounds.

3.3. *The operator $(F, G) \mapsto \int F\, dG$.* Gill [(1989), pages 110–111] showed that (CD) applies in this case, while (FS) fails. Now, (FP) also holds, much as in 3.2, and gives error bounds stronger than those from (CD) and nearly but not quite the strongest possible bounds [Dudley (1992b), Corollary 3.5]. The conclusion only follows for some (not all) p-variation norms on F and G. It would seem then that the (CR) forms could only apply to one of F, G, while stronger conditions are put on the other; I have at this point no precise result to report for (CR) in this case.

3.4. *The product integral.* This integral had been applied in other fields, especially differential equations [Dollard and Friedman (1979)], but seems to have been almost unknown to statisticians until recent years, when Gill and Johansen (1990) proved that (CD) applies under suitable conditions. As with Reeds' work on the composition and inverse operators, it seems that again statisticians were the first to prove a functional differentiability fact which should be of interest to nonlinear analysts. Gill and Johansen's work was an outstanding success for compact differentiability.

The product integral is related to $\int F\,dG$, where F and G are operator-valued functions, on which see Krabbe (1961). Freedman [(1983), Theorem 5.1] shows that a product integral exists for operator-valued functions of bounded p-variation for $p < 2$. I do not know at this writing whether the (FP) differentiability holds for some p.

In the (many) cases where Fréchet differentiability has not (yet) been proved for suitable norms, there are a number of useful positive results for compact differentiability, such as validity of the bootstrap [van der Vaart and Wellner (1994)] and preservation of asymptotic efficiency of estimators of (infinite-dimensional) parameters [van der Vaart (1991)]. About efficiency, slow remainders (Proposition 2.1) may be a concern.

One useful property of a norm is the central limit theorem (Donsker property) which has been proved for the ordinary empirical process in p-variation norms for some values of p in Dudley (1992a). Vervaat (1972) showed that, for any sequence $\{G_n\}_{n\geq 1}$ of nondecreasing stochastic processes on $[0,\infty)$, $I(t) \equiv t$, and any sequence $[\delta_n]$ of positive random variables converging to 0 in probability, $(G_n - I)/\delta_n$ converges weakly in $D[0,\infty)$ to a continuous stochastic process ξ with $\xi(0) = 0$ if and only if $(G_n^{-1} - I)/\delta_n$ converges weakly in the same space to $-\xi$. Whitt (1980) noted that $\xi(0) = 0$ was needed. Vervaat and later others applied his and further results to processes including partial sum and counting processes. I do not know at this writing what can be done with p-variation in such situations.

4. Proofs. We first have the following.

PROOF OF PROPOSITION 2.1. Since T is not Fréchet differentiable, there are $\delta > 0$ and $u_k \in X$ with $\|u_k\| \leq 1$ such that $u_k \to 0$ as $k \to \infty$ and

$$|T(x + u_k) - T(x) - T'(x)(u_k)| \geq \delta\|u_k\|.$$

We can assume that $\delta < 1$ and $\varepsilon_n \geq 1/n$ for all n. Take a subsequence u_{k_n} such that $\|u_{k_n}\| \leq 1/n^2$ for all n. Let $t_n := \|u_{k_n}\|\delta/\varepsilon_n$. Then $0 < t_n \leq \delta/n < 1/n$. Let $v_n := u_{k_n}/t_n$. Then $\|v_n\| = \varepsilon_n/\delta \to 0$, and

$$\|T(x + t_n v_n) - T(x) - t_n T'(x)(v_n)\| \geq \varepsilon_n t_n. \qquad \Box$$

Next, to help in the proof of Theorem 2.2, we have the following.

LEMMA 4.1. *Let f be a real-valued function on $\mathbb{R}$ and let G be a real-valued function on $[0,1]$. Suppose that, for some $\beta > 0$, $G(y) - G(x) \geq \beta(y - x)$ for $0 \leq x \leq y \leq 1$. Let J be the smallest integer greater than or equal to $1/\beta$. Let m be a positive integer, $k = 1, \ldots, m$, and $I_{m,k} := [(k-1)/m, k/m[$. Then, for any Orlicz function ψ,*

$$\sum_{k=J+1}^{m-J} \sup\left\{\psi\left(f(G(x)) - f(G(x) + g_k)\right): g_k \leq \frac{1}{m}, x \in I_{m,k}\right\} \leq (2J + 1)v_\psi(f).$$

PROOF. If $x_i \in I_{m,i}$, $i = k,j$, and $j - k \geq 2J + 1$, then $x_j - x_k \geq 2J/m$ and $G(x_j) - G(x_k) \geq 2/m$, so the intervals with endpoints $G(x_i)$ and $G(x_i) + g_i$ are nonoverlapping for $i = j,k$. It follows that the sum in the statement restricted to indices $J+1, 3J+2, 5J+3, \ldots$ is at most $v_\psi(f)$. Likewise for the restrictions to indices $J+i, J+i+2J+1, J+i+4J+2, \ldots$, for $i = 1, 2, \ldots, 2J+1$, and the conclusion follows. □

Recall that $v_\psi(f) = v_p(f) = \|f\|_{(p)}^p$ when $\psi(u) \equiv u^p$.

PROOF OF THEOREM 2.2. For $f \in V_p(\mathbb{R})$, f is continuous except for at most countably many jumps, so f is Borel measurable, and functions $f(G)$ and $f(G+g)$ are measurable. Let $\delta := \|g\|_s$ and $\alpha := s/(1+s)$. We can assume $\delta \leq 1$, or take $C = 2$. Then $|g| \leq \delta^\alpha$ on a set $A \subset [0,1]$ whose complement has Lebesgue measure at most $\delta^{s(1-\alpha)} = \delta^\alpha$. The same holds if $\delta = \rho_s(0,g)$. Take a positive integer m such that $1/(2m) \leq \delta^\alpha \leq 1/m$. We can assume that $\beta \leq 1$. Let J be the smallest integer greater than or equal to $1/\beta$ and $B := A \cap [J/m, 1 - J/m]$. Now

$$\int_{[0,1]\setminus B} |f(G+g) - f(G)|^p(x)\, dx \leq 2^{\,p} \|f\|_\infty^p \left(\delta^\alpha + \frac{2J}{m}\right). \tag{4.1}$$

Also, letting $B_{m,k} := B \cap I_{m,k}$, with $I_{m,k} := [(k-1)/m, k/m[$ as before, we have

$$\begin{aligned}
\int_B |f(G+g) - f(G)|^p\, dx &= \sum_{k=J+1}^{m-J} \int_{B_{m,k}} |f(G+g) - f(G)|^p\, dx \\
&\leq \sum_{k=J+1}^{m-J} \frac{1}{m} \sup\left\{|f(G+g) - f(G)|^p(x) \colon x \in B_{m,k}\right\} \\
&\leq \left(\frac{2J+1}{m}\right) \|f\|_{(p)}^p \quad \text{(by Lemma 4.1)} \\
&\leq (2J+1)\, 2\delta^\alpha \|f\|_{(p)}^p.
\end{aligned}$$

It follows that $\int_0^1 |f(G+g) - f(G)|^p\, dx \leq C_1 \|g\|_s^\alpha \|f\|_{[p]}^p$, where

$$C_1 := 4J + 2 + 2^p(1 + 4J) \leq \frac{4}{\beta} + 6 + 2^p\left(5 + \frac{4}{\beta}\right) \leq 2^p\left(11 + \frac{8}{\beta}\right),$$

so the first inequality in the statement holds, with $C := 22 + 16/\beta$.

Next, for any $A, B, \rho > 0$, we have $AB^\rho \leq A^{1+\rho} + B^{1+\rho}$ since $1 \leq (A/B)^\rho + B/A$. The second inequality in the statement then follows. Also, we have $\rho_p(0,\eta) \leq \|\eta\|_p$ for any L^p function η.

Now in the converse direction, take $f = th$ as $t \to 0$. For fixed g, we get $\gamma \leq 1$. Then let $g = -\delta 1_{[0,\delta]}$ as $\delta \downarrow 0$, so $\|g\|_s = \rho_s(0,g) = \delta^{1+1/s}$ and

$$\rho_p(f \circ (U+g), f \circ U) = \|f \circ (U+g) - f \circ U\|_p = \delta^{1/p}.$$

The conclusions follow. □

PROOF OF THEOREM 2.3. For any real function f on an interval $[a,b]$ let $\text{osc}_{[a,b]}f := (\sup - \inf)_{[a,b]}(f)$. For a given y let $\xi := G^{-1}(y)$. Take a positive integer m such that $1/(2m) \le \gamma := \|g\|_\infty \le 1/m$. The assumptions imply that $\beta \le 1$. Let $a := \xi - 1/(m\beta)$ and $b := \xi + 1/(m\beta)$. If $a \ge 0$ and $b \le 1$, then by Dudley [(1991), Lemma 4.2],

$$|R_g(y)| \le \text{osc}_{[a,b]}(g)/\beta + \text{osc}_{[a,b]}G'/(m\beta^2). \tag{4.2}$$

Again, consider the intervals $I_{m,k} := [(k-1)/m, k/m[$. Let J be the least integer greater than or equal to $1/\beta$. Suppose $J+1 \le k \le m-J$. Then by (4.2), for $y \in I_{m,k}$ and $I(m,k,J) := [(k-1-J)/m, (k+J)/m[$, we have $a \ge 0$, $b \le 1$ and

$$\begin{aligned} |R_g(y)| &\le \beta^{-1}\text{osc}_{I(m,k,J)}\, g + \text{osc}_{I(m,k,J)}G'/(m\beta^2) \\ &\le \beta^{-1}\text{osc}_{I(m,k,J)}\, g + \omega((2J+1)/m, G')/(m\beta^2), \end{aligned} \tag{4.3}$$

and we note that $(2J+1)/m \le (3+2/\beta)/m$. Recall that ψ satisfies Δ_2 and is therefore quasi-subadditive. Then as in the proof of Lemma 4.1, there is a $K < \infty$ such that

$$\sum_{k=J+1}^{m-J} \sup\{\psi(|R_g(y)|): y \in I_{m,k}\} \le K\left\{(2J+1)v_\psi\left(\frac{g}{\beta}\right) + m\psi(\Delta)\right\},$$

where $\Delta := \omega((3+2/\beta)/m, G')/(m\beta^2)$. Thus

$$\int_{J/m}^{1-J/m} \psi(|R_g(y)|)\,dy \le Km^{-1}(2J+1)v_\psi(g/\beta) + K\psi(\Delta).$$

Noting that $m^{-1} \le 2\gamma$, the latter integral is bounded as in the statement of the theorem. Next, for $0 \le y \le J/m$, we have $0 \le (G+g)^{-1}(y) \le (J+1)/(\beta m)$ since $G(x) \ge \beta x$, so $(G+g)(x) \ge \beta x - 1/m \ge J/m$ for $x \ge (J+1)/(\beta m)$.

Also, $0 \le G^{-1}(y) \le J/(\beta m)$, so $|(G+g)^{-1}(y) - G^{-1}(y)| \le (J+1)/(\beta m)$, while $\sup|(g \circ G^{-1})/(G' \circ G^{-1})| \le 1/(m\beta)$, so $|R_g(y)| \le (J+2)/(m\beta) \le (\beta^{-1}+3)/(m\beta)$. Thus

$$\int_0^{J/m} \psi(|R_g(y)|)\,dy \le \left(\frac{J}{m}\right)\psi\left(\frac{\beta^{-1}+3}{m\beta}\right) \le \frac{4\|g\|_\infty}{\beta}\psi\left(\left(\frac{6}{\beta}+\frac{2}{\beta^2}\right)\|g\|_\infty\right).$$

Since $1/m \le 2\gamma$ and a similar bound holds for the interval $[1-J/m, 1]$, the first statement in the theorem follows. Then for $\psi(u) \equiv u^p$, taking pth roots and since $(A+B)^{1/p} \le A^{1/p} + B^{1/p}$ for $A, B \ge 0, 1 \le p < \infty$, Theorem 2.3 is proved. □

PROOF OF PROPOSITION 2.5. It is easy to check that, for $0 < y < 1$ and $0 < t < \min(1-b, b-a)$,

$$\text{Rem} := (U+tf)^{-1}(y) - y + tf(y) = \begin{cases} -t, & b < y < b+t, \\ a-y+t, & a < y < a+t, \\ 0, & \text{otherwise} \end{cases}$$

(except at finitely many endpoints). Thus

$$\|\text{Rem}\|_p = \left(t^{p+1} + \int_0^t u^p\,du\right)^{1/p} = \left(1 + \frac{1}{p+1}\right)^{1/p} t^{1+1/p}.$$

The second part of the proposition then follows directly. □

PROOF OF THEOREM 2.6. Apply (4.3), where now $\beta = J = 1$ and G' is constant. So we have, for $2 \le k \le m-1$ and $y \in [(k-1)/m, k/m[=: I(m,k)$, that $|R_g(y)| \le \text{osc}_{I(m,k,1)} g$. Let $g = F_n - F$ and square both sides. I claim that for the uniform distribution as here, $F(x) \equiv x$ for $0 \le x \le 1$, the distribution of $\text{osc}_{[a,b]} g$ for $0 \le a < b \le 1$ depends only on $b-a$. The number s of observations X_i in $[a,b]$ is binomial$(n, b-a)$. Given s, the observations are distributed uniformly and independently in $[a,b]$. F_n has a jump of height $1/n$ at each X_i, and $-F$ is decreasing linearly with slope -1 on the interval. The claim follows.

So, consider $k = 2$, and set $[a,b] = A := I(m,2,1) = [0, 3/m]$. We have

$$\begin{aligned}(\text{osc}_A(F_n - F))^2 &= \left(\text{osc}_A\left(F_n - \frac{ms}{3n}F + \left(\frac{ms}{3n} - 1\right)F\right)\right)^2 \\ &\le 2\left(\text{osc}_A\left(F_n - \frac{sm}{3n}F\right)\right)^2 + 2\left(\frac{ms}{3n} - 1\right)^2\left(\frac{9}{m^2}\right).\end{aligned}$$

The expectation of the latter term is $(6/mn)(1-3/m)$. Now $F_n - msF/(3n)$ can be written as $(s/n)(G_s - G)$, where G is the uniform distribution on A. Thus, by the Dvoretzky–Kiefer–Wolfowitz inequality [Dvoretzky, Kiefer and Wolfowitz (1956); see also Shorack and Wellner (1986), page 354], there is an absolute constant C such that the conditional expectation of the former term given s is at most Cs/n^2, so its expectation is less than $3C/(mn)$. So for a constant C', $E((\text{osc}_A(F_n - F))^2) \le C'/(mn)$ and the sum over $m-2$ intervals is at most C'/n. We can deal with the intervals $[0, 1/m[$ and $[1-1/m, 1[$ as in the proof of Theorem 2.3, getting an upper bound of $K\|g\|_\infty^3$ for a constant K. So

$$E\left(\int_0^1 R_{g_n}(y)^2\,dy\right) = O(n^{-3/2}) \quad\text{and}\quad \|R_{g_n}\|_2 = O_p(n^{-3/4}). \qquad \square$$

PROOF OF PROPOSITION 2.7. Let $b := \frac{1}{2}$. We have for $t, u > 0$,

$$(U + tf)^{-1}(y) = \begin{cases} y, & b + t \le y \le 1, \\ y - t, & t \le y < b + t, \\ 0, & 0 \le y < t; \end{cases}$$

$$(U + (t+u)f)^{-1}(y) - (U + tf)^{-1}(y) = \begin{cases} 0, & b + t + u < y \le 1, \\ -t - u, & b + t < y < b + t + u, \\ -u, & t + u < y < b + t, \\ t - y, & t < y < t + u, \\ 0, & 0 \le y < t. \end{cases}$$

As $u \downarrow 0$ for fixed $t > 0$, we see from the range $b + t < y < b + t + u$ that the operator is not continuous in the $\|\cdot\|_\infty$ norm. The difference equals $-u 1_{\{t<y<b+t\}}$, which is linear in u, plus a term $\eta(t,u,y) = (-t-u)1_{\{b+t<y<b+t+u\}}$, plus another term on a disjoint interval. For $1 \le p < \infty$ we then have $\|\eta(t,u,\cdot)\|_p \ge tu^{1/p}$ (so the "remainder" is larger than the "derivative"!). Thus, for $p > 1$, the inverse operator is not Lipschitz and is not Gâteaux differentiable at $U + tf$. For $p = 1$, suppose it were Gâteaux differentiable. Then, for some function $\gamma(t,\cdot)$,

$$\|\eta(t,u,\cdot) - u\gamma(t,\cdot)\|_1 = o(|u|) \quad \text{as } |u| \to 0.$$

Then $\gamma(t,y) = 0$ (almost everywhere) for $y > b+t$, but this yields a contradiction.

Lastly, $(U + tf)^{-1}\left(\frac{1}{2}\right) \equiv \frac{1}{2}$ for $t < 0$ and $\frac{1}{2} - t$ for $t > 0$, showing non-differentiability at $t = 0$. □

PROOF OF PROPOSITION 2.8. (a) Let $f \equiv 1_{[0,1/2]}$, $g \equiv 1$ and $s \downarrow 0$. Then $|f(G+s)(x) - f(x)| = 1$ for $\frac{1}{2} - s < x < \frac{1}{2}$.

(b) For $n = 1, 2, \ldots$, let $g_n(x) = 1/(2n)$ for $j/n \le x < (2j+1)/(2n)$, $j = 0, \ldots, n-1$, and $g_n(x) = 0$ otherwise. Let $f_n = g_n$. Then $\|f_n\|_\infty = \|g_n\|_\infty = 1/(2n)$, $f_n \circ (G + g_n) \equiv 0$ and $\|f_n \circ G\|_2 = 1/(2^{3/2}n)$.

(c) For the same g_n and for $(2j+1)/(2n) < y < (j+1)/n, j = 0, \ldots, n-1$, $(G + g_n)^{-1}(y) = y - 1/(2n)$ while $g_n(y) = 0$, so

$$\|(G + g_n)^{-1} - G^{-1} + g_n\|_2 \ge 1/(2^{3/2}n). \qquad \square$$

5. Orlicz variation and martingales. This section will give a proof [quite different from the one in Dudley (1992a), Corollary 3.8] that $n^{1/2}\|F_n - F\|_{(p)}$ is bounded in probability uniformly in n for $p > 2$, and we also mention an Orlicz function to come as close as possible to 2-variation. [In the proof of Theorem 2.6, a kind of 2-variation was finite because the lengths of the intervals, $1/m = O(n^{-1/2})$, went to 0 fast enough.]

Let $\psi_1(u) := u^2/\log\log(1/u)$, for $0 < u \le e^{-e}$. It can be checked by derivatives that ψ_1 can be defined for $u > e^{-e}$ to be an Orlicz function satisfying Δ_2. For the Brownian process $X_\bullet: t \mapsto X_t$ on a bounded interval $0 \le t \le T < \infty$, Taylor (1972) showed that $v_{\psi_1}(X_\bullet) < \infty$ a.s., while if $\psi_1(u)/\psi(u) \mapsto 0$ as $u \downarrow 0$, for example, $\psi(u) \equiv u^2/(\log\log(1/u))^\alpha$, $\alpha < 1$, then $v_\psi(X_\bullet) = +\infty$ a.s.

Monroe (1972) showed that any right-continuous martingale process M_t having left limits can be written as X_{T_t} for an increasing family of stopping times $T_t < \infty$. Thus any such M_t has bounded ψ_1-variation on bounded intervals $0 \le t \le s$ a.s. [Monroe (1976), page 134]. If, moreover, the martingale M_t has mean 0 and $EM_s^2 < \infty$, then $ET_s = EM_s^2$ [Monroe (1972), Theorems 5 and 11]. Thus for martingales $M^{(n)}$ on $0 \le t \le s$ having $E((M_s^{(n)})^2)$ uniformly bounded, the stopping times T_s will be bounded in probability, uniformly in n, and likewise for $T_t \le T_s, 0 \le t \le s$. So the ψ_1-variations of $M^{(n)}$ will be bounded in probability uniformly in n.

Let F_n be empirical distribution functions of the uniform distribution $F(t) = t, 0 \le t \le 1$. For each n, $M^{(n)}(t) := n^{1/2}(F_n(t) - t)/(1-t), 0 \le t < 1$, is a martingale [e.g., Shorack and Wellner (1986), page 4 and page 133, Proposition 1]. Then $M^{(n)}$ has mean 0 and variances bounded for $0 \le t \le s := \frac{1}{2}$, uniformly in n. Likewise, symmetrically, $n^{1/2}(F_n(t) - t)/t, \frac{1}{2} \le t \le 1$, are reversed martingales with uniformly bounded variances in the given range.

Now the following will be helpful. Krabbe (1961) treats the p-variation case. Lacking a reference for general ψ-variation, I will sketch a proof:

LEMMA 5.1. *Let ψ be any Orlicz function satisfying Δ_2. Then the functions of bounded ψ-variation on an interval form an algebra, and there is a $K < \infty$ with*

$$v_\psi(gh) \le K(v_\psi(\|g\|_\infty h) + v_\psi(\|h\|_\infty g)),$$

for any functions g and h. For $\psi(u) \equiv u^p, 1 < p < \infty$, we have

$$\|gh\|_{(p)} \le \|h\|_\infty \|g\|_{(p)} + \|g\|_\infty \|h\|_{(p)},$$

so that $\|gh\|_{[p]} \le \|g\|_{[p]} \|h\|_{[p]}$, for any two functions g and h.

PROOF. For any points x and y, we have

$$\begin{aligned}|(gh)(y) - (gh)(x)| &= |h(y)(g(y) - g(x)) + g(x)(h(y) - h(x))| \\ &\le \|h\|_\infty |g(y) - g(x)| + \|g\|_\infty h(y) - h(x)|.\end{aligned}$$

Since the Δ_2 condition implies quasi-subadditivity, ψ can be distributed over the sum, with a constant K. Then, taking sums of ψ of such increments over nonoverlapping intervals, the result for v_ψ follows.

For $\psi(u) \equiv u^p, 1 < p < \infty$, the result follows instead from Minkowski's inequality. □

Now apply Lemma 5.1 to $g(t) = M^{(n)}(t)$ and $h(t) = 1 - t$, and set $\alpha_n(t) := n^{1/2}(F_n(t) - t) \equiv (gh)(t)$. For any function f on an interval $[a, b]$ and Orlicz function ψ, denote the ψ-variation of f on $[a, b]$ by $v_\psi(f)_{[a,b]}$. Then

$$v_\psi(f)_{[0,1]} \le v_\psi(f)_{[0,1/2]} + v_\psi(f)_{[1/2,1]} + \psi((\sup - \inf)_{[0,1]} f),$$

and $v_{\psi_1}(\alpha_n)_{[0,1/2]}$ is uniformly bounded in probability. Symmetrically, the same is true for $[\frac{1}{2}, 1]$. Thus $v_{\psi_1}(\alpha_n)$ are uniformly bounded in probability. For $r > 2, |u|^r = o(\psi_1(u))$ as $u \to 0$. It follows that, for each $r > 2$, $\|\alpha_n\|_{[r]}$ are also uniformly bounded in probability, completing the alternate proof of Dudley [(1992a), Corollary 3.8].

I have a proof that $v_{\psi_1}(F_n - F) = O_p(n^{-3/2})$ and, consequently, that in the situation of Theorem 2.6, $\|R_{g_n}\|_2 = O_p((\log \log n)^{1/2}/n^{3/4})$. The proof is omitted since this is a little weaker than Theorem 2.6, which in any case is essentially a known fact.

Lepingle (1976) also treated r-variation of martingales.

Acknowledgment. I thank Gilles Pisier for pointing out the possibility of martingale proofs of the boundedness in probability of $\|\alpha_n\|_{[p]}, p > 2$. He has told me of another proof which uses the notion of "Type 2," as mentioned in Dudley (1992a) and references there.

REFERENCES

APPELL, J. and ZABREJKO, P. P. (1990). *Nonlinear Superposition Operators*. Cambridge Univ. Press.

BAHADUR, R. R. (1966). A note on quantiles in large samples. *Ann. Math. Statist.* **37** 577–580.

BICKEL, P. J., KLAASSEN, C. A. J., RITOV, Y. and WELLNER, J. A. (1993). *Efficient and Adaptive Estimation for Semiparametric Models*. Johns Hopkins Univ. Press.

BROKATE, M. and COLONIUS, F. (1990). Linearizing equations with state-dependent delays. *Appl. Math. Optim.* **21** 45–52.

CSÖRGŐ, M. and RÉVÉSZ, P. (1978). Strong approximations of the quantile process. *Ann. Statist.* **6** 882–894.

DEHEUVELS, P. and MASON, D. M. (1990). Bahadur–Kiefer–type processes. *Ann. Probab.* **18** 669–697.

DOLLARD, J. D. and FRIEDMAN, C. N. (1979). *Product Integration with Applications to Differential Equations*. Addison-Wesley, Reading, MA.

DUDLEY, R. M. (1989). *Real Analysis and Probability*. Wadsworth and Brooks/Cole, Pacific Grove, CA.

DUDLEY, R. M. (1991). Differentiability of the composition and inverse operators for regulated and a.e. continuous functions. Unpublished manuscript.

DUDLEY, R. M. (1992a). Fréchet differentiability, p-variation and uniform Donsker classes. *Ann. Probab.* **20** 1968–1982.

DUDLEY, R. M. (1992b). Empirical processes: p-variation for $p \leq 2$ and the quantile–quantile and $\int F\, dG$ operators. Unpublished manuscript.

DVORETZKY, A., KIEFER, J. and WOLFOWITZ, J. (1956). Asymptotic minimax character of the sample distribution function and of the classical multinomial estimator. *Ann. Math. Statist.* **27** 642–669.

ESTY, W., GILLETTE, R., HAMILTON, M. and TAYLOR, D. (1985). Asymptotic distribution theory of statistical functionals. *Ann. Inst. Statist. Math.* **37** 109–129.

FERNHOLZ, L. T. (1983). *Von Mises Calculus for Statistical Functionals. Lecture Notes in Statist.* **19**. Springer, New York.

FILIPPOVA, A. (1961). Mises' theorem on the asymptotic behavior of functionals of empirical distribution functions and its statistical applications. *Theory Probab. Appl.* **7** 24–57.

FREEDMAN, M. A. (1983). Operators of p-variation and the evolution representation problem. *Trans. Amer. Math. Soc.* **279** 95–112.

GILL, R. D. (1989). Non- and semi-parametric maximum likelihood estimators and the von Mises method (Part 1). *Scand. J. Statist.* **16** 97–128.

GILL, R. D. and JOHANSEN, S. (1990). A survey of product-integration with a view toward application in survival analysis. *Ann. Statist.* **18** 1501–1555.

GRAY, A. (1975). Differentiation of composites with respect to a parameter. *J. Austral. Math. Soc. Ser. A* **19** 121–128.

HAMPEL, F. R., RONCHETTI, E. M., ROUSSEEUW, P. J. and STAHEL, W. A. (1986). *Robust Statistics: The Approach Based on Influence Functions*. Wiley, New York.

HOLT, R. J. (1986). Computation of gamma and beta tail probabilities. Technical report, Dept. Mathematics, MIT.

HUBER, P. J. (1981). *Robust Statistics*. Wiley, New York.

KENDALL, M. G. and STUART, A. (1977). *The Advanced Theory of Statistics* **1**. *Distribution Theory*, 4th ed. Macmillan, New York.

KIEFER, J. (1967). On Bahadur's representation of sample quantiles. *Ann. Math. Statist.* **38** 1323–1342.

KIEFER, J. (1970). Deviations between the sample quantile process and the sample df. In *Nonparametric Techniques in Statistical Inference* (M. L. Puri, ed.) 299–319. Cambridge Univ. Press.

KRABBE, G. L. (1961). Integration with respect to operator-valued functions. *Bull. Amer. Math. Soc.* **67** 214–218.

KRASNOSEL'SKII, M.A. and RUTICKII, YA. B. (1961). *Convex Functions and Orlicz Spaces*. Noordhoff, Groningen. (Translated by L. F. Boron.)

LEPINGLE, D. (1976). La variation d'ordre p des semi-martingales. *Z. Wahrsch. Verw. Gebiete* **36** 295–316.

MOLENAAR, W. (1970). *Approximations to the Poisson, Binomial and Hypergeometric Distribution Functions*. Math. Centrum, Amsterdam.

MONROE, I. (1972). On embedding right continuous martingales in Brownian motion. *Ann. Math. Statist.* **43** 1293–1311.

MONROE, I. (1976). Almost sure convergence of the quadratic variation of martingales: A counterexample. *Ann. Probab.* **4** 133–138.

MUSIELAK, J. and ORLICZ, W. (1959). On generalized variations (I). *Studia Math.* **18** 11–41.

PRATT, J. W. (1968). A normal approximation for binomial, F, beta, and other common, related tail probabilities II. *J. Amer. Statist. Assoc.* **63** 1457–1483.

REEDS, J. A., III (1976). On the definition of von Mises functionals. Ph.D. dissertation, Dept. Statistics, Harvard Univ.

RÉNYI, A. (1953). On the theory of order statistics. *Acta Math. Acad Sci. Hungar.* **4** 191–227.

SEBASTIÃO E SILVA, J. (1956). Le calcul différentiel et intégral dans les espaces localement convexes, réels ou complexes I, II. *Rend. Accad. Lincei Sci. Fis. Mat.* (8) **20** 743–750; **21** 40–46.

SHORACK, G. R. (1982). Kiefer's theorem via the Hungarian construction. *Z. Wahrsch. Verw. Gebiete* **61** 369–373.

SHORACK, G. R. and WELLNER, J. A. (1986). *Empirical Processes with Applications to Statistics*. Wiley, New York.

TAYLOR, S. J. (1972). Exact asymptotic estimates of Brownian path variation. *Duke Math. J.* **39** 219–241.

VAN DER VAART, A. (1991). Efficiency and Hadamard differentiability. *Scand. J. Statist.* **18** 63–75.

VAN DER VAART, A., and WELLNER, J. (1994). *Weak Convergence and Empirical Processes*. IMS, Hayward, CA. To appear.

VERVAAT, W. (1972). Functional central limit theorems for processes with positive drift and their inverses. *Z. Wahrsch. Verw. Gebiete* **12** 245–253.

VON MISES, R. (1936). Les lois de probabilité pour les fonctions statistiques. *Ann. Inst. H. Poincaré* **6** 185–212.

VON MISES, R. (1947). On the asymptotic distribution of differentiable statistical functions. *Ann. Math. Statist.* **18** 309–348.

WHITT, W. (1980). Some useful functions for functional limit theorems. *Math. Oper. Res.* **5** 67–85.

WONG, W. H. and SEVERINI, T. A. (1991). On maximum likelihood estimation in infinite dimensional parameter spaces. *Ann. Statist.* **19** 603–632.

YOUNG, L. C. (1936). An inequality of the Hölder type, connected with Stieltjes integration. *Acta Math.* **67** 251–282.

DEPARTMENT OF MATHEMATICS
ROOM 2–245
MASSACHUSETTS INSTITUTE OF TECHNOLOGY
CAMBRIDGE, MASSACHUSETTS 02139

13
Empirical Processes and p-variation

R. M. Dudley[1]

ABSTRACT Remainder bounds in Fréchet differentiability of functionals for p-variation norms are found for empirical distribution functions. For $1 \leq p \leq 2$ the p-variation of the empirical process $n^{1/2}(F_n - F)$ is of order $n^{1-p/2}$ in probability up to a factor $(\log \log n)^{p/2}$. For $(F, G) \mapsto \int F dG$ and for $(F, G) \mapsto F \circ G^{-1}$ this yields nearly optimal remainder bounds. Also, p-variation gives new proofs for the asymptotic distributions of the Cramér-von Mises-Rosenblatt and Watson two-sample statistics when the two sample sizes m, n go to infinity arbitarily.

13.1 Introduction

This paper is a continuation of Dudley (1994), in which specific remainder bounds are found for the differentiation of operators on distribution functions with special regard for empirical distribution functions.

Let F be a probability distribution function and F_n an empirical distribution function for it. Then $n^{1/2}(F_n - F)$ is an *empirical process*. The p-variation of the process (defined in Section 2) is bounded in probability as $n \to \infty$ if and only if $p > 2$: Dudley (1992, Corollary 3.8; 1994). Theorem 2 below shows that for $1 \leq p \leq 2$, the p-variation of the empirical process is at least $n^{1-p/2}$ (for F continuous) and is at most of the order $n^{1-p/2}(\log \log n)^{p/2}$ in probability.

Then, Section 3 treats the operator $(F, G) \mapsto \int F dG$ and finds a bound in probability for the integral of one empirical process based on observations i.i.d. with distribution F with respect to another such process for a distribution G (where the two processes have an arbitrary joint distribution). Section 4 treats composition $G \mapsto F \circ G$, for fixed F, where $(F \circ G)(x) \equiv F(G(x))$. The operator $(f, g) \mapsto (F + f) \circ (G + g)$, with p-variation norms on f, was considered in Dudley (1994). Section 5 treats the operator $(F, G) \mapsto F \circ G^{-1}$ and Section 6, the two-sample Cramér-von Mises statistics of Rosenblatt (1952) and Watson (1962).

The inverse operator $F \mapsto F^{-1}$ was known to be compactly but not Fréchet differentiable with respect to the sup norm on distribution functions

[1]Mathematics Department, Massachusetts Institute of Technology

E. Giné et al. (eds.), *Selected Works of R.M. Dudley*, Selected Works in Probability and Statistics, DOI 10.1007/978-1-4419-5821-1_23, © Springer Science+Business Media, LLC 2010

F and L^p norm on the range, and likewise for the composition operator $(F, G) \mapsto F \circ G$ with respect to F, for compact differentiability with respect to the L^p norm also on G: Reeds (1976), Fernholz (1983, Props. 6.1.1, 6.1.6). The use of p-variation norms yields Fréchet differentiability and explicit bounds on remainders which for empirical distribution functions F_m and G_n give for these operators, as shown in Dudley (1994), the correct powers of m and n in remainder bounds, although not necessarily the right logarithmic factors or the strongest norms on the ranges of the operators. The present paper does the same for the operator $(F, G) \mapsto \int F dG$. Recall that as shown in Dudley (1994, Proposition 2.1), when Fréchet differentiability fails, as for the sup norm, compact differentiability yields no such remainder bounds.

When an operator is a composition of other operators, the best possible remainder bounds may not follow from the chain rule. For example, as will be seen at the end of Section 5, to represent $\int F dG$ as $\int_0^1 F \circ G^{-1} dt$ gives a non-optimal remainder bound.

13.2 Empirical Processes' p-variation

Let ψ be a convex, increasing function from $[0, \infty)$ onto itself and let f be a function from an interval J into $\mathbf{R}$. Recall that the ψ-variation of f is defined by

$$v_\psi(f) := \sup\{\sum_{k=1}^{m} \psi(|f(t_k) - f(t_{k-1})|): \ t_0 \in J,$$
$$t_0 < t_1 < \ldots < t_m \in J, \ m = 1, 2, \ldots\}.$$

Let $Ly := \max(1, \log y)$. Recall that as noted in Dudley (1994, Section 5), there is a convex, increasing function ψ_1 with $\psi_1(x) = x^2/LL(1/x)$ for $0 < x \leq e^{-e}$. Letting

$$\psi_1(x) \ := \ e^{-2e} + C(x^2 - e^{-2e}) \ \text{ for } \ x \geq e^{-e}$$

for a large enough constant C, we will have $\psi_1(x) \geq x^2/LL(1/x)$ for all $x > 0$. Taylor (1972) showed that for the Brownian motion process $x(\cdot) : t \mapsto x_t$ on $0 \leq t \leq 1$, almost surely $v_{\psi_1}(x(\cdot)) < \infty$, while if $\psi_1(x) = o(\psi(x))$ as $x \downarrow 0$, then $v_\psi(x(\cdot)) = +\infty$ almost surely. So for the Brownian bridge process $y_t = x_t - tx_1, \ 0 \leq t \leq 1$, the same holds. It follows that if $\psi_1 = o(\psi), \ \ v_\psi(n^{1/2}(F_n - F))$ is not bounded in probability as $n \to \infty$, while $v_{\psi_1}(n^{1/2}(F_n - F))$ is bounded in probability: Dudley (1994, Section 5).

For any ψ, the space of all functions f on J such that $v_\psi(cf) < \infty$ for some $c > 0$ will be called $W_\psi(J)$. For $\psi(x) \equiv x^p, \ 1 \leq p < \infty$, the ψ-variation is called p-variation and denoted $v_p(f)$. Then $W_p(J) := W_\psi(J)$ is a Banach space with the p-variation norm defined by $\|f\|_{[p]} \ := \ v_p(f)^{1/p} + \|f\|_\infty$ where $\|f\|_\infty \ := \ \sup\{|f(x)| : \ x \in J\}$.

It follows from the results about ψ_1 stated above that

$$\|n^{1/2}(F_n - F)\|_{[p]} \text{ is bounded in probability as } n \to \infty \text{ if and only if } p > 2. \tag{1}$$

("If" was also proved in Dudley (1992, Corollary 3.8).) For $p \le 2$, although the p-variation norm of the empirical process will not be bounded in probability, its order of growth in n can be bounded as follows.

(2) Theorem *For any distribution function F on $\mathbf{R}$ and $1 \le p \le 2$, $v_p(n^{1/2}(F_n - F)) = O_p(n^{1-p/2}(LLn)^{p/2})$ as $n \to \infty$. Conversely if F is continuous, then almost surely for all n, $v_p(n^{1/2}(F_n - F)) \ge n^{1-p/2}$.*

Proof. To prove the first statement, let U be the uniform $U[0,1]$ distribution function, $U(x) \equiv \max(0, \min(x,1))$, and let U_n be empirical distribution functions for it. Then for any distribution function F, $U \circ F \equiv F$ and $U_n \circ F$ have all the properties of the F_n. So we can take $n^{1/2}(F_n - F) = n^{1/2}(U_n - U) \circ F$. Since F is non-decreasing, we get $v_p(F_n - F) \le v_p(U_n - U)$ for any p. So we can assume F is continuous.

First let $p = 2$. For a given n let $f := n^{1/2}(F_n - F)$, and for any $t_0 < t_1 < \ldots < t_m$ for any positive integer m let $\Delta_i := f(t_i) - f(t_{i-1})$, $i = 1, \ldots, m$. In the supremum of 2-variation sums we can assume that the Δ_i are alternating in sign, since $(A+B)^2 > A^2 + B^2$ for $AB > 0$, so any adjoining differences of the same sign should be combined.

Claim In the supremum we can assume

(a) $\Delta_i > 2/(3n^{1/2})$ whenever $\Delta_i > 0$, and

(b) If $\Delta_i < 0$, then either $i = 1$ or m or $|\Delta_i| \ge 1/(3n^{1/2})$.

Proof of (a). Since $\Delta_i > 0$, there is at least one X_j (in the sample on which F_n is based) with $t_{i-1} < X_j \le t_i$. For the largest such X_j, we can take $t_i = X_j$ since this can only enlarge both Δ_i and $|\Delta_{i+1}|$. Likewise, we can let t_{i-1} increase toward (but not quite equal) some $X_k \le X_j$. If $X_k = X_j$ then (a) holds. Otherwise $X_k < X_j$ and if $\Delta_i \le 2/(3n^{1/2})$ we can move t_{i-1} to a point just less than X_j. This will increase both $|\Delta_{i-1}|$ and Δ_i and make $\Delta_i > 2/(3n^{1/2})$, so (a) is proved.

Proof of (b). If $\Delta_i < 0$ and $i \ge 2$ the sum can only be increased by letting t_{i-1} decrease until $t_{i-1} = X_j$ for some j, and if $i < m$, we can also let t_i increase nearly up to another observation X_k. Suppose (b) fails for Δ_i and there is another r with $X_j < X_r < X_k$. Then by inserting new division points at X_r and just below it, we can enlarge the 2-variation sum. If there are then adjoining positive Δ_j, they can be combined as before. Then (a) is preserved. Iterating, we can assume that there is no such r. Then, I claim the sum is increased by deleting t_{i-1} and t_i: by (a), we have $A := \Delta_{i-1} > 2/(3n^{1/2})$ and $B := \Delta_{i+1} > 2/(3n^{1/2})$, while $0 < C := -\Delta_i < 1/(3n^{1/2})$. Then $(f(t_{i+1}) - f(t_{i-2}))^2 = (A+B-C)^2 > A^2 + B^2 + C^2$, in other words $AB > C(A+B)$ since $C < \frac{1}{2}\min(A,B)$ and $A + B \le 2\max(A,B)$. Here

$A + B - C > 2/(3n^{1/2}) > 0$, so again (a) is preserved, and, iterating, (b) is proved.

Now, continuing with the proof of Theorem 2, since

$$\psi_1(x) \geq x^2/(LL(1/x)) \text{ for all } x > 0,$$

we have for any set $I \subset \{1, \ldots, m\}$,

$$\sum_{i \in I} \psi_1(|\Delta_i|) \geq \sum_{i \in I} \Delta_i^2/(LL(1/|\Delta_i|)).$$

Taking I to be the set of all i for which $|\Delta_i| \geq 1/(3n^{1/2})$, which is all $i = 1, \ldots, m$ except possibly for $i = 1, m$, we get

$$v_2(f) = O\left(v_{\psi_1}(f)LLn + \frac{1}{n}\right) \quad \text{as } n \to \infty.$$

Since $v_{\psi_1}(f)$ is bounded in probability the case $p = 2$ follows.

Now for $1 \leq p < 2$, the p-variation sums can be bounded by the Hölder inequality as follows:

$$\sum_{k=1}^{m} |f(t_k) - f(t_{k-1})|^p \leq \left(\sum_{k=1}^{m} |f(t_k) - f(t_{k-1})|^2\right)^{p/2} m^{1-p/2}.$$

The n observations $X_1, \ldots, X_n$ divide the line into at most n left closed, right open intervals and one interval $(-\infty, \min_i X_i)$. In each such interval f is nonincreasing, and in the supremum of the sums we can assume there are at most two values of t_k in each of the $n+1$ intervals, from the inequality $(A + B)^p \geq A^p + B^p$, $p \geq 1$, $A, B \geq 0$. Thus we can take $m \leq 2n + 2$. Then from the $p = 2$ case, the upper bound follows.

In the other direction, if F is continuous, then for a given n, almost surely $f := n^{1/2}(F_n - F)$ will have jumps of height $n^{-1/2}$ at n distinct points. For $m = 2n$, putting t_k at and just below each jump shows that $v_p(f) \geq n^{1-p/2}$. □

Since $(F_n - F)(t) \to 0$ as $t \to -\infty$, $\|F_n - F\|_\infty \leq v_p(F_n - F)^{1/p}$ and $\|F_n - F\|_{[p]} \leq 2v_p(F_n - F)^{1/p}$. Thus we have

(3) **Corollary** *For* $1 \leq p \leq 2$,

$$\|n^{1/2}(F_n - F)\|_{[p]} = O_p(n^{(2-p)/(2p)}(LLn)^{1/2}) \qquad \text{as } n \to \infty.$$

13.3 The operator $(F, G) \mapsto \int F dG$

This operator is bilinear in F and G. Specifically, we have, if all the integrals are defined,

(4) $$\int (F + f)d(G + g) = \int FdG + \int fdG + \int Fdg + \int fdg.$$

Here if we think of F and G as fixed and f, g as small, approaching 0, then $\int FdG$ is the value of the operator at the point F, G, while $\int fdG$ is a partial derivative term in the direction f, $\int F dg$ is a partial derivative term in the direction g, and $\int f dg$ is the remainder. To apply the operator to empirical distribution functions F_n, G_m we have $f = F_n - F$, $g = G_m - G$.

Young (1936, (10.9)) proved a basic inequality, given here in a slightly different form as in Dudley (1992, Theorem 3.5), for $\int F dG$ in terms of p-variation norms:

$$(5)\qquad \begin{aligned} |\int F dG| &\leq C_{p,q} \|F\|_{[p]} \|G\|_{[q]} \text{ if } \frac{1}{p} + \frac{1}{q} > 1 \\ \text{where } C_{p,q} &:= 1 + \zeta\Big(\frac{1}{p} + \frac{1}{q}\Big) < \infty \end{aligned}$$

and $\zeta(s) \equiv \sum_{n \geq 1} n^{-s}$. Here $\int F dG$ is an extended Riemann-Stieltjes integral defined by Young (1936); for some small clarifications see Dudley (1992, Section 3). Inequality (5) provides a duality between p-variation spaces. But, by (1), the empirical process $n^{1/2}(F_n - F)$ has p-variation norms bounded in probability as $n \to \infty$ only for $p > 2$, and we can't take $p > 2$ and $q > 2$ simultaneously in Young's bound. To bound the remainder here we can apply Theorem 2 to make $p < 2$ or $q < 2$. We have from (4)

$$(6)\qquad \begin{aligned} &\int F_n dG_m = \\ &\int F dG + \int (F_n - F) dG + \int F d(G_m - G) + \int (F_n - F) d(G_m - G). \end{aligned}$$

Here if F_n is based on observations $X_1, \ldots, X_n$ and G_m on observations $Y_1, \ldots, Y_m$, and if $X_i \neq Y_j$ for all i and j, then $mn \int F_n dG_m$ is the number of pairs (i, j) such that $X_i < Y_j$, which is the well-known Mann-Whitney statistic, e.g. Randles & Wolfe (1991, (2.3.8)). The remainder term can be bounded as follows:

(7) **Proposition** *For any $\epsilon > 0$, for any distribution functions F, G and any possible joint distribution of F_n and G_m, as $m, n \to \infty$,*

$$(mn)^{1/2} \int (F_n - F) d(G_m - G) = O_p(\min(m, n)^{\epsilon}).$$

Proof. Choose $1 < p < 2$ such that $1 - p/2 < \epsilon$, then $r > 2$ such that $\frac{1}{p} + \frac{1}{r} > 1$. Apply Corollary 3 to bound the p-variation norm of $F_n - F$ by $O_p(n^{\epsilon - 1/2})$. The r-variation norm of $G_m - G$ is of order $O_p(m^{-1/2})$ since $r > 2$: Dudley (1992, Corollary 3.8). So the Young duality inequality (5) gives the conclusion for $n \leq m$. Otherwise, we can consider $\|G_m - G\|_{[p]}$ and $\|F_n - F\|_{[r]}$ and so multiply by m^{ϵ} rather than n^{ϵ}. □

Proposition 7 implies the following asymptotic behavior of $\int F_n dG_m$: in (6), $\int (F_n - F) dG$ and $\int F d(G_m - G)$ are asymptotically normal and of orders $O_p(n^{-1/2})$ and $O_p(m^{-1/2})$ respectively. If $0 < \epsilon < 1/2$, $m \to \infty$ and $n \to \infty$, then Proposition 7 implies that the remainder is

$$O_p(\max(m, n)^{-1/2} \min(m, n)^{\epsilon - 1/2}) = o_p(\min(m, n)^{-1/2}).$$

The o_p bound on the right, which is enough to imply asymptotic normality of $\int F_n dG_m$ when F_n and G_m are independent, also follows from results of

Gill (1989, pp. 110-111), proved by way of compact differentiability of (a modification of) $(F, G) \mapsto \int F dG$ in the supremum norm.

The stronger bound for the remainder gives directly the following:

(8) **Corollary** *For any possible joint distributions of F_m and G_n, if $m, n \to \infty$ in such a way that $m/n \to c$ where $0 < c < \infty$, then $n^{1/2}(\int F_n dG_m - \int F dG)$ converges in distribution to a sum of two normal variables. The remainder term $n^{1/2}(\int F_n - F\, d(G_m - G))$ is $O_p(n^{\epsilon - \frac{1}{2}})$ for any $\epsilon > 0$.*

If F_n and G_m are independent, then the normal variables are independent, so their sum is normal.

If F_m and G_n are independent and G is continuous, then (as R. Pyke kindly pointed out to me) the expression in Proposition 7 has mean 0 and finite variance: the variance is $\int F - F^2 dG - \int G^2 dF + (\int G dF)^2$, for all m and n, so the expression is $O_p(1)$. So Proposition 7, although not quite optimal, at least indicates the correct powers of m and n. When F_n and G_m have an arbitrary joint distribution, $\min(m,n)^\epsilon$ can at any rate be replaced, not surprisingly, by a logarithmic factor, using Orlicz-variation duality (Young 1938, Theorem 5.1). The details are not carried through here since it isn't clear that the result would be optimal.

13.4 The operator $G \mapsto F \circ G$

In this section we consider differentiability of the operator $h \mapsto F \circ (H+h)$ at $h = 0$ from L^s to L^p, $1 \le p < s$, when F is twice continuously differentiable on a bounded interval $[a, b]$ including the range of H. Such operators have been much studied and are special cases of what are often called Nemitsky operators, e.g. Appell & Zabrejko (1990). We get a bound for the remainder in case $H + h$ also has values in $[a, b]$, which will be applicable when F, H and $H + h$ are all distribution functions. In a more general case, when F (like the uniform $U[0, 1]$ distribution function) is non-differentiable at the endpoints a, b and this turns out to contribute the leading error term in the remainder, we get a larger bound.

A function F on a closed interval $[a, b]$ will be called C^k if it is continuous on $[a, b]$ and has derivatives through kth order continuous on the open interval (a, b) which extend to continuous functions on $[a, b]$.

(9) **Theorem** *Let F be a function from $\mathbf{R}$ into $\mathbf{R}$ whose restriction to a bounded interval $[a, b]$ is C^2.*

(i) Let $(X, \mathcal{A}, \mu)$ be a finite measure space and H a measurable function from X into $[a, b]$. For $1 \le p < s$ and $h \in L^s(X, \mu)$, define the remainder

$$R(h) \ := \ F \circ (H + h) - F \circ H - (F' \circ H)h.$$

If $\|h\|_s \to 0$ for h such that $H + h$ also takes values in $[a, b]$, then

$$\|R(h)\|_p \ = \ O(\|h\|_s^{\min(2, s/p)}).$$

Here $O(\cdot)$ cannot be replaced by $o(\cdot)$.
(ii) Suppose that $F(y) = F(a)$ for $-\infty < y \le a$, $F(y) = F(b)$ for $b \le y < \infty$, F is nondecreasing, $X = [c,d]$ for some $c < d$ with $\mu =$ Lebesgue measure, and H is an increasing C^1 function from $[c,d]$ onto $[a,b]$ with $H'(x) \ge \delta > 0$ for $c < x < d$. Then the operator $h \mapsto F \circ (H+h)$ is Fréchet differentiable at $h = 0$ from L^s to L^p with usual derivative $h \mapsto (F' \circ H)h$, and remainder of order $\|R(h)\|_p = O(\|h\|_s^\zeta)$ where $\zeta = s(p+1)/(p(s+1))$. Again, $O(\cdot)$ cannot be replaced by $o(\cdot)$.

Proof. (i) Since $H + h$ and H take values in $[a,b]$ and F is C^2 there, $|R(h)| \le C|h|^2$ as $|h| \to 0$ for some constant $C < \infty$, and $|h| \le b - a$. If $s \ge 2p$, then $(\int |h|^{2p} d\mu)^{1/p} = O(\|h\|_s^2)$. If $s < 2p$, then the inequality $\int |h|^{2p} d\mu \le \int |h|^s d\mu (b-a)^{2p-s}$ implies that $\|R(h)\|_p = O(\|h\|_s^{s/p})$ and the first conclusion follows.

To see that $O(\cdot)$ cannot be replaced by $o(\cdot)$, let $F(x) = x^2$, $X = [a,b] = [0,1]$, $\mu =$ Lebesgue measure, $H(x) \equiv x$. Then $R(h) \equiv h^2$. For $s \ge 2p$, take $h = c1_{[0,1-c]}$ as $c \downarrow 0$. Then $\|h^2\|_p \sim c^2 \sim \|h\|_s^2$, so $O(\cdot)$ cannot be replaced by $o(\cdot)$. Or if $p < s < 2p$, let $h = (1-c)1_{[0,c]}$. Then as $c \downarrow 0$, $\|h^2\|_p \sim c^{1/p}$ and $\|h\|_s \sim c^{1/s}$, so again $O(\cdot)$ cannot be replaced by $o(\cdot)$. In both cases $H + h$ takes values in $[0,1]$. (If desired, F can be a probability distribution function with $\inf_{[0,1]} F' > 0$: let $F(x) = \frac{1}{2}(x + x^2)$ for $0 \le x \le 1$, and the same examples apply up to constants.) So (i) is proved.
(ii) Here $H + h$ can take values outside of $[a,b]$. Let's assume without loss of generality that $a = c = 0$, $F(0) = 0$ and $b = d = 1$. Then since H is onto, $H(0) = 0$ and $H(1) = 1$. Let A be the measurable set of $x \in [0,1]$ where $(H+h)(x) < 0$. Now since $H' \ge \delta > 0$, $H(x) \ge \delta x$ for $0 \le x \le 1$. So on A, $h(x) < -\delta x$ and $F(H+h)(x) = 0$. Also on A, $R(h)(x) = -F(H(x)) - F'(H(x))h(x)$, $0 \le F(H(x)) \le Cx$ for some C since F and H are C^1, and $0 \le -F'(H(x))h(x) \le D|h(x)|$ for some constant D. Then since $|h(x)| > \delta x$, there is a constant $M < \infty$ ($M = D + C/\delta$) such that $|R(h)| \le M|h|$ on A. By Hölder's inequality

$$\int_A |R(h)(x)|^p dx \le M^p \int_0^1 |h|^p 1_{\{|h| > \delta x\}} dx \le M^p \|h\|_s^p (\lambda(|h| > \delta x))^{1-p/s}$$

where λ is Lebesgue measure. For fixed $\|h\|_s$, $\lambda(|h| \ge \delta x)$ is maximized when $|h| = \delta x$ on an interval $[0,\alpha]$ and $h(x) = 0$ otherwise, when it is $O(\|h\|_s^{s/(s+1)})$. Combining terms and taking pth roots gives, on A, the stated bound for the p-norm of the remainder. By symmetry, we get a bound of the same order for the remainder on the set where $H + h > 1$. Now $p < s$ implies $s(p+1)/(p(s+1)) < \min(s/p, 2)$, so that the remainder bound coming from $H + h$ outside $[a,b]$ dominates the bound from $H + h$ in $[a,b]$. To see that $O(\cdot)$ cannot be replaced by $o(\cdot)$, consider the functions $F(x) = H(x) = x$, $0 \le x \le 1$, and let $h(x) = -2x$ on an interval $[0,\alpha]$, $h(x) = 0$ for $x > \alpha$, and let $\alpha \downarrow 0$. □

Andersen, Borgan, Gill & Keiding (1993, Proposition II.8.8) consider (compact) differentiability of composition with sup norm on the range. Then in the situation of the second half of Theorem 9, differentiability fails, so H is required to have range in a proper subinterval $[s,t]$, $a < s < t < b$.

13.5 The operator $(F, G) \mapsto F \circ G^{\leftarrow}$

Given an interval $[a,b]$, and a real-valued function G on $[a,b]$, and any real y let

$$G^{\leftarrow}(y) \;:=\; G^{\leftarrow}_{[a,b]}(y) \;:=\; \inf\{x \in [a,b] : \; G(x) \geq y\},$$

or $G^{\leftarrow}_{[a,b]}(y) \;:=\; b$ if $G(x) < y$ for all $x \in [a,b]$. The notation $G^{\leftarrow}$ is used instead of G^{-1} to specify a particular definition of "inverse," e.g. Beirlant & Deheuvels (1990).

Suppose that inverses are taken with respect to a bounded interval $[a,b]$. If G is C^2 from $[a,b]$ onto an interval $[c,d]$, $G' \geq \delta$ for some $\delta > 0$, and $1 \leq s < \infty$, then by Corollary 2.4 of Dudley (1994), the inverse operator taking g to $(G+g)^{\leftarrow}$ restricted to $[c,d]$ is Fréchet differentiable at $g = 0$ from $(W_s([a,b]), \|\cdot\|_{[s]})$ into $(L^s[c,d], \|\cdot\|_s)$ (for the L^s norm with respect to Lebesgue measure on $[c,d]$), with derivative $g \mapsto -(g \circ G^{\leftarrow})/(G' \circ G^{\leftarrow})$. The remainder

$$R_g \;:=\; (G+g)^{\leftarrow} - G^{\leftarrow} + (g \circ G^{\leftarrow})/(G' \circ G^{\leftarrow})$$

satisfies

$$\text{(10)} \qquad \|R_g\|_s \;=\; O(\|g\|_{[s]}^{1+1/s}) \;\text{ as }\; \|g\|_{[s]} \to 0.$$

Let U be the $U[0,1]$ distribution function, $U(x) = \max(0, \min(x,1))$ for all x. In the following $\|\cdot\|_\infty$ denotes the sup norm $\|h\|_\infty \;:=\; \sup_{0<t<1} |h(t)|$. On $[0,1]$ let

$$R_{f,g} \;:=\; (U+f) \circ (U+g)^{\leftarrow} - U - f \circ U + g,$$

the remainder in differentiating $(f,g) \mapsto (U+f) \circ (U+g)^{\leftarrow}$ at $f = g = 0$. Let $\|\cdot\|_p$ be the L^p norm with respect to Lebesgue measure on $[0,1]$.

(11) **Proposition** *For some constant $C < \infty$, $1 \leq p < \infty$, any measurable real functions f on $\mathbf{R}$ and g, h on $[0,1]$, let inverses be taken with respect to $[0,1]$. Then*

(a) $\|f \circ (U+h) - f \circ U\|_p \;\leq\; C\|f\|_{[p]} \|h\|_\infty^{1/p}$.

(b) $\|(U+g)^{\leftarrow} - U\|_\infty \;\leq\; \|g\|_\infty$.

(c) $\|f \circ (U+g)^{\leftarrow} - f \circ U\|_p \;\leq\; C\|f\|_{[p]} \|g\|_\infty^{1/p}$.

(d) $\|R_{f,g}\|_p \;=\; O(\|f\|_{[p]}^{1+1/p} + \|g\|_{[p]}^{1+1/p})$ as $\|f\|_{[p]} + \|g\|_{[p]} \to 0$.

Proof. For (a), although Theorem 2.2 of Dudley (1994) is stated only for $s < \infty$, it actually holds (the proof only gets a little easier) for $s = \infty$, so (a) holds.

For (b), let $\delta := \|g\|_\infty$. If $x+g(x) \geq y$ then $x+\delta \geq y$, so $(U+g)^{\leftarrow}(y) \geq y-\delta$. Conversely if $y+\delta \leq 1$ then $y+\delta+g(y+\delta) \geq y$ so $(U+g)^{\leftarrow}(y) \leq y+\delta$, which also holds if $y+\delta > 1$, and (b) follows.

Then (c) follows from (a) and (b). For (d),

$$R_{f,g} = f \circ (U+g)^{\leftarrow} - f \circ U + R_g.$$

So from (10), (c), and the inequality $AB^\alpha \leq A^{1+\alpha}+B^{1+\alpha}$ for $A, B, \alpha > 0$, (d) follows. □

Now consider the case when F, $F+f$, G and $G+g$ are all probability distribution functions, and specifically $F+f = F_m$ (an empirical distribution function) and $G+g = G_n$. Then $F \circ G^{\leftarrow}$ is a procentile-procentile function (P-P plot, e.g. Beirlant & Deheuvels (1990)) and $F_m \circ G_n^{\leftarrow}$ its empirical counterpart. The special case where both F and G equal U is in fact not so special: if F_m and G_n are both empirical distribution functions from the same continuous distribution function F, strictly increasing on an interval $[a,b]$ with $F(a) = 0$ and $F(b) = 1$, then we can write $F_m = U_m \circ F$ and $G_n = V_n \circ F$ where U_m and V_n are empirical distribution functions for $U = U[0,1]$. So $F_m \circ G_n^{\leftarrow} \equiv U_m \circ V_n^{\leftarrow}$. Thus, as is known, any results for $U_m \circ V_n^{\leftarrow}$ yield conclusions for such $F_m \circ G_n^{\leftarrow}$.

(12) **Proposition** *If U_m and V_n are empirical distribution functions for U, which can have any joint distribution, and $m \wedge n := \min(m,n)$,*

$$\|U_m \circ V_n^{\leftarrow} - U_m + V_n - U\|_2 = O_p([(LL(m\wedge n))/(m\wedge n)]^{3/4}) \quad \text{as } m,n \to \infty.$$

Proof. Apply Proposition 11(d) and Corollary 3 for $p = 2$. □

In the situation of Proposition 12, the exponent $-3/4$ for $m \wedge n$ is correct, for example when $m >> n$ by Theorem 2.6 of Dudley (1994) and the comment after it. If we apply the latter Theorem instead of (10) in the proof of Proposition 11(d) in the case of Proposition 12, the bound in the latter is replaced by

$$O_p((LLm)^{1/2}m^{-1/2}n^{-1/4} + n^{-3/4}), \tag{13}$$

which is better by a factor $(LLm)^{1/4}$ if m and n are of the same order of magnitude and still better if they are not. I don't know whether the $(LLm)^{1/2}$ factor is needed.

As in the case of the inverse operator $g \mapsto (G+g)^{\leftarrow}$ [Dudley (1994, Theorem 2.6 and the discussion after it)] it seems that to treat L^p norms of remainders for $2 < p \leq \infty$, one must go beyond the kinds of proofs treated so far in this paper. From Beirlant & Deheuvels (1990, (1.9), (2.1), (2.2), (2.4), (2.6) and (2.7)) it follows that if $m/(n^{1/2}\log n) \to \infty$ and $n^{3/2}/(m \log n) \to \infty$ then

$$\|U_m \circ V_n^{\leftarrow} - U_m + V_n - U\|_\infty = O_p\left(\frac{(\log n)^{1/2}}{m^{1/2}n^{1/4}} + \frac{\log m}{m} + \frac{\log n}{n}\right).$$

(For this conclusion, independence of U_m and V_n is not needed.) If m and n are of the same order of magnitude, the bound for the sup norm becomes $O_p((\log n)^{1/2}/n^{3/4})$, larger than (13) for the L^2 norm.

Suppose $(X_1, \|\cdot\|)$, $(X_2, \|\cdot\|)$ and $(X_3, \|\cdot\|)$ are normed spaces and for $i = 1, 2$, T_i is a map from an open set $U_i \subset X_i$ into X_{i+1}, Fréchet differentiable at a point $x_i \in U_i$ with derivative $T_i'(x_i)$. Suppose $T(x_1) = x_2$ and $T(x) \in U_2$ for all $x \in U_1$. Then by a well-known chain rule, $T_2 \circ T_1$ is Fréchet differentiable at x_1 from U_1 into X_3. Suppose that for some $\alpha > 1$ we have remainder bounds

$$\|T_i(x_i + u_i) - T_i(x_i) - T_i'(x_i)(u_i)\| \;=\; O(\|u_i\|^\alpha)$$

as $\|u_i\| \to 0$, $u_i \in X_i$, for $i = 1, 2$. Then one can get such a remainder bound also for $T_2 \circ T_1$. But such general chain rule remainder bounds may not give bounds of the best possible order for the composition, as in the following example.

It is well known, and not hard to check, that we have the probability integral transformation (for a Lebesgue-Stieltjes integral) $\int_{-\infty}^{\infty} F dG = \int_0^1 (F \circ G^{\leftarrow})(t) dt$ for any two distribution functions F, G of probability measures. The results for $F \circ G^{\leftarrow}$ from this section could then be applied to $\int F dG$, as follows: $\int_0^1 U_m dV_n = \int_0^1 U_m \circ V_n^{\leftarrow} dt$. The integral from 0 to 1 is a linear operation and so differentiable with no remainder. Composing and applying Proposition 12 yields

$$\left|\int_0^1 U_m dV_n - \frac{1}{2} - \int_0^1 U_m - V_n dt\right| \;=\; O_p\left(\left[\frac{LL(m \wedge n)}{m \wedge n}\right]^{3/4}\right).$$

But Proposition 7 gives the stronger bound $O_p((m \wedge n)^\epsilon (mn)^{-1/2})$.

13.6 The 2-sample Cramér-von Mises-Rosenblatt and Watson statistics

Suppose again that F, G are distribution functions. Let F_m be an empirical distribution function for F and G_n an empirical distribution function for G, independent of F_m. Then the two-sample empirical process will be defined by

$$(F, G)_{m,n} \;:=\; \left(\frac{mn}{m+n}\right)^{1/2} (F_m - G_n).$$

Let $N := m+n$ and $K_N := K_{m,n} := (mF_m + nG_n)/N$. Lehmann (1951, p. 174) and Rosenblatt (1952) proposed a test for $F = G$ based on the statistic $\int (F, G)_{m,n}^2 d(F_m + G_n)/2$. Later authors (mentioned in the Notes below) have preferred the closely related statistic

$$W^2 \;:=\; W_{m,n}^2 \;:=\; \int (F, G)_{m,n}^2 dK_{m,n},$$

which has the same asymptotic distribution by nearly the same proof. Earlier A. M. Mood, according to Dixon (1940), suggested for $m = n$ the statistic with F_m (or G_n) instead of K_N.

(14) **Theorem** *(Rosenblatt-Fisz-Kiefer) For any continuous distribution function $F = G$, the statistic $W^2_{m,n}$ converges in distribution as $m, n \to \infty$ to $\int_0^1 y_t^2 dt$, where y_t is the Brownian bridge.*

Notes So, the asymptotic distribution is the same as for the one-sample Cramér-von Mises statistic $n\int(F_n - F)(t)^2 dF(t)$. Rosenblatt (1952) proved the conclusion under the further assumption

$$\text{(15)} \qquad \text{as } m, n \to \infty, \ m/n \to \lambda \ \text{ where } \ 0 < \lambda < \infty.$$

Fisz (1960) gave a correction to the proof. Aki (1981) gave another proof, also assuming (15), based on functional differentiability methods, extending the work of Filippova (1961). Darling (1957, p. 827) indicated that the weaker assumption

$$\text{(16)} \qquad 0 < \liminf m/n \leq \limsup m/n < \infty$$

would suffice, even for weighted forms of the statistic. Theorem 14 holds as stated, with $m \to \infty$ and $n \to \infty$ with no restriction such as (15) or (16) on m/n, as was first shown apparently by Kiefer (1959) and will also follow from the proof below by way of p-variation.

Proof. As before let U be the uniform $U[0, 1]$ distribution function. Let U_m, V_n be independent empirical distribution functions for U. Then as is well known, we can set $F_m = U_m(F)$, $G_n = V_n(G) = V_n(F)$. If X_i are i.i.d. with distribution F, then since F is continuous, $F(X_i)$ are i.i.d. $U[0, 1]$. It follows by these transformations that we can assume $F = G = U$. It's easy to check that

(17)
$$(F, G)_{m,n} = \left(\frac{n}{m+n}\right)^{1/2} m^{1/2}(F_m - F) - \left(\frac{m}{m+n}\right)^{1/2} n^{1/2}(G_n - F).$$

A sequence Y_n of possibly non-measurable functions from a probability space into a metric space (S, d) is said to converge *in outer probability* to a random variable Y_0 if for every $\epsilon > 0$, $P^*(d(Y_n, Y_0) > \epsilon) \to 0$ as $n \to \infty$, where

$$P^*(A) := \inf\{P(B) : B \text{ measurable}, A \subset B\}.$$

By the Donsker property of $\{f \in W_p(\mathbf{R}) : \|\mathbf{f}\|_{[\mathbf{p}]} \leq \mathbf{1}\}$ for $p < 2$, the Love-Young duality for q-variation spaces converse to (5) above (Dudley 1992, Theorem 2.2 and Theorem 3.6), and existence of almost surely convergent realizations, see Dudley (1985, Theorem 4.1), for any $r > 2$, taking $(1/p) + (1/r) = 1$, there exist on some probability space processes α_m having all the properties of $m^{1/2}(F_m - F)$ for each fixed m (but not the same joint distribution for different m) and a Brownian bridge α such that $\|\alpha_m - \alpha\|_{[r]} \to 0$ in outer probability. By right continuity, the r-variation is determined by restriction to rational $t \in [0, 1]$, so we have convergence in probability (outer probability is unnecessary here). On another copy of the probability space, let α_n be called β_n and let α be called β. Taking a product space we have (α_m, α) independent of (β_n, β). Then

$$C_{m,n} := \left(\frac{n}{m+n}\right)^{1/2} \alpha - \left(\frac{m}{m+n}\right)^{1/2} \beta$$

is a Brownian bridge since α and β are independent Brownian bridges and the squares of the coefficients add up to 1. Also, by (17),

$$\gamma_{m,n} := \left(\frac{n}{m+n}\right)^{1/2} \alpha_m - \left(\frac{m}{m+n}\right)^{1/2} \beta_n$$

has all the properties of $(F, G)_{m,n}$. We have

$$\|\gamma_{m,n} - C_{m,n}\|_{[r]} \to 0$$

in (outer) probability as $m, n \to \infty$.

Let $J_N := F + (m^{1/2}\alpha_m + n^{1/2}\beta_n)/N$. Then $W^2_{m,n}$ is equal in distribution to $\int \gamma^2_{m,n} dJ_N$. Now, J_N is an empirical distribution function for F for sample size N (just as K_N is). For any $1 < q < 2$ there is an $r > 2$ such that $(1/q) + (1/r) > 1$. Then $\|\gamma_{m,n}\|_{[r]}$ is bounded in probability by (1), and

$$\|\gamma^2_{m,n}\|_{[r]} \leq \|\gamma_{m,n}\|^2_{[r]}$$

by Lemma 5.1 of Dudley (1994). We have $J_N = F + (J_N - F)$ where

$$\|J_N - F\|_{[q]} = O_p(N^{(1-q)/q}(LLN)^{1/2}) \text{ as } N \to \infty$$

by Corollary 3. Thus by the inequality of Young (1936) ((5) above) again,

$$\int \gamma^2_{m,n} d(J_N - F) = O_p(N^{(1-q)/q}(LLN)^{1/2}) \to 0$$

in probability as $N \to \infty$. Now $\|\gamma_{m,n} + C_{m,n}\|_2$ is bounded in probability and

$$\|\gamma_{m,n} - C_{m,n}\|_2 \leq \|\gamma_{m,n} - C_{m,n}\|_\infty \to 0$$

in probability, so $\int_0^1 \gamma^2_{m,n} - C^2_{m,n} dt \to 0$ in probability. Since $\int_0^1 C^2_{m,n}(t)dt$ has for all m, n the distribution of $\int_0^1 y_t^2 dt$, W^2 has the same limit distribution. □

For the convergence of the uniform empirical process to the Brownian bridge in r-variation norm for $r > 2$, used in the above proof, Huang (1994, Section 3.7; 1995) gave rates: he showed that from the construction of Komlós, Major & Tusnády (1975) one obtains Brownian bridges $B_{(m)}$ such that for some constant $C(r)$,

$$E\|\alpha_m - B_{(m)}\|_{[r]} \leq C(r) m^{\frac{1}{r} - \frac{1}{2}}$$

where the exponent of m is best possible since for any Brownian bridge (or sample-continuous process) B, almost surely

$$\|\alpha_m - B\|_{[r]} \geq m^{\frac{1}{r} - \frac{1}{2}}.$$

Watson (1962) introduced a statistic similar to Rosenblatt's, suitable for use on the circle since it doesn't depend on the choice of initial angle, namely

$$U^2 := U^2_{m,n} := \frac{mn}{m+n}[\int(F_m - G_n)^2 dK_N - \{\int(F_m - G_n)dK_N\}^2]$$
$$= \int(F,G)^2_{m,n}dK_N - (\int(F,G)_{m,n}dK_N)^2.$$

In the same way as above it can be seen that the asymptotic distribution of $U^2_{m,n}$ for $m, n \to \infty$ arbitrarily, as Persson (1979) and Janson (1984, p 504) showed, and as Watson (1962) had shown under (15), is that of

$$U^2_\infty := U^2_{\infty,\infty} := \int_0^1 y_t^2 dt - (\int_0^1 y_t dt)^2.$$

The Brownian bridge has a Fourier series representation

$$y_t = \frac{1}{\pi 2^{1/2}} \sum_{k=1}^{\infty} \frac{1}{k}\{X_k(1 - \cos(2\pi kt)) + Y_k \sin(2\pi kt)\}$$

where $X_1, X_2, \ldots, Y_1, Y_2, \ldots,$ are i.i.d. $N(0, 1)$. Then

$$y_t - \int_0^1 y_s ds = \frac{1}{\pi 2^{1/2}} \sum_{k=1}^{\infty} \frac{1}{k}\{-X_k \cos(2\pi kt) + Y_k \sin(2\pi kt)\}, \text{ so}$$

$$(18) \qquad U^2_\infty = \int_0^1 y_t^2 dt - (\int_0^1 y_t dt)^2 = \frac{1}{4\pi^2} \sum_{k=1}^{\infty} \frac{1}{k^2}(X_k^2 + Y_k^2).$$

Each term $X_k^2 + Y_k^2$ has the exponential distribution with density $\frac{1}{2}e^{-x/2}$ for $x \geq 0$. A convolution of exponential densities with different scale parameters is a linear combination of those densities. It turns out that

$$(19) \qquad P(U^2_\infty > x) = 2\sum_{i=1}^{\infty}(-1)^{i-1} \exp(-2\pi^2 i^2 x), \quad x \geq 0$$

(Watson 1962, (2)). A curiosity here is that $\pi^{-2} \sup_{0 \leq t \leq 1} y_t^2$ has the same distribution [e.g. Dudley (1993, (12.3.4))], while for the Bahadur-Kiefer remainder $\mathcal{K}_n := U_n^{\leftarrow} + U_n - 2U$, $n^3\|\mathcal{K}_n\|_\infty^4/(\pi(\log n))^2$ has the same asymptotic distribution (Kiefer 1970, Theorem 1), where $\|\mathcal{K}_n\|_\infty$ can also be replaced by $\sup \mathcal{K}_n$ or $-\inf \mathcal{K}_n$. The series in (19) converges rapidly, so only a few terms give a fine approximation unless x is small, while the series in (18) converges slowly.

Acknowledgments: This research was partially supported by National Science Foundation Grants.

I'd like to thank Peter Bickel, Richard Gill, Ron Pyke, Jinghua Qian and Jon Wellner for helpful discussions.

13.7 References

Aki, S. (1981), 'Asymptotic behavior of functionals of empirical distribution functions for the two-sample problem', *Annals of the Institute of Statistical Mathematics* **33**, 391–403.

Andersen, P. K., Borgan, Ø., Gill, R. D. & Keiding, N. (1993), *Statistical Models Based on Counting Processes*, Springer-Verlag, New York.

Appell, J. & Zabrejko, P. P. (1990), *Nonlinear Superposition Operators*, Cambridge University Press.

Beirlant, J. & Deheuvels, P. (1990), 'On the approximation of P-P and Q-Q plot processes by Brownian bridges', *Statistics and Probability Letters* **9**, 241–251.

Darling, D. A. (1957), 'The Kolmogorov-Smirnov, Cramér-von Mises tests', *Annals of Mathematical Statistics* **28**, 823–838.

Dixon, W. J. (1940), 'A criterion for testing the hypothesis that two samples are from the same population', *Annals of Mathematical Statistics* **11**, 199–204.

Dudley, R. M. (1985), 'An extended Wichura theorem, definitions of Donsker class, and weighted empirical distributions', *Springer Lecture Notes in Mathematics* **1153**, 141–178.

Dudley, R. M. (1992), 'Fréchet differentiability, p-variation and uniform Donsker classes', *Annals of Probability* **20**, 1968–1982.

Dudley, R. M. (1993), *Real Analysis and Probability*, Chapman and Hall, New York. Second printing, corrected.

Dudley, R. M. (1994), 'The order of the remainder in derivatives of composition and inverse operators for p-variation norms', *Annals of Statistics* **22**, 1–20.

Fernholz, L. T. (1983), *von Mises Calculus for Statistical Functionals*, Vol. 19 of *Lecture Notes in Statistics*, Springer, New York.

Filippova, A. A. (1961), '[the von] mises theorem on the asymptotic behavior of functionals of empirical distribution functions and its statistical applications', *Theory of Probability and Its Applications* **7**, 24–57.

Fisz, M. (1960), 'On a result by M. Rosenblatt concerning the von Mises-Smirnov test', *Annals of Mathematical Statistics* **31**, 427–429.

Gill, R. D. (1989), 'Non- and semi-parametric maximum likelihood estimators and the von Mises method', *Scandinavian Journal of Statistics* **16**, 97–128.

Huang, Y.-C. (1994), Empirical distribution function statistics, speed of convergence, and p-variation, PhD thesis, Massachusetts Institute of Technology.

Huang, Y.-C. (1995), Speed of convergence of classical empirical processes in p-variation norm, preprint, Academica Sinica, Taipei, Taiwan.

Janson, S. (1984), 'The asymptotic distributions of incomplete U-statistics', *Zeitschrift für Wahrscheinlichkeitstheorie und Verwandte Gebiete* **66**, 495–505.

Kiefer, J. (1959), 'K-sample analogues of the Kolmogorov-Smirnov and Cramér-v. Mises tests', *Annals of Mathematical Statistics* **30**, 420–447.

Kiefer, J. (1970), Deviations between the sample quantile process and the sample df, *in* M. L. Puri, ed., 'Nonparametric Techniques in Statistical Inference', Cambridge University Press, pp. 299–319.

Komlós, J., Major, P. & Tusnády, G. (1975), 'An approximation of partial sums of independent RV's, and the sample DF. I', *Zeitschrift für Wahrscheinlichkeitstheorie und Verwandte Gebiete* **32**, 111–131.

Lehmann, E. L. (1951), 'Consistency and unbiasedness of certain nonparametric tests', *Annals of Mathematical Statistics* **22**, 165–179.

Persson, T. (1979), 'A new way to obtain Watson's U^2', *Scandinavian Journal of Statistics* **6**, 119–122.

Randles, R. H. & Wolfe, D. A. (1991), *Introduction to the Theory of Nonparametric Statistics*, Krieger, Malabar, FL. Reprinted with corrections.

Reeds, J. A. (1976), On the definition of von Mises functionals, PhD thesis, Statistics, Harvard University.

Rosenblatt, M. (1952), 'Limit theorems associated with variants of the von Mises statistic', *Annals of Mathematical Statistics* **23**, 617–623.

Taylor, S. J. (1972), 'Exact asymptotic estimates of Brownian path variation', *Duke Mathematical Journal* **39**, 219–241.

Watson, G. S. (1962), 'Goodness-of-fit tests on a circle, II', *Biometrika* **49**, 57–63.

Young, L. C. (1936), 'An inequality of the Hölder type, connected with Stieltjes integration', *Acta Mathematica (Djursholm)* **67**, 251–282.

Young, L. C. (1938), 'General inequalities for Stieltjes integrals and the convergence of Fourier series', *Mathematische Annalen* **115**, 581–612.

Part 6

Miscellanea

Introduction

This chapter contains five papers on a variety of topics including random walks in locally compact groups, approximation theory, stochastic analysis, learning theory and Bayesian statistics. This shows the breadth of Dudley's interests ranging from very pure areas of probability and analysis to much more applied areas including statistics and even computer science.

The first paper deals with the problem of existence of a recurrent random walk in discrete locally compact abelian groups. A random walk in a countable abelian group G is a Markov chain whose transition function is invariant with respect to translations. It is called recurrent if it visits any open set $U \subset G$ infinitely often with probability 1. Dudley proved that such a random walk exists in a locally compact abelian group if and only if the factor group G/H is countable for any open subgroup H of G and there is no discrete subgroup that is free abelian on three generators. The study of this problem started in an earlier paper by Dudley (Proc. Amer. Math. Soc. 13, 1962, 447–450) and it continues a classical line of research on random walks in discrete additive subgroups of finite dimensional vector spaces (e.g, Chung, K.L. and Fuchs, W.H.J., Mem. Amer. Math. Soc., 1951, n. 6).

The subject of the second paper is bounding metric entropies of some classes of sets in $\mathbf{R}^k$. The notion of metric entropy of compact sets was studied by Kolmogorov and Tihomirov in the 50s (Uspehi Mat. Nauk, 1959, 14, 2, 3–86). In part, they were motivated by Hilbert's 13th problem. They proved a number of bounds on the ε-entropy of various classes of smooth functions equipped with classical metrics. In the 60s, Dudley and Sudakov developed a metric entropy method of bounding sup-norms of Gaussian processes. Dudley's interest in studying entropies of classes of sets was apparently related to the development of empirical processes theory where such bounds have become one of the main tools. The paper deals with classes of subsets of $\mathbf{R}^k$ with boundaries of given "smoothness" $\alpha > 0$ and also of closed convex subsets of the unit ball in $\mathbf{R}^k$. The metrics used in computation of their ε-entropy are the Hausdorff metric and the Lebesgue measure of symmetric difference of two sets. In particular, it is shown that for the classes of sets with boundary of smoothness α the entropy is of the order $O(\varepsilon^{-(k-1)/\alpha})$ as $\varepsilon \to 0$ and for the convex sets it is of the order $O(\varepsilon^{-(k-1)/2})$ as $\varepsilon \to 0$.

A short note on Wiener functionals as Itô integrals deals with a basic question in stochastic analysis: given a standard Wiener process $W(t)$, $0 \le t \le 1$ and the corresponding filtration $\mathcal{F}_t^W := \sigma\{W(s) : 0 \le s \le t\}$, is it possible to represent an $\mathcal{F}_t^W$-measurable random variable X as an Itô stochastic integral of an adapted process? It is proved in the paper that it is possible under the assumption that X is finite a.s., strengthening the previously known result for integrable X.

The fourth paper in this chapter is joint with Kulkarni, Richardson and Zeitouni. It deals with a problem in learning theory posed by Benedek and Itai as to whether uniform bounds on the metric entropy of a class of sets ("concepts") would suffice for the probably approximately correct (PAC) learnability of a concept from the class. It happens that the metric entropy is not sufficient for this, which is proved in this paper by constructing a rather subtle counterexample.

The last paper is joint with Haughton and it deals with classical problems in Bayesian statistics about asymptotic normality of posterior probabilities of half-spaces. Namely, let $\mathcal{P}$ be a statistical model with parameter space Θ that is an open subset of $\mathbf{R}^d$ and let π be a prior law on Θ with a continuous, strictly positive density. Let A be a half-space of $\mathbf{R}^d$ that does not contain the maximum likelihood estimator of parameter θ (otherwise A is replaced by its complement). Finally, let 2Δ denote the likelihood ratio statistic. It is shown, under some conditions, including a Glivenko-Cantelli condition on the second partial derivatives of the log likelihood function with respect to θ, that uniformly over all half-spaces A, either the posterior probability $\pi_n(A)$ is asymptotic to $\Phi(-\sqrt{2\Delta})$, or both $\pi_n(A)$ and $\Phi(-\sqrt{2\Delta})$ converge to zero exponentially fast.

E. Giné et al. (eds.), *Selected Works of R.M. Dudley*, Selected Works in Probability and Statistics,
DOI 10.1007/978-1-4419-5821-1_24,

PATHOLOGICAL TOPOLOGIES AND RANDOM WALKS ON ABELIAN GROUPS

R. M. DUDLEY[1]

It is shown in this paper that a part of the structure theory of locally compact abelian groups, which does not explicitly involve compactness, still fails to apply to complete separable metric abelian groups. Specifically, Theorem 2 asserts that any countable abelian group may be discretely imbedded in a metrizable abelian group with a dense cyclic subgroup. This result is independent of the rest of the paper.

Theorem 1 is a natural generalization of the result of [**4**], on the existence of recurrent random walks on discrete abelian groups, to locally compact abelian groups. Theorem 2 shows that it cannot be applied verbatim to complete separable metric groups. However, a statement in terms of dense subgroups rather than discrete ones is formulated as a conjecture.

We call a nonnegative finite measure μ defined on the Borel sets (the σ-field generated by the open sets) of a topological space *regular* if for every Borel set B and $\epsilon>0$ there is a compact set $K\subset B$ such that $\mu(B)-\mu(K)<\epsilon$. It is known that any finite Borel measure on a complete separable metric space is regular, and that assuming the generalized continuum hypothesis "separable" is unnecessary (see [**6**]). Also, a bounded linear functional on the continuous functions on a compact Hausdorff space may be uniquely represented as the integral with respect to a regular Borel measure (see [**5**, §56]).

In order to define random walks on topological groups, we must define convolution of measures. Let G be a Hausdorff topological group and let μ and ν be finite regular Borel measures on G. Then there are disjoint compact subsets $C_1, C_2, \cdots$, of G such that both μ and ν are concentrated in $\bigcup_{n=1}^{\infty} C_n$.

To obtain a product measure for μ and ν on $G\times G$, defined on all Borel sets, we may proceed as follows. On each compact set $C_m\times C_n$ we have a positive linear functional

$$f \rightarrow \int_{C_n}\int_{C_m} f(x, y)\,d\mu(x)\,d\nu(y)$$

Received by the editors August 21, 1962 and, in revised form, January 11, 1963.

[1] Some of the work on this paper was done while the author held a National Science Foundation Cooperative Fellowship in 1961–1962.

E. Giné et al. (eds.), *Selected Works of R.M. Dudley*, Selected Works in Probability and Statistics, DOI 10.1007/978-1-4419-5821-1_25, © Springer Science+Business Media, LLC 2010

on the continuous functions on $C_m \times C_n$. (It follows from the Stone-Weierstrass theorem that the order of integration can be reversed.) This linear functional can, as mentioned, be written as

$$f \to \int f d\rho_{mn},$$

where ρ_{nm} is a regular Borel measure of total mass $\mu(C_m)\nu(C_n)$ on $C_m \times C_n$. The measure $\sum_{m,n=1}^{\infty} \rho_{mn}$ is then a regular Borel measure on $G \times G$ which is easily seen by mutual refinement to be independent of the choice of the sets C_n. We call this measure $\mu \otimes \nu$. It is easily seen to be an extension of the ordinary product measure $\mu \times \nu$.

If there is no countable base for the topology of G, it seems unclear whether $\mu \times \nu$ itself is defined on all Borel sets in $G \times G$, even after completion of the measure, a point not explained in [8]. Another way of obtaining a sufficiently widely defined product measure $\mu\nu$ is given in [1]. Since the $\mu\nu$ measure of any Borel set B is the supremum of the measures of closed sets included in B,[2] and for any closed set F

$$\mu\nu(F) = \sum_{m,n=1}^{\infty} \mu\nu(F \cap (C_m \times C_n)),$$

$\mu\nu$ is regular. $\mu\nu$ has the property that

$$\int f d\mu\nu = \int\int f(x, y) d\mu(x) d\nu(y)$$

for any Borel function f on $G \times G$. Applying this to functions equal to continuous functions on $C_m \times C_n$ and zero elsewhere, it follows that $\mu\nu$ and $\mu \otimes \nu$ agree on all Borel sets.

If ψ is a continuous mapping of a topological space A onto another space B and μ is a regular Borel measure on A, we can define a regular Borel measure μ_ψ on B by $\mu_\psi(M) = \mu(\psi^{-1}(M))$ for any Borel set M in B (see [8]). If g is a bounded Borel function on B and M is a Borel set in B, then

$$\int_M g d\mu_\psi = \int_{\psi^{-1}(M)} g(\psi(x)) d\mu(x)$$

since the equality holds for functions with finitely many values.

Now, if $p(x, y) = xy$ on $G \times G$, we define the convolution $\mu * \nu$ by

$$\mu * \nu = (\mu \otimes \nu)_p = (\mu\nu)_p.$$

[2] This property is stated in [1] only for B open, but since it also holds for B closed and for countable unions and intersections of sets where it holds, it is true of all Borel sets.

Then $\mu * \nu$ is a regular Borel measure on G satisfying, and characterized by,

$$\int f d(\mu * \nu) = \int f(xy) d\mu\nu(x, y) = \int\int f(xy) d\mu(x) d\nu(y)$$
$$= \int\int f(xy) d\nu(y) d\mu(x)$$

for any bounded Borel function f. It follows directly that convolution is associative and is commutative if G is abelian. We also have

$$(\mu * \nu)(B) = \int \mu(By^{-1}) d\nu(y) = \int \nu(x^{-1}B) d\mu(x)$$

for any Borel set B, as in [8]. If the topology of G has a countable base, all these facts can be proved without recourse to [1] since we need only apply Fubini's theorem for the ordinary product measure.

Here is a fact relating convolution and the map $\mu \rightarrow \mu_\psi$:

LEMMA 1. *If G and H are topological groups, ψ is a continuous homomorphism of G into H, and μ and ν are regular Borel measures on G, then*

$$(\mu * \nu)_\psi = \mu_\psi * \nu_\psi.$$

PROOF. If B is a Borel set in H,

$$\begin{aligned}(\mu_\psi * \nu_\psi)(B) &= \int_H \mu_\psi(Bh^{-1}) d\nu_\psi(h) \\ &= \int_G \mu_\psi(B\psi(y)^{-1}) d\nu(y) \\ &= \int_G \mu(\psi^{-1}(B\psi(y^{-1}))) d\nu(y) \\ &= \int_G \mu(\psi^{-1}(B) y^{-1}) d\nu(y) \\ &= (\mu * \nu)(\psi^{-1}(B)) = (\mu * \nu)_\psi(B), \qquad \text{q.e.d.}\end{aligned}$$

To define a "random walk" on an abelian topological group G, we will have regular Borel probability measures μ and ν on G, where ν is an "initial distribution" and μ defines a "transition probability." Thus we want to define a probability measure Pr on the infinite product

$$G^\omega = G \times G \times G \cdots = G_0 \times G_1 \times G_2 \cdots$$

or space of sequences $\{x_n : n=0, 1, 2, \cdots, x_n \in G\}$ such that the variables $x_0, x_1 - x_0, \cdots, x_{n+1} - x_n, \cdots$ are all independent and $\Pr(x_0 \in B) = \nu(B)$, $\Pr(x_{n+1} - x_n \in B) = \mu(B)$ for any $n \geqq 0$ and Borel set $B \subset G$.

To do this, we first define a product measure $\nu \otimes \mu \otimes \mu \otimes \cdots$ on G^ω. If $H_n = G_0 \times G_1 \times \cdots \times G_n$, then we have a regular Borel measure

$$\nu \otimes \mu^{(n)} = \nu \otimes \mu \otimes \cdots \otimes \mu$$

on H_n for each n (it is easy to check that the operation $\otimes$ is associative). For each n there is a natural mapping ψ_n of H_{n+1} onto H_n, and clearly

$$(\nu \otimes \mu^{(n+1)})_{\psi_n} = \nu \otimes \mu^{(n)}.$$

Thus all the hypotheses of a theorem of Bochner [**2**, Theorem 5.1.1, p. 120] on inverse limits of measures are satisfied. Hence there is a probability measure P on G^ω such that if ϕ_n is the natural mapping of G onto H_n for any n, $P_{\phi_n} = \nu \otimes \mu^{(n)}$. It is clear that for any closed set $F \subset G^\omega$ (for the product topology),

$$F = \bigcap_{n=1}^{\infty} \phi_n^{-1}(\phi_n(F)).$$

Thus P is defined on all Borel sets in G^ω. Since there are compact sets $\bigcap_{n=1}^{\infty} \phi_n^{-1}(K_n)$ in G^ω, where K_n is compact in H_n, with measure arbitrarily close to 1, P is regular on closed sets. It follows easily that P is also regular on open sets and hence on all Borel sets.

Finally, let RW be the mapping of G^ω into itself defined by

$$RW(y_0, y_1, y_2, \cdots) = (y_0, y_0 + y_1, y_0 + y_1 + y_2, \cdots).$$

RW is clearly continuous, so that $\Pr = P_{RW}$ is a regular Borel measure on G^ω. The measure space $(G^\omega, \Pr)$ will be called the random walk on G with initial distribution ν and transition probability μ. (If G has a countable base for its topology, P is the usual infinite product measure and our other definitions also coincide essentially with the usual ones.) A random walk will be called *recurrent* if for every open set $U \subset G$,

$$\Pr(x_n \in U \text{ for infinitely many } n) = 1.$$

THEOREM 1. *A locally compact abelian group G has a recurrent random walk if and only if* (a) *for every open subgroup H of G, G/H is countable, and* (b) *G does not have a discrete subgroup free abelian on three generators.*

To prove Theorem 1 we shall first generalize Lemma 1 of [2]. If $(G^\omega, \Pr)$ is a recurrent random walk on G and U is an open subgroup of G, let t_0 be the least n such that $x_n \in U$, t_1 the next least, and so on; each t_j, $j=0, 1, 2, \cdots$, is defined with probability one and equal to a Borel function on its domain. Let $x_n = x(n)$ and $y_n = x(t_n)$ for any n.

Now y_0 has a distribution ν_U and $y_1 - y_0$ has a distribution μ_U, where ν_U and μ_U are regular Borel measures on G. For given values of $t_0, \cdots, t_n$, $t_{n+1} = t_n + k$ if and only if k is the least positive integer such that $x(t_n + k) - x(t_n) \in U$, since U is a subgroup. The set of variables $x(t_n + k) - x(t_n)$, $k > 0$, is independent of $x(t_0), \cdots, x(t_n)$, by a standard argument. Thus $y_{n+1} - y_n$ is independent of $y_0, \cdots, y_n$ and has distribution μ_U for all n. Thus if μ_n is the distribution of $(y_0, \cdots, y_n)$ on H_n, the measures μ_n have an inverse limit $\Pr_U$ on G^ω, concentrated in U^ω, so that $(U^\omega, \Pr_U)$ is a well-defined random walk on U with initial distribution ν_U and transition probability μ_U. The map $\{x_n\} \to \{x(t_n)\}$ of G^ω into U^ω obviously carries Pr into $\Pr_U$.

LEMMA 2. *$(U^\omega, \Pr_U)$ is recurrent.*

PROOF. For any open set $V \subset U$, V is open in G, so that

$$\Pr\{x_n \in V \text{ infinitely often}\} = 1,$$

and since if $x_n \in V$, $n = t_m$ for some m,

$$\begin{aligned} &\Pr\{x(t_m) \in V \text{ infinitely often}\} \\ &\quad = \Pr_U\{y_m \in V \text{ infinitely often}\} = 1, \qquad \text{q.e.d.} \end{aligned}$$

Let us now prove Theorem 1, beginning with "only if." Suppose G has a recurrent random walk. G cannot have an open subgroup H with G/H uncountable since for each n at most countably many cosets of H can be reached with positive probability by time n. To prove (b), we use the fact that G has an open subgroup H of the form $C \oplus V$ where C is compact and V is a vector group (see [7, pp. 160–162]; the restriction to second-countable topologies is removed in the second edition). Suppose J were a discrete subgroup of G, free on three generators. The sum of the ranks of G/H and $J \cap H$ is then at least 3, and the rank of $J \cap H$ is less than or equal to the dimension of V as a real vector space, since J is discrete. Let F be a free abelian subgroup of G/H of maximal rank, and let U be the union of all cosets of H belonging to F. Then U is an open subgroup of G of the form $C \oplus R^k \oplus Z^m$, where R and Z are the additive groups of the real numbers and the integers respectively and $k + m \geqq 3$. (The relative topology of U is the product topology indicated by the direct sum since G/H is discrete.) By Lemma 2, there exists a recurrent random walk on U. Applying the natural homomorphism ψ of U onto $R^k \oplus Z^m$ and using Lemma 1, we obtain a recurrent random walk on $R^k \oplus Z^m$, $k+m \geqq 3$, which is impossible (see [**3**, p. 31]).

To prove "if," again let $H = C \oplus V$ be an open subgroup of G with C compact, V a vector group, and G/H countable. Since G/H is discrete, (b) implies that the sum of the rank of G/H and the dimension of V is at most 2.

Thus G/C has a countable dense subgroup M of rank at most 2. According to [**4**], there is a random walk on M recurrent on M as a discrete group and hence also with its relative topology from G/C. Thus there is a recurrent random walk on G/C with its transition probability μ concentrated in countably many points. We may assume μ has no mass at 0, according to the proof in [**4**].

Let ρ be a normalized Haar measure on C and each of its cosets. Let ξ be the "product" of μ and ρ defined by

$$\xi(B) = \int_{G/C} \rho(B \cap x)\, d\mu(x)$$

for each Borel set B. (Since μ is purely atomic, any function is measurable for it.) Probably ξ is the transition probability of a recurrent random walk, but the proof seems to be easier if we modify it slightly, letting $\sigma = (\xi + \rho_0)/2$, where ρ_0 is ρ confined to C.

It is easy to see that if a transition measure α defines a recurrent random walk on a group K for some initial distribution, and δ is the unit mass at the identity of K, then $\lambda\alpha + (1-\lambda)\delta$ is also the transition probability for a recurrent random walk with the same initial distribution if $0 < \lambda \leqq 1$. Thus $(\mu+\delta)/2$ defines a recurrent random walk on G/C for any initial distribution, recurrent on M as a discrete group.

Let U be any open subset of G, and let A be a coset of C such that $\rho(A \cap U) > 0$ and $A \in M$. Then a random walk on G with initial distribution concentrated at 0 and transition probability σ has probability 1 of being in A infinitely often since it induces a random walk on G/C with transition probability $(\mu + \delta)/2$. If $x_n \in A$, then $\Pr(x_{n+1} \in A \cap U) = \frac{1}{2}\rho(A \cap U) > 0$ since μ has no mass at 0. The events that $x_{n(m)+1} \in A \cap U$, where $n(m)$ is the mth integer n such that $x_n \in A$, are independent for different m. Thus

$$\Pr(x_n \in U \text{ infinitely often}) \geqq \Pr(x_n \in A \cap U \text{ infinitely often}) = 1, \qquad \text{q.e.d.}$$

The following result shows that "locally compact" cannot be replaced by "complete separable metric" in Theorem 1.

THEOREM 2. *Let G be any countable abelian group. Then there is a Hausdorff topology T on the group $A = Z \oplus G$ with a countable base for which it is a topological group, Z is dense, and G has discrete relative topology.*

PROOF. Let $\{g_n\}_{n=1}^{\infty}$ be an enumeration of G in which every element occurs infinitely many times and let u_n be the element $(2^{2^n}, g_n)$ of A, $n=1, 2, \cdots$. For each n, let U_n be the set of all elements of A which can be written as finite sums $\sum a_i u_i$ where the a_i are integers such that $|a_i| \leqq 2^{i-n}$ for all i. (Thus $a_i = 0$ for $i < n$.)

Then we have $U_n = -U_n$ for all n and $U_{n+1} + U_{n+1} \subset U_n$. Thus the U_n form a neighborhood-base at 0 for a topology $\mathfrak{J}$ on A making it a topological group (see [7, p. 55]). If $u \in U_n$ for some n, then $u = \sum a_i u_i$ with $a_i = 0$ for $i \geqq k$ for some k. Thus $U_k + u \subset U_n$, so that the U_n are actually open sets (belong to $\mathfrak{J}$).

The sets $U_n + a$, $a \in A$, $n = 1, 2, \cdots$, form a base for $\mathfrak{J}$. Hence to show that Z is dense in A for $\mathfrak{J}$ it suffices to show that for any n and $a = (z, g)$ there is an integer w such that $(w, 0) \in (z, g) + U_n$, or $(z - w, g) \in -U_n = U_n$. This is clear since $g = g_r$ for some $r \geqq n$ and $u_r = (2^{2^r}, g_r) \in U_n$; let $w = z - 2^{2^r}$.

If $\sum_{j=1}^{N} a_j 2^{2^j} = 0$ with $|a_j| \leqq 2^{j-1}$ for all j, a_j integers, then $a_j = 0$ for all j since

$$\left| \sum_{j=1}^{n} a_j 2^{2^j} \right| \leqq n \cdot 2^{n-1} \cdot 2^{2^n} < 2^{2n+2^n} \leqq 2^{2^{n+1}}$$

while

$$\left| \sum_{j=n+1}^{N} a_j 2^{2^j} \right| \geqq 2^{2^{n+1}} \quad \text{if } a_{n+1} \neq 0.$$

Thus the only element of U_1 of the form $(0, g)$ is $(0, 0)$ so that the subgroup G of elements of this form has discrete relative topology. Finally if $u = (z, g) \neq (0, 0)$, then U_n and $U_n + u$ are disjoint if $0 < |z| < 2^{2^n}$ or if $z = 0$, $g \neq 0$, so that $\mathfrak{J}$ is Hausdorff. This completes the proof.

COROLLARY. *There is a topology for the free abelian group Z^4 on four generators for which it has a recurrent random walk, and in which there is a subgroup free on three generators with discrete relative topology.*

PROOF. Let $G = Z^3$ in Theorem 2 and recall that Z has a recurrent random walk.

To see that Theorem 1 fails for complete separable metric groups it suffices to take a completion of the group described in the corollary with its topology as given by Theorem 2. The uniformity of this topology is metrizable, and the completion has a natural structure of complete metric group with a dense cyclic subgroup and a discrete subgroup free on three generators.

One might conjecture, however, that a complete abelian topological group has a recurrent random walk if and only if it has a compact subgroup whose quotient has a dense subgroup of rank at most 2. "If" follows from the proof of Theorem 1.

I would like to thank the referee for suggesting a number of improvements and corrections in the proofs of both theorems.

References

1. W. W. Bledsoe and A. P. Morse, *Product measures*, Trans. Amer. Math. Soc. **79** (1955), 173–215.

2. S. Bochner, *Harmonic analysis and the theory of probability*, Univ. of California Press, Berkeley, Calif., 1955.

3. K. L. Chung and W. H. J. Fuchs, *On the distribution of values of sums of random variables*, Mem. Amer. Math. Soc. No. 6 (1951), 12 pp.

4. R. M. Dudley, *Random walks on abelian groups*, Proc. Amer. Math. Soc. **13** (1962), 447–450.

5. P. R. Halmos, *Measure theory*, Van Nostrand, Princeton, N. J., 1950.

6. E. Marczewski and P. Sikorski, *Measures in nonseparable metric spaces*, Colloq. Math. **1** (1948), 133–139.

7. L. S. Pontrjagin, *Topological groups*, Princeton Univ. Press, Princeton, N. J., 1946.

8. Karl Stromberg, *A note on the convolution of regular measures*, Math. Scand. **7** (1959), 347–352.

University of California, Berkeley

Reprinted from JOURNAL OF APPROXIMATION THEORY
All Rights Reserved by Academic Press, New York and London

Vol. 10, No. 3, March 1974
Printed in Belgium

Metric Entropy of Some Classes of Sets with Differentiable Boundaries*

R. M. DUDLEY

Mathematics Department, Massachusetts Institute of Technology, Cambridge, Massachusetts 02139

Communicated by G. G. Lorentz

Let $I(k, \alpha, M)$ be the class of all subsets A of R^k whose boundaries are given by functions from the sphere S^{k-1} into R^k with derivatives of order $\leqslant \alpha$, all bounded by M. (The precise definition, for all $\alpha > 0$, involves Hölder conditions.) Let $N_d(\epsilon)$ be the minimum number of sets required to approximate every set in $I(k, \alpha, M)$ within ϵ for the metric d, which is the Hausdorff metric h or the Lebesgue measure of the symmetric difference, d_λ. It is shown that up to factors of lower order of growth, $N_d(\epsilon)$ can be approximated by $\exp(\epsilon^{-r})$ as $\epsilon \downarrow 0$, where $r = (k-1)/\alpha$ if $d = h$ or if $d = d_\lambda$ and $\alpha \geqslant 1$. For $d = d_\lambda$ and $(k-1)/k < \alpha \leqslant 1$, $r \leqslant (k-1)/(k\alpha - k + 1)$. The proof uses results of A. N. Kolmogorov and V. N. Tikhomirov [4].

1. INTRODUCTION

We consider classes of subsets A of R^k whose boundaries ∂A are defined by maps of the sphere S^{k-1} into R^k with bounded derivatives of order $\leqslant\alpha$ for some $\alpha < \infty$. Using Hölder conditions, such classes are defined for all $\alpha > 0$ (not necessarily integral). Given k, α, and a uniform bound M on derivatives of orders $\leqslant\alpha$ (for more detailed definitions see Section 2 below), we have a class $I(k, \alpha, M)$ of subsets of R^k. We ask: given $\epsilon > 0$, how many sets are needed to form an ϵ-dense set in $I(k, \alpha, M)$, i.e., to approximate each set within ϵ, for the Hausdorff metric h or for the metric d_λ which is the Lebesgue measure of the symmetric difference. We find that as $\epsilon \downarrow 0$, the required number $N(\epsilon)$, of sets, is approximated by $\exp(\epsilon^{-r})$ for a suitable exponent r depending on k, α, M and the choice of metric h or d_λ; we write $r = r_h$ or $r = r_\lambda$, respectively. The approximation is proved only in the sense that given $t < r < s$, $\exp(\epsilon^{-t}) < N(\epsilon) < \exp(\epsilon^{-s})$ for ϵ small enough.

* This research was partially supported by National Science Foundation Grant GP-29072.

E. Giné et al. (eds.), *Selected Works of R.M. Dudley*, Selected Works in Probability and Statistics,
DOI 10.1007/978-1-4419-5821-1_26, © Springer Science+Business Media, LLC 2010

Thus $N(\epsilon)$ is asymptotic to $\exp(\epsilon^{-r})$ with suitable factors of lower order of growth inserted, but here we do not find these factors precisely.

Theorem 3.1 below relates r to the given k and α. This result extends previously known theorems on metric entropy of classes of functions (Kolmogorov–Tikhomirov [4, Theorem XV]; Clements [2, Theorem 3]; Lorentz [6, Theorem 10]). The main difficulty in the extension results from the fact that boundaries of sets in $I(k, \alpha, M)$ are not restricted except for differentiability and may intersect themselves in complicated ways. For example, there is a set in $I(2, 17, 1)$ with infinitely many components. We have

$$r_h(I(k, \alpha, M)) = (k-1)/\alpha;$$
$$r_\lambda(I(k, \alpha, M)) = (k-1)/\alpha \quad \text{if} \quad \alpha \geqslant 1;$$
$$r_\lambda(I(k, \alpha, M)) \leqslant (k-1)/(k\alpha - k + 1) \quad \text{if} \quad (k-1)/k < \alpha \leqslant 1,$$

where I conjecture that the last inequality for r_λ is also an equality.

The exponent $(k-1)/\alpha$ for classes of functions goes back to Kolmogorov and Tikhomirov [4]. Relations between sets and boundary functions are developed in the preliminary Section 2. The boundary functions on spheres need not be one-to-one.

In Section 4 we consider the class $C(U)$ of all convex closed subsets of any fixed bounded open set $U \subset R^k$. We find

$$r_h(C(U)) = r_\lambda(C(U)) = (1/2)(k-1).$$

Thus convex sets behave like sets with exactly twice differentiable boundaries, as is perhaps not surprising. (On R^1, a convex function f has a second derivative f'' which is a positive Radon measure; even when f'' is a function, it need not satisfy any Hölder condition.) The proof in Section 4, however, uses convexity rather than second derivatives *per se*.

While the results of this paper were found with probabilistic applications in view [3, Theorems 4.2 and 4.3], it seemed appropriate to give them a separate presentation.

2. Preliminaries: Boundaries

Let (S, d) be any metric space. Given $\epsilon > 0$, let $N(S, \epsilon)$ be the smallest number of sets of diameter $\leqslant 2\epsilon$ which cover S. The *exponent of entropy* of S is defined by

$$r(S) \equiv r_d(S) \equiv \limsup_{\epsilon \downarrow 0} [\log \log N(S, \epsilon)]/|\log \epsilon|.$$

(If $r(S) < \infty$, (S, d) must be totally bounded.)

For any two subsets A, B of S, we have the Hausdorff distance $h(A, B)$ defined as follows:

$$h_1(A, B) = \sup_{x \in A} \inf_{y \in B} d(x, y),$$

$$h(A, B) = \max[h_1(A, B), h_1(B, A)].$$

For closed sets, h is a metric. A set $A \subset S$ is called *ϵ-dense* iff $h(A, S) \leqslant \epsilon$.

For measurable subsets of a measure space (S, μ) modulo null sets, there is a metric d_μ defined by $d_\mu(A, B) = \mu(A \triangle B)$, where $A \triangle B$ is the symmetric difference $(A \sim B) \cup (B \sim A)$.

In the following, λ denotes Lebesgue measure. Exponents of entropy for d_λ will be written r_λ.

Now we define spaces of functions on spheres "with bounded derivatives of orders $\leqslant \alpha$" for any $\alpha > 0$. Let β be the greatest integer $< \alpha$ and $\gamma = \alpha - \beta > 0$. For any open set $U \subset R^k$, let $F(U, \alpha)$ be the set of all real functions f on U such that:

(a) the partial derivatives $D^p f = \partial^{|p|} f / \partial x_1^{p_1} \cdots \partial x_k^{p_k}$ exist for $|p| \equiv p_1 + \cdots + p_k \leqslant \beta$;

(b) $\|f\|_\alpha < \infty$ where

$$\|f\|_\alpha \equiv \sup\{|D^p f(x) - D^p f(y)|/|x - y|^\gamma + |D^q f(x)|: \\ |q| \leqslant |p| = \beta, x \neq y \in U, x \in U\}.$$

Let S^{k-1} be the unit sphere in R^k:

$$S^{k-1} = \{x \in R^k: |x|^2 \equiv x_1{}^2 + \cdots + x_k{}^2 = 1\}.$$

We can cover S^{k-1} by finitely many coordinate patches V_j so that there are C^∞ isomorphisms $\Phi_j: U \to V_j$ where U is the open ball $\{y: |y| < 1\} \subset R^{k-1}$. We can assume that Φ_j is actually a C^∞ isomorphism from a neighborhood W of the closure of U into S^{k-1}. Then each partial derivative of Φ_j is uniformly bounded and the vectors $\partial\Phi_j/\partial x_i$ for $i = 1, \ldots, k - 1$ are linearly independent on W.

We define $F(V_j, \alpha)$ as the set of all real-valued functions f on V_j such that $f \circ \Phi_j \in F(U, \alpha)$. Let $F(S^{k-1}, \alpha)$ be the set of all real-valued functions f on S^{k-1} such that the restriction of f to V_j is in $F(V_j, \alpha)$ for each j. Then let $\|f\|_\alpha = \sup_j \|f \circ \Phi_j\|_\alpha$. This norm $\|\cdot\|_\alpha$ depends on the choice of V_j and Φ_j but is topologically equivalent to the norms defined by other allowed choices of V_j and Φ_j.

Taking the k-fold Cartesian product of copies of $F(S^{k-1}, \alpha)$, we obtain a Banach space $(F^{(k)}(S^{k-1}, \alpha), \|\cdot\|_\alpha)$ of functions from S^{k-1} into R^k, where $\|(f_1, \dots, f_k)\|_\alpha = \max_j \|f_j\|_\alpha$. Now for $M > 0$, let

$$G(k, \alpha, M) = \{f \in F^{(k)}(S^{k-1}, \alpha): \|f\|_\alpha \leqslant M\}.$$

Here we recall a basic definition from algebraic topology. Let f and g be two maps from a topological space S into another space T. Then f and g are called *homotopic* iff there is a continuous F from $[0, 1] \times S$ into T such that $F(0, \cdot) \equiv f$ and $F(1, \cdot) \equiv g$. F is called a *homotopy* of f and g.

Next we shall define an "interior" $I(f)$ for each $f \in G(k, \alpha, M)$, so that, e.g., if f is the identity on S^{k-1}, $I(f)$ is the usual open unit ball. The following definition was kindly suggested to me by J. Munkres.

DEFINITION. For any continuous map f of a topological space S into another space T, let $I(f)$ be the set of all $x \in T \sim \text{range}(f)$ such that in $T \sim \{x\}$, f is not homotopic to any constant map of S into a point $t \in T \sim \{x\}$.

The proof of the following fact was also told me by J. Munkres.

LEMMA 2.1. *Suppose F is a homotopy of f and g. Then $I(f) \triangle I(g) \subset \text{range}\, F$.*

Proof. Suppose $x \in I(f) \sim I(g)$. If $x \notin \text{range}(F)$, then f and g are homotopic in $T \sim \{x\}$. Clearly homotopy is transitive. Since g is homotopic to a constant map in $T \sim \{x\}$, so is f, a contradiction. The proof is complete.

If f is the identity map of S^{k-1} into R^k, then $I(f)$ is the usual open unit ball by well-known theorems of algebraic topology. Also, if f and g are homotopic in $R^k \sim \{0\}$, then $0 \in I(g)$. Thus the above definition seems broad enough to cover cases of interest.

Let $I(k, \alpha, M) = \{I(f): f \in G(k, \alpha, M)\}$.

3. THE EXPONENTS OF ENTROPY OF $I(k, \alpha, M)$

In the following, I conjecture that equality holds in (3.4). It seems that a proof might require construction of some rather pathological sets.

THEOREM 3.1. *Let $0 < \alpha < \infty$ and $0 < M < \infty$. Then*

$$r_h(I(k, \alpha, M)) = (k-1)/\alpha; \tag{3.2}$$

If $\alpha \geqslant 1, \quad r_\lambda(I(k, \alpha, M)) = (k-1)/\alpha;$ (3.3)

If $(k-1)/k < \alpha \leqslant 1, \quad r_\lambda(I(k, \alpha, M)) \leqslant (k-1)/(k\alpha - k + 1).$ (3.4)

If $0 < \alpha < 1$, $r_\lambda(I(k, \alpha, M)) \geqslant (k-1)/\alpha$.

Proof. Let $F(U, \alpha, \gamma) = \{f \in F(U, \alpha): \|f\|_\alpha \leqslant \gamma\}$, using the definitions in Section 2. Kolmogorov and Tikhomirov [4, Sect. 5, Theorems XIII–XV] have shown that for any bounded open $U \subset R^k$ and $0 < \zeta < \infty$, $r_s F(U, \alpha, \zeta) = (k-1)/\alpha$ where s is the supremum metric, $s(f, g) = \sup\{|f(x) - g(x)|\}$. By definition of $G(k, \alpha, M)$ it follows that $r_s G(k, \alpha, M) \leqslant (k-1)/\alpha$, proving $r_h(I(k, \alpha, M)) \leqslant (k-1)/\alpha$.

Now suppose $f, g \in G(k, \alpha, M)$ and $s(f, g) \leqslant \epsilon$, where $\epsilon > 0$. Let $F(t, x) \equiv (1-t)f(x) + tg(x)$ for $0 \leqslant t \leqslant 1$, $x \in S^{k-1}$. By Lemma 2.1, $d_\lambda(I(f), I(g)) \leqslant \lambda(\text{range } F)$.

If $\alpha \geqslant 1$, the maps in $G(k, \alpha, M)$ are uniformly Lipschitzian. Thus $\lambda(\text{range } F) = O(\epsilon)$ as $\epsilon \downarrow 0$, uniformly for $f \in G(k, \alpha, M)$. Hence $r_\lambda(I(k, \alpha, M)) \leqslant (k-1)/\alpha$.

Next let $(k-1)/k < \alpha \leqslant 1$. There is a $K < \infty$ such that for $0 < \delta \leqslant 1$, there is a set $E_\delta \subset S^{k-1}$ such that for all $x \in S^{k-1}$, $|x - y| \leqslant \delta^{1/\alpha}$ for some $x \in E_\delta$, where E_δ has at most $K\delta^{(1-k)/\alpha}$ elements. Then for any $f \in G(k, \alpha, M)$ and $z \in \text{range} f$ there is an $x \in E_\delta$ with $|f(x) - z| \leqslant N\delta$ for some $N > M$.

Let c_k be the volume of the unit ball in R^k. Given $\epsilon > 0$ let

$$\delta = [\epsilon/Kc_k 4^k N^k]^{\alpha/(k\alpha - k + 1)}.$$

Then $\lambda\{x : \exists y: |f(y) - x| < 3N\delta\} \leqslant 4^k N^k K c_k \delta^{(k\alpha - k + 1)/\alpha} = \epsilon$ if $\delta \leqslant 1$, as is true for ϵ small enough. To obtain a $3N\delta$-dense set in $G(k, \alpha, M)$ it suffices to approximate functions within $N\delta$ at each point of E_δ. Hence for ϵ small,

$$\begin{aligned} N(I(k, \alpha, M), \epsilon, d_\lambda) &\leqslant \exp\{K\delta^{(1-k)/\alpha} \log[(2k+1)^k/\delta^k]\} \\ &\leqslant \exp\{C_k \epsilon^{(1-k)/(k\alpha - k + 1)} \,|\log \epsilon|\} \end{aligned}$$

for some constant C_k, so (3.4) follows.

To prove $\geqslant$ and hence equality in (3.2) and (3.3) we use the following fact, due to G. F. Clements [2, Theorem 3]. The proof here is different and seems simpler.

LEMMA 3.5 (Clements). *Let V be a bounded open set in R^{k-1}, $k \geqslant 2$, $\alpha > 0$, and $0 < \gamma < \infty$. Then $r_1(F(V, \alpha, \gamma)) \geqslant (k-1)/\alpha$ where r_1 is the exponent of entropy for the L^1 metric $d_1(f, g) = \int_V |f - g|\, d\lambda$.*

Proof. We can assume V is the open cube $\{x: 0 < |x_j| < 1, j = 1,\ldots, k-1\}$. Let f be a positive C^∞ function with support in V. Let $\|f\|_\alpha = N < \infty$. For $Q \geqslant 1$ and $t \in R^{k-1}$ let $g(x) = f(Qx + t)$. Then for some $Z < \infty$, $\|g\|_\alpha \leqslant ZQ^\alpha$ for all $Q \geqslant 1$.

For each positive integer Q there exist Q^{k-1} such functions g_j with disjoint support, $j = 1,\ldots, Q^{k-1}$. For each set $A \subset \{1,\ldots, Q^{k-1}\}$, let $g_A = \sum_{j \in A} g_j$. We shall show that there are many such sets A, different in many places. This type of result seems to be known, but the following proof seems short enough to include, and I know no explicit references for the result.

LEMMA 3.6. *For any positive integer n and any set B with n elements, there is a collection of sets $E_i \equiv E(i) \subset B$, $i = 1,..., m$, such that $m \geqslant e^{n/8}$ and such that for $i \neq j$, $E_i \triangle E_j$ has at least $n/5$ elements.*

Proof. Given any set $E \subset B$, the number of sets $F \subset A$ such that $E \triangle F$ has at most $n/5$ elements is $2^n B(n/5, n, 1/2)$ where $B(r, n, p)$ is the probability of at most r successes in n independent trials with probability p of success in each trial. According to Kolmogorov's exponential bound [5, p. 254],

$$B(n/5, n, 1/2) \leqslant \exp(-.126n) < \exp(-n/8).$$

Thus we can inductively choose the sets E_i with $m \geqslant e^{n/8}$, proving Lemma 3.6.

Now the functions $h_A \equiv \gamma g_A / Q^\alpha Z$ all belong to $F(V, \alpha, \gamma)$. Let $\kappa = \int |f| \, d\lambda > 0$. Then for $i \neq j$,

$$\int | h_{E(i)} - h_{E(j)} | \, d\lambda \geqslant Q^{k-1} \gamma\kappa / 5ZQ^{k-1+\alpha} = \gamma\kappa / 5ZQ^\alpha.$$

Let $\epsilon = \gamma\kappa/5ZQ^\alpha$. Then Q is proportional to $\epsilon^{-1/\alpha}$. Letting $Q \to \infty$ and applying Lemma 3.6 yields, for some constant $\beta > 0$,

$$N(F(V, \alpha, \gamma), \epsilon) \geqslant \exp\{\beta\epsilon^{(1-k)/\alpha}\}.$$

Thus Lemma 3.5 is proved.

There is a one-to-one C^∞ map $G = (G_1, ..., G_k)$ of S^{k-1} into R^k with a flat face. Here "flat face" means there is an open set $U \subset S^{k-1}$ such that $G_1(U) = \{0\}$, and for some $\delta > 0$ and all t such that $| t | < \delta$ and $x \in U$, $G(x) + (t, 0,..., 0) \in I(G)$ iff $t > 0$. Let $H = (G_2, ..., G_k)$. Then $H(U)$ is an open set $V \subset R^{k-1}$. For some $M_0 < \infty$, $G \in G(k, \alpha, M_0)$. Given any $M > 0$, we can replace G by a small multiple of itself and assume $M_0 < M/2$. We can also assume $V = \kappa C$ where $\kappa > 0$ and C is the open unit cube in R^{k-1}. Then for some small enough $\zeta > 0$, with $\zeta < \delta$, all the following functions $\varphi_A \in G(k, \alpha, M)$:

$$\begin{aligned} \varphi_A(x) &= G(x) \qquad \text{for} \quad x \notin U \\ &= G(x) + (\zeta h_A(H(x)/\kappa), 0,..., 0) \qquad \text{for} \quad x \in U, \end{aligned}$$

where h_A is as in the proof of Lemma 3.5, with $\gamma \leqslant \min(1, M_0)$.

For any sets A and $B \subset \{1,..., Q^{k-1}\}$,

$$d_\lambda(I(\varphi_A), I(\varphi_B)) = \zeta \int_V | h_A - h_B | \, d\lambda,$$

for Q large enough. Thus by Lemma 3.5 and its proof, we have equality in (3.2) and (3.3) for all $M > 0$ and Theorem 3.1 is proved.

4. Convex Sets

Let $C(U)$ denote the class of all convex closed subsets of U. It turns out that the exponent of entropy of $C(U)$, for U bounded, is $(1/2)(k-1)$ although second derivatives of boundaries of polyhedra in $C(U)$ are only measures, not functions.

Theorem 4.1. *Let U be a bounded open set in R^k. Then* $r_\lambda(C(U)) = r_h(C(U)) = (1/2)(k-1)$.

Proof. We choose a fixed point $\zeta \in U$. Let $s = h(U, \{\zeta\})$. We have for any $C, D \in C(U)$ by [1, p. 41, 5]:

$$d_\lambda(C, D) \leqslant 2c_k[-s^k + (s + h(C, D))^k] \leqslant Nh(C, D) \tag{4.2}$$

where N depends on k and s but not on C, D. Thus to prove $r(C(U)) \leqslant (1/2)(k-1)$ we need only consider the Hausdorff metric.

Lemma 4.3. *Suppose given vectors x, y, u, v in R^k such that $(x-y, u) \geqslant 0$ and $(x-y, v) \leqslant 0$. Then*

$$|x + u - y - v| \geqslant \max(|x-y|, |u-v|).$$

Proof.

$$\begin{aligned} |x+u-y-v|^2 &= |x-y|^2 + |u-v|^2 + 2(x-y, u-v) \\ &\geqslant |x-y|^2 + |u-v|^2. \end{aligned}$$

Q.E.D.

A convex set C will be called *analytic* iff there is an entire analytic function f such that $C = \{x \in R^k : f(x) \leqslant 1\}$, and the gradient of f is nonzero on the boundary ∂C. It is known that analytic convex sets are h-dense in the class of all bounded convex sets [1, pp. 36–37]. If C is analytic and $p \in \partial C$, let $\varphi(p) = \operatorname{grad} f(p)/|\operatorname{grad} f(p)|$. Then φ is a continuous 1–1 map of ∂C onto S^{k-1}. Let $e(p, q)$ be the (smallest nonnegative) angle between $\varphi(p)$ and $\varphi(q)$. Then $0 \leqslant e(p, q) \leqslant \pi$. Let $d(p, q) = |p - q|$.

Lemma 4.4. *Given a bounded open $U \subset R^k$, there is an $M < \infty$ such that whenever $0 < \delta < 1$, and C is any analytic convex subset of U, there is a set $A \subset \partial C$ with* $\operatorname{card}(A) \leqslant M\delta^{1-k}$ *such that A is δ-dense in ∂C for $d + e$.*

Proof. Let B be a fixed ball such that $x + y \in B$ whenever $x \in U$ and $|y| \leqslant 1$. Then there is a constant $S < \infty$ such that whenever $0 < \epsilon < 1$ there is an ϵ-dense set $B_\epsilon \subset \partial B$ with $\operatorname{card}(B_\epsilon) \leqslant S\epsilon^{1-k}$.

Let C be convex and analytic, $C \subset U$. Then for every $p \in \partial B$, there is a unique nearest point $n(p) \in \partial C$, with $|p - n(p)| \geqslant 1$. The function $n(\cdot)$ maps ∂B 1–1 onto ∂C. Suppose $q \in \partial B$ and $|p - q| < \epsilon$. Let $u = p - n(p)$, $v = q - n(q)$. Then we can apply Lemma 4.3 with $x = n(p)$ and $y = n(q)$ to conclude $|n(p) - n(q)| < \epsilon$ and $|u - v| < \epsilon$. Let θ be the angle between u and v, so that $e(n(p), n(q)) = \theta$. Let $u_1 = u/|u|$, $v_1 = v/|v|$. Since $|u| \geqslant 1$ and $|v| \geqslant 1$, we have $|u_1 - v_1| < \epsilon$. Also $|u_1 - v_1| = 2\sin(\theta/2)$. We know $\theta \leqslant \pi \sin(\theta/2)$ for $0 \leqslant \theta \leqslant \pi$ by concavity. Thus $e(n(p), n(q)) \leqslant \pi\epsilon/2 < 2\epsilon$. Hence we can let $M = 2^k S$, $A = \{n(p): p \in B_\epsilon\}$, proving Lemma 4.4.

LEMMA 4.5. *Let C be an analytic convex set and $0 < \delta \leqslant \pi/4$. Let A be a δ-dense set in ∂C for $d + e$. Let C_A be the intersection of all half-spaces which include C and are bounded by hyperplanes supporting C (tangent to ∂C) at points of A. Then $h(C, C_A) \leqslant 2\delta^2$.*

Proof. Clearly $C_A \supset C$. Conversely let $x \in \partial C$ and choose $y \in A$ with $(d + e)(x, y) \leqslant \delta$. Let T_x be the tangent hyperplane to ∂C at x. Let u be the unit outward normal vector to ∂C and T_x at x. Then $x + \gamma u \in T_y$ for some $y > 0$. To maximize γ, we may assume $y \in T_x$ (this particular argument does not use analyticity). Now $\gamma \leqslant \delta \tan \delta \leqslant 2\delta^2$ since $\tan \theta \leqslant 2\theta$ for $0 \leqslant \theta \leqslant \pi/4$. For every $z \in C_A$ there is a nearest point $x \in C$, and $|z - x| \leqslant 2\delta^2$. Q.E.D.

Proof of Theorem 4.1. First we prove $r_h(C(U)) \leqslant (1/2)(k - 1)$. We can assume U is a cube. Let t be the diameter of U. We may assume $t \geqslant 2$. There is an $N < \infty$ such that $N \geqslant 1$ and whenever $0 < \epsilon \leqslant \pi/4$ there is an $\epsilon/2$-dense set $U_\epsilon \subset B$ with $\operatorname{card}(U_\epsilon) \leqslant N\epsilon^{-k}$ (where B is a fixed large ball $\supset U$ as in Lemma 4.4), and such that there is a $\tan^{-1}(\epsilon/3t)$-dense set $V_\epsilon \subset S^{k-1}$ for the angular metric e with $\operatorname{card}(V_\epsilon) \leqslant N\epsilon^{1-k}$.

Let W_ϵ be the set of all convex polyhedra $P \subset U$ formed by intersections of at most $M\epsilon^{(1-k)/2}$ half-spaces H_j (here M is as in Lemma 4.4) such that each hyperplane ∂H_j contains a point of U_ϵ and is orthogonal to a vector v in V_ϵ, and v is directed outward from H_j. Then

$$\operatorname{card}(W_\epsilon) \leqslant \exp\{[M\epsilon^{(1-k)/2}] \log[N^2\epsilon^{1-2k}]\}.$$

Hence

$$\limsup_{\epsilon \downarrow 0} (\log \log \operatorname{card} W_\epsilon)/|\log \epsilon| \leqslant (1/2)(k - 1).$$

Now we show that W_ϵ is 12ϵ-dense in $C(U)$ for h. To approximate a set $C \in C(U)$, we may assume C is analytic. We take the set $A \subset \partial C$ provided by Lemma 4.4 for $\delta = \epsilon^{1/2}$. At each $x \in A$ let T_x be the tangent hyperplane to ∂C. Let v_x be the unit outward normal vector at x. Choose $p_x \in U_\epsilon$ with

$| p_x - x - \epsilon v_x | \leqslant \epsilon/2$. Let J_x be a hyperplane passing through p_x, orthogonal to a vector in V_ϵ, and forming an angle with T_x less than $\tan^{-1}(\epsilon/3t)$. Let H_x be the half-space on the side of T_x containing x. Then $H_x \supset C$ since $h(\{p_x\}, C) \geqslant \epsilon/2$ and $(t + \epsilon/2)(\epsilon/3t) \leqslant \epsilon/2$. Let $C_\epsilon = \bigcap_{x \in A} H_x \supset C$.

Now take any $y \in \partial C$ and v_y as above. Take $x \in A$ such that $(d + e)(x, y) < \epsilon^{1/2}$. Then $| y - p_x | < 3\epsilon^{1/2}$ while T_y and T_x form an angle less than $2\epsilon^{1/2}$. We have $x \in C$ and $y \in H_x$. As in the proof of Lemma 4.5, it follows that $y + \gamma v_y \notin C_\epsilon$ for $\gamma \geqslant 12\epsilon$, so that $h(C, C_\epsilon) \leqslant 12\epsilon$. Since $C_\epsilon \in W_\epsilon$, we have proved $r(C(U)) \leqslant (1/2)(k - 1)$.

For the converse inequality, by (4.2) it suffices to consider the metric d_λ.

There is a $c > 0$ such that whenever $0 < \epsilon < 1$, there is a set $A_\epsilon \subset S^{k-1}$ with $\operatorname{card}(A_\epsilon) \geqslant c\epsilon^{1-k}$ such that $| x - y | \geqslant 4\epsilon$ for any distinct x and y in A_ϵ. For each $x \in A_\epsilon$, let C_x be the solid spherical cap cut from the unit ball $B_1 = \{y : | y | \leqslant 1\}$ by the hyperplane orthogonal to x and passing through $(1 - \epsilon^2/2)x$. For some constant $\alpha_k > 0$, $\lambda(C_x) \geqslant \alpha_k \epsilon^{k+1}$.

The caps C_x are disjoint. For an arbitrary set $E \subset A_\epsilon$, let

$$D_E = B_1 \sim \bigcup_{x \in E} C_x .$$

Each D_E is convex. We have $h(D_E, D_F) = \epsilon^2/2$ for $E \neq F$ so the proof is easily completed for h. For d_λ we apply Lemma 3.6; taking the sets $E_i = E(i)$ for A_ϵ, we have

$$\lambda(D_{E(i)} \triangle D_{E(j)}) \geqslant \alpha_k \epsilon^{k+1} c \epsilon^{1-k}/5 = \beta_k \epsilon^2$$

for some constant $\beta_k > 0$. Letting $\delta = \beta_k \epsilon^2/3$ we have

$$N(C(U), \delta) \geqslant \exp\{-\gamma_k \delta^{(1-k)/2}\}$$

for some constant $\gamma_k > 0$. Letting $\delta \downarrow 0$, Theorem 4.1 is proved.

Acknowledgment

I thank J. Munkres for showing me the relation between sets and boundaries developed in §2.

References

1. T. Bonnesen and W. Fenchel, "Theorie der Konvexen Körper," Springer Verlag, Berlin, 1934.
2. G. F. Clements, Entropies of several sets of real valued functions, *Pacific J. Math.* **13** (1963), 1085–1095.

3. R. M. Dudley, "Sample functions of the Gaussian process," *Annals of Probability* **1** (1973), 66–103.
4. A. N. Kolmogorov and V. M. Tikhomirov, ϵ-entropy and ϵ-capacity of sets in functional spaces, *Amer. Math. Soc. Transl.* (*Ser. 2*) **17** (1961), 277–364 (from *Uspekhi Mat. Nauk* **14** (1959), 3–86).
5. M. Loève, "Probability Theory," Van Nostrand, Princeton, N.J., 1963.
6. G. G. Lorentz, Metric entropy and approximation, *Bull. Amer. Math. Soc.* **72** (1966), 903–937.

Printed by the St Catherine Press Ltd., Tempelhof 37, Bruges, Belgium.

The Annals of Probability
1977, Vol. 5, No. 1, 140-141

WIENER FUNCTIONALS AS ITÔ INTEGRALS[1]

By R. M. Dudley

Massachusetts Intstitute of Technology

Every measurable real-valued function f on the space of Wiener process paths $\{W(t): 0 \leqq t \leqq 1\}$ can be represented as an Itô stochastic integral $\int_0^1 \varphi(t, \omega)\, dW(t, \omega)$ where φ is a nonanticipating functional with $\int_0^1 \varphi(t, \omega)^2\, dt < \infty$ for almost all ω.

Let $W(t, \omega)$ be a standard Wiener process, $W_t \equiv W(t) \equiv W(t, \cdot)$. A function $\varphi(t, \omega)$ is called nonanticipating iff for all $t \geqq 0$, $\varphi(t, \cdot)$ is measurable with respect to $\{W_s: 0 \leqq s \leqq t\}$. The Itô stochastic integral

$$f(\omega) \equiv \int_0^1 \varphi(t, \omega)\, d_t W(t, \omega)$$

is defined for any jointly measurable, nonanticipating φ such that for almost all ω, $\int_0^1 \varphi^2(t, \omega)\, dt < \infty$ (Gikhman and Skorokhod (1968), Chapter 1, Section 2). It is known that if $E \int_0^1 \varphi^2(t, \omega)\, dt < \infty$, then $Ef = 0$ and $Ef^2 < \infty$. Representation of an arbitrary measurable f as a stochastic integral was stated, but later retracted, by J.M.C. Clark (1970, 1971).

To illustrate our method, we will first show that for an arbitrary probability law P on R, there is a stochastic integral f with law P. Indeed there is a measurable g such that $g(W_{\frac{1}{2}})$ has law P. Let $\varphi(t, \omega) = 0$, $0 \leqq t \leqq \frac{1}{2}$. Let $\varphi(t, \omega) = 1/(1-t)$ for $\frac{1}{2} < t < \tau(\omega)$, the least time such that

$$\int_{\frac{1}{2}}^{\tau} 1/(1-t)\, dW(t, \omega) = g(W_{\frac{1}{2}})\,.$$

Then $\tau < 1$ a.s. since $\int_{\frac{1}{2}}^1 (1-t)^{-2}\, dt = +\infty$. Let $\varphi(t, \omega) = 0$ for $t \geqq \tau(\omega)$. This yields the desired result.

To prove the theorem stated in the abstract, let $g = \arctan f$. Then $|g| < \pi/2$ everywhere. For a sequence $t(n) \uparrow 1$ to be specified later, let B_n be the smallest σ-algebra with respect to which $W(t, \cdot)$ are measurable for all $t \in [0, t(n)]$. Sample continuity of $W(t)$ at 1 implies that g is measurable with respect to the σ-algebra generated by the union of the B_n. Thus, by martingale convergence, $g_n \equiv E(g \mid B_n) \to g$ almost surely. So $f_n \equiv \tan g_n \to f$ a.s., with f_n measurable (B_n).

Now, beginning with any sequence $s(n) \uparrow 1$ such as $s(n) = 1 - 1/n$, we choose $t(n)$ as a subsequence with $\Pr\{|f_n - f| > 1/n^3\} < 1/n^2$, so that if $x_n \equiv x_n(\omega) \equiv (f_{n+1} - f_n)(\omega)$, then

$$\Pr\{(n+1)|x_n(\omega)| > 4n^{-2}\} < 2n^{-2}\,. \tag{1}$$

Now we consider integrals of the form $X_t \equiv \int_a^t v(s)\, dW(s, \omega)$ for $0 \leqq a \leqq t$ and nonrandom v. Then X_t is a Gaussian process with mean 0 and covariance

Received January 19, 1976.

[1] This research was supported in part by National Science Foundation Grant MPS71-02972 A04.

AMS 1970 *subject classifications.* Primary 60H05; Secondary 60G15, 60G17, 60G40, 60J65.

Key words and phrases. Stochastic integral, Wiener process.

E. Giné et al. (eds.), *Selected Works of R.M. Dudley*, Selected Works in Probability and Statistics, DOI 10.1007/978-1-4419-5821-1_27,

$EX_tX_u = \min(h(t), h(u))$ where $h(t) = \int_a^t v^2(s)\,ds$. Thus, X_t has the same law as $W_{h(t)}$. If $\int_a^b v^2(s)\,ds = +\infty$, then for any x, a.s. there is some $t < b$ with $X_t = x$, $t > a$. Taking τ as the least such t, let G be the random variable $\int_a^{\tau(\omega)} v^2(s)\,ds = h\{\tau(\omega)\}$. Then G has the distribution of the least time $T = T(x, \omega)$ such that $W_T = x$ (starting at $W_0 = 0$, as usual). This law has density, for any $u \geqq 0$ (e.g., Itô–McKean (1965), page 25),

$$\Pr\{u \leqq T \leqq u + du\} = (2/\pi)^{\frac{1}{2}} \exp(-x^2/2u)\{|x|/2u^{\frac{3}{2}}\}\,du < (|x|/2u^{\frac{3}{2}})\,du$$

so $\Pr(T \geqq u) \leqq |x|/u^{\frac{1}{2}}$. Hence

$$\Pr(G \geqq u) \leqq |x|/u^{\frac{1}{2}}\,. \tag{2}$$

Now we define φ. Let $\varphi(t, \omega) = 0$ for $0 \leqq t \leqq t(1)$. For $n = 1, 2, \cdots$, let $v_n(s) = 1/\{t(n+1) - s\}$. Let $\tau_n(\omega)$ be the least $t > t(n)$ such that $\int_{t(n)}^t v_n(s)\,dW(s, \omega) = f_n(\omega) - f_{n-1}(\omega)$ (letting $f_0(\omega) \equiv 0$). Define

$$\begin{aligned}\varphi(s, \omega) &= v_n(s)\,, \quad t(n) < s \leqq \tau_n(\omega)\,;\\ &= 0\,, \qquad \tau_n(\omega) < s \leqq t(n+1)\,.\end{aligned}$$

This defines a nonanticipating function φ such that for each n,

$$\int_0^{t(n+1)} \varphi(s, \omega)\,dW(s, \omega) = f_n(\omega)\,. \tag{3}$$

We have

$$\int_0^1 \varphi^2(s, \omega)\,ds = \sum_{n=1}^\infty \int_{t(n)}^{\tau_n(\omega)} v_n(s)^2\,ds = \sum_{n=1}^\infty G_n$$

where by (2), $\Pr(G_n \geqq n^{-2} \mid B_n) \leqq n|x_{n-1}(\omega)|$. Thus by (1),

$$\Pr(G_n \geqq n^{-2}) \leqq 2(n-1)^{-2} + 4(n-1)^{-2} = 6(n-1)^{-2}\,,$$

so $\sum G_n < \infty$ a.s. and $\varphi(\cdot, \omega) \in \mathscr{L}^2[0, 1]$ a.s.

Thus $\int_0^t \varphi(s, \omega)\,dW(s, \omega)$ is a.s. continuous in t (Gikhman and Skorokhod (1968), Chapter 1, Section 3, Theorem 2). This and (3) give the desired result since $f_n \to f$. □

Acknowledgment. I am very grateful to L. A. Shepp for an introduction to the problem and the idea of using integrands such as $1/(1 - t)$ up to stopping times.

REFERENCES

[1] CLARK, J. M. C. (1970, 1971). The representation of functionals of Brownian motion by stochastic integrals. *Ann. Math. Statist.* **41** 1282–1295; correction, *ibid.* **42** 1778.

[2] GIKHMAN, I. I. and SKOROKHOD, A. V. (1968). *Stochastic Differential Eqautions.* Naukova Dumka, Kiev, (in Russian); Akademie-Verlag, Berlin, 1971 (in German); Springer-Verlag, New York, 1972 (in English).

[3] ITÔ, K. and MCKEAN, H. P., JR. (1965). *Diffusion Processes and their Sample Paths.* Springer-Verlag, New York.

ROOM 2-245
MASSACHUSETTS INSTITUTE OF TECHNOLOGY
CAMBRIDGE, MASSACHUSETTS 02139

Correspondence

A Metric Entropy Bound is Not Sufficient for Learnability

R. M. Dudley, S. R. Kulkarni, T. Richardson, and O. Zeitouni

Abstract—We prove by means of a counterexample that it is not sufficient, for probably approximately correct (PAC) learning under a class of distributions, to have a uniform bound on the metric entropy of the class of concepts to be learned. This settles a conjecture of Benedek and Itai.

Index Terms—Learning, estimation, PAC, metric entropy, class of distributions.

I. Introduction

Let $(\mathcal{X}, \mathcal{B})$ be a measurable space. Let $\mathcal{P}$ be a class of probability measures on $(\mathcal{X}, \mathcal{B})$. Let $\mathcal{C}$ (the "concept class" in the language of learning theory, as introduced in [1]) be a subset of $\mathcal{B}$. Suppose one is given a sequence of independent and identically distributed (i.i.d.), $\mathcal{X}$-valued random variables $X_1, \cdots, X_n$ distributed according to P^n, where $P \in \mathcal{P}$. In addition, for some unknown $c \in \mathcal{C}$, one is given data $(X_1, I_c(X_1)), \cdots, (X_n, I_c(X_n))$ which we henceforth denote by $\mathcal{D}_n(c)$. The problem of learning consists roughly of the question "given $\mathcal{C}$, $\mathcal{P}$, how large should n be for approximating c with high accuracy and low probability of error based on the data $\mathcal{D}_n(c)$?" In mathematical terms, assume that $(\mathcal{X}, \mathcal{B})$ is a Borel space, and define on $\mathcal{B}$ the pseudometric $d_P(c_1, c_2) = P(c_1 \triangle c_2)$. Let $\mathcal{T}$ be the algebra of all four subsets of $\{0, 1\}$. A learning rule is a map $T^n: (\mathcal{X} \times \{0, 1\})^n \to \mathcal{C}$ such that, for any $c \in \mathcal{C}$, any $P \in \mathcal{P}$, and any $\epsilon > 0$,

$$\{(X_1, \cdots, X_n, i_1, \cdots, i_n): d_P(c, T^n((X_1, i_1), \cdots, (X_n, i_n))) > \epsilon\} \in \mathcal{B}^n \otimes \mathcal{T}^n. \quad (1)$$

It follows that for any $c, d \in \mathcal{C}$,

$$\{(X_1, \cdots, X_n): d_P(d, T^n(\mathcal{D}_n(c))) > \epsilon\} \in \mathcal{B}^n. \quad (2)$$

We say that the concept class $\mathcal{C}$ is probably approximately correct (PAC) learnable under the class of probability measures $\mathcal{P}$ (in short: $\mathcal{C}$ is PAC learnable under $\mathcal{P}$) if, for every $\epsilon > 0$, $\delta > 0$, there exist an integer $n = n(\mathcal{P}, \mathcal{C}, \epsilon, \delta)$ and a learning rule T^n such that, for any $P \in \mathcal{P}$ and $c \in \mathcal{C}$,

$$P^n(\{(X_1, \cdots, X_n): d_P(c, T^n(\mathcal{D}_n(c))) > \epsilon\}) < \delta. \quad (3)$$

The notion of learnability in the form (3) has recently received much attention (e.g., see [2], [3], [1]), and in the learning literature is referred to as probably approximately correct (PAC) learning, for reasons obvious from its definition. Intuitively, in PAC learning one attempts to achieve a good prediction on future samples, after seeing some finite number of samples, uniformly in $P \in \mathcal{P}$ and $c \in \mathcal{C}$.

Sufficient and necessary conditions for PAC learnability are by now well known for some cases. Let $B(c, \epsilon) = \{\bar{c} \in \mathcal{B}: d_P(c, \bar{c}) < \epsilon\}$, and define the *ε-covering number* of $\mathcal{C}$ with respect to P by

$$N(\epsilon, \mathcal{C}, P) = \inf\left\{N: \exists c_1, \cdots, c_N \in \mathcal{B} \text{ such that } \mathcal{C} \subset \bigcup_{i=1}^{N} B(c_i, \epsilon)\right\}.$$

The balls $B(c_i, \epsilon)$ above are said to form an *ε-cover* of $\mathcal{C}$, and $\log N(\epsilon, \mathcal{C}, P)$ is often referred to as the *metric entropy* of $\mathcal{C}$ with respect to P. A necessary and sufficient condition for PAC learnability of $\mathcal{C}$ in the special case where $\mathcal{P}$ is a singleton, namely $\mathcal{P} \equiv \{P\}$, is that $N(\epsilon, \mathcal{C}, P) < \infty$ for all $\epsilon > 0$ (see [4] and, in greater generality, [5], pp. 149–151). Moreover, if $\mathcal{P} = M_1(\mathcal{X})$, the space of Borel probability measures on $\mathcal{X}$, then (under suitable measurability conditions) a well-known necessary and sufficient condition for PAC learnability of $\mathcal{C}$ under $\mathcal{P}$ is that the Vapnik–Chervonenkis (VC) dimension of $\mathcal{C}$ be finite, which turns out to be equivalent to the condition that, for all $\epsilon > 0$, $\sup_{P \in M_1(\mathcal{X})} N(\epsilon, \mathcal{C}, P) < \infty$ (see [2], [6], [3], [5], [7], [8] for proofs and additional background on the VC dimension and metric entropy). The similarity between these two extreme cases led Benedek and Itai to conjecture in [4] that the condition

$$\forall \epsilon > 0, \ \sup_{P \in \mathcal{P}} N(\epsilon, \mathcal{C}, P) < \infty \quad (4)$$

is necessary and sufficient for the PAC learnability of $\mathcal{C}$ under $\mathcal{P}$. While necessity is fairly obvious, the sufficiency part is less so because of the difficulty in simultaneously approximately determining $c \in \mathcal{C}$ and $P \in \mathcal{P}$. (We mention that if (4) is replaced by the stronger condition that there exists a fixed finite ε-cover of $\mathcal{C}$ under all $P \in \mathcal{P}$, then the sufficiency is just a standard extension of the single measure case. Some cases where (4) is sufficient are described in [9].) It is the purpose of this note to show, by a counterexample, that (4) is not sufficient in general for learnability. The question of finding a necessary and sufficient condition for PAC learnability of $\mathcal{C}$ under $\mathcal{P}$ remains open.

II. A Counterexample

Let $\Omega = \mathcal{X} = \{0, 1\}^\infty$, let X^i denote the coordinate map of $X \in \mathcal{X}$, and let $\mathcal{B}$ be the Borel σ-field over $\mathcal{X}$. Let $(p_1, p_2, \cdots) \in [0, 1]^\infty$ be defined by $p_i = 1/\log_2(i + 1) \le 1$, and note that for every finite n, $\sum_{i=1}^\infty p_i^n = \infty$. Identifying $p_i = P(X^i = 1)$, the

Manuscript received Nov. 9, 1992; revised Sept. 22, 1993. The research of R. M. Dudley was partially supported by National Science Foundation grants. The work of S. Kulkarni was supported in part by the Army Research Office under Grant DAAL03-92-G-0320 and by the National Science Foundation under Grant IRI-92-09577. The work of O. Zeitouni was done while visiting the Center for Intelligent Control Systems at M.I.T. under support of the U.S. Army Research Office Grant DAAL03-92-G-0115.

R. M. Dudley is with the Dept. of Mathematics, M.I.T., Cambridge, MA 02139.

S. Kulkarni is with the Dept. of Electrical Engineering, Princeton University, Princeton, NJ 08544.

T. Richardson is with AT & T Bell Labs, 600 Mountain Ave., Murray Hill, NJ 07974.

O. Zeitouni is with The Dept. of Electrical Engineering, Technion, Haifa 32000, Israel.

IEEE Log Number 9401814.

0018-9448/94$04.00

vector $p_1, p_2, \cdots$ induces a product measure P_I on the product space $\mathscr{X}$. For any measure P on $\mathscr{X}$, P^n denotes the product measure on $\mathscr{X}^n$ obtained from P.

Let σ denote a permutation (possibly infinite) of the integers, i.e., $\sigma: N \to N$ is one to one and onto, and define P_σ as the measure on $\mathscr{X}$ induced by $(p_{\sigma^{-1}(1)}, p_{\sigma^{-1}(2)}, \cdots)$. The ensemble of all permutations is denoted Σ. Thus, $P_\sigma(X^{\sigma(i)} = 1) = p_i$ and, if σ is the identity map, then P_σ equals the P_I defined above.

Now let $\mathscr{P} \equiv \{P_\sigma, \sigma \in \Sigma\}$, let $c_i \equiv \{X \in \mathscr{X}: X^i = 1\}$, and let $\mathscr{C} \equiv \{c_i, i \in N\}$. It is easy to check that for any $P \in \mathscr{P}$, $N(\epsilon, \mathscr{C}, P) < \infty$. Since any c_i with $p_{\sigma^{-1}(i)} < \epsilon$ satisfies $d_{P_\sigma}(c_i, \emptyset) < \epsilon$, we have that for any $P \in \mathscr{P}$,

$$N(\epsilon, \mathscr{C}, P) < 2^{1/\epsilon}.$$

It follows that $\sup_{P \in \mathscr{P}} N(\epsilon, \mathscr{C}, P) < \infty$. We now claim

Theorem 1: $\mathscr{C}$ is not PAC learnable under $\mathscr{P}$.

Proof: We use a random coding argument. Suppose that the theorem's assertion is false. Then, for each $\epsilon > 0$, $\delta > 0$, it is possible to find an $n = n(\epsilon, \delta)$ and a learning rule T^n which satisfy (3) for all $c \in \mathscr{C}$ and $P \in \mathscr{P}$. In particular, for any finite k, it satisfies (3) for $c \in \mathscr{C}^k$ and $P \in \mathscr{P}^k$, where $\mathscr{C}^k = \{c_i, i = 1, \cdots, k\}$, $\Sigma^k = \{\sigma: \sigma(i) = i, \forall i > k\}$, and $\mathscr{P}^k = \{P_\sigma, \sigma \in \Sigma^k\}$, i.e., $\mathscr{P}^k$ are all possible permutations of the vector $(p_1, p_2, \cdots)$ which involve only the first k coordinates. Let the error event be defined as

$$\mathrm{er}_\sigma^c = \{(X_1, \cdots, X_n): d_{P_\sigma}(c, T^n(\mathscr{D}_n(c))) > \epsilon\}.$$

(It follows from (2) that er_σ^c is a measurable event.) Then, for each $c \in \mathscr{C}^k$ and $P_\sigma \in \mathscr{P}^k$,

$$P_\sigma^n(\mathrm{er}_\sigma^c) < \delta.$$

In particular, if Q is any probability measure on the finite set $\{(\sigma, c): \sigma \in \Sigma^k, c \in \mathscr{C}^k\}$, then

$$E_Q(P_\sigma^n(\mathrm{er}_\sigma^c)) < \delta. \tag{5}$$

Now choose Q such that $Q|_\Sigma$ is uniform over Σ^k while $c = c_{\sigma(1)}$ (i.e., $Q(\sigma, c) = 1/k!$ if $\sigma \in \Sigma^k$ and $c = c_{\sigma(1)}$, and $Q(\sigma, c) = 0$ otherwise). This Q forces the true concept to involve the coordinate of maximal probability (where in fact the probability is 1) in P_σ. Note that by our choice of Q, if $\epsilon < 1 - 1/\log_2(3) = \min_{j>1} d_{P_I}(c_1, c_j)$, then, when (σ, c) are distributed according to Q,

$$d_{P_\sigma}(c, \bar{c}) < \epsilon \Rightarrow c = \bar{c} = c_{\sigma(1)}\ Q \text{ a.s.}.$$

Thus, in this set-up, Q a.s.,

$$\mathrm{er}_\sigma^c = \{(X_1, \cdots, X_n): c \neq T^n(\mathscr{D}_n(c))\}.$$

Using the notation σX to denote the element of $\mathscr{X}$ with coordinates $(\sigma X)^i = X^{\sigma^{-1}(i)}$ and $\sigma\mathscr{D}_n$ to denote the corresponding permutation on $\mathscr{D}_n(c)$ when $c = c_{\sigma(1)}$, i.e.,

$$\begin{aligned}\sigma\mathscr{D}_n &= \left(\left(\sigma X_1, I_{c_{\sigma(1)}}(\sigma X_1)\right), \cdots, \left(\sigma X_n, I_{c_{\sigma(1)}}(\sigma X_n)\right)\right)\\ &= ((\sigma X_1, I_{c_1}(X_1)), \cdots, (\sigma X_n, I_{c_1}(X_n))),\end{aligned} \tag{6}$$

we have

$$\begin{aligned}E_Q(P_\sigma^n(\mathrm{er}_\sigma^c)) &= E_Q(P_\sigma^n(c \neq T^n(\mathscr{D}_n(c))))\\ &= E_Q(P_\sigma^n(c_{\sigma(1)} \neq T^n(\mathscr{D}_n(c_{\sigma(1)}))))\\ &= E_Q(P_I^n(c_{\sigma(1)} \neq T^n(\sigma\mathscr{D}_n)))\\ &= E_{P_I^n} E_Q(1_{c_{\sigma(1)} \neq T^n(\sigma\mathscr{D}_n)}).\end{aligned} \tag{7}$$

For given vectors $\vec{x} = (x_1, \cdots, x_n) \in \mathscr{X}^n$ and $\vec{X} = (X_1, \cdots, X_n) \in \mathscr{X}^n$, denote by $S(\vec{X}, \vec{x})$ the set of permutations $\sigma \in \Sigma^k$ such that $\sigma\vec{X} = \vec{x}$. (Note that for many pairs $(\vec{X}, \vec{x})$, $S(\vec{X}, \vec{x})$ is empty.) It follows from the definition that, for $\sigma \in S(\vec{X}, \vec{x})$,

$$\sigma\mathscr{D}_n = ((x_1, I_{c(1)}(X_1)), \cdots, (x_n, I_{c(1)}(X_n))).$$

By the construction of Q, the distribution of σ conditioned on $S(\vec{X}, \vec{x})$ is uniform there. Let now

$$J^{\vec{X}} = \{i \leq k: X_j^i = 1, \forall j = 1, \cdots, n\},$$

and

$$J^{\vec{x}} = \{i \leq k: x_j^i = 1, \forall j = 1, \cdots, n\}.$$

$S(\vec{X}, \vec{x})$ is nonempty only if $|J^{\vec{x}}| = |J^{\vec{X}}|$. When $\vec{X}$ has distribution P_I^n, we have $1 \in J^{\vec{X}}$ almost surely, so $|J^{\vec{x}}| \geq 1$. Let $\sigma_c \in \Sigma^k$ be a fixed permutation such that $\sigma_c(i) \in J^{\vec{x}}$ if $i \in J^{\vec{X}}$. Decompose each permutation $\sigma \in S(\vec{X}, \vec{x})$ into $\sigma = \sigma_c \circ \sigma_b \circ \sigma_a$, with $\sigma_a: J^{\vec{X}} \to J^{\vec{X}}$, and σ_a equals the identity on $\{1, \cdots, k\} \setminus J^{\vec{X}}$ while $\sigma_b: \{1, \cdots, k\} \setminus J^{\vec{X}} \to \{1, \cdots, k\} \setminus J^{\vec{X}}$ and σ_b equals the identity on $J^{\vec{X}}$. This is always possible because all permutations in $S(\vec{X}, \vec{x})$ must satisfy $\sigma\vec{X} = \vec{x}$. Note that whenever $S(\vec{X}, \vec{x})$ is nonempty then $|\sigma_A| = |J^{\vec{X}}|!$, where

$$\sigma_A \triangleq \{\sigma_a: \sigma \in S(\vec{X}, \vec{x})\}, \qquad \sigma_B \triangleq \{\sigma_b: \sigma \in S(\vec{X}, \vec{x})\}.$$

Using now (7),

$$\begin{aligned}&E_Q(P_\sigma^n(\mathrm{er}_\sigma^c))\\ &= E_{P_I^n}\left(\sum_{\vec{x}} E_Q\left(1_{T^n(\sigma\mathscr{D}_n) \neq c_{\sigma(1)}} \,|\, \sigma \in S(\vec{X}, \vec{x})\right) Q(S(\vec{X}, \vec{x}))\right)\\ &= E_{P_I^n}\left(\sum_{\vec{x}} Q(S(\vec{X}, \vec{x})) \frac{\sum_{\sigma_b \in \sigma_B} \sum_{\sigma_a \in \sigma_A} 1_{T^n(\sigma\mathscr{D}_n) \neq c_{\sigma(1)}}}{\sum_{\sigma_b \in \sigma_B} \sum_{\sigma_a \in \sigma_A} 1}\right),\end{aligned} \tag{8}$$

where in the last equality we have used the uniformity of the conditional distribution over $S(\vec{X}, \vec{x})$, and the sum over $\vec{x}$ is taken over all *different* vectors in $\mathscr{X}^n$. By (6), $\sigma\mathscr{D}_n$ is constant for $\sigma \in S(\vec{X}, \vec{x})$, so

$$T^n(\sigma\mathscr{D}_n) = c_T$$

for some $c_T = c_T(\vec{X}, \vec{x}) \in \mathscr{C}$ not depending on $\sigma \in S(\vec{X}, \vec{x})$. Here $c_T(\cdot, \cdot)$ is measurable by (2). Thus, since the number of permutations $\sigma \in \sigma_A$ for which $T^n(\sigma\mathscr{D}_n) = c_{\sigma(1)}$ is at most equal to the number of permutations in σ_A which have a prescribed index in $J^{\vec{X}}$ unchanged,

$$\sum_{\sigma_a \in \sigma_A} 1_{T^n(\sigma\mathscr{D}_n) \neq c_{\sigma(1)}} \geq (|J^{\vec{X}}| - 1)(|J^{\vec{X}}| - 1)!,$$

whereas

$$\sum_{\sigma_a \in \sigma_A} 1 = |J^{\vec{X}}|!.$$

It follows that, for any $\eta > 1$,

$$\begin{aligned}E_Q(P_\sigma^n(\mathrm{er}_\sigma^c)) &\geq E_{P_I^n} \frac{(|J^{\vec{X}}| - 1)(|J^{\vec{X}}| - 1)!}{|J^{\vec{X}}|!} = \left(1 - E_{P_I^n} \frac{1}{|J^{\vec{X}}|}\right)\\ &\geq \left(1 - \frac{1}{\eta} - P_I^n(|J^{\vec{X}}| \leq \eta)\right).\end{aligned}$$

It remains therefore only to show that $|J^{\vec{X}}|$ may, with high probability, be made arbitrarily large by choosing a k large enough. But this is obvious because, by the Borel–Cantelli lemma, using $\vec{X}^i \triangleq (X_1^i, \cdots, X_n^i)$,

$$P_f^n\left(\vec{X}^i = (1, \cdots, 1) \text{ infinitely often}\right) = 1,$$

since $\sum_{i=1}^{\infty} P_f^n(\vec{X}^i = (1, \cdots, 1)) \geq \sum_{i=1}^{\infty} p_i^n = \infty$. Thus, for any η, one may find a k large enough such that $P_f^n(|J^{\vec{X}}| \leq \eta)$ is arbitrarily small. □

Remark: Note that we have actually shown that, for any fixed n and any $\epsilon < 1 - 1/\log_2(3)$, one may construct a $\mathscr{P}$ and a $\mathscr{C}$ such that the probability of error is arbitrarily close to 1. By defining p_i, $i \geq 2$, to be smaller, we could also take any $\epsilon < 1$.

References

[1] L. G. Valiant, "A theory of the learnable," *Commun. ACM*, vol. 27, no. 11, pp. 1134–1142, 1984.

[2] A. Blumer, A. Ehrenfeucht, D. Haussler, and M. Warmuth, "Learnability and the Vapnik–Chervonenkis dimension," *J. ACM*, vol. 36, no. 4, pp. 929–965, 1989.

[3] D. Haussler, "Decision theoretic generalizations of the PAC model for neural net and other learning applications," *Inf. Comput.*, vol. 20, pp. 78–150, 1992.

[4] G. M. Benedek and A. Itai, "Learnability with respect to a fixed distribution," *Theor. Comput. Sci.*, vol. 86, pp. 377–389, 1991.

[5] V. N. Vapnik, *Estimation of Dependences Based on Empirical Data.* New York: Springer-Verlag, 1982.

[6] R. M. Dudley, "A course on empirical processes," *Lecture Notes in Math. Vol. 1097.* New York: Springer, 1984, pp. 1–142.

[7] V. N. Vapnik and A. Ya. Chervonenkis, "On the uniform convergence of relative frequencies of events to their probabilities," *Theory Probab. Its Appl.*, vol. 16, no. 2, pp. 264–280, 1971.

[8] V. N. Vapnik and A. Ya. Chervonenkis, "Necessary and sufficient conditions for the uniform convergence of means to their expectations," *Theory Probab. Its Appl.*, vol. 26, no. 3, pp. 532–553, 1981.

[9] S. R. Kulkarni, "Problems of computational and information complexity in machine vision and learning," Ph.D. thesis, Dep. Elec. Eng. Comput. Sci., M.I.T., June 1991.

Non White Gaussian Multiple Access Channels with Feedback

Sandeep Pombra and Thomas M. Cover

***Abstract*—Although feedback does not increase capacity of an additive white noise Gaussian channel, it enables prediction of the noise for non-white additive Gaussian noise channels and results in an improvement of capacity, but at most by a factor of 2 (Pinsker, Ebert, Pombra, and Cover). Although the capacity of white noise channels cannot be increased by feedback, multiple access white noise channels have a capacity increase due to the cooperation induced by feedback. Thomas has shown that the total capacity (sum of the rates of all the senders) of an m-user Gaussian white noise multiple access channel with feedback is less than twice the total capacity without feedback. In this paper, we show that this factor of 2 bound holds even when cooperation and prediction are combined, by proving that feedback increases the total capacity of an m-user multiple access channel with non-white additive Gaussian noise by at most a factor of 2.**

Manuscript received Oct. 1, 1991; revised Sept. 1, 1993. This work was partially supported by NSF Grant NCR-89-14538, DARPA Contract J-FBI-91-218, and JSEP Contract DAAL03-91-C-0010.

S. Pombra is at 413 Mountain Laurel Court, Mountain View, CA 94043.

T. M. Cover is with the Departments of Electrical Engineering and Statistics, Stanford University, Stanford, CA 94305.

IEEE Log Number 9401815.

***Index Terms*—Feedback capacity, Capacity, Multiple-access channel, Non-white Gaussian noise, Gaussian channel.**

I. Introduction

In satellite communication, many senders communicate with a single receiver. The noise in such multiple access channels can often be characterized by non-white additive Gaussian noise. For example, microwave communication components often introduce non-white noise into a channel.

In single-user Gaussian channels with non-white noise, feedback increases capacity. The reason is due solely to the fact that the transmitter knows the past noise (by subtracting out the feedback) and thus can predict the future noise and use this information to increase capacity. A factor of 2 bound on the increase in capacity due to feedback of a single-user Gaussian channel with non-white noise was obtained in [1], [2], [10]. Ihara [9] has shown that the factor of 2 bound is achievable for certain autoregressive additive Gaussian noise channels.

Unlike the simple discrete memoryless channel, feedback in the multiple access channel can increase capacity even when the channel is memoryless, because feedback enables the senders to cooperate with each other. This cooperation is impossible without feedback. This was first demonstrated by Gaarder and Wolf [5]. Cover and Leung [6] established an achievable rate region for the multiple access channel with feedback. Later, Willems [7] proved that the Cover–Leung region is indeed the capacity region for a certain class of channels including the binary adder channel. Ozarow [8] found the capacity region for the two-user Gaussian multiple access channel using a modification of the Kailath–Schalkwijk [4] scheme for simple Gaussian channels. Thomas [11] proved a factor of 2 bound on the capacity increase with feedback for a Gaussian white noise multiple access channel. Keilers [3] characterized the capacity region for a non-white Gaussian noise multiple access channel without feedback. Coding theorems for multiple access channels with finite memory noise are treated in Verdú [14].

The case of non-white Gaussian multiple access channel with feedback combines the above two problems. Here feedback helps through cooperation of senders, as well as through prediction of noise. If we simply use the factor of 2 bounds derived by Cover and Pombra [10] and Thomas [11] for the single-user Gaussian channel with non-white noise and the Gaussian multiple-access channel with white noise, respectively, we might expect feedback to quadruple the total capacity of a non-white m-user Gaussian multiple access channel. However this reasoning is misleading due to the following reasons: Prediction of noise by the receiver and cooperation between the senders are not mutually exclusive events. Also the factor of 2 bound on the feedback capacity of a non-white Gaussian channel has been shown to be tight for the case of only one sender, where there is no interference among the senders. If we have more than one sender, the interference among the senders may diminish the feedback capacity gain due to the prediction of noise.

In this paper, we establish a factor of 2 bound on the increase in total capacity due to feedback for an m-user additive Gaussian non-white noise multiple access channel. Throughout this paper, we define the total capacity of the multiple access channel to be the maximum achievable sum of rates of all the senders.

The paper is organized as follows. In Section II (Theorem 2.1), we prove an expression for the total capacity C_n in bits per

0018-9448/94$04.00

The Annals of Statistics
2002, Vol. 30, No. 5, 1311–1344

ASYMPTOTIC NORMALITY WITH SMALL RELATIVE ERRORS OF POSTERIOR PROBABILITIES OF HALF-SPACES

BY R. M. DUDLEY[1] AND D. HAUGHTON[2]

Massachusetts Institute of Technology and Bentley College

Let Θ be a parameter space included in a finite-dimensional Euclidean space and let A be a half-space. Suppose that the maximum likelihood estimate θ_n of θ is not in A (otherwise, replace A by its complement) and let Δ be the maximum log likelihood (at θ_n) minus the maximum log likelihood over the boundary ∂A. It is shown that under some conditions, uniformly over all half-spaces A, either the posterior probability of A is asymptotic to $\Phi(-\sqrt{2\Delta})$ where Φ is the standard normal distribution function, or both the posterior probability and its approximant go to 0 exponentially in n. Sharper approximations depending on the prior are also defined.

1. Introduction. Some examples of half-spaces of interest in parameter spaces are, in a clinical trial of a treatment versus placebo, the half-spaces where the treatment is (a) helpful or (b) harmful. Thus one may not only want to test the hypothesis that the treatment (c) makes no difference, but to assign posterior probabilities to (a), (b) and (c), under conditions as unrestrictive as possible on the choice of prior probabilities [e.g., Dudley and Haughton (2001)]. More generally, we have in mind applications to model selection as in the BIC criterion of Schwarz (1978) and its extensions [Poskitt (1987); Haughton (1988)], specifically to one-sided models and multiple data sets [Dudley and Haughton (1997)].

We will describe our results, omitting some details and conditions until Section 2. Let Φ be the standard normal distribution function and ϕ its density. Let Θ be an open subset of a Euclidean space $\mathbf{R}^d$ and let A vary over half-spaces in $\mathbf{R}^d$. Let A have boundary hyperplane ∂A. Let $\mathcal{P} = \{P_\phi : \phi \in \Theta\}$ be a family of laws having densities $f(x, \phi)$ with respect to some μ. Given n observations $X_1, \ldots, X_n$ i.i.d. with a law P not necessarily in $\mathcal{P}$, let $\hat{\phi} = \hat{\phi}_n$ be the maximum likelihood estimate of $\phi \in \Theta$. If $\hat{\phi}_n$ is in A, replace A by its complement. Let $\tilde{\phi} = \tilde{\phi}_n$ be the maximum likelihood estimate of ϕ in ∂A. Let $LL(\phi) := LL_n(\phi)$ denote the log likelihood $\sum_{j=1}^n \log f(X_j, \phi)$ and let $\Delta := \Delta_n := LL(\hat{\phi}) - LL(\tilde{\phi})$. Then 2Δ is a likelihood ratio statistic. We will show in Theorem 1, under conditions to be given in Section 2, that for any prior law π on Θ having a continuous, strictly positive density, for $A = A_n$ depending on $n \to \infty$, if ∂A_n approaches, however slowly,

Received December 1999; revised October 2001.

[1]Supported in part by NSF Grants DMS-97-04603 and DMS-01-03821.

[2]Supported in part by an NSF grant.

AMS 2000 subject classifications. Primary 62F15; secondary 60F99, 62F05.

Key words and phrases. Bernstein–von Mises theorem, gamma tail probabilities, intermediate deviations, Jeffreys prior, Mills' ratio.

E. Giné et al. (eds.), *Selected Works of R.M. Dudley*, Selected Works in Probability and Statistics, DOI 10.1007/978-1-4419-5821-1_29, © Springer Science+Business Media, LLC 2010

the true or pseudo-true (as defined in Section 2, Assumption A_5) parameter point ϕ_0, then the posterior probability $\pi_n(A_n)$ is asymptotic to $\Phi(-\sqrt{2\Delta_n})$. Or if ∂A_n remains bounded away from ϕ_0, then both $\pi_n(A_n)$ and $\Phi(-\sqrt{2\Delta_n})$ approach 0 exponentially in n.

The classical normal approximation to posterior laws π_n was by normal laws with their means at the MLE (maximum likelihood estimate) $\hat{\phi}$ and their covariance matrix equal to the inverse of the observed Fisher information matrix, for example, in one dimension [Walker (1969) and Johnson (1970)] and for multidimensional ϕ [Hipp and Michel (1976), Chen (1985) and Pauler, Wakefield and Kass (1999)]. Thus the posterior law becomes close to a standard normal law after a linear transformation. On the other hand, Le Cam (1953) proved asymptotic normality of posteriors without assuming second derivatives of the likelihood functions, so that the Fisher information matrix may not be defined.

Here, as mentioned, we will instead use nonlinear transformations by way of likelihood ratio statistics, as did, for example, Bickel and Ghosh (1990), who gave a set of precise assumptions, some of which (e.g., that the prior has compact support) are stronger than needed for proving our theorem. (Bickel and Ghosh's Bayesian results may be viewed as technical lemmas for proving frequentist theorems.) Likelihood ratio statistics give much smaller relative errors of the normal approximation in the tails (see Section 10), and the approximation is also rather easy to compute, requiring only the values of the likelihood at the unrestricted MLE $\hat{\phi}$ and at the MLE $\tilde{\phi}$ over the boundary ∂A of the half-space A, and the knowledge of whether $\hat{\phi} \in A$. We do assume second derivatives of the likelihood functions and use them in showing that relative errors are small. Some sharpened approximations depending on the prior, defined in Section 3, are still more accurate in examples (Section 10). After Theorem 1 we give a corollary showing that under our assumptions the same conclusions hold for the first sharpened Laplace approximation as for the simpler likelihood root approximation; stronger assumptions (e.g., beyond mere continuity of the prior density) would apparently be needed to prove the order of improvement that can be given by the sharpened approximations.

With notation as in the abstract, let $B_\Delta := \Phi(\sqrt{2\Delta})/\Phi(-\sqrt{2\Delta})$ if the MLE $\hat{\theta}_n$ is in the half-space A, or $\Phi(-\sqrt{2\Delta})/\Phi(\sqrt{2\Delta})$ otherwise. Then our result as stated in the abstract, or in more detail in Theorem 1, shows that B_Δ gives an asymptotic approximation to the Bayes factor $\pi_n(A)/(1-\pi_n(A))$ for a half-space relative to its complementary half-space, for a wide class of proper priors, unless $\pi_n(A)$ or $1-\pi_n(A)$ is exponentially small. Yet B_Δ does not depend on the prior. Thus B_Δ might be viewed as a "default Bayes factor" if no prior is given. Berger and Mortera (1999) treat alternative possibilities for such factors when priors may be improper.

Let $a_n \ll b_n$ mean that $a_n = o(b_n)$ as $n \to \infty$. For a fixed half-space A, let d_n be the Euclidean distance from $\hat{\phi}_n$ to ∂A. Let us say we have a case of Laplace deviations if $d_n = O(n^{-1/2})$, of intermediate deviations if $n^{-1/2} \ll d_n \ll 1$, and large deviations if d_n converges to a positive constant d. Let us further say that deviations are positive if $\hat{\phi}_n \notin A$ and negative if $\hat{\phi}_n \in A$.

Asymptotic normality of the posterior distribution holds classically for Laplace deviations [e.g., Hipp and Michel (1976)] and, as the present paper shows, for positive intermediate (and Laplace) deviations in terms of the likelihood root. In the analysis literature, asymptotics for integrals of functions $e^{-ah(x)}$ over half-lines with boundary point near the minimum of h as $a \to +\infty$ had been treated in terms of normal distribution functions by Bleistein [(1966), Section 5], Wong (1973), Skinner (1980) (in the multidimensional case), Temme (1982) and Wong [(1989), Section VII.3]. In Temme (1982), $h = h(x, a)$ can depend on a, as for posterior distributions with $a = n$, and asymptotic expansions giving small relative errors are proved under some (analyticity) assumptions.

In a statistical problem one presumably does not know in advance (before observations are taken) whether the deviation will be Laplace, intermediate or large. If one uses the asymptotic normality approximation we propose then one has small relative errors for posterior probabilities of Laplace or intermediate deviations.

Pauler, Wakefield and Kass (1999) also indicate some cases where posterior probabilities of subsets A which may be half-spaces, or orthants, are useful. They give in their Section 4 a "boundary Laplace approximation" which involves the normal probability of A based on the observed Fisher information matrix at $\hat{\phi}$ times the prior density at the maximum $\tilde{\phi}$ of the likelihood on A. Pauler, Wakefield and Kass (1999) prove under a set of assumptions that as $n \to \infty$ the (relative) error of their approximation approaches 0. Their approximation has the advantages of applying when A is not necessarily a half-space and the data are not necessarily i.i.d. On the other hand, their assumption C1, that the approximating normal probabilities are bounded away from 0, is quite restrictive relative to our situation.

We will treat in Section 9 a multivariate normal location family and present in Section 10 some numerical results for beta distributions.

2. Statements of assumptions and the main theorem. First we give some notation. Let $P_n := \frac{1}{n}(\delta_{X_1} + \cdots + \delta_{X_n})$, with $\delta_x(A) := \mathbb{1}_A(x)$. The log likelihood $LL_n(\phi)$ then equals $LL_n(\phi) = n \int \log f(x, \phi)\, dP_n(x)$, with $-\infty \leq LL_n(\phi) < +\infty$ for all ϕ. We will need the following assumptions.

(A$_1$) The set $\mathcal{P} = \{P_\phi : \phi \in \Theta\}$ is a family of laws dominated by a σ-finite measure μ on some sample space $(\chi, \mathcal{B})$. Let $f(x, \phi)$ be the density $(dP_\phi/d\mu)(x)$, $x \in \chi$, $\phi \in \Theta$. We take $0 \leq f(x, \phi) < \infty$.

(A$_2$) Θ is an open set in a Euclidean space $\mathbf{R}^d$.

(A$_3$) The observations $X_1, X_2, \ldots$ are i.i.d. random variables with values in χ and with some law P (not necessarily in the family $\mathcal{P}$).

(A4) A prior probability $\pi_0(\phi)\,d\phi$ is given on Θ with a continuous density $\pi_0(\phi) > 0$ for all ϕ, so $\int_\Theta \pi_0(\phi)\,d\phi = 1$.

(A5) There is a $\phi_0 \in \Theta$, to be called the pseudo-true value of ϕ, such that for every neighborhood $\mathcal{N}$ of ϕ_0 there is a $\kappa > 0$ such that almost surely for n large enough, $\sup_{\phi \notin \mathcal{N}} LL_n(\phi) < \sup_{\phi \in \mathcal{N}} LL_n(\phi) - n\kappa$.

(A6) For ϕ in a small enough neighborhood W of ϕ_0, the function $f(x, \cdot)$ is strictly positive and C^2 in ϕ for P-almost all x, and the P-Fisher information matrix $E(\phi) := \{E_{ij}(\phi)\}_{i,j=1}^d := \{E_P(-\partial^2 \log f(\cdot, \phi)/\partial\phi_i \partial\phi_j)\}_{i,j=1}^d$ exists and is finite, strictly positive definite and continuous in ϕ.

(A7) For some neighborhood W of ϕ_0, the class $\mathcal{F}_W := \{-\partial^2 \log f(\cdot, \phi)/\partial\phi_i \partial\phi_j\}$ of functions for $\phi \in W$, $i, j = 1, \dots, d$, is a Glivenko–Cantelli class for P; in other words we have $\sup_{g \in \mathcal{F}_W} |\int g\, d(P_n - P)| \to 0$ almost surely as $n \to \infty$.

REMARKS. Sufficient conditions for (A5) to hold can be found from sufficient conditions for consistency of approximate maximum likelihood estimators; see Huber (1967), and for further extensions, Dudley (1998). Berk (1966) and Poskitt (1987) have considered the limiting behavior of posterior distributions for observations with a law P not in a parametric family.

Both (A6) and (A7), if true for one neighborhood W of ϕ_0, are also true for all smaller neighborhoods, so we can take W to be the same in both. Since P will usually be unknown to the statistician, one will need to check that assumptions (A6) and (A7) hold for all P in a class including at least all P_ϕ and preferably all P in some large class. Assumptions (A6) and (A7) hold for an exponential family in standard form, where $f(x, \phi) = e^{x\cdot\phi - b(\phi)}$, since the second derivatives of $\log f$ are constant in x.

Talagrand (1987) gave a characterization of Glivenko–Cantelli classes for a given P. Dudley (1998) applies Talagrand's theorem to extend the Glivenko–Cantelli property to transformed classes; see also van der Vaart and Wellner (2000). Sufficient conditions are known for a class $\mathcal{F}$ to be a Glivenko–Cantelli class for all P, and a criterion is known up to measurability for $\mathcal{F}$ to be a Glivenko–Cantelli class uniformly in P [Dudley, Giné and Zinn (1991)]. See also Alon, Ben-David, Cesa-Bianchi and Haussler (1997).

Note that (A6) and (A7) are preserved by nonsingular linear transformations of $\phi_1, \dots, \phi_d$. Let I denote the $d \times d$ identity matrix.

Let A be a half-space $\{\phi \in \Theta : \phi \cdot v_A \geq M\}$ for some $v_A \in \mathbf{R}^d$, $|v_A| = 1$ and $M \in \mathbf{R}$. Then ∂A is the boundary hyperplane $\partial A := \{\phi \in \Theta : \phi \cdot v_A = M\}$. Let

$$\Delta := \Delta_{A,n} := \left(\sup_{\phi \in \Theta} - \sup_{\phi \in \partial A}\right) LL_n(\phi),$$

where if the suprema are attained at unique points these are $\hat{\phi}_n$ (unrestricted MLE) and $\tilde{\phi}_n$ (MLE in ∂A), respectively. Here 2Δ is known as the (usual) likelihood ratio (test) statistic, and $\pm(2\Delta)^{1/2}$ as a likelihood root [defined, e.g., as m_1 for $p = 1$, $q = 0$ in Lawley (1956), Section 4].

When a statement is made "a.s. for n large enough" it will mean that it holds with probability 1 for $n \geq n_i$ where n_i is a random variable, *not depending on* A, but possibly depending on other quantities, as will be made explicit in the proofs.

THEOREM 1. *Suppose assumptions* (A_1)–(A_7) *hold, for a neighborhood* W *of* ϕ_0 *in* (A_6) *and* (A_7). *Let* $\pi_0(\phi)\,d\phi$ *be a prior law on* Θ *satisfying* (A_4). *For* $x = (X_1, \dots, X_n)$ *let* $\pi_{x,n}$ *be the posterior distribution on* Θ. *Then almost surely* $\hat{\phi}_n$ *exists for n large enough.*

For any $\varepsilon > 0$ *there exists* $\kappa' > 0$ *such that almost surely for n large enough, for all half-spaces* A, *if* $\hat{\phi}_n \notin A$, *then either*:

(a) $\Phi(-\sqrt{2\Delta})/(1+\varepsilon) \leq \pi_{x,n}(A) \leq (1+\varepsilon)\Phi(-\sqrt{2\Delta})$, *or*

(b) *both* $\pi_{x,n}(A) \leq 3e^{-n\kappa'}$ *and* $\Phi(-\sqrt{2\Delta}) \leq e^{-n\kappa'}$.

Or, if $\hat{\phi}_n \in A$, *then the same holds for* $B = \overline{A^c}$ *in place of* A.

Specifically, if U *is a neighborhood of* ϕ_0, *then for some* $\alpha > 0$, $\pi_{x,n}(U^c) \leq 3e^{-n\alpha}$ *almost surely for n large enough, and for small enough* $\kappa' > 0$ (b) *holds for any half-space* $A \subset U^c$. *Or, if* A_n *is a sequence of half-spaces such that* $d(A_n, \phi_0) := \inf\{|\phi - \phi_0| : \phi \in A_n\} \to 0$ *as* $n \to \infty$, *and* $\hat{\phi}_n \notin A_n$ *for each n, then almost surely* $\pi_{x,n}(A_n)/\Phi(-\sqrt{2\Delta_{A_n,n}}) \to 1$ *as* $n \to \infty$.

The proof will be given in Sections 4 through 8. The main steps in the proof are as follows: Assume that $\hat{\phi}_n \notin A$, the case $\hat{\phi}_n \in A$ following by symmetry. For any neighborhood V of ϕ_0, if A is disjoint from V then (b) in the Theorem holds for n large enough. We will find a small enough V so that if A intersects V we can make a linear change of coordinates from ϕ into y so that $\hat{y} = 0$, the boundary ∂A is the hyperplane where the first coordinate y_1 is some constant, a unique MLE $\tilde{y}$ in ∂A exists and is the point of A closest to 0, and the empirical Fisher information matrix $E_n(y)$ with respect to y is close to the identity matrix when y is close to 0.

We then obtain bounds on integrals $I_n(B) = \int_B e^{-nh(y)}\pi(y)\,dy$, where $\pi(\cdot)$ is the prior density in y coordinates, $nh(y)$ is the difference in log likelihoods at the MLE $\hat{y}$ and y, so that $\pi_{x,n}(y \in C) = I_n(C)/I_n(\Theta)$. Outside of a neighborhood $U(\rho)$ of the MLE, $\pi_{x,n}$ is exponentially small, so $\pi_{x,n}(A)$ equals $I_n(A \cap U(\rho))/I_n(U(\rho))$ up to an exponentially small quantity. Propositions 2 and 3 give asymptotic evaluations of $I_n(U(\rho))$ and $I_n(A \cap U(\rho))$, which rely on Lemma 5, relating to Mills' ratios and their extension to ratios of gamma tail probabilities to densities.

3. Sharper approximations. Extending multidimensional Laplace approximation formulas of Hsu (1948) and Fulks and Sather (1961) [see Wong (1989), Section IX.5], approximation formulas for integrals of likelihood functions over hyperplanes or manifolds have been proved under different conditions by Haughton (1984, 1988) and Poskitt (1987). Shun and McCullagh (1995) give further asymptotic expansion terms and consider high dimensions relative to n; see also Barndorff-Nielsen and Wood [(1998), Section 4]. Leonard (1982), Tierney and Kadane (1986), Tierney, Kass and Kadane (1989), DiCiccio and Martin (1991) and DiCiccio and Stern (1993) applied the Laplace method to approximate a marginal posterior density in case of a parameter $\phi = (\psi, \eta)$ where $\psi = (\psi_1, \dots, \psi_p)$ is a parameter of interest and $\eta = (\eta_1, \dots, \eta_q)$ is a nuisance parameter. In our case, where we are interested in posterior probabilities of a half-space A which can be written $\psi_1 \geq \psi_1'$, we have $p = 1$, $\psi \equiv \psi_1$, $q = d - 1$, $\eta_j = \phi_{j+1}$ for $j = 1, \dots, d - 1$. Let $\pi(\phi)$ be the prior density, $Y = (Y_1, \dots, Y_n)$ the observations, $\tilde{\phi}(\psi) = (\psi, \tilde{\eta}(\psi))$ the MLE of ϕ for fixed ψ, if it exists, $\hat{\phi} = (\hat{\psi}, \tilde{\eta}(\hat{\psi}))$ the overall MLE, ℓ the log likelihood, $m(\psi) := \ell(\tilde{\phi}(\psi))$ the profile log likelihood, and $\ell_{\eta\eta} := \{\partial^2 \ell / \partial\eta_i \partial\eta_j\}_{i,j=1}^q$ the Hessian of ℓ with respect to η. Let $\pi_{\psi|Y}(\psi)$ be the marginal posterior density of ψ. Then the approximation in the form stated by DiCiccio and Stern (1993) is

$$\pi^*_{\psi|Y}(\psi) := c^* \left(\frac{\det[-\ell_{\eta\eta}(\tilde{\phi}(\psi))]}{\det[-\ell_{\eta\eta}(\hat{\phi})]} \right)^{-1/2} \frac{\pi(\tilde{\phi}(\psi))}{\pi(\hat{\phi})} e^{m(\psi) - m(\hat{\psi})}, \tag{1}$$

where c^* is a normalizing constant. (Note that functions of $\hat{\phi}$ and $\hat{\psi}$ do not depend on the argument ψ and in that sense are also constants.) Let z be the likelihood root

$$z := \operatorname{sgn}(\psi - \hat{\psi}) \sqrt{2(m(\hat{\psi}) - m(\psi))}. \tag{2}$$

If there is a pseudo-true ψ_0 and a 1–1, mutually C^1 relationship between z and ψ for ψ in some neighborhood of ψ_0, as will be shown in the Appendix, we get an approximate posterior density for z,

$$\pi^{**}_{z|Y}(z) := c^* \left(\frac{\det[-\ell_{\eta\eta}(\tilde{\phi}(\psi(z)))]}{\det[-\ell_{\eta\eta}(\hat{\phi})]} \right)^{-1/2} \frac{\pi(\tilde{\phi}(\psi(z)))}{\pi(\hat{\phi})} e^{-z^2/2} \frac{d\psi}{dz}. \tag{3}$$

We then define an approximation for the posterior probability of a half-space $A : \psi \geq \psi' > \hat{\psi}$ or equivalently $z \geq z_1 > 0$, say, by

$$\pi^{**}(z \geq z_1) := \Phi(-z_1) \left(\frac{\det[-\ell_{\eta\eta}(\tilde{\phi}(\psi(z_1)))]}{\det[-\ell_{\eta\eta}(\hat{\phi})]} \right)^{-1/2} \frac{\pi(\tilde{\phi}(\psi(z_1)))}{\pi(\hat{\phi})} \frac{d\psi/dz|_{z=z_1}}{d\psi/dz|_{z=0}}. \tag{4}$$

The approximation given by Theorem 1 is just $\Phi(-z_1)$. We can now state the following corollary.

COROLLARY 1. *Under the hypotheses of Theorem* 1, *if* A_n *are half-spaces with* $\hat{\phi}_n \notin A_n$ *and* $d(A_n, \phi_0) \to 0$ *as* $n \to \infty$ *where for each* n, A_n *is written as* $\{\psi \geq \eta_n\}$ (*where the parameter* ψ, *a linear function of* ϕ, *can also depend on* n), *then the posterior probability* $\pi_{x,n}(A_n)$ *is asymptotic as* $n \to \infty$ *to its sharpened Laplace approximation* $\pi^{**}(z \geq z_n)$, *which is well defined almost surely for* n *large enough*.

A proof will be given in the Appendix.

To improve the approximation of small posterior probabilities, one can consider π^{**} and the following approximations, but the rest of this section is nonrigorous.

The approximation $\pi^{**}(z \geq z_1)$ can be sharpened further. We will define approximations in the one-dimensional case $p = d = 1$, $q = 0$, assuming that the prior density π is differentiable at least once, or has higher derivatives for higher-order approximations. The ratio of determinants is replaced by 1 and $\tilde{\phi}(\psi) \equiv \psi$. Let τ be the prior density as a function of z, so $\tau(z) = \pi(\psi(z))\, d\psi/dz$. Let $z_1 \geq 0$. The posterior probability $\pi_Y(z \geq z_1)$ equals

$$\int_{z_1}^{\infty} \tau(z) e^{-z^2/2}\, dz \Big/ \int_{-\infty}^{\infty} \tau(z) e^{-z^2/2}\, dz, \tag{5}$$

exactly if the 1–1, increasing, C^1 relationship between z and ψ holds for all $|z| < \infty$ (as in the beta case in Section 10), or up to an error exponentially small in n under broader conditions, with ∞ replaced by $K\sqrt{n}$ for some $K < \infty$.

Let $\pi_Y^{k,*}(z \geq z_1)$ denote the approximation to $\pi_Y(z \geq z_1)$ where τ is replaced by partial Taylor expansions through order k, around 0 in the denominator and around z_1 in the numerator. The integrals of polynomials in z times $\exp(-z^2/2)$ can be found explicitly. In the denominator, odd powers of z yield 0 integrals. We get $\pi^{**}(z \geq z_1) = \pi_Y^{0,*}(z \geq z_1)$ and for $\zeta := z_1$,

$$\pi_Y^{1,*}(z \geq z_1) = \tau(0)^{-1}\big[\Phi(-z_1)\tau(z_1) + \tau'(z_1)(\phi(z_1) - z_1\Phi(-z_1))\big],$$

$$\pi_Y^{2,*}(z \geq \zeta)$$
$$= \frac{\Phi(-\zeta)[\tau(\zeta) - \zeta\tau'(\zeta) + \tau''(\zeta)(1+\zeta^2)/2] + \phi(\zeta)[\tau'(\zeta) - \zeta\tau''(\zeta)/2]}{\tau(0) + \tau''(0)/2}.$$

DiCiccio and Martin (1991) gave a related approximation. In one dimension, letting $l(\psi)$ denote the log likelihood function and $l^{(k)}$ its kth derivative, for a given ψ_1, the posterior tail probability $\pi_n(\psi \leq \psi_1)$ is approximated by

$$\pi_Y^{\mathrm{DM},*} := \Phi(z_1) + \phi(z_1)\left[\frac{1}{z_1} + \frac{\sqrt{-l^{(2)}(\hat{\psi})}\pi(\psi)}{l^{(1)}(\psi)\pi(\hat{\psi})}\right], \tag{6}$$

where $\psi(z_1) = \psi_1$ and we have in mind $\psi_1 \le 0$ in this case. These approximations will appear in the examples in Section 10.

Following a method going back at least to Bleistein (1966), Temme (1982) in effect expands an integral $\int_{-\infty}^{w} \phi(z) f(\delta z)\, dz$ for a smooth function f as

$$f(0)\Phi(w) + \int_{-\infty}^{w} \frac{f(0) - f(\delta z)}{z}\, d\phi(z),$$

integrating by parts and iterating the process to obtain an asymptotic expansion in powers of $\delta \downarrow 0$. Here Temme (1982), going beyond earlier work by analysts as far as we know, allows f to depend separately on δ if it satisfies suitable conditions. In our case $\delta = 1/\sqrt{n}$. Although the resulting expansions appear at first sight different from ours, we have verified that: if $\delta w \to 0$, while possibly $w \to -\infty$, the terms of orders δ^j, $j = 0, 1, 2$, are the same in our expansions of the numerator and denominator of $\pi_Y^{*,2}$ as in those of Temme (1982). Or, if $\delta \downarrow 0$, $w \to -\infty$ and $\delta w \to u < 0$ (a large deviation case), then our expansion and Temme's have different coefficients of $\Phi(w)$ and $\phi(w)$, but via

$$\Phi(w) = \phi(w)\left[-\frac{1}{w} + \frac{1}{w^3} + O(w^{-5})\right]$$

as $w \to -\infty$, Temme's series, and ours except for the error bound, give

$$\phi(w)\left[-\frac{\delta f(u)}{u} + \delta^3\left\{\frac{f(u) - uf'(u)}{u^3}\right\} + O(\delta^5)\right], \tag{7}$$

where the $O(\delta^5)$ term also depends on f and u.

When applying Temme's expansion method to a beta probability

$$I_x(a, b) = \int_0^x t^{a-1}(1-t)^{b-1}\, dt \Big/ B(a, b), \qquad 0 < x < 1,\ a > 0,\ b > 0,$$

there are different possible choices of δ such as $1/\sqrt{a+b}$ [Temme (1982)] or $1/\sqrt{a+b-2}$ for $a + b > 2$ or $1/\sqrt{a+b-1}$ for $a + b > 1$ (Section 10 below). Temme (1982, 1987) evaluates the complete beta function $B(a, b)$ directly, whereas we apply approximations to it as examples of how the approximations might work more generally.

There is an evident similarity between the sharpened Laplace approximations to posteriors just given and the "saddlepoint" approximation to tail probabilities for sample means due to Lugannani and Rice (1980); see also Daniels (1987) and Jensen [(1995), Chapter 3]. One contrast is that the Lugannani–Rice and related approximation formulas (e.g., for distributions of maximum likelihood estimators) are based on moment generating functions, sometimes for approximating distributions [e.g., Fraser, Reid and Wu (1999)], and do not require explicit knowledge of the densities of the sample means. Conversely, $\pi_Y^{i,*}$ and $\pi_Y^{\mathrm{DM},*}$ do not involve moment generating functions, but they do use the likelihood functions and prior densities (which are usually given explicitly in the Bayes case) and their derivatives.

In applications of the Laplace method to posterior probabilities, DiCiccio and Martin (1991) and Fraser, Reid and Wu (1999) gave error rates of $O(n^{-3/2})$ but only for Laplace deviations. We will see in Table 3 of Section 10 that $\pi_Y^{i,*}$ and $\pi_Y^{\mathrm{DM},*}$ also work well in a large deviation example.

4. Preliminary transformations. Here we begin the proof of Theorem 1. Let $L_n(\phi) := \exp(LL_n(\phi))$ be the likelihood function. We first note that the posterior law is eventually well defined under our assumptions: it follows from (A_5) that for a compact neighborhood $\mathcal{N}$ of ϕ_0, a.s. for each n large enough, $L_n(\phi)$ is bounded outside of $\mathcal{N}$. On $\mathcal{N}$ it is bounded by continuity (A_6). Then $\int L_n(\phi)\pi_0(\phi)\,d\phi < \infty$. Also by (A_5), for the same n, $L_n(\phi) > 0$ for some $\phi \in \mathcal{N}$ and so by continuity $L_n(\phi) > 0$ on some nonempty open set. Then by (A_4) $\int L_n(\phi)\pi_0(\phi)\,d\phi > 0$. Thus a.s. for n large enough, $L_n(\phi)\pi_0(\phi)/\int L_n(\phi')\pi_0(\phi')\,d\phi'$ is a well-defined posterior probability density.

We will assume first and for most of the proof that $\hat{\phi}_n \notin A$. The case $\hat{\phi}_n \in A$ will follow by symmetry. We will make a linear change of coordinates in Θ so that we can assume $\hat{\phi}_n = 0$, $E(\phi_0)$ is the identity matrix, A is a half-space with boundary where the first coordinate ϕ_1 is some constant and $\tilde{\phi}$ is the point of A closest to 0. Moreover, we need to make these changes "without loss of generality," which requires proof since A varies over all half-spaces and κ' must not depend on A. Also, our informal statements so far have assumed that unique maximum likelihood estimators $\hat{\phi}_n$ in Θ and $\tilde{\phi}_n$ in ∂A exist, which is true for $\hat{\phi}_n$ for n large enough, but not necessarily true for $\tilde{\phi}_n$ simultaneously for all half-spaces A. It is true if ∂A is close enough to $\hat{\phi}_n$ in a sense to be made precise as Case I (28) after some coordinate transformations. The full transformation involving $\tilde{\phi}_n$ will be made in Case I.

We can take $0 < \varepsilon < 1$. Take $\delta > 0$ small enough so that

$$(1+\delta)^{3d+9} < 1+\varepsilon. \tag{8}$$

Then $\delta < 1/10$. Let $\delta_1 := \delta^2/100$. By a fixed linear transformation of ϕ (not depending on A) we can obtain coordinates θ in which $E(\theta_0) = I$, the identity matrix, where θ_0 is ϕ_0 expressed in θ coordinates. By assumption (A_4) there exists a neighborhood V_0 of θ_0 such that, for the prior density $\pi_0(\phi)\,d\phi$ now expressed as $\pi_1(\theta)\,d\theta$, and $\eta := \inf_{\theta \in V_0} \pi_1(\theta)$,

$$\eta \le \pi_1(\theta) \le \eta(1+\delta), \qquad \theta \in V_0. \tag{9}$$

To recall some facts about matrix norms, let K and J be $d \times d$ real matrices. For the Euclidean norm $\|t\| := (t_1^2 + \cdots + t_d^2)^{1/2}$, we use the matrix norm $\|K\| := \max_{\|t\|=1} \|Kt\|$. Let $K \le J$ mean that $J - K$ is nonnegative definite and symmetric. For any symmetric real matrix K, $\|K\| = \max_i |\lambda_i|$ for the eigenvalues λ_i of K. Thus if $-cI \leq K \leq cI$ and $c \geq 0$, then $\|K\| \leq c$. Conversely if J is symmetric and $\|J - I\| < 2c/3$ where $0 < c < 1/2$ then

$$I/(1+c) \leq J \leq (1+c)I. \tag{10}$$

Then, there is a $\gamma_1 > 0$ such that for W as in (A_6) and (A_7) and

$$V := \{\theta : \|\theta - \theta_0\| < 4\gamma_1\}, \tag{11}$$

$V \subset W \cap V_0$ and $\|E(\theta) - I\| < \delta_1/3$ for all $\theta \in V$.

By a rotation which does not change distances, and keeps $E(\theta_0) = I$, we can assume that $v_A = (1, 0, \ldots, 0)$, so A is a set

$$A = \{\theta : \theta_1 \geq M\}. \tag{12}$$

Let $E_n(\theta)$ be the empirical Fisher information matrix,

$$E_n(\theta) := \left\{ -\int (\partial^2 \log f(x,\theta)/\partial\theta_i \partial\theta_j)\, dP_n(x) \right\}_{i,j=1}^{n}. \tag{13}$$

Almost surely for n large enough, by (A_7),

$$\|E_n(\theta) - I\| < 2\delta_1/3 \tag{14}$$

for all $\theta \in V$, and by (A_5), a.s. for n large enough the supremum of the likelihood will be attained at some point θ where $\|\theta - \theta_0\| \leq \gamma_1$. Specifically, this will occur for $n \geq n_0(\xi, \gamma_1)$ where ξ is a point of the probability space on which X_j are defined. At any such point, the gradient of $LL_n(\theta)$ exists and is 0. By (14) and (10) we have

$$I/(1+\delta_1) \leq E_n(\theta) \leq (1+\delta_1)I \tag{15}$$

for all $\theta \in V$. By (15), E_n is strictly positive definite on V, so almost surely,

(16) for $n \geq n_0(\xi, \gamma_1)$ a unique MLE $\hat{\theta}_n$ exists with $\|\hat{\theta}_n - \theta_0\| < \gamma_1$.

By a translation we can assume

$$\hat{\theta}_n = 0. \tag{17}$$

Then from (12),

$$A = \{\theta : \theta_1 \geq \zeta\} \tag{18}$$

for some ζ depending on $\hat{\theta}_n$, A and the choice of coordinates. The translation preserves distances and the fact that $E(\theta_0) = I$.

By (18), $\zeta > 0$ since $\hat{\theta}_n = 0 \notin A$. Also by (15), LL_n is strictly concave on V. Thus if $A \cap V$ is nonempty, and since we assume $\hat{\theta}_n \notin A$, we have

$$\sup_{A \cap V} LL_n = \sup_{\partial A \cap V} LL_n. \tag{19}$$

For $\theta \in \Theta$, let

$$H(\theta) = H_n(\theta) = \frac{1}{n}\sum_{i=1}^{n}[\log f(X_i, 0) - \log f(X_i, \theta)]. \tag{20}$$

Along every line segment $\theta = tw$ in V, say, $a < t < b$, where $\|w\| = 1$, we have by (15) that $1/(1+\delta_1) \leq d^2H(tw)/dt^2 \leq 1+\delta_1$. A Taylor expansion around $0 = \hat{\theta}_n$, where $\nabla H = 0$, then gives $dH(tw)/dt|_{t=0} = H(0) = 0$,

$$1/(1+\delta_1) \leq [dH(tw)/dt]/t \leq 1+\delta_1 \qquad \text{for } t \neq 0, \tag{21}$$

$$t^2/[2(1+\delta_1)] \leq H(tw) \leq t^2(1+\delta_1)/2. \tag{22}$$

In what follows we let $H(\zeta, X) \equiv H(\zeta)$ if $d = 1$. For $\zeta \leq \gamma_1$ we have

$$H(\zeta, 0) \leq \zeta^2(1+\delta_1)/2 \tag{23}$$

[here 0 is the $(d-1)$-dimensional zero vector]. Also, for $(\zeta, X) \in V$,

$$H(\zeta, X) \geq (\zeta^2 + \|X\|^2)/[2(1+\delta_1)]. \tag{24}$$

Let

$$\rho := \gamma_1/(1+6^{-1}\delta) \leq \gamma_1/\big(1+(\delta^2/36)\big)^{1/2}. \tag{25}$$

Let $r := \rho/2$. Let Ω_{d-1} be the total $(d-1)$-dimensional area of the sphere $S^{d-1} := \{\omega \in \mathbf{R}^d : \|\omega\| = 1\}$. Then $\Omega_0 := 2$, $\Omega_1 = 2\pi$, $\Omega_2 = 4\pi, \ldots$, $\Omega_k = 2\pi^{(k+1)/2}/\Gamma((k+1)/2)$ for $k = 0, 1, \ldots$. Let $C_1 := 1/2$, $C_2 := \pi/9$ and $C_d := \Omega_{d-1}/(2\Omega_{d-2})$ for $d \geq 3$. Let, for κ in (A_5) with $\mathcal{N} = V$,

$$\zeta_0 := \zeta_0(d, r, \delta) := \min(C_d r\delta, \sqrt{\kappa}). \tag{26}$$

Since $\Omega_d = 2\int_0^1 \Omega_{d-1}(1-x^2)^{(d-2)/2}\,dx \leq 2\Omega_{d-1}$ for $d \geq 2$, we have for all $d = 1, 2, \ldots$,

$$C_d \leq 1. \tag{27}$$

5. Cases of the proof. Recall ζ_0 as defined in (26). The proof of Theorem 1 will be divided into two cases:

(28) Case I: $$0 < \zeta < \zeta_0,$$

where the boundary $\partial A = \{\theta_1 = \zeta\}$ is not too far from the MLE $\hat{\theta}_n = 0$, and

(29) Case II: $$\zeta \geq \zeta_0,$$

where ∂A is bounded away from $\hat{\theta}_n$ by a fixed amount.

Here, we will begin the main proof in Case I, where either (a) or (b) in Theorem 1 may hold. We will find a unique MLE $\tilde{\theta}_n$ in A or equivalently by (19), (23) and (26) in ∂A. The following arguments through (38) are immediate if $d = 1$ since (ζ, X) is replaced by ζ and X by $0 \in \mathbf{R}^{d-1}$.

For fixed ζ with $0 < \zeta < \zeta_0$, $\inf_{(\zeta,X)\in V} H(\zeta, X)$ can by (23) and (24) be restricted to (ζ, X) such that $(\zeta^2 + \|X\|^2)/(1+\delta_1) \le \zeta^2(1+\delta_1)$, or equivalently $1 + \|X\|^2/\zeta^2 \le (1+\delta_1)^2$, and thus since $\delta < 1/10$ and $\delta_1 = \delta^2/100$,

$$\|X\|^2/\zeta^2 \le 2\delta_1 + \delta_1^2 \le \delta^2/36. \tag{30}$$

This then implies $\zeta^2 + \|X\|^2 \le \gamma_1^2$ by (25) since $\zeta < \rho$. Thus $(\rho, X) \in V$ by definition (11) of V and (16) for $n \ge n_0(\xi, \gamma_1)$. On the compact, convex set of (ζ, X) satisfying (30) for fixed $\zeta > 0$, H is a C^2 function, strictly convex by (15), so H attains its minimum on $\{\theta_1 = \zeta\} \cap V$ at a unique point $\tilde{\theta} := (\zeta, \tilde{X})$, where $\tilde{X} = \tilde{X}(\zeta)$.

Define new coordinates y by the linear transformation $y = T(\theta)$ such that for $u^{(1)} := (u_2, \ldots, u_d)$, where u may be y or θ,

$$y_1 = \theta_1, \qquad y^{(1)} = \theta^{(1)} - \theta_1 \tilde{X}/\zeta. \tag{31}$$

We have $\zeta < \rho$, so that $(\zeta, 0) \in U(\rho) := \{y : \|y\| < \rho\}$. In the y coordinates, A still is $\{y_1 \ge \zeta\}$ and $\hat{y} = 0$. Now $\tilde{y} = (\zeta, 0)$. Define $h = h_n$ so that

$$h(y) := h_n(y) = H_n(\theta) = H_n(y_1, y^{(1)} + y_1\tilde{X}/\zeta). \tag{32}$$

Then since $\zeta < \rho$, $\|T - I\| \le \delta/6$ by (30). Since T^{-1} and its transpose $(T^{-1})^t$ have the same form as T and T^t, respectively, in this case, replacing $\tilde{X}$ by $-\tilde{X}$, also

$$\|T^{-1} - I\| = \|(T^{-1})^t - I\| \le \delta/6. \tag{33}$$

Thus since $\zeta < \zeta_0$, if $\|y\| < \rho$ then $\|\theta\| = \|\theta - \hat{\theta}_n\| \le (1 + (\delta/6))\|y\| < \gamma_1$ by (25), so $\theta \in V \subset V_0$ by choice of γ_1 and (16), for $n \ge n_0(\xi, \gamma_1)$. We have $D^{(2)}H_n = E_n$ by (13) and (20), and $D^{(2)}h_n = (T^{-1})^t(D^{(2)}H_n)T^{-1}$. By (14) and (33) and a short calculation we then have $\|D^{(2)}h_n - I\| < 2\delta/3$. Thus by (10), if $\|y\| < \rho$,

$$I/(1+\delta) \le D^{(2)}h_n \le (1+\delta)I. \tag{34}$$

Note that the determinants (Jacobians) of an orthogonal transformation, translation and T (31) are all ± 1, so that the volume element $dy = d\theta$ is unchanged. Then by (9), for $\|y\| < \rho$,

$$\eta \le \pi(y) \le \eta(1+\delta), \tag{35}$$

where $\pi(y)$ is the density of the prior π_1 for θ expressed in the y coordinates.

Recall that $U(\rho) := \{y : \|y\| < \rho\}$. Note that $U(\rho)$, by (31), depends on $\hat{\theta}_n$, $\zeta > 0$, $\tilde{X}$, and its radius $\rho > 0$ for y, $U(\rho) = U(\rho, \zeta, \tilde{X}, \hat{\theta}_n)$.

Reviewing then, in Case I, since $0 < \zeta < \rho < \gamma_1$, in the y coordinates, H_n is expressed as $h_n \equiv h_n(y)$, the maximum likelihood estimate $\hat{y}$ is equal to 0, and the maximum likelihood estimate on the set $\partial A = \{y_1 = \zeta\}$ equals $(\zeta, 0)$.

6. Bounds for the likelihood function and integrals. It follows from (20), (32), (34) and a Taylor expansion around $y = 0$ as in (22) that

$$\|y\|^2/[2(1+\delta)] \le h(y) \le (1+\delta)\|y\|^2/2 \tag{36}$$

for all y in $U(\rho)$. We have $A \cap U(\rho) = \{y_1 \ge \zeta\} \cap U(\rho)$. Since h is minimized at $(\zeta, 0)$ on $\{y_1 = \zeta\} \cap U(\rho)$, it follows that $\partial h/\partial y_j = 0$ at $(\zeta, 0)$ for $j = 2, \ldots, d$. By a Taylor expansion around $(\zeta, 0)$, we have by (34) for $(\zeta, X) \in U(\rho)$ and

$$\tau := h(\zeta, 0), \tag{37}$$

$$\tau + \frac{1}{2(1+\delta)}\|X\|^2 \le h(\zeta, X) \le \tau + \frac{(1+\delta)}{2}\|X\|^2. \tag{38}$$

Note that

$$n\tau = \Delta \tag{39}$$

by (17), (20), (32) and (37). Assume that $d \ge 2$ (the case $d = 1$ is easier and will be omitted). Define

$$I_n(B) := \int_B e^{-nh(y)}\pi(y)\,dy \tag{40}$$

for any measurable set B. Note that by (20) and (32), the likelihood function equals $ML\, e^{-nh}$, where ML is the maximum likelihood on Θ.

Recall that $\pi_{x,n}$ is the posterior distribution on Θ, so a.s. for n large enough, for any measurable set $C \subset \Theta$ in y coordinates, $\pi_{x,n}(y \in C)$ equals

$$ML \int_C e^{-nh(y)}\pi(y)\,dy \Big/ \left(ML \int_\Theta e^{-nh(y)}\pi(y)\,dy\right) = I_n(C)/I_n(\Theta). \tag{41}$$

We will prove Theorem 1 by way of three propositions. Proposition 1 will say that $\pi_{x,n}(U(\rho)^c)$ is exponentially small in n; Proposition 2 gives an asymptotic evaluation for $I_n(U(\rho))$ and Proposition 3 does so for $I_n(A \cap U(\rho))$. For any $\rho' > 0$ not depending on n let $U' := U'(\rho') := B(\hat{\theta}_n, \rho') = B(0, \rho')$ in the θ coordinates as ultimately chosen, ending with (17).

PROPOSITION 1. *For some $\nu > 0$ not depending on the observations, a.s. for n large enough, $\pi_{x,n}(U'(\rho/2)^c) < e^{-n\nu}$ and whenever $0 < \zeta \le \rho$, $\pi_{x,n}(U(\rho)^c) < e^{-n\nu}$.*

PROPOSITION 2. *We have almost surely for n large enough,*

$$(1+\delta)^{-(3d+10)/2} \le \frac{1}{\eta} I_n(U(\rho))(n/2\pi)^{d/2} \le (1+\delta)^{(3d+6)/2}. \tag{42}$$

PROPOSITION 3. *In Case* I, *for* $d \geq 2$, *we have almost surely for* n *large enough, not depending on* A,

$$\eta(2\pi/n)^{d/2}(1+\delta)^{-(3d+10)/2}\Phi(-\sqrt{2n\tau}) \tag{43}$$

$$\leq I_n(A \cap U(\rho))$$

$$\leq \eta(2\pi/n)^{d/2}(1+\delta)^{(3d+6)/2}\Phi(-\sqrt{2n\tau}). \tag{44}$$

The three propositions will be proved in Section 8 after Lemma 5. Recall that the orthogonal transformation just before (12) and the translation just after (16) did not change distances in the θ coordinates.

Let $\omega = (\omega_1, \ldots, \omega_d)$, $S(d^+) := \{\omega \in \mathbf{R}^d : \|\omega\| = 1,\ \omega_1 > 0\}$. Let $d\omega$ be the surface element on $S(d^+)$. For any Borel set $S \subset S(d+)$, the surface area measure is $\Omega_{d-1}(S) := \int_S d\omega$. Let $D_\omega := \zeta/\omega_1$, and $h_\omega := D_\omega\|\omega^{(1)}\|$, where $\omega^{(1)} = (\omega_2, \ldots, \omega_d) \in \mathbf{R}^{d-1}$. For $0 \leq a \leq b \leq \infty$ let $[[a, b)) := \{\omega : a \leq D_\omega < b\}$. Then

$$D_\omega^2 = \zeta^2 + h_\omega^2. \tag{45}$$

Let the spherical coordinates of y be (t, ω) where $\omega := y/\|y\|$ and $t = \|y\|$. Then $dy = t^{d-1}\, d\omega\, dt$. Let $S(\zeta, \rho) := \{\omega \in \mathbf{R}^d : \|\omega\| = 1,\ \omega_1 \geq \zeta/\rho\}$. We have

$$I_n(A \cap U(\rho)) = \int_{S(\zeta,\rho)} \int_{D_\omega}^{\rho} e^{-nh(t\omega)}\pi(t\omega)t^{d-1}\, dt\, d\omega. \tag{46}$$

If $\pi(\cdot)$ were constant and h linear then the integrals $\int_{D_\omega}^{\rho} = \int_{D_\omega}^{\infty} - \int_{\rho}^{\infty}$ would be proportional to gamma probabilities. If h is quadratic without a linear term they reduce to gamma probabilities or, if $d = 1$, to normal probabilities. We will bound the corresponding integrals above and below by gamma probabilities. Thus we consider gamma and normal probabilities in the following section.

For $x > 0$, $[[x, \infty)) = \{\omega : \zeta/\omega_1 \geq x\} = \{\omega : 0 \leq \omega_1 \leq \zeta/x\}$. Recalling ζ_0 (26) and Ω_d, C_d as defined just before (26), we have the lemma.

LEMMA 1. *For any* ζ *with* $0 < \zeta < \zeta_0$, $\Omega_{d-1}([[r, \infty))) < \delta\Omega_{d-1}/2$, *where* $\Omega_0(\cdot)$ *is a sum of unit point masses at* 1 *and* -1.

PROOF. For $d = 1$, $S(1^+) = \{1\}$, so $[[r, \infty))$ is empty by (26) and (27) since $\zeta < \zeta_0 < r$. For $d = 2$, $\Omega_1([[r, \infty))) = 2\arcsin(\zeta/r)$. Since arcsin is a convex function on $[0, 1]$, $\arcsin x \leq 2x$ for $0 \leq x \leq 1$. So since $\zeta/r < 1$, and by (26), $\Omega_1([[r, \infty))) \leq 4\zeta/r < 4C_2\delta \leq \delta\pi/2 = \delta\Omega_1/4$. This proves the lemma for $d = 2$.

Let now $d \geq 3$. We have, again by (26), and since $0 < \zeta < \zeta_0$,

$$\Omega_{d-1}([[r, \infty))) = \Omega_{d-1}\{\omega : 0 \leq \omega_1 \leq \zeta/r\} = \int_0^{\zeta/r} \Omega_{d-2}(1 - x^2)^{(d-3)/2}\, dx$$

$$< \Omega_{d-2}\zeta_0/r \leq \delta\Omega_{d-1}/2$$

by definition of C_d above (26). Lemma 1 is proved. □

7. Gamma and normal tail/density ratios. We begin with normal probabilities. Let $M(x) = \Phi(-x)/\varphi(x)$ for $x \geq 0$ (M is sometimes called Mills' ratio).

LEMMA 2. *For all $x \geq 0$ and $0 \leq \delta \leq 1$, $e^{-\delta} \leq M(x(1+\delta))/M(x) \leq 1$.*

PROOF. For $x = 0$ the result is clear. It is easily shown that

$$M'(x) = -1 + xM(x) \tag{47}$$

since $\varphi'(x) = -x\varphi(x)$. By, for example, Lemma 12.1.6(a) in Dudley (1993), we have for all $x > 0$ that $\Phi(-x) \leq \varphi(x)/x$ and so $M(x) \leq 1/x$. It follows that $M'(x) \leq 0$ for all $x > 0$. So the second inequality is proved. For the first, and $x > 0$, $M(x)$ can be written as Laplace's continued fraction [Wall (1948), 92.15],

$$M(x) = 1/(x + 1/(x + 2/(x + 3/(x + \cdots.$$

So again $M(x) < 1/x$, and now we get

$$M(x) > 1\bigg/\left(x + \frac{1}{x}\right) = x/(x^2+1). \tag{48}$$

It follows that $[\log M(x)]' = M'(x)/M(x) = [-1 + xM(x)]/M(x) \geq -1/x$, so

$$[\log M(\cdot)]_x^{x+\delta x} \geq -\int_x^{x+x\delta} \frac{1}{t}\,dt = -\log[1+\delta] \geq -\delta.$$

This completes the proof of the first inequality, and the proof of Lemma 2. $\square$

For $\alpha > 0$ and $s > 0$, let

$$M_\alpha(s) := \int_s^\infty x^{\alpha-1}e^{-x}\,dx \bigg/ (s^{\alpha-1}e^{-s}). \tag{49}$$

LEMMA 3. (a) *For any $\alpha \geq 1$, $s > 0$ and $\delta > 0$, we have $M'_\alpha(s) \leq 0$, $M_\alpha(s) \geq 1$ and $e^{-\delta(\alpha-1)} \leq M_\alpha(s(1+\delta))/M_\alpha(s) \leq 1$.*

(b) *For $\alpha = 1/2$, $s > 0$ and $0 < \delta \leq 1$, we have $M'_{1/2}(s) \geq 0$ and $1 \leq M_{1/2}(s(1+\delta))/M_{1/2}(s) \leq (1+\delta)^{1/2}$.*

PROOF. For (a), note first that $M_1 \equiv 1$. So assume $\alpha > 1$. Using $x^{\alpha-1} \geq s^{\alpha-1}$ for $x \geq s$ it follows that $M_\alpha(s) \geq 1$ for all $s > 0$. We have $M'_\alpha(s) = -1 + M_\alpha(s)[1 - (\frac{\alpha-1}{s})]$. An integration by parts shows that $M_\alpha(s) \leq 1 + (\frac{\alpha-1}{s})M_\alpha(s)$. So $M'_\alpha(s) \leq 0$. This proves the first part of (a). Now $d\log(M_\alpha(s))/ds = -\frac{1}{M_\alpha(s)} + 1 - (\frac{\alpha-1}{s}) \leq 0$. So, $\frac{d}{ds}[\log(M_\alpha(s))] \geq -(\frac{\alpha-1}{s})$ since $M_\alpha(s) \geq 1$. Thus,

$$\int_s^{s+\delta s} \frac{d}{dt}[\log(M_\alpha(t))]\,dt \geq -(\alpha-1)\log(1+\delta) \geq -(\alpha-1)\delta.$$

So $M_\alpha(s+\delta s)/M_\alpha(s) \geq e^{-\delta(\alpha-1)}$. This completes the proof of (a).

For (b), we have $M_{1/2}(s) \equiv \sqrt{2s}M(\sqrt{2s})$. So by (47), with $u := (2x)^{1/2}$, $M'_{1/2}(x) = [M(u)(1+u^2) - u]/u$. Since $M(u) > u/(1+u^2)$ by (48), $M'_{1/2}(x) \geq 0$. This proves the first part of (b) and the first inequality of the second part. For the last inequality, Lemma 2 gives $M_{1/2}((1+\delta)s) \leq [2s(1+\delta)]^{1/2}M([2s]^{1/2})$, so $M_{1/2}((1+\delta)s) \leq (1+\delta)^{1/2}M_{1/2}(s)$. □

Next, let $a > 0$, and let X_a be a random variable distributed according to a Γ_a distribution with density $x^{a-1}e^{-x}/\Gamma(a)$ for $x > 0$. We have:

LEMMA 4. *For any $a > 0$, (a) $M_a(s) \to 1$ as $s \to \infty$.*

(b) *For α fixed in $(0, 1)$ and $0 \leq x \leq \alpha y$, as $y \to \infty$, $P(X_a \geq y)/P(X_a \geq x) \to 0$ uniformly in such x.*

PROOF. (a) The definition of M_a (49) and integration by parts give

$$\int_s^\infty x^{\alpha-1}e^{-x}\,dx = s^{\alpha-1}e^{-s} + (\alpha-1)\int_s^\infty x^{\alpha-2}e^{-x}\,dx,$$

and (a) follows since $\int_s^\infty x^{\alpha-2}e^{-x}\,dx \leq s^{-1}\int_s^\infty x^{\alpha-1}e^{-x}\,dx$.

(b) By (a) it suffices to note that as $y \to \infty$, $y^{a-1}e^{-y}/[(\alpha y)^{a-1}e^{-\alpha y}] = \alpha^{1-a} \times e^{(\alpha-1)y} \to 0$. □

8. The rest of the proof. In this section we will finish the proof of Theorem 1. The following bounds the inner integral in (46) via gamma probabilities:

LEMMA 5. *Let $U_v := (0, v)$ with $v > 0$ and $g : U_v \mapsto \mathbf{R}$. Suppose that for some $0 < \delta \leq 1$, for all t in U_v, $1/(1+\delta) \leq g''(t) \leq 1+\delta$ and assume that g'' and thus g and g' can be extended to be continuous on $[0, v]$, with $g(0) = g'(0) = 0$. Let $\pi(\cdot)$ be measurable on U_v and such that for some $\eta > 0$, we have $\eta \leq \pi(t) \leq \eta(1+\delta)$ for all $t \in U_v$. For $0 \leq a < v$, $n = 1, 2, \ldots$ and $k = 0, 1, 2, \ldots$, let*

$$I_{n,a,k} := I_{n,a,k,v} := \int_a^v e^{-ng(t)}\pi(t)t^k\,dt, \tag{50}$$

$$\Gamma_{n,a,k} := \eta e^{-ng(a)}a^{k-1}M_{(k+1)/2}(na^2/2)n^{-1}.$$

We then have $I_{n,a,k,v} \leq \Gamma_{n,a,k}(1+\delta)^{k+2}$, and $I_{n,a,k,v}$ is bounded below by

$$\frac{\Gamma_{n,a,k}}{(1+\delta)^{k+1}} - \frac{\eta e^{-ng(a)}}{n\sqrt{1+\delta}}v^{k-1}\exp\left[\frac{n}{2}(1+\delta)(a^2-v^2)\right]M_{(k+1)/2}\left(\frac{nv^2}{2}\right).$$

PROOF. We have that $t/(1+\delta) \leq g'(t) \leq t(1+\delta)$, thus $t^2/[2(1+\delta)] \leq g(t) \leq t^2(1+\delta)/2$ for all t in U_v. Now for $0 \leq a \leq t \leq v$,

$$g(t) = g(a) + \int_a^t g'(s)\,ds \geq g(a) + \int_a^t \frac{s}{1+\delta}\,ds = g(a) + \frac{(t^2-a^2)}{2(1+\delta)},$$

so

$$I_{n,\alpha,k,v} \le \eta(1+\delta) \int_a^\infty e^{-ng(a)} \exp\left[-\frac{n(t^2-a^2)}{2(1+\delta)}\right] t^k\, dt.$$

For any $\lambda > 0$ let

$$J_{a,\lambda,k} := \int_a^\infty \exp(-\lambda t^2) t^k\, dt. \tag{51}$$

Then

$$J_{a,\lambda,k} = (2\lambda)^{-1} a^{k-1} \exp(-\lambda a^2) M_{(k+1)/2}(\lambda a^2) \tag{52}$$

by the change of variables $u = \lambda t^2$. Set $\lambda_1 := n/[2(1+\delta)]$. It follows that

$$I_{n,\alpha,k,v} \le \eta(1+\delta) \exp\left[-ng(a) + \frac{na^2}{2(1+\delta)}\right] J_{a,\lambda_1,k} \tag{53}$$

and

$$J_{a,\lambda_1,k} = \frac{(1+\delta)}{n} a^{k-1} \exp\left[-\frac{na^2}{2(1+\delta)}\right] M_{(k+1)/2}\left(\frac{n}{2(1+\delta)} a^2\right). \tag{54}$$

Next,

$$e^\delta \le (1+\delta)^2 \qquad \text{for } 0 \le \delta \le 1, \tag{55}$$

as is easily checked. Set

$$s := \lambda_1 a^2, \qquad \alpha := (k+1)/2 \tag{56}$$

and recall that $\lambda_1 = n/[2(1+\delta)]$. Then $s(1+\delta) = na^2/2$.

By Lemma 3(a), for $k \ge 1$ [so that $\alpha \ge 1$ by (56)], then by (55), we have

$$M_\alpha(s) \le M_\alpha(na^2/2) e^{\delta(\alpha-1)} = M_\alpha(na^2/2) e^{\delta(k-1)/2} \le M_\alpha(na^2/2)(1+\delta)^{k-1}.$$

If $k = 0$, $M_{1/2}(s) \le M_{1/2}(na^2/2)$ by Lemma 3(b). So, in all cases, for $k = 0, 1, 2, \dots,$

$$M_{(k+1)/2}(\lambda_1 a^2) \le M_{(k+1)/2}(na^2/2)(1+\delta)^k,$$

and by (54),

$$J_{a,\lambda_1,k} \le (1+\delta)^{k+1} n^{-1} a^{k-1} \exp\left[-\frac{na^2}{2(1+\delta)}\right] M_{(k+1)/2}(na^2/2).$$

Thus by (53) and (50),

$$I_{n,a,k,v} \le \eta(1+\delta)^{k+2} \exp[-ng(a)] n^{-1} a^{k-1} M_{(k+1)/2}(na^2/2) = (1+\delta)^{k+2} \Gamma_{n,a,k},$$

proving the first statement in Lemma 5.

For the lower bound, we have by the first line of the proof,

$$g(t) = g(a) + \int_a^t g'(x)\,dx \le g(a) + \left(\int_a^t x\,dx\right)(1+\delta)$$
$$= g(a) + (t^2 - a^2)(1+\delta)/2.$$

So

$$I_{n,a,k,v} \ge \eta \exp[-ng(a) + na^2(1+\delta)/2] \int_a^v t^k e^{-(1+\delta)nt^2/2}\,dt$$

or equivalently, recalling (51) and setting $b := (1+\delta)n/2$,

$$I_{n,a,k,v} \ge \eta \exp[-ng(a) + na^2(1+\delta)/2][J_{a,b,k} - J_{v,b,k}]. \tag{57}$$

By (52) with $\lambda = b$, we have

$$J_{a,b,k} = \frac{1}{n(1+\delta)} M_{(k+1)/2}((1+\delta)na^2/2)a^{k-1}\exp[-na^2(1+\delta)/2]. \tag{58}$$

For $k \ge 1$, $(k+1)/2 \ge 1$, so by Lemma 3(a),

$$M_{(k+1)/2}((1+\delta)na^2/2) \ge M_{(k+1)/2}(na^2/2)e^{-\delta(k-1)/2}.$$

Since $0 < \delta < 1$ by (8), we have $e^\delta \le (1+\delta)^2$ by (55), and so

$$M_{(k+1)/2}((1+\delta)na^2/2) \ge M_{(k+1)/2}(na^2/2)\frac{1}{(1+\delta)^{k-1}}$$

for $k \ge 1$. For $k = 0$, by Lemma 3(b), $M_{1/2}((1+\delta)na^2/2) \ge M_{1/2}(na^2/2)$. So,

$$J_{a,b,k} \ge (\exp[-(1+\delta)na^2/2])a^{k-1}\frac{1}{n}M_{(k+1)/2}(na^2/2)(1+\delta)^{-k-1} \tag{59}$$

for all $k = 0, 1, \ldots$. Now, by (58) with v in place of a, and by both parts of Lemma 3, for $k = 0, 1, \ldots,$

$$J_{v,b,k} \le \frac{v^{k-1}}{n(1+\delta)}(1+\delta)^{1/2}M_{(k+1)/2}(nv^2/2)\exp[-(1+\delta)nv^2/2].$$

By (57), (50) and (59), it follows that $I_{n,a,k,v}$ is bounded below by

$$\frac{\Gamma_{n,a,k}}{(1+\delta)^{k+1}} - \frac{\eta}{n}e^{-ng(a)}v^{k-1}\exp[n(1+\delta)(a^2-v^2)/2]M_{(k+1)/2}(nv^2/2)/(1+\delta)^{1/2}.$$

This completes the proof of Lemma 5. □

PROOF OF PROPOSITION 1, ASSUMING PROPOSITION 2. We have for $0 < \zeta \le \rho$ that $\|y\| \le (1 + (\delta/6))\|\theta\| \le 2\|\theta\|$ as noted just after (32). So $U(\rho)^c \subset U'(\rho/2)^c$ and it will suffice to prove the statement about $U'(\rho/2)^c$. Almost surely $\|\hat{\theta}_n - \theta_0\| < \rho/8$ for $n \ge n_0(\xi, \rho/8)$ by (16), so that in the θ coordinates as finally chosen, $\|\theta_0\| < \rho/8$. Let $\mathcal{B} := B(\theta_0, \rho/4)$ in the θ coordinates. Then

$U'(\rho/2)^c \subset \mathcal{B}^c$. By assumption (A5), there is a $\kappa > 0$ such that a.s. for n large enough, specifically $n \geq n_2(\xi, \rho)$, where we can take $n_2(\xi, \rho) \geq n_0(\xi, \rho/8)$, $\sup_{\theta \notin \mathcal{B}} LL_n(\theta) < \sup_{\theta \in \mathcal{B}} LL_n(\theta) - n\kappa$. Thus by (20), $\inf_{\theta \notin \mathcal{B}} H(\theta) \geq \kappa$, and by (32) and (40), $I_n(\mathcal{B}^c) \leq e^{-n\kappa}$. Also, $\|\theta\| \leq 2\|y\|$ for $0 < \zeta \leq \rho$ by (33) so $\mathcal{B} \supset U'(\rho/8) \supset U(\rho/16)$. Thus we have a.s. for $n \geq n_4$ for some $n_4(\xi, \rho, \gamma_1, \delta, d)$,

$$I_n(\mathcal{B}) \geq I_n(U'(\rho/8)) \geq \eta(2\pi/n)^{d/2}(1+\delta)^{-(3d+10)/2}$$

by (42) for $U(\rho/16)$. Thus by (41) for $C = \mathcal{B}$,

$$\begin{aligned}\pi_{x,n}(U(\rho)^c) \leq \pi_{x,n}(\mathcal{B}^c) &= \frac{I_n(\mathcal{B}^c)}{I_n(\Theta)} = \frac{I_n(\mathcal{B}^c)}{I_n(\mathcal{B}) + I_n(\mathcal{B}^c)} \\ &\leq \frac{e^{-n\kappa}}{I_n(\mathcal{B})} \leq e^{-n\kappa}(n/2\pi)^{d/2}(1+\delta)^{(3d+10)/2}\eta^{-1}.\end{aligned}$$

We have $(1+\delta)^{(3d+10)/2} < 1 + \varepsilon < 2$ by (8), and $2(n/(2\pi))^{d/2} \leq \eta e^{n\kappa/2}$ for $n \geq n_5$ for some $n_5 = n_5(d, \eta, \kappa)$. Proposition 1 then holds with $\nu = \kappa/2$ and $n_3 := \max(n_2, n_4, n_5)$. □

PROOF OF PROPOSITION 2, ASSUMING PROPOSITION 3. For $n \geq \max(n_0(\xi, \gamma_1), n_1(\delta, \rho, d))$, the bounds (43) and (44) hold uniformly in $0 < \zeta \leq \zeta_0$ and then we can let $\zeta \to 0$ to obtain the two inequalities when $\zeta = 0$ and $\hat{\theta}_n \in \partial A$. Consider the half-space $A_1 := \{\theta_1 \geq 0\} = \{y_1 \geq 0\}$ by (31). Adding inequalities for A_1 and its complement then gives (42) (which could be proved directly without half-spaces), proving Proposition 2. □

PROOF OF PROPOSITION 3. Recall ρ as defined in (25) and D_ω and h_ω as defined before (45). By Lemma 5, applied to $k = d - 1$, with $g(t) = h_n(t\omega)$ in light of (34), since $\eta \leq \pi(t\omega) \leq \eta(1+\delta)$ for $t\omega \in U(\rho)$ by (35), we have, for ω such that $D_\omega \leq \rho$,

$$\begin{aligned}&\int_{D_\omega}^{\rho} \exp[-nh_n(t\omega)]\pi(t\omega)t^{d-1}\,dt \\ &\quad \leq \eta \exp[-nh_n(D_\omega\omega)](D_\omega)^{d-2}M_{d/2}(nD_\omega^2/2)n^{-1}(1+\delta)^{d+1}.\end{aligned} \tag{60}$$

Since $h_n(D_\omega\omega) \geq \tau + h_\omega^2/[2(1+\delta)]$ by (38), we have by (46) and (60) that $I_n(A \cap U(\rho))$ is less than or equal to

$$\frac{\eta}{n}(1+\delta)^{d+1}e^{-n\tau}\int_{S(\zeta,\rho)} \exp\left[-\frac{nh_\omega^2}{2(1+\delta)}\right]D_\omega^{d-2}M_{d/2}\left(\frac{nD_\omega^2}{2}\right)d\omega. \tag{61}$$

Let $Y(\omega) := Y(\omega, n, d, \delta) := D_\omega^{d-2}M_{d/2}(nD_\omega^2/2)/(1+\delta)^d$ and

$$\begin{aligned}Z(\omega) &:= Z(\omega, n, d, \delta, \rho) \\ &:= \rho^{d-2}\exp[n(1+\delta)(D_\omega^2 - \rho^2)/2]M_{d/2}(n\rho^2/2)/(1+\delta)^{1/2}.\end{aligned}$$

Then by Lemma 5 and (32),

$$\int_{D_\omega}^{\rho} e^{-nh(t\omega)}\pi(t\omega)t^{d-1}\,dt \geq \frac{\eta}{n}\exp[-nh(D_\omega\omega)]\{Y(\omega)-Z(\omega)\}. \tag{62}$$

Recalling $r := \rho/2$ after (25) and $U(\rho) := \{y : \|y\| < \rho\}$ after (35), we have $I_n(A \cap U(\rho)) \geq I_n(A \cap U(\rho) \cap \{D_\omega \leq r\})$, and for $S(\zeta, \rho)$ defined before (46) and D_ω before (45), $S(\zeta, \rho) \cap \{D_\omega \leq r\} = S(\zeta, r)$, so by (46) and (62),

$$I_n(A \cap U(\rho)) \geq \eta n^{-1}\int_{S(\zeta,r)} \exp[-nh(D_\omega\omega)][Y(\omega)-Z(\omega)]\,d\omega. \tag{63}$$

We have $D_\omega \geq \zeta > 0$ for all $\omega \in S(d+)$. Next,

$$\frac{Z(\omega)}{Y(\omega)} = \frac{\rho^{d-2}\exp[-\frac{n}{2}(1+\delta)\rho^2]}{D_\omega^{d-2}\exp[-\frac{n}{2}(1+\delta)D_\omega^2]}\frac{M_{d/2}(n\rho^2/2)}{M_{d/2}(nD_\omega^2/2)}(1+\delta)^{d-1/2}, \tag{64}$$

which from the definition (49) of M_α is easily seen to equal a product $T_1T_2T_3$ where $0 < T_1 := \exp[n\delta(D_\omega^2 - \rho^2)/2]$, $T_2 := (1+\delta)^{d-1/2}$ and since $D_\omega < \rho$ in (63),

$$T_3 := \frac{P(X_{d/2} \geq n\rho^2/2)}{P(X_{d/2} \geq nD_\omega^2/2)} < 1, \tag{65}$$

where $X_{d/2}$ is a gamma random variable with density $x^{(d-2)/2}e^{-x}/\Gamma(d/2)$ for $x > 0$. Then for $0 < D_\omega \leq r = \rho/2$ we have $T_1T_2 \leq 1$ for n large enough, $n \geq N_0(\delta, \rho, d)$, and then $Y(\omega) \geq Z(\omega)$.

Also, $h(\zeta, X) \leq \tau + (1+\delta)\|X\|^2/2$ by (38) for $(\zeta, X) = D_\omega\omega$. So for $n \geq N_0(\delta, \rho, d)$, by (45) and (63), $I_n(A \cap U(\rho))$ is greater than or equal to

$$\eta n^{-1}e^{-n\tau}\int_{S(\zeta,r)} \exp[-n(1+\delta)h_\omega^2/2][Y(\omega)-Z(\omega)]\,d\omega. \tag{66}$$

By Lemma 4(b), we have, for any fixed $d \geq 1$, uniformly for $D_\omega \leq r = \rho/2$,

$$(1+\delta)^{d-1/2}\frac{P(X_{d/2} \geq n\rho^2/2)}{P(X_{d/2} \geq nD_\omega^2/2)} \to 0$$

as $n \to \infty$, recalling $0 < \delta < 1$. So for $n \geq n_1(\delta, \rho, d) \geq N_0(\delta, \rho, d)$ large enough,

$$(1+\delta)^{d-1/2}\frac{P(X_{d/2} \geq n\rho^2/2)}{P(X_{d/2} \geq nD_\omega^2/2)} < \frac{\delta}{2}$$

for $0 < D_\omega \leq r$. It follows by (66), (64), (65) and the definition of $Y(\omega)$ after (60) that $I_n(A \cap U(\rho))$ is bounded below by

$$\eta n^{-1}e^{-n\tau}\int_{S(\zeta,r)} \exp[-n(1+\delta)h_\omega^2/2]\frac{D_\omega^{d-2}M_{d/2}(nD_\omega^2/2)}{(1+\delta)^d}\left(1-\frac{\delta}{2}\right)d\omega.$$

As is easily shown, for $0 \le \delta < 1$, we have $1 - (\delta/2) \ge 1/(1+\delta)$, so for $n \ge n_1(\delta, \rho, d)$, $I_n(A \cap U(\rho))$ is greater than or equal to

$$\eta n^{-1} e^{-n\tau} \int_{S(\zeta,r)} \exp[-n(1+\delta)h_\omega^2/2] \frac{D_\omega^{d-2} M_{d/2}(nD_\omega^2/2)}{(1+\delta)^{d+1}} \, d\omega. \tag{67}$$

Recalling again (45) that $D_\omega^2 = \zeta^2 + h_\omega^2$, the integrand in (67) equals

$$I(\zeta, \omega) := I_{n,d}(\zeta, \omega) := (1+\delta)^{-d-1} \exp[n\zeta^2(1+\delta)/2] f(D_\omega) \tag{68}$$

with $f(x) := \exp[-nx^2(1+\delta)/2]x^{d-2}M_{d/2}(nx^2/2)$. Then

$$f(x) = \exp[-nx^2\delta/2]\Gamma(d/2)(2/n)^{(d-2)/2} P(X_{d/2} \ge nx^2/2),$$

so $f(x)$ is a decreasing function of x. Let $\omega(r)$ be a direction ω such that $D_{\omega(r)} = r$ (choose one such direction). Recall that $[[r, \infty)) = \{D_\omega \ge r\}$, $[[0, r)) = \{D_\omega < r\}$. The function $I(\zeta, \omega)$ depends on ω only through D_ω and is decreasing in D_ω since $f(x)$ is decreasing, so

$$\int_{[[r,\infty))} I(\zeta, \omega) \, d\omega \le I(\zeta, \omega(r)) \Omega_{d-1}([[r, \infty))) \tag{69}$$

and

$$\int_{[[0,r))} I(\zeta, \omega) \, d\omega \ge I(\zeta, \omega(r)) \Omega_{d-1}([[0, r))). \tag{70}$$

Now by Lemma 1 and (69),

$$\int_{S(d+)} I(\zeta, \omega) \, d\omega \le \int_{[[0,r))} I(\zeta, \omega) \, d\omega + I(\zeta, \omega(r)) \delta \Omega_{d-1}/2.$$

So by (70),

$$\begin{aligned} &\int_{S(d+)} I(\zeta, \omega) \, d\omega \\ &\quad \le \int_{[[0,r))} I(\zeta, \omega) \, d\omega + \frac{\delta}{2} \Omega_{d-1} \int_{[[0,r))} I(\zeta, \omega) \, d\omega \Big/ \Omega_{d-1}([[0, r))) \\ &\quad < \int_{[[0,r))} I(\zeta, \omega) \, d\omega \left(1 + \frac{\delta}{1-\delta}\right), \end{aligned}$$

since $\Omega_{d-1}([[0, r))) > (1-\delta)\Omega_{d-1}/2$ by Lemma 1. Since $\delta < 1/10$, it is clear that $1 + \delta/(1-\delta) = 1/(1-\delta) < (1+\delta)^2$. So

$$\int_{S(d+)} I(\zeta, \omega) \, d\omega \le \left[\int_{[[0,r))} I(\zeta, \omega) \, d\omega \right] (1+\delta)^2.$$

Thus, returning to (67), then since $I(\zeta,\omega)$ is the integrand in it,

$$I_n(A\cap U(\rho)) \geq \eta n^{-1}e^{-n\tau}\int_{[[0,r))} I(\zeta,\omega)\,d\omega$$

$$\geq \eta n^{-1}e^{-n\tau}\int_{S(d+)} I(\zeta,\omega)\,d\omega\Big/(1+\delta)^2 \tag{71}$$

$$= \frac{\eta n^{-1}e^{-n\tau}}{(1+\delta)^{d+3}}\int_{S(d+)} \exp[-n(1+\delta)h_\omega^2/2]D_\omega^{d-2}M_{d/2}(nD_\omega^2/2)\,d\omega.$$

Then by (49), (45) and a few calculations, the integral in (71) equals

$$ne^{[n\zeta^2(1+\delta)/2]}\int_{S(d+)}\exp[-n\delta D_\omega^2/2]\int_{D_\omega}^\infty u^{d-1}e^{(-nu^2/2)}\,du\,d\omega. \tag{72}$$

Returning now to Cartesian coordinates, we have $u^{d-1}\,du\,d\omega = dy$ with $u=\|y\|$, so the integral in (71) equals, where $dy := dy_d\cdots dy_1$,

$$n\exp[n\zeta^2(1+\delta)/2]\int_\zeta^\infty\int_{-\infty}^\infty\cdots\int_{-\infty}^\infty \exp[-n\|y\|^2/2]\exp[-n\delta\|y\|^2\zeta^2/(2y_1^2)]\,dy$$

$$\geq n\exp[n\zeta^2(1+\delta)/2]\int_\zeta^\infty\int_{-\infty}^\infty\cdots\int_{-\infty}^\infty \exp[-n(1+\delta)\|y\|^2/2]\,dy_d\cdots dy_1$$

$$=(1+\delta)^{-d/2}n^{(2-d)/2}(2\pi)^{(d-1)/2}M\big(\zeta\sqrt{n(1+\delta)}\big)$$

(after a few calculations). So from (71),

$$I_n(A\cap U(\rho)) \geq \eta n^{-d/2}e^{-n\tau}(1+\delta)^{-(3d+6)/2}(2\pi)^{(d-1)/2}M\big(\zeta\sqrt{n(1+\delta)}\big). \tag{73}$$

It follows from (36) and the definition (37) of τ that $\zeta^2/[2(1+\delta)]\leq\tau = h(\zeta,0)\leq(1+\delta)\zeta^2/2$ (since $(\zeta,0)\in U(\rho)$), and therefore

$$\sqrt{n}\zeta/\sqrt{1+\delta}\leq\sqrt{2n\tau}\leq\sqrt{n(1+\delta)}\zeta. \tag{74}$$

So by Lemma 2,

$$M(\zeta[n(1+\delta)]^{1/2})\geq e^{-\delta}M(\sqrt{n}\zeta/\sqrt{1+\delta})\geq e^{-\delta}M(\sqrt{2n\tau}),$$

since M is a nonincreasing function, so

$$I_n(A\cap U(\rho))\geq\eta n^{-d/2}e^{-n\tau}(1+\delta)^{-(3d+6)/2}(2\pi)^{(d-1)/2}e^{-\delta}M(\sqrt{2n\tau}),$$

thus by definition of $M(\cdot)$ before Lemma 2,

$$I_n(A\cap U(\rho))\geq\eta(2\pi/n)^{d/2}(1+\delta)^{-(3d+6)/2}\Phi(-\sqrt{2n\tau})e^{-\delta}.$$

By (55), for $0\leq\delta\leq 1$, $e^{-\delta}\geq 1/(1+\delta)^2$; thus (43) follows.

Turning to the upper bound (61), consider the integral

$$I := \int_{S(\zeta,\rho)}\exp[-nh_\omega^2/[2(1+\delta)]]D_\omega^{d-2}M_{d/2}(nD_\omega^2/2)\,d\omega.$$

A few calculations and (49) show that as in (72),

$$I = n\exp\left[\frac{n\zeta^2}{2(1+\delta)}\right]\int_{S(\zeta,\rho)}\exp\left[\frac{\delta}{1+\delta}D_\omega^2\frac{n}{2}\right]\int_{D_\omega}^{\infty}u^{d-1}\exp[-nu^2/2]\,du\,d\omega.$$

Returning to Cartesian coordinates, we have $dy = u^{d-1}\,du\,d\omega$ and $u = \|y\|$. Then for $dy := dy_d\cdots dy_1$,

$$I \le n\exp\left[\frac{n\zeta^2}{2(1+\delta)}\right]\int_\zeta^\infty\int_{-\infty}^\infty\cdots\int_{-\infty}^\infty\exp\left[\frac{\delta}{1+\delta}\frac{n}{2}\frac{\|y\|^2\zeta^2}{y_1^2}\right]\exp\left[-\frac{n}{2}\|y\|^2\right]dy$$

$$\le n^{(2-d)/2}(1+\delta)^{(d-1)/2}(2\pi)^{(d-1)/2}M(\zeta\sqrt{n})$$

after a few calculations, using $y_j^2\zeta^2/y_1^2 \le y_j^2$ for $j = 2,\ldots,d$. Altogether, using (61),

$$I_n(A\cap U(\rho)) \le n^{-1}\eta(1+\delta)^{d+1}e^{-n\tau}n^{(2-d)/2}(1+\delta)^{d/2}(2\pi)^{(d-1)/2}M(\zeta\sqrt{n}).$$

Now, since $\sqrt{2n\tau} \le \sqrt{n}(1+\delta)\zeta$ by (74) and by Lemma 2, we have

$$M(\zeta\sqrt{n}) \le M(\zeta\sqrt{n}(1+\delta))e^\delta \le M(\sqrt{2n\tau})e^\delta.$$

So again by (55), (44) follows and Proposition 3 is proved. □

PROOF OF THEOREM 1. Continuing with Case I, we have $\hat\theta_n \notin A$ and $\zeta < \zeta_0 < \rho$. By Proposition 1, there exists $\nu > 0$ such that by (41), for $n \ge n_3$,

$$\pi_{x,n}(A) = \pi_{x,n}(A\cap U(\rho)) + \pi_{x,n}(A\cap U(\rho)^c) \le \pi_{x,n}(A\cap U(\rho)) + e^{-n\nu}$$
$$= I_n(A\cap U(\rho))/I_n(\Theta) + e^{-n\nu} \le \frac{I_n(A\cap U(\rho))}{I_n(U(\rho))} + e^{-n\nu}.$$

Since $\zeta < \rho$ then by Proposition 3 (44) and Proposition 2 (42), for $n \ge n_3$,

$$\pi_{x,n}(A) \le (1+\delta)^{3d+8}\Phi(-\sqrt{2n\tau}) + e^{-n\nu}. \tag{75}$$

We now distinguish two subcases, subcase (i) where $e^{-n\nu} < \delta\Phi(-\sqrt{2n\tau})$ and subcase (ii) otherwise. In subcase (ii), $\Phi(-\sqrt{2n\tau}) \le \delta^{-1}e^{-n\nu}$ and

$$\pi_{x,n}(A) \le [(1+\delta)^{3d+8}\delta^{-1} + 1]e^{-n\nu} \le e^{-n\nu/2}$$

for n large enough so that $e^{n\nu/2} \ge (1+\delta)^{3d+8}\delta^{-1} + 1$, in addition to $n \ge n_3$. Thus in subcase (ii), we have (b) in Theorem 1 for any $\kappa' \le \nu/2$.

In subcase (i), by (75), (8) and (39),

$$\pi_{x,n}(A) \le (1+\delta)^{3d+9}\Phi(-\sqrt{2n\tau}) \le (1+\varepsilon)\Phi(-\sqrt{2\Delta}), \tag{76}$$

so that the upper bound in (a) of Theorem 1 holds.

Now to treat lower bounds for $\pi_{x,n}(A)$, by (41),

$$\pi_{x,n}(A) \ge \pi_{x,n}(A\cap U(\rho)) = I_n(A\cap U(\rho))/I_n(\Theta).$$

Note that a.s. for $n \geq n_3$ by Proposition 1 and (42),

$$I_n(\Theta) = I_n(U(\rho)) + I_n(U(\rho)^c) \leq I_n(U(\rho)) + e^{-n\nu}$$
$$\leq \eta(2\pi/n)^{d/2}(1+\delta)^{(3d+6)/2} + e^{-n\nu} \leq \eta(2\pi/n)^{d/2}(1+\delta)^{(3d+7)/2},$$

for n large enough, say, $n \geq \max(n_3, n_6)$ for some $n_6(\nu, \eta, d)$. Then by (43), (39) and (8),

$$\pi_{x,n}(A) \geq (1+\delta)^{-(3d+9)}\Phi(-\sqrt{2n\tau}) \geq \Phi(-\sqrt{2\Delta})/(1+\varepsilon). \tag{77}$$

Thus by (76) conclusion (a) of Theorem 1 is proved in Case I(i).

To prove Theorem 1 when $\hat{\theta}_n \in \partial A$ we can, as in the proof of Proposition 2, let $\zeta \downarrow 0$ in Proposition 3, then apply the same argument as for Case I(i).

In Theorem 1, the statement about U and $A \subset U^c$ follows from Proposition 1. For the statement about A_n, since $d(A_n, \phi_0) \to 0$ and $d(\hat{\phi}_n, \phi_0) \to 0$ we have by (37), (20), (23), noting that $h(\zeta, 0) \leq H(\zeta, 0)$, and (32) that $\tau = \tau_n \to 0$ as $n \to \infty$. Thus we have subcase (i) [after (75)]. We apply Propositions 1–3 and (8) to get that $\pi_{x,n}(A_n)/\Phi(-\sqrt{2\Delta_n}) \doteq 1$ within a fixed power of $1+\varepsilon$ for n large enough. Letting $\rho \downarrow 0$ and thus $\varepsilon \downarrow 0$, we get $\pi_{x,n}(A_n)/\Phi(-\sqrt{2n\tau_n}) \to 1$ as $n \to \infty$. This finishes the proof of Theorem 1 in Case I for $d \geq 2$.

In Case II (29), $\zeta \geq \zeta_0$. Let $B := \{\theta : \theta_1 \geq \zeta_0\}$ in the coordinates as in (18). Recalling that, for B, by (37) and (39) for $\zeta = \zeta_0$, $\Delta_{B,n} = nh(\zeta_0, 0)$, we have by (A5), $\Phi(-\sqrt{2\Delta_{B,n}}) \leq \exp(-n\kappa)$ for some $\kappa > 0$ and n large enough. Then also for A, $\Phi(-\sqrt{2\Delta}) \leq e^{-n\kappa}$, uniformly for $\zeta \geq \zeta_0$. B is disjoint from $U'(\zeta_0)$, so by Proposition 1, for some $\nu' > 0$, a.s. for n large enough, $\pi_{x,n}(A) \leq \pi_{x,n}(U'(\zeta_0)^c) \leq e^{-\nu' n}$ uniformly over all A with $\zeta \geq \zeta_0$. So in Case II we have conclusion (b) for $\kappa' = \min(\nu', \kappa)$. The proof of Theorem 1 for $d \geq 2$ is complete.

For $d = 1$ the proof is somewhat simpler and is omitted because there is no transformation to spherical coordinates and we do not need Lemma 1. This completes the proof of Theorem 1. □

9. A multivariate normal location family. For this family, explicit calculations can be done. We will see how (a) for a normal prior, slowness of approach of the half-space to ϕ_0 can cause slow convergence to 0 of the relative error of our simplest approximation (to some small probabilities), and (b) for a double-exponential prior, the absolute error for some half-spaces is no smaller than $O(1/\sqrt{n})$, even for the approximation sharpened via (1).

Consider the location family $N(\mu, I)$, $\mu \in \mathbf{R}^d$, on $\mathbf{R}^d$ with, first, a prior $N(0, \sigma^2 I)$ for μ. (In this case the Jeffreys prior, Lebesgue measure, is improper.) Let $x = (X_1, \dots, X_n)$ be observed i.i.d. $N(\mu, I)$. Let $S_n := X_1 + \cdots + X_n$ and $Q_n := \sum_{j=1}^n |X_j|^2$. Let A be a half-space $A := \{\mu \cdot t \geq c\}$ with $|t| = 1$. A brief calculation shows that the posterior probability is $\pi_{x,n}(A) = \Phi(\tau_n S_n \cdot t - c\tau_n^{-1})$ where $\tau_n := (n + \sigma^{-2})^{-1/2}$. Our simple likelihood root approximation to the posterior, provided that $S_n/n \notin A$, is $\Phi(-\sqrt{2\Delta}) = \Phi(n^{-1/2}S_n \cdot t - c\sqrt{n})$. As

$\sigma \to \infty$, so that the prior becomes more and more diffuse, and converges in a sense to the Jeffreys prior, the exact posterior distribution converges to the likelihood root approximation. If we instead fix, for example, $\sigma = 1$, $\mu = \mu_0 = 0$ and $t = (1, 0, \ldots, 0)$ and let $c = c_n = n^{-\alpha}$ for some $\alpha \in (0, 1/2)$, we obtain $\pi_{x,n}(A_n) = \Phi(S_{n,1}/\sqrt{n+1} - n^{-\alpha}\sqrt{n+1})$ and $\Phi(-\sqrt{2\Delta_n}) = \Phi(S_{n,1}/\sqrt{n} - n^{-\alpha}\sqrt{n})$. For $a > 0$ and $0 < \delta < 1$ a short calculation gives

$$\Phi(a+\delta) - \Phi(a) = \phi(a)(1 - e^{-a\delta})(1 + O(\delta^2))/a. \tag{78}$$

We will use the fact that if Z_n are any standard normal variables then $|Z_n| = O(\sqrt{\log n}) = O(n^{\varepsilon})$ a.s. as $n \to \infty$ for any $\varepsilon > 0$. Let $Z_n := S_{n,1}/\sqrt{n}$, $a := a_n := n^{0.5-\alpha} - Z_n$ and $\delta := \delta_n := n^{-\alpha}\sqrt{n+1} - Z_n\sqrt{n}/\sqrt{n+1} - a_n$. Then we have a.s. $\delta_n = \frac{1}{2}n^{-0.5-\alpha} + \frac{Z_n}{2n} + O(n^{-\alpha-3/2})$. Since by l'Hospital's rule and facts stated after (47), $\Phi(-x) \sim \phi(x)/x$ as $x \to +\infty$, it follows from (78) that the relative error in the likelihood root approximation of the posterior is almost surely asymptotic to $a_n\delta_n \sim n^{-2\alpha}/2$ as $n \to \infty$, which converges to 0 slowly for small $\alpha > 0$. The probabilities being approximated converge to 0 rather rapidly in this case.

For the approximation π^{**} (4) the factors in terms of $\ell_{\eta\eta}$ and $d\mu_1/dz \equiv n^{-1/2}$ are constants and divide out, so the interesting factor is $\rho_n := \pi_1(\tilde{\theta})/\pi_1(\hat{\theta})$, which a brief calculation shows to be $\exp(-a_n\delta_n + O_p(\delta_n^2))$. The relative error in this case is $r_{\mathrm{SL}} := [\Phi(-a_n)\rho_n - \Phi(-a_n - \delta_n)]/\Phi(-a_n - \delta_n)$. By (78) we have

$$\Phi(-a_n - \delta_n) = \Phi(-a_n) - \phi(a_n)(1 - \exp(-a_n\delta_n))(1 + O(\delta_n^2))/a_n,$$

and so the numerator of r_{SL} is asymptotic to

$$[\phi(a_n)a_n^{-1} - \Phi(-a_n)](1 - \exp(-a_n\delta_n)) + O(\delta_n^2)\Phi(-a_n).$$

Since $\Phi(-x) - \phi(x)/x \sim -\phi(x)/x^3$ as $x \to +\infty$, also by l'Hospital's rule, the numerator is asymptotic to $\phi(a_n)\delta_n/a_n^2$. Since the denominator is asymptotic to $\phi(a_n)/a_n$, we have $r_{\mathrm{SL}} \sim \delta_n/a_n \sim 1/(2n)$, not depending on α. But in the examples in the next section, the relative error of π^{**} goes to 0 at a slower rate.

Often, rates of convergence for approximations of posteriors have been stated in the literature without precise assumptions. Under certain conditions on the likelihood functions, it is sufficient for convergence without rates that the prior $\pi(\cdot)$ should be continuous and strictly positive [e.g., Walker (1969), Theorem 1 and Corollary 1 above]. For faster rates such as $O(1/n)$ or $O(n^{-3/2})$, beside stronger assumptions on the likelihood functions, it has been assumed that $\pi(\cdot)$ has continuous partial derivatives through order 4 [e.g., Bickel and Ghosh (1990), Erkanli (1994)]. We will see that for such rates, even for the very smooth normal location likelihood, some smoothness of the prior is needed.

A strictly positive, continuous, in fact Lipschitz, but not differentiable (at 0) prior density on $\mathbf{R}^1$ is the double-exponential $\pi(\theta) = e^{-|\theta|}/2$, $-\infty < \theta < \infty$. One can do closed-form calculations with this prior and the normal location family as noted by Pericchi and Smith (1992) and Choy and Smith (1997). Let $d = 2$ and take $N(\phi, I)$, $\phi = (\psi, \eta)$, $S_n = (S_{n1}, S_{n2})$. The profile log likelihood is

$$m(\psi) = -n\log(2\pi) - \frac{1}{2}Q_n + \psi S_{n1} - \frac{n}{2}(\psi^2 - \overline{y}^2),$$

where $\overline{y} := S_{n2}/n$. For the log likelihood ℓ, $\partial^2\ell/\partial\eta^2 \equiv -n$.

For any product prior density $\pi(\psi, \eta) = \pi_1(\psi)\pi_2(\eta)$, the approximation (1) to the marginal posterior density is exact here. Instead we take the 45 degree rotated double-exponential product prior $\pi(\psi, \eta) = \frac{1}{2}\exp(-|\psi + \eta| - |\psi - \eta|)$. Then π is not differentiable where $\psi = \pm\eta$. The approximation (1) is given by

$$\pi^*_{\psi|Y}(\psi) = c^{**}\exp(-|\psi + \overline{y}| - |\psi - \overline{y}| + \psi S_{n1} - n\psi^2/2),$$

where the normalizing constant c^{**} may depend on $\overline{y}$, S_{n1} and n but not on ψ. The exact marginal posterior can be evaluated straightforwardly, say, for $\psi \geq 0$, setting

$$\int_{-\infty}^{\infty} d\eta = \left(\int_{-\infty}^{-\psi} + \int_{-\psi}^{\psi} + \int_{\psi}^{\infty}\right) d\eta,$$

as $c\exp(\psi S_{n1} - n\psi^2/2)\tau_n(\psi, \overline{y})$ where c is a constant with respect to ψ and

$$\begin{aligned}\tau_n(\psi, \overline{y}) := {} & e^{-2\psi}\left[\Phi(\sqrt{n}(\psi - \overline{y})) - \Phi(-\sqrt{n}(\psi + \overline{y}))\right] \\ & + e^{2\overline{y}+2/n}\Phi\left(-\sqrt{n}\left(\psi + \overline{y} + \frac{2}{n}\right)\right) \\ & + e^{-2\overline{y}+2/n}\Phi\left(-\sqrt{n}\left(\psi - \overline{y} + \frac{2}{n}\right)\right).\end{aligned}$$

For $u > 0$ we have $1 - \Phi(\sqrt{n}u) = \Phi(-\sqrt{n}u) \to 0$ exponentially as $n \to \infty$. Thus the relative error in the approximation of $\tau_n(\psi, \overline{y})$ by $\exp(-|\psi - \overline{y}| - |\psi + \overline{y}|)$ for $\psi = a|\overline{y}|$, $a \geq 0$, $\overline{y} \neq 0$, as $n \to \infty$, is asymptotic to $-2/n$ for $0 \leq a < 1$ and to $\sqrt{2/(\pi n)}$ for $a = 1$ (and is exponentially small for $a > 1$). Taking a in a $1/\sqrt{n}$-neighborhood of 1, if for the true parameters $\psi_0 = \eta_0 \neq 0$, then with probability not approaching 0 we have an absolute error in the approximation of posterior probabilities of some half-spaces via (1) of order $1/\sqrt{n}$.

10. Beta numerical results. As an example, we considered the sample space of two points $\{0, 1\}$ and the family of binomial probabilities with $p \in \Theta = (0, 1)$, and the fixed half-space (segment) $p \leq x = 0.7$. We consider in Tables 1 and 3 the Jeffreys prior, which is the $\beta(1/2, 1/2)$ distribution with density $p^{-1/2}(1 - p)^{-1/2}/\pi$, $0 < p < 1$, and in Table 2 the uniform prior. If in n independent trials, there are k successes (1's) and $n - k$ failures (0's) then the posterior is the $\beta(k + 1/2, n - k + 1/2)$ distribution in Tables 1 and 3, $\beta(k + 1, n - k + 1)$ in Table 2. Thus the posterior probability of the interval $[0, x]$ in Tables 1 and 3 is

$$J_{x,k,n} := I_{x,k+1/2,n-k+1/2} := \int_0^x t^{k-1/2}(1-t)^{n-k-1/2}\,dt/B(k+1/2, n-k+1/2),$$

TABLE 1
Jeffreys prior beta examples

n	k	$J_{0.7,k,n}$	r_{cl}	r_Δ	r_0	r_1	r_{DM}	r_2
50	44	$1.3735 \cdot 10^{-3}$	−0.940	−0.091	−0.0103	$-8.6 \cdot 10^{-3}$	$-2.8 \cdot 10^{-3}$	$3.05 \cdot 10^{-5}$
100	85	$2.5483 \cdot 10^{-4}$	−0.932	−0.068	−0.0054	$-3.7 \cdot 10^{-3}$	$-8.5 \cdot 10^{-4}$	$3.30 \cdot 10^{-6}$
200	165	$2.5816 \cdot 10^{-5}$	−0.927	−0.053	−0.0030	$-1.7 \cdot 10^{-3}$	$-2.8 \cdot 10^{-4}$	$5.3 \cdot 10^{-7}$
500	394	$4.6722 \cdot 10^{-6}$	−0.834	−0.034	−0.0016	$-6.2 \cdot 10^{-4}$	$-5.8 \cdot 10^{-5}$	$3 \cdot 10^{-8}$
1000	767	$1.1125 \cdot 10^{-6}$	−0.752	−0.025	−0.0010	$-2.9 \cdot 10^{-4}$	$-1.9 \cdot 10^{-5}$	$\lvert\cdot\rvert \preceq 10^{-9}$
2000	1501	$2.6838 \cdot 10^{-7}$	−0.661	−0.018	−0.0007	$-1.4 \cdot 10^{-4}$	$-6.3 \cdot 10^{-6}$	$\lvert\cdot\rvert \preceq 10^{-9}$
5000	3671	$4.7368 \cdot 10^{-8}$	−0.534	−0.012	−0.0004	$-5.6 \cdot 10^{-5}$	$-1.5 \cdot 10^{-6}$	$\lvert\cdot\rvert \preceq 10^{-9}$

where $B(\cdot,\cdot)$ is the beta function. The MLE for p is $\hat{p} = k/n$. Thus, for the MLE to be outside the half-space (in this case, interval) $[0, x]$ we consider $k \geq nx$. For such n, x and k we have the exact beta probability, which we computed by the algorithm of Holt (1986), and our approximation $\Phi_\Delta := \Phi_{x,k,n} := \Phi(-\sqrt{2\Delta})$. We evaluated the relative error $r_\Delta := \Phi_{x,k,n}/J_{x,k,n} - 1$. For comparison, we took a classical approximation by a normal distribution with mean $\hat{p}$ and variance such that the second derivative of the log of its density (a constant) equals the second derivative of the log of the posterior density at $\hat{p}$. This gave an approximation we call Φ_{cl} with a relative error r_{cl}. [Taking the posterior mode $(k - 1/2)/(n - 1)$ in place of $\hat{p}$ or the second derivative of the log likelihood gave worse approximations in Table 1.] Also, in all three tables, we consider the sharpened approximations $\pi_Y^{i,*}$ and $\pi_Y^{DM,*}$ defined in Section 3 to the posterior probabilities and give their relative errors r_i for $i = 0, 1, 2$, and r_{DM}. The relative error r_2 is the smallest in each row and r_{DM} is usually next smallest.

Table 3 treats a "large deviation" case in which $x = 0.7$ and $k/n = 0.74$ both remaining constant as n becomes large. Here the relative error r_Δ of the simple likelihood root approximation may approach a nonzero constant, which would not contradict our theorem. For all the sharpened approximations, the relative errors

TABLE 2
Uniform prior beta examples

n	k	$I_{0.7,k+1,n-k+1}$	r_{cl}	r_Δ	r_0	r_1	r_{DM}	r_2
50	44	$1.9625 \cdot 10^{-3}$	−0.977	−0.364	−0.0232	$-3.1 \cdot 10^{-3}$	$8.5 \cdot 10^{-3}$	$2.6 \cdot 10^{-4}$
100	85	$3.3070 \cdot 10^{-4}$	−0.960	−0.282	−0.0164	$-3.2 \cdot 10^{-3}$	$2.7 \cdot 10^{-3}$	$7.8 \cdot 10^{-5}$
200	165	$3.1374 \cdot 10^{-5}$	−0.948	−0.221	−0.0106	$-2.1 \cdot 10^{-3}$	$9.4 \cdot 10^{-4}$	$2.1 \cdot 10^{-5}$
500	394	$5.2625 \cdot 10^{-6}$	−0.860	−0.143	−0.0061	$-1.0 \cdot 10^{-3}$	$2.1 \cdot 10^{-4}$	$4.0 \cdot 10^{-6}$
1000	767	$1.2096 \cdot 10^{-6}$	−0.777	−0.103	−0.0040	$-5.6 \cdot 10^{-4}$	$7.0 \cdot 10^{-5}$	$1.1 \cdot 10^{-6}$
2000	1501	$2.8479 \cdot 10^{-7}$	−0.684	−0.075	−0.0026	$-2.9 \cdot 10^{-4}$	$2.4 \cdot 10^{-5}$	$3.1 \cdot 10^{-7}$
5000	3671	$4.9194 \cdot 10^{-8}$	−0.553	−0.049	−0.0015	$-1.2 \cdot 10^{-4}$	$5.9 \cdot 10^{-6}$	$6 \cdot 10^{-8}$

TABLE 3
Jeffreys prior large deviation beta examples

n	k	$J_{0.7,k,n}$	r_{cl}	r_Δ	r_0	r_1	r_{DM}	r_2
50	37	0.27308	−0.037	−0.0269	−0.0134	$-2.5 \cdot 10^{-3}$	$-3.4 \cdot 10^{-4}$	$-5.5 \cdot 10^{-5}$
100	74	0.19233	−0.049	−0.0220	−0.0085	$-1.5 \cdot 10^{-3}$	$-1.4 \cdot 10^{-4}$	$-1.5 \cdot 10^{-5}$
200	148	0.10740	−0.074	−0.0188	−0.0052	$-9.3 \cdot 10^{-4}$	$-6.1 \cdot 10^{-5}$	$-3.8 \cdot 10^{-6}$
500	370	0.02432	−0.141	−0.0162	−0.0026	$-4.5 \cdot 10^{-4}$	$-2.1 \cdot 10^{-5}$	$-4.8 \cdot 10^{-7}$
1000	740	0.00261	−0.241	−0.0151	−0.0015	$-2.5 \cdot 10^{-4}$	$-9.9 \cdot 10^{-6}$	$-8 \cdot 10^{-8}$
2000	1480	0.00004	−0.405	−0.0144	−0.0008	$-1.4 \cdot 10^{-4}$	$-4.7 \cdot 10^{-6}$	$-1 \cdot 10^{-8}$
5000	3700	$2 \cdot 10^{-10}$	−0.713	−0.0140	−0.0003	$-5.7 \cdot 10^{-5}$	$-1.8 \cdot 10^{-6}$	$\lvert \cdot \rvert \preceq 10^{-9}$

do appear to approach 0. For the sharpest approximation $\pi_Y^{2,*}$ the relative error is quite small and becomes smaller very rapidly. We caution, however, that this is a one-dimensional case and the approximation of a marginal of a multidimensional posterior as in (1) and (3) could result in larger relative errors of order $O(1/n)$ [Tierney, Kass and Kadane (1989)]. Alternatively, if the likelihoods or priors are not smooth enough, some of the approximations involving derivatives may not be defined, and others may be defined but less accurate, as seen in Section 9 for a double-exponential prior.

For Table 1, we wrote a computer program which, given n, x, and $\alpha > 0$, starts with the smallest integer $j \geq nx$ and considers $k = j, j+1, j+2, \ldots$ as long as $J_{x,k,n} \geq \alpha$, and finds the k in that range with the largest relative error r_Δ in absolute value. We chose $\alpha = 1/n^2$. Then $k/n \to x$ as n becomes large. Some terms in approximations vanish when $n = 2k$ (so, near $x = 0.5$) and one vanishes for the uniform prior when $k/n \doteq 0.916$. Thus we chose $x = 0.7$. As is perhaps not surprising, we found the largest relative errors when the probabilities $J_{x,k,n}$ were as small as allowed, with $J_{x,k,n} > \alpha > J_{x,k+1,n}$. We used the same k for each n in Table 2 as in Table 1. Since $\alpha \to 0$ as $n \to \infty$ at a slower than exponential rate, by Theorem 1, the relative error r_Δ in the approximation Φ_Δ should approach 0, which fits with Tables 1 and 2. The notation $\lvert \cdot \rvert \preceq 10^{-9}$ indicates that the relative error (of the approximation $\pi_Y^{2,*}$) is less in absolute value than $1.5 \cdot 10^{-9}$; we do not give more exact values since the Holt algorithm by which we computed beta probabilities was constructed to give a relative error less than 10^{-9}. For the same reason, we round off possibly unreliable digits of r_2 when $|r_2| < 10^{-6}$.

The relative errors r_{cl} for the classical approximation Φ_{cl} based on the second derivative at the MLE decrease slowly in magnitude in Tables 1 and 2, even increase in size with n in Table 3, and they are quite large, except in the upper rows of Table 3. The Φ_{cl} approximation can work rather well in the middle of the posterior distribution. The smallest relative error r_2 in Table 3 is of order $n^{-5/2}$, in agreement with (7).

The approximation $\Phi(-\sqrt{2\Delta})$ of Theorem 1, which is the same in corresponding rows of Tables 1 and 2, is a better approximation to posterior probabilities for the Jeffreys prior (Table 1) than it is for the uniform prior (Table 2), by a factor about 4 in the relative errors r_Δ. To understand this, observe as noted, for example, by Woodroofe (1992) that if we take as a coordinate parameter $z = -\sqrt{2\Delta}$ for $p \le k/n$ and $\sqrt{2\Delta}$ for $p > k/n$ as in (5) then the likelihood function becomes exactly normal. The uniform prior dp on $0 < p < 1$ and the Jeffreys prior have densities with respect to dz, say, $\tau_u(z)$ and $\tau_J(z)$, respectively. One can evaluate the logarithmic derivatives of both at $z = 0$, where we find

$$\frac{\tau_u'(0)}{\tau_u(0)} = 4\frac{\tau_J'(0)}{\tau_J(0)} = -\frac{2(2k-n)}{3\sqrt{nk(n-k)}},$$

which is not 0 unless $k = n/2$. Thus with respect to dz, the Jeffreys prior is more uniform than the prior dp around the MLE.

For higher order approximations the situation is more complicated. The approximations again tend to be more accurate for the Jeffreys prior, but not necessarily to the extent shown in Tables 1 and 2 for r_2. For example, when $n = 500$, $x = 0.7$ and $k = 394$, for the Jeffreys prior $r_2 \doteq 3 \cdot 10^{-8}$ and for the uniform prior $r_2 \doteq 4 \cdot 10^{-6}$. The Jeffreys r_2 changes sign nearby from $k = 391$ to 392, so it is unusually small in absolute value. We found a similar sign change for $x = 0.8$, $n = 500$, $k = 437, 438$.

For accurate calculation of the approximations when $|x - (k/n)|$ is very small, analytic subtraction is needed, as for the Lugannani–Rice approximations [e.g., Daniels (1987), page 43; Reid (1996), page 143].

APPENDIX

The relationship between a coordinate and a profile likelihood root.

PROPOSITION 4. *Let $\psi(\phi) = a + v \cdot \phi$, $v \neq 0$, so ψ is a nonconstant affine function of ϕ. Under the assumptions made in Theorem 1, let ϕ_0 be the pseudo-true parameter of* (A5). *Then there is an open interval (c, d) containing $\psi_0 := \psi(\phi_0)$ such that almost surely for all n large enough, in* (2) *there is a 1–1, mutually C^1 relationship, depending on n and the observations, between z and ψ for $c < \psi < d$.*

PROOF. As after (8), we can make a linear coordinate change from ϕ to θ with $E(\theta_0) = I$. In the neighborhood V of θ_0 defined by (11), almost surely for n large enough, $\|E_n(\theta) - I\| < 2\delta_1/3$ by (14). By (16), almost surely for n large enough, there is a unique MLE $\hat{\theta}_n$ with $\|\hat{\theta}_n - \theta_0\| < \gamma_1$. Then $\theta \in V$ whenever $\|\theta - \hat{\theta}_n\| < \rho_1 := 3\gamma_1$ by (11). As in (17) and similarly to (12), by a translation and rotation of coordinates which does not change any of the preceding (but depends on n and $\hat{\theta}_n$), we can assume that $\hat{\theta}_n = 0$ and that $\psi = a + b\theta_1$ for some constants a, b with $b > 0$ where θ_1 is the first coordinate of θ. Thus we can assume $\psi = \theta_1$.

Let $\eta := \theta^{(1)} = (\theta_2, \dots, \theta_d) \in \mathbf{R}^{d-1}$. Consider the vector-valued function $\nabla_\eta H_n$

for H_n defined by (20) from $\{\|\theta\| < \rho_1\}$ into $\mathbf{R}^{d-1}$, where $\nabla_\eta = (\partial/\partial\theta_2, \ldots, \partial/\partial\theta_d)$. Then $\nabla_\eta H_n$ is C^1 and $\nabla_\eta H_n(0) = 0$. By (14), the Hessian $\partial^2 H_n/\partial\theta_i\partial\theta_j$ is within $2\delta_1/3$ of the identity on the open set $\{\|\theta\| < \rho_1\}$, which almost surely for all n large enough includes all θ within $2\gamma_1$ of θ_0.

If I_r is the $r \times r$ identity matrix, a matrix B is $d \times d$ and $B_{ij}^{(1)} = B_{ij}$ for $i, j = 2, \ldots, d$, then it is easily seen that $\|B^{(1)} - I_{d-1}\| \le \|B - I_d\|$. The derivative $G_n := \nabla_\eta\nabla_\eta H_n$ is the Hessian of H_n with respect to η, so by (14) $\|G_n - I_{d-1}\| < 2\delta_1/3$ on $\{\|\theta\| < \rho_1\}$. In particular G_n is invertible at $\theta = 0$. Thus by the implicit function theorem [e.g., Rudin (1976), Theorem 9.28], there are an open set $U_1 \subset \mathbf{R}^d$ and an open interval $W_1 \subset \mathbf{R}$, say $(-w, w)$, both containing 0, such that for all $\theta_1 \in W_1$, there is a unique $\eta(\theta_1)$ such that $\tilde{\theta} := \tilde{\theta}(\theta_1) := (\theta_1, \eta(\theta_1)) \in U_1$ and $\nabla_\eta H_n(\tilde{\theta}(\theta_1)) = 0$. Moreover, the function $\eta(\cdot)$ is C^1 on W_1 and $\eta(0) = 0$. We have by (22) that if $\|\theta\| = \rho_1$ then $H(\theta) \ge \rho_1^2/(2(1+\delta_1))$, while if in addition $|\theta_1| < \rho_1/2$ then $H(\theta_1, 0) < \rho_1^2(1+\delta_1)/8 < H(\theta)$. Thus if $|h| < \rho_1/2$ then $\inf\{H(\theta) : \theta_1 = h,\ \|\theta\| \le \rho_1\}$ is attained at a point $\theta = \tilde{\theta}(h)$ at which $\|\theta\| < \rho_1$ and $\nabla_\eta H(\theta) = 0$. By strict convexity of $H(\cdot)$ on $\|\theta\| < \rho_1$, $\tilde{\theta}(h)$ is unique, so $\tilde{\theta}(\cdot)$ is a well-defined function from $(-\rho_1/2, \rho_1/2)$ into $\{\theta : \|\theta\| < \rho_1\}$, we can take $w \le \rho_1/2$ and then the definitions of $\tilde{\theta}$ agree for $|\theta_1| < w$. By the implicit function theorem, $\tilde{\theta}(\cdot)$ is a C^1 function on some neighborhood of each point in $(-\rho_1/2, \rho_1/2)$. So by compactness of any closed subinterval, we can take $w = \rho_1/2$.

By (A_5), there is a $\kappa > 0$ such that almost surely for n large enough,

$$\inf_{\theta \notin V} H_n(\theta) > \kappa. \tag{79}$$

Inequality (22) gives $H_n(\theta) \le \|\theta\|^2(1+\delta_1)/2 \le \|\theta\|^2$ for $\|\theta\| < \rho_1$. Thus $H_n(\theta) \le \kappa$ for $\|\theta\| \le \min(\rho_1, \sqrt{\kappa})$, so $H_n(\tilde{\theta}(\theta_1)) \le \kappa$ for $|\theta_1| < \gamma_2 := \min(\rho_1, \sqrt{\kappa})/2$. By strict convexity, for $|\theta_1| < \gamma_2$, $H_n(\theta_1, \eta)$ has a strict minimum with respect to η at $\eta = \eta(\theta_1)$ for $(\theta_1, \eta) \in V$ and by (79) for all η. In other words, the profile log likelihood $m(\theta_1) = \ell(\tilde{\theta}(\theta_1)) = \ell(0) - nH_n(\tilde{\theta}(\theta_1))$, so by (2),

$$z = (\operatorname{sgn}\theta_1)\sqrt{2nH_n(\tilde{\theta}(\theta_1))}. \tag{80}$$

Let $g(t) := H_n(\tilde{\theta}(t))$ for $|t| < \gamma_2$. For $-\gamma_2 < u < u + v < u/2 < 0$,

$$\begin{aligned} g(u+v) = H_n(u+v, \eta(u+v)) &\le H_n\left(u+v, \frac{u+v}{u}\eta(u)\right) \\ &= H_n(u, \eta(u)) + v\frac{d}{dt}H_n\left(u+t, \frac{u+t}{u}\eta(u)\right)\bigg|_{t=\tau} \end{aligned}$$

for some τ, $0 < \tau < v$, by the mean value theorem. By (21) and since $\|(1, \eta(u)/u)\| \ge 1$, this derivative is less than or equal to $-|u|/[2(1+\delta_1)]$. Letting $v \downarrow 0$ we get $g'(u) < 0$. Likewise we get $g'(u) > 0$ for $0 < u < \gamma_2$. Thus

$$\frac{dz}{d\theta_1} = \frac{\sqrt{n}|dH_n(\tilde{\theta}(\theta_1))/d\theta_1|}{\sqrt{2H_n(\tilde{\theta}(\theta_1))}} > 0$$

for $|\theta_1| < \gamma_2$ and $\theta_1 \neq 0$ since $H_n(\tilde{\theta}(\theta_1)) > 0$ for $\theta_1 \neq 0$. Also, $dz/d\theta_1$ is continuous in θ_1 for $0 < |\theta_1| < \gamma_2$.

Let $a_{ij} := \frac{1}{2}\partial^2 H_n(0)/\partial\theta_i\partial\theta_j$ for $i, j = 1, \ldots, d$, and $b_j := d\tilde{\theta}_j/d\theta_1|_{\theta_1=0}$ for $j = 1, \ldots, d$. Then by Taylor's theorem with remainder,

$$\sqrt{H_n(\tilde{\theta})}/|\theta_1| = \sqrt{\sum_{i,j=1}^{d} a_{ij}\tilde{\theta}_i\tilde{\theta}_j + o(\theta_1^2)}\Big/|\theta_1| \to \sqrt{\sum_{i,j=1}^{d} a_{ij}b_ib_j}$$

as $\theta_1 \to 0$. The limit is strictly positive because the matrix $\{a_{ij}\}$ is positive definite (being close to $\frac{1}{2}I_d$) and the vector b is nonzero since $b_1 = 1$. By the chain rule, we have for $\tilde{\theta} = \tilde{\theta}(\theta_1)$ that

$$\frac{dH_n(\tilde{\theta})}{d\theta_1} = \frac{\partial H_n}{\partial\theta_1}(\tilde{\theta}) + \sum_{j=2}^{d} \frac{\partial H_n}{\partial\theta_j}(\tilde{\theta})\frac{d\tilde{\theta}_j}{d\theta_1} = \frac{\partial H_n}{\partial\theta_1}(\tilde{\theta})$$

(the latter equation will not be used here). We have as $\|\theta\| \to 0$ for $j = 1, \ldots, d$, $\partial H_n(\theta)/\partial\theta_j = 2\sum_{i=1}^{d} a_{ij}\theta_i + o(\|\theta\|)$, and so

$$\frac{dH_n(\tilde{\theta}(\theta_1))}{d\theta_1} = 2\sum_{j=1}^{d}\sum_{i=1}^{d} a_{ij}b_ib_j\theta_1 + o(|\theta_1|)$$

as $\theta_1 \to 0$. It follows that as $\theta_1 \to 0$, $(\operatorname{sgn}\theta_1)d\sqrt{H_n(\tilde{\theta}(\theta_1))}/d\theta_1 \to \sqrt{\sum_{i,j=1}^{d} a_{ij}b_ib_j}$. So by (80), since z is a continuous function of θ_1, and has a derivative approaching a positive limit as $\theta_1 \to 0$, z is a C^1 function of θ_1 with $dz/d\theta_1 > 0$ for $|\theta_1| < \gamma_2$.

Almost surely for n large enough we will have $\|\hat{\theta}_n - \theta_0\| < \gamma_2/2$, or in our eventual θ coordinates $\|\theta_0\| < \gamma_2/2$, so $|\theta_{01}| < \gamma_2/2$. Then $|\theta_1 - \theta_{01}| < \gamma_2/2$ implies $|\theta_1| < \gamma_2$. A C^1 function with a strictly positive derivative has a C^1 inverse. Thus the open interval $(\theta_{01} - \gamma_2/2, \theta_{01} + \gamma_2/2)$ has the properties stated in the proposition for ψ replaced by θ_1, and the proposition is proved. □

PROOF OF COROLLARY 1. By the last statement in Theorem 1, $\pi_{x,n}(A_n)/\Phi(-z_n) \to 1$. Almost surely for n large enough, we have Case I of the proof (28), so the MLE $\tilde{\theta}_n$ in ∂A_n exists and is unique almost surely, as shown just before (80). Write A_n as $\{y_1 \geq \zeta_n\}$ for coordinates y, depending on A_n, as defined in Section 5, equation (31). We have $d(\partial A_n, \phi_0) \to 0$, $\|\hat{\phi}_n - \phi_0\| \to 0$ a.s., and the coordinate transformations $\phi \mapsto \theta \mapsto y$ in Sections 4 and 5 increase distances at most by a constant factor not depending on A_n, so $\|\tilde{\theta}_n - \hat{\theta}_n\| \to 0$ and $\zeta_n \to 0$. Thus by uniform continuity of $\pi_1(\cdot)$ in a neighborhood of θ_0 and $\pi(\cdot)$ in the corresponding open set in y coordinates we have $\pi_1(\tilde{\theta})/\pi_1(\hat{\theta}) \to 1$ and as in (35), $\pi(\zeta_n, 0)/\pi(0) \to 1$ as $n \to \infty$. Similarly, the Hessians $\ell_{\eta\eta}$ are uniformly continuous in a neighborhood of θ_0 by (A_6) and (A_7), and their determinants are bounded away from 0 almost surely as $n \to \infty$ by (15) for θ coordinates or (34) for

y coordinates. Thus $\det[-\ell_{\eta\eta}(\tilde{\theta})]/\det[-\ell_{\eta\eta}(\hat{\theta})] \to 1$ and likewise in y coordinates. Let $\tilde{y} := \tilde{y}(y_1)$ be the MLE of y for given y_1, or $\tilde{\theta}$ in y coordinates, which is well-defined for $|y_1|$ small enough. Then let $(y_1, \tilde{y}_2, \ldots, \tilde{y}_d) := (y_1, \tilde{y}^{(1)}) := \tilde{y}(y_1)$. By (31) and (32),

$$y_1 \equiv \theta_1 \quad \text{and} \quad h_n(\tilde{y}) \equiv H_n(\tilde{\theta}). \tag{81}$$

Recall also that $\tilde{y}(\zeta_n) = (\zeta_n, 0)$ from just before (32) and so

$$\frac{dh_n(\tilde{y})}{dy_1}\bigg|_{y_1=\zeta_n} = \frac{\partial h_n(y)}{\partial y_1}\bigg|_{y=(\zeta_n,0)} =: \frac{\partial h_n(\zeta_n, 0)}{\partial y_1}.$$

Thus

$$\frac{d\sqrt{h_n(\tilde{y})}}{dy_1}\bigg|_{y_1=\zeta_n} = \frac{\partial h_n(\zeta_n, 0)/\partial y_1}{2\sqrt{h_n(\zeta_n, 0)}}. \tag{82}$$

By (36), $\zeta_n^2/[2(1+\delta)] \le h_n(\zeta_n, 0) \le (1+\delta)\zeta_n^2/2$, so $\zeta_n/\sqrt{1+\delta} \le \sqrt{2h_n(\zeta_n, 0)} \le \zeta_n\sqrt{1+\delta}$. Similarly from (34), $y_1/(1+\delta) \le \partial h_n/\partial y_1 \le y_1(1+\delta)$ for $y_1 > 0$ small enough. Thus by (82),

$$2^{-1/2}(1+\delta)^{-3/2} \le d\sqrt{h_n(\tilde{y})}/dy_1|_{y_1=\zeta_n} \le 2^{-1/2}(1+\delta)^{3/2}$$

for n large enough. So $d\sqrt{h_n(\tilde{y})}/dy_1|_{y_1=\zeta_n} \to 2^{-1/2}$ as $n \to \infty$ since we can let $\delta \downarrow 0$. By (80) and (81), and the end of the proof of Proposition 4, as $n \to \infty$

$$dz/dy_1|_{y_1=\zeta_n} \sim \sqrt{n} \quad \text{and} \quad dz/dy_1|_{y_1=0} \sim \sqrt{n}. \tag{83}$$

It follows from this and Proposition 4 that

$$\frac{dz/dy_1|_{y_1=\zeta_n}}{dz/dy_1|_{y_1=0}} \to 1, \qquad \text{so } \frac{dy_1/dz|_{z=z_n}}{dy_1/dz|_{z=0}} \to 1,$$

and Corollary 1 is proved. □

Acknowledgments. We thank the Editors and two referees for a number of helpful comments, Richard Melrose for advice on implicit function domains and Michael Woodroofe for pointing out the paper of Berk (1966) to us.

REFERENCES

ALON, N., BEN-DAVID, S., CESA-BIANCHI, N. and HAUSSLER, D. (1997). Scale-sensitive dimensions, uniform convergence, and learnability. *J. ACM* **44** 615–631.

BARNDORFF-NIELSEN, O. E. and WOOD, A. T. A. (1998). On large deviations and choice of ancillary for p^* and r^*. *Bernoulli* **4** 35–63.

BERGER, J. O. and MORTERA, J. (1999). Default Bayes factors for nonnested hypothesis testing. *J. Amer. Statist. Assoc.* **94** 542–554.

BERK, R. H. (1966). Limiting behavior of posterior distributions when the model is incorrect. *Ann. Math. Statist.* **37** 51–58. [Correction (1996) **37** 745–746.]

BICKEL, P. J. and GHOSH, J. K. (1990). A decomposition for the likelihood ratio statistic and the Bartlett correction—a Bayesian argument. *Ann. Statist.* **18** 1070–1090.

BLEISTEIN, N. (1966). Uniform asymptotic expansions of integrals with stationary point near algebraic singularity. *Comm. Pure Appl. Math.* **19** 353–370.

CHEN, C.-F. (1985). On asymptotic normality of limiting density functions with Bayesian implications. *J. Roy. Statist. Soc. Ser. B* **47** 540–546.

CHOY, S. T. B. and SMITH, A. F. M. (1997). On robust analysis of a normal location parameter. *J. Roy. Statist. Soc. Ser. B* **59** 463–474.

DANIELS, H. E. (1987). Tail probability approximations. *Internat. Statist. Rev.* **55** 37–48.

DICICCIO, T. J. and MARTIN, M. A. (1991). Approximations of marginal tail probabilities for a class of smooth functions with applications to Bayesian and conditional inference. *Biometrika* **78** 891–902.

DICICCIO, T. J. and STERN, S. E. (1993). On Bartlett adjustments for approximate Bayesian inference. *Biometrika* **80** 731–740.

DUDLEY, R. M. (1993). *Real Analysis and Probability*, 2nd ed., corrected. Chapman and Hall, New York.

DUDLEY, R. M. (1998). Consistency of M-estimators and one-sided bracketing. In *High Dimensional Probability* (E. Eberlein, M. Hahn and M. Talagrand, eds.) 33–58. Birkhäuser, Basel.

DUDLEY, R. M., GINÉ, E. and ZINN, J. (1991). Uniform and universal Glivenko–Cantelli classes. *J. Theoret. Probab.* **4** 485–510.

DUDLEY, R. M. and HAUGHTON, D. (1997). Information criteria for multiple data sets and restricted parameters. *Statist. Sinica* **7** 265–284.

DUDLEY, R. M. and HAUGHTON, D. (2001). One-sided hypotheses in a multinomial model. In *Goodness-of-Fit Tests and Model Validity* (C. Huber-Carol, N. Balakrishnan, M. S. Nikulin and M. Mesbah, eds.) 387–399. Birkhäuser, Boston.

ERKANLI, A. (1994). Laplace approximations for posterior expectations when the mode occurs at the boundary of the parameter space. *J. Amer. Statist. Assoc.* **89** 250–258.

FRASER, D. A. S., REID, N. and WU, J. (1999). A simple general formula for tail probabilities for frequentist and Bayesian inference. *Biometrika* **86** 249–264.

FULKS, W. and SATHER, J. O. (1961). Asymptotics. II. Laplace's method for multiple integrals. *Pacific J. Math.* **11** 185–192.

HAUGHTON, D. (1984). On the choice of a model to fit data from an exponential family. Ph.D. dissertation, MIT.

HAUGHTON, D. M. A. (1988). On the choice of a model to fit data from an exponential family. *Ann. Statist.* **16** 342–355.

HIPP C. and MICHEL, R. (1976). On the Bernstein–v. Mises approximation of posterior distributions. *Ann. Statist.* **4** 972–980.

HOLT, R. J. (1986). Computation of gamma and beta tail probabilities. Technical report, Dept. Mathematics, MIT.

HSU, L. C. (1948). A theorem on the asymptotic behavior of a multiple integral. *Duke Math. J.* **15** 623–632.

HUBER, P. J. (1967). The behavior of maximum likelihood estimates under nonstandard conditions. *Proc. Fifth Berkeley Symp. Math. Statist. Probab.* **1** 221–233. Univ. California Press, Berkeley.

JENSEN, J. L. (1995). *Saddlepoint Approximations*. Oxford Univ. Press.

JOHNSON, R. A. (1970). Asymptotic expansions associated with posterior distributions. *Ann. Math. Statist.* **41** 851–864.

LAWLEY, D. N. (1956). A general method for approximating to the distribution of likelihood ratio criteria. *Biometrika* **43** 295–303.

LE CAM, L. (1953). On some asymptotic properties of maximum likelihood estimates and related Bayes' estimates. *Univ. California Publ. Statist.* **1** 227–329.

LEONARD, T. (1982). Comment. *J. Amer. Statist. Assoc.* **77** 657–658.

LUGANNANI, R. and RICE, S. (1980). Saddle point approximation for the distribution of the sum of independent random variables. *Adv. in Appl. Probab.* **12** 475–490.

PAULER, D. K., WAKEFIELD, J. C. and KASS, R. E. (1999). Bayes factors and approximations for variance component models. *J. Amer. Statist. Assoc.* **94** 1242–1253.

PERICCHI, L. R. and SMITH, A. F. M. (1992). Exact and approximate posterior moments for a normal location parameter. *J. Roy. Statist. Soc. Ser. B* **54** 793–804.

POSKITT, D. S. (1987). Precision, complexity and Bayesian model determination. *J. Roy. Statist. Soc. Ser. B* **49** 199–208.

REID, N. (1996). Likelihood and higher-order approximations to tail areas: A review and annotated bibliography. *Canad. J. Statist.* **24** 141–166.

RUDIN, W. (1976). *Principles of Mathematical Analysis*, 3rd ed. McGraw-Hill, New York.

SCHWARZ, G. (1978). Estimating the dimension of a model. *Ann. Statist* **6** 461–464.

SHUN, Z. and MCCULLAGH, P. (1995). Laplace approximation of high-dimensional integrals. *J. Roy. Statist. Soc. Ser. B* **57** 749–760.

SKINNER, L. A. (1980). Note on the asymptotic behavior of multidimensional Laplace integrals. *SIAM J. Math. Anal.* **11** 911–917.

TALAGRAND, M. (1987). The Glivenko–Cantelli problem. *Ann. Probab.* **15** 837–870.

TEMME, N. M. (1982). The uniform asymptotic expansion of a class of integrals related to cumulative distribution functions. *SIAM J. Math. Anal.* **13** 239–253.

TEMME, N. M. (1987). Incomplete Laplace integrals: Uniform asymptotic expansion with application to the incomplete beta function. *SIAM J. Math. Anal.* **18** 1638–1663.

TIERNEY, L. and KADANE, J. B. (1986). Accurate approximations for posterior moments and marginal densities. *J. Amer. Statist. Assoc.* **81** 82–86.

TIERNEY, L., KASS, R. E. and KADANE, J. B. (1989). Approximate marginal densities of nonlinear functions. *Biometrika* **76** 425–433. [Correction (1991) **78** 233–234.]

VAN DER VAART, A. W. and WELLNER, J. A. (2000). Preservation theorems for Glivenko–Cantelli and uniform Glivenko–Cantelli classes. In *High Dimensional Probability II* (E. Giné, D. M. Mason and J. A. Wellner, eds.) 115–133. Birkhäuser, Boston.

WALKER, A. M. (1969). On the asymptotic behaviour of posterior distributions. *J. Roy. Statist. Soc. Ser. B* **31** 80–88.

WALL, H. S. (1948). *Analytic Theory of Continued Fractions.* Van Nostrand, New York.

WONG, R. (1973). On uniform asymptotic expansion of definite integrals. *J. Approx. Theory* **7** 76–86.

WONG, R. (1989). *Asymptotic Approximations of Integrals.* Academic Press, New York.

WOODROOFE, M. (1992). Integrable expansions for posterior distributions for one-parameter exponential families. *Statist. Sinica* **2** 91–111.

DEPARTMENT OF MATHEMATICS
MASSACHUSETTS INSTITUTE OF TECHNOLOGY
ROOM 2-245
CAMBRIDGE, MASSACHUSETTS 02139
E-MAIL: rmd@math.mit.edu

DEPARTMENT OF MATHEMATICAL SCIENCES
BENTLEY COLLEGE
175 FOREST ST.
WALTHAM, MASSACHUSETTS 02154
E-MAIL: dhaughton@bentley.edu

Breinigsville, PA USA
27 August 2010

244317BV00005B/1/P